PEC
ENVIR S

ISSUES AND ENQUIRIES

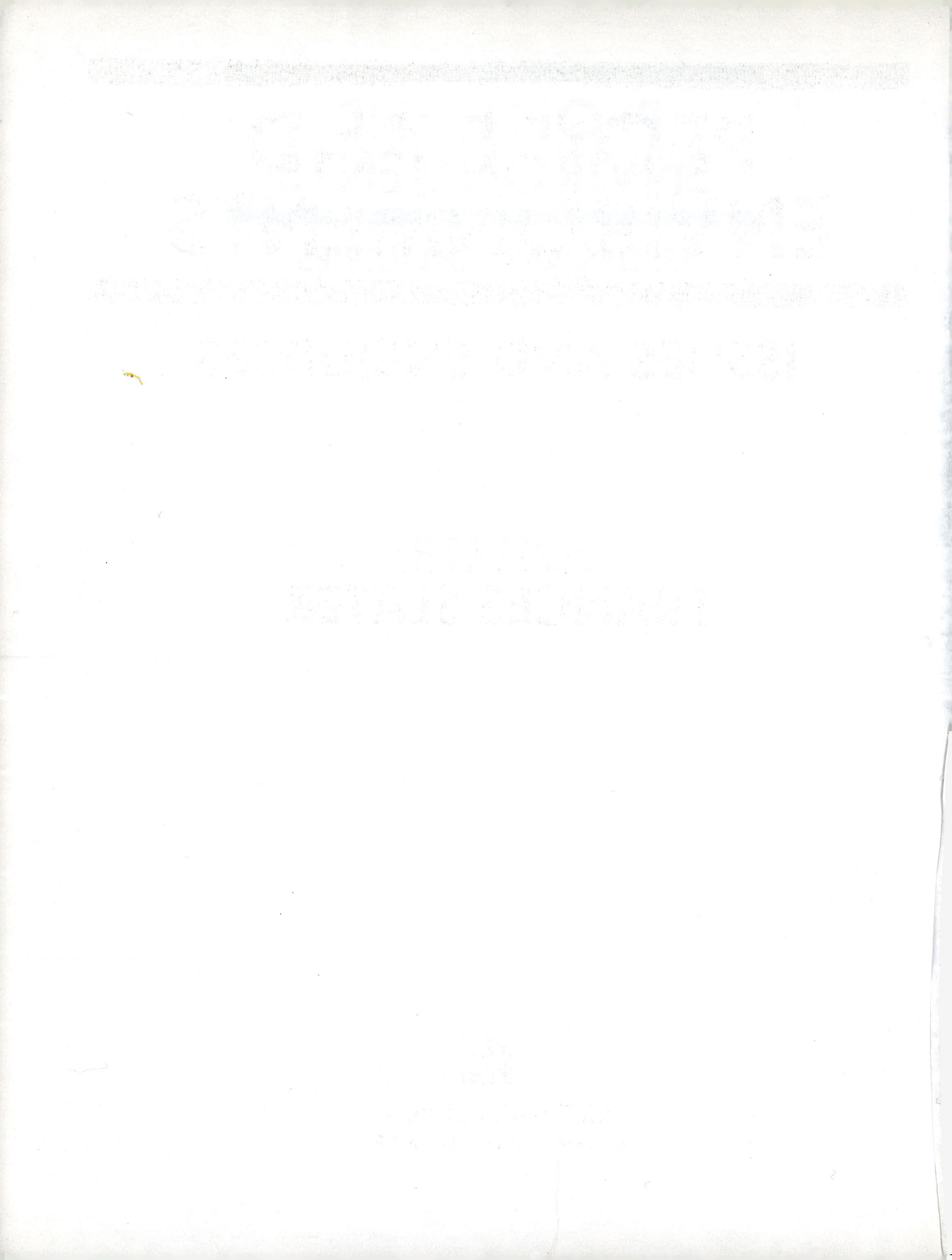

PEOPLE AND ENVIRONMENTS

ISSUES AND ENQUIRIES

EDITOR:
FRANCES SLATER

CollinsEducational
An imprint of HarperCollins*Publishers*

Collins Educational
77-85 Fulham Palace Road
London W6 8JB

An imprint of HarperCollins*Publishers*

First published in 1986 reprinted 1987, 1988, 1989, 1990, 1993, 1994

ISBN 0 00 327402 0

Artwork by Reg Piggott
Cartoons by Peter Shrank

Designed and produced by Sands/Straker Studios Limited
Typeset in Futura by Hope Services (Abingdon) Ltd
Printed and bound in Hong Kong

Editorial Note

The cover photograph shows a *favela* in Rio de Janeiro, Brazil. Poor quality housing, such as this, is not restricted to developing countries — standards of housing vary greatly in all parts of the world.

CONTENTS

PART I PEOPLE, MIGRATION AND URBAN ENVIRONMENTS

PART II POWER, POLICIES AND FOOD PRODUCTION

PART III DEVELOPMENT AND REGIONAL PLANNING

PART IV INDUSTRIAL CHANGE AND RESOURCE DEVELOPMENT

PART V MANAGING AND MISMANAGING THE NATURAL ENVIRONMENT

INTRODUCTION

LANDSCAPES OF CONTRADICTION

'North and South', concepts popularised by the *Brandt Report*, have become a frequently used, shorthand way in which we in the West refer to a simple, two-fold division of our world. We divide the globe perhaps too conveniently into two parts — a more developed, comparatively prosperous, industrialised North, and an underdeveloped, poorer, and as yet less heavily industrialised South.

In this book, a study of over 20 issues enables you to examine and analyse patterns and processes in both North and South and the process links between the two. The issues are loosely grouped into five themes: People, Migration and Urban Environments; Power, Policies and Food Production; Development and Regional Planning; Industrial Change and Resource Development; Managing and Mismanaging the Natural Environment. Exercises, activities and enquiries assist you in analysing each issue so that the gross picture of a North and a South becomes refined and elaborated — the patterns gaining detail while yet being understood in relation to individual, corporate and state decision-making processes.

As you work through the book, a sense of the contradictions inherent in the decision-fuelled processes of industrial, agricultural, urban and resource development should emerge. Development of resources and environments and destruction of resources and environments go hand-in-hand. Who gains in the short term? Who loses in the longer term? We try to manage the various elements of our environment but the landscapes created by the effects of government and company policies and schemes are vivid and contradictory, as the following illustrations and commentaries suggest.

In Mexico many poor peasant farmers are increasingly in debt; 30 million Mexicans are seriously undernourished.

Some very wealthy farmers and multinational companies are exporting crops and cattle for profit.

As the number of landless people increases, homes built from the wastes of cities accumulate as shanty towns.

Exposed to rains by the destruction of forests, millions of tonnes of Himalayan top soil now accumulate as a silt mountain in the Bay of Bengal.

Europe and the USA have an ever-increasing capacity to produce food (resulting in butter mountains, wine lakes, subsidies to farmers, and land lying fallow under grants).

In the South there is insufficient production and people face starvation.

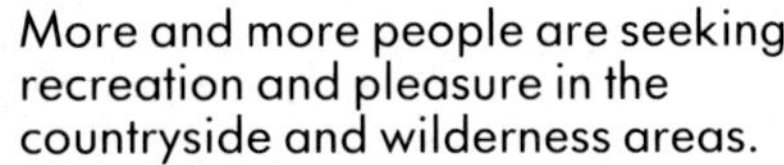

More and more people are seeking recreation and pleasure in the countryside and wilderness areas.

This results in increasing erosion of land and the destruction of the beauty of such areas.

These contradictions and others which you will find later in the book, demonstrate that throughout the world, a wide range of costs and benefits result from people living, working and playing in their various environments. Consequently, it is very important that we try to understand fully how the various parts of these environments fit and work together. In other words, how each part, or variable, impacts on the others, either to maintain a smoothly functioning system or to throw the system out of normal working and thus to change the relationships between variables, possibly for the worse. We are used to thinking of our physical environment in this way. *People and Environments*, encourages you to analyse and understand human systems in the same way — human systems which are full of contradictions and which are affected by our way of looking at things and the way we use our knowledge and understanding. We need to recognise that these human systems are not always managed for the greatest good of the greatest number.

After the chapters in *People and Environments* had been decided on, to illustrate the five main themes and to fill out the detail of 'North and South', a good deal of attention was paid to setting up activities and exercises which would help you to work through and understand the ideas presented by each author. That is the purpose of all the activities — to give you an opportunity to go over the text again, to think for a second, third or fourth time about what is being said, to analyse the main points, and to confront the values and attitudes underlying and fuelling the patterns being described and the decisions being taken. In working through the activities and exercises, you will be developing your knowledge of world issues and your capacity to think about and analyse ideas and make connections between one set of ideas and another. You will also be increasing your ability to problem solve, and to follow through an investigation in a scientific way. The many letter-writing, speech-making, and explaining-to-another-person type tasks, will develop your awareness of the way attitudes, opinions, prejudices and values colour and condition many of our decisions, as well as helping you to work over the ideas and think about them for yourself. Role plays, simulations, debates and conversations are a good way of showing how vested interests operate and contribute to the way our world is patterned.

Major questions about social justice will arise, as will an increasing realisation on your part of the varying interests of different groups in a society and the variations in the way they may perceive and experience their environments. It will become clear over and over again — in the content and through the exercises — that very often powerful decision-makers do not perceive or do not choose to take account of all the necessary and relevant variables in their analysis of an agricultural, industrial, regional or development problem. The cultural blinkers we wear operate to distort views and opinions. Very often we analyse and solve a problem in a one-sided kind of way, from one point of view.

When you have studied the chapters in this book, you should be able to read this introduction again and give several examples for each of the general statements I have made here about the way the world works.

Frances Slater

PART I PEOPLE, MIGRATION AND URBAN ENVIRONMENTS

PEOPLE AND URBAN ENVIRONMENTS

The four chapters in this first section of *People and Environments* are strongly linked by a common emphasis on people in urban environments.

The first chapter, *Population in Europe — Trends and Issues* begins with an overview of population trends in Europe and then introduces a number of issues which have arisen from population processes of key importance:

- Europe's population is ageing (i.e. the number of elderly people as a percentage of the total population is increasing). What problems does this present to policy makers and governments?
- Europe's birth rate has been decreasing. There are fewer young people in the total population. What problems and dislocations does this cause policy-makers in the field of education — schools and teacher supply, for example?
- 'The future of the German people is at stake.' The declining birth rate has many consequences, a number of which are spelt out in relation to the German population. Can you list four in advance of reading the chapter?
- In all European countries, foreign workers form a significant proportion of the population of large cities. Where do they live and with what social and cultural consequences?
- After the Second World War, Europe became an area of net immigration instead of one of net emigration. The 'guest workers' were welcome when jobs were plentiful. In the present recession, jobs are scarce and now the Mediterraneans and non-Europeans are not so welcome. Some people think they should be sent home. What kind of moral dilemma does this pose for people and governments?

The movement of people — the migration of people from one place to another — from country to city and within cities, is most often the result of changes in the economy and in industrial technology, changes in the demand for jobs and changes in opportunities for jobs. The effects of these changes can be seen in cities. Cities show the strain as people adjust to new lives and new circumstances, and the strain is most readily observed in the struggle to provide housing for newcomers to cities. People today pay a high price for changes in the economy and in industrial technology as they struggle to adapt, just as some of our ancestors struggled in the growing industrial towns of Western Europe and North America last century.

All the chapters in this section touch on migration and the social and political issues and demands which arise. Chapters 2 and 3 focus most strongly on the migrant in the developing world. — Chapter 2, *A Tale of Three Cities* questions our stereotyped view of the developing world migrant as someone who makes a basically ill-judged, irrational decision to move from the countryside to a 'shanty settlement' in the city. The authors refute this view with the aid of data gathered on research visits to certain Latin American cities. They show that migrants, faced with a hostile environment, choose sensibly and rationally from a limited range of options. Chapter 3, *Housing in Lusaka — An Example of Self-help* looks closely at the provision of housing in Lusaka, Zambia. The author explores why the term 'shanty town' is really an inappropriate description for this housing — the phrases 'self-help housing' or 'spontaneous urbanisation' capture better the true nature of the process. People do help themselves and yet are still part of a general shift of people from countryside to towns which, while appearing spontaneous, is, in actual fact, a response to economic and technological changes. These migrants in developing countries, like migrants in Europe, are coping with changes which fundamentally affect their lives.

The experience of migration can be seen to link the experiences of people in the developed and the developing worlds, both today and in the past as the extracts on the following pages highlight.

? Read all the extracts — some of them you will study again in the chapters. Make sure that you understand the point of each. Make notes if this helps you to sort out the ideas.

FRANCE ADAPTS TO IMMIGRATION WITH DIFFICULTY

In common with many countries, France has a long history of immigration and is now experiencing repercussions from its eagerness for foreign labour in the 1950s and 1960s.

On the second day of April last year a young Algerian living in Lyons went on hunger strike. He was joined in his protest by a Protestant and a Roman Catholic priest. Three weeks later a similar protest began by 21 Tunisian workers in the Paris suburbs.

They were objecting to a new Act which, from January 1980, had allowed the government to expel immigrants without papers. It highlighted the plight of 300 000 illegal migrants who, with the government's tacit approval, had drifted into the country over the preceding 15 years but whose position was newly challenged by the desire to reduce the size of the immigrant population. Periodic outbursts of violence in cities like Marseille; a protracted strike by immigrant cleaners on the Metro in Paris; the much publicised eviction by the Communist-controlled local council, in December 1980, of 600 African migrants and the destruction of their hostel by bulldozer, added to the increasingly repressive government legislation: all brought the immigrant question to the centre of the political stage at the time of the 1981 Presidential elections.

The events helped to emphasize that there are few groups in society that feel more acutely the effects of economic recession than immigrants. Estimates put the number of foreign workers within the European Economic Community countries at around 15 million, many of whom live in the most disadvantaged parts of cities. France provides a particularly good example: a large immigrant community, of more than four million in 1981, whose diverse nationalities show very different patterns of settlement and adaptation to the urban environment: a long-standing history of immigration, intensified during the 1960s and abruptly reversed in the 1970s; and a range of governmental attitudes from the liberal to the repressive, with substantial changes of approach heralded by the election of the Socialist president François Mitterand in 1981.

French policy shifted sharply against migrants during the last years of the Giscard presidency: the idea of voluntary repatriation was introduced in April 1977 with the provision of grants of 10 000 francs to each worker returning home, with 10 000 francs for a working wife and 5000 francs for each working child. These measures had only limited appeal: between June 1977 and September 1979, 70 000 people sought to return under the scheme; of these, 60 per cent were Portuguese or Spanish. There was considerable sympathy across the political spectrum for halting new immigration and a little less marked, for encouraging repatriation. But great damage was done to the climate of race relations by the government's increasingly rigid attitudes towards illegal migrants: the unedifying spectacle of those with black or brown faces being stopped in the Metro and asked for papers they almost certainly did not possess was an increasingly frequent feature of the Parisian scene.

Many migrants and their families whose illegality in a better economic climate was tacitly accepted came to live in terror of arrest and expulsion. The desultory attitude of police and 'authority' to immigrants was widely resented and much ill-feeling was caused by the case of several youths who, foreign by nationality and migrants with their parents when a few months old, were threatened with compulsory repatriation to a 'home' of which they had no recollection.

The new government of François Mitterand acted swiftly to change policies and attitudes. New migration is strictly controlled but reuniting of families is made easier, expulsion stopped, repatriation grants ended and papers given to most illegal migrants in employment. Details of policy are as yet unclear, but the government appears to recognize that, while return home should not be discouraged, France will retain a large foreign population whose humane treatment is seen as part of a new approach in foreign policy to Third World problems. Discussion within the new *majorité* is, however, just below the surface: the Communists are frankly hostile to immigrant concentration and a recent proposal that foreigners should be given the vote was greeted with cries of outraged nationalism on all sides.

The large *bidonvilles*, the shanty-town refuges of the very poor, have now largely been swept away but *microbidonvilles* were still a feature of cities in the 1970s: 'little tin, or petrol drum shacks deep in the recesses of someone's garden or yard; discreetly hidden from the eyes of the casual passer-by. The more fortunate immigrants have moved to public housing recently erected by government and public authorities. The rest lead a precarious existence in various marginal forms of housing. The transit centres and *foyers*, sometimes run by firms or by private landlords, provide shared accommodation. But many are still to be found in rooming houses in the least salubrious districts: the *marchand de sommeil* ('sleep seller') may still rent beds by the night, or even the hour, in *hôtels meublés* or *hôtels garnis*, overcrowded, badly furnished, insanitary rooms in rundown blocks.

The illegal migrant is particularly vulnerable: underpaid by day and overcharged by night. Poor immigrant zones spread well beyond Paris: in the heart of Marseille, the area known as 'La Cage' is a ghetto: 22 000 North Africans living in slum properties, with a high rate of unemployment, where 15 per cent of children leave school unable to read or write and where the risk of violence has occasionally been given substance. The questions of segregation and housing provision have already proved explosive: Communist local governments, where the immigrant concentrations are naturally very high, have begun to campaign in favour of a more equal geographical spread of immigrants and of the social problems and costs associated with them.

(*Source*: *The Geographical Magazine*, July 1982)

MOVING TO EUROPE

On 15 July, a Saturday, it was exceedingly hot in France. One of the few really hot days France had this summer. The temperature was somewhere in the 90s. A lorry had broken down near the small town of Aix-les-Bains, just a short distance from the Mont Blanc tunnel and the Italian frontier. The French driver, Michel Piteau, was panic-stricken. He'd spent an hour trying to get the lorry going again. But he couldn't. Inside the lorry the shouts and screams of his human cargo were getting louder. Frankly, he didn't know what to do. He'd been paid a lot of money to drive them from Allessandria in North Italy to Paris. On the other hand, with the sun beating down like this on the open stretch of road it was obvious why his cargo was getting hysterical. He couldn't open the lorry himself. It had been sealed in Monza in the same way as the Customs had sealed it in Rome when his cargo had just been sewing machines. It could only be unsealed by his boss in Paris — or the police. But to go to the police meant exposing himself to a heavy prison sentence — and no more big fat payments. It was a good business, even if it was a bit shady, and it had all run so smoothly before. What the hell should he do? In the end Piteau's conscience got the better of him. After all, he didn't want dead bodies on his hands. He plucked up his courage and hitched a lift to the nearest police station. Half an hour later the seals on the lorry were broken, and 59 Malians tumbled out, naked and dizzy with hunger, thirst and lack of oxygen.

And so came into view a vast international cheap labour smuggling racket. Africans from Mali, Mauritania and Senegal were recruited from remote districts by a white man who promised secure, well paid jobs in Europe, with fares, documents, and food and lodging to be provided by the employers. The Africans were moved by road to Tunis, then flown to Palermo in Sicily. Others arrived in fishing boats. They were assembled in a basement flat in Rome and given tourist passports. Next, they were shipped in batches of 50 to 60 by train to Allessandria. Here they were transferred to the lorries and then, 'customs sealed', it was non-stop by road to Paris. With them in the back was a small amount of food and water, and a pile of straw in a corner for a 'toilet'.

Piteau is now in prison awaiting trial; the Africans in a refugee camp near Rome soon to be repatriated; and Interpol are co-ordinating investigations in France and Italy.

The smuggling of cheap labour goes on all over Europe. Employers need labour desperately. However, the smuggling probably represents less than five per cent of the total foreign labour inflow — it serves the more unscrupulous employers who want to avoid union pressure for decent wages and state regulations about health and social insurance. Most employers, however, prefer to do things legally, and by and large the European governments and unions make it fairly easy for them to get the labour they need.

(*Source: The New Proletarians*, Jonathan Power, Community and Race Relations Unit of the British Council of Churches, 1972)

Housing — A Tale of Two Cities

In Calcutta, 30 per cent of the population are either registered *bustee* (or slum) dwellers or have no shelter at all. In Liverpool, as in the rest of Britain, slum dwellers are not registered, although 5.5 per cent of the population is officially designated as living in 'overcrowded conditions' — that is at a density of over 1.5 people per room. In the whole Merseyside area with a population over 1 200 000 under five per cent live in 'officially' overcrowded conditions.

It is interesting to note that the British designation of over 1.5 persons per room constituting overcrowding would be considered spacious, not to say gracious, living in a Calcutta *bustee* where six to eight people sleeping, cooking and eating in a room 200 feet square is commonplace.

Liverpool and Calcutta have one similarity in that their developing years and the prosperity which followed were both masterminded and exploited by the adventurous, ruthless and often avaricious men responsible for building the British Empire. In that sense the responsibility for the state of both cities today lies in Britain.

As Calcutta and Liverpool grew — roughly at the same time — into great 19th century industrial and trading centres, their wealth attracted migrant labour from the surrounding poverty-stricken rural areas and created a need for some form of shelter to be provided for them and in some cases their families.

In Liverpool, for example, the thousands of Irish fleeing the potato famine who flocked to the city for jobs in the flourishing cotton trade and other industries booming in the city found themselves herded in cramped streets lined by row upon row of what we now call jerry-built houses. As more migrants flocked to the city from Wales and other failing agricultural areas to seek work, they too found the only accommodation they could in the crowded, exploited slum areas.

In Calcutta at about the same time the 'sweatshop' jute industry was the front runner in attracting migrant job seekers not only from the agricultural hinterland of Bengal itself but also the neighbouring states of Bihar, Iross and Uttar Pradesh which over the years had been hit by a series of cyclical famines.

The only shelter these migrants could find was in the slums on the marshes fringing the city. These had grown with an ingenious mixture of adhocery and enterprise and to this day are still run in the same way.

Basically, a landlord leases his land to a middle man who builds a number of huts, usually with mud walls reinforced by bamboo and with tin or tiled roofs. In the early days, it was the middleman's responsibility to provide a 'service latrine', usually of the most primitive kind for the use of his tenants, but now this tradition seems to have faded.

Each hut forms a quadrangle and is divided into six or seven rooms. These partitions are rented out to the migrants who usually can only afford to share with six or more living in each. The system suits everyone — industry which has no intention of housing its labour force; the landlord who gets an easy return for the swamps he owns; the middleman hutment owner who takes his profit and the migrant himself who finds some form of shelter readily and cheaply available.

So the *bustee* system proliferated and has continued to do so until today.

(*Source*: *Christian Aid* pamphlet, 1976)

Women on a scrap heap

The world's major cities are exploding. The natural population increase, combined with migrants who swarm in from the surrounding countryside to survive on the margins of existence, a problem which is now recognised as one of the most horrific and all-pervasive of our time.

Guayaquil, Ecuador's largest city, is situated 100 miles upsteam from the Pacific and built on low-lying land subject to flooding. It has a population of over a million with 30 per cent of its annual growth produced by immigration.

In Cisne Dos, one of the largest of the *barrios*, the families living at its furthest edge with up to 15 people to a house — have been there only a year or two, acquiring their 10 by 30 metres plot by personal invasion or more frequently by buying it from professional squatters.

There is no running water, no sewerage, and often catwalks for roads; the logistics of daily living are daunting, and it is the women who bear the brunt. Water, or the lack of it, dominates their lives. It is delivered by tankers at irregular intervals, and distributed into 50-gallon drums put out in the street.

Nobody in Cisne Dos marries in church and free unions, *compromisos*,are the predominant and highly unstable form of relationship. Men start with one girl and after a few years set up a second household with another younger woman, operating both households simultaneously, resulting in a disastrous division of very limited financial resources and no real responsibility for either family.

Men legitimise this by saying: 'In Guayaquil there are seven women for every man and so it's our duty to live like this.' The most 'macho' and successful may end up running four to five households at once. The consequence for the women is frequently abject suffering and heavy economic dependence on men.

The employment situation is so difficult in Guayaquil that poor women, though they may prefer office or shop work, can only find jobs as maids, cooks, street sellers or washerwomen.

Angelita is a *chambera*, a garbage collector and sorter who lives in a hut made of sacks, cardboard boxes and piece of wood and scrap metal, on the city's rubbish tip. Only two of her four children are still alive and she lives with them and her mother among the flies, decaying garbage and thousands of vultures. Her days are spent sorting through the truck-loads of rubbish which arrive from the city centre. She, like the other *chamberas*, makes a living from the collection of cardboard boxes, paper, pieces of aluminium, tin, iron, wire or bottles, anything which can be recycled and resold. She says she enjoys the freedom of working for herself after being a maid.

Politics has given Emma her way out; at 27 she has been married to Eliseo, her tailor husband, for 10 years and has two daughters by him — he has two sons by a previous marriage and is 16 years older than her. Emma has been president of the barrio committee for the past four years and it is largely due to her own and her largely female committee's constant pressurising that the area now has some roads instead of catwalks. She is quite aware that it is only the considerable prestige attached to political work which protects her against male criticisms about going out into the street. But then Emma aspires to bourgeois respectability unlike Angelita who no longer conforms to the traditional subservient role of women.

There are very few ways out of the *barrio*. The more determined and aware of the women clutch at the dressmaking and hairdressing courses provided by the private education industry but for the most part the courses are useless. Community politics is the one real chance of escape.

(*Source*: 'Woman on a scrap heap', Caroline Moser, *The Guardian*, 20 May 1980)

? Working in pairs, complete the following tasks.
1 Imagine your partner is a 10-year-old. Tell the story of Michel Piteau to him/her. Do not omit to say something about Europe's need for immigrants in order to explain why Michel Piteau was involved in smuggling. Then imagine your partner is older than you. Decide how you would now recount the story. How would the emphasis differ? Would it differ again if you were answering an exam question on European migration since 1945?
2 Imagine you are an immigrant worker in France talking to an interested journalist after the unrest in the early '80s. Give your views on French immigration policy and the reactions of yourself and other immigrant workers to it. Your partner should then pretend he/she is a Frenchman/Frenchwoman talking to the same journalist, giving his/her views on the same situation.
3 Work with your partner to draw up a list of the major points being made about housing conditions in Calcutta, Liverpool and Guayaquil. What are the areas of similarity?

The movement of people from frost belt to sun belt in the US, from inner city to suburbs, and the recent movement of young professional people back into decaying inner city areas which were once the industrial warehouse regions are three of the processes discussed in Chapter 4, *The American City Today*. In North America, many people are moving to where the new service-type jobs are — jobs which do not depend on heavy, raw materials but which largely buy and sell information. People working in information-based businesses can choose to locate more freely and often choose environmentally pleasant places.

These processes in the developed world and the developing world can be placed in perspective if we remember that country to city migration dominated the developed world 100 and 200 years ago in response to the industrial revolution and the changes it brought. Now as developing world countries adjust to their industrialisation, and people shift and experience dislocations, the developed world is responding to post-industrial changes which are also shifting people about once more.

An important link between the old processes of industrialisation and the new post-industrial society, is to be found in the case study in Chapter 4 which highlights the way affluent Yuppies (young, urban professionals) are moving back to what were, until recently, undesirable areas of cities.

These once undesirable areas (London's Docklands is a good British example) were once the residential and working localities for those who moved to them in the industrial revolution of last century. The housing was never spacious or affluent and the working conditions were hard and, for most of the period, jobs were poorly paid. Large and often handsome warehouses were a feature of areas such as docklands. Technological and transport changes (containerisation) brought a change in profitability and a massive loss of jobs in such areas. The docks fell into disuse, the warehouses emptied. Often little was done to locate the sort of new industries and businesses in these areas which would provide employment for the old workforce. The people were forced to move to find new jobs in new places. The empty warehouses, however, represent large areas of floor space ripe for redevelopment. Private citizens and large companies have seen an opportunity to change the function of such buildings, to convert them to 'living space', and well-to-do urban dwellers have been attracted to live in such localities with their advantage of proximity to the city centre.

So we can reflect that, in certain areas of cities, the once migrant poor who established strong communities around their place of work are being replaced by a new group of well-to-do people moving in. Will today's shanty towns at some time in the future be subject to redevelopment? Whatever the answer to this question, it is interesting to link migrants and people moving within cities to broad economic changes and to note that the same places can be desirable at different times to very different groups. Who pays the costs of economic and societal change? Who absorbs those costs? You should keep these questions at the back of your mind when reading through the chapters in this section.

1 POPULATION IN EUROPE: TRENDS AND ISSUES

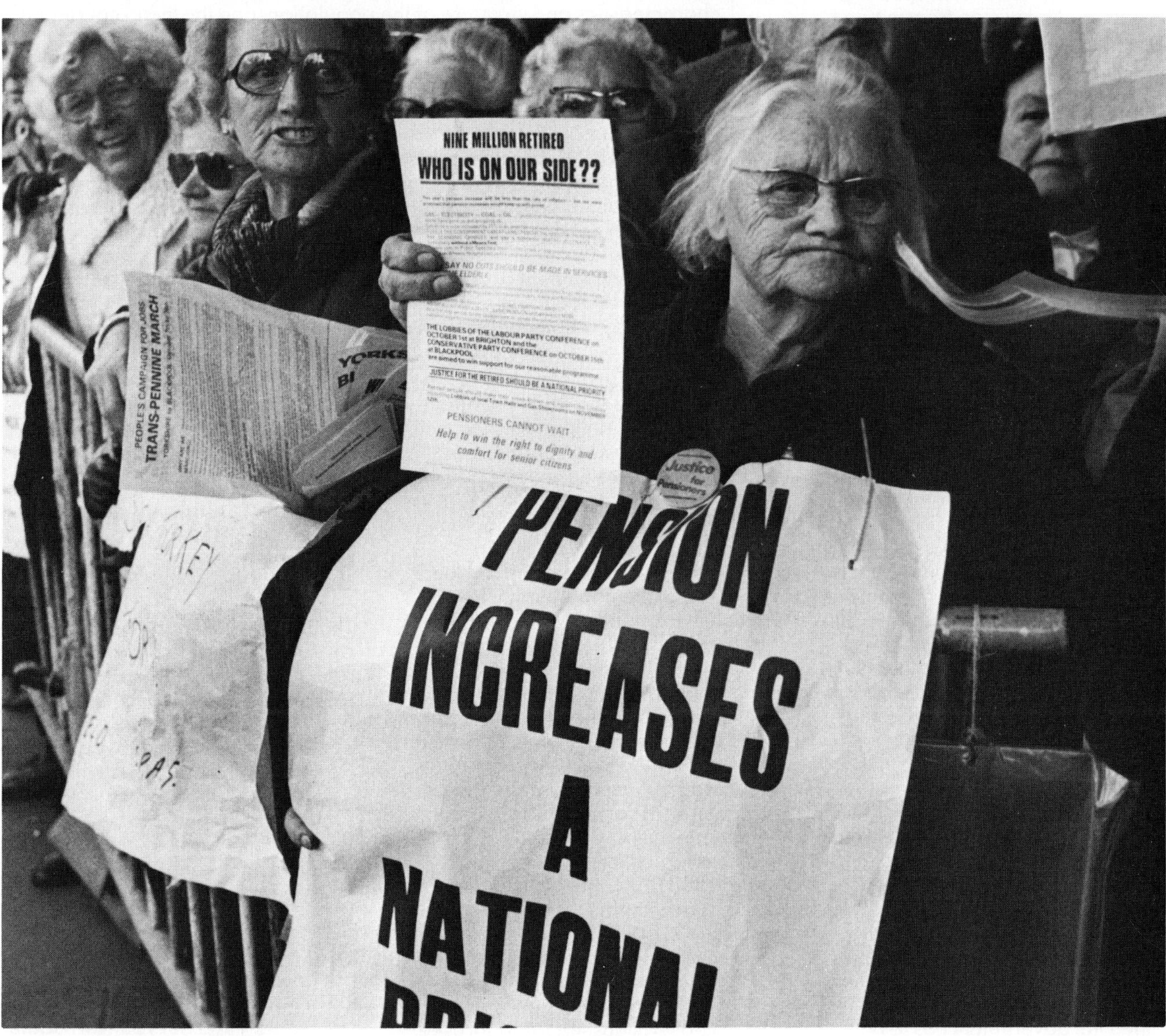

THE NEED FOR DEMOGRAPHIC INFORMATION

Every day a mass of information becomes available about the conditions under which people live all over the world. Newspapers, radio and television transmit reports, and issues arising from them are discussed by journalists, politicians, economists, geographers etc.

The density of the population of an area is an important factor influencing the living conditions in that area. Population size changes continuously and the trends are watched carefully by local, regional and national authorities. The monitoring of these changes is undertaken by the population units of national, regional or local government offices. In addition, national governments take censuses, usually every five or ten years. The children who are born, the people who die, and those who leave or enter an area are carefully documented.

A government is, to a great extent, responsible for the welfare of the population. Not only must there be sufficient food, electricity and public transport; there must also be adequate provision of schools, houses, roads, recreation facilities, etc. If the provision of these services is to be adequate and efficient, a quick reaction to changes in the structure of the population is required. If the number of children increases because of the decrease in infant mortality, then a need to provide more schools and teachers arises. If the number decreases as a result of fewer births, then schools have to be closed down and teachers made redundant. If the number of elderly people increases then the government has to be careful to make the necessary provisions for that group.

In densely peopled areas population increase means, especially in combination with a rise in prosperity, a rapidly growing pressure on the environment. In areas with a decreasing number of people, the quality of living deteriorates if a smaller population cannot make viable the provision of the whole range of facilities (because there is a fall in income from the rates). In turn, the decline in the quality of living conditions promotes an increase in the number of people leaving.

The development of technology and labour saving tools is an important factor in promoting the welfare of the population. The English economist, Malthus, thought that population would outgrow the means of subsistence. Malthus, however, working in the early years of the 19th century, could not foresee the coming technological and social revolution. Advances in technology allowed, among other things a far higher agrarian production per hectare. But how far can technological development be stretched? And, how much more can the environment take? These questions are difficult to answer with any certainty. In any case, governments must watch closely and react early to population changes.

To govern thus means to foresee. In the countries of Western Europe young people have the right to vote, in some cases as early as 18 years of age. So, they have the opportunity to participate in thinking and decision making, about such issues as the way in which a government can maintain a certain quality of living in an area.

It is, at the same time, important to realise that individual behaviour in such matters as family planning, and through that the consumption of goods, services and space, is an important influence on government decision making.

? Divide into small groups to discuss these questions:

1 'The density of the population of an area is an important factor influencing living conditions in an area.' Compare densities of population for different regions or cities you know fairly well. Imagine an area of high density losing its population and an area of low density gaining rapidly. Discuss the statement, then report the outcome of the discussion to the rest of the class.

2 'The children who are born, the elderly who die and those who leave or enter an area are carefully documented'. In addition, governments hold population censuses in order to get a full picture of the numbers, characteristics and structure of the population, and the living conditions. Such a census is normally held every ten years. However, in the early 1970s people began to obstruct censuses.

a Why do you think people obstruct census taking? (Think about the concern being expressed about the storage of information in computers.)

b What possible consequences could result from such obstruction for a government's work and responsibility in relation to the population?

c In what other ways can governments obtain necessary population data? (Think about other forms of data such as that held by the Chamber of Commerce.)

3 'The development of technology and labour saving tools is an important factor in promoting the welfare of the population.' Discuss the pros and cons of technology in relation to the welfare of the population.

EUROPEAN POPULATION TRENDS

The present European population is characterised by its slow growth rate. It has in fact the lowest growth rate in the world — between 1975 and 1980 the increase was only an average of 4 per 1000 a year in Europe, but 17 per 1000 in the world as a whole. Although this slow growth rate appears to be convenient at present, there are also negative implications which will have to be tackled in the future.

? **1** Study Table 1 and Figure 1 and compare the relative shares of the world's population. Using Figure 1, set out your own table expressing the shares in percentages for Europe, the developed world and the developing world.

2 Using Table 1 construct a graph plotting the actual numbers of the world's and Europe's population against time. Then compare the years 1800, 1900 and 2000. Comment upon the change.

THE DEMOGRAPHIC TRANSITION

The slow increase of the European population is not a new phenomenon, as Table 1 illustrates. Population has always tended to increase slowly, but in the past this slow growth was not controlled. It happened naturally as a result of high death rates and high birth rates. Today, however, the slow growth rate is to a large extent due to human control — low death rates because of factors such as improved hygiene, medical care, and increased food production, and low birth rates because of changes in attitudes and conscious birth control.

This change from high birth rates and high death rates to low birth rates and low death rates is known as the *demographic transition*. The course of the demographic transition differs in detail from country to country. However, it is not necessary to investigate every country separately to achieve an overall view for a number of countries have shown more or less the same pattern of birth and death rates.

The UK

During the middle of the 18th century, birth and death rates were close to each other; then the death rate dropped dramatically while the birth rate remained high. By 1930 the birth rate had also dropped. Thus a change took place from a high birth and death rate to a low birth and death rate. This demographic transition occurred between 1750 and 1930 (see Figure 2).

The decrease of the death rate was brought about by a decrease of infectious diseases, such as typhus, smallpox and diphtheria. It is thought that increased prosperity together with improved nutrition helped to give the population greater resistance to disease. The agrarian revolution played an important role, in that the production of more and better plants and livestock helped to improve nutrition.

In addition, improvements in hygiene in the 19th century played an important part in the decline of the death rate. Developments in urban water supply and in the provision of internal washing facilities lessened the risk of infection. Other important

Table 1 *Number and increase of the population of the World and in Europe (excluding the Soviet Union) 1750–2100*

			Increase since previous year; average per year, per 1000 population		*Share of European population in the world population (%)*
	World	*Europe*	*World*	*Europe*	
Year	*Number (millions)*				
1750	771	111	—	—	14
1800	954	146	4.3	5.5	15
1850	1 241	209	5.3	7.2	17
1900	1 634	295	5.5	6.9	18
1950	2 525	392	8.7	5.7	16
1960	3 037	425	18.6	8.1	14
1970	3 695	462	19.8	8.4	13
1975	4 066	474	19.3	5.1	12
1980	4 432	484	17.4	4.2	11
2000	6119	512	16.3	2.8	8
2100	10000	504	4.9	—	5

(*Source*: J. N. Biraben (1979) *Essay sur l'évolution du Nombre des hommes, Population*
UN *World Population trends our policies* 1981 Monitoring Report
UN, Long range global population projections, as assessed in 1980, *Population Bulletin 14*, 1982)

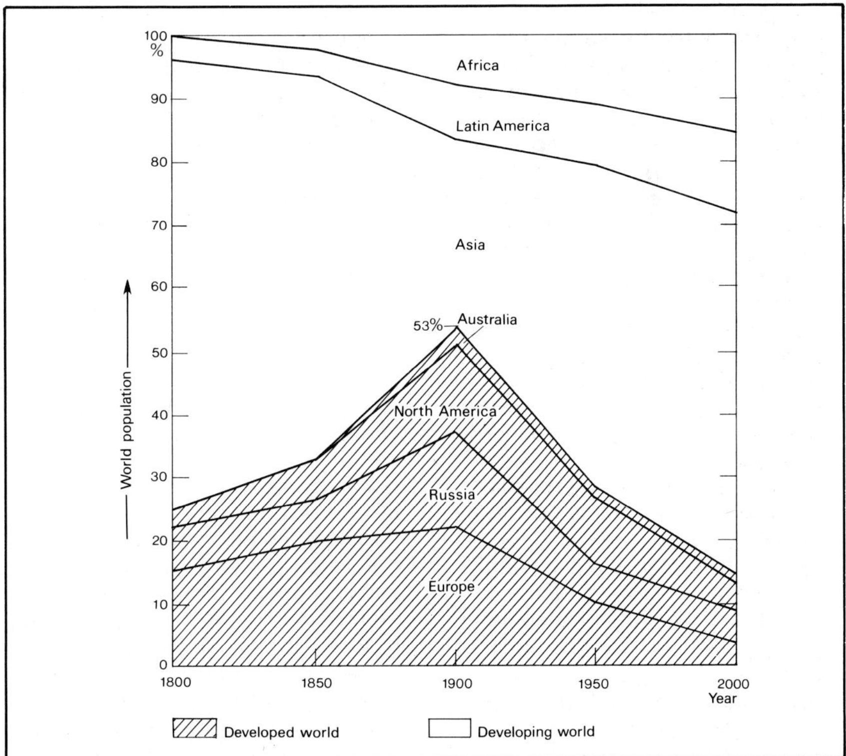

Figure 1 *The developed and developing countries' share of the world's population, 1800–2000*

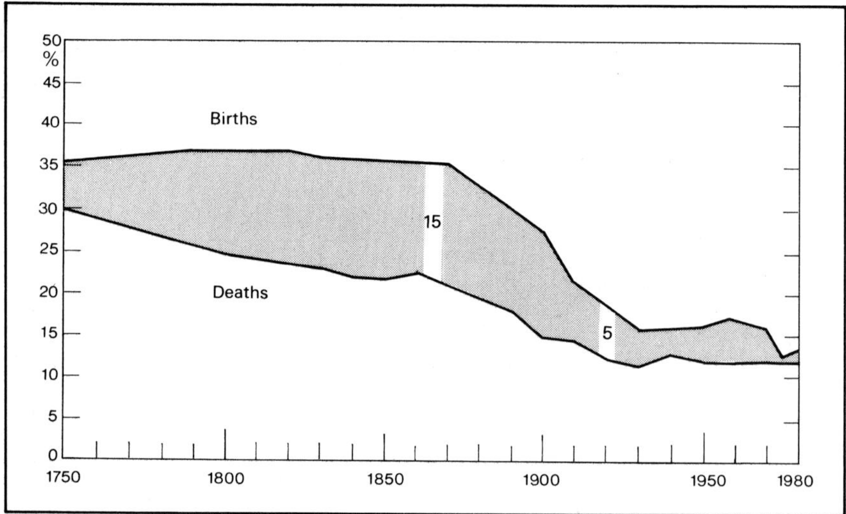

Figure 2 *Natural population increase, England and Wales*

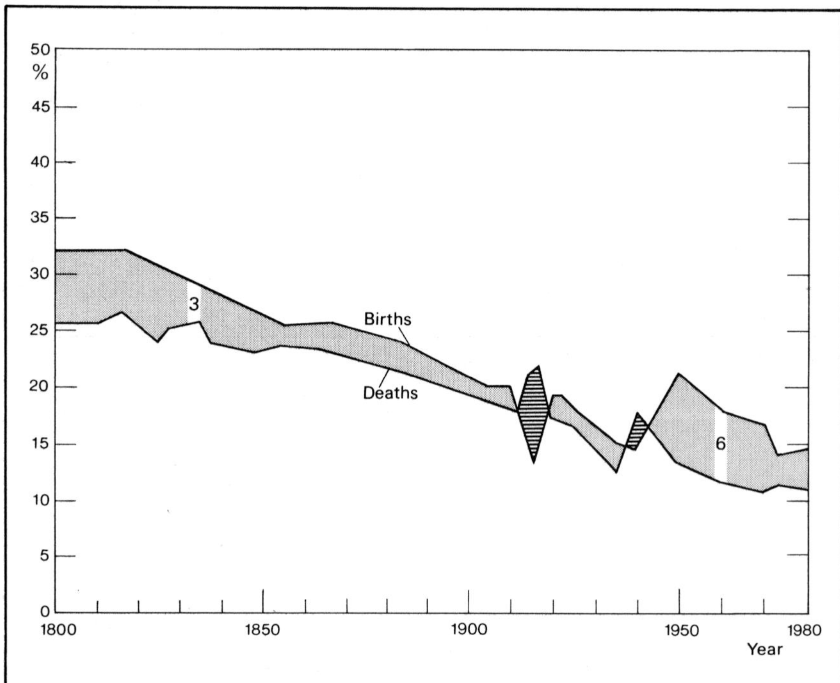

Figure 3 *Natural population increase, France*

measures were the laying down of drainage systems and the paving of streets.

Finally, the progress of medical science also contributed to a constant decline. But it was not until the 1930s that this factor became significant.

The small initial increase in the birth rate was connected with the decline in the death rate and the increase in employment under the influence of the Industrial Revolution. More marriages were contracted, because people did not have to wait so long for a job and thus could marry at an earlier age. In the rest of Western Europe, the custom continued of a couple not marrying until they had saved sufficient money, and could provide their own living accommodation and support their family.

The birth rate remained at a high level for many years. Families were large in size. The reasons for this are simple: children acted as a cheap means of social insurance.

When, in the course of the 19th century, the death rate further declined, and the sense of personal security increased, as child labour was forbidden and compulsory education was introduced, the incentive to bring large numbers of children into the world was lost. Moreover, the striking changes in the socio-economic structure had their effect. The growth of factory production led to increasing demands for skilled labour. This, in turn, exposed the need for better education. Better education made possible an increased social mobility. At the same time, educational costs increased. This and other factors led to the use of birth control techniques, first among the higher income groups, then spreading lower down the social scale.

The rapid decline in the birth rate had to await the time when the great mass of the working classes achieved a degree of material prosperity, and took advantage of better educational opportunities to achieve social mobility.

When we subtract the death rate from the birth rate we have the natural increase in the population. In the UK the demographic transition began with a natural increase of around five per thousand, a normal rate for the pre-industrial period. By the 1920s the natural increase had returned to this level (Figure 2). We could therefore say that at this time the transition was over. During the transition the natural increase achieved a maximum of 15 per thousand, three times the level at the start and finish.

A number of countries have had a transition closely resembling that of the UK.

France

The natural increase per year until about 1950 was around 2 to 3 per thousand and it was only after 1950 that this figure reached 6 to 7 per thousand (Figure 3). Thus in France the course of birth and death rates have been totally different from those of the UK. The most obvious difference is, that both birth and death rates have declined at the same time. The early decline in the birth rate is connected with certain characteristics of French culture. The Revolution of 1789 strengthened the trend towards a decline in the birth rate because people became increasingly prosperous and 'social mobility' was fostered — people with aspirations have fewer children. The small natural increase led to an ageing of the population. Only after the Second World War did the French government promote a more active population policy, increasing the natural rate of growth.

Spain
In 1870 the countries of southern Europe had high birth and death rates. Until 1930 the birth rate was decreasing, but it still remained high. The death rate then declined more quickly and a large natural increase was responsible for a rapid overall population increase (Figure 4). The stage in which the UK found itself in 1930 was only reached by the southern European countries in 1970. The main cause of this lag was industrial underdevelopment and the related social framework.

Ireland
In agrarian Ireland, a rapid population increase was brought to a halt in the 1840s by famine, caused by the decimation of the staple food supply as a result of potato blight. This resulted in massive emigration, which in turn slowed down the natural increase in the population, because many of the emigrants were young people. Of those who remained, many married later, and a far greater proportion than was normal for Europe did not marry at all.

The trend in Ireland was similar to that of France but the causes underlying it were totally different.

The Netherlands
In 1930, the natural increase in the Netherlands was twice as high as in the UK (Figure 5). This was the consequence of a relatively high birth rate and a relatively low death rate, as compared with England. The low death rate was brought about by good medical facilities and a high standard of hygiene. The high birth rate was in large part a result of the restraining influence of the church, preventing the rate from falling because of birth control.

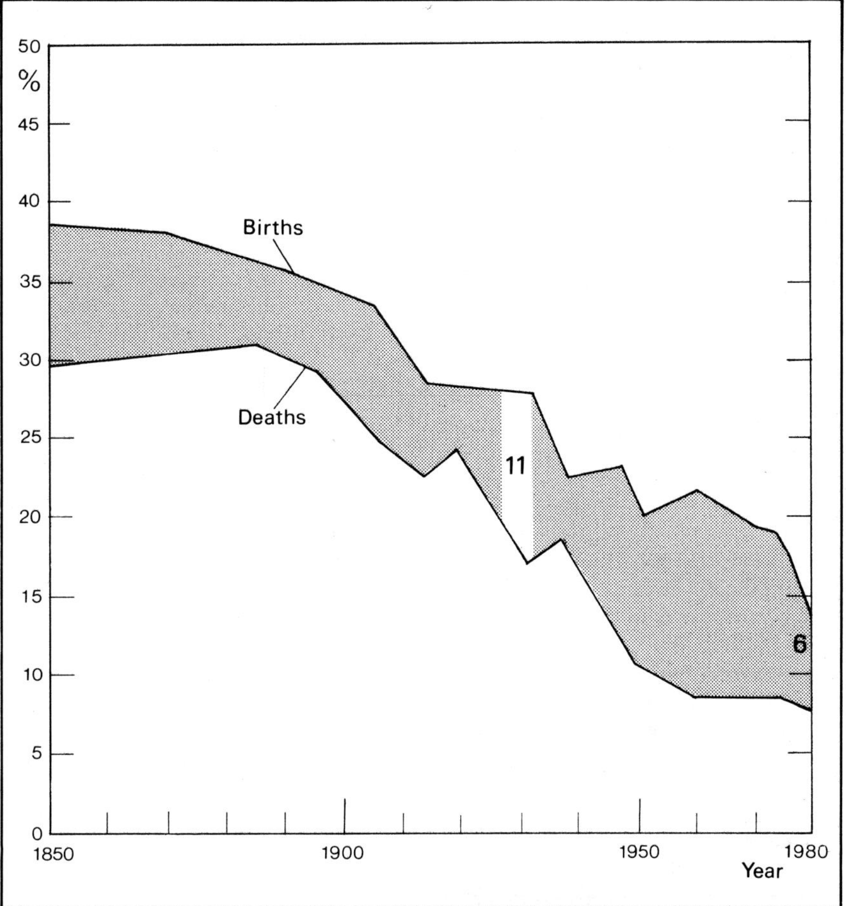

Figure 4 *Natural population increase, Spain*

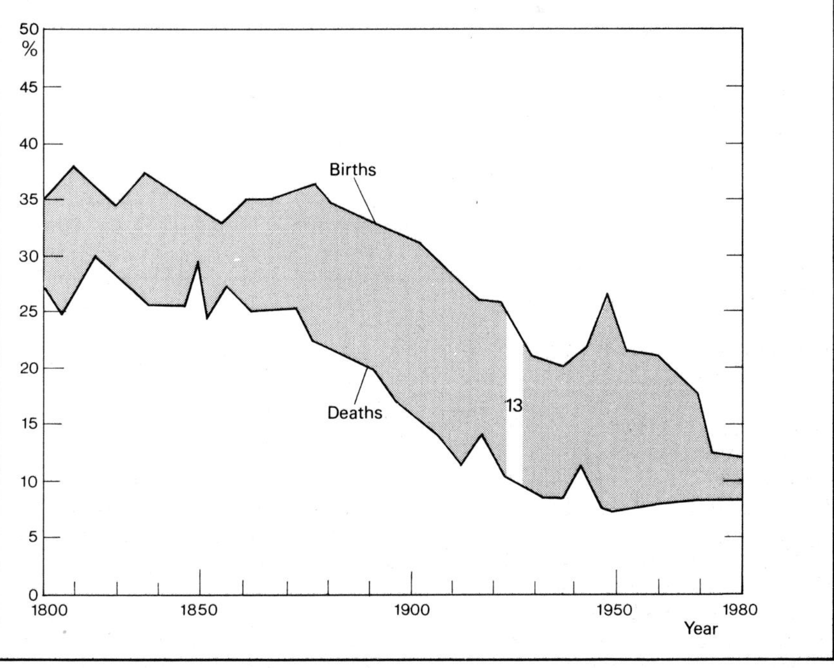

Figure 5 *Natural population increase, the Netherlands*

THE 19TH CENTURY POPULATION 'EXPLOSION'

This is recognisable in Table 1 (*page 9*), but the maximum natural growth rate of +7.2 per 1000 population in 19th century Europe pales into insignificance beside the present growth rates of the world, let alone the individual growth rates of the developing countries. (In Africa, growth is as high as 30 per 1000 population, the birth rate being 50 and the death rate 20.) In 19th century Europe, the excess of births over deaths varied between 15 in England and Wales to less than five in France, where the decline in the death rate was followed immediately by a decline in the birth rate. The overall average excess of births over deaths in Europe was about ten per 1000 population, with a birth rate of around 35 and a death rate of about 25.

Compared with many of the present developing countries, natural population increase in 19th century Europe was actually rather low, due to a relatively low birth rate as well as a relatively high death rate.

Also of importance was the opportunity of emigration for the European population. They could emigrate overseas to practically uninhabited countries and start a new life. An estimated 50 to 60 million Europeans left their continent between 1820 and 1940 (Figure 6).

POST-WAR EUROPE

In the 1950s, 1960s and early 1970s, population growth increased again for a while. Excess of births over deaths increased (Figure 7) and the traditional net emigration fell away. After having been an emigration area for centuries, Western Europe became a modest immigration area (Table 2). The decolonisation process meant that many former emigrants returned home. At the same time, the industrial centres of Europe attracted many European and non-European workers as the extract below highlights.

> The Second World War led to unforeseen new demands. By 1944 Germany had seven million foreign workers. Every fourth German tank, lorry, and field gun was made by a foreigner. Much of this, of course, was forced labour.
>
> But at the end of the war, in 1945, the consensus of expert opinion was that most European countries would not be able to provide enough employment for their own citizens. The big worry was Western Germany which had received eight million refugees from the East. But in fact the reverse happened. The expanding economies of the West European countries soon took up the slack and before long attention was swivelling outwards to foreign sources of labour. In Britain it began with the *Emperor Windrush* in May 1948 — the first ship to dock with a cargo of West Indian immigrants. They were closely followed by Indians and Pakistanis. The numbers of immigrants increased year by year until, in the late fifties and early sixties, Britain's economy went into the permanent doldrums. Ironically, at this time, racism and xenophobia pushed anxious governments into building high walls of protection to limit the immigrants who no longer wanted to come. No one in public life — apart from George Brown in one prophetic outburst — asked the question: how will Britain get future supplies of raw labour assuming the British economy one day regains its form and begins to grow again? All in all, in this time of immigration Britain absorbed 800,000 workers and their families — small fry compared with what was going on in some other European countries. . . .
>
> The British have the feeling that the great influx of Commonwealth immigrants in the last two decades is a special British phenomenon — a legacy of Empire. But this is only true inasmuch as the British tend to have black workers rather than white ones. But *all* European countries have experienced what the British have experienced — a great influx since the war of large numbers of foreign workers. In Germany they are nearly

Figure 6 *European emigration in the 19th century and first half of the 20th century*

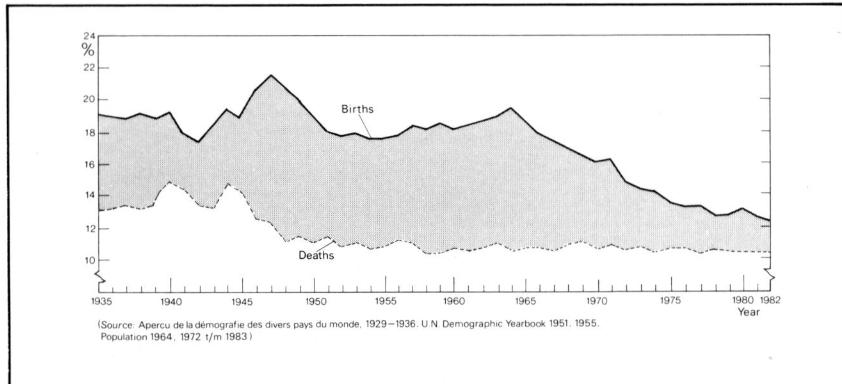

Figure 7 *European birth rates and death rates, 1935–1982*

all white — Greeks, Turks, Yugoslavs, Spaniards, and Italians. In France the principal groups are Portuguese, Spanish and North African. In Belgium they are mainly Italians, Turks, and Greeks. In Holland, more similar to Britain, there are large numbers of Indonesians and people from the Dutch West Indies, although Turks, Greeks and Italians are numerous too.

They are the new proletarians — doing the menial jobs, poorly paid with little job security (unless they are Italians or Greeks, fellow members of the EEC), separated from their families, living in housing that is often appalling.

Willy Brandt succinctly summed up their contribution when he said: 'In every way, foreign workers help us to earn our daily bread . . . although foreign workers are in Germany because at home they live in indigent circumstances, Germany needs them urgently. They're dependent on us. But we are even more dependent on them for otherwise they would not be here.'

President Pompidou put it rather more crudely when he said in September 1963: 'Immigration is a means of creating a certain easing in the labour market, and of resisting social pressure.'

(*Source*: Jonathan Power, 1972, *The New Proletarians*, Community & Race Relations Unit of the British Council of Churches)

THE 1970S AND BEYOND

Since the early 1970s the excess of births over deaths has decreased again and, due to the economic crisis, the economic centres of Europe have lost some of their draw, so that European population growth has declined to an unprecedented level. It is expected that Europe will remain an area of low population growth. The slowing down of population growth in the developing countries will still take a long time. If the present trends continue, the world population could stabilise at approximately 10 billion around the year 2100. The Europeans, while in 1900 still forming 18 per cent of the total world population, will by then represent no more than 5 per cent of the total world population (refer again to Table 1, page 9).

Table 2 *Estimates of net migration in Western Europe, 1950–1976, in 000s*

Region/Country	*50–59*	*60–69*	*70–73*	*74–76*	*77–79*
UK and the Irish Republic					
UK	− 58	− 123	− 167	−157	− 37
Irish Republic	− 397	− 161	− 5	+ 2	+ 21
Total	− 455	− 284	− 172	−155	− 16
Scandinavian countries					
Denmark	− 52	+ 20	+ 22	+ 8	+ 17
Finland	− 73	− 141	− 18	− 15	− 26
Norway	− 14	+ 4	+ 13	+ 15	+ 11
Sweden	+ 93	+ 204	+ 29	+ 46	+ 38
Total	− 46	+ 87	+ 46	+ 54	+ 40
Core area of the continent					
Austria	− 141	+ 38	+ 87	− 12	+ 2
Belgium	+ 59	+ 152	+ 62	+ 58	+ 4
F.R. Germany	+2723	+2057	+1706	−280	+394
France	+1080	+2178	+ 532	+ 50	0
Luxemburg	+ 7	+ 15	+ 14	+ 5	+ 1
Netherlands	− 142	+ 92	+ 98	+126	+ 96
Switzerland	+ 296	+ 334	+ 35	−121	− 24
Total	+3882	+4866	+2534	−174	+473
Mediterranean countries					
Greece	− 196	− 455	− 103	+ 95	+169
Italy	−1166	− 792	+ 217	+251	+ 39
Portugal	− 662	−1290	− 428	+474	− 3
Spain	− 826	− 551	− 243	+143	+ 82
Total	−2850	−3088	− 557	+963	+251
Western Europe, *Total*	+ 531	+1581	+1851	+688	+748

(*Source*: H. Wander (1978) 'The Role of International Migration' *Proceedings of the Third European Population Seminar* and *Recent Demographic Developments*, 1984, Council of Europe)

? Study Table 2.

1 Compare the net migration of the core area with the Mediterranean area. Comment on any differences.

2 Compare the net migration of the core area with the UK and the Irish Republic. Comment upon the deviation of the UK.

3 Compare the net migration of France and Germany with the other core area countries.

4 Net immigration in the core area after 1973 is partly a consequence of women and children joining their husbands and fathers after some years. Compare the Netherlands and Switzerland. Comment upon the difference.

PRESENT DAY POPULATION STRUCTURE

The percentage of elderly, those older than 65, is now much higher than ever before: the European population is an aged population. In France, the percentage of elderly (in this case from 60 years and over) has risen from 7.3 per cent in 1775 to 15.6 per cent in 1980. This percentage is expected to be about 17 per cent in the year 2000 (Table 3). In most European countries, one out of seven inhabitants (in this case of 65 and over) is aged. In the developing countries this figure is only one out of 35. This increase in the percentage of elderly in European society is due, for the largest part, to a decrease in the birth rate and not, as is often thought, to a decrease in the death rate. As the proportion of youngsters declines, the proportion of elderly increases, while the middle group, 20–64 years of age, remains relatively unchanged (Table 3 and Figure 8).

? Using Table 3 and the formula set out below, examine the influence of **a** ageing population on the dependency ratio **b** the index of ageing

$$\text{Dependency ratio} = \frac{(numbers\ 0\text{–}19\ yrs) + (numbers\ 60\ yrs)}{(numbers\ 20\text{–}59\ yrs)} \times 100$$

Table 3 *Division of the population of France into three age categories between 1775 and 2000 (%)*

Age	*Years*					
	1775	*1851*	*1901*	*1959*	*1980*	*2000*[1]
0–19	42.8	38.5	34.3	31.8	30.4	28.1
20–59	49.9	51.3	52.7	51.6	54.0	54.7
60 and over	7.3	10.2	13.0	16.6	15.6	17.2

[1] forecast
(*Source*: R. Pressat (1961) *L'analyse Démographique*, PUF Eurostar, Demographic Statistics, 1981)

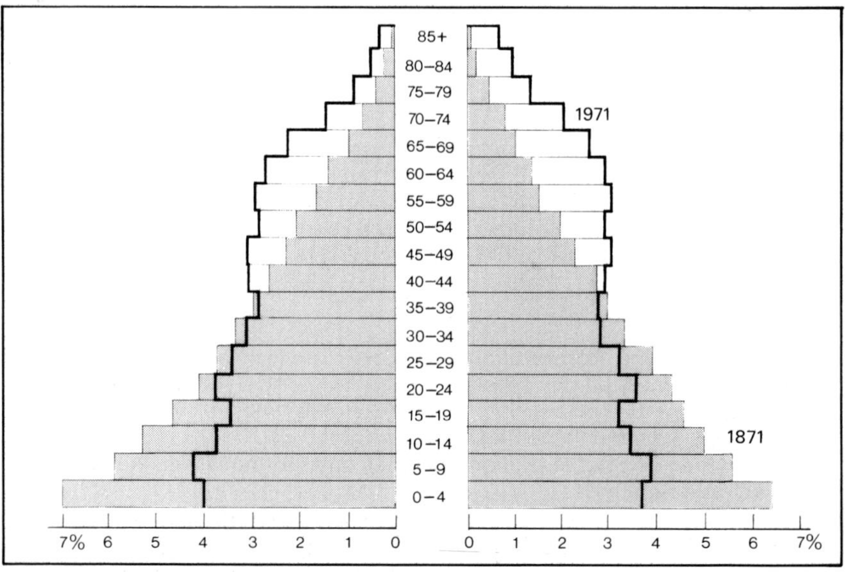

Figure 8 *Population structure, England and Wales, 1871 and 1971*

FUTURE EUROPEAN POPULATION DECREASE?

The most striking development in the years following the beginning of the final stage of the demographic transition (around 1935) and the present day, is the sharp fall of the birth rate since 1964 (Figure 7 *page 12*). The most significant significant causes of this decline have been the introduction of reliable birth control methods and a more tolerant attitude towards birth control.

In 1982, the birth rate was about two per 1000 population higher than the death rate. Will there be a time in which the death rate will be higher than the birth rate, causing population increase to turn into population decrease?

In order to answer this question, we must consider the fertility of the population, or in other words, how many children a woman bears on average during her lifetime. If the population is to remain stationary (at the same level), fertility must be 2.1 (an average of 0.1 child more than 2, as not all women have children). Table 4 shows that fertility in most Western European countries is lower than necessary for the population to remain stationary.

Despite the low fertility, the present birth rate is still higher than the death rate in Europe as a whole. This is due to the fact that many Europeans were born before 1964 (when fertility was much higher than the replacement level), and these people will have their children in the 1980s and 1990s. The European population will decrease only then, when the people born in the low birth rate phases have children. (This does not take into account any high net immigration.) In the population projections for 2100, the decreasing European population has been taken into account (refer to Table 1, *page 9*).

Not all European countries are in the same position when considering future population trends. In the Federal Republic of Germany, the country with the lowest fertility in the world, the birth rate has been lower than the death rate for some years. Should fertility remain as it is, then the population will decrease from today's 62 million to some 45 million in 2030. The percentage of elderly will increase from 15 per cent to 23 per cent.

If, after a period of population decrease, West European countries want to achieve a situation of zero growth or of population increase, fertility will have to increase considerably.

The best alternative, both for Europe and for the rest of the world, would seem to be a situation of zero growth (stationary population). The population explosion started in Europe. Will it also be here that, for the first time ever, a people-controlled population growth will be established, approximating as closely as possible to a stationary population?

?

Look at Table 4.

1 In which countries has fertility declined most and in which countries least between 1964 and 1982?

2 If the total fertility rate declines below 2.1, then in the course of time, the mortality rate will be equal to the birth rate. In which countries is this already or nearly the case?

Table 4 *Birth rates and death rates, birth surplusses and total fertility rates of Western European countries in 1964 and 1982*

Country	*Birth rate*		*Death rate*		*Birth surplus*		*Total fertility rate*	
	1964	*1982*	*1964*	*1982*	*1964*	*1982*	*1964*	*1982*
Norway	17.70	12.50	9.50	10.10	8.20	2.40	2.96	1.71
Sweden	16.00	11.10	10.00	10.90	6.00	0.20	2.47	1.62
Finland	17.60	13.70	9.30	9.00	8.30	4.70	2.40	1.72
Denmark	17.70	10.30	9.90	10.80	7.80	−0.50	2.60	1.60
United Kingdom	18.80	12.90	11.30	11.80	7.50	1.10	2.90	1.84
Ireland	22.30	20.30	11.40	9.40	10.90	10.90	4.08	2.95
Netherlands	20.70	12.00	7.70	8.20	13.00	3.80	3.17	1.49
Belgium	17.20	12.20	11.70	11.40	5.50	0.80	2.71	1.60
Fed. Rep. Germany	18.50	10.10	10.80	11.60	7.70	−1.50	2.55	1.41
France	18.10	14.70	10.70	10.00	7.40	4.70	2.90	1.94
Switzerland	19.20	11.60	9.10	9.00	10.10	2.60	2.68	1.55
Austria	18.50	12.50	12.30	12.10	6.20	0.40	2.77	1.70
Portugal	23.80	15.40	10.60	9.50	13.20	5.90	3.14	2.17
Spain	22.20	13.50	8.70	7.50	13.50	6.00	3.00	1.99
Italy	19.90	11.20	9.60	9.60	10.30	1.60	2.62	1.57
Greece	—	14.00	—	8.80	—	5.20	2.32	2.09

(*Source*: *Population*, various issues)

AN AGEING POPULATION — WHO PAYS?

QUANTITATIVE ASPECTS AND QUALITATIVE ASPECTS

In the near future the elderly will not only form a larger group (Table 5) in the total population but will also be different in a qualitative sense.

It is to be expected that they will show up as a more socio-economic resistant group. Already we can perceive this change. The change is due to better and broader child and adult education, to better public health facilities, to early retirement and to public pensions. All in all the elderly are fast growing into a more independent section of the population. The policy of encouraging the elderly to stay on their own and to take care of themselves as long as possible is widespread. Indeed, most of them have still a household of their own. Others live with their children, and only a small group live in institutions.

Table 5 *The elderly population of England and Wales by broad age groups 1971–2021*

Population at mid-year	*Estimates based on pop. census*		*Mid-1981 based projections*			
	1971	*1981*	*1991*	*2001*	*2011*	*2021*
Total numbers (millions)						
60/65 plus[1]	8.3	9.0	9.4	9.2	9.8	10.8
75 plus	2.3	2.9	3.5	3.8	3.7	4.1
85 plus	0.4	0.5	0.8	1.0	1.1	1.1

[1] 60/65 plus means women aged 60 and over and men aged 65 and over.
(*Source*: J. Craig (1983) 'The growth of the elderly population' *Population Trends* no. 32 p. 30.)

?

1 What is a senior citizen's pass? What do you think is the social and human impact of such a pass on the lives of the elderly? Mention at least three aspects.

2 However active and lively the elderly may be, they do get older and eventually most of them will need facilities particular to their age. What type of services are needed? Give examples. Who is going to provide and to maintain the facilities? Who is going to pay for these services? Read the article 'Healthy and wise' which will give you a particularly provocative perspective on this whole issue.

? **3** The money for state old-age pensions comes from the tax-payer. As fertility decreases, so in the medium- and long-term there will be fewer tax-payers, whereas old-age pensioners will grow in proportion.
Look at Figure 9. Past years of high and low birth rates can be traced in the number of people reaching pension age.

a Which periods of the 20th century would appear to have been peak birth rate years?

b Which periods would seem to have been low birth years?

c When will the cost of pensions for the elderly become a problem for the government? Explain your answer, using Table 5 as well as Figure 9, and also the article 'The dispensable pensioners'.

d What do you think will be the solution to the problem?

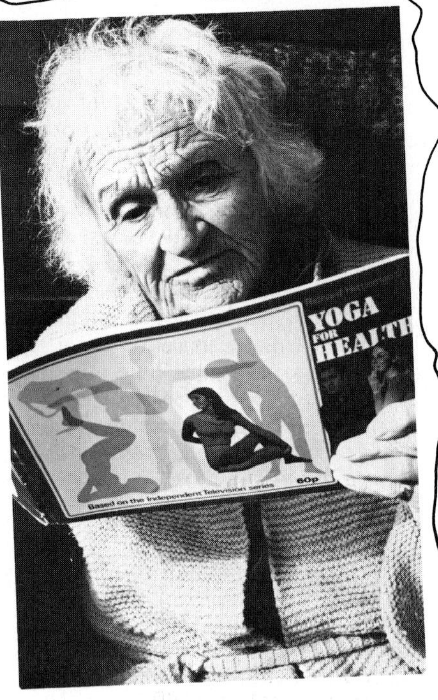

Healthy and wise

The ILO predicts that only 25 per cent of retired men and six per cent of women will be receiving a pension by the year 2000. The rest will have to work as long as they are able and then turn to their children and other relatives for help. Dependent on the state, dependent on the family. It is a miserable prospect — for the up-and-coming old as well as for the younger generations who will have to shoulder their growing greying burden. And, apart from wholesale euthanasia, there doesn't seem to be a solution.

But there is another way of looking at it. The equation: development means lower mortality, means lower fertility, means more old people, means more dependent people means poverty for all — is not entirely logical. It entails the assumption that old people are 'naturally' dependent on the rest of the population as soon as they open their eyes on their 60th birthdays. If that assumption is correct then ageing is indeed something for everyone to fear.

Looking at the reasons why old people become dependent made me realise that there was nothing 'natural' about that state at all. Because, in prolonging life we have prolonged not disabling old age, but vigorous middle-age.

To my astonishment (and relief) I discovered that at least 75 per cent of old people in the developed world are virtually free of disabling disease and that less than five per cent suffer the brain atrophy that leads to senility. Experts on geriatrics point out that ageing is not a disease. If you are decrepit when you grow old it's because you've led an unhealthy life. While this means that the statistics are likely to be much less cheerful in the poor world, where many old people are the survivors of a lifetime of hardship and a host of life-sapping diseases, it does mean that I — and you too — have every chance of becoming spry sexagenarians — nearly as healthy as we ever were and just as capable of learning new skills and holding down a job.

And that means that we will probably be perfectly capable of looking after ourselves for the majority of our old age — if we are allowed to do so. But unfortunately, unless there are some radical changes in current attitudes towards the old, you and I will be forced into dependence whether we like it or not. As one old lady wrote recently to a Sunday newspaper:

> 'We must prepare to do battle to maintain our independence. I am haunted by the fear that unless I can dispel the assumption that I am a senior citizen, the following events may reasonably occur:
> - I shall have a gang of young thugs sent to my home to paint my kitchen instead of going to prison;
> - I shall have patients from the local mental hospital drafted to dig my garden;
> - I may be forced to go to 'suitable' entertainments, drink tea and wear a paper hat;
> - I may receive vast boxes of assorted food to which I feel I am not entitled.
>
> We pensioners are in a terrifying position. We are *recipients*. Hands off please, *I* am in charge of my life.'

That old woman is angry — as are several other old people quoted in this issue of *New Internationalist*. And it is easy to see why. They are the same energetic, talented people as they always were. But suddenly society informs them they have grown old and slams shut its door in their faces.

We don't wear out with use. Given the right opportunities, we can improve with age. At 90, Picasso declared, 'We don't get older, we get riper.' Ripe, not rotten.

The old — like any other disadvantaged minority — are not disabled by old-ness. Prejudice and discrimination are the main reasons for their disability.

In Britain it was decided, in the interests of economy, that an adult education class must have enrolled at least 12 students to qualify for local government subsidy. It was also decided that a student of pensionable age would only count as *half a person*. So a class of 23 pensioners would fail to get off the ground!

Job-sharing, flexible retirement and continuing education are only dreams in most countries. But it is possible to conceive of a system where the boundaries between education and work are blurred, where people can learn a new career three or four times during their lives and become what Alex Comfort calls 'lifelong pilots' — instead of 'work-orientated kamikazes, one-way projectiles, designed to explode at the end of their trajectory'. There are few signs of this dream becoming a reality however. Some old people are angry. But, considering that they are one-third of the electorate in some countries, they are keeping remarkably quiet.

Because of their lack of power, the old have to appeal to society's pity rather than demand what is theirs as of right. And, so unpleasant are the stereotyped images of ageing, that few old people want to identify themselves as members of an 'old' age group.

(*Source*: *New Internationalist*, June 1982)

The dispensable pensioners

In western Europe, more and more people are living to an increasingly ripe old age. That means not just that an increasingly large proportion of each country's population consists of elderly people, though that is true, but it also means that a growing proportion of the elderly are aged over 80 or even 90.

As a result, welfare services are under increasing strain. The amount of money that has to be spent on pensions grows in direct proportion to the total number of elderly people. But the amount of health care, and personal and community services increases not simply with numbers but also with age. It is therefore becoming more and more difficult to keep elderly people in the style to which they are accustomed, let alone in the style to which our growing concern for old people would suggest that they ought to become accustomed.

Sweden, for example, is often held up as the most advanced welfare state in the world. And the Swedish government has summarised its policy towards old people as follows: 'The basic aim of old-age care is to provide elderly people with a secure economic platform, good housing, an opportunity to obtain services and special care, a sense of community with others and meaningful activities.'

Though the ideal is a fine one, Sweden may not be able to go on affording it in years to come. In 1971, 14.1 per cent of all Swedes were in the 65 plus age group. By 1975, that proportion had grown to 15.1 per cent; it has continued growing ever since, and it is expected to go on growing — 16.6 per cent in 1985, 16.8 per cent in 1990. Furthermore, amongst the elderly, those aged 80 and over make up a growing proportion.

The pattern is similar all over Europe, and it is an expensive one. Take the example of Britain, where the proportion of the population in the age group 75–84 increased from the 1973 figure of 2.1 per cent to 2.7 per cent in 1980. In 1977, each Briton aged 65–74 cost the state, in health and community care, on average £160; but those over 75 cost more than twice as much — £365. The jump in the costs of personal social services was over twice as much again proportionately: £30 a head for those aged 65–74; £130 a head for those over 75.

In terms of cash, though, what counts most is pensions. In Britain, most social security payments consist of old age pensions. In 1973, social security payments accounted for £11.5 billion, or just under one-fifth of total government spending. By 1982, that figure rose by £5 billion to £16.5 billion, just over one-quarter of total government spending. Or, to look at it another way, more than twice as much as the government spent on defence and nearly twice as much as it spent on education.

The reasons for this dramatic growth are not just to do with the numbers of old people or how long they are living. When the economy is doing badly as it has been for most of this decade, it is easier to cut back on goods and services, such as health and education, than it is on cash benefits like pensions. Moreover, it has always been more politically popular to increase old age benefits, where the extra money can be seen by everybody, than to expand health care for the aged, where it will be noticed by only a few, even though it may well do more good.

Whatever form it takes, however, the extra money spent on the elderly still has to come from somewhere. And it can only come, in one form or another, from working people — that is to say, those aged between 16 and 65. With high levels of unemployment built in to most Western economies, and with the numbers of the elderly growing at a faster rate than the numbers of those still working, the long range forecast is very bleak indeed.

All in all, paradoxical though it may seem, the best time to be old is probably about now.

(*Source*: *New Internationalist*, 1979)

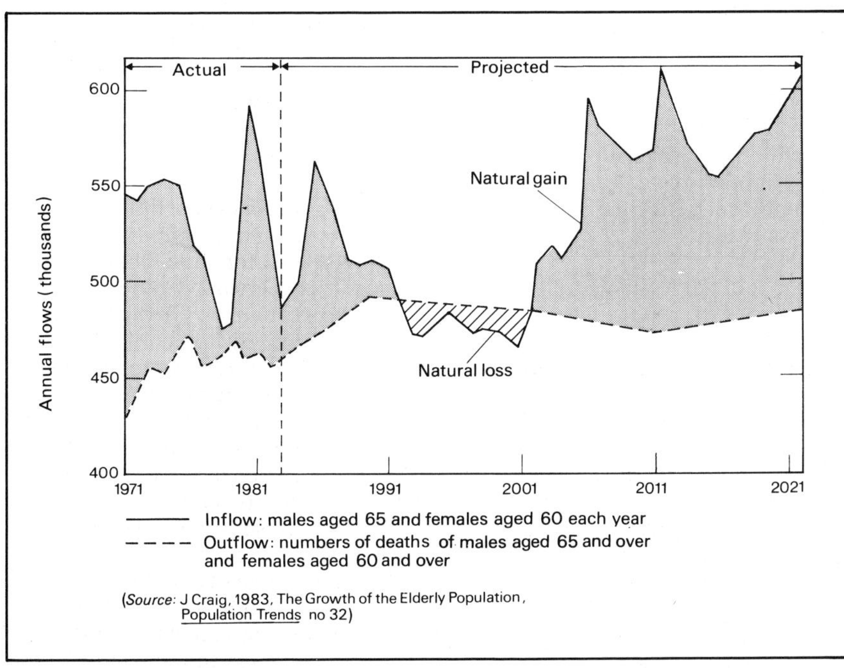

Figure 9 *Annual inflow to, and outflow from, the pensionable ages, 1971–2021, England and Wales*

Table 6 *Number of students admitted to teacher training in the UK*

Year	*1972*	*1975*	*1978*
1 year PGCE* Course	10 365	10 054	9 144
3/4 Year Courses	37 381	27 710	9 887
Total	50 632	40 700	21 672

*PGCE: Post-Graduate Certificate in Education

Table 7 *Number of geography graduates admitted to PGCE geography courses*

Year	*1972–73*	*1980–81*	*1981–82*	*1982–83*	*1983–84*
From University	717	382	351	313	313
From Public Sector	—	212	217	134	65
Change from previous year(s) %		17	4	21	15

(*Source*: (for this table and Table 6): R. E. Hebden (1983) 'Teacher Training and the Future of Geography', *Area*, Volume 15, no. 4)

THE IMPACT OF DEMOGRAPHIC CHANGE ON EDUCATION

In modern western society a number of fundamental changes have taken place at the same time. We are having to deal with socio-economic changes, with technological changes and with changes in population structure. But such changes have taken place all through history. In order to survive, people have to modify their ways.

The changes affecting modern society have important implications for education, teacher training and geography as a subject. The overall number of teachers admitted to initial teacher training in England and Wales very much decreased in the 1970s as is clear from Table 6.

This decline is also evident in the study of geography. Fewer geography students are admitted to PGCE geography courses today as can be seen in Table 7.

Geography has an important role to play in understanding issues in society and there are important new functions which it can serve. Since Western society is increasingly moving towards a reliance on information as a resource (see *pages 283–285*), and since people are gaining more leisure time, geographical education could find an important outlet in adult forms of education.

?

1 Explain the decrease in the admissions to teacher training colleges. (Take Figure 7 into consideration.)

2 Think of the subject of geography as a whole and/or geography as an element of social studies. Discuss ways in which geographical study provides insights into both the immediate and more distant environments — environments which are constantly changing and which in turn are changing people's relationships to them. Give examples of geography's contribution to a better understanding of developments in society.

3 In groups, discuss why people in different sections of society require geographical information. Think about environmental and planning issues, decision making, and leisure activities.

4 For teachers, including geography (and social science) teachers, a whole new field of activities is opening up in adult education in an increasingly information-based society. Why is this so? Give examples of geographical education and geographical information for which adults might be looking.

SPATIAL IMPLICATIONS OF FOREIGN WORKERS IN BRUSSELS

A DIVIDED BRUSSELS?

All big Western European cities lose inhabitants to suburbs, new towns, extended towns or simply to the countryside. Regular contact with the city centre is maintained due to good and new communication facilities, and commuting for work or shopping is for many a daily activity.

As a consequence of various social, economic and political developments many non-Western European workers have settled in the big cities. The combination of the post-war economic boom, the tight labour market in the core area of Western Europe, and the abundance of labour in Mediterranean Europe and in developing countries brought an influx of such workers in the 60s and early 70s when immigration was still fairly easy. They have found places to live in the 19th century housing quarters. In doing so they have changed the structure of the city population and since non-Western Europeans have a different cultural living pattern, they have changed the appearance of those 19th century quarters into which they have moved. We shall now examine the impact of this trend on Brussels.

BRUSSELS

Brussels originated at the intersection of the Zenne river, which runs north–south, and the overland route from Ghent to Cologne, running east–west. On the high, eastern bank of the Zenne river are the centres for administration and commerce, the ecclesiastical institutions (cathedral etc.) and the living quarters of the bourgeois and the nobility. The low, western bank of the river was reserved for industry and associated working class districts.

Brussels, situated right in the heart of Belgium, houses many of the central functions for the whole country. It is also the seat for important international institutions, such as the EEC and NATO. The core of the city is called 'The Pentagoon' because of its shape, and is the area of historical interest. The Brussels agglomeration consists of 25 municipalities. Of the Brussels population 24 per cent are foreigners — out of a total number of 1 million in 1981, 240 000 were foreign. Two types of foreigners can be distinguished: migrant workers and staff of international institutions.

We shall now focus on two municipalities with a high percentage of foreigners:

St Gilles: a municipality of the Brussels agglomeration on the high bank of the Zenne river, situated at the south point of The Pentagoon. The busy south railway station is located in the north of St Gilles. Total population (1981): 46 000 of which 46 per cent, or 21 000 are foreigners (mainly Spaniards).

Tervuren: a south-eastern municipality just outside the Brussels agglomeration, situated in a woody landscape.
Total population (1981): 19 000 of which 18 per cent, or 3 600 are foreigners (mainly English, Dutch, German, French and American).

?

1 Locate on the map in Figure 11 St Gilles and Tervuren and comment on the settlement in both districts: St Gilles for migrant workers and Tervuren for staff members of international institutes.
2 Study the population diagrams of St Gilles and Tervuren (Figures 12 and 13 on next page).
a Give a precise description (at least as precise as possible) of the composition of both population groups in the two areas according to age, sex and marital status.
b Comment on the social consequences of the present structure for both areas, both for the present and for the near future.

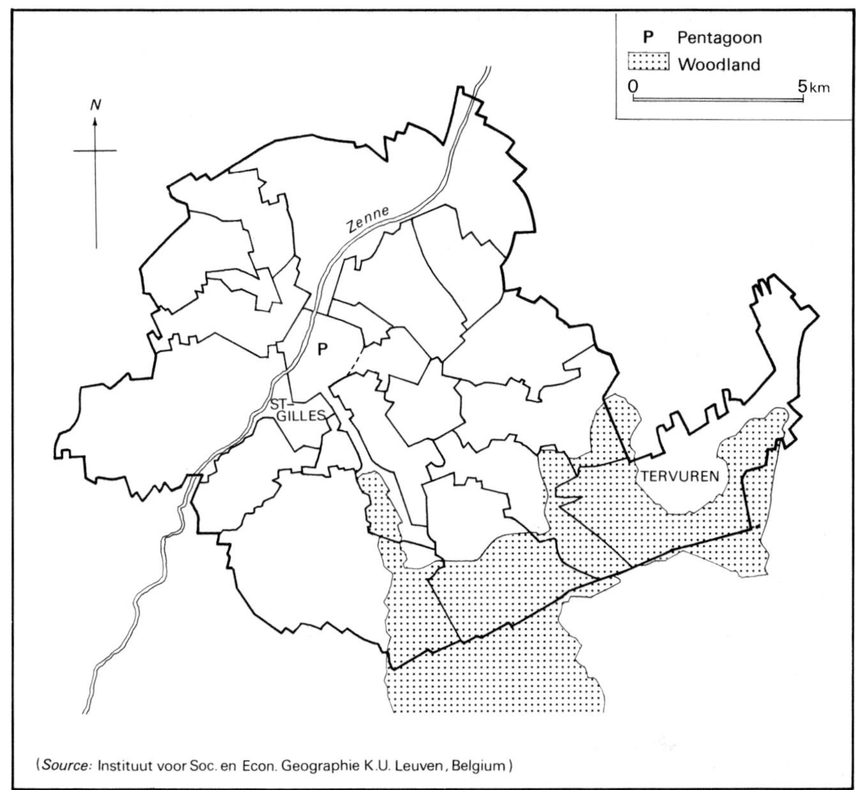

Figure 10 *The Brussels agglomeration — the location of St Gilles and Tervuren*

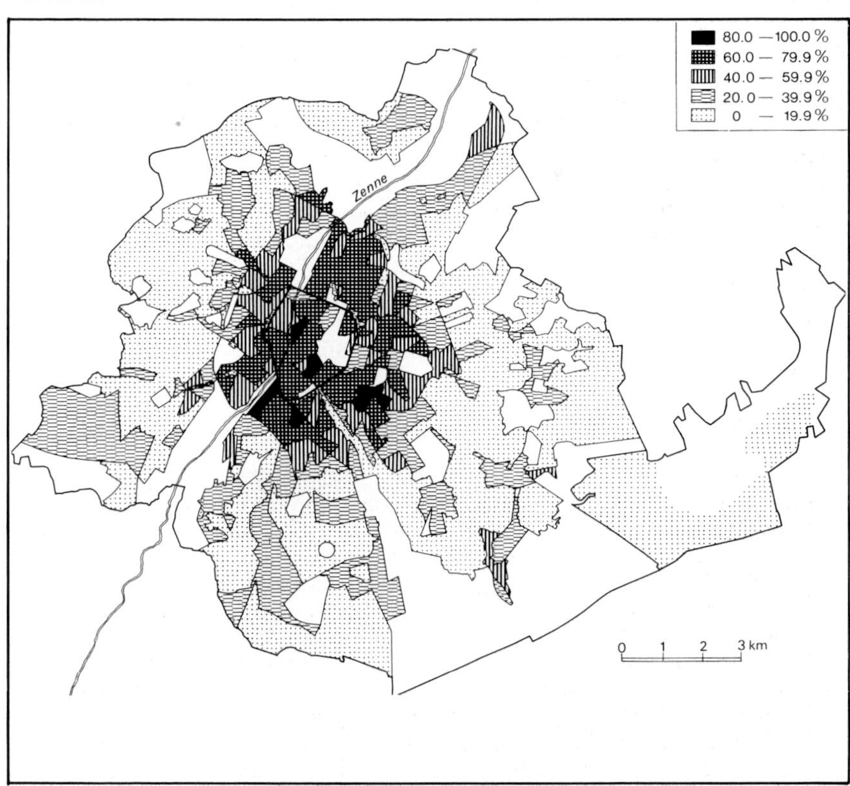

Figure 11 *Proportion of residences within different districts of Brussels built before 1919*

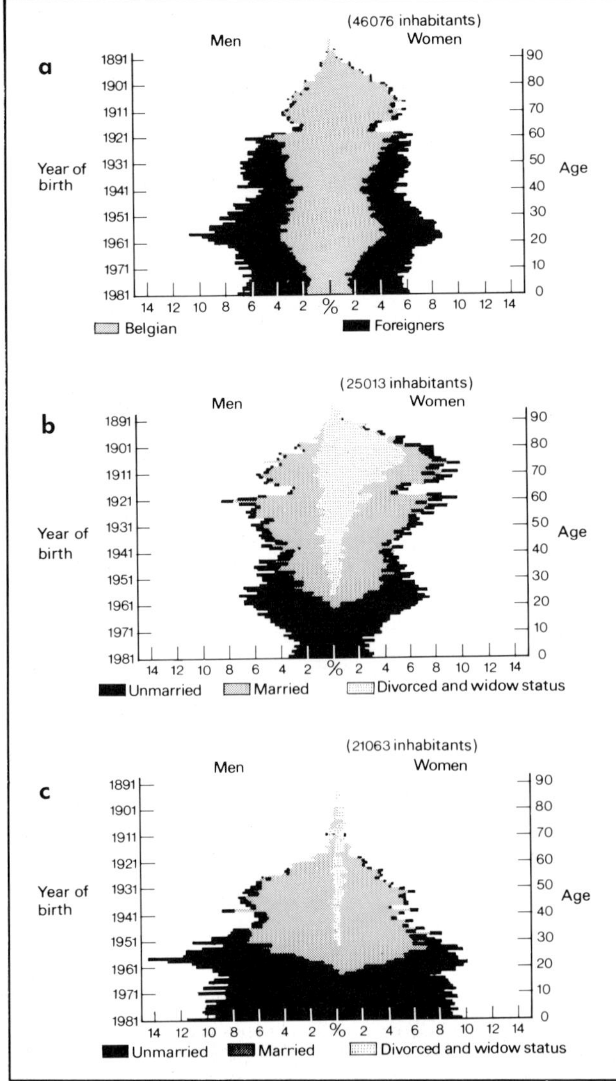

Figure 12 *The population of St Gilles, Brussels*
a *Total population of St Gilles by age and nationality, 1981*
b *Belgian population of St Gilles by age, sex and marital status, 1981*
c *Foreigners in St Gilles by age, sex and marital status, 1981*

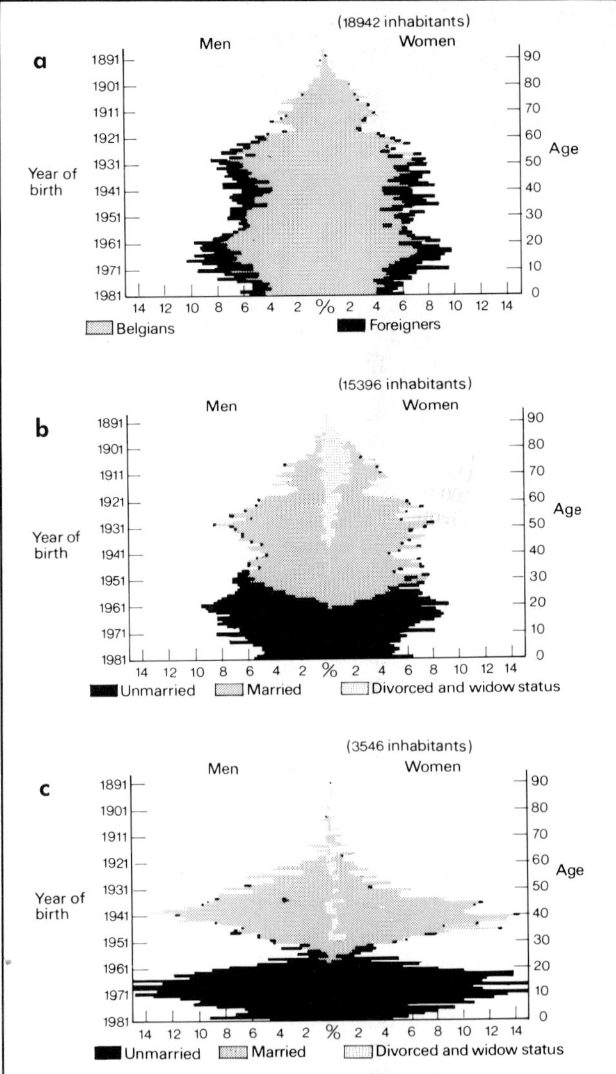

Figure 13 *The population of Tervuren, Brussels*
a *Total population of Tervuren by age and nationality, 1981*
b *Belgian population of Tervuren by age, sex and marital status, 1981*
c *Foreigners in Tervuren by age, sex and marital status, 1981*

Print 1 *Street-corner scene in St Gilles, Brussels*

Print 2 *Tervuren, a well-to-do suburb of Brussels*

THE FUTURE OF THE GERMAN PEOPLE AT STAKE?

Germany given grim warning on birthrate

From Michael Binyon, Bonn

Unless German women produce 200 000 more babies every year, the German population will fall from 56.9m at present to only 38.3m in less than 50 years, with drastic consequences for defence, education and the labour market.

This warning was given yesterday by Herr Horst Waffenschmidt, Parliamentary Secretary in the Ministry of the Interior, after the Cabinet had discussed the second part of a working party report on Germany's falling population.

For the past nine years the Federal Republic has had the lowest birth rate in the world, and the population has been falling steadily since 1972. The birth rate is now one-third below replacement rate, and if this trend continues the present population of ethnic Germans — not including foreign immigrants — will fall to 51 100 000 by the end of the century.

The Government has long been worried by this, and there have been regular warnings and calls for more encouragement for couples to have children. A special inter-ministerial working group was set up to study the problem, and Herr Waffenschmidt said yesterday that its report gave rise to 'serious concern'.

He called on his country to become again a society that welcomed children.

The working party's report points out the burdens the growing proportion of old people will place on the social services. It also notes, ominously for future race relations, that the integration of foreigners living in West Germany — who now number 4 530 000 and have a high birth rate — will become more difficult as their share of the population increases. Indeed, it is only because of their numbers that there has been only a small fall in the total number of people living in the country — now 61.5m — in the past 10 years.

The report outlines the way the balance between young and old people will alter. At present, the proportion of those under 18 is 22.4 per cent, which will fall to 18.9 per cent in the year 2000 and 15.3 per cent in 2030. Those over 63 will rise from 15.1 per cent at present to 15.2 in 2000 and 23.8 in 2030.

Society already had to reckon with the following changes, the report adds: Marriages will be fewer, and some ten per cent of the population will remain single; divorce will be more common, with 25 per cent of all those married after 1970 getting divorced; families will be smaller, with 20 per cent of all couples remaining childless, 19 per cent of all children being only children, 45 per cent having only one brother or sister and 36 per cent growing up in families with three or more children.

The consequences of the falling population will be especially severe in schools. Where the number of pupils will fall by 25 per cent by 2000. Teachers will find it increasingly hard to get jobs, and until 1990 some 150 000 teaching college graduates will be unemployed.

Social services, the police, fire brigades and the medical services will all be short of labour. The situation will be particularly acute in the Army, which needs 225 000 men each year. There will be a shortage from next year onwards so that by the end of the century the army will be severely below strength.

(*Source*: *The Times*, 15 December 1983)

?

1 'The future of the German people is at stake' and 'The population pyramid upside down' are the headlines that emerged from a recently held press meeting in Bonn, following the publication of a report of an inter-governmental working group discussing the newest prognosis on the development of the German population (Germany in this case means the Federal Republic of Germany). Interpret these headlines, in the light of the information given in the section 'European Population Trends' (*page 8*).

2 The inter-governmental working group, in studying and discussing the population decrease, drew attention to four explicit consequences. These consequences are outlined below. Think them over and discuss them, either in groups, or as a class.

- By 1995 there will be a shortage of recruits for military service. The recruitment of women, especially for administrative staffing, might be a solution.
- To continue the present financial level of pension, the premiums to be paid will have to be doubled in the next 30 years (from 18.5 per cent to 35 per cent). (See also the section on the elderly, *page 15.*)
- In the next few years about 150 000 teachers will no longer be needed. (Think also about new functions for education.)
- The proportion of foreigners will rise. In the year 2030 their number is estimated at 7 million. The population of the Federal Republic of Germany by 2030 is estimated to be 45 million of which 38 million will be Germans and 7 million will be foreigners. The working group is afraid of a sub-proletariat being generated especially in big cities. (See Table 2 *page 13* for migration trends during the Seventies, and look again at the section about foreign workers in Brussels, *page 18.*)

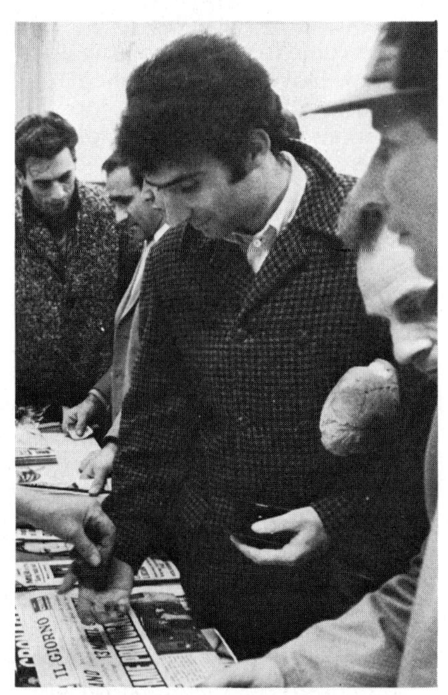

Print 3 *A special Italian newspaper stand outside Volkswagen's factory in Wolfsburg, West Germany*

A MORAL DILEMMA

As we have seen (*page 12*), after the Second World War, Europe became a net *im*migration area instead of a net *e*migration area. The main causes for the influx of foreigners are:

- People from former colonies, so used to the Western socio-economic pattern of living, migrated to their corresponding European country (the UK, Belgium, the Netherlands, France are all good examples).
- 'Guest-workers' from the Mediterranean regions and the West Indies were encouraged by West European countries in the days of economic prosperity and labour shortage.
- Political refugees gained entry to the free 'West' for reasons of political, cultural and religious oppression.

Since the early Seventies, Western Europe has had to cope with very fundamental socio-economic changes. Governments have financial problems, there is less employment, and the tension between western Europeans and non-western immigrants is growing. The mass media (newspapers, radio and television) pay considerable attention to this phenomenon. Although the tension is understandable, it is very undesirable from a humanistic point of view. Governments, political parties, welfare institutions, together with many ordinary people, take notice of the problem and try to find a way out. Can you suggest any solutions?

? THE QUESTION OF GUEST WORKERS

Read the extracts below and then answer the following questions, making use of the extracts and other information you have gained from reading this chapter.

1 Explain the term 'guest-workers'. *What* countries welcomed guest-workers, *why* and *when*?
2 Why has the need for guest-workers declined?
3 What do you think of the research done by the city of Düsseldorf on the effects of a hypothetical exodus of the guest-workers. Are you in favour of such studies, or are they useless in relation to the problem? Explain your answer.
4 Read the following possible solutions to the problem of the surplus of guest-workers. Comment on each of the solutions, state your preference, and justify your choice.

- Order the guest-workers to go home or to return to their country of origin. (Within the EEC workers have complete freedom of movement. Note that Spain and Portugal have now been accepted as members of the EEC and that Turkey will probably join in the next few years.)
- Encourage guest-workers to go home immediately with no right of return by giving them a 'package of financial sweeteners'. (This has happened in Germany and in the Netherlands.)
- Ask the guest-workers to return, and transfer to them each month, as long as necessary, the unemployment benefit and the children's allowance. (Living in Turkey, for example, is less expensive and maintaining facilities in Western countries is costly. A suggestion made by the Dutch.)
- The government, the multinationals and the big industries of the concerned Western countries should set up and finance agrarian and industrial development projects in the guest-workers' countries of origin. (The Dutch have already promoted such a project in the south of Spain and returned guest-workers find jobs there.)
- Seek to integrate the guest-workers into Western society. This, however, poses difficulties in most cases, especially for the first and second generations. For certain nationalities, having to adapt to a totally different culture and religion presents major problems. Comment on these difficulties.

Immigrants

FROM OUR PARIS CORRESPONDENT

Long, hot summers are traditional enemies of good race relations. France is no exception. The past month has seen a series of shooting attacks on immigrants and a tear-gas assault on a block of flats occupied mainly by Turks in the normally peaceful Auvergne town of Bourganeuf. The incident which caused most anger was the shooting, on July 9th, of a 10-year-old Algerian boy, Toufik Ouannès, as he let off fireworks to celebrate the end of Ramadan in a Paris suburb.

The fact that the victims were from Africa and the Caribbean has revived a debate about how France can best cope with its 4.3m immigrants who do not qualify for French nationality — particularly the 1.4m Arabs and the 106 000 from black Africa. The less visible though larger immigrant populations from Portugal, Italy and Spain are rarely mentioned, and understandably keep their heads down.

Extreme right-wingers, who did unexpectedly well in municipal elections in March, say that if France booted out its 2m immigrant workers unemployment would disappear, ignoring the unwillingness of most Frenchmen to take the dirty, poorly-paid jobs immigrants perform. At the other extreme, some of the more emotional mourners at Toufik Ouannès's funeral vowed to turn the housing estate where he died into an immigrant fortress which no policeman would dare enter.

The Socialist government is well aware of the potential explosiveness of the immigrant issue, and is somewhat embarrassed by it. 'Left-wingers would like to see France as an open-hearted country ready to receive the poor and downtrodden. The realities of office have forced them to introduce measures to cut immigration to a trickle and to crack down on illegal immigrants who failed to register during a grace period last year.

(*Source: The Economist*, 6 August 1983)

West Germany tries repatriation

A month ago the government offered some of its jobless foreigners a hearty but not very generous handshake to encourage them to go home. It unveiled a scheme for 'repatriation assistance', which will be available to non-EEC citizens who are made redundant after October 1st this year or who have been working on short time for at least six months. Each worker from this select group who decides to leave West Germany for good will receive DM 10 500 (£2 625) plus DM1 500 (£375) for each homeward-bound child. With such modest incentives, nobody is going to rush back to Anatolia. Besides, the number of those entitled to the money will be tiny — an estimated 7 000 this year, and 12 000 in 1984.

(*Source*: *The Economist*, 6 August 1983)

Too many little Ahmeds

Inevitably some foreigners are more equal than others. Germany's 600 000 Italians and 300 000 Greeks — and perhaps soon the Spaniards and the Portuguese, who between them number 250 000 odd — can come and go as they please, courtesy of the EEC. The recruitment ban was really directed against the Turks, who make up much the largest single group of foreign workers — about 1.6m, treble their 1970 number. They are also much the most obviously foreign. German housewives are apt to notice that Turks eat lamb and garlic where Germans eat pork and cabbage, and that their Turkish neighbours have, by German standards, an inordinate number of children (30 births per 1 000 population, compared with 10) who are swamping German schools. But under German tutelage, and encouraged by strict regulations on living space for residence and working-permit purposes, the Turkish guest workers too are now beginning to appreciate the virtues of birth control. 'Allah cross, but baby away', one young Turkish mother-not-to-be told the pregnancy advisory service. If it is either that, or back to Turkey, many are choosing to face up to Allah's wrath.

What would happen if lots of them did pack their bags? The Social Democrat-governed city of Düsseldorf put its money where its conscience was and commissioned a study to discover the effects of a hypothetical exodus, over two or three years, of three-quarters of its 36 500 foreign workers (out of a total population of 600 000). The study showed that public commuter transport and rubbish-disposal services would probably collapse; that the city would suffer a massive loss in tax revenue and purchasing power; that construction and hospital services would be severely cut back; that many schools and nurseries would have to close, making a good number of (mostly German) teachers redundant; and that whole residential quarters of the city would become derelict; to say nothing of all the pizza parlours and kebab houses which would close their doors.

(*Source: The Economist*, 4 February 1984)

A whiff of xenophobia

Die ich rief, die Geister,
Werd ich nun nicht los.
(The spirits I called
Refuse to depart.)

Like the sorcerer's apprentice, West Germany now complains that it cannot get rid of the army of foreign workers it called in during the labour-starved 1960s. Not that it has not tried: it imposed a ban on recruiting workers from outside the EEC as long ago as 1973, during the first oil crisis. Yet over the past 10 years the number of foreigners living in Germany has risen from under 4m to nearly 4.7m, or 7.6% of the population. All that the recruitment ban achieved was that the once-itinerant guest workers settled down in Germany, got their wives and children to join them and produced yet more children. The number of foreign workers has actually declined, from nearly 2.6m a decade ago to 1.9m now. Times are tough: the unemployment rate among them is over half as high again as the German average.

When the Germans first started wooing large numbers of foreigners to come and work in their country in the 1960s, they — as well as the supplier countries — expected the immigrants to stay around for just a few years, accumulate a nest egg by working even harder than the average German, and then go home. It did not work out that way. At the last count over two-thirds of resident foreigners had been living in Germany for over six years, many of them much longer. Most of them had no intention of returning home in the foreseeable future, if ever. Yet only a tiny proportion of Germany's foreigners — about three in a thousand — have taken German nationality. The vast majority are happy to remain Turkish, Greek, Jugoslav or whatever, and visit the folks back home as often as they can — but return to Germany after every trip home.

While the German economic miracle was in full swing, the *Gastarbeiter* (guest-workers) were regarded as a necessary, if not always desirable, adjunct to economic success, without whom many of the less attractive jobs — in hotels and catering, car production and mining among others — would have been left undone. But with unemployment now over the 2m mark, many Germans have become openly hostile to the foreigners. Some 80% of the natives, according to polls, reckon that the number of foreign residents in Germany is too high.

(*Source: The Economist*, 4 February 1984)

2 LATIN AMERICAN MIGRANTS: A TALE OF THREE CITIES

THE MIGRANT

?

1 Write your own hypothetical account of a migrant moving to a Latin American city. Fill out the portrait as fully as possible. Who are the 'typical' migrants? Where do they come from, how do they travel? On arrival where do they live; what sort of work might they get; what are their housing conditions like? Read and discuss some of the portraits in class. Consider carefully what sources of information have informed your ideas.

2 Now read the first few pages of this chapter. How far does this description conform to your idea of the process of migration? Is it appropriate only to Brazil or also to other Latin American countries?

Her name was Maria. A woman waiting in the long line at the water tap. She was short, and dressed in old clothing which failed to protect her from the damp winter air. She was obviously quite young, but her face, with its deep set eyes and lined cheeks, told of experience of ages. How had she come to be here, in the appalling conditions of this São Paulo *favela* (shanty town). She did not want to talk of the past, only of new hopes for the future; but after some time her story came out, in staccato sentences as though each word was painful to utter:

'It was all right in Pernambuco last year. The rains came in time, and the crops grew. There was even some cotton to sell and we could buy a pair of shoes for Romero — his first. The stores of maize and rice lasted just long enough; until the first crop was due this year.

But the rains failed again. It happens so often in the north-east, and it seems as though everything was against us. This was the third, or maybe the fourth, very bad drought I remember. And it was just like the ones before.

Just like before, someone said, 'There's food at the big farm,' and the men marched off in their bedraggled squad to try to force the rich man to give up some of the stores. It didn't work, it never does. If you're rich you can afford guards and dogs to punish the hungry ones who beat at your doors.

Sometimes the men came back with enough scraps to last for a few days. Then we would be back at the beginning, and no hope left. It was a hard time, last summer, especially for the children. My new baby, Leila, she got sick because I couldn't feed her. There was nothing we could do. Last time we went to the feeding centre in the city. But it's a long way, and the food is not enough — too many people go there.

There wasn't much to pack. Our clothes fitted into two bundles, the cooking things made a third, and the old stool my father gave me made it up.

And so we left our home. It wasn't easy. That board shack was not much, but better than we have now. And the land was good to work — it was a clean land. We didn't have much, but we danced and sang with joy on the Saints' Days. Maybe soon we'll go back.

It took many days, our travelling. I don't know how many. I only know that the road is long and hard. The dust of the trucks stings your eyes until the tears run, and the load on your head becomes heavier as your stomach becomes emptier.

We were stopped once and asked for our papers. But we had none, so they said we would have to turn back. My husband said, 'OK', but as soon as their backs were turned we ran and hid for a day and a night. There were so many on the road that the officials were soon busy with others.

On we went. At one town we were taken to a church and given food and a bed. They said we could stay a little, and they would help us to settle down. But my Manuel is proud. He did not want their help then. Sometimes now he says we should have stayed, but then he said that we were going to São Paulo where he would find a good job. The kind man at the church shook his head wonderingly, but he didn't stop us.

We tramped on through the hills. One day, on the road, we passed a family we knew from home. They had a very sick child, and were resting.

And then, at last, we reached the big city. Oh, it was so beautiful that our hearts sang! Everyone had lovely clothes, there was much food in the shops, piles of rice and bread. All wonderful things.

Manuel, he said, 'I'll get a job, and soon you'll be able to buy good clothes and eat proper food.' So he went up to the factories one by one. And they all asked for his papers. He had none. Asked him to read. He couldn't. Asked him what he was trained to do. Only grow rice. And they all said — those clean men who drive cars, splashing mud on to the bare feet of the poor — that there were no jobs.

We were directed to this place by others who live here. It costs very little, they told us. So Manuel found some planks, some tin and some card and we built our shack in the *favela*. It's only for a while, but until then life is hard.

For a few weeks I worked as a cleaner for one of the rich ladies. She was kind. But the child became sick, and I had to leave. Manuel sometimes works. Once he earned some money shovelling the dirt from a new building area. But I am afraid. He has begun to drink much more than before. In the north he was a proud man. He could work, and we could usually eat. Now he has to humble himself very low to find work, and often there is no food in the house. Our clothes are very poor now, and Romero's shoes fell apart long ago . . . .'

? What elements does this description have in common with your account of a migrant? Are there any differences? How sure do you feel about the evidence on which your description is based and this story of Maria and Manuel? What picture does this give us of the kind of people that migrants in developing cities are?

STEREOTYPES AND REALITY

The quotation was taken from a Christian Aid brochure about Brazil, published in 1971. The intention, reasonably enough, was to evoke our sympathy for the poor of that country; it was an invitation to put our hands into our pocket to help. At this level the quotation works well. But does the description fit the *reality* of migration in Brazil or in other Latin American countries? The author of the booklet has no doubt that it does: 'Maria is fictitious, but her story is true'. We have our doubts. In fact, we believe that few people who move to the Latin American city are like Maria and Manuel; such people are among the poorest minority of migrants. Our major reservation about the description, however, is that it accentuates our prejudices about poor people in Latin America. People in the so-called 'developed world' do not have a flattering picture of those who live in the 'developing world'. Latin Americans are often thought uneducated, irrational and prodigal; what else can be expected of a people living in poor nations containing too few schools and producing too many children? Lack of education and too many children evokes an image of people who cannot bring order to their lives. They lack the ability to choose the sensible course of action.

Our experience is that poor people in Latin America are neither prodigal nor irrational; they are as capable as us of making the right decision about how to order their lives. When they decide to migrate, they are just as likely to have made the correct decision as someone living in, say, Europe. They are certainly faced by a hostile environment, both in their home region and in their urban destinations. But despite the difficulties they face, and the lack of education, they make sensible decisions. The basic flaw in the description is that Maria and Manuel are made to appear both helpless and irrational — little children who do not understand the world.

The basic aim of this chapter is to show how and why that picture is wrong. We will present data on migrants and shanty dwellers from three Latin American cities, Bogotá in Colombia, Mexico City and Valencia in Venezuela (see introductory map). We have chosen these three cities because they are where we carried out a large survey of 13 shanty towns in 1978 and 1979. It may be that migration in Brazil is different from that in Colombia, Mexico and Venezuela but we doubt that there are important differences. Our migrants are likely to be very similar to Brazilians moving to São

Print 1 *Although poor, residents of low income barrios are not destitute, nor do they wear exotic rural costumes*

Paulo. We will show, however, that the individuals and families we met differ markedly from Manuel and Maria. We will discuss how they adapted to the city, where they live and the jobs they do. We will show that most of these people have made sensible decisions about their jobs, houses and life chances. They have chosen rationally from the limited range of options open to the Latin American poor.

SHANTY SETTLEMENTS IN BOGOTÁ, MEXICO CITY AND VALENCIA

? **1** Examine Table 1 and write down what seem to you to be the most important points. Look across the rows. Are there any particular figures in the table that surprise you? Why? Does the data support the idea of large differences between the cities or are they broadly similar? Write a paragraph about the most important differences that you observe between the cities and suggest possible reasons for those differences.
2 Now look down at the columns. Are there any figures that don't seem to be consistent with the impression given by other statistics in the column?

Table 1 *Economic, housing and servicing conditions in the three cities*

	Bogotá	*Mexico City*	*Valencia*
Population in 1981 (millions)	4.5 [1]	15.0 [2]	0.6 [3]
Population per hectare	161.0	177.0	68.0
7–11 year olds attending school (%)	64.5 [4]	84.3 [5]	63.8 [6]
GNP per capita in 1979 (US dollars)	1010.0 [7]	1640.0 [7]	3120.0 [7]
Daily minimum wage in 1979 (US dollars	3.0	6.1	3.5
Irregular settlements (%)	46.0	42.0	47.0
Families renting accommodation (%)	51.0	58.0	29.0
Dwellings without water (%)	8.8 [8]	19.7 [9]	22.4 [10]
Dwellings without inside water connection (%)	23.7 [11]	39.6 [12]	27.9 [10]
Dwellings without drainage connection (%)	10.1 [8]	24.5 [13]	53.5 [10]
Dwellings without electricity (%)	5.3 [8]	8.8 [3]	n.a.

Sources and notes

1 Projected at the 6 per cent 1964–73 population growth rate.

2 Fox, R. (1975) *Urban population growth trends in Latin America*; Washington, Inter-American Development Bank, *page* 19.

3 Provisional 1981 census results including Guacara. *XI Censo Nacional de Población y Vivienda*, Venezuela.

4 Colombian Secretariat of Education. 1970 data, unpublished.

5 Mexican Census 1970, adapted from Tables 4 and 17.

6 Feo-Caballero, O (1979) *Estadisticas Consultativos del Estado Carabobo*, State Government of Carabobo, *page* 172.

7 World Bank (1981), *World Development Report.*

8 Colombia, DANE (Departamento Administrativo Nacional de Estadística), *La vivienda en Colombia (1977) page* 53. Dwellings without water connection.

9 Mexican Census, 1970, tables 32 and 35, our calculation.

10 Venezuelan Census. Dwellings without inside or outside connection or without toilet connected to drainage.

11 Number of dwellings (Colombia, *DANE*, 1977: 52) minus the number of residential accounts by the Bogotá Water Authority for 1973. We assume that the latter registers connections to individual homes.

12 Mexican Census, 1970, tables 32 and 35. Census disaggregates dwellings with a supply inside and outside the home.

13 Mexican Census, 1970, tables 32 and 35. Drainage system means a hygenic system for the removal of waste waters.

THE CITIES

The three survey cities in Table 1 differ from one another in several respects. Mexico City has by far the largest population, but neither Bogotá nor Valencia could be called small. Valencia is the wealthiest city both because it is located in oil-rich Venezuela and because it is a major industrial centre. Bogotá is the least prosperous city, even though it is much more affluent than most other Colombian cities. But although there are differences, the cities share certain important similarities:

1 A large proportion of the population is housed in what most of us would consider to be unsuitable accommodation.

2 A substantial proportion of the population is very poor, living on incomes far below what we would consider to be necessary for survival in this country.

3 Much of the population originated in the countryside.

4 The cities have all been growing very rapidly in recent years.

5 Many parts of the urban area lack adequate services and infrastructure.

In these respects they are little different from other large Latin American cities, including São Paulo.

THE SETTLEMENTS

The data in this chapter were collected between September 1978 and October 1979. They are based on interviews with 360 households in Bogotá, 631 in Mexico City and 178 in Valencia. The settlements involved had been in existence for between four and 15 years; the average age being between eight and nine years. Most of the settlements were close to the edge of the urban area (Figure

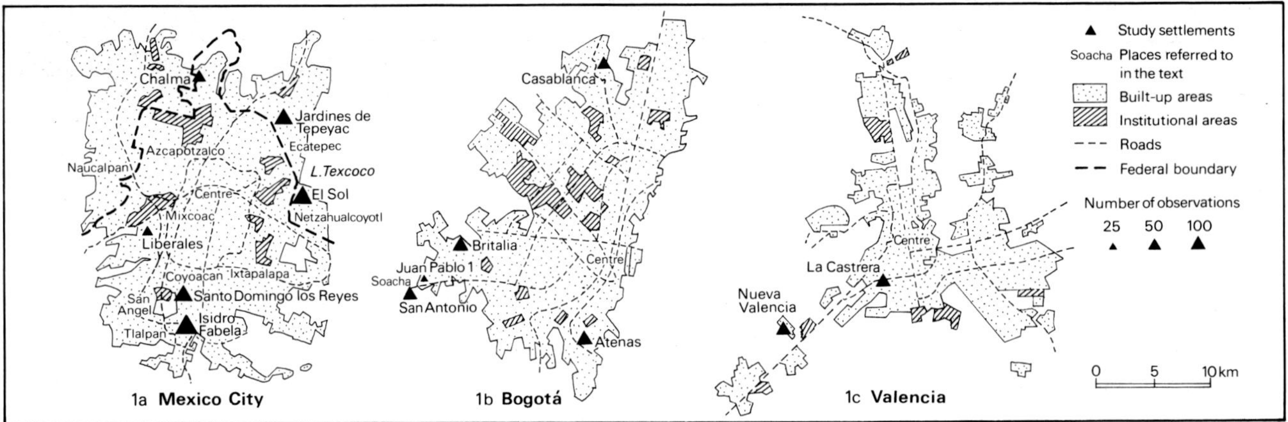

Figure 1 *Base map and settlement locations for the three cities*

1). The settlements contained a mixture of dwelling types, some consolidated, others not. Some services and infrastructure had been provided but few had the complete range of electricity, water, drainage, paved roads, schools and community centres.

We chose the settlements to reflect the different ways in which the poor acquire land in each city. The sample contained cases where land had been invaded, where land had been purchased but planning permission was lacking, and where agricultural communities had illegally sold communal land to settlers. The interviews sought information on the socio-economic characteristics of the inhabitants, their views on housing and servicing problems, their participation in community programmes and politics, their employment and residential histories, and the level of consolidation of their homes. This information should show clearly whether people like Manuel and Maria are typical of these settlements and even whether they are to be found there. Some basic information about these settlements or *barrios* is presented in Table 2 opposite.

Our sample was not designed to be representative of the whole population of each city. Nevertheless, we believe that it covers a good cross-section of those people who have moved to each city in search of jobs, and that the data describe accurately the conditions found in most shanty towns.

?

1 Examine Table 2 and write a paragraph highlighting the significant aspects of each city. Now compare the three cities. Don't worry too much about the specific shanty towns but try to get an overall impression of housing in the settlements. Write another paragraph comparing the three cities.

2 Having noted that our sample survey was not designed to be representative of the whole city population, think about the possible reasons why our survey settlements might give an imbalanced picture of low income communities and of migrant populations. For example, all of the settlements in which we worked are of relatively recent origin: would all low income settlements be as new? Write a paragraph of other possible distortions that might have occurred through our selection of these particular settlements.

3 What reasons might there be to suppose that conditions facing migrants in Brazil are different from those in Colombia, Mexico and Venezuela?

Sources and notes

1 *Age on January 1, 1979. Birth of a barrio* defined as the mean of month in which owners arrived but excluding those who bought from third party. Given that it is a mean value all settlements are likely to have been established a little earlier than the age suggests.

2 Inv. = Invasion; Sub. = Subdivision. Sub-divisions are settlements created by a developer who sells off individual plots, usually without services. Invasions are land captures, usually organized by a group of squatters who settle on the land. This is often done overnight or within a matter of days. These settlements have no services and no papers that indicate their 'ownership'. Invasions are a much more risky process.

3 This comprises a score 0–20 covering the presence/absence and extent of servicing in each settlement for the following:
electricity, water, drainage, roads, telephones, transport, church(es), market(s) and policing. It was based upon detailed observations in each settlement and is a *different* index to the one given below (see note 6) which was derived from household questionnaire surveys.

4 Data derived from a random survey of households in each settlement (PIHLU Survey). The number of questionnaires completed in each case is indicated in brackets below the *barrio* name.

5 Column percentages. Owing to rounding, totals may not always equal 100.

6 'Overall consolidation score' is a composite score combining individual household scores for quality and permanence of dwelling structure, the services to each plot, and the range of household possessions that were present.

Table 2 *Comparative data for the* barrios *sampled in each city*

	Bogotá					*Mexico City*						*Valencia*	
	Casablanca	*Britalia*	*Atenas*	*San Antonio*	*Juan Pablo*	*Isidro Fabela*	*El Sol*	*Santo Domingo*	*Jardines*	*Chalma*	*Liberales*	*Nueva Valencia*	*La Castrera*
Age of settlement (in years)[1]	9.5	2.9	12.1	9.2	3.0	12.0	6.8	7.4	9.7	6.0	2.2	5.2	7.6
Origin of settlement[2]	Sub.	Sub.	Sub.	Sub.	Sub/Inv	Inv.	Sub.	Inv.	Sub.	Sub.	Inv.	Inv.	Inv.
Services and utilities score[3]	12	6	15	11	0	15	14	11	14	11	6	6	13
Density: average lot space per person (m^2)[4]	23	23	21	22	25	29	33	25	52	34	25	99	52
% households that are 'owners'[4]	57	71	56	55	89	63	70	81	66	74	82	97	93
Primary building material used in walls (%)[4,5]													
• Scrap materials or laminated cardboard	3	3	1	6	27	5	4	3	2	20	—	39	19
• Coarse bricks	63	68	45	54	74	59	76	79	76	78	81	50	63
• Prefabricated blocks/'fancy' bricks	34	30	54	41	—	36	20	18	22	2	19	11	18
Primary building material used in roof (%)[4,5]													
• Scrap materials or laminated cardboard	3	3	1	6	27	28	51	41	43	33	19	—	—
• Asbestos/corrugated iron	63	68	45	54	74	23	28	18	21	52	23	95	95
• Concrete or tiles	34	30	54	41	—	49	22	41	36	15	58	5	5
Level of water provision (%)[4,5]													
• Brought in by water lorry/ no service	1	—	4	—	—	—	1	3	—	—	92	—	—
• Carried from public standpipe	1	79	2	2	3	8	1	55	—	92	—	2	—
• Hose from standpipe to plot	5	21	1	—	97	4	20	3	4	8	—	25	4
• Formal connection to plot	10	—	21	4	—	50	70	32	80	—	8	51	27
• Formal connection inside home	82	—	72	94	—	38	8	7	16	—	—	22	69
Level of electricity provision (%)[4,5]													
• None or street lighting only	1	6	6	5	3	—	3	2	3	—	4	—	—
• Illegal hook-up	23	33	10	3	97	2	8	2	1	100	6	99	3
• Legal supply but unmetered	10	11	8	3	—	—	1	—	—	—	3	1	3
• Legal and metered	66	50	76	90	—	98	88	97	97	—	88	—	95
Number of rooms in each dwelling (%)[4,5]													
• Houses with 1 room	40	24	39	46	54	7	4	5	11	15	8	11	6
• Houses with 2 rooms	36	43	32	29	40	26	40	41	30	47	27	34	27
• Houses with 3 rooms	11	22	16	17	6	29	36	35	35	28	36	24	37
• Houses with 4 rooms or more	13	11	13	7	—	38	20	19	24	10	29	31	30
Overall household consolidation score (%)[4,5,6]													
• Houses with less than 14 points	4	14	5	1	50	1	2	4	4	22	10	10	1
• Houses with 15–18 points	7	47	8	7	41	10	11	19	7	27	23	37	4
• Houses with 19–22 points	35	28	43	30	9	28	37	45	35	48	37	38	32
• Houses with 23–26 points	37	10	38	45	—	32	38	23	29	3	26	15	34
• Houses with 27 points or more	16	1	6	17	—	29	13	9	25	—	4	—	29

IS THERE A 'TYPICAL' MIGRANT OR SHANTY TOWN DWELLER?

? Examine Table 3 opposite, and write up a profile of each migrant to each city. Compare the three cities: how do migrants appear to differ? To what extent does your inspection of the data lead you to agree with the passage that follows?

Although many migrants share strong similarities, there is no typical migrant. Some migrants are poor, others not; some are educated and others cannot read; some move to the city for work, others for schooling; some move with their families when children, some move as single adults, some move with their wife and offspring; some work in agriculture, some in industry, some move from towns, others from the countryside. Average figures are always subject to the dangers of misinterpretation and nowhere is this more true than in considering flows of people. Thus, although Table 3 records a young average age for migrants, we found that only 25 per cent moved when they were over 33 years of age compared to 45 per cent who were under 22 years when they arrived in the city.

In addition to these variations among the migrants to particular cities there were certain differences in the patterns between the three cities. More people moved as children in Mexico, and more moved as adults with their own families in Colombia and Venezuela. We do not want to make too much of this point but it is vital to remember that if certain categories of people are more likely to migrate, heterogeneity remains an important feature of the migrant flows. There are migrants like Manuel and Maria, but they do not represent the vast majority of migrants.

If migrants are heterogenous in their age, background and socio-economic circumstances, so too is the population living in the shanty towns. While the majority are migrants, a considerable number came to the city when young and have lived much of their lives in the city. Between one-half and two-thirds of the household heads had migrated to the cities as adults, whereas around one-quarter had migrated to the city when they were children and the rest had been born in the city. Thus the shanty dwellers have a vast range of urban experiences; a few came straight from the countryside, others have lived in the city all their lives. There are various other kinds of difference as well. Some own their houses in the *barrio*, some rent homes, others share homes with kin. Some are employed in factories, some work independently, others are unemployed. There are few millionaires in the shanty towns but there is a handful of very smart homes in many *barrios* — social heterogeneity is an important feature of most low income settlements.

WHERE DO THE MIGRANTS COME FROM?

Maria and Manuel came from the countryside. They were farmers and they had come a long way to live in Brazil's largest city. In the survey settlements in Bogotá and Mexico City we also found many migrants from the countryside. Indeed, between 60 and 70 per cent had worked in agriculture before moving to the city. But the situation in Valencia was different — only 30 per cent had worked in agriculture, prior to moving to the city. Where migrants had worked previously in agriculture, a majority in all three cities had worked their own or their parents' land. Between 30 and 44 per cent had worked as day-labourers or had rented land in the countryside. If it is certain that those who had owned land owned little, this fact still suggests that it is not necessarily the poorest and most destitute who move.

Where our data differ most from the picture of Manuel and Maria is in the distance that they had moved. As Figures 2a–2c show, on *pages 32* and *33*, most of the migrants to Bogotá, Mexico City and Valencia had moved from nearby states. Relatively few moves were of more than 150 kilometres. By contrast, Maria and Manuel had travelled from the north-east to the south-east of the country, a distance of well over 1600 kilometres. We believe Maria and Manuel's journey to be very exceptional. Most migrants are rational in the sense that they move to the nearest large city. If they are looking for urban jobs any largish city will do and there is often one quite close by. Why did Maria and Manuel travel so far and pass through cities as large as Recife or Salvador on the way? Most sensible migrants would have stopped in one of these cities. Our migrants were certainly not going to travel further than they had to.

We did not ask our migrants but we are sure that they did not walk to the city. Maria and Manuel may have liked walking but in most parts of Latin America people take the bus. Few migrants in Latin America would walk 8 kilometres let alone 1000 or more; road communications are adequate and relatively cheap.

IS IT THE POOREST AND LEAST SKILLED WHO MIGRATE?

Manuel could not read and is certainly not the only Latin American with that disability. But most migrants to the cities can read and write. Indeed, many studies have shown that those who move from the countryside are the better trained and educated. The vast majority of the migrants to the survey settlements had followed some form of primary education.

Many migrants also have some skill which they can use in the city. This is not true of all, as the number of agriculturalists would suggest, but many among our sample had worked previously as builders, bricklayers, drivers and itinerant salesmen; skills that would serve them well in the city.

The fact that most migrants have some form of education and possess some marketable skill is strange in so far as the education and job situation in the countryside is poor. Throughout Latin America, with the probable exception of Cuba, rural facilities and services are extremely limited. It is certain, therefore, that those who have moved are untypical of the mass of rural people. The migration process is highly selective, and it is usually those who are most suited to life in the cities who move.

If migration is selective it is not motivated only by an escape from poverty. If rural misery were the only primary factor causing out-migration then the Manuels and the Marias

Table 3 *Some characteristics of migration to Latin American cities*

	Bogotá %	*(No.)*	*Mexico City* %	*(No.)*	*Valencia* %	*(No.)*
Proportion of migrant heads of household in survey settlements[1]	59.2	(213)	54.7	(346)	65.7	(117)
Mean age of migrants on arrival in the city (years)	28.5	(213)	25.1	(346)	29.2	(117)
Proportion of migrants who worked in the agricultural sector prior to moving to the city	62.2	(125)	69.7	(212)	29.6	(32)
Of those working in agriculture, proportion who rented land or were day-labourers	43.8	(61)	31.3	(62)	44.4	(36)
How migrants moved to the city						
• Alone	36.9	(76)	28.5	(97)	26.5	(38)
• With spouse or with children	43.2	(89)	27.6	(94)	47.8	(54)
• With parents or kinsfolk	14.6	(30)	39.7	(135)	15.9	(18)
Assistance received upon arrival						
• None	47.6	(98)	34.2	(117)	48.3	(56)
• Accommodation and/or help in finding work	30.6	(63)	44.4	(152)	31.0	(36)
Who provided assistance?						
• Kin	73.4	(80)	82.1	(184)	63.3	(38)
• Friends	19.3	(21)	13.4	(30)	30.0	(18)
Tenure of first home in which lived for more than 6 months after arrival in the city						
• Owner	5.6	(11)	2.1	(6)	26.4	(23)
• Renter	66.3	(128)	54.6	(153)	28.7	(25)
• With kinsmen	18.7	(36)	30.4	(85)	23.0	(20)

Comparison of educational level of migrants and non-migrants	*Bogotá Non-migrants* %	*(No.)*	*Bogotá Migrants* %	*(No.)*	*Mexico City Non-migrants* %	*(No.)*	*Mexico City Migrants* %	*(No.)*	*Valencia Non-migrants* %	*(No.)*	*Valencia Migrants* %	*(No.)*
No schooling	11.1	(16)	19.2	(41)	12.1	(34)	27.7	(94)	16.6	(10)	34.5	(40)
Incompleted primary education	33.3	(48)	37.1	(79)	31.4	(88)	40.7	(138)	30.0	(18)	25.9	(30)
Completed primary (6 years)	37.5	(54)	31.0	(66)	33.9	(95)	22.1	(75)	36.7	(22)	28.4	(33)
Some post-primary education	18.1	(26)	12.7	(27)	22.5	(63)	12.4	(42)	16.6	(10)	11.2	(13)
Totals	100.0	(144)	100.0	(213)	99.9	(280)	99.9	(399)	99.9	(60)	100.0	(116)

Notes

[1] We excluded those heads of households who were born outside the city and migrated before their sixteenth birthday.

(*Source*: PIHLU Survey, 1978–9)

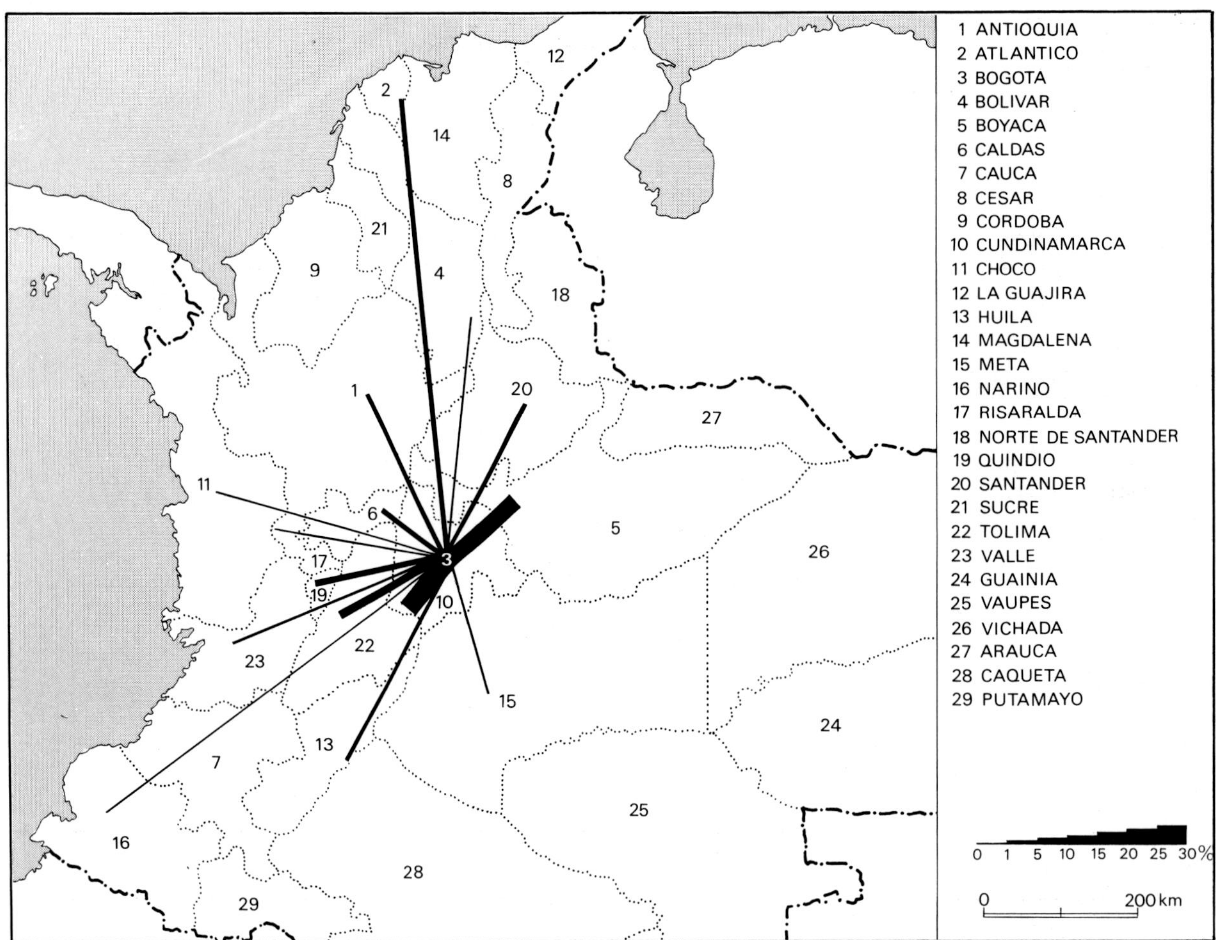

Figure 2a *Flow of migrants to Bogotá by birthplace (Source: PIHLU survey)*

Examine Figures 2a–2c. What principal conclusions do you draw from them? In what single respect do they differ most from the description of Maria and Manuel at the beginning of this chapter?

would be the first to move. Indeed, those with nothing to eat, the poorest and the least educated would be among the first to arrive in the city. In times of disaster this may temporarily be the case, but it is not true of the majority. In normal circumstances it is the poorest and least adaptable who remain in the countryside. This point is not to deny that people in the countryside go hungry, nor does it mean that major improvements in rural conditions are unnecessary. It is merely to say that in response to negative conditions in the countryside and opportunities in the city, only those who think they can adapt to the city tend to move. The rest wisely stay at home.

In general, migrants do not move, like Manuel and Maria, in ignorance. They do not migrate to cities that they have never seen. Most Latin American peasants have been to the city and seen what it has to offer. Regular, cheap bus services allow visits to the cities. Moreover, country people have relations and friends already living there. Migration after all is no longer a new phenomenon in Latin America. It is a process that has been underway in most countries for at least 30 years. Few villages lack former inhabitants now living in the cities. Most families have some friend or acquaintance living in the nearest city. Before moving, therefore, what is more sensible than to consult with the urban dweller, perhaps to arrange to spend a few days there to get to know the city? Those like Manuel and Maria who move without knowledge and information are a tiny minority — those who are forced to move because there is absolutely no alternative. Rural 'push' may operate but it is selective in its effects because the rural poor weigh up the costs and benefits of movement.

Our survey shows how many of the migrants not only sought advice from relations and friends but actually relied upon them once they arrived in the city. Three in ten in Bogotá and Valencia either stayed with contacts on arrival or got help in finding work; in Mexico City the figure was 44 per cent. It is true that a high proportion claimed to have received no help but this does not mean that they had moved in ignorance nor that they had no one

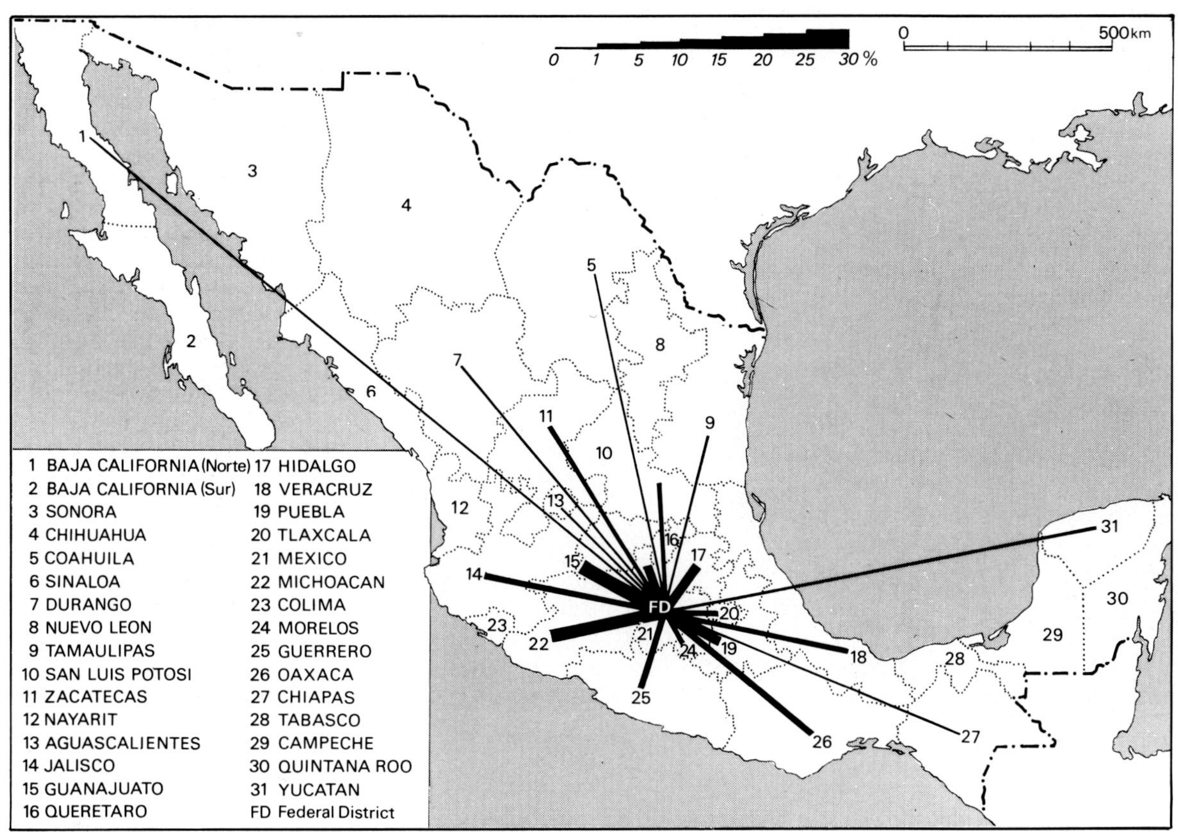

Figure 2b *Flow of migrants to Mexico City by birthplace (Source: PIHLU survey)*

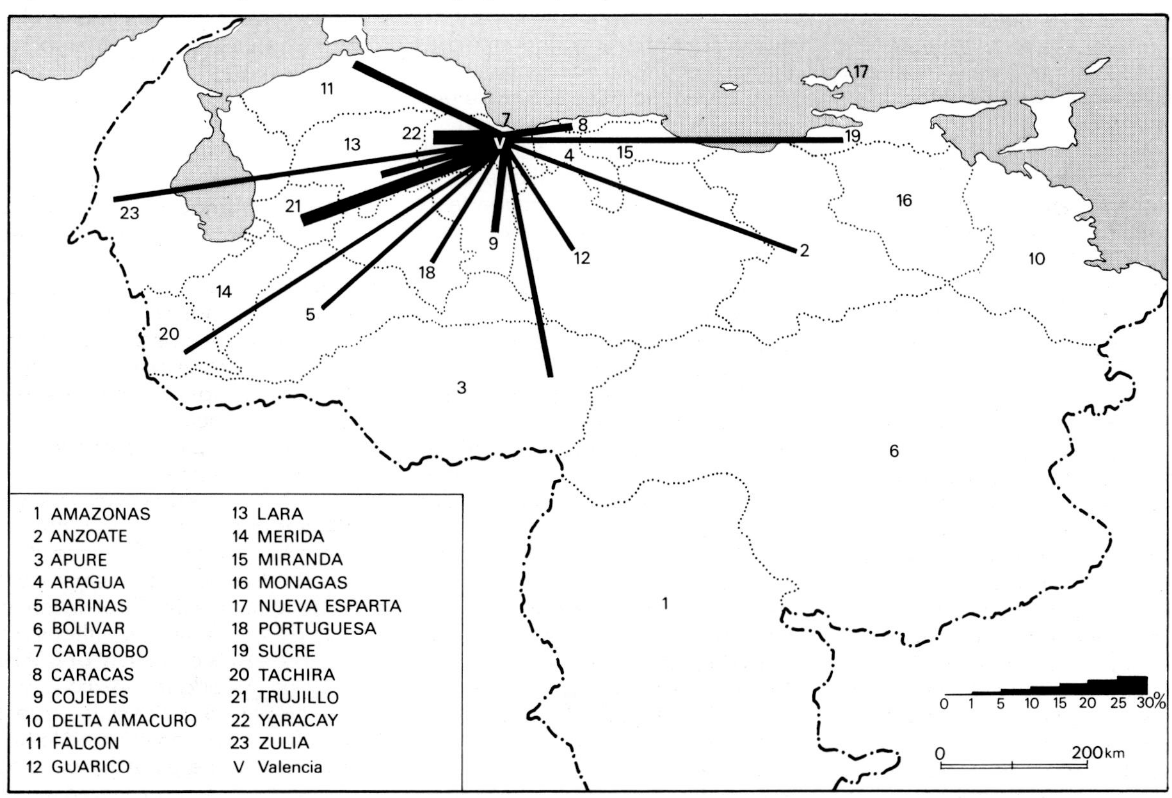

Figure 2c *Flow of migrants to Valencia by birthplace* (*Source*: PIHLU survey)

Print 2 *Most migrants use modern systems of transport to move to the towns*

to consult in times of difficulty. If over half the migrants in Bogotá and Valencia and two-thirds of those in Mexico City stated that they had help on arrival, the rest undoubtedly benefited from prior information about the city. Manuel and Maria clearly knew no one; a situation that arose from moving all the way from Pernambuco to São Paulo. Most migrants have better information and more sense.

HOW DO THEY EARN THEIR LIVING?

> **?**
> **1** Examine Table 4 and look specifically at the employment and income data. Are there any figures in the table that surprise you? Why? Looking across the rows, do the data support the idea of large differences between cities or are they broadly similar? Write a paragraph about the most important differences between the cities giving two or three possible reasons for the differences.
> **2** Looking down the columns, are there any figures that don't seem to correspond to the impression given by other statistics in that column?

Manuel could not get a job and only occasionally worked. In the settlements in Bogotá, Mexico and Valencia he would be the exception. Few poor people in Latin American cities can afford to be unemployed. Since there is no unemployment pay, some form of work is essential if the family is to survive. In our survey very few household heads were without work and most respondents had worked all of the previous year.

The problem of the poor in Latin American cities is not that there is no work but that they earn very little for what they do. The majority work long hours to earn a low wage. In this sense the term commonly used to describe their problem, 'underemployment' is less than apt. The term 'working poor' is much more appropriate; a working week of 50 or 60 hours is common. At work they may not be highly productive; the hawkers will sell little in the long hours that they work, the building labourers will not have the equipment to construct as much per day as a labourer in a developed city. The result is that they earn little after a long week.

However, if most people in shanty towns are poor, they are not all bootblacks and newspaper sellers. The petty services that are so notorious in the literature on 'overurbanisation' and 'excessively large' cities do not figure so prominently in the survey settlements. In fact, only 12–16 per cent of the household heads fell into the private services category (Table 4). More typical occupations in the survey settlements were manual work in factories, labouring in the construction industry, driving buses and working for the council. Bogotá, Mexico City and Valencia are all important manufacturing centres so perhaps it is not surprising that such a high proportion of the sample population was employed in industry. In Bogotá, most received little more than the minimum income (al-

Print 3 *Most people work once they arrive in the city, either in factories or employed as casual labour. These children are working in a brick factory in Bogotá*

Table 4 *Employment characteristics and incomes of heads of households in shanty towns in Bogotá, Mexico City and Valencia*

	Bogotá		*Mexico City*		*Valencia*	
Employment	%	*(No.)*	%	*(No.)*	%	*(No.)*
Manufacturing	23.5	81	31.7	187	19.8	33
Construction	21.2	73	14.9	88	16.2	27
Commerce	13.6	47	14.6	86	14.4	24
Transport	10.1	35	7.8	46	15.6	26
Private services	16.2	56	11.9	70	13.8	23
Public services	8.1	28	13.6	80	10.2	17
Housewives	2.9	11	2.2	13	6.0	10
% in active employment	84.4	304	91.2	571	81.5	145
Proportion of households with more than one wage earner	39.0		31.0		41.0	
Average monthly income of head of household, $US, 1979 (figure in brackets = local currency)	$ 103.2	(4309)	$ 212.1	(4838)	$ 419.8	(1805)
Distribution of incomes for head of household in local currencies	*Colombian pesos* %	*(No.)*	*Mexican pesos* %	*(No.)*	*Bolivares* %	*(No.)*
< 1500	4.5	14	5.0	28		
1501–3000	19.0	59	7.5	43		
3001–4500	41.6	129	40.0	228		
4501–6000	21.9	68	30.1	172		
6001–7500	5.8	18	8.4	46		
> 7500	7.1	22	9.6	55		
< 700					3.4	5
701–1400					41.9	60
1401–2100					30.2	44
2101–2800					13.0	19
2801–3500					6.0	9
> 3501					6.0	9

though one explosives expert was very well rewarded for what was possibly destined to be a short career). There were many construction workers, ranging from a few master builders to unskilled labourers. Such an activity is particularly useful in a shanty settlement where the inhabitants contribute substantially to the construction of their own houses and where building experience can help to supplement the regular income. The construction industry is, however, a highly volatile activity and many of the interviewees had undergone short periods of unemployment. Table 4 shows the range of activities represented in the settlements. We found printers, lawyers, domestic servants, government workers, drivers of buses and of taxis, shop-owners, shop assistants, production-line workers and waiters. What is clear from this table is that the people who live in shanty towns are employed not only in the so-called informal sector but also in modern industry, shops and offices. They are working not in 'marginal' activities but across a wide range of urban activities. What unites them is not their lack of integration into urban life but the limited remuneration that integration affords them.

Most families survive on their incomes, but few have much money to spare. Many households exist only because several people go out to work. Indeed, many Latin American writers now argue that as times get harder the average number of workers per family increases. Our data

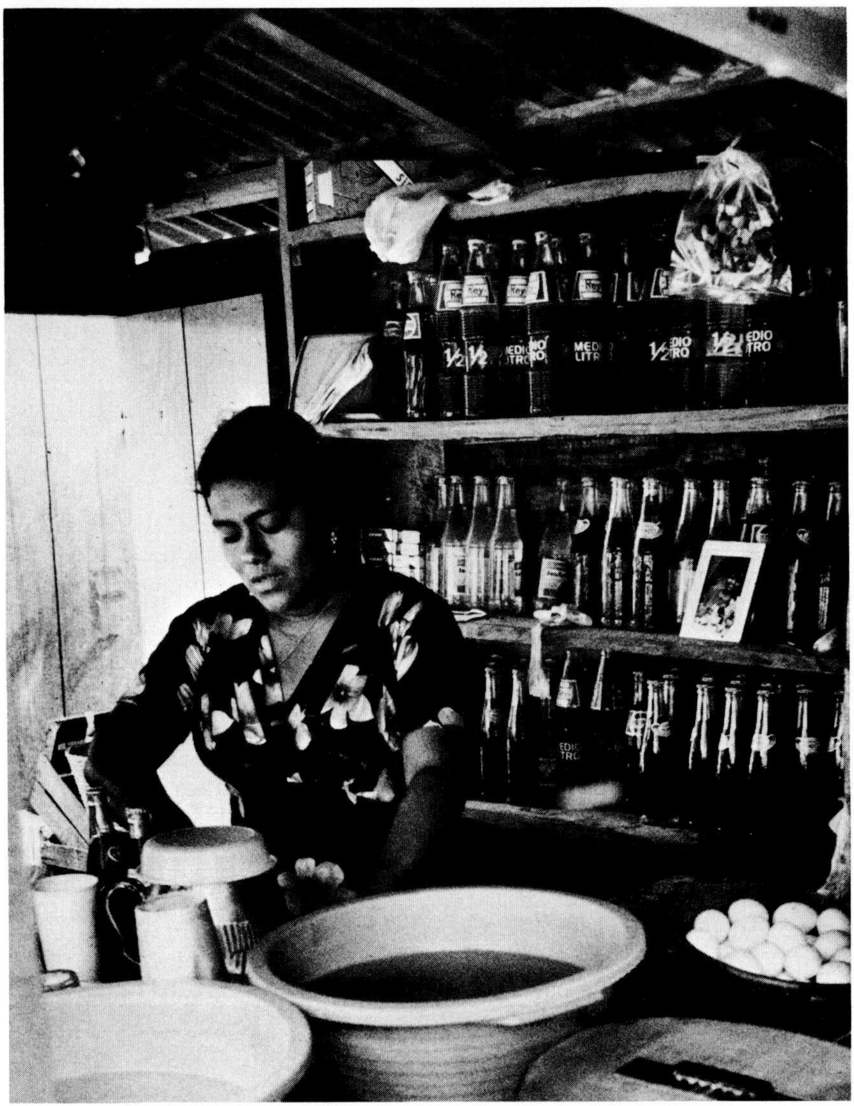

Print 4 *It is common for all the members of a migrant family to have some form of employment. This woman has set up her own roadside snack bar in Mexico City*

cannot demonstrate this point but does underline the fact that many families have several wage earners (Table 4). Low income households undoubtedly make use of a range of workers, some of whom will work in the formal sector, others in the informal. The father may work in a factory with a son, the wife may run a small shop from the house, a daughter may work as a domestic, a younger son may assist a driver learning how to operate and repair a bus. In the sense that both Manuel and Maria were struggling to work, even if they could not because of the lack of opportunities or Maria's pregnancy, they were typical of the survival strategies of many Latin American households. When a family budget is limited it is supplemented in as many ways as possible.

HOW DO THEY DECIDE WHERE TO LIVE?

?

1 In what part of the city would you expect migrants to live when they arrive — the central city area or the periphery?
2 Are migrant arrivals likely to want to rent accommodation ? Why?
3 From whom would they get to know about available accommodation?

Manuel and Maria arrived in the *favela* by chance. Few of the families in our survey arrived in that way. Most were told of vacancies by friends or kin. A few found their house or room by searching in a particular neighbourhood, others heard of the sale of land, others were told about a forthcoming invasion of land. Chance was not a word that we often encountered except in the sense that there were numerous localities where they could have lived — they had looked at several areas and chosen to locate themselves there.

Manuel found some materials and started building the family shack. In Bogotá and Mexico he would probably have been arrested, in Valencia he would have been asked for his party card. Obtaining land is not that easy. In some cities in Latin America new settlements are created by invasions of land (see later extract from *The Guardian*).

In Valencia, 33 per cent of our interviewees had occupied land through invasion. But to participate in this invasion they had to know that it was going to occur. They had to have contacts to enlist their participation. Secrecy is essential in an invasion so that the owner does not find out. Only friends and acquaintances, and often only members of a particular political party are told about the imminent event. Without contacts or a party card Manuel would not have been involved in an invasion in Valencia. In Venezuela, as in certain other Latin American cities, the government permits the invasion of private or more usually public land under certain circumstances — most typically close to elections when political parties or even military rulers are seeking support from the poor. But participation is not indiscriminate; it requires contacts and Manuel would almost certainly have been excluded.

He would have fared still worse in Bogotá and Mexico City where invasions are rarely permitted. Most people in those cities obtain land through purchase. The land lacks services and planning permission but the original owner or an intermediary will divide it into plots and sell it to those among the poor who can afford to pay. Around half of the population of Bogotá live in settlements which began as so-

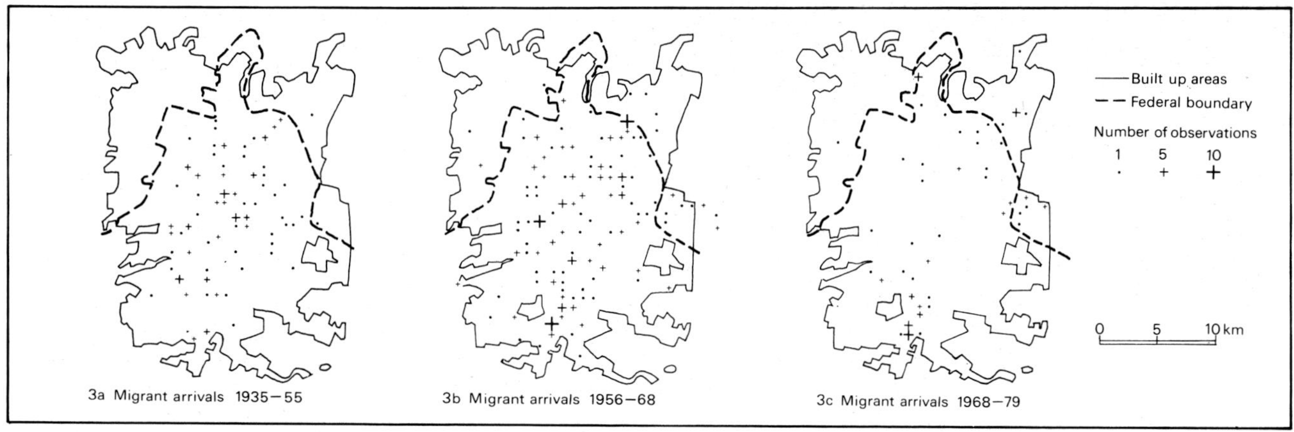

Figure 3 *First place of residence after arrival in Mexico City*

called 'pirate' urbanisations. Eventually they may be serviced and regularised, but initially they suffer from some degree of illegality — this is why the land is cheaper than in other parts of the city. Nevertheless, land is still not cheap and in 1979 it cost the equivalent of a monthly salary to buy ten square metres of land in Bogotá.

Building versus renting

Even if Manuel had managed to obtain a plot, either by joining an invasion or by purchasing land, he would have been very fortunate indeed to have been able to find some planks, tin and card with which to build his shack. With some 16 million people living in Mexico City, half of whom may be engaged in self-help construction, there may well be a shortage of planks left around the streets. If Manuel were to build a shack he would have to buy the materials, little in these cities is free and certainly not those commodities which are in great demand. Indeed, one of the big questions about the ability of poor people to build and consolidate their own homes relates to the difficulty of obtaining sufficient land and materials. Since the real costs of land and materials appear to be rising in many cities, it is probably getting more difficult for the poor to build a self-help home. As such it is the more fortunate, rather than the down and out, who build self-help houses.

Manuel and Maria would probably not have arrived by chance in a *favela* and started to build. It is more probable that they would have found accommodation in the centre of the city or in an already consolidated settlement. They would not have built a house but would have rented a room or shared with kin. In Bogotá and Mexico City, where land is relatively scarce, around half of the population rents accommodation (see Table 3, *page 31*). The poorest people in the city and the young who have not yet accumulated sufficient savings cannot afford to own. It is an interesting point that it may take many years before a family can own its first home. In the settlements in Bogotá, the average age at which people owned their first shack was 35.8 years, a long time after they first established a family. Manuel and Maria's shack was certainly no palace, but it is unlikely that they would have had the opportunity to build anything at all. In our survey only two per cent of migrants in Mexico City lived in their own house on arrival in the city, and only five per cent in Bogotá (Table 3, *page 31*).

The site of the first home

? In what ways do the data about first place of residence displayed in Figures 3 and 4 contradict the pattern described for Maria and Manuel?

The evidence from Bogotá and Mexico City, though not from Valencia, also casts doubt on the idea that Manuel would have moved straight

Print 5 *A migrant's first place of residence in the city is usually a multi-family tenement such as this one in Mexico City*

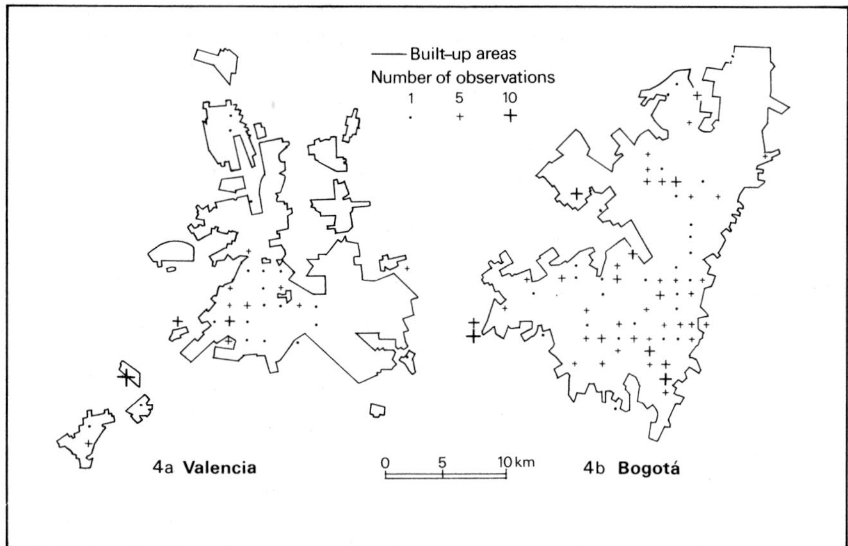

Figure 4 *First place of residence after arrival in **a** Valencia **b** Bogotá*

to a peripheral settlement. Figures 3 and 4 show where most migrants first resided. The vast majority moved first into consolidated older settlements and only later into the newer settlements. If Valencia seems something of an exception this is due to the fact that the original invasion of one survey settlement (Nueva Valencia, see Figure 1c *page 28*) was led by a Peasant League. The organisers of that invasion recruited in the countryside, whereas most organising groups recruit in the town.

HOW BAD ARE HOUSING CONDITIONS?

? Look back at Table 2 (*page 29*) and examine the rows of data about housing conditions. Compare the three cities. Are there any particular figures in the table that surprise you? Does the data support the idea of large differences between the cities or are they broadly similar? Does the overall impression that the data provided correspond to the idea that these are slums? Are there likely to be other low income areas that are worse off? Write a paragraph stating your view and giving your reasons.

CONSOLIDATION OF THE HOME

Manuel built his shack in the *favela* but it was only a temporary solution. Few migrants would adopt such a strategy. Once they have struggled to acquire land, their principal preoccupation is to retain the land and to improve the house. Indeed, one of the most fascinating aspects of shanty town life is how rapidly and successfully the poor improve their housing once they know that they will not be ejected. Gradually, one room shacks expand, horizontally and vertically. Wood and corrugated iron give way to more solid construction materials. The majority of houses in each city contain rooms with brick walls. Many have concrete ceilings which can support a second floor. Our data show clearly how the poor have managed to improve their housing conditions (Table 2). Most dwellings in our sample had at least two rooms (excluding the kitchen). In the older established communities numerous owner occupiers had houses with several rooms. Renters tend to live in one or two rooms in a consolidated house. In Bogotá, where a higher proportion of the survey population was renting accommodation, numbers of rooms per house were somewhat lower.

The degree of consolidation is such that the term 'shanty settlement' is really inappropriate. The word shanty conveys the image of straw huts on hillsides. Many of the settlements are indeed on hillsides but few contain straw huts beyond the initial few weeks. Much more appropriate terms are 'irregular settlement', 'self-help housing' or 'spontaneous urbanisation'. These terms convey the idea of consolidation and improvement even if they also contain implicit value biases. We do not wish to suggest, however, that all settlements consolidate — some are threatened with extinction by the authorities. Sometimes the government does not want a particular piece of land invaded because the

Print 6 *Only several years later do they squat or buy land and build a temporary shack upon it*

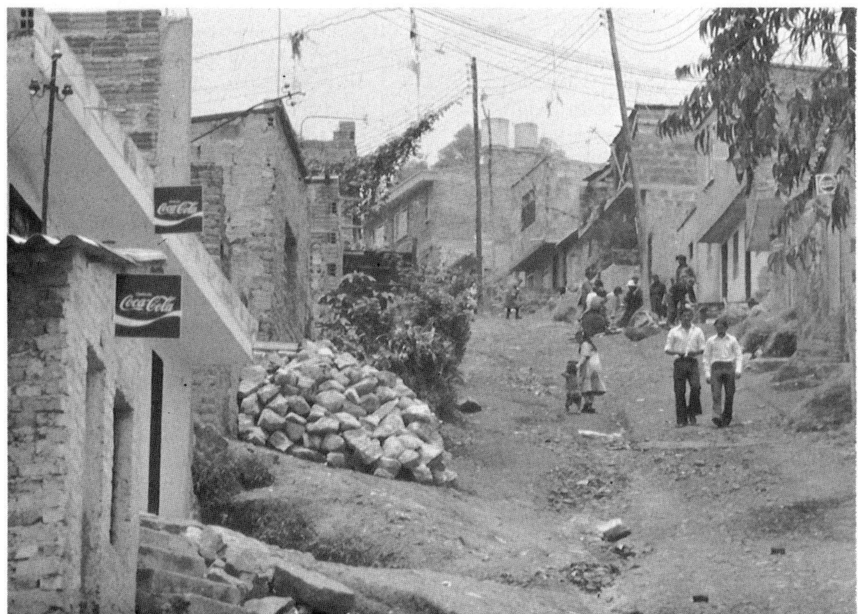

Print 7 *Once they have security of tenure they substitute shacks for permanent dwellings — building their homes with the help of family and occasional paid labour. This settlement began about 12 years ago*

private owners have protested, or because the land is required for public works or urban development. In such circumstances the population will do little more than to establish their presence by occupying the site. They will put up flimsy houses but they will not consolidate the huts in case the police come and knock them down — consolidation costs money. One of the settlements in Bogotá fell into this category. Table 2 shows that the services available in Juan Pablo and the quality of the building materials were far below that of Britalia even though both were three years old. The answer is simple: Juan Pablo was under threat whereas Britalia was secure.

The other principal factor that determines the rate of house consolidation is the affluence of the community. The rate of consolidation obviously depends on the amount of spare cash controlled by the household — it is difficult to buy bricks, cement and glass without savings. The speed of house improvement, therefore, depends upon the incomes of the inhabitants. We found some notable differences both within and between the survey settlement populations. Furthermore, some aspects of home improvement are more difficult and expensive than others. A concrete roof, for example, requires considerable expenditure on materials, scaffolding and labour. Hence many households have to wait several years before they can make such a major investment. But material prosperity also varies considerably between settlements in the same city. For example, more poor people live in Juan Pablo in Bogotá than in the other settlements. Only those with little money 'choose' to live in a settlement under threat. Over and above these differences each community contains its share of very poor and less poor people; heterogeneity is an important feature of low income settlements.

SERVICES

But we should not judge settlements only by their levels of income or by the rates of individual house consolidation. We need also to consider whether or not the settlements have obtained services and infrastructure. In most cases electricity is first supplied by tapping the main power lines and water by splicing into main pipes. Regular servicing often follows much later although the situation of the settlements varies and some cities do a much better job at supplying their inhabitants than others. In Mexico City, for example, 85 per cent of the population has a metered supply of electricity whereas in Bogotá and Valencia the proportion is 64 per cent and 45 per cent. The situation of individual settlements depends greatly on the specific policies of the servicing agencies. In Bogotá most settlements will obtain water if they can collect the money to pay for installation; only the most distant or problematic communities remain unserviced (Table 2).

Print 8 *Twenty years on and the settlement is fully consolidated with services and street paving. The move to the city has paid off handsomely*

In Valencia, settlements formed through invasion obtain services quickly if they occupy public land but not if they are on private land. The authorities claim that servicing invasions on private land both raises the price of that land, for which they may later have to pay compensation to the owner, and encourages further invasions. Whatever the rights and wrongs of that argument it is clear that there are major differences in the servicing levels of invasion settlements on public land or private land. The two settlements where we worked in Valencia were both nine years old. La Castrera, founded on public land, had been supplied with electricity, drainage, roads, telephones and recreation facilities very quickly. Nueva Valencia, built on private land, had waited eight years for water, still had illegal electricity, had no paved roads or drainage.

ARE MIGRANTS WORSE OFF THAN THE CITY BORN?

? Write a paragraph about the ways in which you would expect migrants to be worse off than city born people and give your reasons.

It is unclear from Maria's account whether she and Manuel were worse off than the city born. Many accounts infer that migrants are ill-adapted to the city but our data suggest that they would not have been out of place. A series of variables may be used to show that migrants and the urban born are very similar in their socio-economic characteristics. (In one sense this is obvious; if most urban dwellers are migrants the two populations will be almost identical.) When we compared the differences between migrant and city born heads of household in terms of wage levels, levels of house consolidation, and household size we found no statistical difference. Education levels between migrants and non-migrants, however, were found to differ markedly with city born heads of household or those who arrived as children having had more schooling (Table 3, *page 31*). This reflects the inadequacy of schooling in rural areas in the three countries. Nevertheless, less schooling does not seem to have affected the migrants' achievements relative to urban natives. Broadly it seems that in Bogotá, Mexico City and Valencia, the migrant and city born are very similar.

MIGRANTS *VERSUS* THE ENVIRONMENT

Latin American shanty dwellers are poor and suffer from many disadvantages and problems. Despite their hostile urban environment, however, they are generally rational in their response to the problems it creates. The original move to the city was carefully considered and not undertaken without prior consultation with those already living in the city. Those who move tend to be those who can best adapt to urban life and work. In the city, they obtain jobs in both the formal and the informal sector; they are rarely unemployed. The shanty towns are often notable more for the well-consolidated dwellings than for the shacks. Services and infrastructure are deficient but they are gradually installed.

Manuel and Maria are atypical. It is important to correct this common view of the poor as being innocent and helpless. The urban poor with whom we conversed were realistic, sensible and practical. They took sensible decisions about important matters. If there is to be an answer to the major problems that face most poor families in Latin American cities, it lies not in educating the poor to improve themselves. The poor have no problem adapting to their environment. The problem lies in the poverty and hostility of that environment. If the Christian Aid pamphlet distorts the picture of Manuel and Maria and produces a stereotype, at least it is accurate in portraying an unfair and difficult world. Any solution will require major changes; not in the *favelas* and shanty towns but in the wider urban economy and society. The problems of migrants and the poor are not resolvable by more education, they require major reforms in land holding, tax systems and the distribution of wealth. If the Christian Aid booklet distorts the picture of the poor, its view of the causes of poverty is much closer to the truth.

? Write a new description of the migrant for Christian Aid. It should be about the same length and should accurately convey the reality of being poor, the difficulties of moving to the city and the nature of adjustment to the urban environment.

We should like to thank Christian Aid for the right to quote from their booklet *Brazil: Profile of Poverty*. We would point out that the booklet was written in 1971 and that more recent Christian Aid material such as the film *Exiles in Their own Land* gives a much more accurate portrayal of the lives of most poor people. Other popular accounts, however, still employ similarly evocative and rather misleading stereotypes.

LIMA: A FURTHER CASE STUDY

In this chapter, the authors have contrasted their research results in three Latin American cities, with the portrait of a 'typical' migrant. Use the article from *The Guardian* as another account of the process of migration, to be set against the other two. Compare the processes described in the chapter. Highlight the major areas of agreement and any points of disagreement.

In Peru, as in other developing countries, shanty towns can arise almost overnight. Christopher Roper examines the problems of the Peruvian Government in the face of migration from the interior to the cities.

The most revolutionary fact in Peru today is the growth of its capital city, Lima, whose population has increased from 600 000 in 1940 to 3 000 000, and is still rising at a rate of 7 per cent a year.

The city, which is shaped liked a right-angled triangle with its hypotenuse on the Pacific and its apex in the low sandy foothills of the Andes, absorbs immigrants from the interior in shanty towns, known as *barriadas* or more politely, *pueblos jovenes* (young towns) at a rate of about 20 000 families a year.

The topic is suddenly arousing great interest among town planners, demographers, economists, and architects because unplanned urbanisation is now perceived as a world-wide phenomenon, closely linked to the process of industrialisation, and virtually unstoppable.

The question is not what alternatives can be provided by central governments, nor even how the process can be arrested, but rather how can planners intervene constructively in the process to ensure the medium- to long-term viability of the settlements, and to increase the social options of the settlers?

This pattern of unplanned urbanisation, modified in every case by local geographic and economic circumstances, occurs in places as far apart as Manila, Ankara, Nairobi and virtually every major Latin American city. The process is perhaps more dramatic in Latin America than elsewhere, partly because the process of industrialisation has gone further than other countries in the dependent world, and partly because the traditional structures of rural society in Latin America have been more shattered than in Africa and Asia.

Furthermore, it was in Lima that the first research was done which attempted to emphasise the true nature of the shanty towns, which had usually been portrayed as a disgraceful blight on the capital cities of the Continent. John Turner, a British architect, and William Mangin, a United States sociologist began their research in the late 1950s and came to see the growth of the *barriadas* as a positive response of the most vigorous elements of society to changing economic and social conditions.

The traditional Peruvian economic and social structure, which was geared to the provision of riches to a very few people who controlled the export of raw materials to North America and Europe, simply could not accommodate the revolution in expectations which has been the product of the spread of radio receivers, the construction of air strips in the remotest parts of the country, and road building pro grammes.

The development process by which the more aggressive and intelligent members of Peru's rural population are assimilated through urbanisation into the country's cash economy was first described by Turner and Mangin, who have been followed by many other researchers.

They saw that migration was not a simple move from the rural village to a *barriada* outside Lima. Most of the immigrants, unless they are displaced by an earthquake, have been gradually orientated to urban life in a provincial city before making the move to the capital. On arriving in Lima the immigrants do not move immediately into *barriadas*, but rather into inner city slums, described by Mangin and Turner as reception areas.

It is only after a further period of adjustment to metropolitan life and with a measure of economic security that land invasion and building a house in a *barriada* emerges as an alternative option. The squatters, who have been winners in the selection process all the way down from the Andes to the coast, may spend the next 15 years building their houses.

A conventional house building programme quite simply could not deal with the problems of these people, who have low incomes and whose major investment in their houses will be 15 years' work. The occupation of the Villa Salvador site, which he observed last year and which will eventually accommodate a population of 200 000, may be compared with a United Nations low-cost housing project in Peru, which had spent a million dollars over a three-year period without providing shelter for a single person. Furthermore, in spite of a wildly extravagant competition and any amount of high-minded theorising, the final houses will cost so much to construct that they will only be accessible to middle class families and do nothing to help the city assimilate the rural immigrants.

The initial response of the largely Spanish descended inhabitants of Lima to the invasion of Quechua speaking immigrants from the mountainous hinterland was to imagine that the revolution had at last arrived and that they would soon have their throats cut by the poverty-stricken mob. In a sense, the option offered by the *barriadas* postponed the revolution; potential revolutionaries were too busy building their own houses.

The challenge to existing structures is too massive and too immediate for orthodox planning solutions to cope with. Houghton and others who think like him believe that the *barriadas* process shows very clearly how active participation provides a solution to the problem of shelter in a context of scarce resources and widespread poverty. They believe that only planning based on active participation is likely to succeed and that this must also be reflected in other areas of national activity.

Turner and Mangin's original contribution was to rectify a development process, which had previously been seen merely as a 'social problem'. Today they themselves are reconsidering development policies in the light of that first insight. Houghton, and a Chilean architect, Hans Fox, who also works at the Architectural Association, are looking for points in the *barriadas* process at which Government and professional intervention and support could be necessary and helpful, trying to determine the kind of input which could be constructive.

They are particularly interested by the Cuban experience in the matter of housing policy. After the revolution, the Cubans believed that the answer to the problem of accommodation was to industrialise housing production and to provide completed housing units for all inadequately housed people. This simply did not work and Cuba is now solving its problem by setting architects, civil engineers and planners directly to work with the users, the people who need houses, helping them to make better use of the materials which lie directly to hand.

In this model of housing development an architect's task ceases to be the provision of finished houses and becomes an effective partnership between the designer and the user, who will also be the builder. Amenity ceases to be an artificially imposed set of abstract criteria and becomes the best use of a given urban space, as perceived by its inhabitants.

Houghton believes strongly that there are lessons in this for the developed as well as for the dependent world. He sees low-cost housing projects geared to minimum standards of space and cost as being as evil in Britain where workers are shepherded into council houses, managed by the local authority, as in South Africa where the 'solution' is a township of dehumanising dog-kennels.

The invasion studied by Houghton in Lima last May had elements which were typical of the process described in general terms above. The site chosen — Pamplona Alta on the outskirts of Lima — was not typical as it was partly privately owned, whereas the squatters normally choose publicly owned land.

A typical factor was the organisation of the invasion, which mobilises all kinds of ad hoc leaders — traffickers in building plots, political organisers, people with natural qualities of leadership who happen to need a building site, entrepreneurs in building materials — particularly of the straw mats, which provide shelter during the first weeks. The price of such mats may rise by 15–20 per cent during the period of an invasion. Since it never rains in Lima there is no hurry to put up a permanent roof.

A second factor was the timing of the invasion. The squatters try to pick a moment when the Government is badly placed to take repressive action against them. In this case the invasion followed a major Cabinet reshuffle, with new Ministers as yet not fully in control of departmental machinery. An additional bonus was the fact that a meeting of the governors of the Inter-American Development Bank was due to open in Lima the following week with visitors and journalists from all over the world.

The Government does not encourage new invasions, partly because they upset urban planning procedures and partly because they involve new pressures on fully stretched services. In the case of Pamplona, the Minister of the Interior took action against the squatters and at least one person was killed with 64 injured. Ultimately this cost the Minister his job and gave the squatters increased leverage with the Government, which felt the eyes of the world upon it.

Although the Government was able to persuade the squatters off the land they originally occupied, they had to be granted a rather larger site at Villa Salvador, with hastily installed services — albeit in a rather less desirable sector.

Unless Lima's economy adjusts to the new immigrants, there is every chance that the city will be one of the world's larger urban disaster areas by about 1990. However, the present Government is responding, more or less positively, to the challenge.

Adjustment is not only needed in Latin America, Nick Houghton says. He hopes for change in the attitudes of the developed world — attitudes to foreign aid (most of which he believes to be counter-productive or even destructive in its present form), attitudes to the rôle of cities, and an understanding of the nature of dependence which is shaping and warping the development of countries like Peru.

(*Source*: *The Guardian*, 23 May 1972)

3 HOUSING IN LUSAKA: AN EXAMPLE OF SELF-HELP

HOUSING THE URBAN POOR

Much has been written about the problems associated with urban growth in Third World countries, particularly the impact of squatter housing on the urban landscape. In June 1975, the United Nations Conference of Human Settlements, called *Habitat*, was held at Vancouver. At this conference, governments were able to discuss strategies for dealing with the basic need for shelter in Third World cities. The extracts from articles in the *New Internationalist* indicate the changing attitudes of governments towards the problem of housing the urban poor.

There are some discordant notes in the general harmony of agreement on 'self-help' as the solution to the world's housing problems.

Firstly, relying on 'the efforts of the poor people themselves' could easily become not the solution to the problem but a convenient way of forgetting about it. Only when people have the land, the basic services, the means and the opportunity to create their own communities and their own homes will 'self-help' mean any actual improvement in the living conditions of the poor. These factors cannot be taken for granted, yet without them the self-help approach is not an enlightened step forward but a benighted stride backwards to the Victorian philosophy of 'poverty is the responsibility of the poor'.

Secondly, the new approach could also be an excuse for solidifying dual society — with self-help improvements to crowded slums becoming acceptable for the poor whilst the rich continue to live in houses of a far higher standard and subsidised to a far greater extent by government spending. Self-help only makes sense and justice in the context of the dismantling of the dual society. In the absence of greater equality, the implementation of Habitat's conclusions would be a retrograde step.

CALLING OFF THE BULLDOZERS

To house everybody in the world by the year 2000 would mean building an estimated 75 000 new homes every day for the next twenty-four years. In conventional terms — the brick and steel and glass three-bedroomed home built to national standards borrowed from the West — this simply cannot be done.

In June the United Nations 'HABITAT' Conference on Human Settlements met in Vancouver to pool the experience of over a hundred countries and to discuss new ways and means of meeting people's basic need for shelter.

Ben D'Souza's argument for government-assisted self-help housing won the day at Vancouver, as JEREMY BUGLER reports.

In an extra-ordinary conversion, governments all over the world appear to be revising their ideas of how to house the huge numbers of their poorer people. Suddenly, it seems, the bulldozers have been called off from razing to the ground the shanty-towns around Third World cities. Huge monolithic government house programmes have been shredded or altered drastically. Instead, governments appear to have turned to involving the people themselves in the actual building of their houses and communities.

Habitat, the United Nations Conference on Human Settlements, which ended in Vancouver on 11 June, bore witness to the measure of this change of thinking. The movement, given a seminar to itself under the shorthand name of 'self-help housing', swept Habitat, and was one of its outstanding successes as a conference.

It became apparent that the strategy of laying down simply the bare bones of a new community — the roads, the sites of the houses, the water supply, the sewage and rubbish collection, perhaps electric power — has taken hold in many countries. In the Philippines, in El Salvador, in Francistown in Botswana, in Port Sudan in the Sudan, in Papua New Guinea, in Zambia, in Jamaica and Columbia and Pakistan — and of course in India as Ben D'Souza points out — 'site and services' schemes are under way and in many places completed.

The reason why site and services schemes, using the self-help or community-help abilities of the local people, have caught on in so many places is that government or municipally-built houses are simply too expensive. They are too expensive both for the governments that build them and for poor people to rent them.

PEOPLE KNOW BEST

The official view of Third World cities used to be that their teeming squatter settlements were breeding grounds for disease, crime and social and family disorganisation. An opposite view, whose leading spokesman is the architect John Turner, sees them instead as a triumph of self-help and mutual aid among people who have nothing to gain from grandiose housing schemes. His approach to the problem of housing is parallel to educational views of his close friend, Ivan Illich.

Turner, evolved his ideas when studying illegal squatter settlements in Peru in the early '60s. Since then he has led the assault on the idea that the housing needs of the poor can be met by building huge, costly, standardised estates. Centrally administered housing is characteristically hideous, it defiles personal relations, dirties the environment and alienates the users. It is also uneconomic and extravagant with resources. And they ignore the most vital factor in housing: the involvement of the occupants.

Under the influence of Turner and colleagues, there has been a move away from expensive token projects towards aided self-help.

But allowing rein to people's energy and willingness to invest in their own housing is not to be a cheap way for housing authorities to opt out of their responsibilities. Large organisations have no business building houses any more than the directors of a football club have any business scoring goals. But they should instead be installing infrastructure, supplying tools and materials that people and their own small enterprises can use locally, and fairly distributing scarce resources. This requires what Turner defines as 'decentralising technologies and network structures'.

Turner's ideas on user-control have proved unacceptable to some governments who are still commissioning big prestige projects in the belief that what poor people do for themselves cannot be right and proper. But the lesson from the Third World — that people get more satisfaction at less expense if they are in control — is slowly gaining application in the decaying cities of the West.

?

Read the articles and answer the following questions.

1 Is the 'self-help' approach with government aid for basic infrastructure such as roads, sites and water supply, the best way to deal with the housing plight of the urban poor?

Consider this question from two angles:

a Is it desirable to follow this policy?

b Is it the most practical solution?

2 Discuss your answers with others in your class.

You might feel that you do not have enough background knowledge to give a sound judgement on the issue but it does not matter at present. By looking critically at the resources provided in this chapter you will develop your own ideas further. At the end of the chapter you will be asked to reassess your answers to these questions.

Uncontrolled settlements

The Zambian project

Zambia provides an interesting case study of an evolution in public policy toward self-help, including squatter housing. Prior to independence in 1964, migration to urban areas had been administratively controlled through pass laws which restricted admission to those with official consent. Squatting was therefore a minor problem. With the advent of independence, rural–urban migration accelerated dramatically, and urban population doubled between 1963 and 1969 to reach 1.13 million, or 29 per cent of the total. According to various estimates, squatter population increased by between 19 and 25 per cent per annum over the period. That is, it doubled every three to four years.

In the mid-1960s — ahead of many other governments — Zambia began implementing 'site and services' (serviced lots on vacant land),initially as a means of resettling squatters. By 1969, planned self-help housing was established to accommodate new growth. The Government accepted the fact that available resources were grossly inadequate for conventional contractor-built houses for every urban family, and at the start of the Second National Development Plan (1972–76), it instructed local authorities to provide serviced sites only. A review of the initial attempts to resettle squatters revealed that it was not possible to resettle all squatters who, by 1972, represented about one-third of the total urban population. Therefore, as a corollary to its site and services policy, the Government determined to upgrade all improvable squatter settlements.

In 1973, Zambia requested World Bank assistance in the financing and implementation of its shelter programme. A project was formulated comprising site and services and squatter upgrading affecting some 30 000 households, or about 40 per cent of the total in Lusaka at a cost of about $40 million. Although not catering for all the shelter requirements in the capital, the project was seen as the first of a series within the national programme. As the project was the first systematic attempt at squatter upgrading (designed to improve four squatter complexes comprising nearly 24 000 households), it was also anticipated that the lessons learned in Lusaka would benefit other towns and cities throughout the country. This project got under way at the end of 1974, and the upgrading of the first squatter complex is expected to be completed by mid-1976.

(*Source*: C. Madavo, 1976, 'Uncontrolled Settlements' from *Finance and Development*)

Print 1 *'Site and service' housing in Chawama, Zambia*

? The previous extract is taken from an article by Madavo in which he refers specifically to government policy in Zambia.

1 Draw a flow-line time chart to indicate changes in public policy over time in relation to urban migration and squatter settlements. As you can see, site and service schemes were started after independence, and the upgrading of squatter settlements took place in the early 1970s.

2 Study all three articles below and explain how the changing policies of the Zambian government since the 1960s reflect changes in public attitudes towards squatters.

In the 1960s many people began to be aware of the rapid growth of squatter settlements and became concerned about the welfare of their inhabitants . . .

Meanwhile a Squatter Control Unit was actively trying to persuade or threaten potential squatters not to build, to the extent of demolishing their homes, but pressures for housing were such that their efforts made no tangible impact. In 1969 Doxiadis Associates proposed a 'Special Programme of Action for the Settlement of Squatters' under which every squatter family (estimated by Doxiadis Associates to be 16 000) would be resettled onto a new plot within a year. The first stage of this was Kaunda Square, the speed of development of which proved that this would never have been a realistic course of action. Four years later only just over 1000 of the 2000 plots had been allocated.

In 1971 a new climate of opinion began to develop: a committee, on which the Council was represented, was convened to discuss the future of squatters. It recommended that upgrading of squatter settlements was generally the best approach to adopt, and thus it was that this proposal was incorporated into the Second National Development Plan, published at the beginning of 1972.

(*Source*: R. Martin, 1974, from *Local Government in Zambia*, ed. N. R. Hawkesworth)

These settlements are characterised by a lack of any regulated plot demarcation, uncontrolled housing standards, overcrowding, no pure water supply, and no proper sanitation. Their growth has been spontaneous and unauthorised; they are 'illegal' settlements, although the law has generally turned a blind eye to their existence and has seldom intervened to evict their populations or restrict their growth. Indeed, the major and most common criticism of both central and local government policy towards these settlements has been that it has tended to ignore them, hoping (unsuccessfully) to inhibit their growth by denying them services, unwilling to take positive measures to demolish them, and that it has been dilatory in providing legal alternative housing for their occupants. However, this policy is at last being replaced by a realisation that since the squatter way of life is certain to persist for years to come, amenities such as stand-pipe water and sewerage have got to be provided, and security of tenure introduced for squatter plots.

(*Source*: A. J. F. Simmance, 1972, 'A Working Paper', from *Urbanisation in Zambia*, The Ford Foundation)

'SQUATTER SETTLEMENTS'

The term 'squatter settlements' refers to land which has been taken over illegally by people and used to build their homes. Other terms are also used, e.g. 'uncontrolled settlement', 'unauthorised settlement', 'spontaneous settlement', 'temporary settlement'; or more localised terms such as '*favelos*' in Brazil, or '*barriados*' in Peru. The phrase 'shanty town' is often used as a synonym for squatter settlements.

?

1 List as many words and/or phrases as you can think of which would describe the physical appearance of a squatter settlement as you imagine it.
2 Discuss your images with other members of the class.
3 What are the most common images portrayed by the lists you have made?
4 In what way do they differ?

In reality, squatter settlements differ a great deal in their physical appearance and the stereotyped image often represented by the media is not typical of all settlements. For example, Richard Martin in his article about squatter housing on the outskirts of Lusaka, the capital city of Zambia, describes such as at worst being:
'. . . like a run-down village house, and at best it equals and betters conventionally built housing. In between there is a wide range of well-built, but cheap, and sun-dried brick houses, plastered and painted, standing in a clean and neat environment. Densities are high, relative to other housing in Zambia, at between thirty to forty houses per hectare, but not excessive.'
(*Source*: R. Martin (May 1976) *Geographical Magazine)*

This chapter will look more closely at the demand for low cost housing in Lusaka. The city provides a useful case study of the way in which squatter communities have improved their living environment through self-help. It also offers evidence of the work of a fairly progressive city council in finding partial solutions to the housing shortage. But first, it is necessary to look at the urbanisation process in Zambia as a whole.

23 Existing and future plots need to be controlled by town planning and building by-law requirements. Especially cities and townships selected for industrial locations, or likely to become attractive for extensive settlement, will be town-planned before the construction of industrial establishments in their area is begun.

24 It is recognised that although squatters' areas are unplanned, they nevertheless represent assets both in social and financial terms. The areas require planning and services and the wholesale demolition of good and bad houses alike is not a practical solution.

25 First priority must be given to the acquisition of land when any unauthorised settlement on it is to be upgraded. Strict control of any further development must be enforced both inside and outside the designated area. The following services must then be provided in any upgrading exercise: piped water supplies, sewers and sewage disposals, roads and surface water drainage, street lighting and other communal services.

(*Source*: *The Zambian Second National Development Plan, 1972–1976*, Ministry of Development and National Guidance)

URBAN GROWTH IN ZAMBIA

Apart from South Africa and Zimbabwe, Zambia is the most urbanised country in Southern Africa. The urban centres shown in Figure 1 are concentrated along 'the line of rail' between Livingstone in the south, and the Copperbelt which lies to the north of Lusaka. These centres developed during the colonial period when Zambia was under British rule. The towns to the north of Lusaka grew as townships for the mining communities involved in extracting copper, lead, zinc and other minerals. Copper still forms over 90 per cent of the exports of the country, and lead and zinc are other leading exports (see chapter 16, *pages 328–340*). The towns to the south of Lusaka grew as agricultural centres for the European commercial farming which developed in the early twentieth century. These farms produced food for the mine townships further north. Livingstone was initially the capital city, but the colonial government decided that Lusaka should take over this function in 1931 as it was more centrally located.

Lusaka, like all the other urban centres, has attracted people from the rural areas, and, after independence in 1964, it was the fastest growing urban centre in the country. Its growth is also due to the natural increase within the city itself. Consequently, the question of housing is a serious one as it is estimated that there are at least 250 000 squatters out of a population of over 641 000 (1980 estimate), that is, almost 40 per cent of the total urban population.

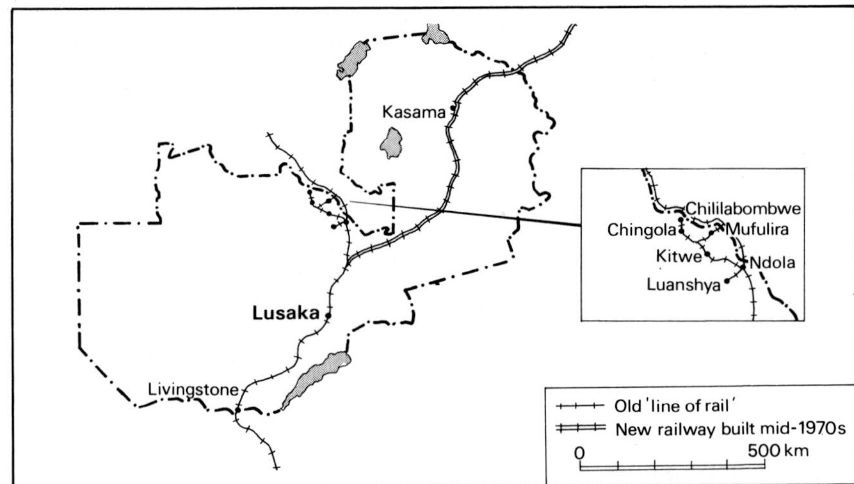

Figure 1 *The location of Lusaka and other major 'line of rail' towns in Zambia*

? Study the information given in Table 1.

1 Rank the towns from 1 to 10 according to their population size in 1980.

2 Comment on the distribution of the largest urban centres in Zambia.

3 Draw a graph to show the change in total population for the five largest towns for 1969, 1974 and 1980.

4 Write a short paragraph on population change in Zambia between 1969 and 1980.

RURAL–URBAN MIGRATION

The migration of people from rural areas to towns has played an important part in urban population growth in Zambia, and a number of conclusions can be drawn from research which has been carried out

Table 1 *Population change in Zambia 1969, 1974 and 1980*

	1980 population[1] *mid-year estimate*	*1974–1980 average annual growth rate %*	*1974 population*[2]	*1969–1974 average annual growth rate %*	*1969 population*[2]
Total Zambia	5 834 000	4.0	4 695 000	3.0	4 056 995
Large Urban Areas					
Chililabombwe	77 000	6.3	56 000	4.5	44 862
Chingola	192 000	7.2	131 000	5.3	103 292
Kabwe	147 000	8.3	98 000	8.2	65 974
Kalulushi	60 000	7.7	41 000	4.9	32 272
Kitwe	341 000	6.0	251 000	4.7	199 798
Livingstone	80 000	6.3	58 000	5.1	45 243
Luanshya	184 000	8.7	121 000	4.7	96 282
Lusaka	641 000	10.0	401 000	8.0	262 425
Mufulira	187 000	6.3	136 000	4.8	107 802
Ndola	323 000	6.8	229 000	7.5	159 786
Percentage urban	43.0[3]		35.3		29.4

Sources:
[1] *The Europa Year Book* 1981, Vol. 2, Europa Publications Ltd.
[2] Sample Census of Population 1974; Preliminary Report February 1975, Central Statistical Office, Republic of Zambia.
[3] World Development Report 1982, The World Bank, Oxford University Press.

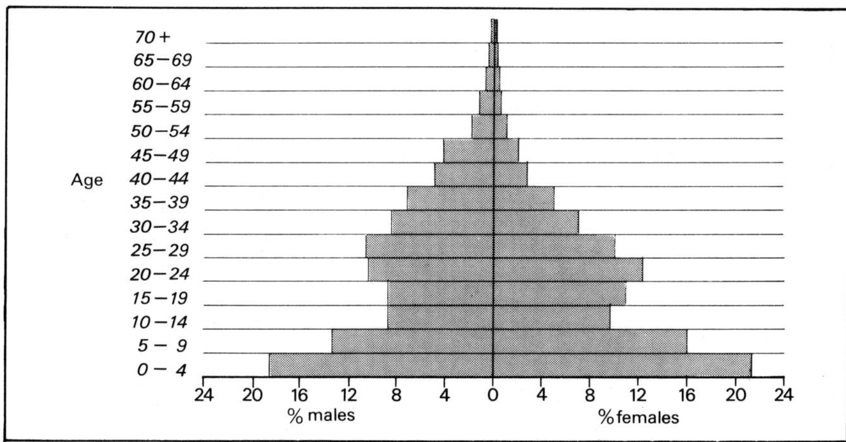

Figure 2 *The age–sex structure of Lusaka, 1969*
(*Source*: National Housing Authority, 1975, *Self-help in Action: a Study of 'Site and Service' in Zambia*)

by Jackman in 1973. Some of these conclusions are summarised briefly below.

1 The Zambian population was highly mobile.
2 Movement patterns were dependant upon social links.
3 The migrants were predominantly young (and well educated) and this distorted the age structure (see Figure 2).
4 Rural migrants often used the rural townships as staging posts before moving on to the larger towns.
5 Improved communications only heightened rural–urban migration.
6 There was a distortion in the sex structure in both rural and urban areas. In the 1960s, most of the migrants were males establishing themselves in the towns. In the 1970s more females and children moved to the towns to join the menfolk.
7 The destination of the migrants was determined by distance, ease of travel and the rural–urban social links.

It is suggested that in Zambia, the 'pull' factors are more important than the 'push' factors. It seems that it is the income differential between the rural and urban areas which is of most significance in affecting migration. In the period between 1965 and 1976, the manufacturing, service and commercial agriculture sectors in the urban and peri-urban areas experienced higher growth than the subsistence agriculture sector in the periphery. Even so, the urban areas have not been able to absorb all the migrants and this has resulted in unemployment, under-employment and the growth of what has come to be known as the *informal sector*. The situation has contributed to urban poverty.

?

1 What are the 'push' and 'pull' factors which bring about large-scale rural–urban migration?
2 What are the social and economic consequences in the rural areas?
3 What are the social and economic consequences in the urban areas?
4 To what extent does the population pyramid for Lusaka in 1969, shown in Figure 2, support the conclusions, about the effect of migration on the distortion of the age structure of urban areas listed above?

HOUSING IN LUSAKA

One of the main difficulties confronting the urban newcomer is finding a place to live. In Lusaka there is a shortage of low-cost housing, and squatter settlements have mushroomed because there is no alternative accommodation for the low-income groups. These squatter settlements not only house the new arrivals from the rural areas but also many of the people who are resident in the city and who are unable to find a home. This is particularly the case with young newly married couples ready to start their own families.

In order to understand the housing problem it is necessary to look at the growth of the city, its internal structure, the spatial distribution and availability of different types of housing stock. Figures 3, 4 and 5, and Tables 2 and 3 provide information on those features, as does the following extract.

Planning the Capital in 1931

Adshead also presented an outline plan and his basic recommendations laid the planning foundations for the city. He proposed building the capital away from the unattractive existing settlement, and on the Ridgeway, a low, breezy schist ridge a mile or so to the east. Here he designed a 'garden city' on an imaginative and spacious groundplan, in contrast to the original gridiron. Ample space and greenery were planned, with large building plots for European officials centred around the Ridgeway axial road (now Independence Avenue), with its government offices, legislature, High Court and Government House. Plot sizes decreased somewhat away from this 'Snobs Hill', but densities were everywhere low. Adshead also accepted the colonial *de facto* residential segregation of races. A limited African population was envisaged, mostly government employees and domestics. Some servants' families would be housed on European plots, but most Africans would occupy separate locations, mainly south and west of the capital area. These had somewhat higher densities and much lower standards of amenity than European areas. An Asian residential quarter, with a secondary business district, later emerged near the old town, and police and army camps and other institutional areas developed. Segregation was virtually complete until the eve of independence. The development of both industry and services was somewhat inhibited by reliance on Southern Rhodesia, especially during Federation, and by growing dominance of the Copperbelt regional market. Lusaka grew steadily without again changing its character until independence in 1964.

(*Source*: D. H. Davies, 1971, *Zambia in Maps*)

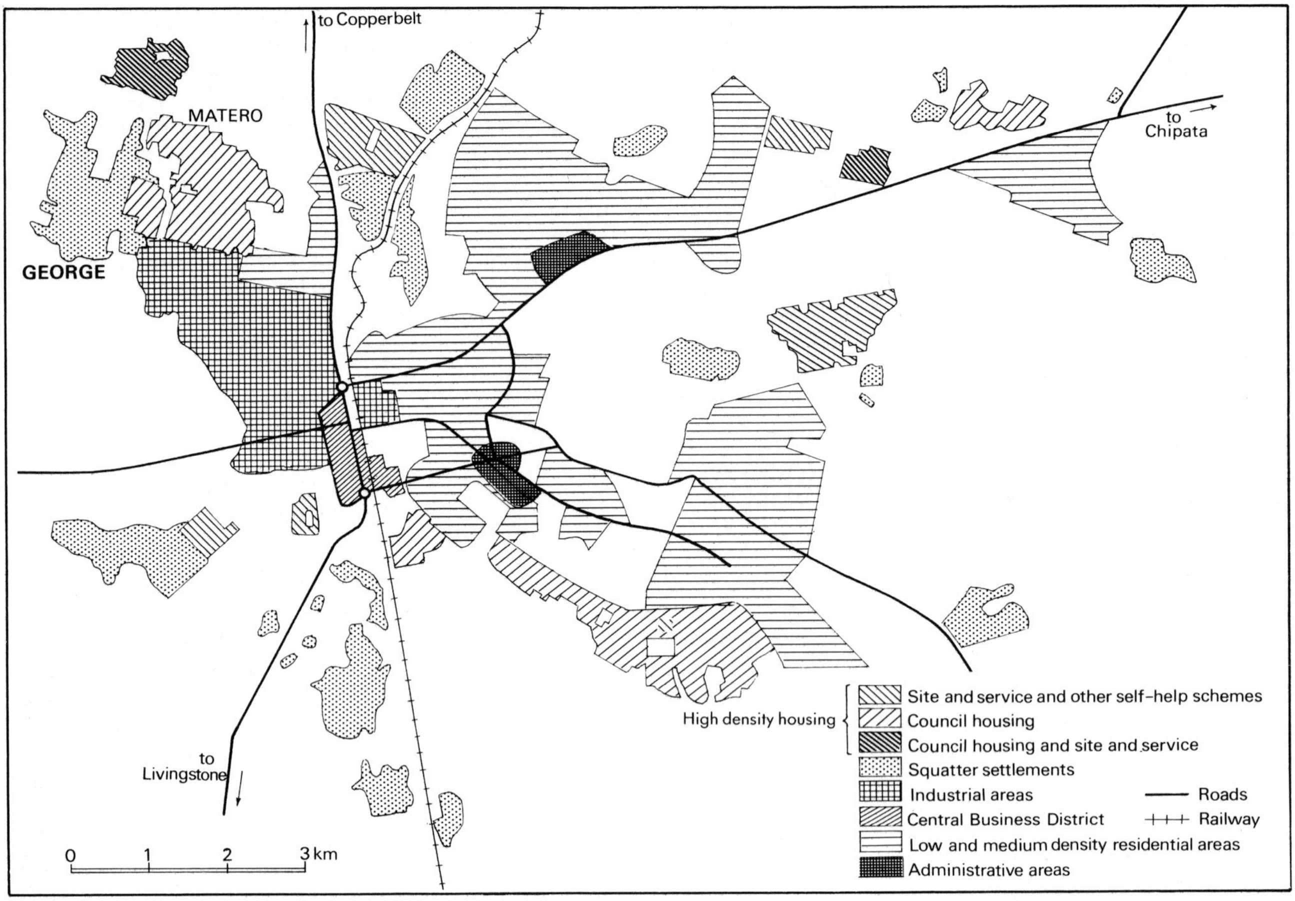

Figure 3 *Present day pattern of land use in Lusaka*

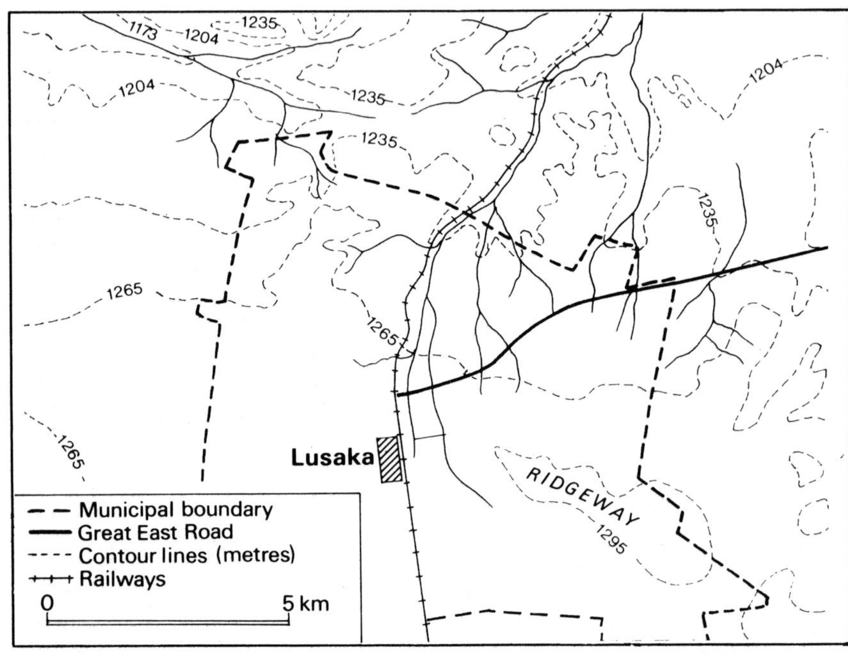

Figure 4 *Relief and drainage*
(*Source*: D. H. Davies, 1971, *Zambia in Maps*, Hodder and Stoughton)

Table 2 *Housing stock in Lusaka* (number of housing units)

High and medium cost housing	
Private	5 413
Government	2 534
Lusaka City Council	367
	8 214
Low cost housing	
Government owned	2 170
Servants' Quarters (estimate)	5 351
Lusaka City Council	14 440
	21 961
Self-help and 'site and service'	4 988
Squatters	28 300
	63 463

(*Source*: see Table 3)

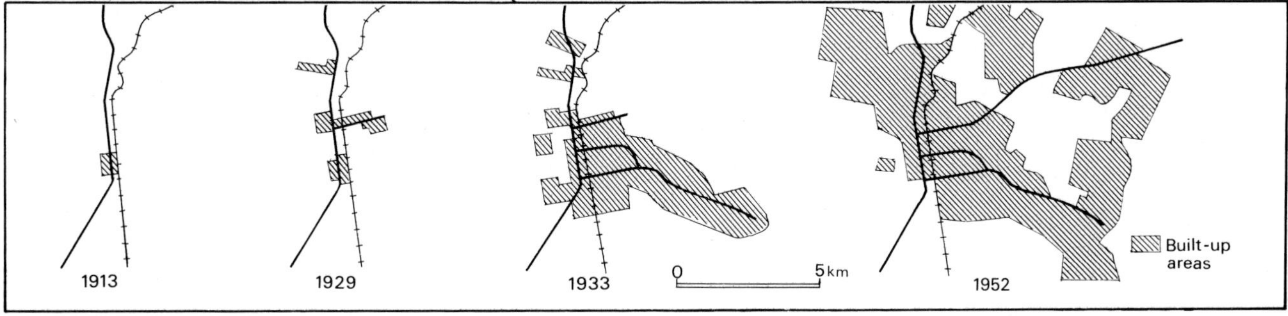

Figure 5 *The planned growth of Lusaka*
(*Source*: D. H. Davies, 1971, *Zambia in Maps*, Hodder and Stoughton)

?

1 Figure 3 shows the present day pattern of land use in Lusaka. Draw a simplified land use model of the city showing the central business district, the industrial zone, the administrative area, the low and medium density residential zone and the high density residential zone. (*Note*: the high density housing has been divided into three sub-groups on the map.)

2 Describe the internal structure of the city.

Cities in Africa which grew during the colonial period are different from Western cities and a full explanation of Lusaka's land use pattern has to take into consideration the effect of the colonial administration. In addition, the physical landscape has had an effect on the city. The maps in Figures 4 and 5 show the local relief and the planned growth of Lusaka between 1913 and 1952.

3 Using those resources, together with the written extract *Planning the Capital in 1931* (*page 49*), by Hywel Davies, give an *explanation* of the city's pattern of growth since 1913. Pay particular attention to the high density zones. (*Note*: Adshead was the architect involved in the planning of Lusaka in 1931 when it became the new capital city.)

4 Information on the Lusaka housing stock is given in Table 2. Draw a pie chart or bar graphs to show the proportion of:
a High and medium cost housing.
b The total low cost housing.
c Site and service housing.
d Squatter housing.
Comment on the graph you have drawn.

5 Table 3 provides a breakdown of the squatter and 'site and service' units according to location in 1972. Each settlement is numbered and Figure 6 shows their location. Trace the map, and use the statistics provided in Table 3 to draw located bar charts to show the total number of housing units. Use different shading to distinguish between the squatter settlements and the 'site and service' schemes.

6 Describe the pattern of distribution you have mapped.

7 What relationship is there between the original city council boundary and the majority of the squatter settlements?

8 What is the average distance of these settlements from the city centre?

The city council boundary was extended in 1970 to form Greater Lusaka. You will have noticed that the largest settlement of all is George, a squatter settlement on the north-western outskirts of the city. A more detailed study of George is given later in the chapter.

Table 3 *Housing stock in squatter settlements and 'site and service' and self-help schemes*

Squatter settlements (October 1972)		
1	George, including Desai, Freedom, New Chibolya, Paradise, Chikolokoso	7 094
2	Chaisa	3 900
3	Nguluwe, Mtengo, Antonio, Howard	1 465
4	Chipata	2 337
5	Old Kanyama	2 841
6	Kalingalinga	1 620
7	Kalikiliki	340
8	Bauleni	518
9	Ngombe	261
10	Chelston (Banda, Kamanga, Roadworks) and (since demolished) Mawuzu	280
11	Chainda	660
12	Chilanga (October 1973)	510
13	Chawama	4 050
14	Quarries	775
15	John Howard	1 297
16	Jack	352
		28 300
'Site and service' and self-help		
1	Marrapodi-Mandevu	1 120
2	Mtendere	1 451
3	Kaunda Square Stage 1	970
4	Chunga	182
5	New Kanyama	720
6	Chawama	373
	Others	172
		4 988

(*Source*: N. R. Hawkesworth, 1974, ed. *Local Government in Zambia*, Lusaka City Council)

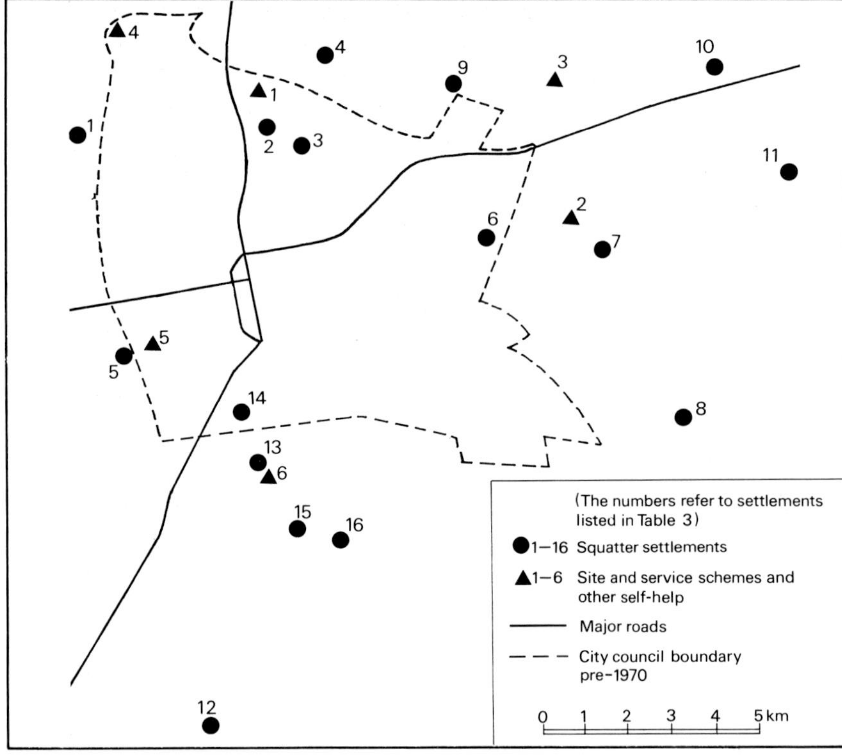

Figure 6 *The location of squatter settlements and 'site and service' schemes*

THE RESETTLEMENT OF SQUATTERS: THE 'SITE AND SERVICE' SCHEMES

The previous exercise will have given you some indication of the shortage of low cost housing in Lusaka. One of the aims of the First Official Housing Programme, introduced in 1965, was to construct as many low cost houses as possible, of which 30 per cent were to be on 'site and service' plots. In 1966, this aided self-help housing scheme was launched. There were six major organisational features:

- Each participant would be eligible for a loan of 72 Kwachas (about £25 at the 1985 rate of exchange) for building materials repayable over a four-year period at an interest of 5 per cent per annum.
- The local authority had to establish stores at which the participants could buy, or obtain against their loan, building materials.
- The local authority had to establish a project management team which would include the mayor and members of each interested department to co-ordinate and guide the development.
- A field team should be established, which would include artisans and community development workers to advise and assist the participants on building their houses.
- Participants would be given a ten-year renewable lease and would be entitled to sell their houses with the approval of the Council. They could, however, be evicted if they failed to pay service or loan charges.
- House plans were provided. The Zambia Housing Board prepared a booklet of five house plans. Each plan was presented in two stages, the core version and the expanded version together with a materials list for each type, and an illustrated guide to house construction. Houses could be built of sun-dried brick, burnt brick, concrete block or cinva ram brick. (The cinva ram is a hand operated mould for making soil-based bricks.)

(*Source*: National Housing Authority, 1975)

At this time most of the squatter areas were under the jurisdiction of the Ministry of Local Goverment and Housing and a 'Squatter Control Unit' was set up to demolish squatter settlements as the 'site and service' schemes were implemented. These early site and service plots were each 12 metres × 27 metres in size and were provided with road access and piped water. At the rear of the plot a shower, latrine and wash area

Print 2 *A concrete block 'site and service' house in Chawama, Zambia*

were built and the houses were left to the householders to build. These early plots were later called *normal* plots.

There were limited public resources available for these 'site and service' schemes and much of the capital that was available was used for high and medium cost housing. In 1968, the Ministry of Local Government and Housing suggested that a different type of service plot should be provided. They were called *basic* plots and remained the same size as the original ones, but the water was to be shared by about 25 households from stand-pipes. Low cost gravel roads were provided and householders had to build their own latrines.

The need to remove the squatter settlements and resettle the inhabitants was reflected in the *Special Programme of Action for the Settlement of Squatters* (1969) which was prepared by the international planning consultants — Doxiadis Associates. At this time, they were in the process of preparing the Lusaka Development Plan which was eventually completed in 1977. The aim was to allocate and service 40 000 plots, and to resettle *all* the squatters. In reality these plans came to very little; only one housing area was built in 1969. This was Kaunda Square, to the north-east of the city. In two-and-a-half years, only 376 houses were finished although 2000 plots were planned for this site. The scheme was not very popular because of its gridiron layout, its very high density of housing (even compared to some squatter settlements), and its distance from the city centre.

The National Housing Authority, created in 1971 to deal with the housing problem, carried out a survey of four 'site and service' schemes between January and April 1973. Three of the schemes, Kaunda Square, Mtendere and Chawama were in Lusaka; the fourth in the survey was in Ndola. A breakdown of the sample they used is seen in Table 4.

Table 4 *Sample size by settlement of the 1973 'site and service' survey*

Area	*No. of plots*	*No. of completed houses*	*No. included in survey*	*Sample percentage*
Chawama	240	172	60	35
Kaunda Square	1 196	530	60	12
Mtendere	2 753	2 751	61	2
Lubuto (Ndola)	844	844	45	5
	5 033	4 297	226	

(*Source:* adapted from National Housing Authority, 1975, *Self Help in Action: a Study of Site and Service Schemes in Zambia*, p. 11)

Table 5 *Reasons for wanting to build on 'site and service' schemes (number of responses in ranking order)*

	Chawama	*Kaunda Square*	*Lubuto*	*Mtendere*	*All*
Own house	11	17	20	22	70
Bigger/better house	11	10	24	9	54
Piped water	9	—	—	24	33
No rent	3	16	5	6	30
Forced to go	11	13	—	3	27
Loan available	12	13	—	—	25
Better services	2	9	11	—	22
Piped water/loan	—	13	—	—	13
No eviction	6	1	—	2	9
Nowhere else to go	1	3	2	1	7
Bigger plots, good roads	2	—	1	3	6
Assistance from employers	—	—	4	—	4
Peaceful living	—	—	—	3	3
Low rent	2	—	—	—	2
Good looks of 'site and service'	—	—	1	1	2
Relatives directed me	1	—	—	—	2
Liked the place	—	—	—	2	2
Good toilet	1	1	—	—	2
People help each other here	2	—	—	—	2
Wanted to live in town	—	—	—	1	1
Places clean	—	1	—	—	1
Did not like squatter compound	1	—	—	—	1

(*Source*: National Housing Authority, 1975, *Self-Help in Action*: A study of 'site and service' schemes in Zambia)

? **1** What do you think are the limitations of the sample used for the survey in drawing fairly accurate generalisations?
2 Use Table 5 to work out what the priorities of the residents were in choosing to go to a 'site and service' scheme.
3 The six major likes and dislikes of the residents brought out in the survey are indicated in rank order below.

Likes	**Dislikes**
1 peaceful	1 lack of shops/markets/butchers
2 houses are big/good/permanent	2 no electricity
3 piped water	3 lack of clinic
4 larger plot	4 no farming plot
5 own house/no evictions	5 bad roads and transport
6 no rent	6 no schools

What kind of image do they give of the 'site and service' schemes? How do these likes and dislikes compare with the expectations that prompted them to go there in the first place?
4 The Chawama self-help project was provided with a lot of technical assistance and the participants worked on a co-operative basis. The work they put in was recorded on time sheets as in Figure 7 below. Draw a graph to show the amount of work done for the weeks shown on the family time sheet. Comment on the division of labour and periods of greatest activity.

? **5** When the survey was carried out, one of the researchers wrote his *own* comments on the people he interviewed. These comments provide a *subjective* view of the people but they were included in the final report because his perceptions seemed to provide a realistic view of the settlement and its inhabitants. A few of the comments are shown in Figure 8. Use these comments as a basis for discussion about the housing, the people, their work, problems and attitudes.

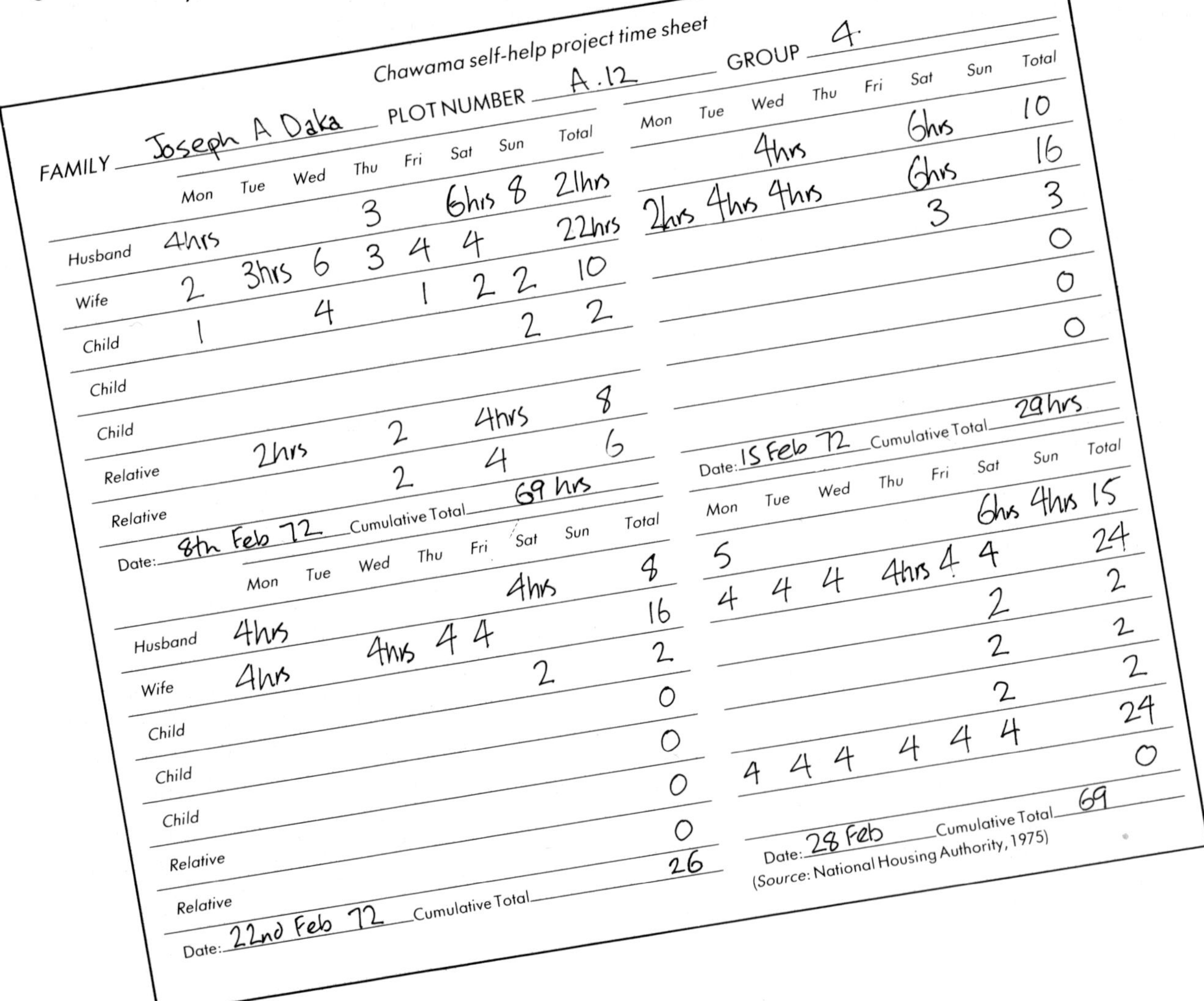

Chawama self-help project time sheet

FAMILY Joseph A Daka PLOT NUMBER A.12 GROUP 4.

	Mon	Tue	Wed	Thu	Fri	Sat	Sun	Total
Husband	4hrs			3		6hrs	8	21hrs
Wife	2	3hrs	6	3	4	4		22hrs
Child	1		4		1	2	2	10
Child							2	2
Child								
Relative		2hrs		2		4hrs		8
Relative				2		4		6

Date: 8th Feb 72 Cumulative Total 69 hrs

	Mon	Tue	Wed	Thu	Fri	Sat	Sun	Total
Husband	4hrs					4hrs		8
Wife	4hrs		4hrs	4	4			16
Child						2		2
Child								0
Child								0
Relative								0
Relative								0

Date: 22nd Feb 72 Cumulative Total 26

	Mon	Tue	Wed	Thu	Fri	Sat	Sun	Total
Husband			4hrs			6hrs		10
Wife	2hrs	4hrs	4hrs			6hrs		16
Child						3		3
Child								0
Child								0
Relative								0
Relative								

Date: 15 Feb 72 Cumulative Total 29hrs

	Mon	Tue	Wed	Thu	Fri	Sat	Sun	Total
Husband	5					6hrs	4hrs	15
Wife	4	4	4	4hrs	4	4		24
Child						2		2
Child						2		2
Child						2		2
Relative	4	4	4	4	4	4		24
Relative								0

Date: 28 Feb Cumulative Total 69

(Source: National Housing Authority, 1975)

Figure 7 *Chawama self-help project time sheet*

A SUBJECTIVE VIEW — THE RESIDENTS OF A 'SITE AND SERVICE' SCHEME

TELEPHONIST —

This young man has been very frank. He is not the one who built the house. It was built by the Company for K500. The Company hired their own labour and paid them. He will own the house after paying the K500 back to the Company. Because of all this — I mean the Company building the house — he does not have any idea about the cost of building materials.

WAITER —

This man has been very reliable — but had plenty of complaints. He says that he was forced to come here by the Council officials and the land resettlement officials. He was promised a lot of help — such as students and other voluntary organisations to help him build the house. But to his surprise he says that he was only dumped here and left to do all things on his own. He says that he is not happy with this place at all — because he just doesn't have the money to complete his house. He was alright where he was — he says. His house is very far from complete, and the walls don't look strong. Whoever the bricklayer was, he didn't know his work.

BRICKLAYER —

This man is an unemployed bricklayer. At the moment he does some part time jobs in the compound building other people's houses. His monthly income in this case is not known as he is only paid after the completion of the work which in most cases would take more than a month. His wife sells some fruits, tomatoes and vegetables at the market. Her income is not fixed as she gets between K3 to K4 per day depending on how many customers she has . . .

SAND DEALER —

This is one of the most beautiful houses in Lubuto. It has good wide windows with well plastered and painted walls. He owns a lorry in which he transports sand for sale to the other people who haven't completed their houses within the community. He paid nothing for the sand used for his own house as he gets his supply free from the bush. For the rest of the materials he did not have to go far to buy them, for the Site Store is very near to his house.

SELF-EMPLOYED BRICKLAYER —

This man wants to summon the people who were in his work group to court. He says that he laboured and built their houses — but they never helped him in building his house, so he had to build it alone. The only people who helped him were two neighbours. They only helped him for 4 days and then stopped. There were 9 people in his group and he helped them all build their houses. His house was the last on the list to be built by the group. By the time they came to it, everyone was tired and left all the work to him.

RAILWAY WORKER —

He was a very co-operative and understanding man. He and his wife contributed a lot to the group work. The husband admits, however, that the wife did more of the group work than he himself, as he was always at work during the day time and did a few hours group work in the evenings, 6 pm to 9 pm and 7 am to 12 am on Sundays. The wife worked from 7 am to 1pm nearly everyday of group work. She brews *kachasu* as an additional income for the family.

Figure 8 *A subjective view — the residents of a 'site and service' scheme*
(*Source*: National Housing Authority, 1975)

GEORGE AND ITS UPGRADING — A CASE STUDY

It was recognised in the early 1970s that it would not be possible to resettle all the squatters in Lusaka. Consequently, a second strategy was used as a corollary to the 'site and service' scheme. This involved the upgrading of some of the squatter settlements like George.

George lies about 6 kilometres north-west of Lusaka city centre, and west of the heavy industrial area and the council housing area of Matero (see Figure 3 *page 50*). The settlement itself has had several name changes. It was initially called George it seems, because the land was owned by a European settler of that name who left the country at independence. In 1968 it was called Kapwepwe Compound after the then vice-president, and later it was officially changed to Mwaziona after a change of government. This name was not accepted by the local people and it now has been officially changed back to its original name of George.

GROWTH AND DEVELOPMENT

Many squatter settlements in Lusaka originated as contractors' camps. George, on the other hand, was deliberately started by a businessman who illegally rented out small plots of poor quality land for about 1 Kwacha a month. In 1976 there were over 56 000 people living in George and the settlement looked very different to the way it did in 1958 when it had a total population of 570 (see Figure 9).

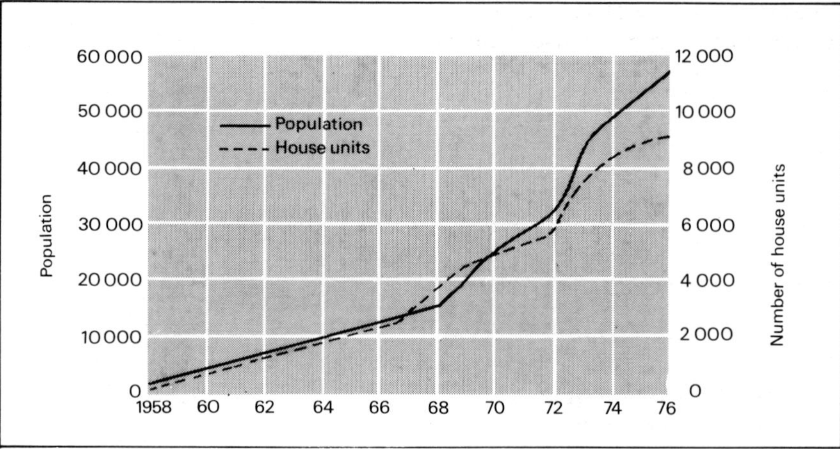

Figure 9 *Population change in George*

The early growth of the settlement took place along a low ridge which slopes down to the north-east, but since the early 1960s, the settlement has spread out from this ridge to the lower wetter lands to the north and north-east and towards the stream to the south-west. These low lying areas are susceptible to floods during the rains. From Figure 10 you should be able to see the relationship between the physical landscape and the stages of growth.

The early houses were made of poles and dagga (a type of hemp) and had grass thatched roofs. By the late 1960s many buildings had galvanised iron roofs and the walls were made of sun-dried bricks plastered with cement. Ten years later many of the earlier houses

Print 3 *Central part of George, 1977*

have been replaced by bigger and better concrete block houses built on the same site. Many of these houses have been extended, and this has contributed to an increase in the density of the built-up area which in George can be as many as 70 houses per hectare. Housing structures have improved over time but the problem of servicing them still exists. In the rains, many roads are impassable and a few people still obtain their water from polluted wells.

SOCIAL ORGANISATION — THE ROOTS OF SELF-HELP

At first sight, the physical layout of the houses appears disorganised but this is not the case. Richard Martin, writing in the *Zambian Geographical Journal* about what appears to be superficially a disorganised layout says:

'. . . the physical organisation is extraordinarily similar to many villages. The common ingredients are size (20 houses), use of space (a common open space in the centre, with private and semi-private open spaces off it and between the huts), physical groupings approximate a circle and there is general absence of cultivation within the village. In the squatter areas this physical separation of one village unit from another is not immediately obvious. In particular the position of the door is significant.'

The extended family, common in the villages, has been replaced by the nuclear family in the settlement, but there is still a strong sense of community. The residents have mobilised themselves to run the community and improve the environment. There is a sizeable community of business people in George — shops vary from being very large, modern, well-built structures with a wide range of goods down to the mobile tea cart (see Figure 11 on next page).

The local officials of the United National Independence Party (UNIP), now the only political party since the country became a one-party state in 1972, has played an important part in the mobilisation of the squatters. A newcomer has first to see the local UNIP official before he is allocated a house plot. The new applicant must have a sponsor from within the community, and if possible, a space is found somewhere near the sponsor.

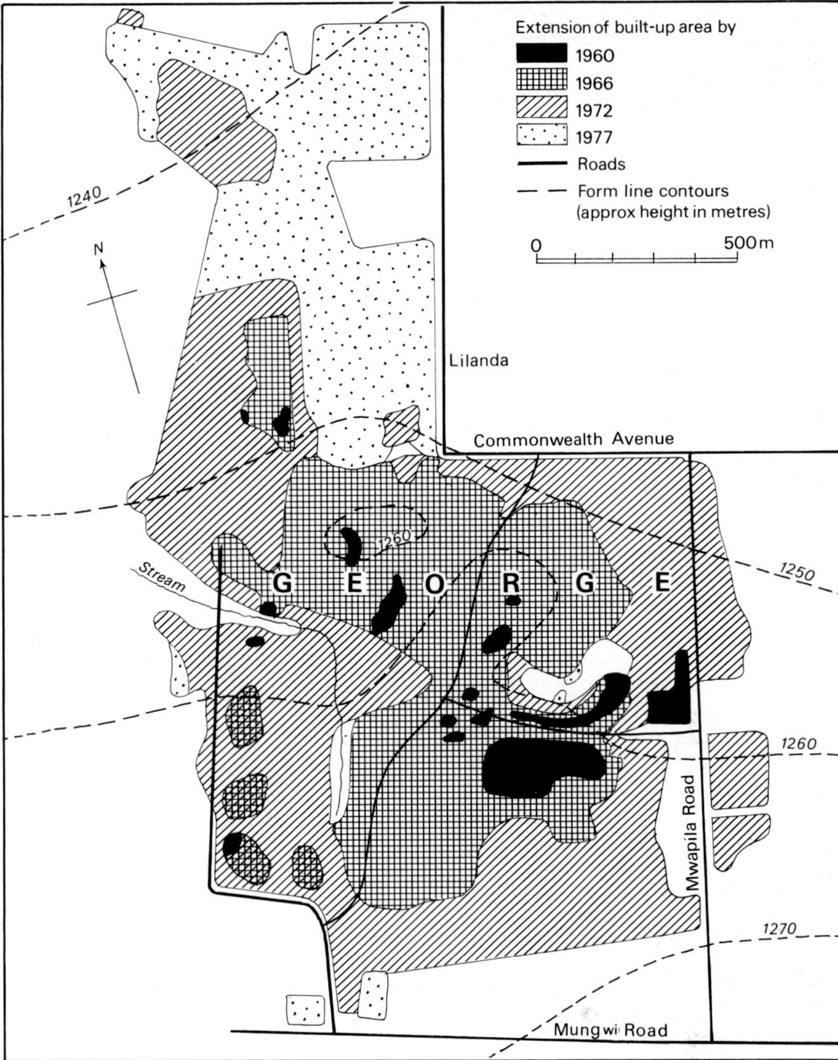

Figure 10 *Map of areal growth of George, 1960–77*
(*Source*: adapted and simplified from Schlyter and Schlyter, 1979)

Richard Martin describes the political organisation of the community in this way:

'. . . it is thanks to their exclusion from the law by their first gesture of breaking it that the people have had to declare a sort of unilateral declaration of independence. In so doing they have no longer to depend on the bureaucracy; instead they can evolve a relevant machinery. This machinery — the revolutionary council as it were — is the Party, already well established on the cell principle and with virtually universal membership. Through the party the people can decide what should and what should not be controlled, and through the party unacceptable members of society, "trouble makers", are expelled from the area: strict hierarchical rules apply. But it is a peoples' party, and it is remarkable to witness the speed with which those within it who do not perform satisfactorily are replaced in the annual elections.'

Each 'village' community of about 20 families form one *section* of UNIP and 2000 families form one *branch*. In 1960, the settlement had one branch of the party but by 1973 there were three, and in 1977 there were twelve.

THE EFFECT OF THE UPGRADING PROGRAMME ON GEORGE

The City Council became responsible for George when it was incorporated into the City of Lusaka in 1970, after the enlargement of the city boundary. George was one of the four squatter complexes included in the upgrading policy set out in the *Second National Development Plan* in 1972. The two main features of the programme are referred to in the extract below from a 1974 City Council publication.

The policy also involved the acquisition of the land on which the settlements had developed, control of further development, and provision of basic amenities such as water standpipes, drainage channels, street lighting, road improvement, refuse removal, clinics, schools and community centres.

Upgrading started in George in 1977, and the project was partly financed by the World Bank. One of the main problems was in the laying down of the infrastructure, because of the layout of the houses. For example, the main water pipe into George was fixed according to the *Lusaka Development Plan* and 130 houses were affected and had to be demolished. These householders were given plots in the overspill area to the north of George. The routes of the minor water trenches were adapted to serve the existing houses and the standpipes located in a central position at a density of about 25 houses per unit.

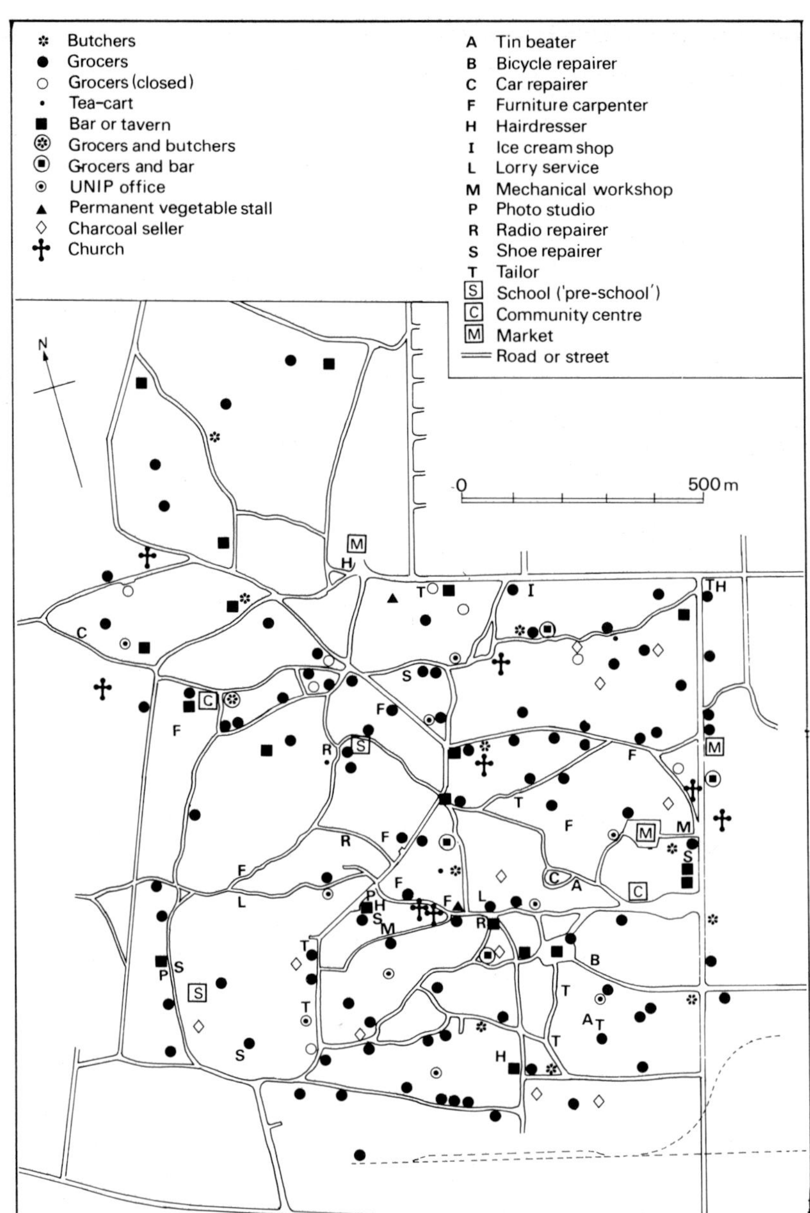

Figure 11 *Commercial and community facilities in George, 1977*
(*Source*: Schlyter and Schlyter, 1979)

The housing programme for Lusaka

The first of these features is that the exercise is being seen as a social as much as a physical one. There are certain systems already at work in the squatter settlements which have already been described. There is a sort of local government and there are well-developed mechanisms for mobilising labour for self-help work. There are positive qualities which are not developed to anything like the same extent elsewhere in the city, and it is intended to reinforce them, rather than supersede them, in the upgrading process.

The second feature is the principle that continuous improvements should be allowed for, so that as aspirations rise individuals have the possibility to fulfil them. Squatter areas and 'site and service' schemes will be the main areas of owner-occupied housing in Lusaka; not only that, a comparatively large proportion of the residents will do some or all of the work themselves. Thus the water pipes are sized so that although they only serve standpipes initially, individual houses can have a connection at a later date. Densities are being examined with a view to ensure that every house can be expanded to a suitable size, because if the present rate of improvement of houses continues they will all be two or three bedroom dwellings within a relatively short period. It will also be possible for groups of houses to collaborate in the installation of a septic tank so that they may have waterborne sanitation should this facility not be available by that stage.

So the general effect will be to make it possible for houses to be improved on their existing sites up to a level comparable with housing elsewhere. Thus it will not be necessary for the better-off members of society to leave the community if they require a better standard of service: the means will be there for them to install it.

(*Source*: R. Martin, 1974, from *Local Government in Zambia*, ed. N. R. Hawkesworth)

THE CHANGING ENVIRONMENT

Detailed research was carried out at George by Swedish workers. This was in the form of a longitudinal survey. They carried out fieldwork in 1969, 1973 and 1977 and their findings were published in 1979, *George — the Development of a Squatter Settlement in Lusaka* (Schlyter and Schlyter). This was followed by further research between 1977 and 1980 with a second publication in 1981 *Upgrading a Squatter Environment* (Schlyter and Rakodi). The change in the landscape was carefully monitored and the study included a random sample survey of six house groups, *A* to *F* (see Figure 12) in the settlement. Some of the changes can be seen in Tables 6 and 7 and Figures 13 and 14.

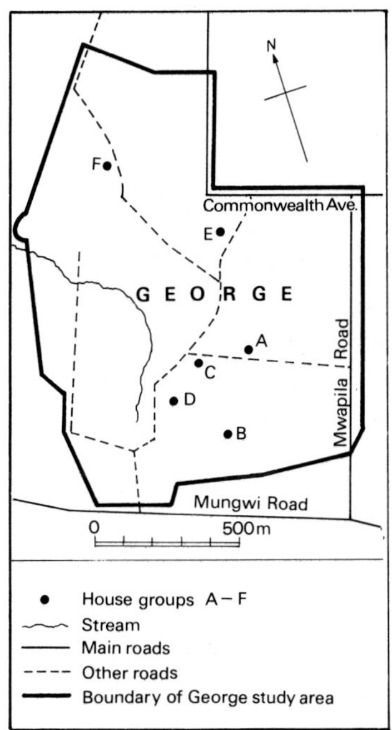

Figure 12 *Map of George study area showing the location of the six house groups used in the 1969, 1973 and 1977 surveys*

(*Source*: adapted and simplified from Schlyter and Schlyter, 1979)

Table 6 *Changes in land use for a random sample of six house groups (A–F) in George*

Type of area	*1969 % of surveyed area*	*1973 % of surveyed area*	*1977 % of surveyed area*
Plot area		75.1	75.5
Overlapping plot area		0.9	0.2
Roofed area – houses	16.0	18.8	20.8
Roofed area – latrines	1.0	1.2	1.2
Roofed area – Stores, animals, etc.	0.4	0.2	0.5
Swept area	8.0	9.8	12.5
Vegetable garden	11.5	7.0	10.0
Flower garden	0.6	1.5	0.8
Grass	11.8	8.8	6.0
Public communication	8.7	9.5	9.2
Enclosure for animals	0.5	0.4	0.1
Stores or car wreckage	0.0	0.0	0.3
Garbage holes	1.2	1.0	0.4
Wells	0.4	0.4	0.3
Collapsed or unfinished structures	0.3	0.6	0.1

(*Source*: A. Schlyter and T. Schlyter, 1979, *George — The Development of a Squatter Settlement in Lusaka, Zambia*)

?

1 Table 6 shows the land use changes of the six house groups in the sample. Draw divided rectangles for the three years surveyed to show the change in the land use pattern. What changes have taken place? Suggest reasons for these changes.

2 Figures 13 and 14 show the pattern of land use for house group *C* in 1969 and 1977. What major changes have taken place in the eight-year period?

3 One of the things you will have noticed in answering the previous question is the change in the density of the built-up area through house extensions. Table 7 shows changes in house size for all the house groups in the sample survey. Draw a frequency graph (histogram) to show the distribution of houses of different sizes for 1969, 1973, and 1979. The *Y* axis should show the number of houses and the *X* axis, the number of rooms. For simplicity, group 'one room' and 'one room and verandah/kitchen' together, and so on. Comment on the changes which have taken place.

4 George's commercial and community facilities in 1977 are shown in Figure 11 (*page 58*). Draw up a table to show the total number of each type of facility. Add a third column to show the population served by each type of facility. Do any facilities seem over- or under-represented, given that the total population of George at the time was 56 000?

5 Use the description of George as a basis for class discussion on the values reflected in the physical and social organisation of the community.

6 Imagine that you are taking photographs for an article to be written in the *National Geographic Magazine* on the changes which have taken place in George between 1969 and 1977. Take a walk through house group *B*, following the route given in Figures 15 and 16 and give details of what your photographs would show, pointing out the changes that have taken place in the eight-year period.

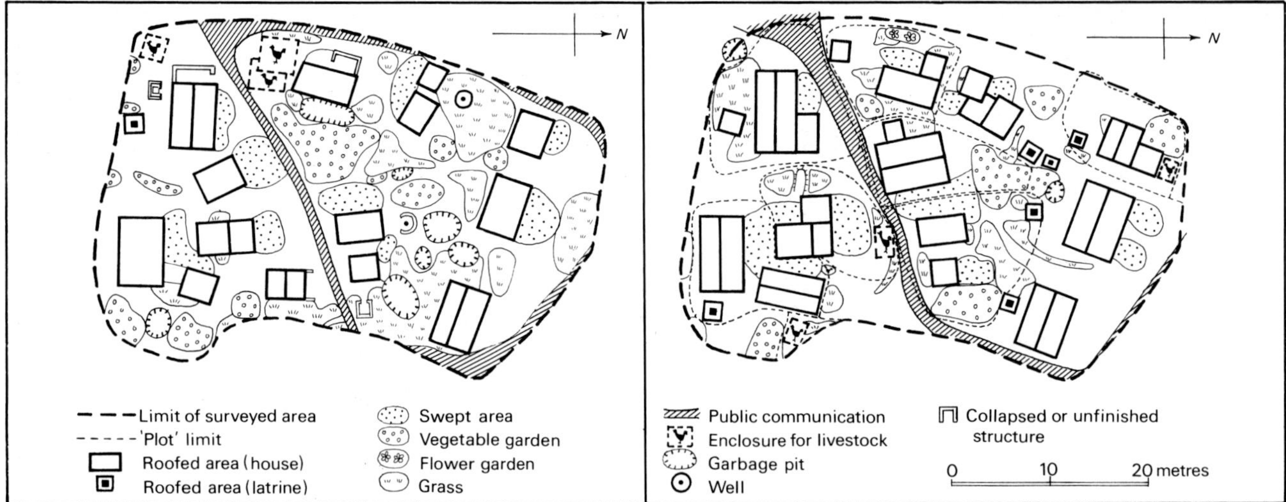

Figure 13 *House group C, 1969*

Figure 14 *House group C, 1977*
(*Source*: for Figures 13, 14, 15 and 16, Schlyter and Schlyter, 1979)

Table 7 *Changes in house sizes in George*

Houses with	*1969*	*1973*	*1977*
One room	16	7	6
One room + verandah/kitchen	3	1	—
Two rooms	11	14	11
Two rooms + verandah/kitchen	5	5	8
Three rooms	8	9	10
Three rooms + verandah/kitchen	6	5	4
Four rooms	4	10	9
Larger houses	3	11	12
Number of houses	56	62	60

(*Source*: Schlyter and Schlyter, 1979)

EVALUATING THE LUSAKA HOUSING PROJECT

Attitudes which different groups adopt towards squatter settlements vary markedly. Much depends on the observer's class and social position, political persuasion and self-interest in relation to the settlements.

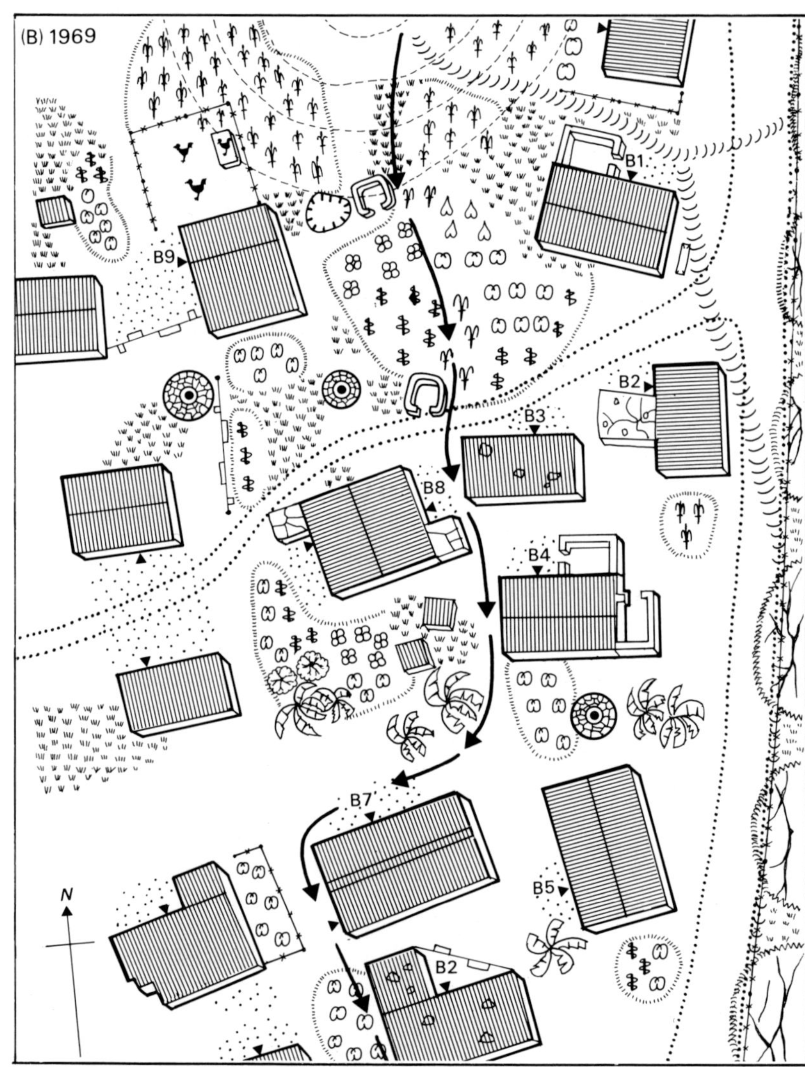

Figure 15 *House group B, 1969*

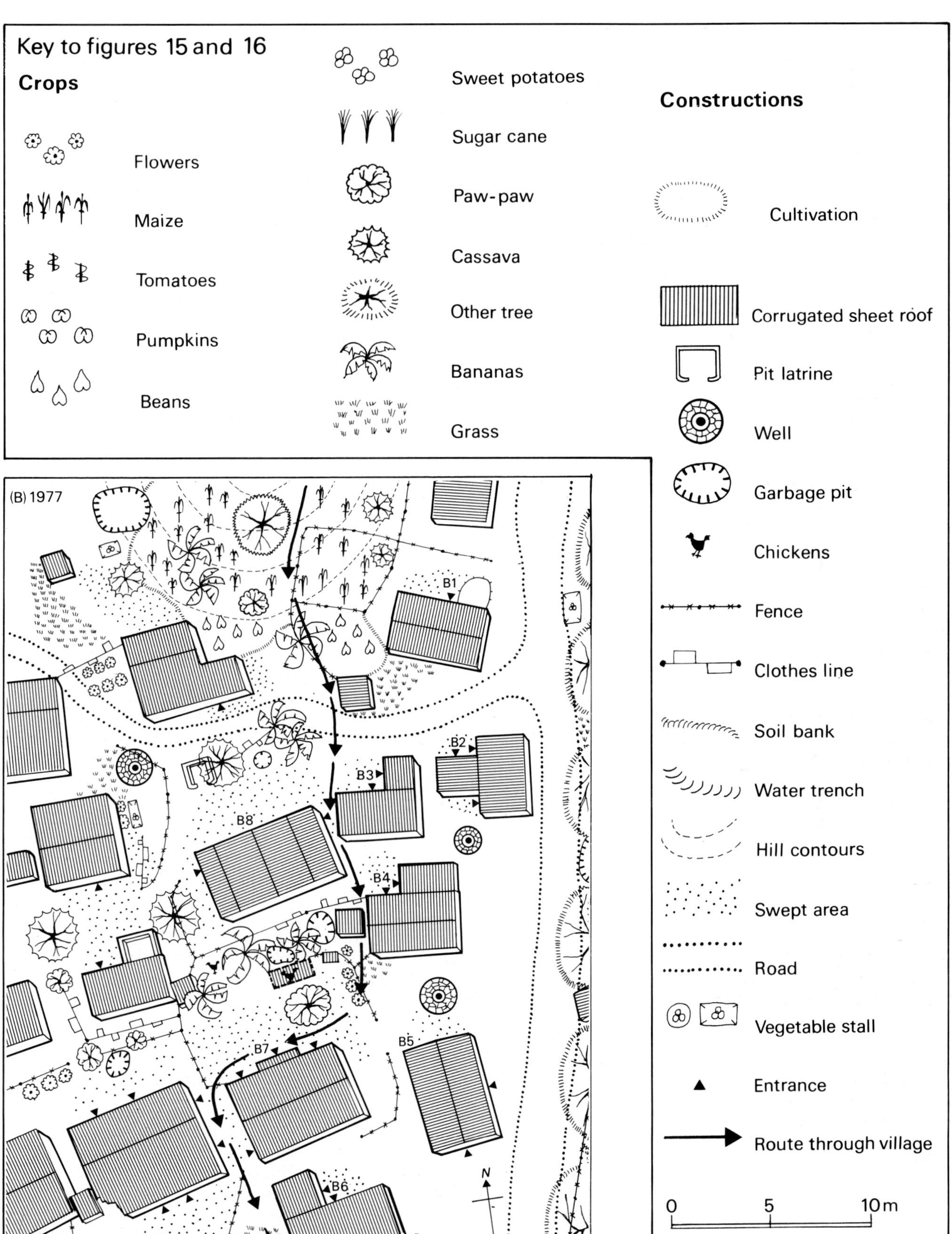
Key to figures 15 and 16
Crops
Flowers
Maize
Tomatoes
Pumpkins
Beans
Sweet potatoes
Sugar cane
Paw-paw
Cassava
Other tree
Bananas
Grass
Constructions
Cultivation
Corrugated sheet roof
Pit latrine
Well
Garbage pit
Chickens
Fence
Clothes line
Soil bank
Water trench
Hill contours
Swept area
Road
Vegetable stall
Entrance
Route through village
0
5
10 m
(B) 1977
B1
B2
B3
B8
B4
B7
B5
B6
N

Figure 16 *House group B, 1977*

?

1 Bearing in mind what you have read about the Lusaka Housing project imagine yourself in the roles of the following observers. Describe the kind of attitudes you are likely to hold if you are:
a A middle-class American journalist.
b A local upper- or middle-class individual living in the wealthier part of Lusaka, e.g. the Ridgeway.
c An administrator, trained in Europe and now in charge of solving housing and planning problems.
d An architect on the local council, disappointed with large-scale projects and keen on community participation.
e An international planning consultant brought in to advise on solving the housing and planning problems.
f A local landowner whose land has been invaded by squatters.
g The owner of a construction and building materials firm.
h A factory owner.
i A health worker.
j An academic researcher from the university.
k A local MP.
Discuss your views with the rest of the class.

2 What sort of questions are raised which are concerned with the conflict of interest associated with urban housing development?

3 Write a short account on the likely interests and motivation of the squatters. Consider the following questions in your answer:
a Why do they live in these areas?
b Why are they involved in the practice of self-help?
c What are their options for improving their living conditions and for obtaining housing outside the self-help process?
d Do they have a free choice in becoming illegal squatters or is it out of necessity?
e What effect might 'site and service' schemes have on attitudes, and are they likely to be divisive?

HOUSING SCHEMES IN OTHER PARTS OF THE WORLD

Similar features to the Lusaka housing programme can be seen in schemes in other parts of the Third World, although different environmental, social, and cultural circumstances mean that each scheme has its own individual characteristics.

NALEDI, BOTSWANA

The resettlement scheme and upgrading project which has taken place at Naledi, a squatter settlement outside Gaberone, the capital city of Botswana, has many features in common with the Lusaka project, but on a smaller scale. Gaberone has a population of only 34 000, which is less than the population of George. Even so, the rapid urban growth which has taken place since independence in 1966 has brought the same problems of unemployment and illegal squatting. Between 1966 and 1975,the annual average population growth rate of the city was 21 per cent.

Gaberone was designated a capital city in 1963 when it was no more than a village of 600 people. Naledi grew up as a camp for construction workers who were needed for building the new city. In 1973 the local authority took action to deal with the squatter problem: by bulldozing and resettlement. People were moved to a new low-cost housing area called New Naledi and they were charged subsidised rent. The 'site and service' scheme could not keep up with the numbers and consequently, there was a change in policy in 1975. Naledi became legal and plans were made to upgrade it. The residents gained security of tenure through a *Certificate of Rights* over house plots on a 99-year lease. The residents themselves were involved in the upgrading and the city council provided loans for building material and advice on building methods. By 1980, Naledi was a fairly self-contained community with many services, such as a health centre, market place, schools and community centre.

SOUTH-EAST ASIA

As it did in Lusaka, the World Bank has provided substantial aid for housing projects in South-east Asia. The BLISS Project (Bagang Lipunan Improvement of Site and Service) in the Tondo district of Manila in the Philippines, and the KIP Project (Kampung Improvement Programmes) in the kampungs of the major cities of Indonesia are examples. The housing programmes are based on similar principles to those of Lusaka and Naledi. The projects are expected to:
1 use the social organisation and leadership already existing within the settlements,
2 provide support for the existing tradition of self-help,
3 be available to all income groups,
4 provide security of tenure for squatters in the upgrading programmes,
5 provide basic 'site and service' provision for those who are to be resettled,
6 bring about community development.

The main difference between these schemes and the African examples relates to house density. Overcrowding is far more common in South-east Asian cities, and the health problems associated with poor water supplies and sanitation are intensified and far more critical than in Lusaka. South-east Asian housing programmes, therefore, have to give very high priority to the provision of health clinics and to informal health and nutrition education.

COLOMBO

The emphasis on healthier living is well-illustrated by the Colombo Project in Sri Lanka — the official title is 'Environmental Health and Community Development in the Slums and Shanties of Colombo'. This project was partly financed by UNICEF and partly by the government.

The city of Colombo, like Lusaka, is a colonial creation. The port acts as the economic centre for the country's major exports of tea, rubber and coconut. The port is encircled by a spacious administrative area, and that in turn by a third ring of growth with crowded streets, small businesses and houses (Figure 17). As this area became more congested, the upper and middle classes began to migrate out to the new suburbs on the city's periphery. The inner area was left to the poorer classes who divided up the large colonial houses into tenements, and built the present slums in the large gardens of these houses. It is for this reason that the slums are called the 'Gardens'. There are 1200 of them altogether.

The Colombo Project was set up to upgrade the slum 'Gardens' and the squatter settlements on the marshy eastern edge of the city, but in practice it has been restricted to 300 slum 'Gardens' — housing 15 per cent of the total slum and shanty population. Its tangible aims are to provide 1400 latrines, 875 washrooms and 700 clean water standpipes. Three thousand, two hundred bucket latrines were to be converted into water-sealed toilets. Apart from providing these basic amenities, the Project has been concerned with health and nutrition, education and community development. Nearly 100 primary health care workers were trained to work in the 'Gardens'. In addition, each 'Garden' was encouraged to be involved in the development process through the election of its own Community Development Council. The Council's function was to liaise between the residents and outside organisations, for example the Health Department, the various municipal services, the Common Amenities Board, the Women's Bureau and the Youth Training Services Council.

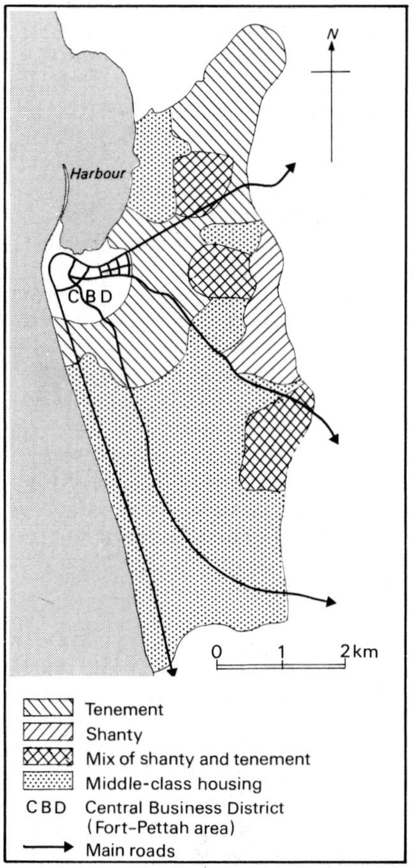

N
Harbour
CBD
0 1 2km

Tenement
Shanty
Mix of shanty and tenement
Middle-class housing
CBD Central Business District (Fort-Pettah area)
Main roads

Figure 17 *Colombo housing pattern*
(*Source*: adapted from map and data in a study prepared by ILO, Geneva, 1978, *The Informal Sector of Colombo City*, report published by Marga Institute, Colombo, 1979)

? A number of case studies on housing programmes can be followed up through your own research. Those already mentioned can be looked at in greater detail from sources referred to in the reading list at the back of the book.

Find out what similarities and differences there are between the Lusaka housing project and any of the case studies you select.

INEQUALITY AND THE CITY

Peter Adamson, in his report on the Colombo Project (commissioned by UNICEF) attempts to look at the project from a wider perspective. He suggests that the provision of basic services in the 'Gardens' of Colombo is an improvement on no services at all but asks whether they are:

'. . . intended as the beginning of a development process which could lead to greater opportunity for personal and community development — or whether they are merely the cheapest possible way of containing the problems of poverty at the minimum possible cost and without necessitating any change in the society of which that poverty is a part?'

In his view, the success of the project will depend on the people's own organisations — the Community Development Councils — continuing their work after the formal project has ended.

A radical viewpoint would tend to favour the explanation that self-help and 'site and service' schemes only *contain* the problems of poverty and do not necessarily break down the dual society of the rich and the poor. The schemes are seen as palliative and unlikely to redress structural inequality. The urban poor are existing in a relationship of dependence with regard to other powerful interest groups and are forced by necessity to provide for themselves.

The counter argument, from what could be called a liberal viewpoint, would suggest that structural inequalities do not easily disappear and therefore self-help, by improving material standards, cannot be objected to *unless* it prevents a more equitable solution. Might not material improvement in the conditions of the poor be more likely to radicalise them, in any case, than if they were left alone?

THE URBAN HOUSING PROBLEM IN THE CONTEXT OF NATIONAL DEVELOPMENT

THE WIDER CONTEXT

The housing issue is involved with the immediate practical problems in the urban environment, but the city is not an independent variable and the issue has to be looked at in a much wider context. A government has to consider cities in the context of national planning for development. In general, most governments have to balance the *equitable* distribution of resources with the efficient working of the economy — that is social justice with economic returns on investment.

In Zambia there is a distinctive core area along the old 'line of rail' between the Copperbelt and Livingstone. The core attracts investment, raw materials and population to the detriment of the large periphery to the west and the east. The First National Development Plan (FNDP) and the Second National Development Plan

(SNDP) put very little emphasis on rural development as a means of slowing down the migration to the towns. More recently, there has been a change in government policy with greater emphasis on food production and rural development. This is reflected in the Third National Development Plan (TNDP), launched in 1980.

In Tanzania, Zambia's northern neighbour, a grassroots approach to development has been followed which has emphasised the need for rural development through a 'villagisation' (*ujamaa*) programme and self-reliance. This strategy for rural development has reduced the pull of the towns. Rural people in dispersed settlements have been regrouped into villages in order that they can benefit from communal facilities. This regrouping has not been without its problems. Even so, the country does seem to be in a better position to check the process of urbanisation.

A useful theoretical housing policy framework is shown in Figure 18. The diagram indicates that there are several levels within the decision making process — each with its own set of influences. If we apply the model to Zambia, its political philosophy of 'Humanism', a form of socialism, would put it mid-left of the political spectrum. On the whole, the effectiveness of government economic policies has resulted in a positive urban bias, although attempts to redress this balance can be seen in all three national development plans. The development goals are both social and economic — perhaps it is the economic ones which have been most effective as development has continued to take place along the line of rail. Projects in the rural areas have not always been successful owing to a lack of infrastructure, poor planning and execution of policies. It has been suggested that these difficulties are associated with the original colonial structure. Regional imbalance was beginning to break down in the areas of health and education as a result of the priority given to them in the FNDP. Housing for lower income groups was given greater priority in the SNDP. It is the ineffectiveness of rural development in Zambia which has intensified the problems associated with urbanisation, although

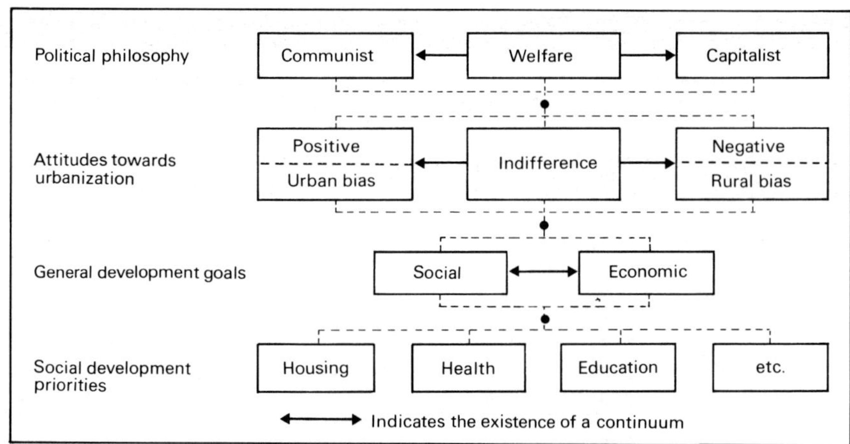

Figure 18 *Housing policy framework*
(*Source*: D. Drakis-Smith, 1979, *Housing in Third World Countries*, Murison and Lea)

migration has slowed down slightly in recent years.

There are two opposing views of urbanisation in the context of development. In the first view, where development is equated with modernisation, cities are seen as centres of innovation. They act as growth poles, and modernisation occurs through a process of diffusion originating at the top of the settlement hierarchy and spreading downwards and into the rural periphery. In the second view, urbanisation is interpreted as a process which creates under-development — the city is seen as parasitic, exploiting the rural hinterland.

These views, perhaps, provide a useful reminder that the pressure on housing in cities like Lusaka needs to be considered, not just at the city level, but from this wider context.

HOUSING PROBLEMS IN THE DEVELOPED WORLD

It is very easy to see the problems of the Third World as being 'out there', but similar housing problems do exist in our own society, though not on the same scale. There are many people who live in sub-standard housing in inner city areas, and others who have resorted to squatting in vacant property. It is important to consider to what extent the explanation for housing problems in our own society is similar to or different from that for cities of the Third World.

In tackling this exercise it might be worth referring to specific cities for comparison (e.g. compare one city in Britain with one in Africa). Alternatively, if you have less time, you could make a general comparison between countries. Helpful references are given in the Further Reading section.

? Now that you have had an opportunity to look in more depth at self-help housing, you should return to the questions asked at the beginning of the chapter.

1 Is self-help housing a desirable policy?

2 Is it the most practical solution?

Try to look at these questions critically, bearing in mind not just the immediate problem of housing, but the wider context of national development. Are there any provisos to your answer? Make sure that you support your opinion with evidence.

4 THE AMERICAN CITY TODAY

URBAN CHANGE IN THE UNITED STATES

We are all familiar with the image of mobility that characterises North America. From the period of initial colonisation, individuals, families and many different groups of people have settled, moved and resettled with a frequency unknown in Europe. From the era of the Gold Rush, through the mass movements away from the land depicted in John Steinbeck's *The Grapes of Wrath*, to the restless exploitation of the motor car for every kind of journey as described in Jack Kerouac's *On the Road* and Hunter Thompson's *Fear and Loathing in Las Vegas*, the message has been: keep moving. This is still the case, and is the message that underlies this chapter. Americans are still moving — from city to city and within their cities. As a result, a whole series of problems exists within the North American city, and we shall examine some of them in the chapter.

MAJOR POPULATION MOVEMENTS

Initially, let us examine the most recent population changes that are taking place within the USA as a whole. Firstly, we can state that the country is still growing rapidly, with an 11.5 per cent increase between 1970 and 1980. The total population is now 226 million.

This figure hides some major transformations, however. Some states, and by no means the smallest ones, have grown far more quickly than that. Nevada increased its population by 63.8 per cent from 1970 to 1980, and even the large state of Texas showed a 27.1 per cent expansion from 1970 to 1980. When we examine the top ten states in terms of growth, we find consistent *types* of states experiencing population expansion, and this has occurred at the expense of particular *types* of declining states (see Table 1). We can see that people have been particularly attracted to the so-called sunbelt states, where lower taxes, attractive environmental conditions, pleasant climates and a reputation for high-growth industrial opportunities have acted as powerful magnets. (Florida is a little different, as retirement has prompted migration.) Many migrants have left the so-called frostbelt, particularly states like Ohio, Illinois and New York. The latter alone has lost 3–8 per cent of its 1970 population in the last ten years.

Print 1 *Oklahoma, 1939, a migrant family on the move*

? **1** On an outline map of the USA, traced from an atlas, shade in red the top ten growth states in 1970–80, and in blue the bottom ten growth states (Table 1). Write a paragraph commenting on the distribution.
2 How do the fastest growing and fastest declining cities fit within this pattern (see Table 2 below)?

Table 2 *Fastest growing and fastest declining urban areas, 1970–80*

Fastest growing cities	*%*	*Fastest declining cities*	*%*
San José	+36.1	St Louis	−27.9
Austin	+35.4	Buffalo	−22.9
Albuquerque	+34.5	Detroit	−21.8
Phoenix	+33.7	Pittsburgh	−18.5
El Paso	+31.7	Louisville	−17.6

(*Source*: extracted from 1980 Census)

Table 1 *Patterns of growth and decline, USA, 1970–80: top ten and bottom ten states*

State	*Population increase and decrease (000s)*
1 California	3697
2 Texas	3029
3 Florida	2948
4 Arizona	942
5 Georgia	876
6 North Carolina	790
7 Washington	716
8 Virginia	694
9 Colorado	679
10 Tennessee	664
41 Nebraska	84
42 Connecticut	75
43 Vermont	66
44 Pennsylvania	65
45 Massachusetts	47
46 Delaware	47
47 North Dakota	34
48 South Dakota	23
49 Rhode Island	−2
50 New York	−684

(*Source*: extracted from 1980 Census)

Print 2 *Intel Plant No 4 at Santa Clara, Silicon Valley. New industries such as this have attracted people to the region from all over the United States*

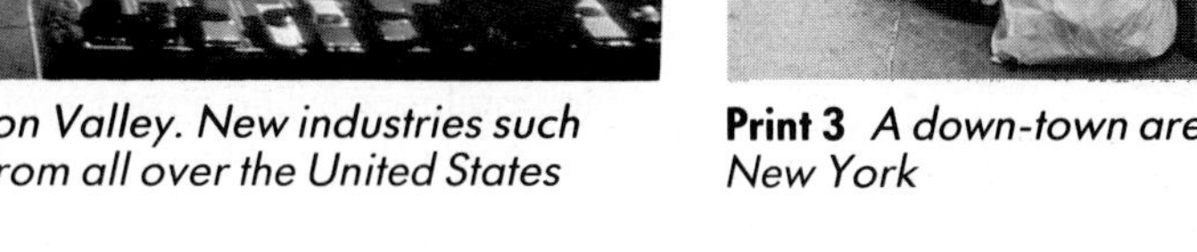

Print 3 *A down-town area of New York*

Movements between cities
Many migrations take place from city to city. Consequently, when we examine specific cities such as those in Table 2, we find similar patterns of growth and decay. Thus the success-stories are places like San José, California (the heart of Silicon Valley and the home of the country's computer industry), and the cities which stretch across the growing states: Phoenix, Albuquerque, Houston, San Antonio, Charlotte. In these urban areas, the only problems are those associated with growth — how to build enough roads, homes and schools to deal with the immigrants.

By contrast, the cities of the frostbelt are facing quite different problems. As industries (like Ford and General Motors in Detroit, and steel producers in Pittsburgh) contract, and the most mobile members of society begin to move away, the urban remnants suffer a shrinking tax base and a proportionately larger population of unemployed and elderly persons. It is little surprise that several of these cities have hovered close to bankruptcy for the last ten years.

Movements within cities
We have seen that new cities are emerging within the country, and the number of metropolitan areas (i.e. those containing an excess of 50 000 people) has risen from 243 to 284 between 1970 and 1980. At the same time, however, many of these metropolitan areas have experienced an *internal* redistribution of population. At the end of the Second World War, nearly two-thirds of people lived 'downtown', and only one-third in the suburbs (only half the country lived in metropolitan areas at that time). This changed as families were lured to the suburbs by attractive financial packages on the homes being built there in their thousands, and by the prospect of a 'new' life, away from congestion, crime and, later, the perceived threat of racial integration in the urban school system. In 1980, three-quarters of the country lived in metropolitan areas, but the downtown/suburban ratio was almost exactly reversed. The expansion of the suburbs has been made possible by the progressive spread of freeways out from the downtown areas; the cheapness of both cars and petrol (prices are far below those experienced in the UK); and, most recently, by the continued movement of industries and services out to suburban locations, which has reduced commuting costs for many workers. This has all added to the problems faced by the city centres, and has simply made the suburbs increasingly attractive.

It could easily be argued that the USA has always led the world in terms of showing how the urban future is likely to look and we shall examine this idea again towards the end of the chapter. However, we should not forget that this evolution has been guided by the demands of the consumer and the interests of the market. The strict development control, which ensures that the views of the élite tend to predominate in Britain, is completely alien to the USA. Thus sprawl, clutter and decay are common, as M. Thomson describes:
'In Chicago, as in Detroit, Los Angeles, and New York, one can drive through miles and miles of uniformly anonymous strip development — an endless succession of cheap hamburger cafés, hot-dog stalls, motels, bowling alleys, gasoline stations, doughnut counters, parking lots, workshops, truck pull-ups and a variety of other hard-sell operations, all heavily interlaced with enormous billboards, telegraph posts, power lines, TV lines, gaudy name signs, bunting and a welter of street furniture. . . . The main roads are wide and heavily painted. The pavements are commonly cracked, ill-stained and strewn with litter, the kerbs are crumbling. . . . A feature of

these main roads is that they never seem to reach a centre. They go on and on, never arriving anywhere.'

(*Source*: M. Thompson, 1978, *Great Cities and their Traffic*)

This means that the American city (with some notable exceptions, such as San Francisco and Boston) is rarely anyone's idea of an ideal, liveable unit. Problems abound, and this is one reason why suburban flight, or attempts to move to newer cities in the south or west, are so popular. As we shall see in the next sections, particularly serious problems relate to transport, service provision and inner city decay.

? **1** Draw a diagram showing all the variables influencing the intra-urban redistribution of city populations.
2 Make lists of what you consider to be the desirable and undesirable effects/consequences of such redistributions. Add to this list as you work through the chapter.

TRANSPORTATION ACROSS THE FRAGMENTED METROPOLIS

The US was the nation of the drive-in movie theatre in the 1950s and 60s, and although videos are now more popular than films, this is not a result of decreased mobility. Consider some of the following statistics (figures for 1980):

- There are 145 million registered drivers, approximately 88 per cent of the population aged between 14 and 75.
- There are 161 million motor vehicles, 121 million of which are private cars. This means that there are approximately 1.5 cars per household, something like three times the comparative figure for the UK.
- About 83 per cent of Americans travel to work by private vehicle, and a miniscule 6 per cent uses public transport.
- Studies undertaken by the Department of Transportation indicate that the average journey-to-work distance is approximately 9.2 miles.
- All this adds up to something like 108 000 *million* car trips every year.

THE AMERICAN CITY — A MOTOR CITY

Growth has been predicated upon an assumption of car ownership, with the result that most suburban residential locations can only be reached by private vehicles. Americans regard their vehicles as a symbol of independence, and rarely share rides in a car pool. This means that, despite the fact that virtually all highways have at least four lanes, and many freeways can be as wide as ten or twelve lanes in total, there are frequently serious problems of traffic congestion which would make rush hour in London, Leeds or Newcastle seem mild by comparison. No one who has ever travelled in Los Angeles between 4 and 6 p.m. in the evening will forget the sight of hundreds of thousands of automobiles crawling nose-to-tail along every available road — not just away from the downtown area, but in every conceivable direction. The freeways vibrate to the sound of traffic late into the night, and the periodic chaos of over-use is such that a radio station exists whose only function is to relay traffic information and advice.

Los Angeles is, along perhaps with Houston, the archetypal product of the motor age. Even in the 1930s and 40s, it was a sprawling collection of communities (as depicted in any of the Raymond Chandler novels, such as *The Big Sleep*). This sprawl originally developed because of street cars and railway links, but was encouraged with the building of the highway system, which began in 1941.

The results of this growth can be interpreted in several ways. To many observers, the urban sprawl is unpleasant. The endemic pollution problem is in no sense aided by the carbon monoxide levels, although California does lead the country in automobile emissions control equipment, which has to be fitted to every vehicle. Fatalities and injuries from accidents are high.

Probably the biggest problem, however, is actually organising the transport system, and this is true for many cities besides Los Angeles. Because growth has spread across so many counties (an extent of approximately 3 500 square kilometres), it is virtually impossible to integrate the many local governmental agencies responsible for traffic problems.

Print 4 *The high-tech look of a BART passenger train*

In an attempt to cut through this labyrinth of government, some metropolitan areas have set up single agencies, with their own finances, to develop some form of co-ordinated public transport. Obviously, the assumption is that increasing congestion can only be slowed by some transport alternative. Los Angeles has been attempting to develop a new integrated system for some years without success. We shall now go on to examine the experiment undertaken by its northern Californian rival, San Francisco, and see how successful such developments can be.

THE BART EXPERIMENT

Although San Francisco does not appear as scattered as Los Angeles, it is, in fact, a very widely dispersed urban area. Commuters travel an average of 15.8 miles, due in the main to the topography, which makes some locations impossible to develop. There are three main employment concentrations: San Francisco itself, a major service and tourist centre; Oakland, a predominantly black city where much industry and most port facilities are located; and San José, which was mentioned previously. Residential development has spread all along the East Bay, from Richmond to Freemont, and inland

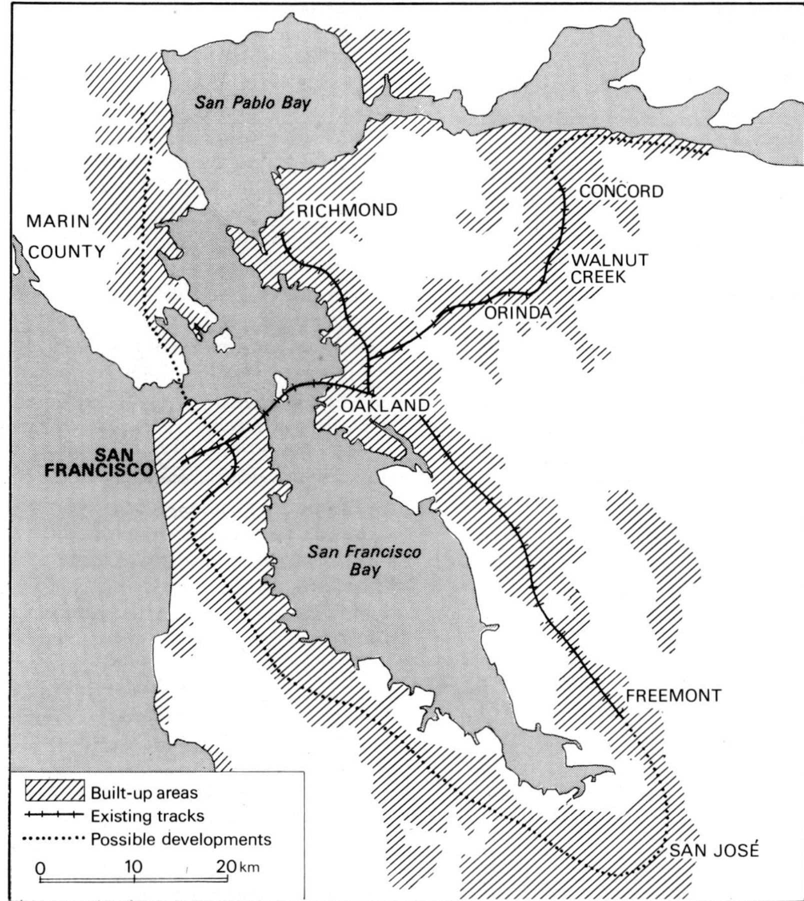

Figure 1 *The Bay Area Rapid Transit system, San Francisco, California*

to Walnut Creek and Orinda (Figure 1).

The Bay Area Rapid Transit System (BART) was first promoted in the 1950s. The intention was to develop an alternative to suburban sprawl by promoting the redevelopment of San Francisco and Oakland. It was assumed that new business expansion would halt the continued sprawl of both homes and businesses out to the north, south and east. To make this possible, a fast, light railway system was envisaged, which would allow commuters to travel in from the existing suburbs at high speed. The result, it was hoped, would be a strengthening of the city centres and a halt to suburban drift (see Figure 1).

It was originally assumed that the system would cost about $790 million, and local governments set out to borrow that amount in 1962. Forecasts suggested that approximately one-quarter of a million passengers would be expected each weekday, and that the system would make an $11 million surplus every year. The system opened in 1972, at a total cost of $1585 million dollars, and runs at a net loss. More importantly, many of the expected transport advantages have not materialised.

? Below, you are provided with information on BART's impacts. While reading this evaluate the mistakes that have been made, and note down suggestions for how they could have been avoided.

BART: patterns of use

The system is comfortable and fast, although each train stops at all stations, which are 2.5 miles apart. Most journeys are relatively long ones: three-quarters of journeys are over 7 miles, half are over 12 miles, and a quarter are over 19 miles. Some people have been attracted from the roads, but approximately half the users have been attracted away from existing public transport, particularly buses.

BART: who pays?

The system has proved a great deal more expensive than predicted. Furthermore, because use has been lower than anticipated, fares have had to be heavily subsidised, in order to attract passengers. In 1976 the average subsidy was $3.75 per journey. This support comes from three major sources: tolls on the many bridges in the area; a tax on motorists; property taxes (which are paid by residents in proportion to the size of their homes; and sales tax, paid by consumers (rather like British VAT). All these taxes are *regressive*, in that they tend to be more of a burden on the relatively poor than on the rich. This is compounded by the fact that most users are relatively affluent suburban commuters: in other words, these regressive taxes are going to subsidise high-income passengers.

Table 3 *Costs and subsidies for families with different incomes using BART*

Annual household income ($)	*Proportion of income paid in BART taxes %*	*Proportion of Bay Area residents with this income using BART %*
5 000	0.65	3.50
10 000	0.40	4.50
15 000	0.35	6.50
20 000	0.30	9.00
25 000	0.25	15.50

(*Source*: Webber, 1976)

? Examine Table 3 and assess to what extent it confirms this last sentence.

BART: the outcomes
Despite the fact that passengers are getting an enormous subsidy, usage has not been as high as expected, and as we have seen, BART has tended to attract bus passengers rather than automobile users. Only 35 per cent of BART users formerly used their cars, and they have already been replaced on the roads. One of the more amazing findings is that:

'In accordance with the Law of Traffic Congestion (which holds that traffic expands to fill the available highway space until just tolerable levels of congestion are reached), other people began driving their cars on trips they would not otherwise make. Perhaps they are suburban spouses who now have the family car during working hours. Perhaps people are visiting friends, now that highways are less crowded.'

(*Source*: 'The BART Experience: What Have We Learned?' *The Public Interest*, Webber, 1976)

Whoever they are, the net effect is an increase in mobility, but with no net decrease in congestion, reduction in petrol usage, or reduction in pollution output — and at a cost of 1.6 billon dollars!

? For the purposes of this exercise you are the temporary Director of the Bay Area Rapid Transit System. Your predecessor has resigned because she is facing political pressure to make BART pay, to reduce traffic congestion, and to make the service costs more equitable for all income groups. You will face election soon, and you must prepare a convincing political statement. Write, in 500 words, what you intend to do with BART and how that is to be achieved. You should bear in mind the following points:
a Rapid transit is supposed to ease congestion and make movement throughout the area possible.
b If fares are high, people will not use the service.
c If fares are low,the rich travel but the poor pay proportionately more through taxes.
d You cannot be too ruthless in your measures: Americans depend upon their cars and don't like to be parted from them. You should aim to be politically expedient.
Below are some possible strategies you might consider:
e Increasing bridge tolls on car use, especially for single passengers rather than those who 'car pool'.
f Higher parking costs in downtown areas.
g Cheaper fares for low-income neighbourhoods.
h Cheaper fares for *reverse* commuter journeys and off-peak use.
i Higher fares for express, suburban trains.
You should also examine the distribution of lines (Figure 1) and consider whether a more complete service would be attractive, despite the fact that this will involve more expensive construction.

SERVICE PROVISION IN CITY AND SUBURB

In the last section we examined the way in which many American cities have grown, and in the process have become fragmented. It is important to bear in mind that although such cities have *functional interconnections*, they also have clear *social and political barriers* within them. Put more simply, the cities have many internal links, involving commuters, the movement of goods and materials, flows of people seeking recreation and leisure, children and young people attending schools and colleges, and other types of cohesion. Conversely, racial barriers and income levels are frequently manifested in terms of separate communities, and these have by no means harmonious relationships. One of the ways in which this disharmony manifests itself is in terms of service provision, and this is examined in detail in this section.

FISCAL ISSUES IN CITIES

As mentioned in the first section, many North American cities have hovered near bankruptcy throughout the 1970s and early 1980s. Detailed studies have shown that costly attempts to improve housing conditions (as a means of encouraging population to stay) have a negative effect on this indebtedness, and that militant municipal labour unions have also pushed up the wages of city employees (police, transit workers, garbage disposal employees). On the whole, there is a downward cycle at work, whereby downtown areas require more and more investment and have to borrow more and more to stay afloat, while the suburban political areas attract higher-income families. This is illustrated in Table 4, which shows the distribution of incomes in school districts in Texas in 1982. (School districts are separate political areas which raise taxes and run the local school systems.) Table 4 shows that some of these districts can only draw on downtown low-income residents, whereas other districts clearly have numerous wealthy residents. This means that wealthy school districts can afford lavish schools, well-qualified teachers and costly equipment, whereas low-income areas must depend upon local fund-raising to make ends meet.

Services like transport, education, water and sewerage, police protection and public housing are important parts of any city's liveability. It is important that such services exist, are properly paid for and that they are of the necessary quality. It is also important that all

Table 4 *Distribution of rich and poor school districts in Texas*

Total property value in the school district ÷ no of pupils $	*No of districts in each category*	*Tax yield per pupil ($)*
100 000+	10	585
50 000–100 000	26	262
10 000–50 000	70	187
below 10 000	4	60

(*Source*: adapted from Johnston, 1982, *The American Urban System*, N.Y. St Martins)

residents have a share in both the positive and the negative aspects of service provision.

Positive and Negative Externalities
Most services have positive and negative overtones. A school within the neighbourhood cuts down on students' travel times, but may mean increased noise levels for local residents. A nearby park may be a recreational advantage *or* a location for crime and violence. Most services have this quality of being both positive and negative *externalities*: so-called because they are external to the individual's control, except by political action.

Studies have shown that residents are very clear about what services they want in their neighbourhood, and what they want elsewhere, in order that they can increase their share of the benefits available and decrease their share of the disbenefits. A particularly clear instance of this is the way in which communities treat the mentally-ill, as the following example indicates.

Geographer Michael Dear has written extensively about the ways in which the mentally-ill have been treated historically. Originally, the 'insane' were incarcerated in institutions, primarily for the supposed safety of the community at large. In recent years, it has become more usual for those undergoing successful treatment to go through a series of institutional settings — hospital, clinic, group home — with the intention of progressively releasing the individual back into the social mainstream.

As Dear shows however, the social mainstream does not often regard this process as desirable, and may try to resist the setting up or expansion of services which allow former patients to be half in care, half within the community. For the ex-patients, there are real positive externalities in the existence of patient concentrations. In San José for example, there are 2000 ex-patients in an area housing a total of 20 000 residents. This concentration can support job referral programmes, and a community mental health centre. For residents, this however all poses a negative externality: a possible disruption of the community and a threat to property values.

Opposition resulted in political activity and a change in the land-use regulations, which placed a ceiling on the numbers of group patient homes. Dear summarises these tensions as follows, and indicates the resulting geography of services for the mentally-ill, which he describes as the emergence of a new 'asylum without walls':
'The burden of care seems to fall on a few communities. The developing asylum seems to be strongly influenced by three forces which are peripheral to the problem of care-giving: (i) the final assignment of patients to aftercare facilities which tend to proliferate in downtown locations because of planners' actions; (ii) community opposition to facilities in other neighborhoods; and (iii) an informal filtering process by which some patients gravitate towards transient areas of rental accommodation in the inner city.'
(*Source*: Dear and Wittman, 1980)

In short, controversial services tend to become overconcentrated in downtown areas because of suburban political opposition, which influences planners' actions, such as the way in which they zone the city, allowing particular developments only in less-sensitive locations. In addition, many of the mentally-ill are constrained to live in the inner city because of low housing costs.

LAND-USE DECISION MAKING

Placing services in the right place is not an easy task. As we have seen, it will probably have political overtones. In addition, there are usually different strategies available to a planner: they can be summarised in terms of efficiency, equality and equity. Let us examine these in turn.

Efficiency
This is a basic consideration in any public arena — the tax-payer's dollar has to be spent in an effective manner. One way that this is frequently achieved is by exploiting economies of scale, that is by building one very large facility. In such situations, building costs may be lower and it is possible to use workers for a range of tasks.

Equality
The problem with an efficient facility is that it may be a long way away from some people. (There are obvious reasons why there is usually only one government centre in a county, due to the desirability of integrating the process of management. Conversely, that one centre may be remote from some voters and tax-payers.) The opposite possibility is to distribute services equally — one goal might be to provide one school for every 1000 families, and to place these at equal distances throughout the city.

Equity
A third possibility is to provide services in relation to *need*. One might, for example, argue that health services should be placed where ill health is greatest and where life expectancy is lowest.

?

1 In this exercise, you are aiming to solve a land-use problem. You must act as a planner and decide where the facilities outlined in Figure 3 should be located (Figure 2). There are two types of service provision under discussion (education after high school and health care) and you should consider what kinds of solution would be produced using the different goals.
2 Some of the blanks in the chart (Figure 3) have already been filled in by way of an example. What you have to determine is: What is the best financial solution likely to be? What is the most politically advantageous to a city government, and what is political suicide? After filling in all the blanks, you should opt for a strategy and justify your choice of provision for both education and health.
3 Now consider your choices again, and suggest which strategy or strategies you think carry most weight in American cities — efficiency, equality or equity? Does this exercise perhaps go some way towards accounting for the continued variations in terms of health and educational attainment that exist between rich and poor, white and black (and in future, white, black and hispanic)?

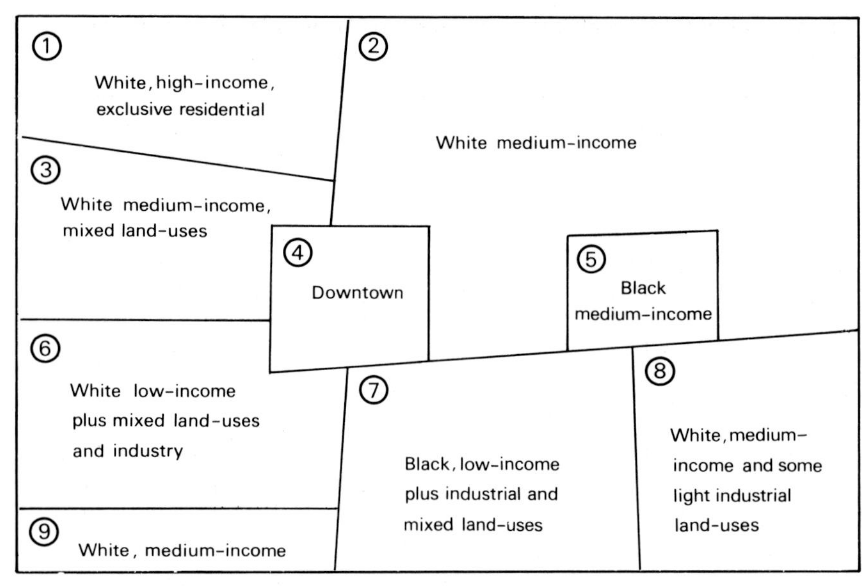

Figure 2 *Generalised distribution of neighbourhoods in the American city, on income and racial bases*

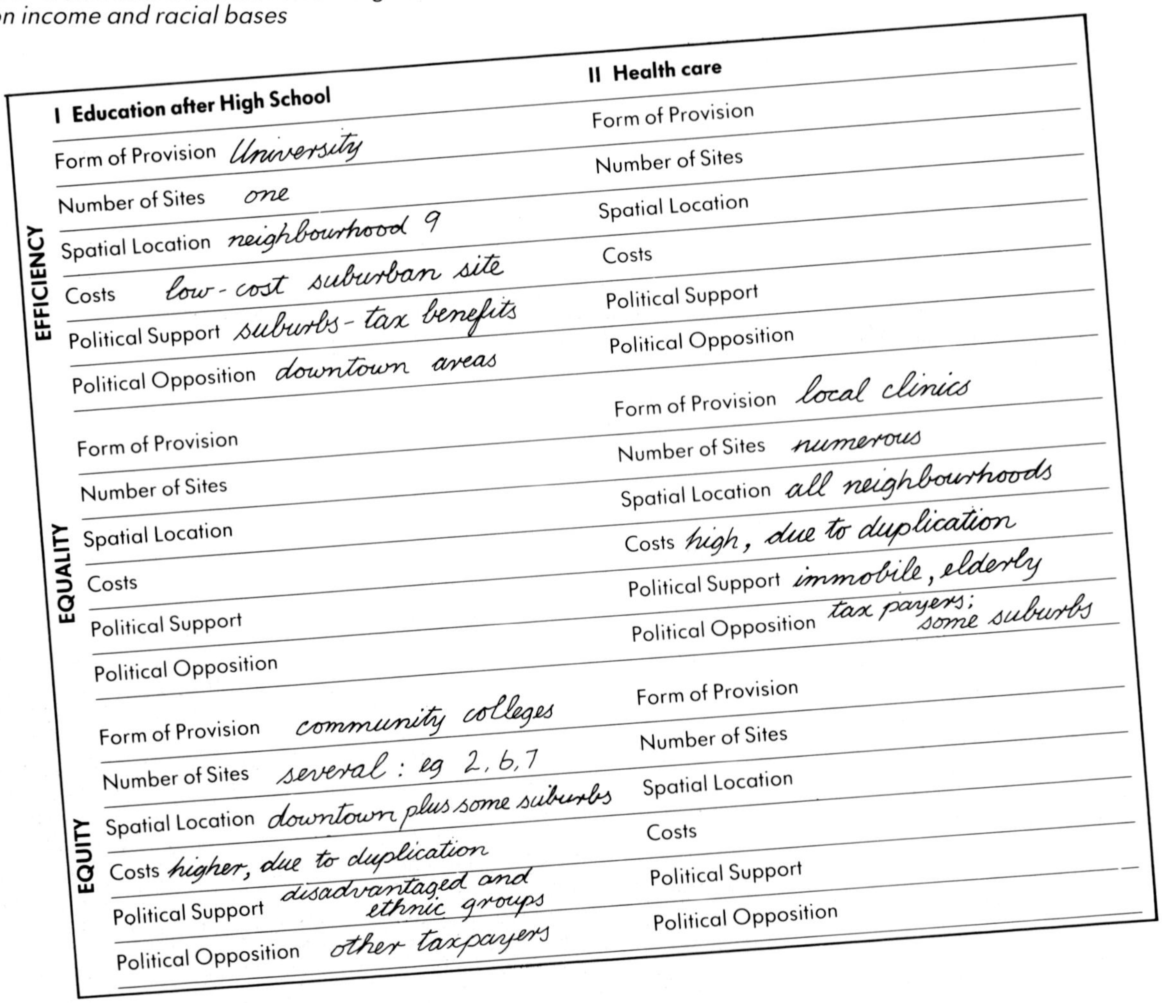

		I Education after High School	II Health care
EFFICIENCY	Form of Provision	University	
	Number of Sites	one	
	Spatial Location	neighbourhood 9	
	Costs	low-cost suburban site	
	Political Support	suburbs - tax benefits	
	Political Opposition	downtown areas	
EQUALITY	Form of Provision		local clinics
	Number of Sites		numerous
	Spatial Location		all neighbourhoods
	Costs		high, due to duplication
	Political Support		immobile, elderly
	Political Opposition		tax payers; some suburbs
EQUITY	Form of Provision	community colleges	
	Number of Sites	several: eg 2, 6, 7	
	Spatial Location	downtown plus some suburbs	
	Costs	higher, due to duplication	
	Political Support	disadvantaged and ethnic groups	
	Political Opposition	other taxpayers	

Figure 3 *Exercise chart*

REVITALISING THE CITY

So far, we have contrasted the central city with suburban areas, and indicated the advantages enjoyed by the latter. Suburbs, as we have seen, have enjoyed transport subsidies in the form of public transport or, more likely, expensive freeway developments. They are able to maintain exclusivity by politically opposing unwelcome developments, but are usually close enough to cities to enjoy metropolitan facilities like theatre, opera and restaurants.

Central city areas, in contrast, have been associated with poverty and racial minority growth. Between 1960 and 1978, city centres lost nine per cent of their white populations but increased black proportions by 40 per cent — resulting in, on average, hispanic or black concentrations of approximately one-third of the total population. (There are of course several cities where this proportion is far higher — Gary, Indiana has 71 per cent, Detroit 63 per cent, compared with a national average of blacks, 12 per cent, and hispanics 6 per cent.)

In the past, these imbalances in population have caused researchers to liken the downtown area to a 'reservation', in which the ethnic poor are trapped. If this analogy ever had any validity, it has largely disappeared. To begin with, city centres have in recent years attracted large quantities of federal dollars to improve the quality of life and to renew job skills. Table 5 shows the amounts spent in just one year for selected cities.

The uses to which these funds have been put have varied, but increasingly they have gone towards attracting and/or creating new economic activities. Small businesses have been nurtured, but a lot of investment has gone into large prestigious developments, like the Renaissance Center in Detroit. The name is not coincidental, for the huge concrete edifice, which contains numerous shops, restaurants, offices, plazas and car parks, is specifically designed to be a symbol of regrowth. Many of the larger cities are now competing for new hotels, convention centres and

Table 5 *Federal Government aid to selected US cities in 1978, in dollars per capita*

	GRS[1]	*CDBG*[2]	*LPW*[3]	*CETA*[4]	*ARFA*[5]	*Total*	*% Black*[6]
Atlanta	17.60	32.69	16.91	66.36	4.90	133.56	66.6
Baltimore	33.21	34.15	23.67	36.76	11.50	127.79	54.8
Boston	37.87	39.84	25.83	52.26	8.83	174.82	22.4
Chicago	42.29	37.18	12.01	33.79	4.65	125.27	39.8
Cleveland	25.11	55.18	15.26	38.82	6.29	140.66	43.8
Dallas	17.84	17.56	—	14.44	—	49.84	29.4
Denver	26.92	24.54	17.76	29.38	5.22	103.82	*
Detroit	30.63	42.88	20.39	44.97	12.35	151.22	63.1
Houston	14.64	16.62	5.95	17.16	1.04	55.41	27.6
Los Angeles	17.00	18.18	16.69	43.21	4.93	100.01	17.0
Memphis	17.48	22.93	—	16.47	1.10	57.98	47.6
New Orleans	36.23	34.45	16.36	34.38	11.17	132.59	55.3
New York	41.11	30.13	25.85	40.13	16.07	153.29	25.2
Oakland	17.53	34.22	33.64	65.31	8.59	159.29	46.9
Philadelphia	28.90	35.03	30.46	37.95	11.84	144.18	37.8
Pittsburgh	27.44	51.20	36.17	44.02	8.47	167.30	24.0
Portland	27.15	27.62	32.06	46.83	6.17	139.83	*
St Louis	26.82	61.35	29.05	46.45	10.40	174.07	45.6
San José	10.55	11.05	17.08	31.71	3.01	73.40	*
Washington DC	40.95	46.40	42.74	38.06	12.59	180.74	70.3

Notes:
1 General Revenue Sharing: basic source of Federal aid for social services.
2 Community Development Block Grant: urban renewal and housing expenditure.
3 Local Public Works Program: new infrastructure support.
4 Comprehensive Employment Training Act: retraining support.
5 Anti-recession fiscal assistance: targeted to areas with high unemployment.
6 Percentage of city population Black, 1980 Census.
* Black population less than 15 per cent

Print 5 *The Renaissance Center, Detroit — symbol of rejuvenation*

leisure areas, in the hope that these will attract tourist and business revenues. The kinds of developments that were once confined to London (St Katharine's Dock *page 77*) and San Francisco (Fisherman's Wharf) are being replicated in many locations from Portland to Boston.

Leverage

In these instances, the aim is so-called leverage, by which public money is spent in order to attract private investment. In general, this is being successful. There seems to be an increasing tendency for corporations to allow themselves to be lured back into the city. Approximately $5 billion has been invested in Manhattan, New York this decade, increasing office space by 25 million square feet. Thus the city skylines familiar from the opening credits of television programmes like *Dynasty* (Denver) and *Dallas* are being replicated throughout the country.

Gentrification

One of the unexpected results of this corporate/service growth is a new demand for medium- to high-cost residential space within the city. A new breed — the YUPPIE or *young, urban professional* — is attracted to a city existence. This is not the same life style as that enjoyed by their grandparents, however. The new residential developments are high-rise condominiums — blocks of owned apartments, usually containing a concierge service, a swimming pool and a garage. Again, the development is fuelled by real estate speculation and the possibility of high profits for investors.

These kinds of development have come to be called gentrification, an extension of a term first used in Britain. The incomers are not 'gentry', but they usually have a high income, and the life style they seek effectively displaces the poorer residents who live in inner city neighbourhoods. The speed with which this happens can be unexpected. In San Francisco, the location of the BART stations within the city originally avoided low income hispanic areas in a neighbourhood known as the Mission, in order to preserve the image of the service. Stations were ultimately built in the Mission district, and immediately served as nodes around which gentrification began. Again, the unplanned nature of much of what happens in the American city is emphasised.

In his study of the American City, Eliot Hurst writes about the issues which are important:

'In the best of all possible worlds, in which no-one starves and small children do not go without food and medical treatment, in which there is no injustice, no mechanised oppression, no racist exploitation of one group of people by another, it would be an interesting exercise to study hierarchies of shopping centers, central place models, and pure city forms; to write about the 'quality of life', to analyse the distribution of widgets and gizmos. . . .'

(*Source*: M. Eliot Hurst, 1975, *I Came to the City*, Houghton Mifflin)

This perspective is correct, but only half so. We do need to know about the poor and the exploited. It is also the case, however, that these are not the people who make political or investment decisions, or who determine the shape and form of the city. The young urban professional or the investment banker is more likely to do that.

Print 6 *Gentrification in Boston — these houses have been converted into condominiums*

?

1 Figure 4, *page 75*, shows a recent advertisement taken from *Vogue* magazine. It offers condominiums in the downtown area, in the price range of 1.5 to 11 million dollars, a figure high even for Los Angeles. Clearly, for that price the buyer could purchase a home anywhere in the United States. So what is the attraction of an urban existence?

a Why would anyone choose to live in the downtown area, and pay so much to do so?

b Are the new urban dwellers — the YUPPIES — attracted by what the city has to offer?

c Are they repelled by the idea of a suburban existence?

2 Clearly, not all those moving into condominiums are fabulously wealthy; nevertheless, if the rich, who have freedom of location, do it, then we must assume that it is a popular choice, and that there are attractions worth paying for. For a glimpse of what these might be, examine the extract above the advertisement taken from a recently published book which is an 'exposé' of the YUPPY lifestyle. Here the authors have changed the more usual term 'gentrification' to 'yupification'.

Even in this short piece, the authors identify a cast of characters who all have a role to play in this process: most of them will benefit from yupification, some — like the previous residents — will be ejected. Try to place yourself in each of these roles, and examine the motivations and particularly the locational decisions that are being made by these different characters. They are:

a **Investment banker**

b **Gourmet food store owner**

c **Bar owner**, specialising in jazz and expensive drinks

d **Owner of laundrette**, about to lose lease and business; cannot afford new rents

e **Real estate developer**

f **Former resident**, evicted from single men's hotel; now homeless

g **Former dress factory worker**, now unemployed after conversion of factory

Yupification

The Yupification of a neighbourhood has been compared to termite infestation. Yuppies descend in swarms and leave nothing behind but dumpsters filled with discarded linoleum. Sure signs of impending Yupification include:

- forced relocation of candy stores and laundromats
- proliferation of gourmet food stores, outdoor cafes, and historical society plaques
- disoriented 'bums' trying to figure out why their favourite bar now has asparagus ferns in the window

Residences are no longer referred to as houses or apartments; they are now called 'living spaces'. The term is necessarily vague. Factories and garages are being converted into living spaces. Condominium conversion is rampant. Here is a not unlikely urban scenario:

Struggling immigrant opens a tailor shop in the local ghetto. By working 18-hour days and eating nothing but rye bread, he manages to save enough to open a small dress factory. He rents a floor in a cast iron building and hires a few employees. He marries one of his stitchers, moves to a better neighbourhood and sends both his children to city colleges. His son, the accountant, marries a schoolteacher, buys a house with a two-car garage and an acre of wooded property, and sends both of his children to private colleges. The son of the accountant, an investment banker, marries another investment banker and buys Grandpa's dress factory, which has been converted into a 'loft living space' and is going for a quarter of a million.

Moral of the Story: Yuppies will live anywhere, as long as the floors are genuine parquet and there's another Yuppie on the block.

(*Source*: Prisman and Hartley, 1984, *The Yuppie Handbook*, Pocket Books)

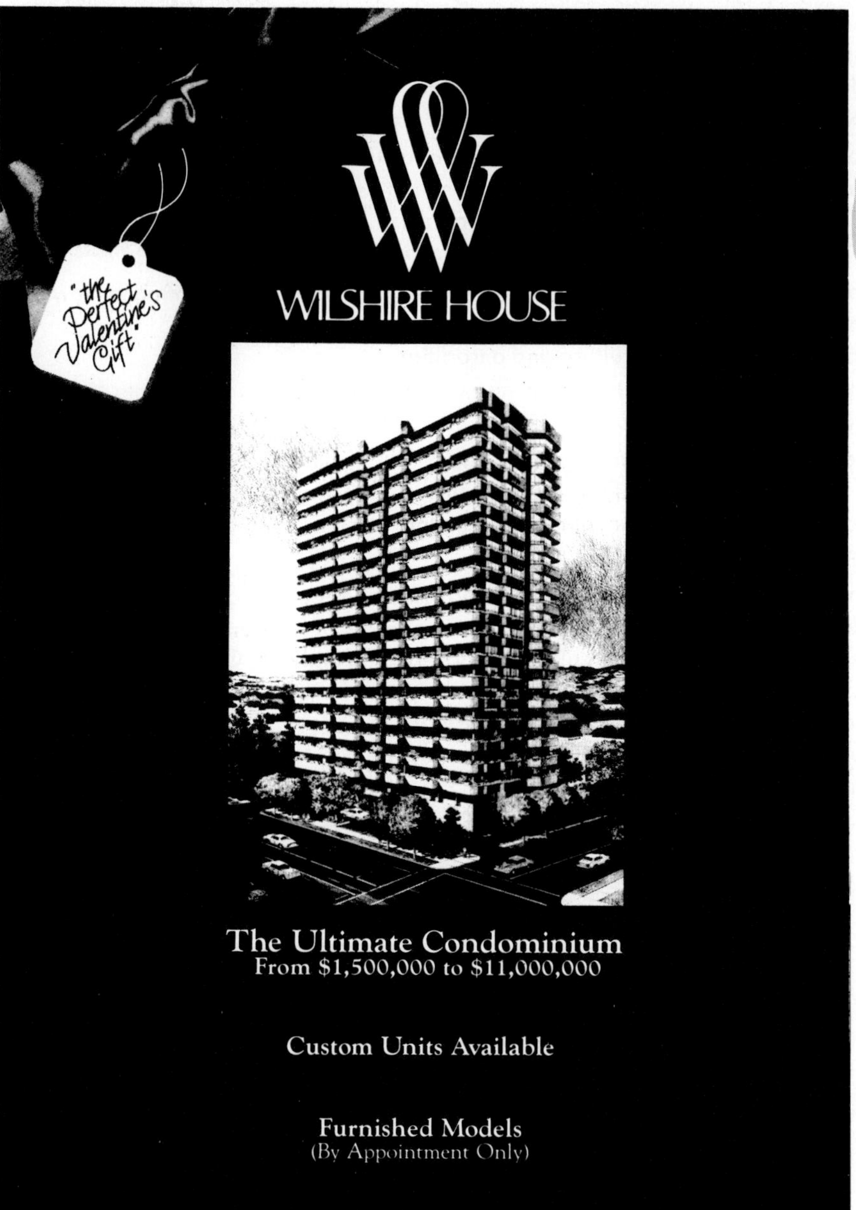

Figure 4 *Gentrification in practice*

THE AMERICAN URBAN FUTURE

One of the aims of this chapter has been to emphasise the mobility and change which exists within American cities. Many of the stereotypes are altering, and population is still mobile. This all adds up to a complex urban future: indeed, given the sheer size of the country, it is likely that there will be various urban futures.

Seven different types of urban development are shown in Table 6, including the mechanisms at work and some examples. What is clear from this crystal-gazing is that different forces are at work in different parts of the country, and that different tastes and possibilities can produce different city forms. It is unlikely that there will be any real convergence towards a typical 21st century urban model, and far more likely that a diversity will continue to offer opportunities for individual choice (for those at least who are mobile and affluent).

Table 6 *Seven urban futures*

	Urban futures	*Stimuli*	*Contemporary examples*
I	Conflict City	Achieved by drift, inaction, and crisis; opposed by pro-action residents, but supported by civic cynicism, lethargy, lack of political vision, and class and race hostility.	Detroit; Youngstown; any large Frostbelt city
II	Wired City	Promoted by commercial pressure on competitive firms and the lure of vast profits in new information services; opposed by technophobes and companies vulnerable to telecommunication charges.	Manhattan in New York City; Washington, D.C.
III	Neigbourbood City	Promoted by historic appreciation for small-scale community advantages; opposed by modernists who perceive the city as a single operational entity; opposed also by those who see neighbourhoods as narrow bastions of ethnic and racial isolation.	Santa Barbara, California; Seattle, Washington; Boulder, Colorado
IV	Conservation City	Promoted by ecology and environmental conservation pressures and partisans; opposed by cultural hostility to 'beehive' living arrangements, and by traditional pride in the right to wide options in land development.	Habitat (Montreal); Manhattan in New York City
V	International City	Promoted by commercial pressure on competitive firms and the need to secure new jobs for local labour force; opposed by a provincial culture and a tradition of urban insularity and ethnocentrism.	Toronto; Manhattan in New York City
VI	Regional City	Promoted by recognition of inadequacy of city resources to meet problems of larger scope, and by the need to achieve economies of scale and central political direction; opposed by boosters of traditional political boundaries.	Greater Miami; Greater Toledo; Greater Toronto; Minneapolis-St Paul
VII	Leisure City	Promoted by a shorter work year, and by a steady rise in joblessness, and in dual income, smaller, better-off households, and by pro-recreation culture; opposed by a Calvinist Work Ethic culture, by those who see leisure undermining the productivity of the local work-force and by those who fear 'the devil will make work for idle hands'.	Reno, Nevada; Palm Beach, California; Orlando, Florida; Anaheim, California

(*Source*: adapted from Shostak, 1982)

? In this exercise, you will evaluate the implications of the different urban futures outlined in Table 6. Use your skill and imagination to fill out a copy of Table 7. Then, compute the scores (5 is high, 1 is low) for each type of urban future, and decide on that basis which scenarios are the most attractive and which are the most unattractive (with the implication that perhaps some action should be taken by government to change the future).

There are various things you may want to discuss in groups first before you fill in the table: what *kind* of employment (for the skilled, the unskilled), for example, and just what determines the quality of life — is it the climate, is it a low crime rate, or is it a combination of a number of factors?

In conclusion, you should also ask: what does this suggest may happen here say, in Europe? Often American trends become European trends too. What would you most like to see happen to, for example, Britain's cities, and which of these futures would you least like to see when you are 40 years old?

Table 7

Scenario 1–7	*Employment opportunities* High = 5 Low = 1	*Racial harmony* High = 5 Low = 1	*High quality of life* High = 5 Low = 1	*High cost of amenities* High = 5 Low = 1	*Overall evaluation* Total Score
1					
2					
3					
4					
5					
6					
7					

POSTSCRIPT: REDEVELOPING LONDON'S DOCKLANDS

Angry council tenants in East London have vowed to prevent their homes being sold off to wealthy 'trendies' seeking to snap up riverside homes.

Once a slum area, Wapping is now rapidly becoming the fashionable place to live. Property prices are soaring with a three bedroom Thames-view flat now commanding well over £150 000.

The tenants fear they might have to move out while refurbishment is being carried out. Then they would not be allowed back, with their homes instead offered on the open market.

Mr Patrick Massett, 57, of Willoughby House, a former docker and one of hundreds of GLC tenants in the area, said: 'There are lots of rumours that we might have to move out while renovation takes place. But if we leave we fear that we will never get back in.

Our homes will go to the rich so we are not going anywhere. We'll fight to stay put and I'll barricade myself in if necessary.'

But a council spokesman said today: 'We certainly have no plans to move tenants out. The idea of selling off properties to wealthy purchasers is quite abhorrent to us.'

(*Source*: *Evening Standard* 10 January 1985)

The following brief outline and exercise on London's Docklands are intended to illustrate some parallels between North America and Britain as regards the renewed interest being displayed in the redevelopment of urban areas.* Is yupification a feature of Docklands redevelopment?

ST KATHARINE'S DOCK

The Docklands are a narrow belt of residential, port, warehousing and industrial land that extend over 20 kilometres east of the City of London along the Thames River. This large area contains some of the worst examples of poor living conditions and blight of any urban area in the industrialised world. This has been caused by the downstream movement of port activities to the deeper, wider Thames estuary.

Extensive studies of conditions in the Docklands were carried out in the 1970s. These produced many ideas but little agreement between the five dockland councils and the GLC (known as the Joint Docklands Committee), and central government.

In July 1981 the London Docklands Development Corporation was formed by government charter and given the job of revitalising the area. One of its first moves was to establish a Free Enterprise Zone on the Isle of Dogs, offering small-scale businesses up to ten rate-free years if they moved into the zone. The LDDC has powers above those of the docklands local councils and has been able to implement many new commercial developments with great rapidity. This has been bitterly resented by both local councils and residents who feel that their communities are being threatened and that the LDDC is no better than a private property developer with widesweeping powers.

One area that the JDC oversaw the redevelopment of was that of St Katharine's Dock in the borough of Tower Hamlets. The location of this area is unique as it is adjacent to the City of London and the tourist attractions of the Tower of London and Tower Bridge. Consequently, it had enormous tourist and commercial development potential after the docks closed in 1968.

Print 7 ***A view of St Katharine's Dock showing the marina***

* This activity is based upon 'St Katharine's Dock — a model for docklands' redevelopment', by J. Fien et al, *Classroom Geographer*, May 1981.

What sort of redevelopment?

1 Divide into groups of four, each member of the group adopting one of the roles listed below. Study your role carefully, working out how you would have liked to see St Katharine's Dock redeveloped.

Role 1: The Developer — Mr J. Fisher of McDonalds Construction Ltd
You represent a large successful firm which will carry out the development. The firm's main interest is in making profit, perhaps at the expense of providing things which the local people really need. Your firm has the finance needed to develop the area. Your firm wants to use the land for the type of buildings that can be leased for high rents, such as commerce, tourist facilities and certain types of housing. However, you must get planning permission for your schemes, so they must fit in with the desires of the Greater London Council, the body that will give planning permission. (But the GLC needs your resources to carry out the plan.)

Role 2: Mr Henry Orson-Young of the Royal Yachting Association
You are a spokesman for the Royal Yachting Association, a group concerned with putting the case of boat owners and the boating industry. Your association needs a site for mooring boats owned by Londoners, and a permanent site for the collection of historic boats and ships. You would also like to see accommodation built for boat owners to give them access to their boats. You are a boat owner yourself and would like to be able to live and work in the City, yet have easy access to your boat.

Role 3: Mr Fred Bates of the Greater London Council
The Council are the owners of the land. They have purchased the dock from the Port of London Authority by compulsory order. You represent the GLC, who sees the St Katharine's Dock area as part of London: you want to attract money into London, but you also want to provide for the people of London as a whole. You believe the site provides a good opportunity for business development as it is near the City and has great potential for tourism — another good way of earning money. However, the docks were previously important for employment of local people, so new forms of employment are needed for them.

Role 4: Mrs Joyce Lyons of the local Tower Hamlets Council
You represent the working people of the docklands area and are very concerned that local people should benefit from the development, not just the more affluent members of society. You would like to see developments such as housing and entertainment which are accessible in price to everyone.

2 Each member of the group should select three land uses which he/she considers appropriate according to his/her role and without reference to others in the group. Nine possible uses include: offices, private housing, marina/open space, shops, GLC housing, industry, hotels, entertainment facilities, car parks.

3 Again without reference to the other members of your group, select 7 of the 16 area units shown on the planning map (Figure 5) to plan where land uses should be sited.

4 Planning Meeting — redevelop the dock area. Acting your roles, the members of your group now meet to decide on the land uses for all of the 16 units on the base map. A proposed land use map for the area should be prepared during the meeting (trace off Figure 5) and notes should be taken on how the decisions were made. (For example did some members make a 'deal' and join forces? Did others have to make compromises?)

5 Your final task is to produce a ten minute report which includes:
a A map of the plan decided on by the group.
b The reasons behind the final form of the plan.
c A brief account of the way the decisions were arrived at.
d A brief account of whether each member was satisfied by the final result.

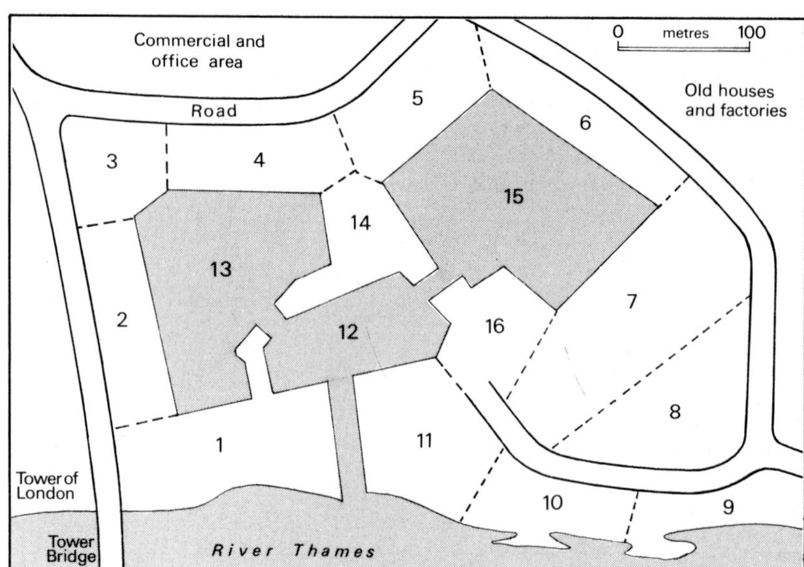

Figure 5 *Base map for replanning St Katharine's Dock*

The plan that went ahead

The firm of Taylor Woodrow was prepared to redevelop the area. Their plan (Figure 6), now largely completed, provided for commercial interests (a first class hotel, offices and a marina), tourists (the hotel, a maritime museum, restaurants and an historical inn) and, to a lesser extent, local residents (GLC housing). This plan has been criticised for being socially and economically elitist and for not providing adequately for the working class inhabitants of the surrounding area.

Many Tower Hamlets councillors and residents have been fiercely opposed to the Taylor Woodrow plan. As one of their leaders, Paul Beasley, has said, 'It is simply not for East Enders. It is too crowded, too expensive. . . . You cannot proceed down here as if it was a green field. You cannot operate against the hostility of the local people. . . . We don't want any more developments like it down here.'

Peter Drew, a spokesman for Taylor Woodrow, acknowledges this view but claims in reply, 'We have provided jobs, over 1700 of them. We pay Tower Hamlets rates of £1.25 million a year and have generated economic activity where before there was none. We are also building council housing around the East Dock in an attractive setting. I know they don't like the private sector, but it is the private sector which is going to get Docklands moving.'

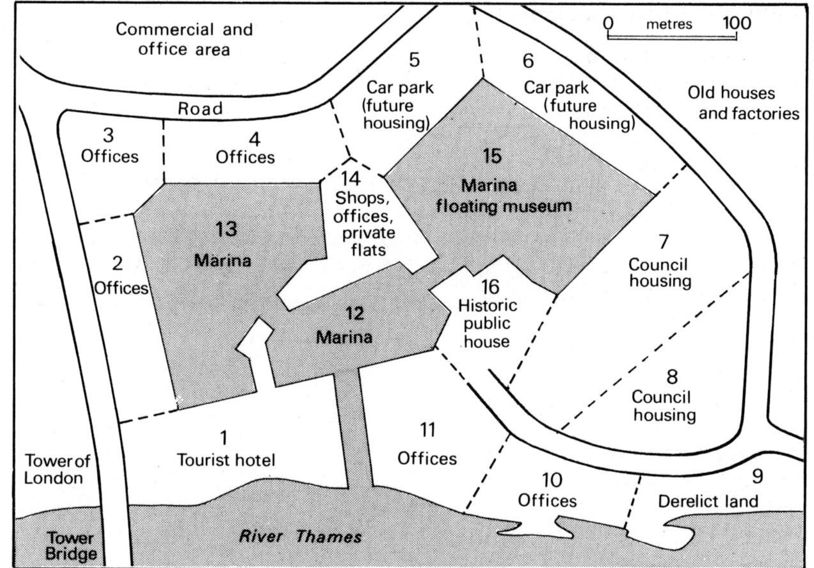

Figure 6 *The Taylor Woodrow Plan for St Katharine's Dock*

?

1 Who has benefited from the Taylor Woodrow redevelopment plan? Use the following scoring system to rate the benefit of each of the development land uses to the various groups listed in Table 8 below:
very beneficial — 2   of no benefit — 0   somewhat beneficial — 1
Enter the scores on the table, total the scores and then answer the following questions.

a Who are the two highest scoring groups? How does the plan benefit them?

b Who are the two lowest scoring groups? What criticisms would they make of the plan?

c Why were these groups unable to get the developments they would have desired?

d Why did the highest scoring groups get the developments they wanted?

e What could be done with the still vacant Area 9 and possibly Areas 5 and 6 to make for a more socially just environment?

f Evaluate your group's plan on the matrix in Table 8. Would your plan have been a fairer one? Why?

2 Who do you think is right, Paul Beasley, the residents' leader, or Peter Drew, the Taylor Woodrow spokesman? Which view favours yupification?

Table 8 *Matrix for evaluating the Taylor Woodrow Plan*

Land use	*Taylor Woodrow*	*Other Companies*	*Tourists*	*GLC*	*London as a whole*	*Local residents*
Offices						
Shops						
Hotels						
Private Housing						
GLC housing						
Leisure						
Marina/open space						
Car parks						
Total						

PART II POWER, POLICIES AND FOOD PRODUCTION

FEEDING THE WORLD

'The Committee sees no more important task before the world community than the elimination of hunger and malnutrition in all countries.' *(Source: Brandt Report)*

There has been much controversy over recent years — controversy in which Oxfam has joined — about the food problems of poor countries and the best way of meeting them, especially about the value — or otherwise — of food aid. The whole question was looked at again in the Brandt Report.

Everybody agrees that the best long-term solution is for food production within poor countries to be increased so that they can meet most of their own needs without outside assistance and without using previous foreign exchange to pay for food imports. External aid should thus obviously be mainly devoted to improving the capacity for local food production, rather than to shipping in food supplies which, whether free or subsidised, compete with local production and may actually discourage it by depressing prices. This is why donor Governments agreed in Rome in 1974, after the last major world food crisis, that the prime aim of international policy must be to improve food production within poor countries. The controversy concerns the interim period before this is achieved. If local food production is insufficient so that mass hunger results, and if Western countries, such as the USA, have large surpluses which might otherwise be wasted, should the latter be used to fill the gap, and if so how and on what terms?

The Brandt Commission, like all others who have examined the problem, accepted that the main need is to increase domestic food production within poor countries. But they noted that at present, with rapid increases in their populations and only slow increases in agricultural productivity, the trend is for developing countries to be come *more* dependent on outside supplies. While over the last 25 years or so, food *production* has risen by more than 2½% a year in such countries, *demand* for food, with increases in population and incomes, has gone up by more than 3%. As a result, imports of food have increased from low levels in the early 1950s, to 20 million tonnes in 1960–61, 5 million in the early 1970s and nearly 8 million tonnes in 1978–79. By 1990 such imports could be 147 million tonnes a year, more than half of which would go to the poorer countries of Africa and Asia. Poor countries are unable to finance such imports from the proceeds of their own exports, so some regular outside assistance will continue to be necessary (in addition to emergency aid in disaster situations, which the Report sees as a much worse way of helping, being far the most expensive in terms of cost and lives lost).

Since the first priority is to increase food production in poor countries, the governments of those countries, the Report concludes, must devote a large part of their development effort to increasing agricultural production (at present in more than half the poor countries food production is growing more slowly than population). This will require that such countries should devote more of their development funds to irrigation of all kinds, storage facilities, fertilisers and agricultural research. It is also of great importance that measures of land reform should be introduced to increase productivity and the purchasing power of the very poor. Farming systems that are suited to local circumstances, making use of ample local labour supplies, and not the blind transfer of capital-intensive western methods, should be used. Efforts should be made also to develop local fishing industries to enable them to exploit to the full the fish stocks in the large Exclusive Economic Zones now under the control of many poor countries.

But these efforts will need international support. External aid to develop agriculture in low-income countries was about 3 billion dollars a year at the end of the 1970s; but the report estimates that this should go up to somewhere between 4 and 8 billion dollars a year within the next decade (in addition to local investment) if substantial improvment is to be made. And if food deficits are to be eliminated altogether, 12 billion dollars a year might be required. The International Fund for Agricultural Development (IFAD), which has 124 member Governments and a relatively democratic management structure, could well prove to be the most efficient channel for disbursing such funds. The use of these funds for investment in irrigation and water management, rural electrification, roads and other communications, and draught power, as well as in agricultural research, especially on plant strains and stock development, might eventually enable poor countries to meet most of their food requirements from local production and should at least reduce the need for emergency food operations which only demonstrate, at heavy cost, that other efforts have been inadequate.

Even so, there is need for some security in food supplies to guard against shortages, whether local or international, because of unfavourable conditions.

- Does food aid benefit poor countries or only discourage local production?
- Can food production in poor countries be increased as fast as population?
- If food imports remain inevitable in poor countries for the next decade, how can they best be organised and on the basis of what criteria?
- Should poor countries adopt agricultural methods that bring big increases in productivity (such as the Green Revolution) if they also increase inequality and landlessness?
- Should poor countries introduce desirable social reforms, such as land reform, if the effect is to reduce agricultural production, or seek to boost production whatever the social costs?

? Read the extract taken from an Oxfam bulletin and then read the following statements and decide whether you consider them to be true or false. Each of the statements (or ones very like them) are taken from *World Hunger 10 Myths* by Lappé and Collins, Institute for Food and Development Policy.

1 People are hungry because of scarcity. T F

2 Hunger results from overpopulation: there are just too many people for food-producing resources to sustain. T F

3 To solve the problem of hunger the top priority must be for growing more food. T F

4 If countries where so many go hungry did not produce agricultural exports, then the land now growing food for foreign consumers would nourish local people. Export agriculture, therefore, is the enemy. T F

5 To help the hungry we should improve and increase our foreign aid programmes. T F

The above activity has introduced you to some of the main issues and problems examined in the chapters in this section. As you work through the chapters you should assess the truth or falsity of the above statements again and make notes and record examples of evidence which support or refute the statements.

POWER AND FOOD PRODUCTION: PERU AND THE US

The chapters in this section, taken together, highlight the awful dilemma we face of too much food for some people and not enough food for others. Each chapter also raises other issues not touched on in the five statements, such as the nature of export agriculture and the management and mismanagement of land reform schemes, both of which are first examined in *Agricultural Production and Land Reform — Peru*. The next chapter, *US Agriculture — the Problem of Overproduction* provides a stark contrast to Peru. But after you have worked through the chapter and begun to realise the connection between the political power of US farmers and the implementation of policies which *protect* the farmer, look back to the Peruvian study and pick out phrases like '. . . but the implications of this fact for rural development depend, of course, on who controls the government bureaucracy'. Think about who controls Peruvian society, what power farmers have there in contrast to farmers in the US and consider the connection between power and food production.

FURTHER CONTRASTS IN MEXICO, EUROPE AND CHINA

The dilemmas and realities of the first two chapters are further explored and reinforced in the next two which look at Mexico and Europe.

An extract from a newspaper article you will read in *Government Policy and the Rural Environment — Mexico* highlights the relationships between food production and government agricultural policy. It also points out, like the cartoon on *page 136*, the relationship between food exports from developing countries to wealthy countries with the money to pay for them and the problem of reaching self-sufficiency and an adequate diet within the less wealthy countries themselves.

The theme of *overproduction* is taken up again in *Peasants, the Environment and the Common Agricultural Policy*. Significant to our understanding of European agricultural production is the central position given to agriculture and the living standards of peasants and farmers in the EEC. The CAP may be seen as a social policy encouraging the modernisation of agriculture and maintaining and increasing the living standards of workers in agriculture. The French policy of self-sufficiency and protection of the market by a system of price maintenance policies is to be understood in the context of the social aim of protecting French workers in agriculture. The burden which this now imposes and the debate it causes is often reported in our newspapers — changes in the structure of the CAP is history and geography in the making. 'Snatches from a Rural Congress' (*page 145*) gets to the heart of the CAP debate. It is interesting that massive subsidies to agriculture have still left problem regions. How successful indeed are any government policies?

In China too, top priority is given to agriculture and self-sufficiency and the last chapter in this section, *China — Can Self-Sufficiency be Achieved?* evaluates the pattern and process of China's agricultural policy. The Chinese situation should be contrasted with the other countries studied, especially Peru and Mexico.

NATURAL DISASTERS AND WORLD HUNGER — THE BIGGEST MYTH OF ALL?

One of the most widespread beliefs about the causes of world hunger is that it is precipitated by droughts, hurricanes, floods and similar natural disasters. The article 'Nature pleads not guilty', will help you to think about the validity of beliefs such as drought causing starvation and studying this section will reveal the role of power and policy in arranging the relationships between people and their environment — often, of course, to the detriment of some people and some environments. It would be so much easier if we could blame nature entirely; unfortunately we cannot and there is still much to be achieved through policy making.

Nature pleads not guilty

Every year the Sahara swallows up acres of farmland as it marches indomitably towards the south and west. As human settlements get drier and drier, pastoralists dependent on tiny patches of grasslands give up their herds and migrate to the cities. Agriculturalists abandon their villages to hunt for water or food. In Mauritania's capital, Nouakchott, migrants have swelled the population to 500 000, about a third of the country's total.

Two thousand miles to the south, the Kalahari desert gobbles up the semi-arid lands of Lesotho, Botswana, Zambia and Zimbabwe, forcing farmers as far away as Tanzania to crowd together on the remaining good land, putting more and more pressure on exhausted soil. To make matters worse, fertile forests are being cut and cleared at a rate as high as one million acres per year in Zambia and other southern African countries. All that remains is poor soil to bake and harden in the sun, making it nearly impossible to till.

African rains are often either inadequate or torrential intensifying problems of poor soil and desertification. In Mozambique the two years of drought which had crippled production in 1982–83 were swept away by rains when cyclone Domoina lashed the southern provinces in January this year. After drought robs the soil of vital nutrients, it becomes looser, more vulnerable to landslides and wind. If, in turn, the rain comes too fast, the soil cannot contain it. Silt-thick African rivers yearly sweep tonnes of topsoil into the sea.

Poor soil, erratic rainfall, accelerating population growth, blatant overuse of land: the environmental cards appear to be stacked against Africa. And the results seem obvious. Africa's total cereal production has declined by one per cent annually since 1970. In the 1930s Africa was a food exporter; in the 1950s it was self-sufficient. But by 1980 sub-Saharan Africa was importing 8.5 million tonnes of cereals annually.

To what extent are the climate and 'poor' African soils responsible for food shortages in Africa? In the mid-70s a group of meteorologists and other academics organised a project to study the effect of climate on the great Sahel famine. After a few months an entirely different picture began to emerge. They found that the role of drought was much smaller than assumed and there was no simple cause-and-effect link between drought and famine. 'In 1976', their report argues, 'there was also a drought in Britain. We believe that nobody would have thought it 'natural' for thousands of British children to die *because of the drought*. The loss of even a few dozen children would have been nothing less than a scandal.'

Significantly, their report is titled *Nature Pleads Not Guilty*. People are to blame. The spreading of the Sahara and Kalahari deserts can be linked directly to overgrazing and overuse of land. Even shortages of rain, foresters speculate, are caused not by natural fluctuations in climate but by the rapid clearing of rain forests. Blaming the weather, moaning over acts of God and accusing poor farmers are superficial responses to a complex problem. It is more revealing to examine the policies which starve the poor, pressure the land and make entire countries vulnerable to drought.

Growing food for local consumption, for example, receives low priority from many governments. In Upper Volta last year drought devastated millet production in the northern provinces yet farmers in the south — with bumper surpluses of millet to sell — were discussing planting cotton instead. The reason is simple. The farmers know that cotton for export will be quickly collected and paid for by the government. Thus cotton production has shot up over 20 times since independence while yields of sorghum and millet — the major food crops for the region — have stagnated.

This trend is typical for many African countries. In Mali, during the great drought between 1976 and 1982, while food production plummetted, cotton production increased by 400 per cent. During the drought of 1973–74 in Tanzania, sales of maize fell by a third while the output of tobacco continued to grow. Both crops need about the same amount of rain. The difference comes in inputs available and the incentives to the growers. Sixty-two per cent of money loaned by the Tanzanian Rural Development Bank between 1978 and 1979 went for tobacco and only 19 per cent for maize.

Donors such as the World Bank and the US Agency for International Development (USAID) have been pouring in millions of dollars for development over the last decade. The rhetoric is of improving food production for local consumption but the reality has proven different. By 1975 the World Bank had invested over $200 million in Tanzania without supporting a single project designed to produce basic foodstuffs. Things haven't changed too much since then either. A recent survey by USAID of 570 projects in Africa found that only 22 were directly related to food crop production.

The entire food distribution and storage system in many countries also causes a gap: not in food production but in the number of people who have access to food. In Zimbabwe the fruits of the 1981–82 bumper harvest were nearly lost because the government didn't have the facilities to store maize.

Often the government simply can't purchase or transport grain, and private traders find a way to profit through illegal grain sales. In Mali an estimated 20 per cent of the government millet stocks disappear into the hands of the black marketeers who sell it across the border where prices are higher. Development workers in Niger — one of the few countries self-sufficient in food production — tell of a regular system of withholding grain until the price for food in Nigeria rises and then selling the surplus there.

For the government the priority is to keep prices low for urban consumers with little thought to the effect on rural producers. An extreme case comes from Mali, where in 1980–81 it cost farmers about ten cents to produce a kilo of rice while the official price paid by the government was only 6 cents.

Such policies are a recipe for low-production and a boost to cross-border smuggling, not an encouragement for local self-sufficiency. And it is man-made policies just as much as god-given forces that are keeping Africa hungry.

(*Source*: *New Internationalist*, September 1984)

GOVERNMENT POLICIES

Once you have worked through all the chapters in this section you should take a second look at the political policies described in them and compare and contrast them. How successful are government policies anywhere? Decide by what criteria success is to be measured — in global terms, national and regional terms or from the viewpoint of the individual farmer. Rule up a table like this one and jot down your carefully considered assessments. You should also be able to support or defend the statements about food production taken from *World Hunger 10 Myths* made at the beginning of this section. Have the government policies been successful?

Perspective	Peru	US	Mexico	Europe	China
Global Criteria					
1					
2					
3					
National Criteria					
1					
2					
3					
Regional Criteria					
1					
2					
3					
Individual Criteria					
1					
2					
3					

5 FOOD SUPPLIES, AGRICULTURAL PRODUCTION AND LAND REFORM: THE CASE OF PERU

FOOD SUPPLIES IN THE WORLD CONTEXT

> Putting an end to hunger is a challenge to the world's economic system, requiring complementary national and international measures. Only major efforts of investment, planning and research can make enough food available for the six billion people the world will probably hold by the year 2000. But not only must the food be there; the people who need it must be able to buy it. The reduction of poverty itself is equally essential for abolishing hunger . . . . To conquer hunger, every family must have a reliable livelihood, which means much greater gainful employment in both agriculture and manufacturing.

The extract quoted above is taken from the Report of the Independent Commission on International Development Issues, *North–South: a Programme for Survival.* The Report was the result of a series of conferences of an authoritative group of commissioners under the chairmanship of Willy Brandt (it is often known as the *Brandt Report*), former Chancellor of the Republic of West Germany, which included a majority of representatives from the Third World. The Report continues:

> Food production in all the developing countries rose by two and a half per cent annually between 1950 and 1975; but demand for food has grown by well over three per cent a year as population and incomes have gone up. As a result, the developing countries have rapidly increased their imports of cereals, from relatively low levels in the 1950s to 20 million tonnes in 1960 and 1961, to over 51 million tonnes in the early 1970s, and nearly 81 million by 1978–79. On current trends, the Third World could be importing 147 million tonnes of food by 1990, 81 million of which would be needed by the poorer countries of Africa and Asia. It is unlikely, within the prevailing economic climate, that these countries' own exports, or even additional aid, can finance such massive food imports. And even if the financing problems can be solved, there are doubts whether the major grain producers could supply the amounts needed. The suffering, unless something is done, will be appalling.

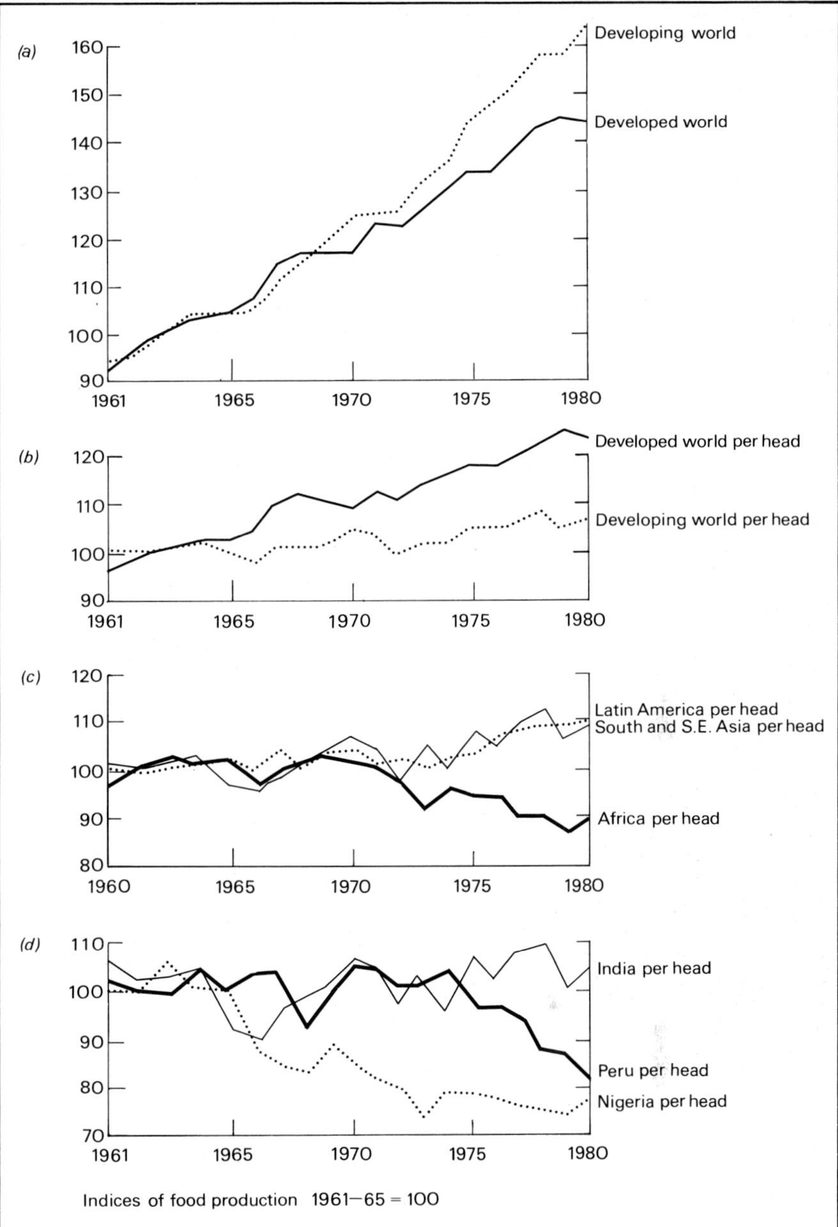

Figure 1 *Trends in food production and food production per head, 1961–80* (*Source*: *FAO Production Yearbook*, 1972 and 1980)

More information about trends in agricultural production can be gleaned from the statistics published annually by the United Nations Food and Agriculture Organisation in the *FAO Production Yearbook*. Figure 1 draws on these statistics to show, in a and b, the trends of production and of production per head of population in the advanced industrial world and in the Third World; in c, the trends per head in the major regions in the Third World; and in d, trends in individual countries, including the case of Peru.

DIFFERENCES AND TRENDS IN WORLD FOOD PRODUCTION

As Figure 1a shows, in the period 1961–80 total world food production increased more rapidly, in absolute terms, in the developing world than in the advanced industrial world. The two graphs were more or less parallel up to 1974, by which date production had increased by 35 per cent in the developing world compared with the average for 1961–65, and by 29 per cent in the advanced industrial world. Since 1974, however, food production in the developing world has increased much more rapidly than in the advanced industrial world. But account must be taken of the fact that the population of the Third World is increasing at a much faster rate than that of the industrial world, so Figure 1b, showing indices of food production per head, is a much more useful guide. Per head of the population, 20 per cent more food is produced in the developed world than in 1961–65, but only 5 per cent more in the developing world. Much of this modest improvement has been gained since 1974 largely as a result of some success in the application of 'Green Revolution' techniques in Asia and Latin America.

But these are overall average figures which conceal many differences among individual regions and countries. Africa, for example, has had a particularly poor record, and it is clear from Figure 1c that food production per head in developing Africa is less than it was 20 years ago. In South and South-east Asia, progress in food production per head has been made since the early 1970s. In individual countries, however, there are much greater fluctuations in food production, and they do not always follow the trends for the major regions in which they are located (see Figure 1d). There are many reasons for such fluctuations: drought, for example in the Sahel in Africa, in Peru in the late 1970s, or in Ethiopia in recent years; flooding or unseasonable rains and other natural hazards may affect levels of production. But other major factors are the success or failure of government policies, political disturbances and the state of international trade.

For many countries, then, food production per head is no greater, or only slightly greater, than it was 20 years ago, and even in the early 1960s food supplies were by no means adequate for a large percentage of the populations of many Third World countries. Many millions then, as now, suffered from malnutrition. Table 1 lists countries, as reported by the FAO Production Yearbook, which are producing less than they formerly did.

Table 1 *Countries producing less food per head in 1980 than in 1969–71 and producing less in 1970 than in 1961–65*

Africa	ALGERIA	GAMBIA	MALI	Sierra Leone
	Angola	GHANA	MAURITANIA	Somalia
	Benin	Guinea	MAURITIUS	Sudan
	Central Afr. Rep.	GUINEA-BISSAU	Morocco	Tanzania
	CHAD	Kenya	Mozambique	Togo
	Congo	LESOTHO	NAMIBIA	UGANDA
	Egypt	LIBERIA	NIGER	Upper Volta
	ETHIOPIA	Madagascar	NIGERIA	ZAIRE
	Gabon	Malawi	SENEGAL	ZAMBIA
				ZIMBABWE
Latin America and the Caribbean	***Argentina***	El Salvador	Haiti	Panama
	Barbados	***Guadeloupe***	Honduras	Peru
	Chile	Guatemala	JAMAICA	PUERTO RICO
	Ecuador	GUYANA	MARTINIQUE	Trinidad
				Uruguay
Asia	AFGHANISTAN	HONG KONG	LEBANON	***Syria***
	BANGLADESH	IRAQ	NEPAL	***Vietnam***
	Bhutan	JORDAN	Saudi Arabia	YEMEN
	Burma	Kampuchea		

Countries producing less food per head:
- in 1980 than in 1969–71 are shown in normal type: Angola
- in 1970 than in 1961–65 are shown in italics: ***Argentina***
- in 1980 than in 1969–71 *and* in 1970 than in 1961–65 are shown in capitals: ALGERIA

(Source: FAO Production Yearbooks 1980, Tables 4 and 6, and 1972, Tables 7 and 9)

Table 2 *International trade in cereals and cereal products, 1979, as percentages of world trade (excluding the Eastern bloc)*

Commodity	Value (million $ US)	*Imports* To developed countries (%)	To Third World countries (%)	*Exports* From developed countries (%)	From Third World countries (%)	Net exports from developed countries to Third World countries (%)
Wheat and wheat flour	13 222	39.9	60.1	92.9	7.1	53.0
Rice	3 407	20.8	79.2	46.3	53.7	25.5
Barley	2 039	75.1	24.9	98.3	1.7	23.2
Maize	9 735	73.8	26.2	88.5	11.5	14.7
Other cereals and flour	1 920	63.2	36.8	68.9	31.1	5.7
Cereal products	2 748	64.9	35.1	91.1	8.9	26.2
Total	33 072	53.6	46.4	85.6	14.4	32.0

(Source: based on UN Yearbook of International Trade Statistics, 1980, vol. 2, pp. 25–32)

THE CONSEQUENCES OF INSUFFICIENT FOOD PRODUCTION

Clearly, the deterioration or the lack of significant improvement in food production has meant suffering, especially in those countries where income is unequally distributed among differing sectors of the population. It has also meant that food supplies must be imported, as is indicated in the extract from the Brandt Report quoted on *page 85*. Table 2 gives some indication of the scale of this trade. In 1979, total world trade in cereals (excluding the trade of the Eastern bloc) and in cereal products amounted to 33 072 million US dollars. Of this total, 10 586 millions (or 32 per cent) represented *net* exports from the developed world to the Third World. Rice and maize were the only cereals producing significant exports from Third World countries, but both of these trades were much less important than the import into the Third World of wheat and wheat products.

?

1 How far do the indices of food production indicated in Figure 1 bear out the comments made in the Brandt Report?

2 What factors may help to explain:
a the variations in food production in major world regions as shown in Figure 1c and
b the variations in food production at the level of individual countries?

3 On a map of the world locate the countries listed in Table 1. Use three different coloured circles to indicate periods of declining food production in the countries shown, according to the key given in the table. What suggestions do you have about the factors which may in part be responsible for low rates of food production in three of the countries you have marked?

4 What other conclusions can be drawn from Table 2 besides those indicated in the text? Why do you suppose that wheat is the most important of the major cereals to be imported into the Third World from developed industrial countries?

PERU: THE ORGANISATION OF AGRICULTURE

Peru may be taken as an example of a country in the developing world which has found it increasingly difficult to feed its own population from domestic output, in spite of the fact that it is a significant exporter of agricultural produce. In recent years, agricultural products, chiefly essential foodstuffs, have made up 15 to 20 per cent of the total import bill, yet agricultural products have also constituted about the same proportion of total exports. Table 3 gives more specific details on the composition of these imports and exports in relation to total agricultural production.

THE NATURE OF IMPORTS

By far the largest component in food imports are cereals: chiefly wheat and wheat products. The country is self-sufficient in the production of root crops such as potatoes — an important element in the staple diet of the sierra, and to some extent in the coastal region — and of manioc, a staple crop in the eastern regions of Peru (see Figure 2, *page 89*). But substantial imports of maize and rice are surprising in view of the fact that areas in northern Peru have developed rice cultivation as a speciality since the Second World War. Maize is a traditional staple crop which is widely cultivated, both on the coast and in the sierra, up to about 3000 m above sea level. Animal products, especially butter, cheese and dried milk, are imported despite the extent of natural pastures in the sierra and the expansion of cultivated fodder crops in the coastal region and parts of northern Peru. Certainly, land suitable for wheat production is limited to pockets of moderately level land in the intermontane basins of the sierra, but much more could be produced. However, as in other countries of the Third World, the consumption of wheat products, such as bread, pasta, cakes and biscuits, appeals particularly to the increasingly large proportion of the population which has come to prefer, and to attach prestige to, the kind of diet common in the advanced industrial world of Western Europe and North America. Consumption of wheat products was once a mark of social status and mainly an urban practice, but it has spread widely and cases have been noted in rural, mountain areas where traditional local staples such as potatoes and maize have suffered from the competition of imported flour, bread and pasta when these become available as a result of the improvement of roads. This fact illustrates the important point that the improvement of communications and reduction of costs of transport are *not* always followed by new agricultural opportunities for the communities involved.

THE REASONS FOR EXPORTS

At first sight, it seems paradoxical that although Peru must import basic staple foodstuffs, it also exports a substantial volume of agricultural products. Coffee is a recent success story, but sugar, cotton and wool have been long-standing staples of Peru's export trade over the last hundred years or more. Their significance demands further elaboration in a more general context. Peru's trade in agricultural products, indeed, is similar to that of many other countries of the Third World — exports of food and raw materials destined for the temperate countries of the advanced industrial world which either cannot produce them for climatic reasons or can only do so more expensively — rubber, palm-oil, coffee, cocoa, cane sugar, bananas and cotton are examples.

The concentration on growing agricultural products to export to the industrial world has a long history — associated from the 16th to the 19th centuries with slavery, and later with imports of cheap labour where local communities were too sparse or too unwilling to abandon their traditional farming systems. Because of the orientation towards exports, plantations had to be easily accessible to coastal areas

Print 1 *Cut sugar-cane being tranported to a processing plant before export*

Table 3 *Peru: production, imports and exports of some major agricultural products, 1979*

Commodity	*Imports* Value (million $US)	Quantity (tonnes)	Total national production (tonnes)	*Exports* Quantity (tonnes)	Value (million $US)
Wheat	114.2	828 100	105 700		
Rice	47.6	152 800	565 900		
Maize	18.2	130 800	656 300		
Sorghum	3.5	25 400	54 900		
Soybeans	6.3	22 900	7 100		
Beef	1.3	1 600	90 400		
Milk products	17.2	19 800	145 200		
Sugar			714 200	194 100	34.3
Cotton			81 300	20 000	49.3
Coffee			105 700	70 600	244.8
Wool			12 200	5 800	33.5

(*Sources*: FAO Production Yearbook, 1980. A World Bank Country Study, 1981, *Peru: Major Development Policy Issues and Recommendations*)

? **1** Look at Table 3 and
a work out what percentage of each food crop consumed within the country has to be imported and
b calculate the percentages of total national production of sugar, cotton, coffee and wool that are exported
2 Draw pie charts for
a commodities imported and
b for commodities exported, according to value. The size of the circles should be proportional to the total value of exports and imports. What do these statistics tell you about the balance of payments in terms of agricultural products?

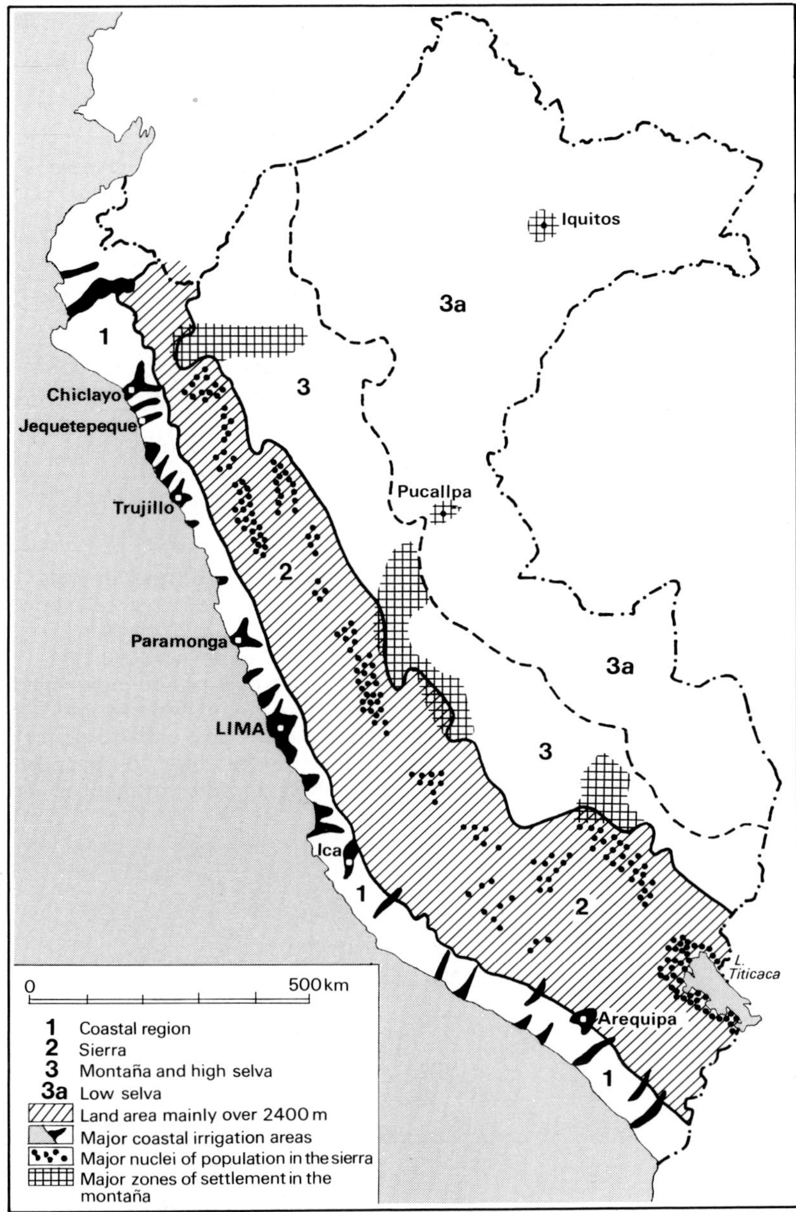

Figure 2 *The major regions of Peru*

where ports were built and which were linked by roads and railways. Above all, plantations were associated with monoculture on large estates, most often owned and managed by foreign entrepreneurs, usually from the metropolitan countries. There was, therefore, a tendency for the development of dual economies in many colonial or ex-colonial territories. One sector was characterised by 'modern' or capitalistic forms of development, notably in export-orientated agriculture and associated commercial activities. The other sector was characterised by traditional farming systems geared to subsistence and the support of a very modest urban sector. But the last 50 years have seen a profound change in this simple pattern — the rapid growth of population, and more particularly of urban populations.

THE EFFECT OF POPULATION GROWTH

In the countryside, traditional farming systems have been strained to the limit by the growth of population, which has brought about the fragmentation of holdings so they are no longer big enough to support peasant families. It has also caused a vast increase in the volume of migration to the expanding towns, either for seasonal or permanent labour in the expanding capitalistic sector, and an internal social differentiation within peasant communities which itself threatens the survival of traditional ways of life. Peasant systems are changing rapidly under pressures to switch from subsistence to commercial farming in a context of population growth. Farmers have often found it impossible to feed both themselves and the burgeoning urban sector, which buys in cheap, imported food from elsewhere. The problem tends to be compounded by the geographical fact that the urban sectors, particularly the larger cities, have usually grown in locations with easy access to the sea links which bound them to their former colonial masters and to the export–import trades. The cost of importing grain in bulk to such coastally orientated cities is often less than the cost of assembling and transporting grain from the relatively remote areas in which traditional farming sectors have survived.

THE INFLUENCE OF THE LARGE ESTATES

The preoccupation with export orientated agriculture for sugar, cotton and wool has been closely bound up with the structure of Peruvian society, particularly since the late 19th century. The export sector in many colonial countries in Africa or South-east Asia was dominated by individuals or companies from the metropolitan countries. But Peru gained its independence from Spain in 1828 and the social consequences of the development of an export economy were not quite so starkly obvious as in other parts of the Third World, but there were similar features. With the expansion of exports, particularly of sugar, cotton and wool, there emerged a powerful group of large landowners/exporters — an agriculturally-based bourgeoisie. They were able to control land and water (the latter fundamentally important on the arid Peruvian coast) and emerge as a dominant force in Peruvian politics, pursuing policies favourable to the development of the export economy

Some of the families and individuals involved in the export economy were of old, colonial origin, others were immigrants who became Peruvian nationals, though Peru never attracted European immigration on the scale of, for example, Argentina and Brazil. Fernandini and Gildemeister, formerly two very important landowning family groups, were of Italian and German origin. Some were foreign-based, like the American Grace Company which owned important sugar estates (see Figure 3, *page 91*), or the Cerro de Pasco Mining Corporation, which owned vast pastoral estates in central Peru. However, foreign ownership in export agriculture was never dominant, and much of the profit from agricultural exports remained within Peru, unlike the situation in some colonial countries elsewhere in the Third World. The coastal oligarchy, on the other hand, invested some of their wealth in different sectors of the economy to diversify their interests into mining, commerce, banking and industry.

Nevertheless, until the land reform of 1969–75, the concentration of landownership in large estates was extreme, even by Latin American standards. Table 4 illustrates this concentration, and also the number of smallholdings incapable of supporting a family.

Large estates dominated coastal export agriculture and wool production in the sierra, but there were significant differences in the organisation of production according to the nature of the product and the conditions under which it was produced.

Table 4 *Number and size of agricultural holdings in Peru, 1961*

	Agricultural units				
	Number		*Total*	*Area*	*Average*
	Total	*%*	*(ha)*	*%*	*size (ha)*
Large multi-family estates	33 172	3.9	15 001 000	80.6	1 300.0
Family farms	98 370	11.6	876 000	4.7	8.9
Sub-family farms	719 110	84.4	1 124 000	6.0	1.6
Communities	808	0.1	1 604 000	8.6	1 985.0
Total	851 460	100.0	18 605 000	100.0	21.8

(*Source*: Inter-American Committee for Agricultural Development, Washington, 1966, *Tenecia de la Tierra y Desarrollo Socio-económico del Sector Agricola*)

AGRICULTURE ON THE COAST

SUGAR

In many of the northern valleys of the coastal region, physical conditions favoured the expansion of sugar cultivation. There were moderately high temperatures throughout the year with a high degree of insolation favouring the growth of sugar cane with a high sugar content. There were also adequate supplies of irrigation water from streams originating in the relatively well-watered high Andes of northern Peru, some of them fed by glacial meltwaters and snowmelt, and maintaining their flow throughout the year. Various factors encouraged the amalgamation of smaller units into large estates for the production of sugar (see Figure 3, *page 91*). Economies of scale increased the optimum size of sugar-refining mills, which in turn meant that factories needed to draw on a wider area for supplies of sugar cane. To transport it cheaply required investment in

Print 2 *This aqueduct carries water over an intervening valley to irrigate land for sugar on the Paramonga estate*

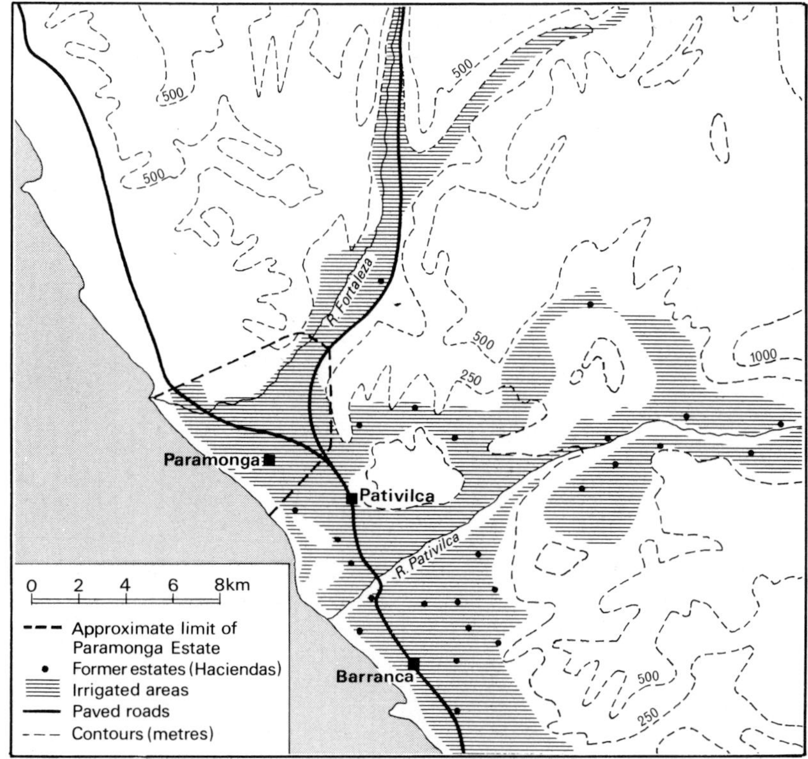

Figure 3 *Coastal valleys and the Paramonga sugar plantation. Until the agrarian reform of 1969, agriculture was dominated by the* haciendas, *of which one of the largest, Paramonga, with 7000 ha amalgamated from four smaller* haciendas, *was owned by Grace Co., an American-owned enterprise with interests in shipping, air transport, and in Peru, the manufacture of paper, chemicals, textiles and food products, including sugar, rum and biscuits. The town of Paramonga, with approximately 10 000 people, grew around the sugar refining plant under the aegis of the company. The estate was transformed into a co-operative in 1969*

efficient systems of light railways, or later of tractor-drawn wagon trains on good roads. Furthermore, in the arid, sunny climate of northern Peru, sugar cane can be brought to maturity at most times of the year by the careful control of irrigation water in relation to planting and harvesting. This fact ensures a more continuous use of sugar-crushing and -refining plants than is possible in parts of the world where sugar cultivation is dependent on seasonal rainfall. The need for investment in transport, but above all the need for careful scheduling in planting, irrigation and harvesting encouraged the concentration of land under sugar into large estates, which also controlled the sugar mills. The demand for labour in sugar growing is high, but less seasonal than in other branches of agriculture. Sugar estates tended to become agro-industrial complexes with a large, permanent, wage-paid labour force with some short-term and seasonal labour, and with a skilled management and salaried staff. Some of the sugar estates were run on highly paternalistic lines and incorporated company towns with health and education facilities provided by the company. Paramonga was an excellent example.

COTTON

In the central coastal valleys, and also in the far north, cotton was formerly the main crop. In the central coastal region, a high incidence of cloudiness during the cool season discouraged sugar cultivation, but both here and in the Piura valley the river regimes tend to be very seasonal, dependent solely on rain-fed streams, from the western slopes of the Andes, which become increasingly dry towards southern Peru. Conditions suited cotton cultivation with its highly seasonal need for water during the growing season, and for labour, especially during the harvest period. Some of the cotton estates were operated by wage-paid labour and imported temporary seasonal labour from the sierra. But cotton cultivation also lent itself to systems of share-cropping contracts whereby the landowner dictated land use and methods of cultivation, and provided land and

Print 3 *Workers on a co-operative harvesting cotton*

some of the agricultural inputs, including credit, in return for a share of the crop.

Other elements in coastal agriculture are more closely orientated to the domestic market. In the north, the water-retentive soils of the Jequetepeque valley and adequate supplies of water for irrigation lent themselves to the commercial cultivation of rice, which has also spread across the north towards valleys draining into the Amazon basin. In the south, notably in the Ica valley, concentration on viticulture (primarily for the domestic market) has a long history dating back to the 16th century. Dairying, with the intensive cultivation of fodder crops, is highly specialised around the city of Arequipa in the south, and is increasingly important, with intensive poultry farming and market gardening, in the neighbourhood of Lima.

Print 4 *Irrigation channels through the desert between Ica and Nasca in the southern coastal strip of Peru*

AGRICULTURE IN THE MONTANA AND THE SELVA

The lower eastern slopes of the Andes are hot and humid with high rainfall, dense montane forest, and intricately dissected landscapes. They descend to the low plains of the selva, the Amazonian rainforest, where there is an important distinction between old higher terraces of poor soils and the seasonally inundated lands of the floodplains.

Still sparsely populated, it is, however, a region in which colonisation and settlement have been taking place steadily since the early part of the century. Settlement is focussed along the lines of access roads, most of which have been constructed or greatly improved in the last 30 years. Colonisation has been of various kinds. Some *haciendas* (large estates) have early origins which go back to colonial times, but most of the larger holdings were the product of land concessions made in the last hundred years. Medium-size holdings were established as a result of formal, planned settlement schemes in the 1960s and more recently. Land is, however, abundant and 'spontaneous' colonisation on quite a large scale has been undertaken by Indian peasant farmers of the sierra.

Various forms of shifting cultivation are normal on small and medium holdings, mainly for subsistence crops of manioc, maize, sweet potatoes and the like, and though methods of cultivation are poor, the subdivision of landholdings is nothing like so extreme as in the highlands, and most farms are adequate to support a family. Coffee has come to be the main export crop from medium-sized plantations, but a great variety of other products are grown — bananas, tea, cocoa, palm-oil in an experimental way, fruits for juice and for canning, and in recent years, there has been a rapid expansion in the cultivation of coca. Coca has long been grown on the slopes of the montaña as a crop exported as dried leaves to the Indians in the highlands who chew them with lime as a stimulant and narcotic, but in recent years, the montaña has become a centre for the production of coca on a large scale for the illegal extraction of cocaine for the international drug traffic.

In the north, rice has extended inwards from the coastal region, but in many parts of the montaña and selva, colonisation has been followed by the destruction of forest and its replacement by pasture for cattle and beef production. Soil erosion and soil exhaustion are increasing problems, but the eastern regions of the country do hold out considerable potential for further development, provided that communications with the coast and the major markets can be improved and maintained.

AGRICULTURE IN THE SIERRA

It is in the highly complex and varied environments of the sierra that the most intractable problems of development occur. In part, these are problems of distance, remoteness from markets, lack of suitable land for arable farming and the unpredictability of rainfall. This is both highly seasonal and very variable from one year to another, with hazards at higher altitudes from hail and frosts as well as drought.

HACIENDAS

In many ways it is the social environment of agriculture which is problematic. Landholdings have long been polarised between the *haciendas* and the smallholdings of predominantly Indian populations, many of whom were organised into communities which have pastures in common. *Haciendas* originated in the late 16th and 17th centuries as a consequence of the Spanish conquest, but they increased in numbers and size during the 19th and early 20th centuries with the growth in value of natural pastures for the sheep and wool they could produce, quite apart from meat, hides, dairy produce etc. for the domestic market. The following extract summarises many of the characteristics of the haciendas in the sierra in the 1960s:

The most important tenure system of the region is, undoubtedly, the large traditionally-operated estate, which has been called the *latifundio–minifundio* complex. Next in importance is the transitional large estate, because the sierra does not have any really modern commercial operations like those on the coast. The *latifundios* in the sierra, in general, make their influence felt not only in the region where they are located, but also in other areas of the country, and these *haciendas* are the key to numerous economic, political and social problems at a national level.

In the sierra, the traditional *hacienda* is often worked without a resident master. The owner rarely lives there, and arrives only for the harvest, the profits of which he will invest far from the *hacienda*.

As a result, in the Peruvian sierra, the *latifundio–minifundio* complex exists as a result of the combination of the following factors: **a** a large expanse of lands in the hands of a proprietor with an almost feudal mentality; **b** small plots of land which are worked by Indians who are not landowners; and **c** the institution of servitude adapted to the prevailing circumstances.

The most typical characteristic of the *latifundio–minifundio* complex is its work relationships. The labour structure determines the distribution of small plots in exchange for services rendered. The complementary aspects of the problem are: **a** payment of rent in products demanded by the large landowner for the plot he gives to the *colono*; and **b** the token wages given to the latter as payment for his obligatory work on the lands which the owner exploits directly on the *hacienda*. The landowner maintains 'his Indians' at a bare subsistence standard of living in order to keep them on the farm.

Print 5 *Quechua Indians tending their sheep and goats*

There are two principal types of arrangement between the *patron* or master, and the *colono*, as the worker who lives on the hacienda under such an agreement is called. If it is a sheep-raising *latifundio*, as are the majority, the *campesino* (peasant) is a shepherd and the concession is that he is allowed to graze his sheep or llamas and alpacas. If it is agricultural, which is less frequent, the *campesino* has to do obligatory and unpaid work on the lands of the *hacienda*. Sometimes there is a combination of these two systems.

The *colono* also has a varied series of obligatory duties to perform. Among these is *ponguaje*, which is very common and consists of the domestic service of the family of the *colono* in the landowner's *hacienda* house. Another is *mitani*: transporting the harvests of the *latifundio* on their own animals or their backs. The *campesino* calls his *patron* '*taita*' . . . '*papay*' or '*papacito*', Quechua and Spanish terms for father or little father, which clearly demonstrate the paternalistic nature of the sierra *latifundio*.

Corporal punishment of the Indians still exists in the sierra, and is inflicted both by the landowner or by his men and by those who exercise authority. The *campesino* often becomes indebted to his *patron* for the sheep losses of the hacienda herd. As the *campesino* has no way of compensating for the losses, the debt is noted and the ties of the submission or dependence are strengthened. . . . In some cases the children of the *campesinos* must repay the debts of their parents.

(*Source*: ed. Solon Barraclough, 1973, *Agrarian Structure in Latin America*, Lexington Books, Lexington, Mass.)

Nevertheless, *haciendas* were beginning to change in the 1950s and early 1960s away from the neo-feudal patterns described above. Money wages, though minimal, were becoming more normal. Some *haciendas* had been turning towards a more intensive use of land for dairying in the north and centre, for example. In the south, on pastoral estates, however, the tendency was more towards the eviction of Indian shepherd families on some of the more so-called progressive *haciendas* where the introduction of improved breeds of sheep required the separation of *hacienda* flocks from the Indian shepherds' own flocks. This involved the fencing of pastures which made it possible to dispense with the labour of Indian shepherds and their troublesome flocks. But eviction of the shepherds exacerbated the contrast between the underused and extensively grazed lands of the *haciendas* and the overpopulated and overcultivated lands of the Indian communities nearby.

PEASANT SMALL HOLDINGS

In many parts of the sierra, however, predominantly Indian and *mestizo* (mixed blood) populations held land independently of the large estates, and many of them were loosely organised into communities with or without legal title. Communities sometimes retained their common pastures and complex systems of individual holdings and land rights over a wide range of altitudes and ecological niches appropriate for the cultivation of a wide variety of subsistence crops. A single family might hold a patch of irrigated land on the valley bottom for the cultivation of maize and fodder crops, and assume the right to take in land for temporary cultivation on hillsides for crops such as barley, potatoes and other root crops which could flourish up to 3500 m. In addition, they may have the right to graze sheep or llamas on the high pastures which are too cold for arable farming.

On a larger scale, systems of barter within the community, or among neighbouring communities, established traditional values for the exchange of high altitude products for wheat, maize or fodder crops.

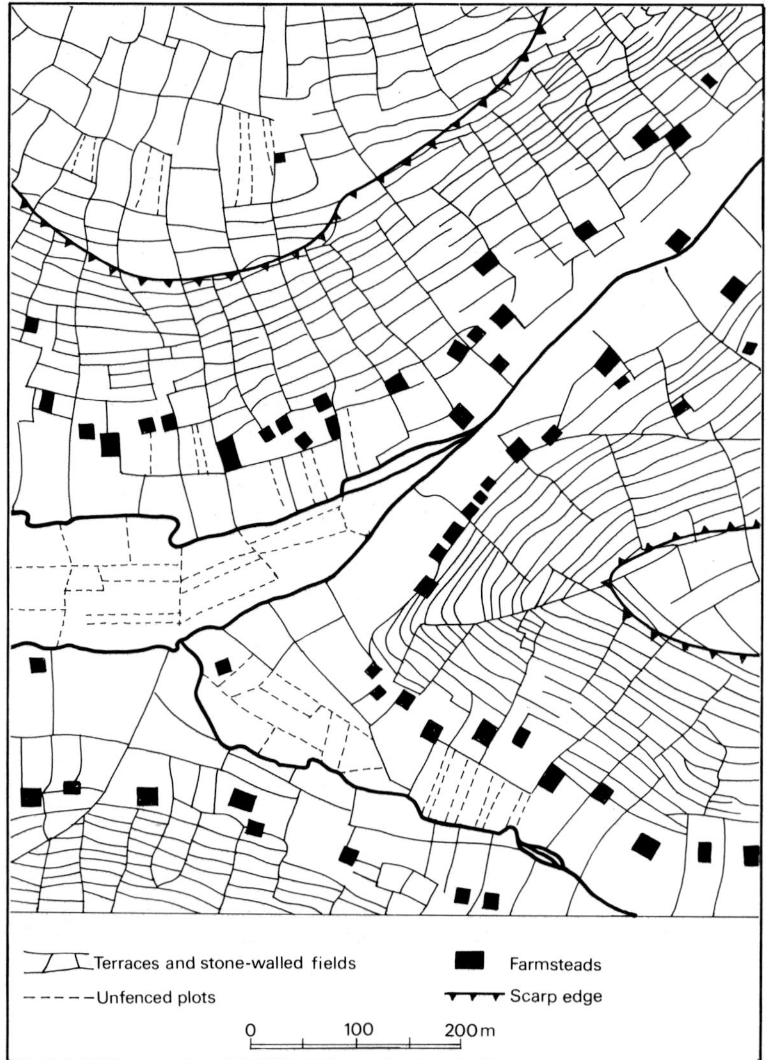

Figure 4 *Peasant small holdings and settlement near Lake Titicaca. A plateau surface at 3300 m above sea level and capped with basalt is deeply dissected by rivers with level flood plains. The slopes are terraced and the plateau surface divided into small stone-walled fields. These and the valley floor are occasionally subdivided into open plots*

Increasing commercialisation has threatened and distorted traditional exchange values of this kind, but the major problems of the small, landowning peasant farms of the sierra have stemmed to a large extent from the growth of population, which has put increasing pressure on scarce land resources. Fragmentation of holdings of arable land in a difficult mountainous environment with limited potentially cultivable land has reduced peasant holdings to the point at which they cannot provide adequate subsistence for a family. This problem has often been aggravated by the resultant overstocking of limited pasture and unwise cultivation of steep slopes which leads to severe soil exhaustion and erosion. In 1973 there were 650 000 holdings in the sierra with less than 5 hectares (5–6 hectares of arable land are regarded as the minimum-sized holding in the sierra capable of supporting a family at an acceptable level of living with existing methods of cultivation), see Figure 4.

? 1 Draw an annotated east–west section of Peru to show the major regions referred to in the text and in Figure 2. (A good atlas should also be used for this exercise.) Annotate the sketch map to indicate the contrasts between the regions in terms of
a climate
b subsistence and commercial farm products and
c the type and size of landholding.

PEASANT RESPONSES TO THE PROBLEMS

Peasant reaction to the problems posed by the pressure of population, decreased holdings and the increasing concentration of land into large holdings has been of various kinds:

Commercial farming

Some peasant farmers, particularly those more fortunately located with respect to communications and with adequate land to undertake new ventures, have involved themselves in commercial farming for the urban markets, producing potatoes, milk, vegetables or fruit.

Migration

Many peasants have responded to the problems by migration in order to earn cash income by selling their labour rather than crops from their land. Migration has taken various forms. Seasonal migration from the sierra to coastal cotton plantations and for short-term contracts to the sugar plantations has been a feature of rural development in Peru since the late 19th century. Nearby *haciendas*, construction work in mines, on the roads and in urban building have all attracted temporary migrants from their home areas, particularly if seasonal labour elsewhere can be fitted into the seasonal fluctuations of labour needs on family farms. But increasingly, it is migration to the cities, and especially to Lima, which has attracted migrants on a more permanent basis.

Many of these migrants retain links with their home villages, returning for family gatherings and to celebrate local fiestas. Some send money back to their villages helping to supplement inadequate family incomes from the land. But one other effect is also important. Returned migrants bring with them a wider knowledge of the world; they may initiate improvements in their communities of origin; but above all, they bring an awareness and an expectation of standards of material welfare far beyond those of the older generation. A 'revolution of rising expectations', as this process has been called, spreads into remote Andean villages, creating discontent with the existing conditions of poverty, and often fostering conflict between the older generation and the younger, returned migrants, challenging time-honoured concepts of seniority rule in the village community.

Protest

Rural rebellion is nothing new in Peru, but land invasions of *haciendas* became common in the 1960s and again in the early 1970s as peasant communities attempted to regain lands held in *haciendas*. The following extract gives further details:

> The generic term 'invasions' has come to be given to the socio-economic-political phenomenon signified by the *de facto* occupation of *hacienda* lands of which the peasantry has been dispossessed. This phenomenon, as a massive process of national importance, is about ten years old. It has been significant for about ten years, but the greatest activity occurred between August 1963 and August 1964.
>
> The peasants have never considered the invasions as illegal, and have generally spoken of the 'recovery of lands' or 'the restitution of their own rights', but it is undoubtedly the case that there have been many invasions of lands in which there was no legal problem of boundaries, or opposing and overlapping claims, or any question of this kind. The fact that a large indigenous population is restricted to a small area, while largely unused or abandoned but potentially cultivated land surrounds them, and the general discontent with conditions in the sierra, have been important motivations taking this process beyond the scale of simple 'recovery of lands'. Indeed, it is often difficult to say whether one is dealing with a true invasion or whether the *colonos* of the *haciendas* are simply refusing to provide obligatory labour to the *hacienda*, or whether it is a combination of these factors. In general, the immediate motivation is perhaps of less importance than the fundamental cause, which is the injustice of the agrarian structures in the sierra.
>
> (*Source*: IDA, 1966, *Peru: Tenencia de la Tierra*, Unión Panamericana)

Writing of the invasions of 1963–4, the same report continues:

> The most important feature of this period is the fact that no less than 3000 peasants have been involved, mainly members of the communities (*comuneros*), *hacienda colonos*, and landless labourers. An important aspect of this stage is that there is no longer much interest in legal argumentation before proceeding to the occupation of lands. It is equally characteristic that they are not only occupying natural pastures in the sierra as heretofore, but also cultivated lands in crops and fallow. The motives which drive the peasant groups are the same as always, but the reasons they offer and make public have changed. Instead of legal documentation they present the slogan 'land or death', and they argue that lands have been occupied because they have been paid for by long years of unpaid or badly paid labour.

Land invasions were not organised on anything but a local scale, though they had support from the peasants' trades unions. They were usually *ad hoc* affairs; some were successful, some were repelled, not without brutality and the occasional loss of life. They were, however, symptomatic of peasant discontent and of a land hunger which could only be satisfied on a large scale by the redistribution of land from the large estates for the benefit of those who actually worked the land, either as tenants on the estates themselves or in neighbouring communities.

Print 6 *Near Lake Titicaca, peasant farms cultivating potatoes, barley and other grain crops are fragmented into tiny plots as seen here*

?

1 **a** Taking sample areas from Figure 4; estimate the average size of agricultural plots. What problems does the subdivision of land, to the extent shown in Figure 4, present in relation to the modernisation of agricultural techniques?
b In this area of Lake Titicaca, peasant holdings, even though small in total, are usually subdivided into several plots which may be scattered in different places. Why do you think this occurs? Does this parcellisation of holdings have any advantages from the point of view of the peasant farmers?

2 Adopt the role of the head of a peasant family with insufficient land either to feed yourself and your family or to provide a small cash income for the purchase of essentials by the usual methods of farming in the Peruvian Andes. Consider what options are open to you. What are the advantages and disadvantages to you of adopting one of the options?

3 Imagine you have been a member of a television crew in Peru for a month (before land reform) making a programme about food production. Write imaginary transcripts of your interviews with one farmer from each of the regions — the coastal region, the montaña and selva region, and the sierra region. Then, as a summary, make suggestions about which areas offer the most potential for increasing food production.

LAND REFORM

Land reform has meant many things at different times and in different places. It has sometimes meant no more than the abolition of unjust and inequitable tenancies, like the labour-service tenancies described in the last section. But the central and most important feature is the redistribution of rights on land, involving the breakup of large estates into small or medium holdings, and sometimes the consolidation of excessively small and fragmented holdings into adequate farms. Pressures working for land reform gathered pace in the 1960s, and not only in Latin America. Apart from peasant communities and peasant unions which obviously stood to gain, the political left saw land reform as a means of breaking the power of the landlords. But other views in favour of some measure of land reform were gaining circulation in the interest of economic efficiency and modernisation as well as social justice. At the international level, the 'Alliance for Progress', an important development programme for Latin America, instigated by the USA and with, at first, considerable support from Latin American countries, put forward the idea that rural reform was a pre-requisite for the granting of aid. The *Declaration of Punta del Este* in 1961, at the foundation of the Alliance for Progress, stated that governments would:

'. . . encourage, in accordance with the characteristics of each country, programmes of integral agrarian reform, leading to the effective transformation, where required, of unjust structures and systems of land tenure and use; with a view to replacing latifundia *and dwarf holdings by an equitable system of property, so that, supplemented by timely and adequate credit, technical assistance and improved market arrangements, the land will become for the man who works it the basis of his economic stability, the foundation of his increasing welfare, the guarantee of his freedom and dignity.'*

(*Source*: Doreen Warriner, 1966, *Land Reform in Principle and Practice*, OUP)

The Brandt Report takes up a similar theme in the late 1970s:

But the conquest of hunger calls for much broader international and domestic efforts to ensure that additional food reaches those who need it, and that it is bought, either by individual families or by governments for subsidised distribution. But governments cannot sustain subsidised schemes for long, and an end to hunger can only be foreseen if there are more wage-earners, and a more equitable distribution of income — a challenge for many developing countries where growth has bypassed the poor. Agrarian reform is a critical means to benefit the poor — though naturally the measures needed differ from country to country. In some areas the key issue is reform of tenancy to give greater security of tenure. In others it is to divide large parcels of land among those who can farm it more intensively. Yet others require consolidation measures to overcome the excessive fragmentation of holdings which has already occurred. All these can increase the incentive for farmers' investment.

PRESSURES FOR LAND REFORM

In Peru, the pressures for land reform came from various directions. Peasant pressure, described earlier, was supported by peasant trade unions and the trade union movement as a whole in Peru. Left-wing political parties, though not very powerful in the country, wanted land reform as a means of securing greater equality of opportunity in the countryside, and also as a means of breaking the power of the landed oligarchy, so that land reform could be seen as a key to the restructuring of society along socialist lines. There were also other, more qualified and ambivalent voices in favour of some degree of agrarian reform. In Peru, as in other Latin American countries, agrarian reform has a wider meaning than land reform alone, embracing not only the redistribution of land in favour of peasant or medium-scale farmers, but also the provision of roads, electricity, rural education, agricultural extension services and rural credit. Opinion among liberal political parties could therefore support a programme of agrarian reform which stressed the need to improve the productivity of agriculture without radical redistribution of the large estates, particularly the highly productive coastal plantations. Many economists took the view that agriculture would be more efficient in the hands of small- or medium-scale proprietors, especially if there were to be a parallel improvement in the rural infrastructure. Domestic food supplies would be increased and higher incomes for rural producers would provide demand for the manufactured goods which Peru's industrialisation was creating. In addition, under the provisions of the 'Alliance for Progress' charter, foreign aid for rural development along these lines could be that much more easily forthcoming.

Above all the prospect, held out to the peasantry of benefits in landownership and to tenants on large estates of independence, or at least of an end to the neo-feudal and socially oppressive tenancy systems under which they worked, would halt the spread of communism and support for the extreme revolutionary left at a time, in the early 1960s, when rural guerrilla movements were still very active in Venezuela, Colombia, Peru and Bolivia. If the rhetoric about agrarian reform could achieve this aim, then the actual execution of a real and effective land reform could be delayed indefinitely. Indeed, it was the *promise* of agrarian reform that took the steam out of the peasant invasions of 1963–64.

Print 7 *Cotton workers, protesting at the government's slowness in implementing land reform, are dispersed by riot police, Peru 1964*

As for the landowners, most were implacably hostile, particularly those involved in the export-orientated farming of the coastal zone. Some responded to the threat of land reform by parcelling out their estates, occasionally to existing tenants, more commonly as a subterfuge to keep land in the family's hands. Some suspended judgment until they knew of the compensation they might receive in return for the expropriation of their land and stock. Favourable terms would give them capital to invest in the booming urban property market or in foreign investments.

RESULTS OF AGRARIAN REFORM

In the event, an agrarian reform law was passed in 1964, but remained largely ineffective. The government of the day responded to internal pressures and the encouragement of the 'Alliance for Progress' (as did some other Latin American countries) by legislation rather than action. Very little land reform was achieved and only a few estates in the sierra were expropriated and allocated to peasant beneficiaries. Some attempts were made to improve the rural infrastructure, but the major thrust of rural development was towards the colonisation of the montaña and the selva for which grandiose plans were made. Pressures for land reform were thus diverted towards infrastructural development and eastern colonisation, and the interests of large landowners were hardly threatened. Coastal export agriculture was, in any case, excluded from the provisions of the land reform law.

1968 — FURTHER REFORM

In October 1968, however, a military coup put General Velasco into power at the head of a government which called itself revolutionary and put into effect a series of more or less radical reforms. Interpretations of the political character and ambitions of the military regime have varied widely. It *was* nationalistic, and it certainly sought to extend the power of the State by the control of heavy industry, banking and the press. Nationalist sentiment, for example, was initially expressed by the expropriation of American oil interests and the establishment of a 200 mile limit of territorial waters to protect the highly important fishing industry. Policies were also aimed at the mobilisation of patriotic sentiment by programmes of worker participation in the control of industry, and also by a programme of agrarian reform which was much more radical and effective than that of the 1964 law as the newspaper extract below describes:

Print 8 *A group of former tenants, now members of a co-operative, are gathered at the old* hacienda *headquarters for discussion*

Agrarian Reform 1969

A Descriptive Analysis and Review of Intentions and Possible Effects

The Military Government's Agrarian Reform Law and the spectacular take-over of the coastal sugar estates is widely regarded as a positive contribution to the very complex problems involved in 'modernising' a country like Peru. Naturally it is not a simple piece of legislation and many important questions remain to be answered. The *Peruvian Times* has consulted leading economists and legal commentators in preparing this report on the law, of which the following is the first, introductory section. In general the whole agrarian reform programme is seen to be 'ambitious but rational' though major problems, especially in preparing adequately trained numbers of technicians, will certainly occur in the short term.

A Peruvian Times Report

The 1969 Agrarian Reform Law lays down the broad lines of future policy for the country's agricultural sector and has set up a large number of instruments for putting this policy into effect. On a broader front it is also intended to bring about important social, economic and political changes over the country as a whole.

As far as the stated objectives are concerned the new law is to a large extent a reformulation of its 1964 predecessor, which it repeals. There are, however, significant changes, both in scope and emphasis. The addition of the large coastal sugar estates to the area affected by the reform is of fundamental importance, and the range of policy instruments is much wider than before.

Altogether the law assumes very wide powers. All rural holdings within Peruvian territory are subject to its legislation and very few exceptions are permitted (the low jungle is for the moment one of these).

The law sets out, in its 196 articles, a series of economic, social and political objectives which are not easy to separate for analytical purposes, and individual provisions often have wider implications — for instance, the constantly reiterated aim of effecting a basic transformation of 'structures' clearly has repercussions in all three spheres.

Balanced Growth

The principal economic aim is to change tenure and property arrangements in the agrarian sector, with a view to promoting 'balanced growth' — that is, to increase the contribution of the rural sector to the country's general economic development, by means of promoting increased production and productivity. The law contemplates that this will involve increasing rural per capita income, which will have the effect of stabilising the *campesino* on his land, giving him a basic minimum economic security and tending to remove rural instability — such as that which occurred in La Convención in the early 1960s — and effecting large-scale redistribution of rural wealth in the direction of small- and medium-size proprietors.

Equal if not greater emphasis has been placed on the purely social objectives of the reform: hence, for instance, the stress laid on the 'social function' of land ownership and the reference to 'extreme conditions of rural social injustice' on the first page of the law. It is assumed that the structural changes being undertaken will tend towards the promotion, not merely of economic, but also of social development in rural areas, particularly the total integration of the predominantly Indian rural masses into the national society. President Velasco promised to extend the vote to illiterates as evidence of the seriousness of this proposal.

The general objectives of the reform may appear to be too ambitious, even idealistic. However, studied in detail it shows a reasonable and balanced outlook, especially considering Peruvian historical and political circumstances.

At a very basic level it is simple: give the land to those who actually work on it. Practically no renting or share-cropping (and therefore the possible holding of large quantities of land) is to be permitted. Exceptions can be seen to be highly marginal to the main land-holding situation.

(*Source*: *Peruvian Times*, 14 November 1969)

The sugar estates of the coastal region were the first to be put into the hands of the government, to be reorganised as co-operatives. A regional programme for the execution of land reform was set up and many of the large estates of the sierra, as well as those of the coast, were expropriated as co-operatives In some cases, communities in the neighbourhood of reformed estates were to participate in the co-operative structure (these were known as SAIS — agrarian societies of social interest). Both the previous newspaper extract and Table 5 give information as to the scale and location of reform activity and the organisation of newly reformed lands, in which, it will be noted, the formation of co-operatives took priority over the allocation of land into small- or medium-sized peasant holdings. A substantial proportion of Peru's agricultural land on the coast and in the sierra was thus profoundly affected by land reform, which was brought to a halt abruptly in 1975 when an internal coup within the military government put into power a group of generals who brought Peru's radical experiments to an end.

HOW SUCCESSFUL HAS LAND REFORM BEEN?

THE SOCIAL EFFECTS

The results of Peru's land reform programme have been mixed. In general terms, it has broken the power over land formerly exerted by a small minority of powerful landowners, and it has benefited a large number of former tenants or wage-earners on the former estates whose incomes and status have been substantially improved. It has not, however, resulted in a basic restructuring of social relationships. Former landowners have shifted their interests to urban investment, commerce, industry and other activities. Indeed, the expansion of industry, of the bureaucracy, and of urban occupations in general has meant that ownership of land, as such, no longer confers the social and political prestige that it did 30 years ago. Reform has resulted in an extension of the power of government over land on an unprecedented scale, but the implications of this fact for rural development depends, of course, on who controls the government bureaucracy.

Table 5 *Agrarian reform: adjudications to 1976*

	Area (hectares)	*Families affected*	*Number of enterprises*
Individuals	137 702	19 097	—
Co-operatives	2 131 802	101 034	478
Communities	652 527	67 995	235
SAIS	2 592 217	59 764	57
Other	1 123 058	31 705	495
TOTAL	6 637 306	279 595	1 265

(*Source*: Mariano Valderrama, 1969–76, *Siete Años de Reforma Agraria Peruana*, Universidad Católica del Perú)

? **1** Consider the *attitudes* likely to be adopted by the following groups in relation to land reform:

a The owner of a large sugar plantation on the coast.

b An absentee landowner of a pastoral *hacienda* in the sierra who lives in Lima and needs capital for investment in a company he owns concerned with urban construction.

c A share-cropping tenant on a cotton estate.

d A peasant member of a village community who owns 2 hectares of land.

e The owner of a small company making paraffin cookers with a market mainly in rural areas.

THE EFFECT ON AGRICULTURAL PRODUCTION

Land reform has been followed by a substantial *drop* in agricultural production. Certainly, the late 1970s were years of drought for much of Peru, and it is impossible to disentangle completely the effects of drought, international market conditions, internal pricing policies and land reform, but there is no doubt that agricultural production both for export and for the domestic market, suffered a drastic decline in the 1970s, and put even greater strain on imports, in spite of legislation requiring 40 per cent of land in agriculture to be used for the production of crops for the home market. Figure 1 (*page 85*) shows the downward trend quite clearly.

What happened? Why were the consequences of reform apparently so disastrous as far as production is concerned? Other factors were certainly important.

Low prices

Apart from drought, government pricing policies for foodstuffs throughout the period tended to maintain low prices for the benefit of the urban consumer. This was politically necessary in view of the size and volatility of the Lima population, but it was a policy which discouraged commercial agriculture.

The preoccupation with land reform had meant, among other things, that trained personnel and the resources of the Ministry of Agriculture were almost wholly concerned with land reform at the expense of agricultural extension services, which were virtually abandoned.

Insufficient land

The transformation and regrouping of former *haciendas* into co-operatives involved drastic reorganisation, and disruption was perhaps inevitable. But the way in which land reform was carried out, with a considerable emphasis on co-operatives, had social effects which were not foreseen, and which again tended to reduce the level of agricultural production for the market.

First of all, the expropriation of the large estates did not provide enough land to satisfy the needs of the rural population. In the event, the main beneficiaries of land reform were the wage-paid labourers, tenants and share-croppers who formerly worked the estates, not the peasants in the overcrowded communities subsisting on small areas of arable land alongside the estates. According to the agricultural census of 1972, 85 per cent of holdings in the sierra, most of them small, were worked and occupied by their owners and only 15 per cent of the peasants were tenants of one kind or another. The majority of peasants, then, were not affected by the land reform at all, except in a few areas where SAIS (agrarian societies of social interest) were created. There was considerable resentment and rural unrest and the invasion of estates recurred on a large scale in 1973 when the application of land reform was in full flow. Tension built up between the small farmers, often organised in communities, and the former tenants and labourers on the estates who were now members of the co-operatives and materially much better off than the independent peasants.

Problems within the co-operative system

The co-operatives themselves were a source of new tensions. Beneficiaries were fully-fledged members, but the management of the co-operatives was entrusted quite often to men who had been administrators of the old *haciendas*, or to a new breed of agricultural bureaucrats. The social gulf between them and ordinary members was not easily crossed. On former coastal sugar estates, wage-paid labour and salaried officials alike were drawn into membership of the co-operatives, with its privileges of profit-sharing and voting. Productivity of labour fell drastically on some of the sugar co-operatives, and many of the skilled workers and salaried officials left because of the new problems, to the detriment of former standards of production which had been very high.

In the sierra, response to the formation of co-operatives varied. In some cases, where former *haciendas* were operated by wage-paid labour rather than tenancy systems, the new co-operatives have been moderately successful. In many others, the co-operatives exist almost in name only as peasants have increasingly claimed plots of land for themselves rather than contribute their labour to the co-operative effort. Some have virtually disintegrated into small- or medium-sized holdings, and in some the co-operative function has been restricted to the provision of agricultural inputs and the sale of produce.

THE END RESULT

In sum, land reform in Peru has profoundly changed the old order of *hacienda* control of land resources. It has done little to change the basic structure of society, because the expansion of industry and of urbanisation has effectively displaced the levers of power from the old landed oligarchy towards an industrial, commercial and bureaucratic elite. Land reform has massively extended the power of the state into the agricultural sector. It has benefited a substantial number of former tenants and agricultural labourers, but has had little impact on the majority of peasant small-holders, who still remain as a marginal element in the Peruvian population as a whole. It has been followed by a substantial decline in agricultural production, contrary to the expectations aroused by the prospect of land reform in the 1960s.

What options remain?

How can standards of production be raised, agricultural exports be revived and the national bill for imported foodstuffs be reduced? The World Bank Report on Peru in 1981 suggested the following:

'The revival of agricultural production depends on adequate incentives and the appropriate application of technology. Adequate incentives essentially mean remunerative producer prices in line with international prices and more efficient marketing channels (with increased involvement of private enterprise) supported by an adequate market news service. The appropriate application of technology calls for an effective combination of extension, research and credit supply.'

This further comment was also made:

'The development and application of improved agricultural technology through research, extension and credit need firm guidance from the Ministry of Agriculture towards the country's production priorities, with a strong commitment to make agrarian reform eventually work. In line with the present state of production and nutrition, priority should be given to export crops (particularly sugar) and food crops (particularly low-cost animal proteins).'

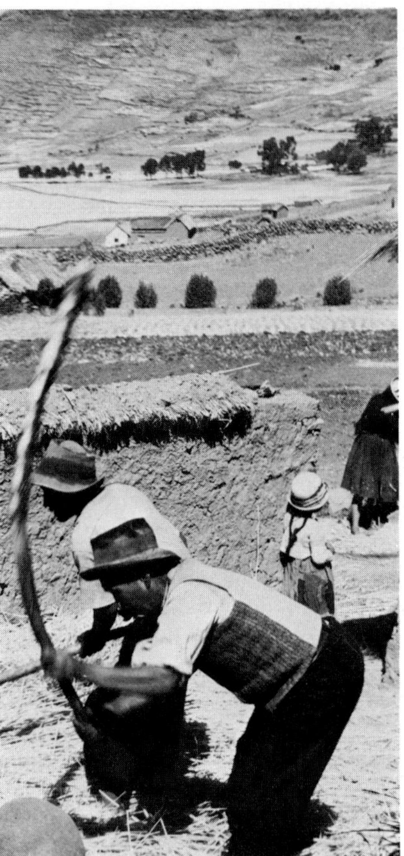

Print 9 *Traditional farming techniques — grain threshing in the higher Andes*

The issues involved are, indeed, extremely difficult to resolve. Stability in the social and economic environment as well as reasonable pricing structures are necessary if private investment in agriculture in the sierra is to be forthcoming on anything like the scale needed to introduce new techniques, new strains of high-yielding crops and adequate use of fertilisers etc.

Agriculture on the coast is already very highly commercialised; mechanised farming and the use of rural credit are understood and there is recognition of the ways in which farming may be made more intensive. In the montaña, transport to the coast is bound to be costly, but there are substantial opportunities for the production of rice, beef, bananas and other fruits for the domestic market as well as opportunities for the production of coffee.

The major intractable problem still lies in the sierra, where land reform has by no means solved the problem of access to land by the overcrowded and predominantly Indian peasantry. Agricultural techniques remain relatively backward, though some progress has been made in the use of better seed and fertilisers. Low levels of education, lack of land and poor communications make it very difficult for peasant farmers to commit themselves to commercial farming, the hazards of accepting rural credit, the discipline of co-operative farming and the trial of new techniques which are not always appropriate to their mountainous environment.

?

1 As a new Minister of Agriculture, consider the following policies you may wish to adopt, and put them in order of priority, bearing in mind that the budget for agricultural affairs is a limited one, and that some policies will conflict with those of other government departments:
a Subsidised rural credit for agricultural improvements.
b Programmes of plant and stock breeding in universities.
c Agricultural advisory and extension services.
d The consolidation of excessively small farms.
e Support for, or dismantling of, the co-operatives created by land reform.
f Improvement of roads in rural areas.
g A regional policy focussing agricultural assistance on the coast, *or* the sierra, *or* the montaña and the selva.
h Better controlled prices for agricultural staples in the urban markets.
i The abandonment of controls over prices and of subsidies to rural credit.
2 Should poor countries introduce desirable social reforms, such as land reform, if the effect is to reduce agricultural production, or should they seek to boost production whatever the social costs? Justify your answer in relation to what you have read in this chapter and in other chapters in this section.
3 Read the briefing sheet below, taken from the Oxfam package on the Brandt Report. Take the commentary paragraph by paragraph and decide whether what it says is substantiated in the account of Peru in this chapter. Give examples and illustrations so that it could be rewritten specifically as an account of Peru.
4 To what extent do physical factors interplay with human factors to heighten the problems of agriculture in Peru?

The Report of the Brandt Commission raises issues of great importance for Latin America. Nowhere are the opportunities and dilemmas seen more starkly than in a region where, on the one hand, countries have struggled since the 19th century for a fairer deal from the international economy and, on the other, internal pressures are increasing for more just societies.

Few would disagree that the Report's recommendations on trade, investment, aid, and technology address themselves to the main mechanisms inhibiting sustained and balanced growth in Latin America. Yet many Latin Americans, with their long experience of attempting unsuccessfully to bring about reforms in these spheres, may wonder how they are to be brought about. Some may feel that the solution lies in building up the bargaining power of the region as a whole, though, as the shaky history of the Andean Pact shows, there is no 'One Latin America', just as 'One World' is at present sadly a long way off: Brazil and Mexico are knocking on the doors of the club of rich Western countries; Venezuela guards its position in OPEC; and Cuba is tied to COMECON. The weaker Latin American countries have learned from experience to expect little generosity from their more influential neighbours.

This raises the central question of the quality of growth and the distribution of economic benefits within the underdeveloped world. Over the last 15 years, Latin America — to the surprise of many — has experienced a respectable annual per capita growth rate of about three per cent. Yet this 'progress' has brought mixed social benefits. There has been little advance in employment opportunities: a quarter of the labour force (some 30 million people) is without fully productive work. Statistics for housing, education, health and nutrition show little change for the vast majority of the population. The gap between rich and poor is in fact widening: while Brazil has leapt ahead of countries like Haiti and Honduras, the distance within Brazil itself between living standards in the rich south and the poor north-east has similarly increased. In the cities, the contrasts grow between the shanty-towns on the one hand, and on the other the rich suburbs and business centres.

Broadly speaking, this distorted growth has to do in part with the form of economic development chosen, which in turn is influenced by the international economy in which the south must compete.

Beyond this, however, both economic policy and the distribution of income and wealth are matters of political choice. An explanation of poverty in Latin America must to a large extent take account of who takes decisions over the allocation of resources. It is no accident that the greatest inequalities are to be found in countries governed by small, frequently authoritarian élites, usually backed by the armed forces who control the government. Every such case is basically a conflict between two sets of competing values: justice, freedom and equity on the one hand; and responsibility, hierarchy and efficiency, on the other.

This raises perhaps the most important challenge posed by Brandt: how to bring those working for better societies in the countries of the south together with the international decision-makers to whom the Report is principally addressed. This will be a difficult task. Some in Latin America may think that in their struggle for freedom, where inequality and injustice have become deeply entrenched, it is now too late to speak of accommodation. But there are ways in which, at the very least, channels of communication can be kept open.

6 US AGRICULTURE: THE PROBLEM OF OVERPRODUCTION

US AGRICULTURE AND THE PROBLEM OF OVERPRODUCTION

Cheese is driving the United States Department of Agriculture mad. They give it away, they store it, freeze it and destroy it — and still it keeps on coming. In 1983 alone the beleaguered Department will receive 14 000 million pounds of cheese, butter and dried milk and, horror of horrors, dairy farmers are continuing to increase production.

(*Source*: *The Times*, 16 February 1983)

THE SOMBER WAVES OF GRAIN

The crop glut prompts the administration to reconsider its hands-off farm policy.

The ground is frozen and the fields are bare, but January is normally a productive month on Carlyn Johanningmeier's farm in Postville, Iowa. Like thousands of others across the Midwest, the 46 year old farmer and his family tinker with machinery, shop for seeds — and start plowing a harvest of profits back into their business. But this winter, the Johanningmeier spread is at a near standstill. More than 90 000 bushels of the farm's corn crop were unloaded last year at a loss; 60 000 more are stored in grain bins around the property. Instead of going shopping for farm supplies, the Johanningmeiers have gone heavily into hock. Says US Department of Agriculture spokesman Herb Jackson: 'This is the year many farmers are asking themselves whether they should plant or get out of the business.'

Johanningmeier hopes to plant again, but right now he and other farmers are waiting on Washington.

(*Source*: *Newsweek*, 17 January 1983)

Most of us would agree that the Food and Agriculture Organisation's (FAO) estimate that there are 500 million seriously undernourished people in the world is a greater horror than the US cheese mountain or corn mountain but they are both aspects of our common experience and reflect fundamental differences in the nature of the people–environment relationship that exists between countries and between societies.

Traditionally it has been believed that the fundamental problem facing humankind is the essentially Malthusian one of human reproduction outstripping material production, but there are some societies which can organise production at such levels that their very success creates its own problems. These problems raise a series of economic, social, political, ethical and physical-resource management issues which are complex to the point of intractability but which have to be analysed and understood.

In the US, an enormously inventive and productive society is having to come to terms with the consequences of its success in increasing agricultural production, but although short-term administrative measures may provide a temporary respite in some areas of pressing difficulty, the real problems of overproduction (if problems they are?) are long-term and are deeply rooted in both the American economy and society, and in present systems of international trade and relations.

AGRICULTURAL PRODUCTION IN THE USA

It is clear from Figure 1 that total farm output has been increasing at least since the end of the Second World War, although there have been minor fluctuations in livestock production.

?

1 Look at Figure 1. What are the principal changes in output that have taken place? Describe the differences before and after 1970.

2 What are the advantages and disadvantages of using an index as a measure of change? Production figures are given for

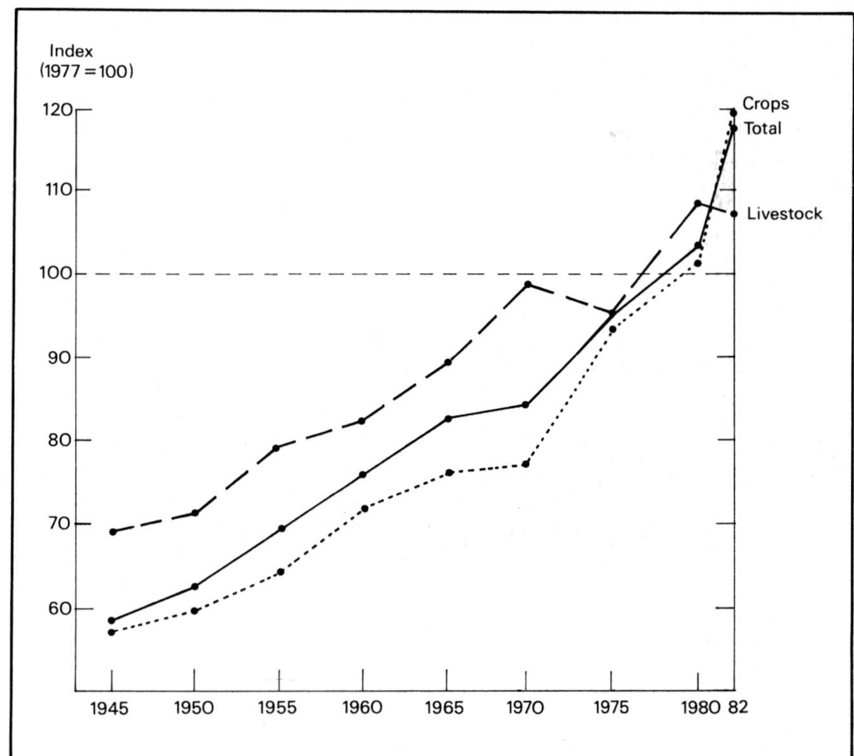

Figure 1 *The growth of agricultural output in the USA, 1945–82*

selected crops in Figure 2 and for selected animal products in Figure 3.

a Measure the scale of agricultural production in the USA.

b Identify the general trends of production, both overall, and for specific products.

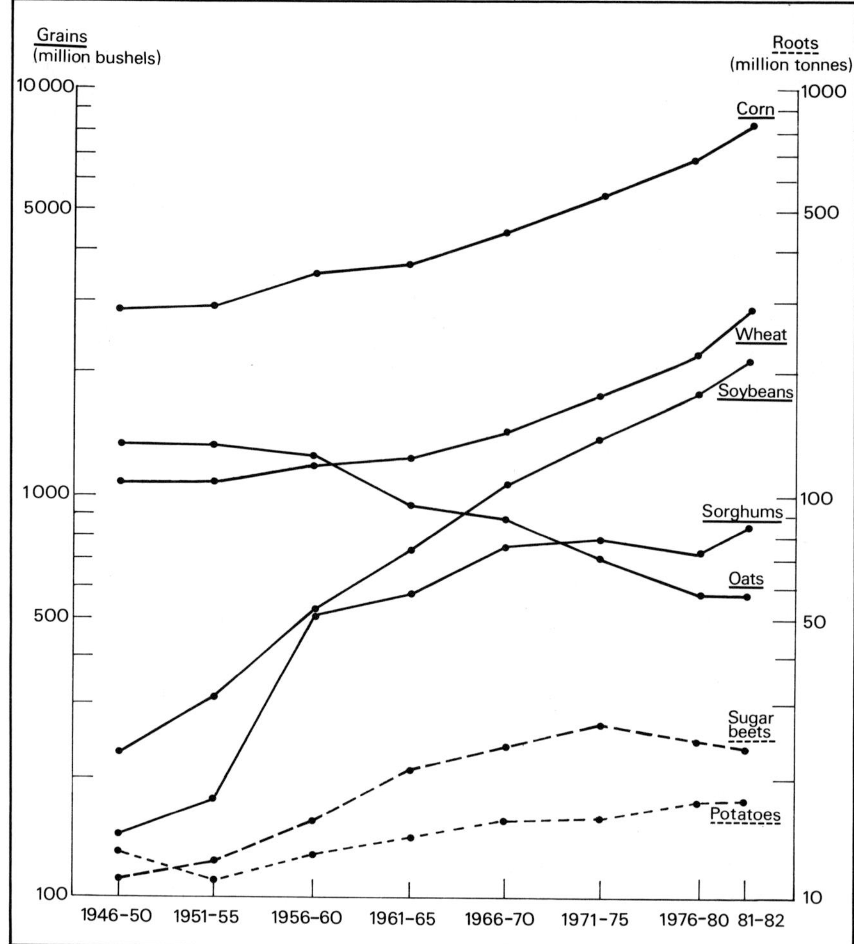

Figure 2 *The production of principal crops: quinquennial averages 1945–80, biennial average 1981–82*

Extension and intensification

As a result of the increases in farm output, and despite the fact that the population of the USA grew by over 75 million from 1950 to 1980, the US was able to export over $42 billion of agricultural products in 1981. In fact, each farmer was then able to feed 75 fellow Americans and substantial numbers in the world outside.

Increases in production can be brought about by the extension and/or the intensification of the cultivated area. Historically, the cultivated area of the US has been extended by the process of Westward Expansion, and in more recent years, new areas have been brought into use as a result of the extension of irrigation, particularly in the western states where the irrigated area increased from 25 to 45 million acres (10 to 18 million hectares) between 1950 and 1980. A comparison of the acreage of principal crops harvested at the beginning and end of the post-war period is given in Table 1.

? Calculate the percentage change for each crop shown in Table 1.

American agriculture has also long been characterised by the more intensive use of already cultivated land and by the intensification of animal husbandry through the application of science and technology. For example, it took from 12 to 14 weeks to produce a 4 lb chicken in 1940 and the bird ate 4 lbs of feed for each pound of meat it gave. By 1980 a similar chicken could be produced in only 8 weeks at a cost of 1.9 lbs of feed per pound of meat. In crop farming, yields are about 20 per cent lower than those achieved in Europe because of the more extensive nature of American farming. Trends, including milk yields, are given in Figure 4.

Table 1 *Acreage of principal crops harvested in the USA: 1946–50 and 1978–82* (*Figures in brackets — 000 hectares*)

	1946–50 av.		*1978–82 av.*	
Corn	75 711	(30 639 ha)	73 042	(29 559 ha)
Wheat	70 312	(28 454 ha)	69 957	(28 310 ha)
Soybeans	11 263	(4 558 ha)	67 847	(27 456 ha)
Hay	73 639	(29 800 ha)	60 802	(24 605 ha)
Sorghums	7 283	(2 947 ha)	13 359	(5 406 ha)
Cotton	21 421	(8 669 ha)	12 403	(5 019 ha)
Oats	39 409	(15 947 ha)	9 046	(3 660 ha)
Rice	1 718	(695 ha)	3 239	(1 311 ha)
Potatoes	1 992	(806 ha)	1 258	(509 ha)
Sugarbeets	797	(323 ha)	1 167	(472 ha)
US crop total	353 600	(143 094 ha)	342 423	(138 572 ha)

(*Source*: Statistical Abstract of the US)

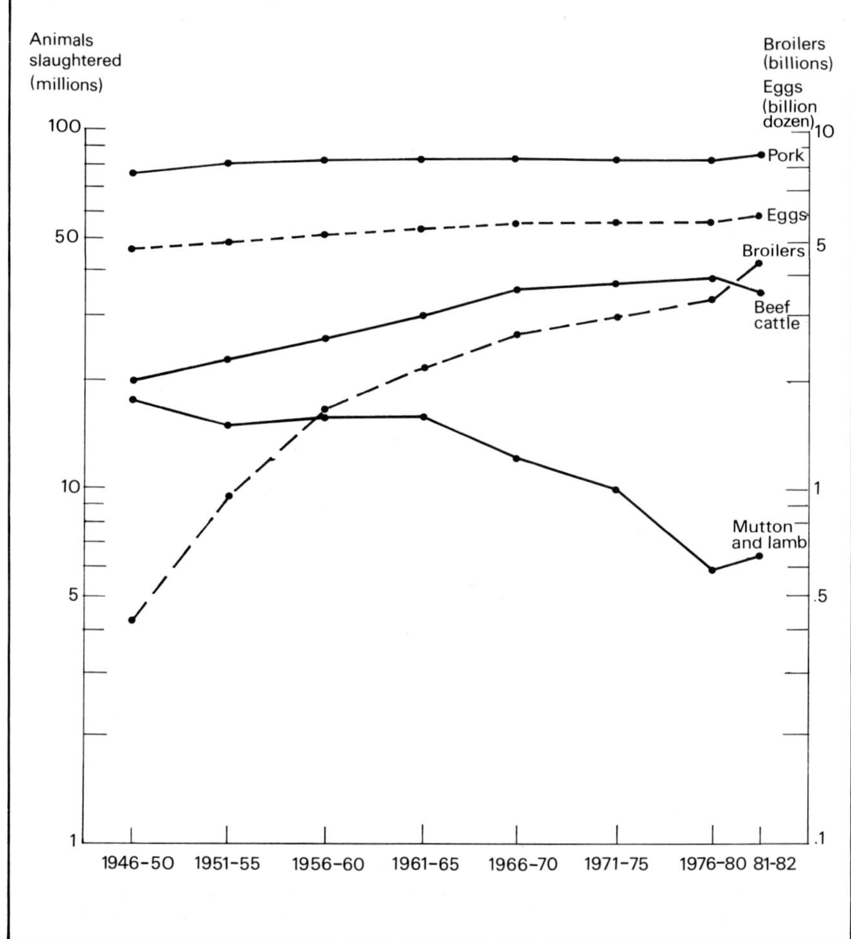

Figure 3 *The production of principal animal products: quinquennial averages 1945–80, biennial average 1981–82*

? **1** Compare Table 1 and Figure 4 and, with a partner, explain the mechanism of increasing production in the USA.
2 Assess the relative importance of the two processes of extension and intensification in these increases.

It must be remembered that the two processes of extension and intensification do not operate independently and the complex interrelationship that exists between them can be illustrated by the fact that mechanisation, which is central to the intensification process, has effectively extended the cultivated area by releasing millions of acres of land that were formerly used to grow crops to feed draught animals such as horses and mules.

One of the less desirable consequences of this phenomenal productive power is that American agriculture has always been susceptible to market changes and when demand has periodically fallen and supply has been maintained, the inevitable surplus has to be stored, given away, or even destroyed.

In the late 1970s there were, once again, large surpluses of wheat and feed grains, dairy products, tobacco, sugar and so on, which had become 'mountains and lakes' by late 1982 as supply greatly outstripped demand. Once again, government action was taken to attempt to balance supply with demand (see stocks listed in Table 4) and once again a short-term solution was used to treat an essentially long-term problem.

? **1** Read the following extract and then in a group, debate the issues raised in the chapter so far.

All California quickens with produce, and the fruit grows heavy, and the limbs bend gradually under the fruit so that little crutches must be placed under them to support the weight. Behind the fruitfulness are men of understanding and knowledge and skill . . . and men are proud, for of their knowledge they can make the year heavy. They have transformed the world with their knowledge.

And first the cherries ripen. Cent and a half a pound. Hell, we can't pick 'em for that . . . . And the yellow fruit falls heavily to the ground and splashes on the ground. The yellow-jackets dig into the soft meat, and there is a smell of ferment and rot.

The decay spreads over the State, and the sweet smell is a great sorrow on the land. Men who can graft the trees and make the seed fertile and big can find no way to let the hungry people eat their produce. Men who have created new fruits in the world cannot create a system whereby their fruits may be eaten. And the failure hangs over the State like a great sorrow.

The works of the roots of the vines, of the trees, must be destroyed to keep up the price, and this is the saddest, bitterest thing of all. Carloads of oranges dumped on the ground. The people came for miles to take the fruit, but this could not be. How would they buy oranges at 20 cents a dozen if they could drive out and pick them up? And men with hoses squirt kerosene on the oranges, and they are angry at the crime, angry at the people who have come to take the fruit. A million people hungry, needing the fruit — and kerosene sprayed over the golden mountain.

(*Source*: John Steinbeck (1939) *The Grapes of Wrath*)

THE PIK POLICY: THE SHORT-TERM RESPONSE TO OVERPRODUCTION

IMMEDIATE CAUSES

The year 1983 opened with massive surpluses of which 1 billion bushels of wheat, 2.5 billion bushels of feed grains (corn, oats, barley, sorghums), and 2.4 billion pounds of

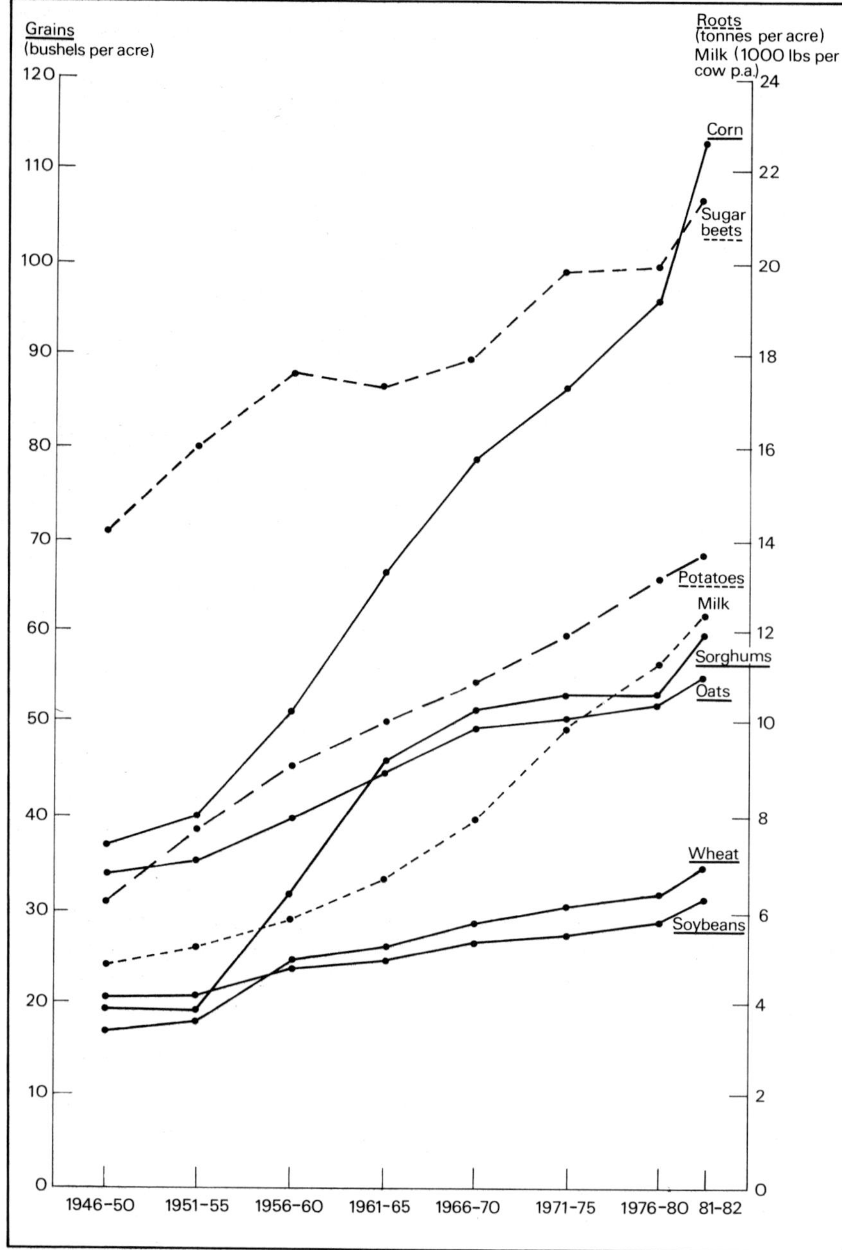

Figure 4 *Average yields of selected crops and milk: quinquennial averages 1945–80, biennial average 1981–82*

dairy products were just the largest.

The causes are not difficult to find, and as you read through this paragraph attempt to draw a diagram of the variables and their interrelationships. The supply of food and feed grains, for example, had accelerated during the record harvests of 1981 and 1982 which were created by a combination of good weather, the high and growing efficiency of American farmers, and government policies which are designed to protect farmers' incomes. Unfortunately, demand had fallen for a complex variety of reasons which illustrates the fact that agriculture, whether in developed, developing or planned economies cannot be considered in isolation from its total environment. At home, economic depression and changing food fashions meant that red-meat consumption had fallen by 20 per cent since 1976. Abroad, where surpluses were expected to be sold, demand was also low because of the world economic recession and because harvests in many countries benefited from the good weather. Moreover, the dollar was very strong and traditional importers such as Mexico and Poland could not, for very different reasons, afford large grain imports. The embargo on extra grain sales to the USSR, imposed by President Carter in 1980 following the invasion of Afghanistan, had also tarnished America's reputation as a reliable supplier. This created especially serious problems because the USSR was the largest single customer for US agricultural exports in 1979, taking products worth $2.9 billions, but the Russians were very cautious about ever becoming dependent upon so easily withdrawable US supplies.

THE PAYMENTS IN KIND POLICY

Faced with surpluses on this scale, and the massive costs of both storage and its agricultural policies in general, the Federal Government decided that it must take action. On the one hand it was eager to cut government spending as part of its general political philosophy, but on the other it was under pressure from the farming lobby who, faced with falling prices as a result of a glutted market, were demanding even more Federal expenditure in the form of higher crop-price supports and protectionism. Anxious not to lose the farm vote, President Reagan announced an emergency Payments In Kind (PIK) scheme for wheat, corn, sorghums, rice and cotton in January 1983. It was based on a scheme first devised in the Depression years of the 1930s and tried again, none too successfully, in the 1960s.

In essence, participating farmers would agree to take some or even all of their crop land out of production and would receive from existing stocks up to 80 or 90 per cent of what they would have produced, to either sell or to feed to their animals. The scheme would thus have the admirable triple aim of raising farm prices by cutting production, of reducing stocks and of preventing additions to them. Moreover, it would be almost cost-free.

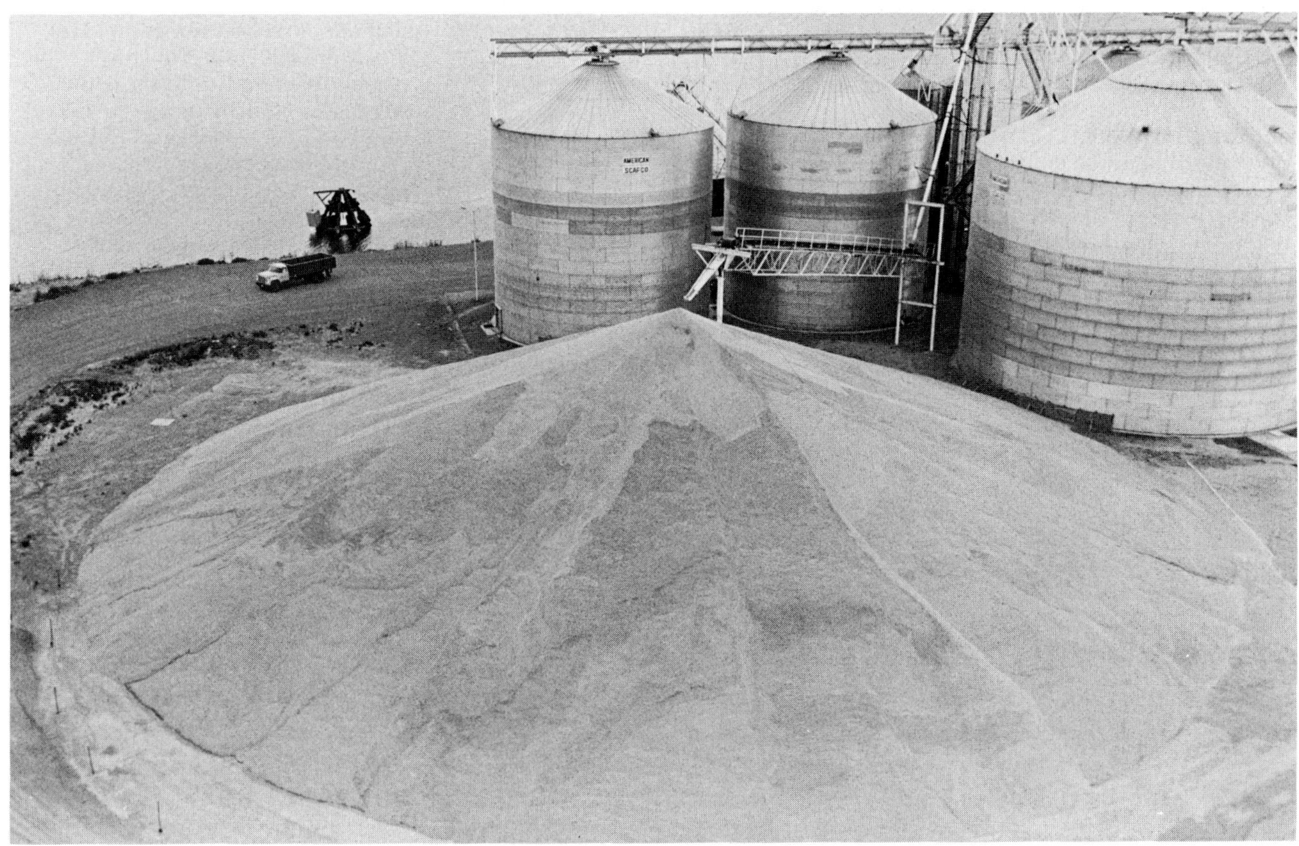

Print 1 *Piles of wheat at a grain terminal on the Snake River*

Problems arising from the PIK policy

Unfortunately, reality proved to be more complex and a number of problems arose which made the PIK policy rather less successful than had been hoped:

1 The number of farmers participating in the scheme was much larger than expected on the basis of participation rates in previous crop reduction schemes. Forty-three per cent of corn farmers and 35 per cent of wheat farmers actually took part. Instead of an estimated 23 million acres (9.2 million hectares) lying fallow as predicted by the US Department of Agriculture, 46.3 million acres (18.5 million hectares) were taken out of production and such was the enthusiasm for the scheme that about one-third of the winter wheat crop was destroyed after the autumn planting, by farmers who wished to take part.

2 Consequently, the stocks of some crops actually owned by the government were not large enough to cover reimbursement to farmers, and the government therefore had to buy additional supplies in a market where prices were rising because of the shortages created by government policy and compounded by drought. As a result, the 'cost-free' scheme was estimated to have cost from $10–16 billion, and there were political repercussions to be faced because no limits were put on the amount an individual farmer could receive and the richest farmers received the largest amounts.

3 The drought severely affected the corn harvest since it came at a critical stage in the pollination process, and for much of Illinois it was the worst since 1936. As a result, it was estimated that one billion bushels of corn had been lost to the drought to add to the 2.2 billion bushels not produced under PIK, and some parts of the corn belt were declared agricultural disaster areas. On the other hand, those farmers in the PIK scheme were protected from the drought and happily received their allocations in a rising market.

4 The 'swings and roundabouts' situation that prevails in agriculture saw the weather giving an almost record wheat harvest despite the acreage cuts (which were usually made on the marginal land anyway) and the surpluses of wheat were little affected, with prices consequently remaining low. Just to illustrate the difficulty of it all, the hoped-for rise in corn prices reduced the farmers' potential for exports since corn became more expensive to buy from the USA and so the problems were compounded.

5 There were also repercussions in associated activities. Worries that the farm machinery, fertiliser and seed industries would be adversely affected by an agricultural contraction were not realised because of increased production on land kept in cultivation. In fact, PIK stipulations that land must be protected against erosion led to a tripling of grass and clover seed prices. As a response, farmers put an increased number of feeder cattle onto their grassed fields with obvious consequences for future supplies of meat.

What this brief examination of the PIK policy clearly illustrates is the fascinatingly interconnected and interrelated nature of the people–environment relationship and the difficulty of decision making in a situation characterised by past consequence, present complexity and future uncertainty.

? List the circumstances, just described, which were actually within the government's control when it formulated its policy and consider what this shows about the nature of the problems facing government decision makers.

LONG-TERM CAUSES OF OVERPRODUCTION

AGRICULTURAL DECISION-MAKING IN THE US CONTEXT

In the American context, overproduction means a surplus that cannot be sold, which in a needy world may be a debatable definition. Of course, given the size and variety of American agriculture and the wide range of variable factors affecting supply and demand that are not within the farmers' control, the notion that farmers can balance their supply to meet demand must be seen as an ideal that is probably far from any achievable reality. Nevertheless, what is being achieved in the US are large and persistent surpluses that represent a serious imbalance in the people–environment relationship and the reasons for this great capacity to overproduce need to be understood.

This problem of overproduction may seem a surprising state of affairs given the quantity and quality of information available to literate and responsive farmers from both government and private sources. A great amount of research is undertaken to give farmers more control over nature, to increase their production and to reduce their production risks. As a result, a great amount of technical information and advice is available and is disseminated by the US Department of Agriculture Extension Service and by private agencies such as chemical companies, seed merchants and machinery and equipment manufacturers. Since 1976 a commercial earth-satellite company has been offering a 'Cropcast' service which can predict yields halfway through the growing season. Government agencies also provide information such as weather forecasting and soil and water management advice to assist the farmer in maintaining and improving output.

Help to market this output is also available through the US Department of Agriculture's Agricultural Marketing Service which aims to promote an orderly and efficient market at home, and its Foreign Agricultural Service which seeks to maintain and extend its overseas markets. Since the USA is the world's leading exporter of agricultural products with about one-third of all crop land producing for export, the Foreign Agricultural Service plays a central role in maintaining the prosperity of American agriculture. Regular and up-to-date information about market conditions is also available from the US Department of Agriculture and from private agencies such as banks and market analysts.

Nevertheless, the predicament of the individual farmer as decision maker has been stated thus:

Print 2 *A district conservationist discussing autumn tillage plans with a farmer*

'Being in a market so large that he can have no perceptive influence on it, the farmer's only alternative for survival is to cut costs. Thus he has eagerly availed himself of the new technologies that increase his output per unit of input. Yet the resulting expansion of production is confronted by the inelastic nature of the aggregate demand for food, so that prices are driven down and the resulting reduction of farm income sets the stage for still another vain effort by operators to preserve their farms.'

(*Source*: H. F. Gregor, 1982, *The Industrialisation of US Agriculture*)

? Gregor's statement sets out the dilemma for the individual farmer. Simplify his statement using no more than 250 words.

The difficulty of making decisions in a large and complex reality is clearly only part of the explanation of American agricultural over-production. The farmer is an American, operating in an American environment, with all that that involves.

As can be seen from Figure 5, the information that is gathered and used by farmers exists within a decision environment which is created by the interconnection and interaction of the cultural and physical environments of their country. This interconnection and interaction produces a complex and dynamic framework of cultural and physical opportunities and constraints within which the farmer must operate. It is within this framework that the farmer's decisions are formulated and it reflects an American society which values freedom, change, progress and achievement, and which is rooted in a rich and bountiful environment.

? Study Table 2 and contrast the attitudes of Americans and Tanzanians, farming in dry areas, to the drought hazard that they face. In particular, say how the decision environment of dry areas is affected by differing attitudes to technology, the role of magic vis-à-vis science, and the role of people in adapting to or attempting to change nature.

Table 2 *A comparison of possible adjustments to drought suggested by 96 American and 131 Tanzanian farmers*

United States			*Tanzania*		
Adjustment	*No.*	*%*	*Adjustment*	*No.*	*%*
1 Change farming method to conserve moisture	78	40.2	1 Move away from area	51	36.4
2 Introduce irrigation	46	23.7	2 Do nothing, wait	17	12.1
3 Change landscape: add dams, ponds, trees	26	13.4	3 Live on savings of food and money	16	11.4
4 Live on insurance and savings, reduce expenses	16	8.3	4 Rainmaking, prayer	15	10.7
5 Adapt crops	2	1.0	5 Introduce irrigation	15	10.7
6 Rainmaking, prayer	2	1.0	6 Change crops	9	6.4
7 Quit farming	1	0.5	7 Change plot location	4	2.9
8 No suggestions	16	8.3	8 Change farming methods	1	0.7
9 Others	7	3.6	9 Others	12	8.6
TOTAL ADJUSTMENTS	194	100.0	TOTAL ADJUSTMENTS	140	100.0
Adjustments per farmer = 2.02			Adjustments per farmer = 1.07		

(*Source*: T. F. Saarinen, 1976, *Environmental Planning: Perception and Behaviour*)

Of course, decisions are made not only by individual farmers, but by other decision makers, each of whom, as in the case of the government's PIK policy for example, may clearly affect the others. The situation is further complicated by the fact that the consequences of all decisions feed back into the decision environment, to a greater or lesser extent, and make it even more difficult for the individual farmer, already coping with capricious and unpredictable natural forces, to predict future events.

? **1** Use Figure 5 to assess the ways in which the decision environment of an American farmer is different from the decision environment of a farmer in this country.
2 Contrast the balance of opportunity and constraint facing the American farmer, both cultural and physical, with the balance facing a farmer in a developing country which you have studied. Again use Figure 5 as a guide.

ENVIRONMENTAL OPPORTUNITIES AND THE OVERCOMING OF ENVIRONMENTAL CONSTRAINTS

Rich opportunities of the natural environment

The first major factor affecting agriculture is the nature of the physical environment, and the USA is a land rich in physical opportunities for the farmer. This gives high levels of production and forms the basis of American agricultural wealth. For agriculture to take place successfully, there must be a sufficient and reliable supply of water and temperatures must be above specific minima for sufficient lengths of time to enable plants and animals to grow and flourish. Good soil is also required since this provides the support for plant growth and the essential minerals and trace elements which constitute plant food. The quality of the opportunity presented by soil depends upon its composition, texture and depth and this in turn is the product of the type of parent material from which it is formed, the type of relief and vegetation cover, the activity of soil organisms ranging from microscopic bacteria to large mammals, and past and present human use. It is some measure of the richness of the ecological complex in the USA that in 1978, 46.7 per cent of the land area of the country was put to agricultural purposes, of which 471 million acres (20.8 per cent) were in crop land and 587 million acres (25.9 per cent) were in pasture (188 million hectares in crop land and 235 million hectares in pasture).

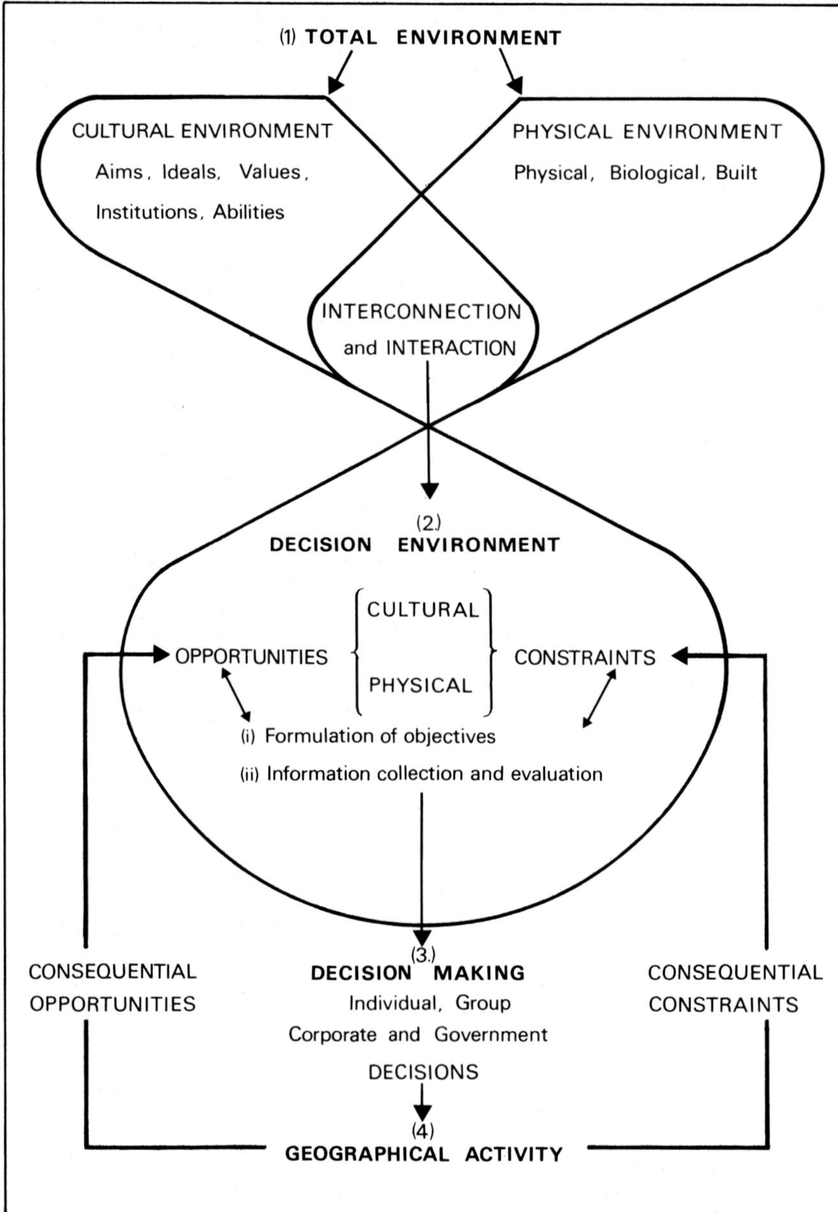

Figure 5 *The process by which the system of interconnection and interaction in the total environment is translated into geographical activity*

Constraints to the ecumene
There are constraints of course. Large areas of the country are hostile environments outside the ecumene where temperature or rainfall deficiencies caused by altitude or location pose a constant threat to humankind. The ecumene is also constrained by environmental limits created by seasonality or the life-cycles of plants and animals. Even within the ecumene, environmental hazards pose an intermittent threat, and to add to such threats as flood or drought there are over 10 000 species of harmful insect, 1800 different weeds, 1500 fungal diseases and 1500 different kinds of nematodes present in the USA.

However, Americans agree with Francis Bacon that 'a man must make his opportunity as oft as find it' and a great deal of effort and energy is put into overcoming these constraints and extending and intensifying the ecumene. Overcoming the constraint of distance, for example, has been achieved by regional specialisation, the integration of crop and animal farming and especially by the development of transport facilities. This illustrates well the role of human culture as a factor in the opportunity/constraint balance and confirms the view that the people–environment relationship can only be understood by an appreciation of the close interaction of physical and cultural circumstance.

THE PHILOSOPHICAL AND TECHNOLOGICAL ENVIRONMENT

Agricultural activity takes place within a cultural environment and is affected by the aims, attitudes, values, beliefs, institutions and abilities of a society which presents opportunities for farmers, but which also puts constraints upon how and what can be produced.

? Muslims and Jews will not rear pigs, nor Hindus rear cows, for food. What do you think are the principal food and social taboos in Anglo-American society that influence how and what farmers can grow?

The technological revolution
Technology represents our abilities to reduce the constraints of nature and to create new opportunities for the utilisation of the earth. Because of this, the concept of the ecumene, or the inhabited earth, is a dynamic one as we extend and intensify our occupancy. The physical environment is constantly being reassessed as technology advances.

Agriculture in the US is a high technology industry geared to produce with a high capital to labour ratio, and it operates in an environment where rapid changes in technology and the acceptance of new ideas are not only possible but are normal. (This contrasts markedly with many developing countries which are constrained by taboo and the dead hand of custom and where changes in methods of production often involve unacceptably high risks.) Indeed:
'It shall be the object and duty of the state experimental stations to

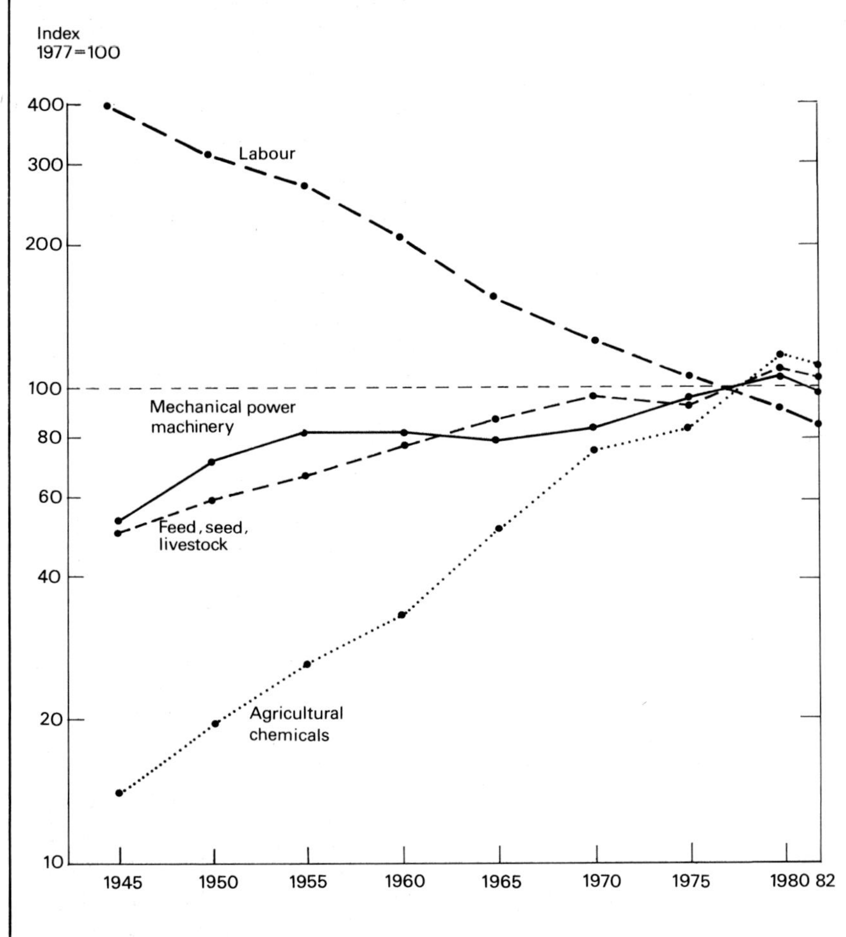

Figure 6 *The changing balance of agricultural inputs 1945–82*

conduct original and other researches, investigations and experiments bearing directly on, and contributing to, the establishment and maintenance of a permanent and effective agricultural industry.'
(*Source*: Hatch Act, 1887)

'A basic goal of agricultural research is to establish a high-yielding agriculture that also supports a quality environment and conserves energy.'
(*Source*: US Department of Agriculture, 1980)

These statements illustrate the long-established importance of public support for the application of science to agriculture beginning with the Morrill Land Grant Act of 1862 which established the state agricultural colleges, continued with the Hatch Act of 1887 which established experimental stations, and was capped by the Smith-Lever Act of 1914 which added the Extension Service to carry their findings directly to the farmers.

Although the Extension Service is very effective, not all taxpayers are happy about it and argue that those who benefit from it ought to pay for the research. Of course, commercial organisations such as chemical and seed companies also undertake research and the high levels of production have been produced by a technological revolution whose basic characteristics are given in Figure 6 and whose effects are illustrated by the quotation below:

'Mr Donald Newby has a farm on the outskirts of Des Moines, Iowa. He can buy during a year 7 000 two month old pigs weighing about 40 lbs and sell them to slaughter-houses when they are six months old and weigh between 220 lbs and 230 lbs and still find time to raise 720 acres (288 hectares) of soyabeans and maize with only a single hired hand to help him.'
(*Source*: *The Economist*, 5 January 1980)

?

1 Look at Figure 6 and then look back at Figure 4 (*page 106*). Now attempt a more detailed explanation of the changes shown in Figure 4.

2 Look at Figure 6, think about what has been written in the chapter so far, and make a list of the characteristics of the technological revolution in farming. Read on and compare your list with the text.

The major components of the technological revolution are as follows:

1 Mechanisation, in a country where, in 1982, 2.4 million farmers had 4.7 million tractors with new ones averaging 100 hp, enabling much larger areas to be farmed. Mechanisation is not only the application of power but also the development of machines such as mechanical harvesters and electronic sorters which can handle crops as fragile as tomatoes and thus cut the need for expensive and possibly unreliable labour.

2 The widespread application of artificial fertilisers to maintain soil fertility and to increase production.

3 The use of improved feedstuffs for animals using bought-in concentrates and formula feed from specialist manufacturers.

4 The overcoming of the limitations of nature by the use of improved strains of plants and animals, giving higher yields and being more resistant to pests and diseases. In this way, for example, the average yield of corn has increased from 30 bushels per acre (74 bushels per hectare) in 1945 to 115 bushels per acre (284 bushels per hectare) in 1982.

5 The increasing protection of growing plants and animals. Various techniques have been developed to either protect them directly through the use of drugs and pesticides, or indirectly by destroying the environments in which pests thrive, by encouraging the pests' natural predators, or by interrupting the pests' life-cycles.

6 The use of improved equipment, principles and practices. Irrigation, for example, has been used not only to extend the ecumene but also to increase yields in traditional farming areas and to provide security of supply in areas of unreliable rainfall.

7 The use of new methods of farming, such as feedlots for cattle or battery units for poultry (factory farming), in which organisational technology plays as important a role as the mechanical technology previously described, and higher production results from improved crop and livestock management.

One of technology's greatest benefits has been to take the drudgery out of the farmer's life. It has been said that the new rotation is wheat–Hawaii–sorghums–Florida, but technology, as we shall see, has not been without its costs.

THE ECONOMIC AND POLITICAL ENVIRONMENT

The second major component of the cultural environment consists of the economic and political frameworks and institutions which are closely interconnected and which play a central role in affecting the nature and volume of agricultural production.

Farming is America's biggest business and employs directly as many people as transport, steel and cars combined. It also creates much subsidiary employment in industries such as farm machinery, fertilisers and food processing and offers pehaps 20 per cent of jobs in the private sector of the economy.

US agriculture operates in an economic environment where attitudes and structures are dedicated to production. As a result, there is an emphasis on efficiency and competition, and farms are operated as 'agribusinesses' characterised by high capital inputs, specialisation and sophisticated organisation. The changing balance of inputs, shown in Figure 6 (*page 111*), illustrates the increasing industrialisation of agriculture in the US as the tool of technology is applied in pursuit of bigger and better production.

These changes have created some problems since, in 1980, a combine harvester cost \$70 000, a 135hp tractor \$35 000 and an oxygen-free silo \$50 000. Income:debt ratios of 1:10 are typical. Consequently, the farmer is placed under considerable financial pressure when prices fall and conflicts arise as, for example, grain farmers demand high prices and animal feeders insist that they be kept low.

In these circumstances, as financial stakes and risks have grown considerably, good management has become essential and five out of six farmers are now members of co-operatives. These buying and marketing organisations give farmers access to cheaper inputs, lower taxes, cheap credit and advice from the US Department of Agriculture's Co-operative Service. Moreover, the need for economies of scale has resulted in a marked decline in the number of farms from 5.6 million in 1950 to 2.4 million in 1982 with a corresponding increase in the average size of a farm from 214 to 433 acres (from 86 hectares to 175 hectares). This and the growth of

Print 3 *Imperial Valley, California, receives water for irrigation from the Colorado river. The valley is intensively farmed with grain and vegetables and supplies the nearby urban centres of San Diego and Los Angeles with fresh produce*

corporate farms has led to concern that the traditional idea of the American family farm is disappearing, but many corporate farms are family owned and the distinction has become far from clear.

Although there has been considerable success in organising production at high levels, inevitably there has been less success in organising markets, whose fluctuations have created many problems. Some purely economic solutions have been sought, such as contract farming in which farmers produce specific quantities and qualities of crops for specific customers such as supermarkets or canneries, or by vertical integration in which, for example, canning firms own farms which produce their fruit or vegetable supplies. These arrangements now account for perhaps 20 per cent of agricultural production, but since one-third of US agricultural production is exported into unpredictable markets where it is often used by foreign buyers as a 'supply of last resort' (basically to top-up their own production deficits), fluctuations in demand can and do create serious surpluses.

However, such is the importance of agriculture economically and strategically (and politically and emotionally) that it has been observed that:

'No policy that leaves the American farmer to the mercy of the market place is ever likely to get through Congress.'
(*Source*: *The Economist*, 14 January 1984)

So, US overproduction cannot be understood without reference to the political environment within which the farmer operates.

Given that 226.8 million people had to be fed in the USA in 1980 and that it is expected this figure will increase to 280 million by the year 2000, and given that self-sufficiency is seen as an economic and strategic necessity, the political importance of agriculture can be understood. Nevertheless, it leads to a very interesting conflict of values because although farmers resent government 'interference' when prices are high, they demand government 'intervention' when they are low. Farming illustrates very well the conflict between political ideals and political realities in the US because the Depression years of the 1930s killed 'laissez-faire' capitalism; the rugged individualism of the American farmer exists as a state of mind rather than a concrete reality, with farmers very dependent upon the government for their prosperity. Since the New Deal legislation of 1933, the government has moved from its very peripheral 19th century role as landlord and advisor to a central role as protector of farmers' incomes, and 50 years on, direct payments to farmers totalled $22 billion in 1983.

Farmers also benefit from government infrastructure investments such as cheap irrigation water, rural electrification and soil conservation programmes, and are aided, advised and cushioned by measures such as cheap finance, a generous tax climate, protection against natural disasters and a whole range of financial and technical services. These contribute indirectly to the problem of overproduction, but more directly it is the policies begun in the 1930s and continued in the Food and Agriculture Act of 1977 which kept prices high and encouraged overproduction, although these were modified in 1985.

The aim of the legislation, backed very strongly by the farming lobby, to protect farm incomes by ensuring that farmers are not exposed to the full rigours of the market when prices fall. This is done through a baffling variety of programmes of which the PIK policy is just one. The principal methods used are price-support programmes in which the government makes direct payments and loans to farmers who choose to take part if price levels fall below specified levels based on costs of production. In 1982, for example, the world price of sugar was 8.5 cents/lb but US growers needed 20 cents/lb to break even and it cost the taxpayer $1 billion to support sugar growers. The complexity of the system can best be seen when applied to food and feed grains and is shown in Figure 7 using the support prices for wheat that were in operation in a typical year, 1980.

? Study Figure 7, read the following text, and explain the diagram to a partner.

PRICE SUPPORTING

In Figure 7, when the market price was above level *A* (equivalent to the government's target price), then the farmer would sell in the open market, but if it fell to between level *A* and *B* (the government's loan price, or guaranteed amount per bushel that the farmer could borrow using the grain as security), then a deficiency payment scheme would come into operation — if the crop was sold by the farmer, the difference between the market and the target price would be received from the government. If the market price fell below *B*, then the farmer need not sell, but could borrow money from the government at the loan price and hand over the grain as collateral, still qualifying for the

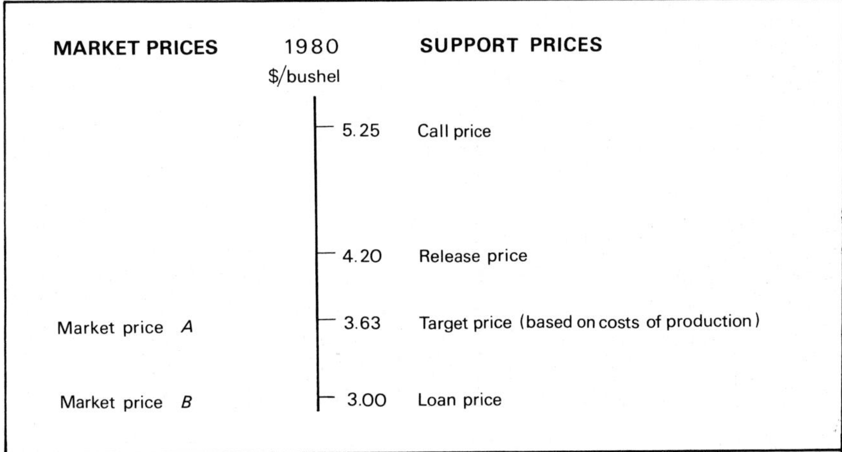

Figure 7 *Price support values for wheat in 1980*

Table 3

	1978	*1979*	*1980*	*1981*	*1982*
Target price (per bushel)	$ 3.40	$ 3.40	$ 3.63	$ 3.81	$ 4.05
Loan price	$ 2.35	$ 2.50	$ 3.00	$ 3.20	$ 3.55
Market price	$ 2.98	$ 3.82	$ 3.91	$ 3.66	$ 3.39
Average yield (bushels per acre)	$31.40	$34.20	$33.40	$34.50	$35.60

deficiency payment. In addition, a grain farmer could put surpluses into the Grain Reserve programme and receive a 3–5 year loan on the grain which would be stored at government expense. The farmer could not, however, get grain out of store until the market price had reached the release price and if it was still in store when it reached the call price, the government could call on the farmer to repay the loan. Furthermore, if a farmer wished to store surpluses at home on the farm, then farm facility loans to build drying and storage facilities were available. In effect the farmer was allowed to operate in a government created economic environment that bore little relation to economic reality, but was made very real politically by the power of the farming lobby.

?

1 Using Table 3, calculate first the amount of deficiency payment a farmer would have received in growing and selling 500 acres of wheat per annum during the period shown. Assume that he decided against the loan option in 1982 because he needed the cash income.
2 Calculate how much this decision cost the farmer in 1982.

Because of the clear need to cut the cost to the taxpayer of deficiency and storage costs, the farmer who wished to receive price supports was, however, required to take part in production adjustment programmes whereby 30 per cent of the farmland was taken out of production. These measures have their origins in the acreage quota and soil bank programmes of the 1930s and 1950s but, like them, they have had only limited success because it has usually been the marginal land that has been 'set aside' and increasing yields have simply meant that more has been produced upon less. With the safety net of target prices and the prospect of good profits if prices rise, farmers have nothing to lose and everything to gain by increasing production and the resulting surpluses can be seen in Table 4.

The substantial stocks shown in Table 4 confirm that similar programmes exist for crops such as rice, cotton, peanuts, tobacco and sugar, and as we have seen, the situation is even more acute for dairy products where the government is committed to buy unlimited amounts at highly artificial prices and is saddled with enormous costs and problems of storage.

Additional incentives for farmers to produce have resulted from the imposition of quotas and tariffs on imported products as a result of political pressure from the farming lobby, and protection from foreign competition has added to the artificiality of the market. At the same time, the farmer is encouraged to produce for export with the aid of export subsidies despite the known volatility of the export market.

Table 4 *Annual production and stocks of principal crops in the USA: 1973–82*

	1973	*1974*	*1975*	*1976*	*1977*	*1978*	*1979*	*1980*	*1981*	*1982*	
Corn	5671	4701	5841	6289	6505	7268	7939	6645	8202	8397	Production million bushels
	708	484	361	400	886	1111	1304	1617	1034	2286	Stocks
Wheat	1711	1782	2127	2149	2046	1776	2134	2374	2799	2809	Production million bushels
	597	340	435	666	1113	1178	924	902	989	1164	Stocks
Sorghums	996	623	754	711	781	731	809	579	879	841	Production million bushels
	73	61	35	51	91	191	160	147	109	297	Stocks
Soybeans	1548	1216	1548	1289	1762	1869	2268	1792	2000	2277	Production million bushels
	60	171	188	245	103	161	174	359	318	266	Stocks
Rice	92.8	112.4	128.4	115.6	99.2	133.2	131.9	146.2	182.7	154.2	Production million cwt
	5.1	7.8	7.1	36.9	40.5	27.4	31.6	25.7	16.5	49.0	Stocks
Sugar	6158	5753	6401	6907	6186	5692	5886	5828	6326	5986	Production 000 tonnes
(Cane + Beet)	2823	2646	2854	2856	3498	4491	5754	3701	3082	3461	Stocks
Cotton	13.0	11.5	8.3	10.6	14.4	10.9	14.6	11.1	15.6	11.9	Production million bales
	4.2	3.8	6.7	3.7	2.9	5.3	4.0	3.0	2.7	6.6	Stocks
Tobacco	1742	1990	2182	2137	1914	2025	1527	1786	2064	1982	Production million lbs
	2943	3006	3297	3540	3560	3601	3259	3260	3555	3790	Stocks

(*Source*: Statistical Abstract of the USA)

?

1 In percentage terms, which of the crops shown in Table 4 are most seriously overproduced, allowing for the fact that some stocks are essential?
2 Assemble in list form the evidence so far presented to support or deny the statement that 'the rugged individualism of the American farmer exists as a state of mind rather than a concrete reality'.

ISSUES ARISING

'The corn is as high as a elephant's eye,
And it looks like its climbing clear up to the sky'
(*Source*: Rogers and Hammerstein, *Oklahoma*)

Although the words of the song celebrate the American farmers' productive power, the application of science and technology in an open, flexible and competitive society has led to the industrialisation of agriculture involving major changes in organisation and methods of production and in the types of crops grown. This industrialisation has produced consequences that go far beyond the immediate problem of selling or storing surplus production and it raises some important issues.

INTERNAL ISSUES

1 Rural depopulation
There have been major consequences for the rural population as the small farmer and the agricultural worker have come under increasing pressure from the need for 'efficient' production. In just over 30 years, from 1950 to 1982, the number of farms decreased from 5.6 to 2.4 million under the impact of the mechanical/chemical revolution in farming and the job of the farm worker became increasingly skilled and decreasingly available. As a result, one of the greatest, and only semi-voluntary, migrations in history has taken place as the farm population has decreased from 23 to 5.6 million.

The complexity of the rural depopulation issue is well-illustrated in the case of tobacco farming where one government agency has issued a warning on all cigarette packets that 'The surgeon-general has determined that cigarette smoking is dangerous to your health', while another government agency pumps money into its growth through price supports and acreage allotments. The problem is that tobacco is a labour-intensive crop and tobacco farms are too small to grow economically other crops such as soybeans or corn, so there are few alternatives open apart from the politically contentious one of (mainly black people) leaving the land.

2 Environmental degradation
The industrialisation of agriculture has also led to increasing levels of environmental degradation, particularly soil erosion and pollution.

Concern about the soil has played an important role in American agricultural thinking since the Soil Conservation Service was established in 1936, following the tragic lessons of the Dust Bowl. However, it is estimated that 4.5 billion tonnes of topsoil are still being lost each year and there is growing concern about the cultivation of marginal and highly erodable land, especially in the western states, Alabama and Georgia, by farmers seeking to reap the benefits of agricultural subsidies.

However, it is not the activities of 'quick-buck' operators that cause most concern, it is the fact that a central feature of the new agriculture is that the farmer has become a unit of production in a complex system, and is often divorced from the immediate environment by bought inputs. The new production methods lead to serious pollution problems and the increasing use of agricultural chemicals (see Figure 6 *page 111*) is especially serious. For example, only ten per cent of nitrogen added to land actually ends up in the crop,

Print 4 *A wheat field damaged by rill erosion*

and the rest is introduced into natural environmental systems where its effects can be devastating. In addition, over-production is creating its own pollution problems since pesticides have to be used on crops in store. The use of ethylene dibromide as an insecticidal fumigant in grain silos, for example, has been banned in Britain and other countries and is now creating worries in the US where, although its direct application to soil is already banned, there are fears that food stocks may have been poisoned.

3 The drain on resources
The new production methods also drain water, soil and energy resources to an alarming degree and a third issue has been raised about the real *efficiency* of US agriculture. Great stress is laid on the productivity of American farm workers and there has indeed been a marked increase in labour productivity per man and per hour resulting in vast increases in production — but this has been at the expense of great increases in non-labour inputs. Although mechanisation has released millions of acres of land previously used to feed horses and mules, the resources then consumed were renewable, but agriculture now uses more non-renewable petroleum than any other single industry in the USA and the farmer is increasingly buying-in services that were formerly tackled by the farm's own workforce. Although it is difficult to measure, US agriculture is not very energy efficient compared with other countries. Thus the energy–output/energy–input ratio for US corn is 2.9 compared with 10.7 and 6.4 for traditional hand-production methods in Mexico and Nigeria respectively. For rice, the figure is 1.6 in California compared with 3.3 in the Philippines and 2.5 in Japan. For peanuts it is 1.4 in Georgia compared with 2.6 in Thailand. For crops such as apples, oranges, spinach, tomatoes and sprouts, more energy is consumed in their production than the food can replace.

4 The political weight of farmers
As in the EEC, there is increasing concern over the power of the farming lobby and the seemingly bottomless demands of farmers on taxpayers' money to provide the array of subsidies, supports and incentive payments which maintain farm incomes. In a country which promotes free markets as the ideal, the artificial economic world of the farmer is a curious anachronism, and can be explained only in political terms. Although farmers constitute only 4 per cent of the population, they are found in every state and therefore every senator is involved to a greater or lesser degree with farm policy matters, giving the farm lobby a powerful lever.

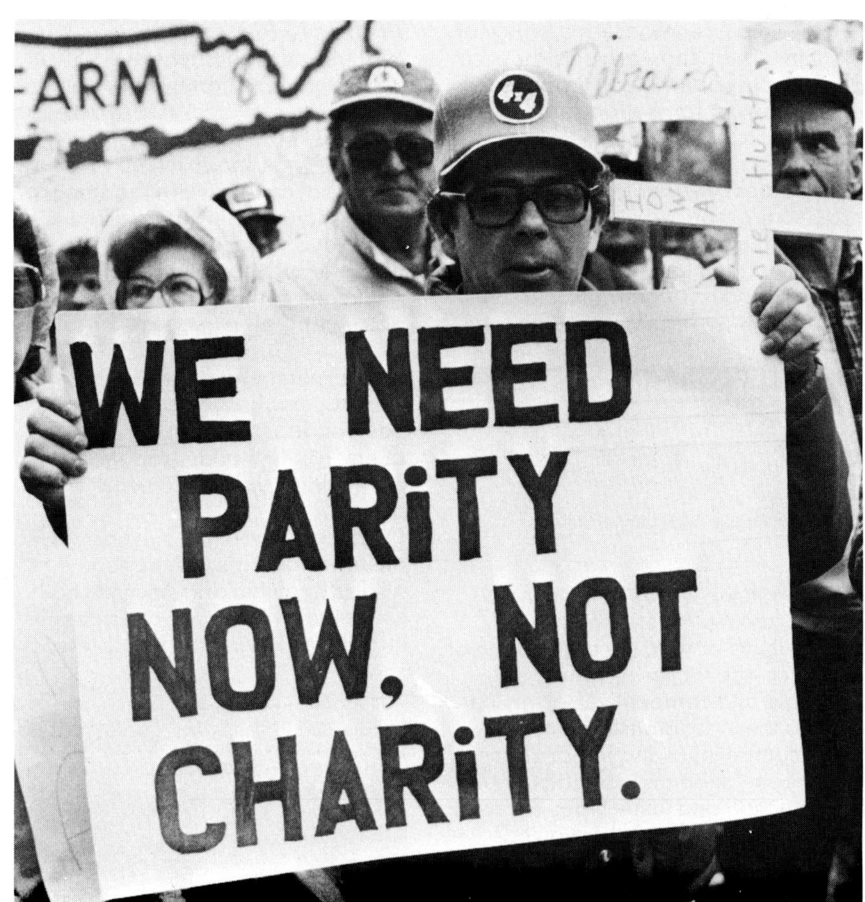

Print 5 *March 1985 — Angry farmers converged on Washington to demonstrate their plight and seek assistance from the Federal Government*

? **1** In a group, debate the extent to which American farming can be considered to be efficient, given its dependence on the large-scale consumption of non-renewable resources.
2 What are the essential differences between production and productivity?

Upon his re-election in November 1984, President Reagan announced his intention to cut Federal spending, and conscious of public concern over subsidies and surpluses, included cuts in spending on agriculture. Predictably, the new farm bill of February 1985, which reduced price supports and was designed to make farming more market-orientated, faced strong opposition from the farming lobby in Congress, especially since the combination of glut, low prices and high indebtedness among farmers threatened that many small- and medium-sized family farms would go out of business without continuing Federal supports. However, it is some measure of the role of government in American farming that even after the proposed cuts, it was intended that direct farm supports should total $12 billion in 1986.

5 Hunger within the US
There is also ethical concern that, despite the general wealth and productive power of the USA, and despite the fact that billions of dollars of public money are given to farmers each year to produce food, there are serious claims of widespread hunger among the poor in the USA. In response, President Reagan set up a commission to investigate such claims in 1983, even though the food stamp programme introduced by President Johnson in 1961 as an anti-poverty measure was costing about $10 billion a year by 1982. Although it reported in 1984 that 'general claims of widespread hunger can neither be positively refuted nor definitively proved', the issue has remained a live one since the chairman announced after publication that 'we find hunger to be a real and significant problem throughout the nation'.

EXTERNAL ISSUES

1 Hunger in Third World countries
There is less doubt about the extent of hunger and malnutrition outside the USA and it might be argued that the obvious solution to the problem of US overproduction would be to use the surpluses to feed the poor and needy in the Third World where different cultural and environmental circumstances have produced even more serious imbalances in the people–environment relationship.

However, American agriculture is not organised primarily for charity and the farmer and taxpayer might justifiably see little profit in giving expensively produced food to overseas governments, some of whom may be corrupt or inefficient. Indeed, many poor people in the USA might argue that charity should begin at home.

In fact, the US government does play a major humanitarian role in alleviating world hunger but such programmes as Food for Peace often create as many problems as they solve, because the dumping of US food surpluses in underdeveloped economies simply serves to keep food prices low and thus discourages local farmers from investing in more productive processes.

It is not only internal markets in Third World countries that might be upset. Any attempt, for example, to unload US dairy surpluses on world markets would present international complications since these dairy markets are presently dominated by New Zealand and the EEC countries whose farmers can only prosper if the markets are carefully managed.

More helpful might be the removal of tariffs and quotas on imports of tropical products such as sugar, peanuts or tobacco which can be produced more cheaply in Third World countries and would give enormous incentives to local farm economies, but the outcry that this would raise from American producers would be difficult if not impossible to overcome.

2 Sales to the Soviet Bloc
A second group of external issues arises from the sale of agricultural products to countries in the Soviet Bloc that are hostile to the USA. Many Americans see little sense in spending vast amounts on defence when it would be much simpler to wait for the chronically inefficient communist economic system to starve itself into extinction.

? Read Frederick Forsyth's novel *The Devil's Alternative* (1979) for a fictionalised account of why this may not be a wise policy.

Moreover, although the Soviets have great difficulty in feeding themselves, capitalism needs markets to sell in and grain sales to the USSR are very popular among producers in the Midwest who, on any other issue, are markedly anti-Russian. In 1972, the USSR bought up huge amounts of subsidised grain and was a major market for American producers until President Carter imposed his partial ban in 1980 as a reprisal for the invasion of Afghanistan. New negotiations were suspended in 1981 when martial law was declared in Poland but despite all these difficulties, President Reagan concluded a new five year agreement in 1983 under pressure from the farm lobby which was concerned about the problem of its massive stocks. Perhaps Lenin was right when he observed that 'the capitalists will sell us the rope with which we will hang them'.

FARMERS AT FORUM RESPOND WARMLY TO MCGOVERN

Ames, Iowa, January 21

If applause could be translated into votes, former Senator George McGovern would have got the largest share today from about 2500 farmers gathered from about a dozen states to hear pledges from six candidates for the Democratic Presidential nomination on measures they might support to help repair a troubled farm economy.

Among the six, it was the South Dakotan who appeared to say what the farmers, gathered for the forum at Iowa State University, most wanted to hear. Mr McGovern promised to support farm prices and income at sharply higher levels, though Senator Gary Hart vied with him for attention on the income-support promises.

'What's wrong with going from $3.50 a bushel to $6 for our wheat in commercial sales in international markets?' Mr McGovern asked. The cheers were thunderous.

But the audience was not stingy with applause either for former Vice President Walter F. Mondale, who is leading the Democratic candidates in the polls, or any of the other candidates here.

Whether the responses would equally please consumers or other farmers who hear a later statewide televised broadcast of the session was a question that could not be answered quickly.

(*Source: The New York Times*, 22 January 1984)

? The Secretary of Agriculture, who is a member of the President's Cabinet, is to appear on a discussion programme on national television partly to answer Mr Mondale's denunciation of the Reagan Administration made at the Iowa forum that 'this crowd has managed to hurt farmers and raise the cost of farm programmes to $22 billion at the same time' and partly to advance the President's case and claim for re-election. Put yourself in his place and answer the following questions on his behalf.

From a wheat farmer in Kansas:
'I have borrowed money on the strength of the value of my land in order to make my farm efficient and I currently owe the bank over $400 000. Last year my income was $60 000 but with the prime interest rate over 11 per cent and the present depressed state of the market I can't sell my wheat at a profit. Why don't you adopt the system that they have in the EEC's Common Agricultural Policy whereby the government sets guaranteed prices for the crops we produce that are high enough to give us a decent living?'

Reply . . .

From a cattle rancher in Texas:
'I agree with my friend from Kansas that we need to make sure that prices are high enough to make farming strong and profitable but you must take into account the fact that I need feed-grain prices to be kept as low as possible to enable me to produce beef without my costs rising too steeply. How will you keep my feed-grain prices down?'

Reply (remembering Vice-President Hubert Humphrey's advice to aspiring politicians, 'never answer a question from a farmer') . . .

From an office worker in St Louis:
'In 1983 I earned $30 000 and lost over a fifth of it to the taxman. I have just read in the newspapers that in 1983 farmers cost the taxpayer $22 billions in direct payments and that total farm support to our 2 million farmers added up to $60 billion in one form or another. I want you to tell me why I should give my money to people who are richer than I am, and what you are going to do to make farmers stand on their own feet and to stop them demanding and getting massive government hand-outs?'

Reply . . .

From a member of the right-wing John Birch Society in California:
'The speaker from Missouri should remember that we must pay our taxes to keep America strong and free, but I cannot understand why the government spends billions of dollars on arms to defend this country and then sells cheap food to our enemies. Why don't you just ban all trade with Russia and so destroy the Communist threat for ever?'

Reply . . .

From an environmentalist in Oregon:
'Every year our country becomes a more dangerous place to live in but the main threat doesn't come from the Communists but from those damn-fool farmers who are intent on turning America into a chemical desert. Why don't you ban the use of all artificial fertilisers and pesticides and go back to natural farming before they kill everything that moves in the countryside, including you and me?

Reply . . .

From a Third World visitor to New York City:
'Would you not agree that the best way to help the poor people of the world would be to send us technical aid and equipment so that we can improve our agriculture and increase our production, and then provide us with an opportunity to sell in your rich market by removing the tariff and quota barriers that you have erected against our produce?'

Reply . . .

7 GOVERNMENT POLICY AND THE RURAL ENVIRONMENT: MEXICO

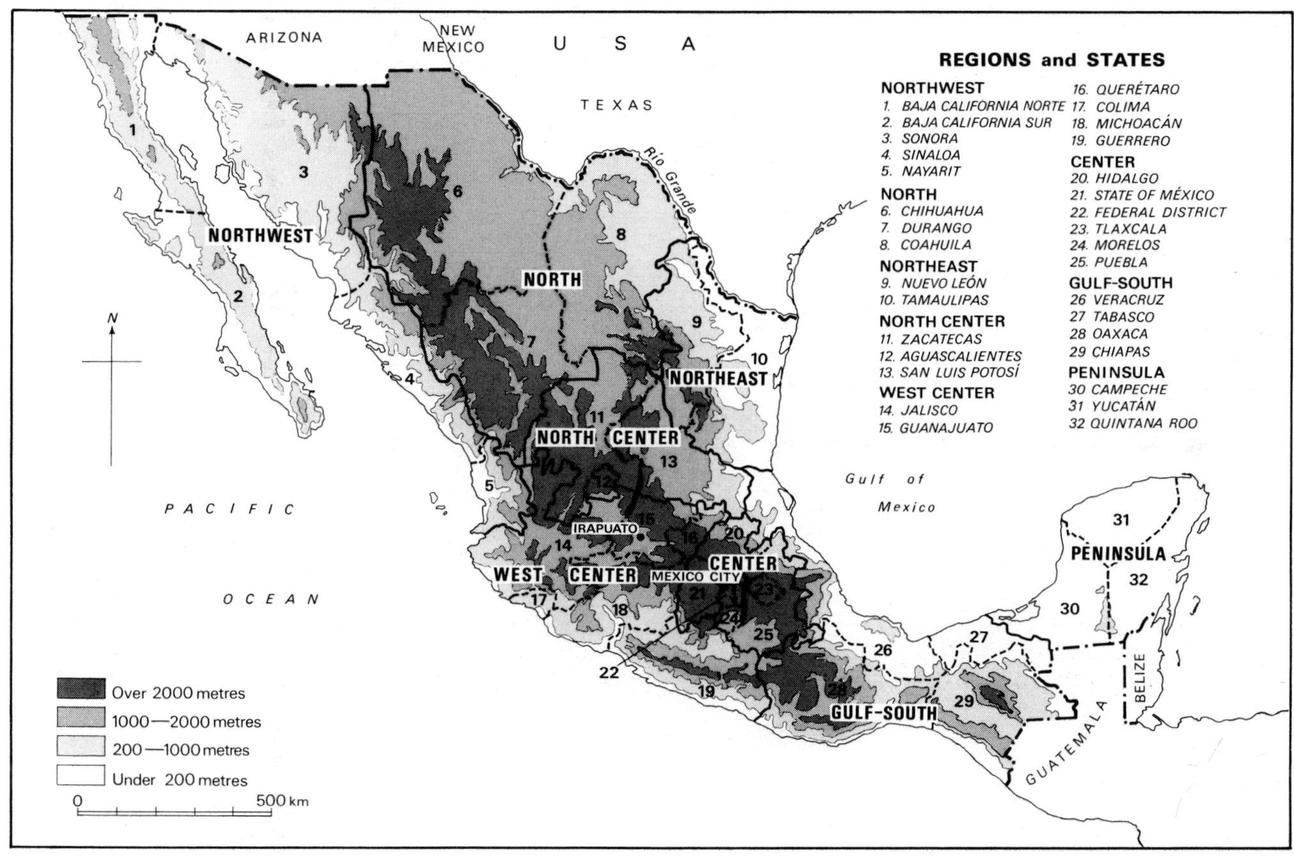

ARIZONA
NEW MEXICO
U S A
TEXAS
Rio Grande
REGIONS and STATES
NORTHWEST
1. BAJA CALIFORNIA NORTE
2. BAJA CALIFORNIA SUR
3. SONORA
4. SINALOA
5. NAYARIT
NORTH
6. CHIHUAHUA
7. DURANGO
8. COAHUILA
NORTHEAST
9. NUEVO LEÓN
10. TAMAULIPAS
NORTH CENTER
11. ZACATECAS
12. AGUASCALIENTES
13. SAN LUIS POTOSÍ
WEST CENTER
14. JALISCO
15. GUANAJUATO
16. QUERÉTARO
17. COLIMA
18. MICHOACÁN
19. GUERRERO
CENTER
20. HIDALGO
21. STATE OF MÉXICO
22. FEDERAL DISTRICT
23. TLAXCALA
24. MORELOS
25. PUEBLA
GULF-SOUTH
26 VERACRUZ
27 TABASCO
28 OAXACA
29 CHIAPAS
PENINSULA
30 CAMPECHE
31 YUCATÁN
32 QUINTANA ROO
NORTHWEST
NORTH
NORTHEAST
NORTH CENTER
WEST CENTER
CENTER
GULF-SOUTH
PENINSULA
IRAPUATO
MEXICO CITY
Gulf of Mexico
PACIFIC OCEAN
GUATEMALA
BELIZE
N
Over 2000 metres
1000—2000 metres
200—1000 metres
Under 200 metres
0 500 km

SAM: A FOOD POLICY FOR MEXICO

This chapter analyses the progress of rural development in one Latin American country, Mexico, by concentrating attention on two closely related policy areas: food policy and the rural environment.

It analyses a new strategy towards food and the environment, which was adopted by the Mexican government in May 1980. This strategy was called the *Sistema Alimentario Mexicano* (Mexican Food System), generally referred to by its acronym, SAM. It was intended to return the direction of Mexico's rural development towards greater self-sufficiency in basic foods and the better conservation of rural resources, and in turn, to help the problems of rural poverty in many areas of the country. This is how the Mexican President, José López Portillo announced the forthcoming policy in March 1980:

'SAM provides Mexico with an opportunity, perhaps unique and unrepeatable, to satisfy our great potential for growth of the food system without making unnecessary concessions to sovereignty, without being strangled by external attachments of financial servitude.'

SAM Two Years Later

Has it Moved Mexico Closer to Self-Sufficiency?

By ANNIE O'CONNOR

In 1979, Mexico imported more than a third of its basic staples from the United States, prompting one agricultural economist to observe, 'One out of every three tortillas eaten in Mexico is made with US corn.'

A year later, many considered the goal of self-sufficiency a bureaucratic daydream when it was proposed by José López Portillo. Launching the *Sistema Alimentario Mexicano* (SAM) in 1980, López Portillo stated that 'Mexico should not give away the independence and power its petroleum reserves have given it by importing large quantities of food.'

SAM was designed as a group of consultants who, with *CONASUPO*, the Ministry of Agriculture, and the National Rural Credit Bank, work as a central planning body for Mexico's agricultural system. Most have excellent credentials; Luiselli, the programme coordinator, has a Ph.D. in agronomy from the University of Wisconsin.

And, successful results are important for SAM, since the programme is not cheap. The SAM bureaucracy alone costs taxpayers $15 million dollars, and together with related government service agencies amounts to about $568 million. In 1982, officials say, the programme will consume $758 million of the federal budget.

But SAM officials, confident of Mexico's ability to meet its basic grain needs without imports, feel that the programme has justified its expense.

The 1981 harvest was the greatest in Mexico's history, according to SAM figures. Of the 14.7 million hectares sown, half was in corn, filling a long neglected gap in domestic food production.

Mexico is now reportedly self-sufficient in the major national staples: corn, beans, and rice.

As Luiselli says, Mexico hasn't always been dependent on the US for domestic food needs.

'Mexico has the capacity to feed itself. Over the years, we have achieved self-sufficiency, lost it, and regained it again, according to the policies of each administration.'

SAM works with both small farmers and agribusinesses, attempting to structure its credit and pricing policies to meet the needs of both.

Basic grains, for instance have, since the late 1950s only been grown by Mexico's small farmers. Due to government imposed price ceilings, crops like corn and beans were not profitable enough for large-scale farmers to produce. But in 1981, through SAM recommendations, the price of corn went up 31 per cent, and beans rose 25 per cent. The raises, said Luiselli, encouraged farmers to raise production in addition to raising the standard of living for small farmers.

A number of SAM's new strategies are directed at the small farmers, known as *ejidos* in Mexico.

One is the new credit card, which is intended to cut down on bureaucracy as well as give the farmers a wider choice of suppliers. Another is better credit terms. Credit can now be obtained at a three per cent interest rate from a new rural bank, freeing farmer's dependence on local credit suppliers, who often ask exorbitant interest rates.

Luiselli admits that crop insurance, which is a prerequisite to the purchase of any credit deal, is still expensive and therefore inaccessible to many small farmers. 'We still have a long way to go,' he said.

Another improvement aimed at small farmers is a shared risk programme, where the government will agree on a probable production and profit level with the farmer, and then split the losses if crop damage sends the actual production level below the mark.

'The idea,' Luiselli said, 'is to make sure that higher production levels result in increased social benefits for the producers themselves.'

Luiselli maintains that a large reason for the loss of Mexico's self-sufficiency in basic foods was the growing demand in Mexico during the late 1960s for processed foods and meat products, which take large amounts of grain to produce, and only feed 'members of a certain social class.'

Another factor, Luiselli said, was the intensive programme of import substitution begun in the 1950s and financed by Mexico's growing agricultural export sector.

Luiselli admits that the SAM plan often runs into problems, though. One of these problems is corruption. Reportedly, some local CONASUPO officials have refused to buy farmers' products. The farmers are then forced to sell to *caciques* — local political leaders in rural areas — who buy the products at well below the market price, and sell it to the CONASUPO dealers. Both parties then split the profit.

'There is a CIA in the Mexican countryside,' Luiselli said. 'The *cacique*, the intermediary merchant, and the *agiotista* (or credit shark).'

SAM has little control over most of these problems, since they occur where much of the production takes place — in isolated rural areas. In addition, many practices — like *caciquismo* — have existed since pre-Colonial times and are thoroughly ingrained in the social system.

'In the Mexican countryside,' admits Luiselli, '*caciques* are the natural political leaders.'

Nevertheless, Luiselli believes, SAM can solve social problems as well.

'They cannot be solved overnight, but through better rural organisation, we can eventually combat them.'

(*Source*: *The News*, Mexico City, 20 July 1982)

Ejido — communally owned land
CONASUPO — the government's food distribution organisation
Sexenio — the six-year term of a government
Cacique — literally, 'coyote' (local political leaders in rural areas)
Campesino — peasant farmer
Campesinistas — pro-peasant political group

? Read through the newspaper extract to get a broad idea of what SAM is about. Jot down notes under the following headings: The purpose of SAM; How Sam is supposed to operate; Potential problems with SAM.

MEXICO'S RURAL SECTOR

The fact that such a policy was launched at all raises some leading questions about Mexican rural development. The Mexican Revolution of 1910, led by Emiliano Zapata and Pancho Villa, succeeded in giving land to the peasants, so why was it necessary, seventy years later, to stimulate basic food production? What had gone wrong with Mexican development? To find the answer to this question we need, in the next section, to undertake a survey of rural Mexico at the beginning of the 1980s.

? Look at Tables 1–5 and Figures 1 and 2 and find answers to the following questions:

1 What proportion of the cultivated land is irrigated? Where is it? What are the main crops grown there?
2 In what areas are the population densities highest?
3 What staple crops do most peasants grow? Where?
4 What are the most valuable crops in market towns?
5 How self-sufficient is Mexico in wheat and maize?
6 What percentage of land is communally owned?
7 What percentage of the rural population is dependent on wages earned away from the land?

Table 1 *Cultivated land, 1980*

	Hectares (millions)
Rain-fed areas	11 262
Irrigated areas	3 176
Total	14 438

(*Source*: Mexican Government)

Table 2 *Principal crops, 1980*

	Percentage distribution of cultivated area
Food staples (corn, wheat, beans)	59% (20% of total value)
Forage crops (sorghum, alfalfa etc.)	11% (16% of total value)
Fruit and vegetables	7% (20% of total value)
Others (coffee, tobacco, oil seeds etc.)	23% (44% of total value)

(*Source*: Mexican Government)

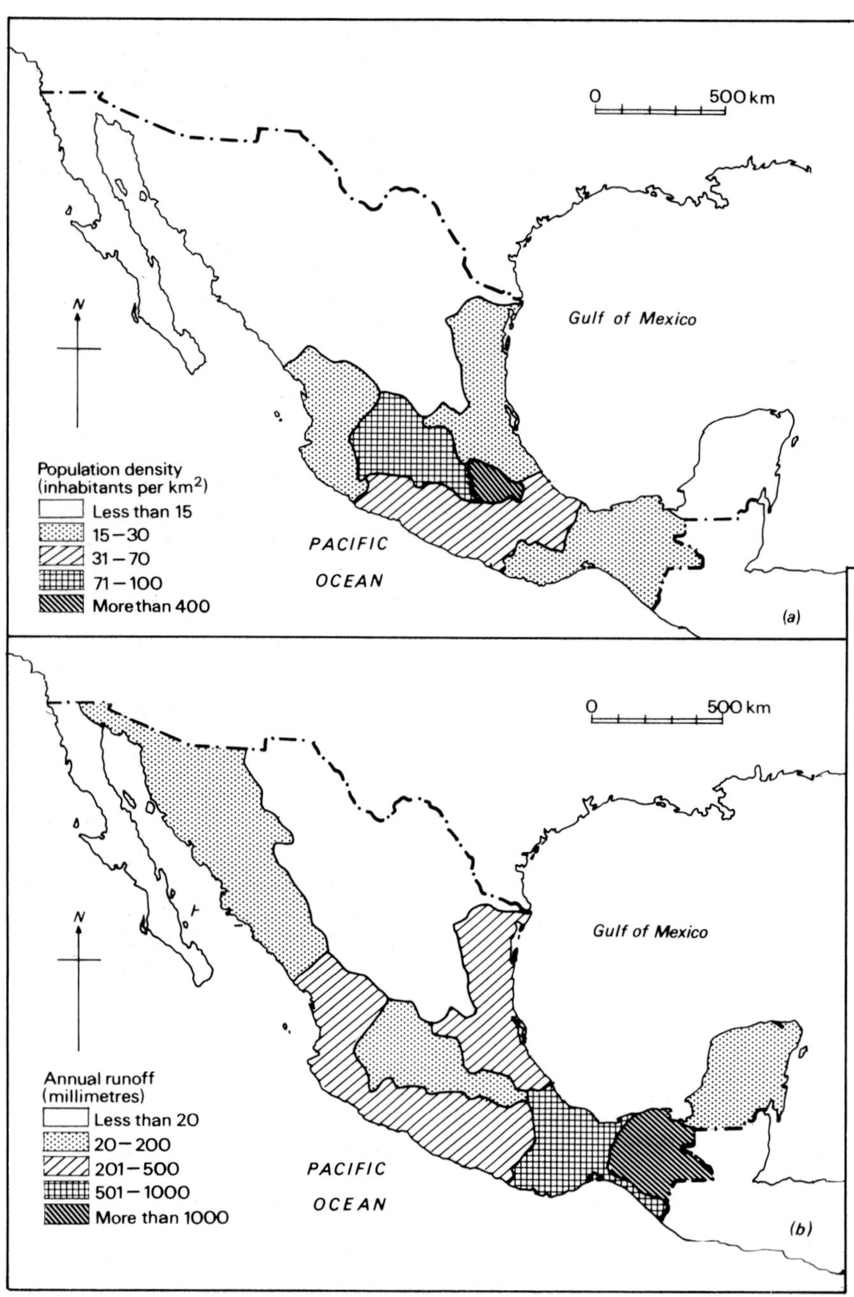

Figure 1 *A basic problem confronting agricultural planners in Mexico is that the distribution of the population (map a) does not coincide with the distribution of the water resources (map b). Thus most of the people (and most of the farms) are concentrated in the drier areas of the country, particularly in the central highlands, which have more than half of the population but only about 10 per cent of the water resources. In contrast, approximately 40 per cent of the country's available water supply is in the humid south-eastern region, where only about 8 per cent of the people live*

The rest of this section puts the figures in Tables 1–5 into an overall context. It is important to work out who is benefiting from changes in Mexican agriculture; who is gaining from increased production and who has obtained most of the credit given by the government under the SAM policy.

THE PHYSICAL ENVIRONMENT

Mexico may be divided into three ecological zones (see Figure 2) in which food and environmental policy may have very different effects.

The highland rain-fed areas

The bulk of the rural population inhabit the central plateau and mountain area, even though these areas are prone to drought and unreliable rainfall. In these regions, over 11 million hectares of land are

Table 3 *Annual rates of growth in foodcrops and forage crops, 1965–1980*

Basic foodcrops	*%*	*Forage crops*	*%*
Maize	1.75	Alfalfa	5.5
Beans	6.15	Oats	26.5
Wheat	2.30	Barley	18.8
		Sorghum	15.0

(*Source*: D. Barkin, 1981, *El uso de la tierra agrícola en México*, Working Papers in US–Mexican Studies no. 17, University of California)

Table 4 *Maize and wheat imports as a percentage of total consumption*

	Maize	*Wheat*
1960	0.5%	—
1970	8.6%	—
1980	34.2%	18.1%

(*Source*: *CONASUPO*, Mexico 1981)

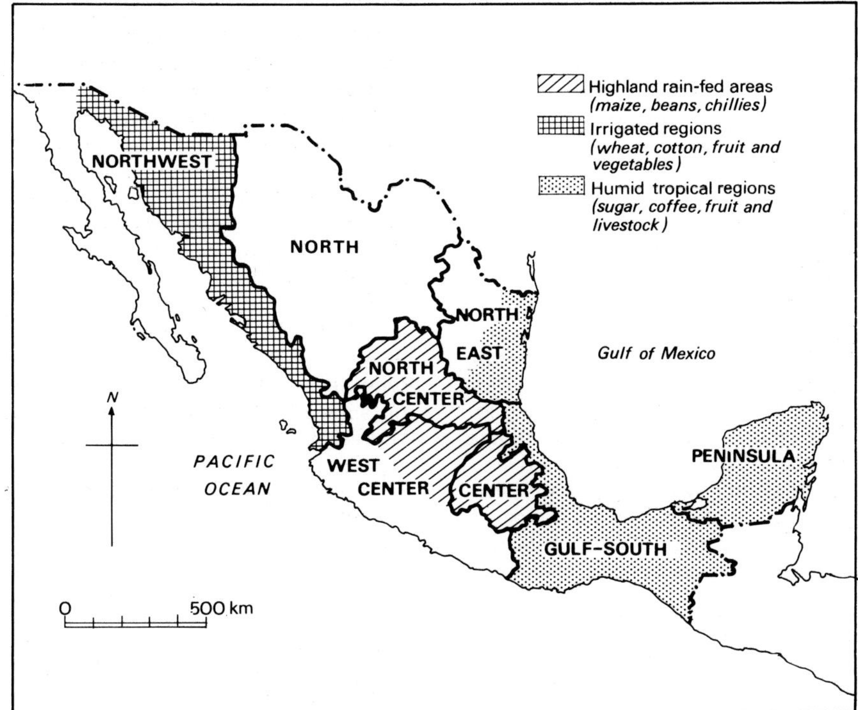

Figure 2 *Map showing the highland, rain-fed areas, the humid tropical regions and the irrigated regions in Mexico*

Table 5

Land tenure	*% of land area*
Ejidos	43
Independent communities	7
Private farms	50
	Number of families 1980
Ejidos	1 600 000
Independent communities (peasant farmer 1)	200 000
Private farmers	600 000
Hired workers, seasonal and full-time	900 000
Total	3 300 000

(*Sources*: Mexican Government 1971 and Yates 1981)

worked by peasant producers and their families. They grow maize, beans and chillies.

The irrigated areas

The three million or so hectares that are irrigated are located primarily in the north-west and north of the country. These areas have benefited from the 'Green Revolution' in wheat production, by using chemical fertilisers and new seed varieties to give increased yields. They also grow cotton and fruit and vegetables for export (mainly to the United States). However, as Figure 1 illustrates, these are not densely populated areas.

The humid tropical regions

These are in the south and south-east and include Yucatán (see location map at beginning of chapter) where the forests are being cleared for cattle ranching. Conflict between peasants and ranchers is acute. The rainfall is reliable, but few people live in these areas. Where agriculture is practised, the people grow rice, sugar-cane and bananas.

THE CROPS

For most peasant producers, the staple crops are maize and beans, and Table 2 shows that these crops, together with wheat (which is grown principally in the irrigated areas) take up almost 60 per cent of the cultivated land area. But, during the last twenty years, and particualrly during the last decade, other crops have assumed importance. In 1980 forage crops, for feeding to animals, covered 11 per cent of the cultivated land area and accounted for 16 per cent of crop values. Other crops, including fruit and vegetables, coffee and tobacco account for 30 per cent of the cultivated land area and almost two-thirds of the market value of all the crops grown. The growth in forage crop production during 1965–1980 is shown in Table 3.

During the 1970s Mexican agriculture showed a decline in basic foodcrops, largely at the expense of crops for animal feeds and for export to the United States. Animal products are consumed only by a minority of relatively affluent people in Mexico, and much of the fruit and vegetables grown are sent directly to California, Texas and other places during the winter period when production falls off there. Cash crops like coffee and tobacco are important income earners for Mexico, but the benefit they provide in terms of the balance of payments is not passed on to the rural poor. As Table 4 shows, in 1980 Mexico had reduced its self-sufficiency in basic foodcrops to a point where a third of its requirements were being met by imports. Even wheat was being imported on a large scale, despite the 'miracle' Green Revolution seeds being used in the irrigated north-west. At the same time Mexico was feeding its middle class rather better, exporting fruit and 'winter vegetables' very successfully, and raising more cattle for the hamburger chains in North America.

THE PEOPLE

Table 5 indicates that over 40 per cent of the cultivated land area in Mexico was worked as *ejidos*, or communally owned land. The *ejido* was introduced after the Revolution to provide a land base for the peasantry, which had largely been dispossessed of land by large farmers and speculators, many of them foreign. *Ejidos* are rarely collectively worked, and most of the land is divided into small plots worked by individual families. During the 1930s a collectivisation programme was begun under President Cárdenas but these policies were reversed by Cárdenas' successors. Today, most peasant farmers on *ejido* land need to earn income from other sources to sustain their families. Consequently, they migrate regularly across the border to the United States, or to other parts of Mexico for seasonal employment in agriculture. Many *ejido* peasants maintain close links with urban centres, where they or their children find casual work for part of the year.

The heading 'private farmers' in Table 5 covers a variety of farming classes. Some work small- or medium-size farms — Mexicans call them *rancheros*. But, the most important in this group are the large commercial farmers, some of them also government officials or beneficiaries of the 1910 Revolution in its earlier phases. The interests of this group are closely and inextricably linked with export production and the United States. Crops such as strawberries and citrus fruits are grown specifically for American canneries and we shall be looking further at the effect this has on Mexican agriculture.

The number of hired workers is probably much greater than suggested by the figures in Table 5. Many *ejido* peasants work away from their land plots. The agricultural census figures also under-represent women and children, who play a large part in seasonal employment, picking coffee and tobacco, cotton and fruit. Women and children are the poorest and most exploited section of Mexico's rural population.

MONEY AND THE PEASANT FARMER BEFORE 1980

At least half of Mexico's rural population is dependent on wages, often earned at some distance from home. The 1910 Revolution that promised land to those who worked it, had succeeded by 1980 in removing a significant section from the land altogether. Those who did work the land were very unlikely to receive outside technical assistance, credit for buying fertilisers and seeds or help with marketing their produce. These important rights were in the power of intermediaries and rural bosses called *caciques* or *coyotes*.

Borrowing money and obtaining credit are of critical importance to a peasant farmer. Three-fifths of the *ejidos* in Mexico received no official government credit in 1980, yet they constituted about 71 per cent of the cultivated land. The Mexican Government's 'National Rural Credit Bank', BANRURAL, had distributed over one billion US dollars' worth of agricultural credit in 1979, an increase of 32 per cent over 1978. Since most of BANRURAL's credit went to smaller producers (the private banks look after the bigger farmers) the bulk of official credit must therefore have gone to a minority of relatively rich peasant farmers. Improving the situation of the poor farmers was one of the goals of SAM.

MEXICO'S DEPENDENCE ON THE UNITED STATES

Mexicans have an expression about what it feels like to live 'cheek by jowl' with the United States: 'Poor Mexico! So near the United States and so far from God!' To understand Mexican feelings about their giant neighbour to the north one should recall that most of the south-west USA and California once belonged to Mexico. Today, Mexicans feel, it is they who are the 'junior partner', valued more for the labour they provide than their culture or distinct identity.

The majority of Mexicans, including those who migrate to the USA, are very much influenced by North American capitalism. They pick fruit and vegetables for agribusiness firms, many of which are American-owned, see extract on 'Canned Imperialism'. They also work in the border industrial plants, or 'component assembly shops', which were established near the US/Mexican border with support from the Mexican Government. The idea behind the border industries was that employment would be created for people and would persuade them not to migrate to the United States.

In fact, the border plants employ young Mexican women, who assemble components for electrical equipment imported from the United States which are then re-exported to the States and sold. Not only has male unemployment remained untouched, but female labour is often grossly exploited in these plants. Furthermore, there is no evidence that the border industries help Mexico to accumulate capital of its own for development. They are mainly a means through which US corporations can utilise cheap and docile Mexican labour.

Many Mexicans also spend all or part of their lives in the United States. The majority go there as migrant workers, or 'wetbacks', as they used to be called (they reached the USA by swimming across the Rio Grande). They work in California and Texas, picking fruit and vegetables. Many have settled in these states; some have moved into higher status occupations. It has been calculated that by 1990 over half the population of California will be of Mexican descent, including *Chicanos* (naturalised US citizens), permanent and seasonal migrants.

Another dimension of Mexico's dependence on the United States is foreign trade. Table 6 shows that over half Mexico's import/export trade is with the USA. In this area the USA is much more important to Mexico than Mexico is to the United States, whose trade is shared among a number of major trading partners. Increasingly, Mexico exports oil to the United States in return for machinery and high technology goods (see Tables 7 and 8). Food tastes, like so much else in Mexico, are dominated by North American

Print 1 *A tale of two civilisations — Mexico to the south, a developing world hispanic nation with 30 per cent unemployment, high birthrate and widespread poverty. To the north, the dream of many Mexicans; America the beautiful, the prosperous, cruising into the post-industrial age . . . in between, the 'tortilla curtain' or the border, America's answer to illegal Mexican immigration, which has increased drastically since the seventies, and continues to trickle steadily into Texas, New Mexico and California . . . The ocean-to-ocean manmade barrier is patrolled night and day. Anything to prevent the 'wetbacks' — guided across the border at a stiff price by 'coyotes' — from coming to the USA. These* Chicanos *in California are the lucky ones for they have managed to secure work permits and become naturalised US citizens*

Table 6 *Direction of Mexico's foreign trade 1980**

	Exports US $ million	*Imports US $ million*
USA	825.0	793.0
Spain	63.0	59.7
Israel	52.0	57.4
France	21.8	29.1
Japan	20.6	24.3
Puerto Rico	19.4	24.2
Brazil	18.7	20.4
West Germany	16.6	20.4
Others	410.9	197.5
Total	1448	1226

* These are monthly averages, by value, for January–March 1980. They show that 72 per cent of Mexico's exports and 65 per cent of imports are to/from the United States (*Source*: Adapted from *Quarterly Economic Review of Mexico*, The Economist Intelligence Unit, 1980)

Table 7 *Mexican exports to the USA, 1980*

	*US $ million**
Petroleum and products	491
Machinery and transport equipment	167
Foodstuffs	133
Minerals and chemicals	70
Clothing and textiles	30
Other exports	159
Total	1050

*These figures are monthly averages, by value. (*Source*: Adapted from *Quarterly Economic Review of Mexico*, The Economist Intelligence Unit, 1980)

Table 8 *Mexican imports from the USA, 1980*

	US $ million
Machinery, transport equipment and scientific instruments	556
Minerals and chemicals	189
Foodstuffs	98
Clothing and textiles	26
Other imports	266
Total	1135

companies and marketing. 'Junk Food' has arrived in Mexico, via the USA, for those who can afford it. As Mexico finds it difficult to produce enough staple foods for its own consumption, maize, wheat and beans are also imported.

FOOD FOR HOME OR ABROAD?

The following extract is from a recent book, *Agribusiness in the Americas* by Roger Burbach and Patricia Flynn. It examines how North American agribusinesses operate in Mexico and other parts of Latin America.

? Read the extract through and then, working in pairs, pretend that one of you is in Mexico making a film strip for the Del Monte Corporation. What slides would you take (limit yourself to 10 or 12) and what commentary would you provide? The other person has been commissioned by the Third World Trust to make a short documentary of Del Monte's operation in Mexico. What slides and commentary would you provide?
Set out the work by drawing a space for each slide, describing what it would show and writing the commentary underneath.

Canned Imperialism: Del Monte in Mexico

Around the world, agribusiness corporations like Del Monte are moving into regions that are struggling with the problems of malnutrition, poverty, and land distribution. The corporations usually pose as saviours, claiming that their fertilisers, tractors, hybrid seeds, and food processing plants will help solve the world's problems by expanding food production and providing employment opportunities. However, these corporations exert enormous control over the economic life of countries where they operate, and often control their most important sources of foreign exchange. They dominate vast tracts of land which are used to produce export crops, instead of being used to produce staple foods or to provide a source of income for the local population. Further, because their profits depend on control of vast landholdings and the exploitation of cheap labour, plantation operators like Del Monte must oppose progressive changes in the Third World, and consistently align themselves with the most conservative forces. A case study of the Bajio Valley in Mexico shows how Del Monte's operations merely accentuate the extremes of wealth and poverty, turn out highly processed foods that are priced beyond the reach of most of the country's population, and force peasants off their lands.

The Bajio Valley is one of Mexico's richest agricultural regions. Located 200 miles north of Mexico City in the state of Guanajuato, the valley is endowed with fertile soils and a mild climate suitable for a wide variety of crops. The valley itself is sizeable — about 100 miles in length and from 20 to 50 miles wide. Bustling cities and towns dot the Bajio, linking the region to greater Mexico and the rest of the world.

The cradle of Mexican independence, the Bajio today finds itself invaded by foreign agribusiness interests. For over a decade and half, US corporations have been at work altering the region's agriculture and integrating the valley into the network of international capitalism. Three multinational food processing corporations — Del Monte,

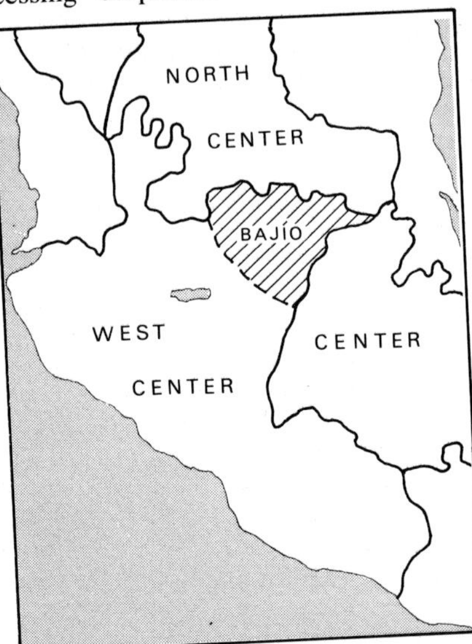

Campbell's and General Foods — operate canning and packing plants in the valley. Ford and John Deere tractors till the land, insecticides from Bayer are used to control plant diseases, and cattle are fed special formula feeds milled by Ralston Purina and Anderson Clayton. As one Del Monte vice-president noted, 'When you go to the Bajio today, it's almost like being in one of California's valleys.'

Del Monte has had a more profound impact upon the people and agriculture in the valley than any other corporation. Its factory in Irapuato at the west end of the valley employs more people than any other food processing company in the region, and the plant turns out the largest variety of fruits and vegetables of any Del Monte facility in the world. The tillers of the land as well as the factory workers have felt Del Monte's presence: agricultural techniques have been altered and crops that were never seen before in the valley are now cultivated on large tracts of land for Del Monte

Contract Farming

When Del Monte first sent its technicians to look at the Bajio in 1959, they found a region ill-suited to the needs of the world's largest canner of fruits and vegetables. Grain production predominated, with corn and beans serving as the mainstays of the local diet. In Del Monte's own words, 'vegetable production was small and limited to a few crops grown exclusively for the local fresh market.'

The Bajio's land tenure system was also incompatible with Del Monte's needs. Due to the valley's population density and the breakup of the large landed estates under Mexico's agrarian reform laws, the average landholding was small, ranging from ten to 20 acres. Some of the land was held in *ejidos*, large, communally owned farms that are subdivided into many small plots and worked by peasants, or *ejidatarios*. Mexican law prohibited the sale of these lands, and it also placed restrictions on land ownership by foreign corporations. For a company used to owning plantations and working with US growers who own hundreds or even thousands of acres, the conditions in the Bajio did not appear auspicious.

But Del Monte found the perfect tool for changing the valley's agriculture — contract farming. Under the contract system, the farmer or grower agrees to plant a set number of acres of a particular crop, and the company in return provides financial assistance which usually includes seeds and special machinery, as well as cash outlays for purchasing fertilisers and hiring farm labour. All these costs are discounted from the farmer's or grower's income when the crop is delivered to the cannery.

In Mexico, where agricultural credit is limited or non-existent, contract farming was an attractive offer. As such, it was influential in changing the structure of agriculture in the Bajio. Del Monte revealed just how influential its crop financing was when it noted that in the early 1960s, Productos Del Monte was practically the only source that many of its growers could turn to for short-term crop loans.' By skillfully using its financial leverage, Del Monte affected the valley in several ways: it introduced crops that had never been grown there, favoured the development of the larger growers at the expense of the smaller, more marginal producers, and gained operating control over large tracts of land.

From the start of each growing season, Del Monte exerts tight control over its contract growers. It specifies seed varieties and fertilisers and often supplies special planting equipment. After the land is sown, frequent visits are made by Del Monte technicians who insist that the company's irrigation and cultivation specifications be strictly followed. If these specifications are not carried out, the contract gives Del Monte the right to take over direct control of the crop. Harvest time also finds Del Monte in the field: the company maintains the largest pool of machinery in the valley, much of which is sent to harvest its crops.

Viewed from a productivity perspective, this paternalistic system of agricultural production has some positive effects. As Del Monte boasts, 'from 1962–1972 yields per acre among Productos Del Monte contract growers rose steadily,' while gross income per acre sometimes rose by 50 per cent. But these results must be placed in the larger context of agricultural changes in the valley. Although the Bajio's landowners are small by US standards, those with more land tend to be favoured by the changes in agricultural production, while the smaller producers or *ejidatarios* are increasingly marginalised.

From the start, Del Monte worked with the larger growers in the valley. In 1964 the company had contracts with 21 growers for 413 acres, or an average of 20 acres per grower. Since most of the growers contracted only a portion of their land to Del Monte, their landholdings were actually much larger than the average 10 to 20 acre farms that predominated in the valley. Since then, Del Monte has tended to work with even larger tracts of land. In 1974 the company had 110 growers with a total of 5000 acres, or an average holding of 45 acres.

Many of the reasons for working with larger growers are implicit in capitalist agriculture. For Del Monte it is easier to supervise a contract with one grower who owns 50 acres than with five growers who each own 10 acres. Ownership of capital assets also influences the company's choices. Although Del Monte provides much of the credit for each vegetable crop, anywhere from 10 to 60 per cent of the actual cash needs must come out of the grower's pocket. Del Monte also insists that the grower put up some agricultural equipment or machinery as collateral. For the small producers who rent most of their machinery, this stipulation automatically prevents them from becoming Del Monte's contract growers. As one agricultural technician in the valley noted, 'the small landowner doesn't have the economic resources to plant vegetables.'

Thus Del Monte's contract farming leads to an increasing concentration of wealth among the valley's larger growers. As the American Chamber of Commerce in Mexico noted in a recent study: 'In the Bajio, as elsewhere in the Republic, there is a sharp split between relatively prosperous commercial farming and the fragmented plots of *ejidal* land starved for investment and technology.' The growers who already have capital or land can enter into new areas of production which will further augment their wealth, while the *ejidatarios* or small producers are increasingly marginalised and often forced out of production. Agricultural experts at several government agencies in the valley noted that a large number of small landowners are already renting or selling their lands to the larger growers. The dispossessed either serve as paid farmworkers for the new owner, or search for other ways to eke out a living in a valley that is already noted for its high rate of unemployment.

Although the growers who work with Del Monte are relatively prosperous, their relationship to the company is a difficult one. Crops delivered to the factory have to be of prime quality or be subjected to price discounts. Sometimes the poor condition of the produce is Del Monte's fault: one grower lost his entire pea crop because the company failed to send out its pea harvester on time. As one grower commented, 'The company does what it wants with the contract.' Throughout the valley Del Monte has a reputation as a tightfisted and manipulative company that drives a hard bargain and takes advantage of the growers whenever it can.

The growers who survive the pressures of capitalist agriculture are driven to produce for the national and international markets instead of raising staples for local consumption. The three largest crops that Del Monte contracts for are sweet corn (1000 acres), peas (1500 acres), and asparagus (2250 acres). None of these crops figures prominently in the diet of the Mexican people. Canned peas and sweet corn are marketed as delicacies and purchased exclusively by the middle and upper classes, while over 90 per cent of asparagus is shipped abroad to markets in the industrialised countries. In 1974 alone, over $US 4 million worth of canned goods from the Irapuato plant were shipped to 20 different countries.

Del Monte's Workforce

Del Monte's employment policies at the Irapuato plant contribute to the economic and social instability of the Bajio. During the course of a year the company employs approximately 1750 workers. But only 120 are permanent workers. The remainder are seasonal employees, 90 per cent of whom work no more than four to six months of the year. Wages for these workers are the minimum required by Mexican law — 61 pesos per day in 1977, or approximately $4.90. Even people who have worked at the cannery for six years receive no more than the minimum wage.

To keep its wages low, Del Monte draws on the valley's large pool of unemployed labourers, using those who most desperately need work. The majority, around 75 per cent, are women. Some are young women, still in their teens, who come to Del Monte looking for their first job, while others are older women who need any kind of work to sustain their families. As one woman said, 'I don't work at Del Monte because I like it — I have to feed my children.' Many of the men and women who work at the plant are the sons and daughters of *ejidatarios* or small landowners who migrated to the valley's cities and towns looking for jobs that did not exist.

Because the workforce needed at the plant varies from day to day, Del Monte has a hiring hall where workers report to find out whether or not they are needed on a given day. There they often sit and wait for hours before being told if there is work. Some spend hours travelling from their homes in the countryside, only to find no work available at Del Monte. For some at the hiring hall the situation is desperate. As one woman said: 'If I don't get work today, how will I manage to eat?'

In theory the workers at Del Monte are represented by a union affiliated with the *Confederacion de Trabajadores Mexicanos*, or the CTM. But the CTM is an official government union that is known more for its efforts to quell labour militancy than for its defence of worker interests. During Del Monte's 14 year history in Irapuato, there has been only one strike, in 1969, which only lasted several hours. One woman said that a couple of years ago 'we wanted to strike for better wages, but the union leaders said no, that people would be brought in from the countryside to take our jobs.' The strike never occurred.

In addition to the 1750 cannery workers, Del Monte says it provides employment for another 3500 people in the Bajio. These are mainly the field hands who work for the company's contract growers. Some of these agricultural workers have been dispossessed of their lands, others maintain small plots that they and their families work to help sustain themselves, and still others form part of the migratory workforce that moves around the valley looking for employment. These workers, like the factory employees, can expect only seasonal employment since the asparagus harvest season lasts only about three months.

As elsewhere, advertising has been the key to Del Monte's marketing expansion in Mexico. McCann Erickson, Del Monte's US advertising agency, works closely with company executives out of its offices in Mexico City. When Del Monte entered Mexico in the early 1960s, it found that the Mexicans had little need or use for its products. The company admitted it confronted a difficult situation:

'Del Monte was an expensive brand in the minds of the consumers in the large population centers where our products could be found, and in the countryside Del Monte Brand awareness was virtually non-existent.'

But Del Monte and McCann Erickson moved aggressively to change this situation, using radio, television, billboards, magazines, and other media to create a public awareness of Del Monte products. McCann Erickson even developed a new symbol for the campaign — a talking parrot with the Del Monte emblem emblazoned across its chest. By 1968, Del Monte propaganda had made an imprint on the public mind: a survey found that 70 per cent of the Mexican people were aware of the Del Monte brand name.

Del Monte makes no secret of the fact that its primary market is not the working masses who make up the majority of Mexico's population, but the new urban middle class and the upper class. The company boasts: *'Canned foods are becoming more and more accepted, and are no longer found only in the homes of the wealthy. The middle class, developing as a result of the jobs created by companies like our own, is a fast growing consumer of our products.'*

Del Monte did not point out that of the more than 5000 people who work for the company in the Bajio, only a small fraction receive wages adequate to fulfill their minimal dietary needs, let alone purchase Del Monte canned foods.

Del Monte is only one of the many multinationals that dominate the Mexican food industry. During the past two decades, foreign food processing companies such as Kraft, General Foods, Carnation, Anderson Clayton, and Nestlé have established new plants and acquired locally owned companies. Del Monte itself has two snack food factories in Mexico, besides its plant at Irapuato. The far reaching impact of these companies in Mexico was summed up by Fernando Camora, a former director of the Economic Research Institute at the National Autonomous University of Mexico: *'The multinational food processing firms . . . act as monopolies, increasing the cost of food, . . . determining the zones of production and the types of crops, and deciding what is to be exported. They also determine what seeds, fertilisers, insecticides, and machinery should be used, and they fix the salaries of the field and factory workers. . . . In the broadest sense Mexican agriculture is victimised and controlled by the foreign firms in the food industry.'*

(*Source*: P. Burbach and P. Flynn, *Agribusiness in the Americas*, Monthly Review Press, NACLA, New York, 1980)

?

1 What do you think is meant by 'food dependency'? Would you say that Mexico was more dependent on the United States than *vice versa*?

2 Considering what you have read about 'border assembly plants' and US agribusiness in Mexico, what do you think could be done for Mexican rural areas to prevent people leaving the land in large numbers?

THE POLICIES AND OBJECTIVES OF SAM

Having looked at the state of Mexican agriculture pre-1980, we can now return to SAM. Basically, SAM was a package of related policies drawn up after initial research by over twenty committees. The proposals were intended to achieve three objectives:

1 To increase domestic production of strategically important foodcrops especially maize, beans and rice;

2 To streamline food delivery systems between the urban and rural poor;

3 To improve the nutrition of the poorest people in rural and urban areas.

Specific measures included raising the price of corn by 31 per cent and beans by 25 per cent, so that peasant farmers would be encouraged to grow and sell these crops rather than turn to other crops, or abandon the land altogether. Improved seeds would be made

Print 2 *Many Mexicans buy food from small outdoor markets. SAM directed food subsidies to such outlets in an attempt to improve the daily diet of Mexico's poor*

available to poor farmers and they were promised the free delivery of 610 000 tonnes of fertiliser at prices 20 per cent below commercial rates. Resources for combating plant diseases were to be increased and the cost of crop insurance reduced for the peasant farmer. Perhaps most important of all, agricultural credit policy was redesigned with the interests of poor maize-producing farmers in mind, freeing them from the necessity of buying through intermediaries, who took money and thus reduced the farmer's profit.

Efforts were also made to improve food delivery systems by modernising and improving retail shops owned by the Mexican Government (*CONASUPO*, see below). Sections of the food industry that collaborated with SAM were given a financial boost by the injection of state funds. Existing plans to stimulate government controlled agribusiness were published.

The nutritional elements in the policy package were in some ways the most innovative. It was calculated that about 35 million Mexicans, more than half the country's population, failed to reach *per capita* daily food intakes of 2750 calories and 80 grams of protein. Of this number, over half — 13 million in rural areas and six million in cities — were estimated to have fallen well below these 'minimum' nutritional levels. The most vulnerable were rural women and children, whose nutritional levels would be improved through subsidising a *Recommended Basic Food Basket*, reducing its cost to poor consumers to about 13 Mexican *pesos* per day per head. (This was about 26 pence in January 1980.)

SAM recognised that to reach the target population it was necessary to increase the number and efficiency of the retail outlets used by poor people, especially those of the government's food distribution organisation, *CONASUPO*. In the cities, poor people tended to use small grocery stores or public markets, many of which were mobile (*mercados sobre ruedas* — 'markets on wheels'). Food sold in such shops and markets received the highest subsidies, so directly benefiting the poor.

Figure 3 *The way the new food policy was portrayed in the Mexican press: 'Uncle' Sam at the bar with the 'Mexican' Sam*
(*Source*: Proceso, 31 March 1980)

HOW WAS SAM TO BE PAID FOR?

Mexico was in the fortunate, and rather unusual, position of having enormous petroleum reserves. By 1980 it was the world's fifth largest oil producer, and the country's potential reserves were ranked second only to Saudi Arabia's. In 1938, under President Cárdenas, the Mexican Government had nationalised the petroleum industry. In 1980 the subsidies received by SAM amounted to almost 4 billion US dollars. This could easily be financed from oil income. After all, if one million barrels of oil were sold each day (which was well within capacity), at a price of $US 40 a barrel, it would yield approximately $US 14.6 billion a year. The Mexicans, following on the heels of the Venezuelans in the early 1970s, were 'sowing their petroleum' by ploughing back oil revenues into rural development. On the face of it such a strategy had everything to recommend it. However, in the following section we shall consider the principal obstacle to its success: the role that the Mexican state had come to play in the rural sector. Mexico alone in Latin America had the means to develop its rural sector rapidly and effectively. But it had also inherited a burden of suspicion and corruption which hampered attempts to increase ordinary people's participation in the implementation of development policy.

? From what you have learned already in this chapter about the relations between those who manage government assistance and the people in the countryside, discuss the meaning of the cartoon in Figure 3. What does this tell you about the attitude of some Mexicans towards government plans?

THE ROLE OF THE MEXICAN STATE IN RURAL DEVELOPMENT

It has been suggested that the Mexican state bureaucracy might not be equipped to deal with the problems of rural development, either technically, managerially or politically. It is important to look now at the extent to which rural development policy *can* or *cannot* be implemented.

The ability and experience of Mexico's public servants is probably greater than in most other Latin

American countries. Mexico possesses many highly trained civil servants and planners, and the scope provided by the country's development has enabled many of these people to gain valuable experience of policy making and policy implementation. Food policy, in particular, had been recognised as being of critical importance by public servants before SAM was introduced.

Within government agencies, personal support from the president is vitally important and the careers of most civil servants follow that of a political patron to whom they became attached early on. The future of the civil servant or professional in the public sector is dictated by the fortunes of 'leaders'. One of the difficulties in assessing the technical ability of the Mexican state to confront development problems is that much of the effort of government employees goes into maintaining their personal position within the bureaucracy, rather than implementing agreed policy.

The degree to which a government is committed to a policy will vary widely in specific cases. In Mexico, presidents frequently support two or more teams of researchers and civil servants within the same policy area. Each team reports directly to the president and he decides which will be given the green light. Frequently the president's support for one team is conditional on its achieving certain goals, and he is liable to change allegiance midway through the six-year presidential term (*sexenio*).

In addition, the development period for government policy in Mexico is determined not by any considerations derived from the policy itself, for example, the need for speed or urgency, but by the sequence of events which make up the *sexenio*. Merilee Grindle describes this time-frame very vividly in her book about *CONASUPO*:

> The influence of the *sexenio* on *CONASUPO* was clear, and the patterns of behaviour it encouraged are repeated in hundreds of other public agencies in Mexico with predictable regularity. At the beginning of each presidential term, bureaucratic agencies are assigned leaders who must then set about learning the intricacies of their new responsibilities. Soon they begin to replace the middle and top level officials who have remained, uncertain virtually inactive, from the previous administration. At the same time, the new managers evaluate the organisations they have acquired and attempt to introduce revised policies and new programmes. This process takes time. A year or more might go by before a satisfactory team has been recruited; another six or twelve months might be devoted to study, reorganistion, and policy development. During this period, the regular functions of the organisations are reduced to a minimal level as 'old' administrators equivocate and 'new' ones acquire experience.
>
> (*Source*: Merilee Grindle, 1977, *Bureaucrats, Politicians and Peasants in Mexico*, University of California Press)

From this account we can gain some idea of the difficulty of changing the course of development policy. Not surprisingly, most professionals and administrators are at pains to demonstrate the success of a policy. This is true anywhere in the world. However, in Mexico, the opportunity to carry out policy is confined, effectively, to about one-third of each six-year presidential term. Within this restricted length of time public policy can rarely work unless the President is swayed by individuals.

As we have seen, problems within the Mexican bureaucracy make it difficult to establish whether there is a sound basis to state assisted rural development. Nowhere is this more important than in the practice of corruption, which plays an important role in the implementation of rural development.

PIDER: A CASE STUDY OF OFFICIAL PATERNALISM IN RURAL DEVELOPMENT

The Mexican government's *Programme for the Integrated Development of Rural Areas* (PIDER) started in 1973 when it covered 43 so-called 'microregions' with a total population of 2.4 million people. The fate of PIDER projects illustrates the practice of corruption and helps to explain the political difficulties of implementing rural development policies.

By 1978, *PIDER* 'microregions' covered almost one-fifth of Mexico's rural population, five million people. Within these 'microregions' *PIDER* attempted to integrate the functions of different government agencies. It was a pioneering effort, which was supported with some enthusiasm by the World Bank. The programme embraced almost any kind of project from rabbit production at the individual family level, to dairy farming units comprising over 300 milk cows. There were irrigation projects of different sizes, nutrition and preventive health courses and support for commercial fruit growing. Apart from 'integrating' what government agencies did in specific targeted areas, *PIDER* sought to gain the support of the local population for its projects. An office was established to evaluate the progress of *PIDER* projects and make recommendations for fuller public participation. However, not until the programme was well established did it become clear that local people often did not want the kind of projects favoured by the government officials.

The buildings shown in Prints 3 and 4 were completed but never used by local people. Why? First, because nobody had asked the local people what they wanted. Second, the budget for construction companies undertaking public works included bribes to public officials who ensured that the work was done. They were thus able to report to their seniors that the project had been completed, regardless of whether or not it was useful, or even being used. Villages exist throughout Mexico where drinking water systems have been built, in line with published policy, but in which the water has never been connected. The bribes or *mordidas* ('little bites') that oil the wheels of bureaucracy do not necessarily distribute resources fairly, and are often wasteful. Not surprisingly, poor rural people often treat government officials with indifference or, by emulating their behaviour, make the problems of implementing rural development even more difficult. Projects in rural Mexico are often planned so as to generate money for political clients or to pay off political patrons. They are rarely responsive to local needs or managed by local people.

Print 3 *This picture was taken near Perote, a cold region to the east of Mexico City, with few natural resources and considerable rural poverty. The grey concrete building is a potato drying shed that was never used. Local potato producers, living in the hills around, preferred to send their potatoes to Mexico City, rather than sell them to local peasant* caciques, *or bosses who operated the shed*

Print 4 *A multi-storey hospital near Perote that was also constructed as part of a PIDER project but again never used*

PEASANT ACCOUNTABILITY

The other side of the coin to the paternalistic attitudes of government and the corrupt behaviour of the officials is the behaviour of the peasantry itself. The livelihoods of many poor rural people are very precarious. Their production strategy is designed to reduce risks, rather than increase profits. This 'mentality' is logical and sensible for them, but its logic often eludes professional experts from the cities. Thus, suspicion towards government officials is often combined with mutual incomprehension. Where a peasant farmer is becoming progressively more dependent on wages than on his land, he may seek a solution to his poverty in demands for more access to land. However, those poor farmers who do have land are more likely to resist interference from the state. This is the face of peasant conservatism so often decried by middle class professionals.

As the state does little to ensure better understanding among poor landholding farmers, it is hardly surprising that the farmers themselves should be suspicious of government attempts to encourage co-operative or collectivist agriculture. The story is the same throughout rural Mexico, from the dry irrigated regions of the north-west to the tropical basins — a collusion of a minority of peasants with government, to the exclusion of the poorest and most vulnerable. The Mexican state has urged 'co-operation' upon a seemingly apathetic peasantry, which is determined, if at all possible, to work land on a family–household basis, free from government controls. Non-government organisations which have worked closely with peasant farmers in Mexico, such as the Mexican Foundation for Rural Development, draw attention to the peasant farmer's lack of 'accountability' — *'He is not used to being accountable for what he does with public money. When the benefits of development reach a rural community it is usually assumed by the peasant beneficiaries that this is a payment for their political support. It is not surprising, then, that poor rural people often regard the money they receive in agricultural credit, as payment for a favour they have performed. They refuse to be any more accountable for it than the public service bureaucrat.'*

It is, therefore, clear that we cannot separate the technical and administrative aspects from the political aspects of rural development in Mexico. It is also clear that the political system within which policy is formulated and implemented, bypasses formal structures in allocating resources — credit, fertilisers and so on. Clearly the relationship between the Mexican state and peasantry has militated *against* the creation of a social base capable of supporting and implementing SAM.

It seems likely, therefore, that the existing social and political alignments in the Mexican countryside are incompatible with the role that SAM expects of state agencies.

?

This exercise should help you to understand the complexity of the technicalities and politics involved in the administration of SAM.

1 You are a capable geographer who has been appointed by a newly formed commission on food production in Mexico to write a report for the president. He wants to know:

a The facts about the present state of rural development, e.g. Who owns the land? What products are being produced? Where? For whom? What major inequalities exist?

b The main features of a possible new initiative the government might undertake. (You will probably make recommendations similar to SAM but you may also add ideas of your own.) Justify all recommendations as economic good sense. Also, identify as precisely as you can how and for what reasons your recommendations may prove difficult to implement.

2 Now throw off the role of geographer, letting your own personal attitudes and values as a sixth-former living in a western democracy come into play. Set out what you see to be possible ways of overcoming the difficulties identified by the geographer.

3 Now look at the proposals put forward in response to **2** and decide whether they could be acceptable to a Mexican President. Why or why not?

4 In groups, discuss the meaning of the term 'double bind' and 'Catch 22'. To what extent do these phrases apply to Mexican agricultural development?

THE EFFECTS OF POLICY CHANGE ON THE RURAL ENVIRONMENT

We have seen that Mexico can be divided into three ecological zones (see Figures 1 and 2 on *pages 122–23*) and, as may be expected, changes in food and environmental policy have different effects in each zone.

THE HIGHLAND RAIN-FED AREAS

As Figure 1 shows, most rural Mexicans live in the highland areas which run through central Mexico. Only about 30 to 40 per cent of these highland, rain-fed agricultural regions receive enough rainfall (750 millimetres per annum and less than 35 per cent probability of drought) for such crops as improved varieties of maize to be grown successfully. But, if soils are good and rainfall relatively reliable, land can nevertheless have enormous production potential. This was the case in the area covered by *Plan Puebla* between 1967 and 1973, an agricultural extension project designed to help peasant farmers acquire credit with which to buy fertilisers and improve seed varieties. According to one authority, if relatively well-endowed areas received similar assistance, Mexican maize production could be trebled.

The peasant farmers who work land in favoured rain-fed regions are not all rich. But many of them can expect to receive increased attention from Mexican government agronomists and lending agencies. Urban growth ensures that the demand for basic foods such as maize, beans and chillies, will continue in Mexico, and the peasant farmer producing for the market is likely to attract more and more assistance from government and private sectors.

Most of Mexico's rural poor also live in these highland rain-fed regions. *BANRURAL* (National Rural Credit Bank) assesses the number of seriously 'under-employed' at about 40 per cent of Mexico's rural population. Many of these people still spend part of the year as migrants, in Mexican cities or across the border in the United States. While they remain in rural areas these peasant farmers are still producers of basic crops, but increasingly they consume more than they can produce themselves. The shortfall has to be made up by working for wages, or in the so-called informal sector of cities where casual employment is to be found. These people can be expected to continue to desert the land for most of the year, while retaining a nominal interest in the continuation of the *ejido* land unit.

Despite initiatives such as SAM, the future for this highland 'peasant' region is bleak. Some peasant farmers may prosper in the more resource rich zones within the region, but the mass of the population cannot make more effective use of the poor land and water.

IRRIGATED REGIONS

The irrigated regions of Mexico produce most of the country's wheat, as well as cash crops such as cotton, tobacco, fruit and vegetables. One of the principal objectives of SAM was to reverse the balance of advantages currently enjoyed by the irrigated zones. Between 1950 and 1970 more than 70 per cent of the government's expenditure on agriculture was devoted to irrigation projects. In 1979, the maintenance and extension of existing irrigation accounted for 41 per cent of the total agricultural budget, and the Mexican state is still heavily committed to supporting the irrigated zones financially, whatever new directions are being launched in food policy.

Irrigation systems suffer from major technical problems, notably salinisation, and their upkeep is expensive. To ensure reliable water supplies irrigation officials need to be bribed, and politicians tend to show favour towards the needs of the larger users. Most of the benefits go to those who can obtain water cheaply, and sell their produce through established marketing channels. So these rich farmers, some of whom work *ejido* land, have prospered.

Many of the products of the irrigated regions, where most of the government money has been spent, are destined for export to the USA or for consumption by the relatively large Mexican middle class. Since the demand for high value foodstuffs is unlikely to fall, and SAM has done nothing to reduce the investment of American owned companies in Mexican food production, there is little likelihood that government money for irrigated agriculture will be curtailed in the near future. As a major employer of labour from the poorer rain-fed regions, the land-owning classes in Mexico's irrigated regions try to ensure that little is done to threaten their supply of cheap labour. In 1976 President Echevarria surprised the country by supporting the demands of peasants in the irrigated north-west for more land. The effect of his support was to destabilise the Mexican presidency and reduce international confidence in his administration. Against this backcloth, it is extremely unlikely that land reform will be recommenced in the irrigated regions, while Mexico faces the vast foreign debt it does today. So it would seem that, despite SAM, the irrigated regions will get richer and the rain-fed areas poorer.

THE HUMID TROPICAL REGIONS

According to Cassio Luiselli, one of the architects of SAM, Mexico's 'agricultural frontier', consists of more than 11 million hectares of land which are suitable for agriculture but still remain unexploited. A more modest estimate, that of Mexico's Global Development Plan, refers to 3.3 million hectares. Most of this land lies in the humid tropics, especially the states of Tamaulipas and Veracruz on the Gulf Coast. In these states, and others in the south and south-east of Mexico, land is often devoted to cattle rearing rather than the production of basic grains.

Since 1937, between six and nine million hectares of cattle land have been 'protected' from the agrarian reform process by decree of the president. Most of this land is in the drier north of the country, but the protection also extends to these more fertile tropical regions. Here, the *campesinos* (peasant farmers) are used as a cheap means of clearing the forest before being ejected by the wealthy ranchers. Although the humid tropics is a relatively fertile region, livestock production is almost as extensive as in the arid and semi-arid north. Cattle raising is a speculative activity in the Mexican tropics, which provides easy gains at low cost, ties up large tracts of land, requires relatively little labour, and makes wasteful use of natural resources.

Resource use in the humid tropics reflects the class interests of the locally powerful ranchers and the government bureaucrats committed to managing land and water resources on a capital-intensive basis. They do not see it as an opportunity for giving land to the rural poor for crop raising. Capital-intensive agriculture of this kind ignores the experience, needs and participatory potential of poor peasant families. However, it is not clear that this policy can be avoided while financial incentives exist to encourage the rapid conversion of virgin forest into extensive grazing land. Clearly the classes which own and control natural resources in the humid tropics cannot be expected to favour different resource uses, as outlined by SAM, such as integrated farming systems. Not surprisingly therefore, SAM policies have failed to be effective in the humid tropical regions where they were, arguably, of greatest relevance.

? Read through the case studies which follow and then complete the exercises to help you understand the interests and points of view of farmers in the different regions.

1 Write notes for an article for *The Economist*, under the headline 'Local farmer outlines his plan for feasible farm policy'. Either be a local peasant farmer in a highland rain-fed area, an irrigated region or a humid tropical region.

2 Get together with people who took the same role as yourself. Go over ideas. Revise your notes. Write up the article individually.

3 After the articles have been written have a class session in which a person from each group explains the problems and dilemmas in his/her region.

HIGHLAND RAIN-FED AREA

Family: Edelberto Cases
Victoria Pasto de Cases
Three children, aged 10, 6, 3 years
Size of holding: 1.5 hectares
Crops grown: maize, fruit
Family income: Production is 1.5 tonnes of maize per hectare and 60 baskets of apples at 1 basket a tree. Although the family grow almost half of the maize they need for making their *tortillas* (maize pancakes), they still need to buy in additional maize for themselves and their livestock. Their only form of income with which to purchase this is derived from selling their fruit. They are left with the equivalent of only £23 a year cash surplus.

1 *Household consumption*

• daily maize consumption of family members	5.0 kilograms
• daily maize consumption by animals (2 pigs and 25 chickens)	3.5 kilograms
• maize grown for family members' consumption	1 500.0 kilograms
• total income (from selling fruit)	£102 per annum

2 *Family expenditure*

• additional maize bought for family members (1 602 kilos)	£ 61 per annum
• cost of fertiliser	£ 18 per annum
• total expenditure	£ 79 per annum

3 *Surplus* (income *minus* expenditure)

£102 *minus* £79	£ 23 per annum

Background information
This family has less than £2 per month to spend on food (other than maize), clothes, transport, health, education etc. Since this is not enough, most families need to earn additional income through migration to towns. This is the case of Edelberto and Victoria and their children. In the village, 87 per cent of families earn what they need from migration, often in the city of Puebla. In the city, Edelberto works in construction and in the street markets as a porter. This work is very casual and unreliable, but it does bring in more cash.

IRRIGATED REGION

Family: Juan Martinez
Maria Galvan de Martinez
Five children between the ages of 9 months and 14 years
Size of holding: 28 hectare share of 1000 hectare farm (owned jointly by 41 families)
Crops grown: cotton and wheat

Background information
About 50 per cent of the co-operative's profit from wheat is distributed to Juan Martinez's family and other families like his. The rest is set aside by the co-operative to pay for office help, community projects, new machinery, irrigation water, fertilisers, seeds and insecticides. Families like Juan's work as day labourers on the co-operative's land, receiving a wage. They have felt no real deprivation in recent years but their lifestyle and expectations have begun to outrun their income. Their central problem is increasing indebtedness to agricultural suppliers and merchants.

HUMID TROPICAL REGION

Family: Jose Piropo
Concepción Lopez de Piropo
Four children between the ages of 3 and 12 years
Size of holding: 10 hectares
Crops grown: rice

Background information:
Families like the Piropos 'settled' land in a frontier zone. Except in small areas of alluvial soils, the decline in fertility was very rapid. Average yields decreased by 60 per cent by the third year after clearing. From then on only three to four hectares were planted in each parcel and the rest left fallow. Credit was provided for the rice and other higher value crops tried, including tobacco.

But vital inputs for rice cultivation always arrive late, payments to farmers are delayed and debts have increased. Local merchants have gradually re-established their hold over marketing arrangements as the government backs out.

(Based on information derived from: (1) PIDER, 1976, Reports from State of Puebla, Mexico. (2) Cynthia Hewitt, 1976, *Modernizing Mexican Agriculture*, UNRISD, Geneva. (3) Peter Ewell and Thomas Poleman, 1980, *Uxpanapa: Agricultural Development in the Mexican Tropics*, Pergamon.)

THE LESSONS OF SAM

What happened to SAM programme once it had been officially introduced? Were its objectives achieved? What lessons can be learned?

The answers to these questions are not as simple as they may seem for a number of reasons. First, SAM was not properly evaluated in Mexico, and many of the lessons of SAM were given little publicity for political reasons. Less than two years after it had been introduced the policy was effectively abandoned, during the presidential campaign of López Portillo's successor, de la Madrid. So an objective appraisal of SAM is difficult.

INCREASED PRODUCTION?

Certainly in the first year of the programme, SAM enabled impressive strides to be made in staple food production. Maize production in Mexico rose by 19 per cent between 1980 and 1981. Production of beans and rice rose by 51 per cent and 41 per cent respectively. Most of these increases were achieved by extending the land area under cultivation, although there was also an increase in the *productivity* of each of these crops, that is, more was grown on each hectare of land. From a production standpoint, then, SAM did seem to work.

Closer scrutiny of the figures, however, should lead us to reconsider even this statement. The 1981 harvest was a particularly good one in Mexico, partly for climatic reasons (just as the 1979 harvest had been a particularly bad one). In 1982 subsidies were cut back in the agricultural sector, as Mexico began to grapple with a full-scale financial crisis. Weather conditions also contributed to a fall in the production of basic crops. According to the Ministry of Agriculture estimates, maize production fell from 14.9 million tonnes in 1981 to 12.4 million tonnes in 1982, and beans from almost 1.5 million tonnes to just over a million tonnes in the same period. Production was not very much better in 1982 than it had been in 1978, the last 'normal' harvest before SAM was announced.

We should also remember that increases in total production might hide changes in production figures between relatively rich commercial farmers and poor peasant farmers. This is what seems to have happened in the case of SAM during its only full year of subsidies. In the advanced irrigated areas (such as Sonora in the north-west) maize production increased by almost 200 per cent in the first year, while in the states of Morelos and Oaxaca, both rain-fed regions (see location map on *page 119*), maize production actually declined by one per cent in the same period. Fertiliser use increased dramatically in the irrigated regions, together with the introduction of heavily subsidised new hybrid seeds. In the poorer rain-fed states the use of both actually declined between 1980 and 1981.

Print 5 *Workers fertilising a rice field as part of a SAM project. Irrigation has also been a priority*

INCREASED GOVERNMENT SUPPORT?

These provisional figures, supplied by the Mexican government itself, are further corroborated by other observers, who witnessed SAM's introduction. For example, the policy of 'sharing risks' between peasant farmers and the Mexican government, by insuring crops against losses, was only applied to a very small proportion of the land under staple crops. It has also been suggested that SAM, far from helping to put capital into the hands of peasant producers actually helped retain it in the hands of Mexican government agencies. *BANRURAL* (National Rural Credit Bank), for example, was supposed to deliver fertilisers to producers at reduced prices under SAM. In fact it delivered them at market prices, promising to pass on the difference in price to the peasant farmers after the harvest. Even then *BANRURAL* made it clear that this 'subsidy' would not be handed over in cash. We do not know for certain how much of the subsidy reached those who needed it most — the poor peasant farmers in the rain-fed regions — but we can assume that it was very little indeed.

The more radical aspects of SAM were never implemented at all. The economic crisis of 1982 and the devaluation of the Mexican peso against the dollar effectively increased the cost of living for poorer people. The government withdrew subsidies from large numbers of foodstuffs in an attempt to arrest a further financial crash. According to a recent study 65 per cent of the Mexican population no longer eat meat and 80 per cent do not eat eggs. Some 70 per cent never eat fish, a remarkable figure in a country both devoutly Roman Catholic and possessed of several thousand miles of rich coastline. Food consumption has dropped sharply since the 1982 crisis and the Mexican Government admits that 30 million Mexicans (out of 70 million) are now seriously undernourished. Imported food cost Mexico one and a half billion US dollars in 1983.

Does this suggest that SAM was not a success, or that it was never given the *chance* of being a success? As a recent Mexican commentary on SAM argued: '. . . it was a programme to stimulate basic crops that ended by helping commercial producers capable of responding with greater agility to the opportunities of the market.' In retrospect perhaps this was inevitable, given the circumstances under which SAM was implemented. The question we might ask, then, is not whether SAM was successful but whether it could have been successful, without, at the same time, much greater changes in Mexico's economy and society. What would be necessary to make a programme like the SAM work better a second time around?

WHAT IS WRONG WITH MEXICO'S RURAL SECTOR? COULD ALTERNATIVE POLICIES BE ADOPTED?

Within Mexico, there are three main contrasting opinions about what has gone wrong and what should be done:

1 The *campesinistas* — the pro-peasant group — believe that poverty in rural Mexico is a result of the way peasant farmers have been treated by the government and by corrupt rural bosses (*caciques*). They believe that feeding the urban population requires a 'return to the land' for the peasant masses.

2 People on the left of Mexican politics believe that Mexico's economic dependency on the United States has impoverished the rural people in favour of the multinational organisations and agribusinesses.

3 The right wing view is that there are too many people with access to land, not too few. They believe that Mexico should concentrate on resources and products which make money, such as oil. In the rural sector this means producing cash crops such as animal feeds, fruit and cotton, or clearing jungle for cattle rearing (with the help of the multinational organisations and agribusinesses), so that the export of these goods brings money into the country with which to pay for imported foodstuffs if necessary. (This view is rarely voiced publicly as it goes against the 'revolutionary' ideals of giving land to the peasants.)

?

1 Which of the views is the right one in your opinion? Are they perhaps all right to some extent?

2 Discuss this issue in class, and then write down the arguments for and against each view in the form of a balance sheet, referring to specific information from this chapter.

3 See if you can devise an overall answer to Mexico's rural problems — something which the Mexican government has consistently failed to do!

The physical environment is perceived through three main 'filters' — those of values, politics and the economic system.

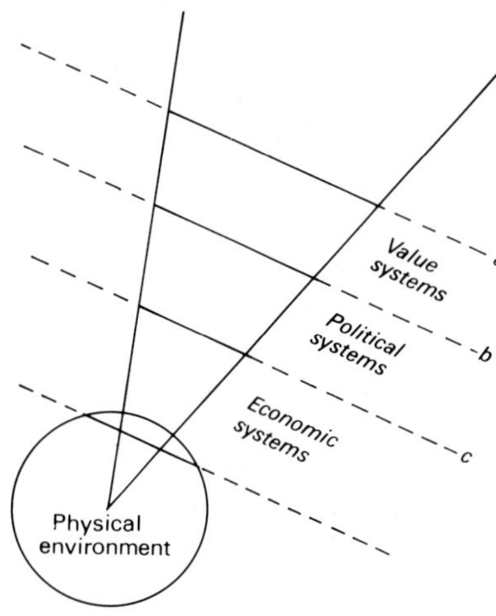

Figure 4 *A possible framework within which to evaluate Mexico's agricultural policies and development*

a The value systems in this chapter include the risk-averting behaviour of poor peasant farmers and the profit maximising behaviour of cattlemen or commercial farmers in irrigated areas. Also, the patronage that characterises personal relationships.

b Political systems which in Mexico are characterised by government bureaucracy, corruption and the bureaucracy behind rural development projects themselves, as well as the political turmoil during each six-year Presidential term.

c Economic systems include the land tenure situation, the markets governing the rural producer's crops, the labour market in which the peasant sells his labour as a migrant, and the country's dependence on the US and foreign aid.

?

If you had been presented with the ideas in SAM before reading this chapter, would you have voted 'yes' or 'no' to the programme? How would you vote now? Explain why and justify your vote.

8 PEASANTS, THE ENVIRONMENT AND THE COMMON AGRICULTURAL POLICY

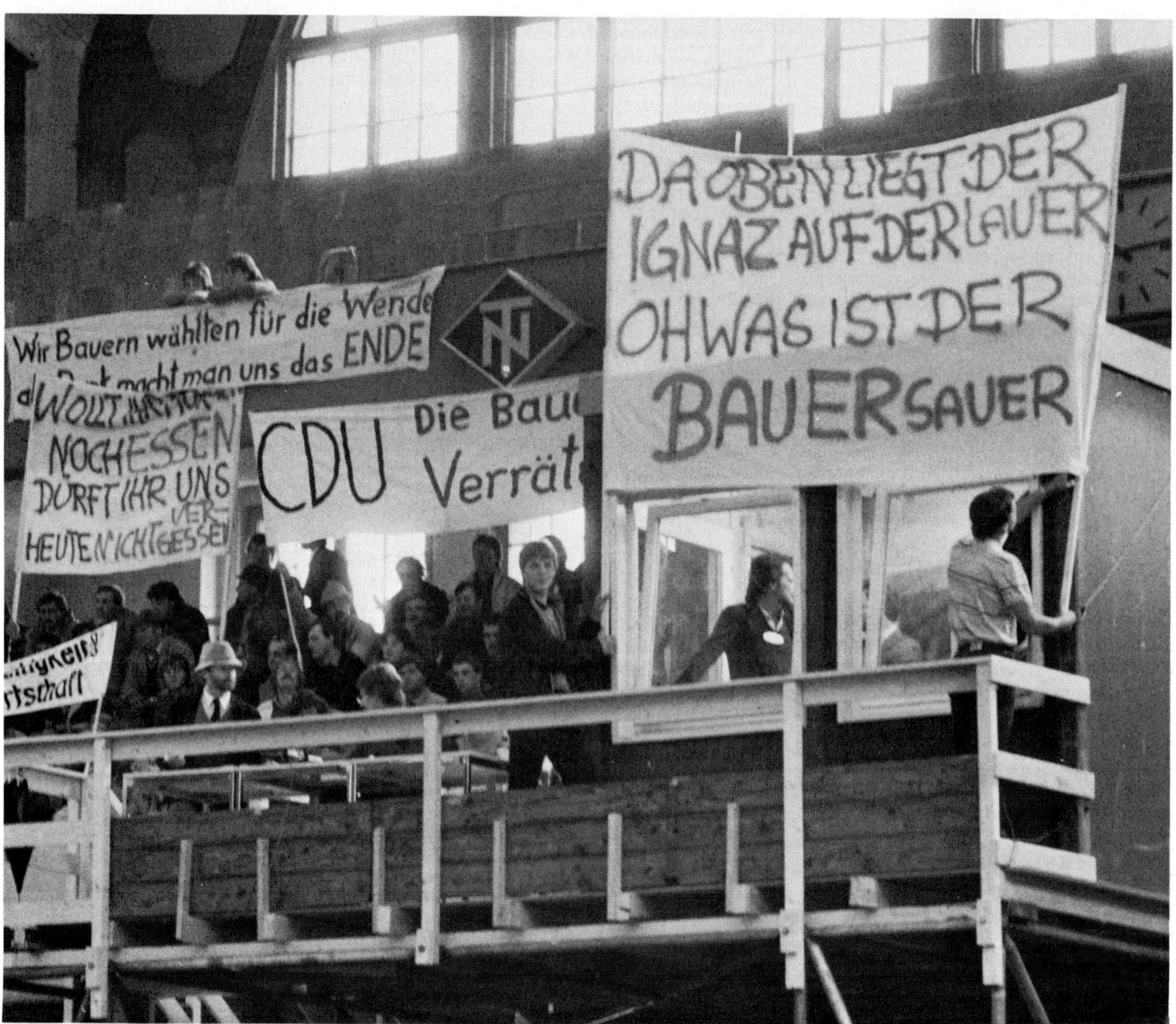

THE FASHIONING OF THE COMMON AGRICULTURAL POLICY

DOWN ON THE FARM

The common agricultural policy (CAP) is Europe's proudest achievement, and its prodigal son. By boosting output too successfully, it has generated surpluses that are costly to store and dump and that make the EEC's trading partners livid. And it now infuriates farmers too, by not propping up their incomes as much as they would like.

BE FAIR

To be fair to the CAP, it has actually achieved some of the targets laid down in the Treaty of Rome. Agricultural productivity in the Nine, thanks to new crops and techniques, has been growing at a rate several points faster than productivity in industry.

But as far as CAP spending goes, there are doubts that it has achieved one of its most important aims: to give farmers a decent income (though remember, as the saying goes, a farmer lives poor and dies rich).

Two tunes in Germany

West Germany, which in its more plaintive moments likes to describe itself as the 'milk cow of Europe', in fact does quite an efficient job milking the CAP system to its own best advantage.

Italian apprehensions

If European enthusiasm survives anywhere in the community, it is in Italy — and against considerable odds. When the EEC was founded, Italy was by far the most heavily agricultural country of the original Six, with more than a quarter of its labour force in farming at the time. But the farm deal which eventually emerged contained little to please the Italians. Under combined pressure from the northern producers, the common agricultural policy was framed to give plenty of protection to northern farm products but only slight help to Mediterranean ones.

Europe's green and expensive land

The EEC's common agricultural policy (CAP) is a spectacular example of a tail wagging a dog. The nine common market member states have 260m consumers, but only 8m farmers; and yet the CAP, still the only common policy the community has managed to devise, eats up three quarters of the EEC's budget.

The EEC and its Common Agricultural Policy (CAP) receive frequent mention in the British press as the above headlines and extracts illustrate. This chapter looks at some of the issues surrounding European agriculture and the CAP. Why was it introduced in the first place? Have subsequent modifications been for the better? How has it affected European peasant agriculture? Has it improved or reinforced rural inequalities within countries and between countries within the EEC? Should the CAP be scrapped, or modified?

The 1958 Stresa Conference

'In this auditorium there is no one whose family tree over many generations does not have its roots in a peasant family. We all know what the rural classes signify for Europe, they embody not only economic values but social and moral values too.'

These words were spoken by Dr Walter Hallstein, President of the EEC Commission, in 1958 at Stresa in Italy, when he addressed the first agricultural conference attended by the signatories of the Rome Treaty. The order of the day for the conference was to determine the aims and long-term orientation of a common agricultural policy in the six countries, France, Germany, Belgium, Italy, the Netherlands and Luxembourg.

Dr Hallstein made his assertion with some confidence. There was no British delegation present. Had there been, his statement would not have rung true. In all of the six original member countries of the EEC, however, the peasantry had remained a significant social group throughout the 19th century and well into the post-war era. One-quarter of the active labour force was still occupied in agriculture, he claimed. Even in Germany — a country with a very high level of technological and industrial development — about one-fifth of the labour force was still employed in farming, and most of it on peasant farms. The proportions were even higher for France (28 per cent) and Italy (37 per cent). Only Belgium retained a notably lower percentage (12 per cent) but, together with the Netherlands (20 per cent), Belgium distinguished herself by a much higher level of productivity per unit of land. It was in the region of twice the productivity levels recorded for France and Italy. Nevertheless, these latter two countries contained the bulk of the Six's farm population — 12 million people — with Germany adding another 4 million.

Politically, the Community was designed to end Franco-German enmity. Without prosperity there would be no peace. Hence a major

Print 1 *Delegates at the Stresa Conference, 1958*

aim of the EEC was to raise the living standards of the West Europeans: by freeing trade between member nations, thereby allowing a better and more rational distribution of the Community's productive forces; by accelerating the structural changes in employment associated with modernisation.

WHY WAS AGRICULTURE SO CENTRAL AN ISSUE?

Agriculture was inevitably involved on many counts. Firstly, the farm sector would supply much of the labour force required by the new urban industries. Secondly, with fewer people engaged in farming, agriculture would nevertheless have to meet the demand for more and better food which would arise as living standards improved. Only an agricultural industry with productive levels approaching the level of the Benelux countries would be in a position to meet the enlarged and shifting demand of a richer population that was also experiencing the consequences of the post-war baby boom. It was a series of simple but fundamental facts like these — often not appreciated by the general public — that gave farming such a central position in the negotiations leading to the establishment of the Community.

The objectives of the Common Agricultural Policy were set out in Article 39 of the Treaty of Rome:

1 To increase agricultural productivity. (This was given pride of place.)

2 To ensure a fair standard of living and an increase of individual earnings for those engaged in agriculture.

3 To provide stable markets, guaranteed regular supplies and reasonable prices to the consumer.

NATIONAL VIEWPOINTS AT STRESA

The speeches delivered by the leaders of each of the national delegations attending the Stresa Conference revealed not only the concerns and interests of each country as it contemplated the inauguration of a Common Agricultural Policy but also highlighted many problems for which no satisfactory answer has yet been found.

The Germans were acutely conscious of their dependence upon international trade, much of it with other nations and the USA in particular, who were not signatories to the Rome Treaty. Their delegate stressed that the creation of the Common Market could not be undertaken in isolation, and should not fail to take account of the relations between its individual members and the rest of the world. The CAP still remains a source of friction between the Community and third parties as the extract at the beginning of the chapter suggests. The Germans pointed out that their own agriculture had displayed a long-term rise in productivity when consumer spending on farm products was slowing down. Hence they feared the appearance of surpluses in the near future and sought to avert the problem by introducing structural reforms and regional specialisation in farming. Nevertheless, they wished to preserve the family farm as the principal element of the farm structure, for social and political reasons.

The French believed their country, with its great diversity and variety of climates, farming systems and regional economies, provided a scaled-down model of the whole Community, implying that this made their own programmes for agriculture particularly suited to the Community's needs. Their programmes had been preoccupied with two issues. Firstly, the improvement of conditions upon family farms (no reference was made to capitalistic farms). Secondly, the maintainance of stable prices and markets for the farmers, adequate provision of farm produce for the home market and industry, and the export of farm produce to assist their trade balance.

The Italian delegate emphasised the unfavourable geographical conditions under which their agriculture operated. Eighty per cent of the country consisted of arid mountains and hills with poor soils. The remaining 20 per cent, composed of plains, had been rendered fertile only through an immense application of human effort. All Italy had to compensate for a poor environment and a lack of capital was an abundance of labour. Better living standards could be achieved only if more capital were available for investment in farming and in industry — to absorb the surplus rural population — and if Italian exports from farm and factory to the Community members were to increase. The remaining members of the Community were left in no doubt regarding the financial and economic implications of Italy's membership.

The Dutch viewpoint was conditioned by their experience as an exporter of farm produce. Their delegate underlined the fact that the common interest of the Community was not the sum of their six particular interests. Somewhere along the line trade-offs were unavoidable if the Community was to be successful. As a basis for settling any differences the Dutch delegate suggested the consumer's interest should have priority. *'There is no economic justification of production for which there is no demand. But too often we reverse the principle. The farmer, the horticulturalist, the fisherman produce, and the consumer is left with the task of absorbing every bit of it.'* Unfortunately, she continued, the conference was a conference of producers. It lacked a competent, well-organised opposition made up of consumers. The value of the Community lay in the dynamics it introduced into European affairs, dynamics that would be dissipated if the Community remained an inward-looking, autarkic body, uninterested in specialisation and the international division of labour. Exporting, she concluded, is unthinkable without importing!

The Belgian delegate was preoccupied with the specific issues and difficulties of his country's agriculture. (The same was true of the speaker for Luxembourg.) Nevertheless, he emphasised the need for the community's policy to assist in the maintainance of family farming, and the need to defend the interests of a class of people whose mode of life was important for the maintainance of internal, political and social equilibrium. He acknowledged that a modernised profitable form of family farming should be the objective of policy making.

THE INITIAL REGULATIONS

Against the more liberal views of the Germans and the Dutch, those of the French were much more protectionist and for that reason (amongst others) they were opposed to the British proposals for a free trade area and to British membership of the Community.

The initial regulations of the CAP which came into operation in 1962 revealed that the French viewpoint had won.

? Read through the previous details again to clarify in your own mind the positions taken by individual countries. Take the role of the spokesperson for one country and in groups of ten (i.e. two per country) put your position and listen to that of the others. You will need to agree on debating rules, and on who is to be first and second spokesperson in each case. Next try to draw up a CAP similar to the three rules shown on *page 140*. Allow up to an hour for this. Your teacher will then carry out a debriefing exercise based on such questions as:

1 What happened in the debate?
2 What additions would have made the debate more realistic?
3 Did the outcome of the debate seem fair?
4 What ideas or hypotheses about reality did the debate suggest?
5 How close to, or distant from the actual policy was your policy? Why?

The Common Agricultural Policy (*first stage*), *1962*

The initial regulations governing the operation of the CAP guaranteed prices and markets for farm products by:
1 Establishing a single market for farm products which can be sold without impediment throughout the Community at the same price.
2 Protecting this single market by tariffs levied on the import of non-member's farm products.
3 Assuming joint financial responsibility for the implementation of the policy.

The Guarantee sector of the *European Agricultural Guidance and Guarantee Fund*, EAGGF (also FEOGA), which absorbs about 95 per cent of the total amount spent on agriculture by the Community, administers the prices set annually by the Council of Ministers; buys up any surplus produce at guaranteed prices; sets the tariffs; charges a variable import levy on imports from third countries (e.g. USA, NZ); and subsidises the export of surplus to non-member countries (e.g. USSR, Poland, Bangladesh, Egypt, India, Somalia and also UNICEF and the International Red Cross).

Variable import levies ensure that produce can be imported into the Community only at the level of the common prices set annually by the Commission. This means countries like New Zealand which could market butter at a lower price, are not allowed to do so within the EEC. These regulations now cover most cereals, sugar, milk, beef and veal, pig and sheep meats, certain fruits and vegetables, table wines and fishery products. The Guarantee Sector of the European Agricultural Guidance and Guarantee Fund (EAGGF) finances the Community's price support policy.

The initial regulations meant that the long-standing practice of all member governments in maintaining the incomes of their farmers by offering them price support and protection was now a Community practice. The immediate result was the creation of surpluses and the disposal of these surpluses, below cost, to Third World and Communist countries. Considerable financial embarrassment to the Community ensued. The price support schemes absorbed most of the EEC budget.

This outcome had been foreseen at Stresa by Dr Mansholt, the vice-president of the EEC Commission. Unless combined with a policy of restructuring farming, price fixing was a dangerous instrument. Not only would it lead to overproduction, it would exacerbate the differences amongst farmers and between the favoured and the less favoured regions (Figure 1).

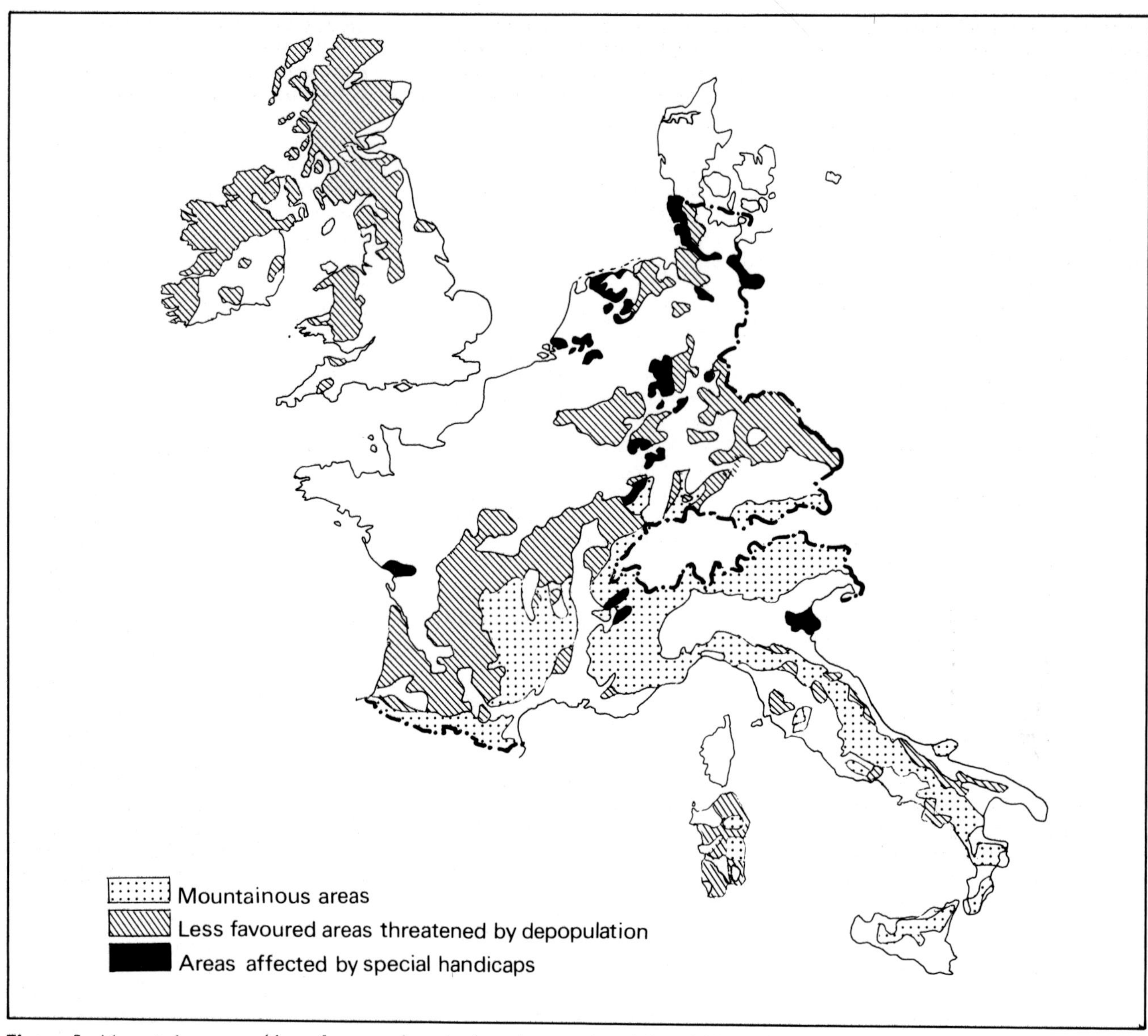

Figure 1 *Mountainous and less-favoured agricultural areas (as defined by EEC Council Directive)*
(*Source*: Agricultural Information Service, European Community Commission, 1980)

THE MANSHOLT PLAN: THE IMPLICATIONS FOR PEASANT AGRICULTURE

The Mansholt Plan, 1968, sought:
1 A levelling off in the Community's expenditure upon price and market support.
2 A reduction by 5 million of the numbers engaged in farming.
3 A reduction of the area farmed by 5 million hectares, with 80 per cent of that area reafforested.
4 Measures to assist those prepared to leave farming with retraining and further educational schemes and pensions, in addition to regional development programmes to provide alternative jobs.
5 The creation of larger farms by a number of farmers to form Units of Production, or one farmer to increase his scale of operation and form a Modernised Holding. In either case a minimum size had to be achieved: 80–120 hectares for crop farms; 40–60 cows for a dairy farm and 150–200 animals for a beef farm.
6 The participation of farmers' organisations in implementing the programme and in forming producer groups to improve marketing managements.
7 A *nationally* implemented programme to take account of regional diversity.

THE MANSHOLT PLAN AND LESS-FAVOURED REGIONS

The District of Queyras: A Case Study

The district of Queyras was selected for investigation under the auspices of the Directorate General of the EEC Commission for Agriculture to evaluate the implications of the Mansholt Plan for regions where agriculture had a distinctly unpromising future. Queyras is located in the French Alps and borders on Italy (see Figure 1). It consists of a score of high valleys (all of the settlements are above 1400 m, the highest at 2000 m) that drain into the Guil, a tributary of the Durance. The winter is hard and long. For six months the average minimum is below zero, the average maxima 6.5°C. Snowfall amounts to an average of 2.7 m at the lower and 4.3 m at the higher settlements. Only July and August are free of frosts. Rainfall ranges between 640 mm and 840 mm but the variations from year to year are considerable and frequent. The critical months are May and June, a good fall will produce good pasture growth; July's rainfall determines the crop of hay. To counter the summer dryness, irrigation channels have been constructed to water the meadows.

In the 'Old World', irrigation works have always spelled vast inputs of human labour. This is true of Queyras. To survive in this difficult environment has required enormously long hours of work for the whole family in the summer months, and survival has established a population threshold below which traditional farming cannot be maintained. Consequently, at the time of the French Revolution, 7200 persons were estimated to live in the district (compared with 3600 in the fifteenth century). The population peaked around the 1930s at 8492, when seasonal migration was still a necessary addition to income for some families. Progressively seasonal migration was replaced by permanent migration: to Paris or Marseilles, even Mexico. But the farm economy depended upon the arrival of Italian workers for the hay harvest well into the present century. By then a proper road had been constructed leading out of the basin and the rural exodus continued, to the industrial cities and to the battlefields of the World Wars. The population fell below its medieval level without any major change in the way of life. Farming still depended upon producing cereals on the restricted areas of flat land and terraced land. Potatoes were rotated with barley or oats and rye was followed by a fallow — a very extensive system of farming as opposed to an intensive four or even eight crop rotation. The upland pastures, the alps, pastured the flocks and herds of animals in the summer months until the hay was cut and laboriously carried down to the barns in the village on the backs of mules.

When the writer René Dumont visited the area in 1949, every quintal (50 kg) of hay stacked in the barns at the end of the harvest required one man's labour for a day. On French farms located on the plains and using machinery drawn by animals, the input of one day's labour was sufficient to produce one tonne of hay. On mechanised American farms of the same era, one day's labour produced 10 tonnes. Faced with this enormous input of labour for such poor returns, many of the farmers of Queyras and their families decided, post-war, to leave farming altogether or to purchase a farm elsewhere. Amazingly, some of them had sufficient savings to do this, having taken advantage of the scarcity and high price of food during the Occupation (a good illustration of the peasants' capacity to save and to take advantage of the slightest opportunity for profit). But they could not save themselves by mechanisation — the relief prohibited it or made it an enormously expensive operation. Those that remained — and by 1972 the number of farms had halved — either developed a secondary source of income via tourism (three-quarters of the families had at least one member with a second occupation) or, continued as marginal units. Only a very few farms remained capable of achieving the standards required by the Mansholt plan for survival as a modernised unit. Eight per cent of the farms fell into this category, but their future was made uncertain by the fact that the land needed to enlarge their farms was hard to obtain. Those who were leaving farming were reluctant to sell their land, it had too great a value as sites for tourist development. Distance and accessibility raised the costs of production and put the modernisers at a disadvantage with respect to the farmers in more favourable locations. The relatively small volume of milk they could make available to the processing stations, and their limited season for production, constituted other disadvantages. A continuing adaptation to changing prices and costs became the condition for survival. One other small but significant matter worked against the modern *chef* — something no city dweller could foresee. The chance of finding a wife grew smaller as the population decreased, as women preferred other occupations and other localities. The threat of remaining celibate was a factor driving both sexes away.

Sixty-seven per cent of the farms in Queyras were categorised as having a dim economic future, and one-third of them were destined to disappear because there was no family member prepared to continue farming. The economic future of the majority of the families (and note it is finally families not farms that have to be considered) would depend upon their drawing income from at least two sources, farming and tourism or other part-time work. The geography of some localities in the district excluded tourism as the secondary occupation. In sum, this amounted to a degree of disguised unemployment and a measure of low productive work for a section of the population. Despite the exodus, the labour force available exceeded by some 60 per cent the actual needs of farming.

Nevertheless, the investigators concluded that farming was essential to the ensemble of activities undertaken by the families. It did not satisfy every need but allowed the families to exist and it provided food and shelter. In the majority of cases neither the farm nor the secondary activity alone was sufficient to maintain the family. The farm required a large amount of work in summer, and little in winter, but the secondary work was also very seasonal. Within the family, both activities were carried on, the division of work often being according to generation. If the father abandoned farming, the economic basis of the family would be sapped. Thus while not meeting the standards of the modern farm, the farms had to be maintained. Without them the whole economy of the district would collapse.

Conclusions to be drawn

People have won a living from this difficult environment by populating — by overpopulating — these valleys so that the huge demands for labour power can be met during the critical summer months. Peasant farming here has rested upon the existence of an underemployed stratum of people being available within the family, or the commune. It has provided nourishment, shelter, at times full employment, the price being low returns per head. If the peasantry abandon the land, desertification will result, and this has happened in some of the communes in Queyras. To survive, the peasantry must acquire alternative sources of income and retain price support from the state. They can modernise if both the natural and human environment is favourable. In Queyras both environments are working against the modernisers.

? Think carefully about the last statement. Define what is meant by the two environments. Explain how they are working against the modernisers.

Print 2 *French agriculture is highly mechanised in places but traditional farming practices are still widely used — this photograph was taken in 1980*

THE CRUCIAL ISSUES SURROUNDING THE MANSHOLT PLAN

The significance of the Mansholt Plan lay in what it acknowledged, not what it achieved. It acknowledged what had long been known but had been very difficult to admit publicly. If the future of Europe's farming lay with the family farm, this did not mean every farm had a future. There was an excess of farms and too many of them were uneconomic.

Put to the Commission in December 1968, ten years after Stresa, six years after the first Community Regulations, the Mansholt Plan created uproar by proposing that five million people would quit farming in the coming decade. (At the time it was not generally realised that as many people had already left the land in the past ten years.) The small farms, the Mansholt argument ran, depended upon livestock production, especially milk, for their income. The herds were too small to provide them with a satisfactory livelihood. Whatever price support they were given would not allow their incomes to keep pace with urban incomes but price support did encourage them to stay in business resulting in enormous surpluses of dairy produce which burdened a market that was slow to grow. A decent standard of living could be achieved only by a smaller number of farmers each managing a larger herd and obtaining a greater share of the market.

The Plan thus sought to establish larger modernised family farms employing the equivalent of two full-time labour units earning a livelihood that matched the incomes of their urban counterparts. The aim was income parity for the rural people. To provide the increased acreage that the modernised farms required, farmers over 55 years of age were offered an annuity provided they left farming and their land was put up for sale. Younger farmers who could not possibly hope to obtain an economic unit were offered retraining grants and unemployment benefits if a suitable job was not to be found. By these means some 20–25 million hectares of farmland were to be released: to be added to the modernised units, to be taken out of production and reafforested, and to be included in national parks and recreational areas. With assistance from the Guidance Section of the EAGGF funds, the implementation of the plan was left in the hands of the national governments, partly on account of the fact that similar and existing schemes were already in operation, partly on account of the fact that existing agrarian structures and specific regional characteristics, like the ones we have looked at in Queyras, prohibited any centralised direction. The preface of the Mansholt plan read as follows:

'European agriculture is on the point of breaking with its time honoured traditional structure so that it can adjust to modern industrial society and the large dynamic market it has been thrust into. There is no more time to lose . . .'

Maybe not, but no one was going to let it happen without a debate first.

Snatches from a Rural Congress

TREASURY OFFICIAL
By offering price support without a structural policy we are maintaining too many people on the land. An excessive amount of public money is going to the farm sector, and within that sector the benefits are not spread evenly. The bigger farms gain disproportionately. We are left with surpluses that we can get rid of only by dumping on the world markets or selling cheaply to the Soviets so that our taxpayers subsidise the Russian consumer.

BRUSSELS TECHNOCRAT
A few statistics would grant us a necessary perspective. Look at who has the cows: Seventy-two per cent of the cows in France are in herds of less than ten. For the Community as a whole the proportion is 80 per cent. Milk sales provide the main source of income for 4 million farms and only 75 000 of these farms have more than 20 cows. No one can make a decent living from a herd that size unless the price of milk is maintained at an exceedingly high level, a practice the Community cannot afford — nor would the taxpayer support over any long period of time . . .

RURAL HISTORIAN
Excuse me, let me have my say. What conclusion can we draw from these simple statistics that the average herd size is only five or ten cows, whereas in America or the Antipodes the figures would be 50 to 100? Unlike these New World countries we didn't begin our Industrial Revolution with a country unsettled and a small population. We began with a society and an economic structure that had evolved over the centuries without the guarantees of wealth and security provided by science and technology. Our farming structure was built — not without a struggle against the remnants of feudalism and ignorance — to provide income and employment for the vast majority of the population on the basis of their own labours, at a standard of living that may look meagre in comparison with our current levels, but which nevertheless supported a rapidly growing population, and which witnessed the spectre of famine finally dismissed from Western Europe. And remember the society we built upon that economy gave us astonishing intellectual achievements in every field.

Remember when half the population was still engaged in agriculture in France and Germany, X-rays had been discovered in Germany; the velocity of electromagnetic waves had been measured; Darwin's *The Origin of Species* had long been published; Marx was writing *Das Kapital*; Wagner had made his synthesis of all the arts; Manet and Monet were both at work.

The society we live in, which is the heir to all these achievements, nevertheless has distinct and different objectives. By means of the managed-welfare economy — the use of Keynesian economic measures — it seeks to abolish unemployment and poverty and to guarantee equal standards of living to all. The disadvantaged of our society are to be found in the rural areas in those farms with their five cows. That historic system of organising people's livelihoods is not designed to fulfill the promises of an industrial technological society. It is destined to disappear.

BRUSSELS TECHNOCRAT
Quite so and therefore . . .

POLITICIAN (with a rural constituency)
The peasantry are and will remain the life blood of the country. Without them the provinces will wither. What is distinctive about our national life will vanish. The peasants incorporate important values — the family is the prime social institution; the self-managed enterprise is the economic basis of a liberal society. Vital to the rural economy and society the self-managed enterprise becomes all the more important in these materialistic times dominated by the faceless, often foreign, corporation, and the sense of alienation that spreads in the city.

CYNIC (a British Journalist)
He fears the city because rural depopulation spells an end to his constituency!

POLITICIAN
I recognise our English friends and welcome them in the hope that they may bring an understanding to their fellow countrymen so that eventually the historic mission of the Community will be strengthened by Britain's presence. My English friend, no one understands the peasantry better than the Englishman Philip Oyler. I always remember what he wrote, *'When I look upon those little farms of the Dordogne valley and compare them with our own acres I feel . . . deeply ashamed of the small amount we obtain per 100 acres compared with the variety and abundance of those peasant holdings. Yet we deride them as being hundreds of years behind.'*

YOUNG FARMER AND ACTIVIST
It is true we peasants exemplify important values which lie at the base of Western civilisation and we wish to protect that institution — the family farm — which above all other

institutions ensures the survival of those values. But we must pass beyond the rhetoric of those in power. For too long the family farm — *l'exploitation familiale* as our French brothers put it — has rested upon the exploitation of the family. No longer are we prepared to have the rich and the influential speak for us. We are devising our own programmes — and the governments must support them — that will produce a structure of farming that will give a fair return to the farmer for his labour, his management and his skills.

INTERJECTION
And recognise the farmers' wives contribution!

YOUNG FARMER
I will come to that later . . .

WOMAN ACTIVIST
No, let me speak!
For too long farming has been a man's world built on the foundations of the labour and the resourcefulness of women. We contribute our work and judgement to the farm and its affairs, and bring up the children. And all too often we have been without — or the last to receive — the elementary services that the urban housewife takes for granted. The new farming structure, in addition to providing us with a standard of life at least equivalent to that of the urban working family, must allow us the time — occasionally at least — to spend weekends together with our spouses, as city couples do, and give us leisure time together.

OLD PEASANT
This is a veritable minefield of opposing views — the old values seem far away.

SECOND YOUNG ACTIVIST
There is no way back grandpa. You must all recognise this: in finding a place in the modern world for the energetic, vocationally well-prepared young farmer you are eliminating thousands upon thousands of rural dwellers, young and old, from the farming industry. They have needs and rights and we must show solidarity. Programmes must be devised to ease the departure of the old without a loss of dignity — to provide a security in their declining years better than the farm itself could provide. And the young who migrate must not arrive in the city unprepared, without skills, to go to the bottom of the pile in the expanding metropolis. There must be solidarity amongst us whilst we create a new rural society. And we should seek the establishment of second chambers where we are represented as economic interests, giving us the full parliamentary weight we possess economically. Only in this way will our real interests be defended.

REGIONAL PLANNER
We are witnessing the decline of old structures, old relationships, and if we are not careful it will end with their destruction, not their reformulation. We must hold the people in the countryside by careful planning that will relocate new industries, that will revive regional structures and regional demographies to halt the tragic agglomeration in the cities. Those of you who know the southern parts of Germany and Switzerland know the countryside is vital where industry is decentralised.

SECOND OLD PEASANT
I sympathise with the young in their attempt to get something better — may their leaders show good judgement. But can we be assured of the point of arrival? Don't forget, young men, René Dumont wrote soon after the war: '*The rural exodus demands a full employment economy, in constant expansion, without the depression of the type we had from 1928–1939*'. You see the future doesn't lie in your hands — and those who are in control, can they assure constant expansion?

ECOLOGIST
The costs for the rural people of the restructuring will, in the end, be far greater and more encompassing than you ever envisage. Desertification will result in some instances. At best the countryside in some regions will become the backyard of the city dwellers who can afford to motor there for the weekend and the holidays. Whilst the cities will be filled with the unemployed who will be mostly young and who have enforced upon them a leisure they cannot afford or enjoy. The countryside we now have is a symbiosis of man and nature created over the centuries, not by profit and loss accounting, but by man labouring with the things of nature. We are disturbing that symbiosis. Go to the modern city — its new suburban quarters. What's the achievement? Look at the costs! Why not spend some of this money on the countryside?

CYNIC
Why not just simply come out and say that all the attainment of modernisation in Western Europe — of which the Common Agriculture Policy is but an instrument — will have done is to have exchanged disguised unemployment in the countryside for undisguised unemployment in the cities.

? Make sure you are familiar with the aims of the Mansholt Plan. Read the *Snatches from a Rural Congress* again. The points of view and ideas are complicated. Discuss these in groups, or as a class, and condense the views into essential points.

Then divide the class in two. One group is to be a party of interested geographers from England. They are to investigate and put the case for the Mansholt Plan. The second group, also interested geographers, are to investigate and put the case against the Plan. When spokespeople from each group have explained their views, your class can vote for or against the Plan.

SOME CONSEQUENCES OF THE CAP

We have dealt so far with the opinions, the policies and the events of the 1960s, a period covering the first decade of Green Europe, a period of initiatives and implementation. One looks to the second decade, the 1970s, for the consequences of the CAP and they are there, of course, but with results somewhat less than conclusive.

From amongst the many issues that commanded the Commission's attention two are of particular interest. Firstly, the ramifications of the Commission's structural policy as it was implemented in the Community's mountainous and Mediterranean regions (see Figure 1 *page 142* and Figure 2 below). Secondly, the nature of the agrarian society that began to emerge as the exodus from agriculture continued, and then showed signs of slowing down.

THE POLICY FOR THE MOUNTAINS

The structural programme acquired a specifically geographic connotation with the Directives concerning mountain and hill farming and farming in certain less favoured areas (May 1975). The accession of Eire and the United Kingdom meant that the problems of the Scottish and Welsh Highlands and of the west of Eire were now within the purview of the Commission, as well as the problems of the German highlands, Italian and French Alps, the Pyrenees, the Appennines and the Massif Central. The fears which prompted this legislation have already been touched on in our study of Queyras: large-scale depopulation arising out of low incomes and limited opportunities raised problems of conservation, even erosion in some districts. The maintenance of the countryside for the city dweller, and the collapse of whole regional economies were other issues addressed by the legislation.

Compensatory allowances were provided for farmers, in some instances for tourist or craft industries, out of the EAGGF funds. Specific programmes for accelerated drainage schemes in the west of Ireland (1978) and the Eire–Northern Ireland border (1979) were to follow. 1979 also saw the publication of a wide-ranging proposal to stimulate the agricultural development in the less favoured districts of western Ireland. EAGGF expenditure marginally reflected these shifts in policy. Ninety-five per cent of its total expenditure ($EEC 14 493 million) still went on price support. Within the remaining five per cent allocated to structural programmes, the largest share, 44 per cent, was devoted to programmes with a regional basis in 1982, in contrast to 1978, when 41 per cent had been diverted to farm projects concerned with improving farm structures.

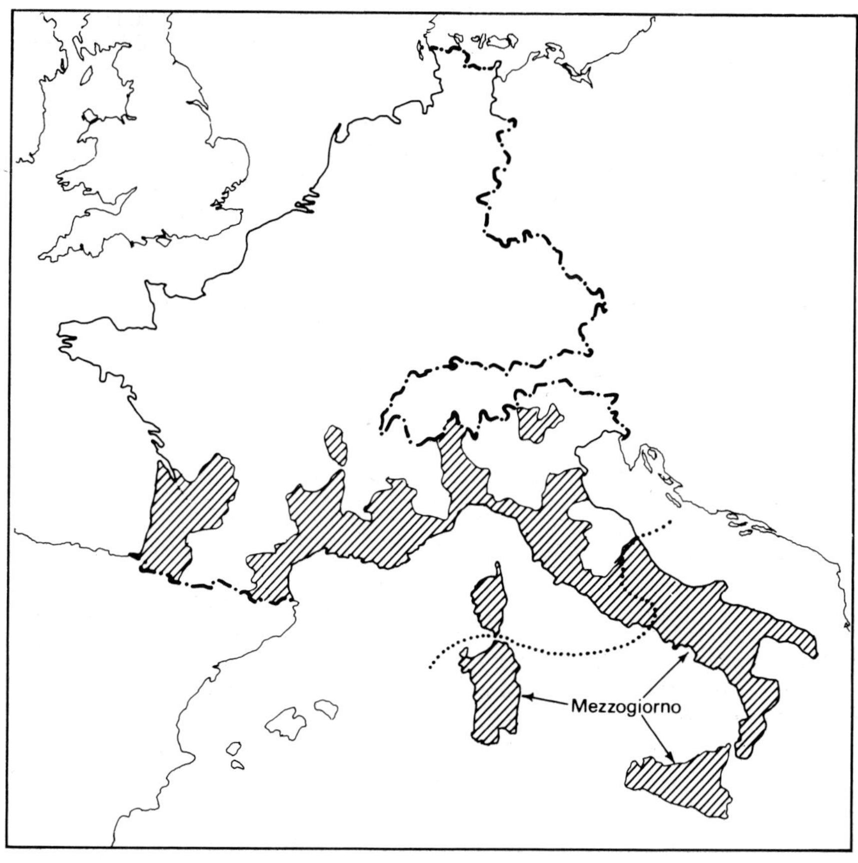

Figure 2 *The CAP's Mediterranean Regions, 1977*
(*Source*: Division for Agricultural Information, ECC, May 1977)

Geographic diversity in the mountain regions

The boundaries of the mountainous areas benefiting from Community aid correspond closely to the 1000 mm isohyet. German agricultural experts recognised this line to be one of the major geographic boundaries within the Community from its inauguration. It contains the bulk of the regions where grassland is the basis of livestock production, which is in itself so fundamental to the peasant farm economy. As the agricultural economy of the Community has moved towards a more capitalist and industrial structure, and as the crop producers have benefited more than other producers from the price support schemes, inevitably, the social problems associated with the new economy have appeared in the wetter, higher upland regions. As so often happens, aspects of the natural environment acquire a new significance when society changes.

Of course, within these very broad limits, as always, a considerable geographic diversity prevails. French geographers have focussed on this issue with respect to their own mountainous areas. Between 2.3 and 2.7 million people live in the mountainous regions, representing about 5 per cent of the total population. Forty-five per cent of them are classified as living in urban and industrialised zones which showed an increase in population of 9 per cent between 1962 and 1975. The total French population increased by 13 per cent during the same period. Rural dwellers who live in communes where tourism is a major industry also increased their numbers by 8½ per cent. In the communes dependent solely upon agriculture, population fell from 1 164 000 in 1962 to 999 000 in 1975 — a 14 per cent decline. Our knowledge of Queyras would lead us to expect such a result.

THE POLICY FOR THE MEDITERRANEAN REGIONS

As the alpine areas become one geographical focus of the Community's structural programme, the Mediterranean parts of the EEC become a second focus on account of their specific problems and their social and economic significance within the whole agrarian economy of the community (Figure 2). As the Directorate-General for Agriculture in the Commission drew up its proposals for the region everyone was aware that they were engaged with a problem that would magnify once the major Mediterranean states of Spain, Portugal and Greece acceded to the Treaty. Within the compass of the nine, the Mediterranean regions accounted for 30 per cent of the holdings and 17 per cent of the utilised farmland. Admittedly these proportions were influenced by the manner in which the Commission defined the area (Figure 2). Their definition had a carefree, indeed almost uninformed touch. For not only were the Mezzogiorno (Italy's southern regions) and Corsica included as might have been expected, but so were Italian provinces like Trento, Novara, Vercelli, Asti, Alessandria. Along with the French regions of Provence, the Cote d'Azur and Languedoc, Aquitaine also found a place. The principle was not a climatic one; it depended upon the particular region having 40 per cent or more of its total agricultural production composed of typically Mediterranean products: wine, fresh and processed fruit and vegetables, durum wheat, tobacco, mutton and lamb. Agriculture in the Mediterranean regions was characterised by the following:

1 Farm incomes were distinctly below the average for the Community, as was the rate of growth of agricultural income and output.

2 The agricultural sector was a diminishing part of a whole economic structure that was depressed and incapable of sustaining its own development. Despite the massive outflow from farming, little corresponding change or renewal had accompanied it.

3 Marketing methods which had been recognised as a contributing cause of backwardness in the 1950s were still archaic.

4 The destruction of topsoil had reached appalling levels in some of the mountainous districts, creating flooding downstream and requiring a combined programme of farming, forestry development and water control to produce, as the Commission put it, an ecological stabilisation of the countryside.

Table 1 *Increase or decrease of population in France's mountainous regions 1962–1975 (per cent)*

	Total population 1975*	*Urban and industrialised zones*	*Rural zones with tourism*	*Rural zones dependent upon agriculture*
Pyrenees	153 000	+3.5	+4.7	−18.6
Massif Central	1 157 000	+4.1	+1.9	−16.9
Vosges	198 000	−1.1	−5.7	−16.6
Jura	249 000	+21.3	+26.6	−4.2
Alpes du Nord	419 000	+20.5	+7.7	−13.4
Alpes du Sud	222 000	+21.3	+8.3	−5.0

(*These figures add up to give the lower total of 2.4 million)

The Mezzogiorno — any improvement?

The Mezzogiorno was the single largest region included within the Commission's Mediterranean programme. Region it may be, but to call it so may be misleading, for its population of 19.8 million people exceeds that of many member states of the EEC: Ireland, Denmark, Belgium, Luxembourg, Greece, and even the Netherlands. The Fund for the South (the Cassa per il Mezzogiorno) had already been in operation then since 1951.

Neither the Fund nor the Mediterranean programme has brought any real reduction in the disparity of regional incomes in the Mezzogiorno, despite a remarkable increase over time in real per capita incomes. This latter achievement was made possible by means of an enormous inflow of capital matched by a large outflow of labour from the region (see Figures 3a, b and c). Disappointingly, an economic structure capable of sustaining its own economic growth has not been created in the South by this investment. The net increase in employment has been almost nil.

Despite the fact that the Mezzogiorno is now far less dependent upon farming than it was 30 years ago, the listing of its problems made recently by the Commission has a familiar ring to it:

1 There are still too many people in farming for everyone to have a decent standard of living.

2 Marketing arrangements are inadequate for the requirements of a modern economy.

3 More investment, better placed, is required.

4 The reforms that have been introduced have been inadequate to overcome the severe limitations of the physical environment and blockages in the human environment.

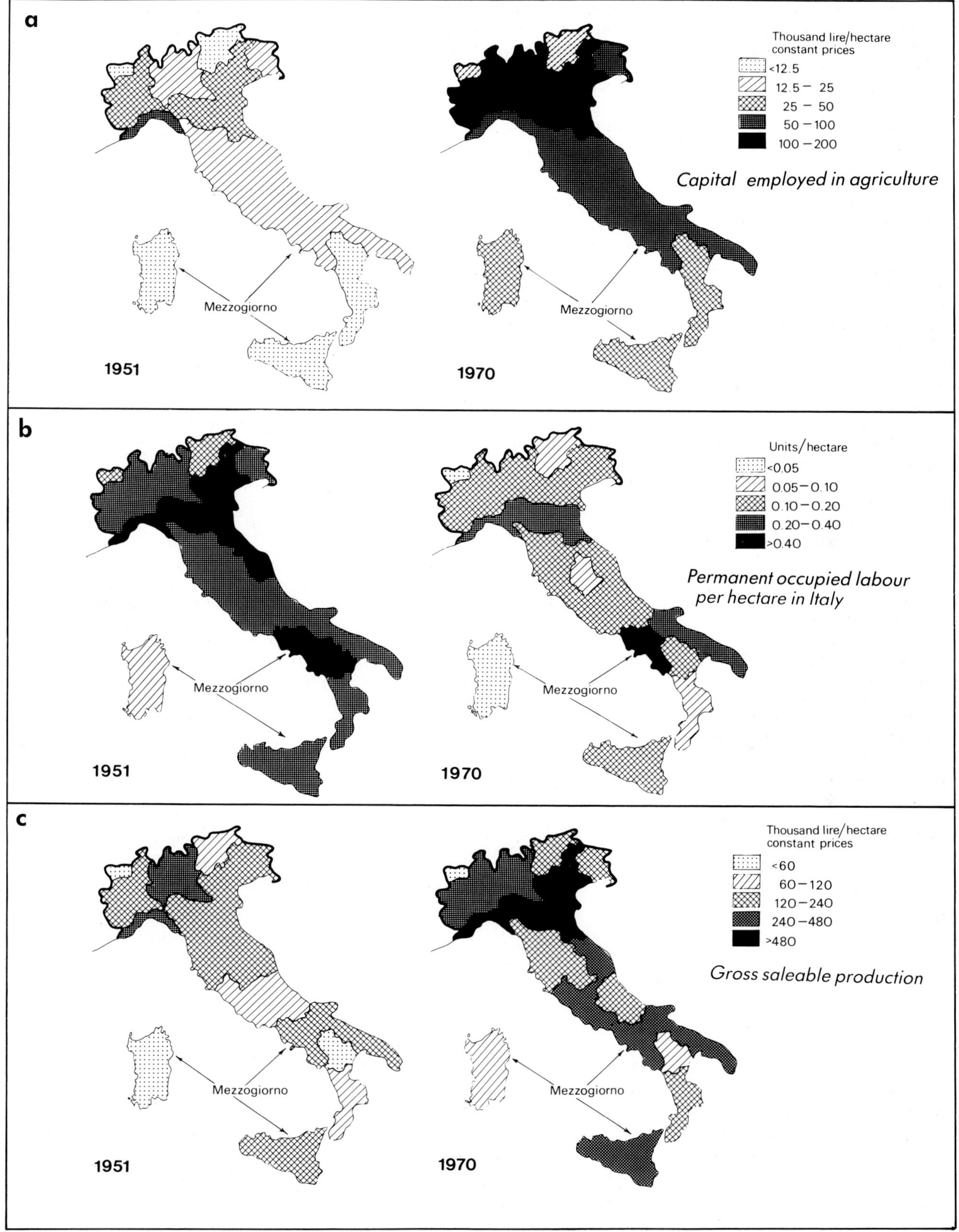
a
Thousand lire/hectare
constant prices
<12.5
12.5 – 25
25 – 50
50 – 100
100 – 200
Capital employed in agriculture
Mezzogiorno
1951
Mezzogiorno
1970
b
Units/hectare
<0.05
0.05–0.10
0.10–0.20
0.20–0.40
>0.40
Permanent occupied labour
per hectare in Italy
Mezzogiorno
1951
Mezzogiorno
1970
c
Thousand lire/hectare
constant prices
<60
60–120
120–240
240–480
>480
Gross saleable production
Mezzogiorno
1951
Mezzogiorno
1970

Figure 3

The latter point, about the limitations of the physical and human environments deserves amplification. Long ago, the famous agrarian economist Rossi-Doria classified the Mezzogiorno in terms of its soft and boney parts. The boney parts are the hills and mountains that erode devastatingly once the topsoil has been broken by cultivation or over-stocking. The winters can be bitterly cold, with snow and frost; the summers gaspingly arid with occasional excessive and damaging downpours of rain. Pressure of population meant this environment, right up until the 1950s, was adopted for cereal cultivation and the planting of poor-yielding vines and olive trees, when ecologically it was suited to a more extensive, less damaging forestry–pastoral system of farming. In their scramble for land the peasantry pulverised the landscape into thousands of inadequate marginal holdings and, although they have subsequently abandoned farming, rarely have they given up the land. Until the units can be reassembled into large farms, a farming system suited to the mountain ecology cannot be instituted.

In the soft parts of the Mezzogiorno — the coastal plains together with the alluvial valleys of the few rivers that penetrate the mountains — things are different, but the end result is not much better. Nature is certainly kinder, and in some places where good soil, drainage and water are combined, the Mezzogiorno resembles the Garden of Eden — but an over-crowded one. Despite the abundance and variety of crops — in fact because of them — the land to person ratio takes on Asian proportions in some rural areas surrounding Naples. Marketing is chaotic, quite unsuited to the demands of the supermarket chains of the industrial North. There is little scope for intensification in these 'Gardens', but away from them there are many plains where irrigation, once provided or enlarged in scope, would allow intensification on a larger scale of operation than is customary, with the labour organised on a capitalistic basis.

The South has had both good and bad experiences with large-scale irrigated farming. In Sicily and the Basilicata, large farms simply wasted the state's investment in irrigation by not intensifying cultivation (and hence increasing the demand for labour). Instead, they continued to raise cereals by mechanised means, using the irrigation system if it proved convenient. Unless governments ensure the controlling institutions they establish are effective, the farming systems will not automatically be adapted to the best of the limited opportunities which climate and soil offer.

Inevitably, we can only assess the prospects for the Community's Mediterranean programme against the experiences of the past. Too little time has elapsed for us to do more. Nevertheless the prospects look far from promising. However, with respect to its structural programmes enough time has elapsed for some results to emerge and for us to review them at the national level solely because only at this level has the programme been enforced or pursued with greater or lesser energy.

CONTEMPORARY TYPES OF FARMS AND FARMERS

During any tour of the European countryside one can expect to come across five basic types of farmers, farms and farm families.

1 Large-scale, highly productive, capitalistic farms

The capitalistic farmer runs a large, heavily mechanised holding, employing a permanent staff of hired labourers, heavily mechanised

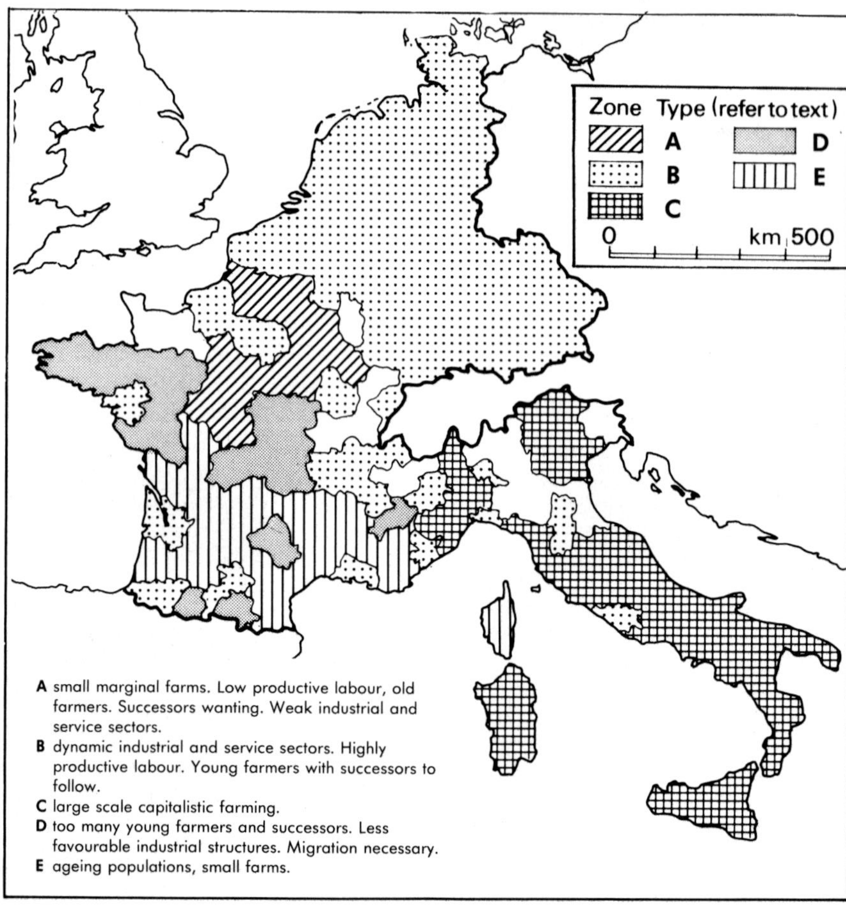

Zone Type (refer to text)
A
B
C
D
E
0
km 500
A small marginal farms. Low productive labour, old farmers. Successors wanting. Weak industrial and service sectors.
B dynamic industrial and service sectors. Highly productive labour. Young farmers with successors to follow.
C large scale capitalistic farming.
D too many young farmers and successors. Less favourable industrial structures. Migration necessary.
E ageing populations, small farms.

Figure 4 *Distribution of farm types within the EEC.* Note: A map including the newer members of the EEC is not available for inclusion here as the necessary information has not yet been collated and published by the EEC Commission

with an above average amount of capital per unit area or per labourer. These farmers are, sociologically speaking, the 'invisible men' of the countryside, ignored by the rural sociologists because numerically they are a relatively insignificant group (though in productive terms they are very important) and because they have posed none of the problems associated with the peasantry. Being large, often rich and politically influential, they have never been popular ideologically speaking, and this is one reason why they have been ignored, even though they have been responsible for innovation and represent the high point towards which the structure of farming is evolving. The Paris Basin with its large cereal and livestock farms is their natural habitat. They can be found in parts of Bavaria, in the plains of Northern Italy,and in the Mezzogiorno too, but at a less advanced stage of organisation. It is extremely difficult to calculate their numbers — the statisticians do not group their data on a sociological basis. However in the mid-1960s, I calculated their numbers in France to be, at a minimum, around 24 000 occupying 12 per cent of the farmland, with more than ten times the capital of the peasant farms, and seven times the income. By 1975 in France, their numbers had increased to 32 000, and the area of farmland worked to 17 per cent; they were responsible for 14 per cent of the total output of farming. If one adopts a less conservative basis for estimation, then in 1975 there were 142 000 big to very big farms controlling 42 per cent of the farmland, providing 34 per cent of the output, and representing only 10 per cent of all farms.

2 Modernised family farms

It is even more difficult to calculate the numbers of *chefs* running these farms who, with their book-keeping practices and now computers, their better education, and larger, more capital intensive farms, have been a principal object of the Commission's care and support. Unlike the capitalist farmer, the *chef* usually does not employ any permanent hired labour — though some can afford to take on one hired man. With a greater or lesser contribution from his wife, he himself undertakes most of the work with the aid of machinery and some participation in a mutual assistance scheme. Livestock play an important role in the economy of the farm. He lives well (better than his ancestors did) in a farmhouse with all the modern amenities, because on the one hand, his (and his wife's) parents have contributed something to the initial land, buildings and livestock of the farm, and because his family is the recipient of social welfare expenditure (child allowances, for instance). He lives better also because he has enlarged the farm and intensified the farming, often incurring considerable debt. His survival depends on keeping the costs of production low enough to meet the debt, investment charges and running costs and still leave enough for his family to live on (see boxed material).

3 The marginal family farm

The modernised full-time family farm grades off into the marginal family holding where the *chef* has to meet the same economic standards as his more fortunate counterpart, but fails to do so because his farm is slightly too small, his organisational capacity a little lacking, his choice of products not the best and his location unfavourable. Everyone, his wife included, works very hard but the net return after meeting all charges is inadequate. These farmers tend to hang on and seek to supplement their income by part-time work for the *chef* himself, his wife, or another member of the family. As they have realised the hopelessness of their future, some of these farmers, in France for instance, have joined radical political movements, such as MODEF (MODEF roughly translates as 'the league for the preservation of the family farm') where communist influences have penetrated. Recognising that many of these peasants will eventually be driven out of farming, the communist strategy is to defend their interests whilst they are still farmers so that when they become workers they will be sympathetic towards the communist strategy for the class struggle in the towns.

The Germans have pursued the modernisation of their peasantry more actively, with greater resources and a better statistical follow-up than any of the other major continental members of the Community. Since the introduction of their Green Plan in 1955, they have sought a parity income for their farmers. They have achieved higher

RECENT HISTORY OF A BRETON FAMILY

In 1948, René Dumont, the French author and ecology party candidate, visited one of the most archaic farms he had ever seen. He described it as being half a century out-of-date. At Pluzunet, in the Department of the Côte du Nord, with 21 hectares, the farm supported a *chef*, his wife, four children and the *chef's* three brothers who had remained unmarried in order to preserve the size of the holding! One could ask for no better illustration of how the kinship order and the economic order are inextricably mixed in peasant life.

At 18 years of age, after a period at a private agricultural college, one of the children took over the farm in 1958. The farm was exactly as it had been ten years previously. It had three wasted horses, six cows, three heifers and the same old equipment. Progressively, the young chef improved the farm, purchasing the first tractor in 1962, inviting ridicule from the district when he applied artificial fertilisers. Taking down the stone walls he reduced the number of *parcelles*.

By 1975 when Dumont revisited the farm, he ran 58 cows, 15 heifers on a rent run, and 1000 battery chickens. From Holstein–Friesian crosses he obtained 5256 litres per cow. Whereas his father and uncles had grown wheat and rye he had introduced rye grass, alternating it with maize for silage and supplementing it with purchased feedstuffs to support the larger herd. As one French geographer put it: farming's image depends upon who undertakes it. If they are young and dynamic it looks progressive. If they are old it has a routine look about it.

Table 2 *Change in farm size in Germany between 1959 and 1981*

Farms of:	Decrease		Increase
< 1 ha	−106 296	20–25 ha	+6 506
1–2 ha	−205 104	25–30 ha	+21 886
2–5 ha	−402 801	30–50 ha	+35 633
5–10 ha	−259 299	50–100 ha	+15 188
10–15 ha	−72 250	100+ ha	1 585
15–20 ha	−7 653		

farm incomes and better distributed incomes (Figure 5), but parity has escaped them as urban incomes have risen faster than rural ones. A better distribution of income has been made possible by the elimination of thousands upon thousands of marginal farms. In 1959 there were 1.7 million holdings; by 1981 only 0.81 million remained. The decreases (and increases) occurred as shown in Table 2. Nevertheless, after a generation of change, 4 million of a total 12 million hectares of farmland were still controlled by marginal holdings of less than 20 hectares.

4 The part-time farmer

Once so characteristic of German agriculture, the part-time farmer has become a common phenomenon throughout the Community (Figure 6). As the size of the holding is insufficient to support a family full-time, the function of the farm is merely to supplement the income of the family. It is worked at weekends by the *chef* who often uses his factory earnings to purchase machinery while the land and the animals are tended daily by his wife. Ever since the last century, particularly in south-west Germany and the Rhinelands (especially the industrialised parts), men have left their farms daily to work in the factories of the adjacent towns. Their sons, their daughters and less frequently their wives, have followed them. Branch factories from the towns have themselves sought out the labour in the villages and, more strikingly, some *chefs* have become entrepreneurs establishing their own factories. At its best part-time farming has produced some of the soundest rural economies to be found in the world, with high standards of living, good schooling and cultural facilities, and avenues for the promotion of the able in an environment that remains rural and unspoiled.

The success of the German examples is, however, based upon the very favourable structure of the Federal Republic's economy, with its orientation towards exports, technology and particularly technical education, which has been introduced into the smaller towns of the regions, and of course the high rates of economic growth the Germans have achieved. Their experience with part-time farming is not consequently easily transferred.

Farms %

1978-79

1969-70

Thousand Dm

(*Source*: Agrarbericht, Bonn 1980)

Figure 5 *This chart shows the effect of the German government's agrarian policy combined with the exodus from farming upon the standard of living of full-time farmers. It plots the distribution of income per family labour unit for full-time farms (i.e. part-time farms excluded). Over time, the distribution of income has become less skewed and a greater percentage of farms are in the higher income brackets*
(*Source*: Agrarbericht, Bonn 1980)

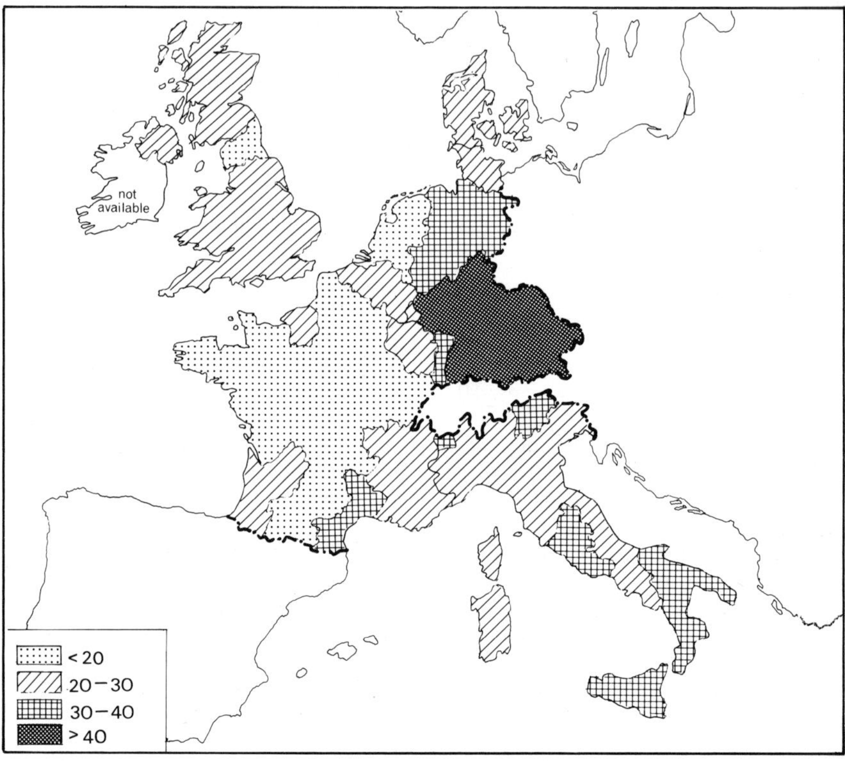

Figure 6 *Farmers with another gainful activity, 1975*
(*Source*: Agricultural Information Service, ECC, 1980)

Print 3 *This mixed arable and dairy farm in Hesse, West Germany also offers 'farmhouse holidays' in the summer months. Its facilities include a sunbathing terrace, pony riding, table tennis and a barbecue site*

In the French Alps, part-time farming is associated with tourism which, as already mentioned, is on the increase, and also associated with craft industries which do not necessarily match the technical calibre of the German worker-peasant villages. Furthermore, in those parts of France and southern Italy where there are still too many marginal small farms — where part-time farming might make a useful addition to income — the structural and dynamic conditions that have favoured its extension in Germany are totally absent.

5 'Pensioner' farms

The fifth type of farm may easily be confused with the part-time farm. Small in area, it is quite unsuited to employing a family full-time. The *chef* may therefore have an additional source of income. You, yourself, may in fact stay on one of these farms as a tourist. If you do, look round for the critical signs: the age of the farmer and his wife, the absence of young members of the family, the distinctly unmodernised farm practices and the small size of the herd of cows. Farms like these are basically pensioner farms for old people who have been in farming all their life and who cannot and do not want to migrate to the towns, their farms providing a source of psychological and economic security. Twenty one per cent of the farmers in the Community are more than 65 years of age. They control an equivalent percentage of the farm numbers and 13 per cent of the farmed area. Often these farms are described as being 'without a successor' which means there are heirs who will not continue farming after the death of the *chef*, but who will hang on to the land — the price of farmland being so high — to make a capital gain, selling it to some rich city dweller, or ambitious modernising *chef*. The Commission admits there has been relatively little success with their programme to encourage the cessation of farming amongst the old and its reallocation to those with drive.

This simple classification of farmers does not allow for the very specialised activities of farmers who concentrate upon particular crops, e.g. flowers and run farms small in area which nevertheless create a lot of employment and employ a lot of capital. The classification also draws attention away from a very important social feature of the countryside, namely the creation of non-farm jobs in the upstream and downstream industries which supply inputs to the farm — fertilisers, pesticides, animal semen, machinery — and process its products in factories and packing plants. These industries are most likely to be located in nearby towns and regional centres where they absorb labour within the region and provide better incomes and better working conditions for those classes who would have been worse off had they been forced to remain in farming. These industries, furthermore, help to create a better regional structure of employment.

SHOULD THE CAP BE REVIEWED?

? The article below, taken from *Europe 82* highlights certain ways in which Sir Henry Plumb felt the CAP might be restructured without being radically altered. List his suggestions and in each case say whether, based on what you have read in this chapter, you think the suggestion is likely to be practical.

What is not highlighted in the article above, is the extent to which disparities still exist in the agricultural sector. In its reflections on the CAP, published in 1980, the Commission of the EEC admitted to a close correlation existing between the level of farm income and the extent of EAGGF's support expenditure. Per labour unit support expenditure exceeds by half as much again (150) the Community index (100) in those regions where agricultural incomes are highest: Paris Basin, Belgium, the Netherlands, Denmark and North Germany. However, it falls below 80 per cent, even 50 per cent of the average in the mountainous regions,

'THE CAP IS NOT SUCH A BAD SYSTEM . . .'

For nine years Sir Henry Plumb was the formidable president of the National Farmers' Union. He is now chairman of the European Democratic Group of the European Parliament.

Does he believe the CAP must be radically altered? His answers venture into a barely-publicised area of future European concern — the restructuring of rural areas with national funding co-ordinated at a European level.

'No. Not radically altered. The CAP is not such a bad system if you analyse it. Let's start with the cost, because most people say it is enormously expensive.

'There are now about six million productive farms in the Community. Divide the total spent on the CAP among them and you'll find (*including* the 14 per cent of the budget spent on food-aid and for export restitutions when there are surpluses) that the cost of support is £272 per year per farm. The cost of supporting one unemployed person is about £6,000!

'One of the reasons we've got over 10 per cent unemployed — 13 per cent in Belgium — is that there has been a drift, or a push, from the land. With new technology, farming has become less labour intensive. There has been de-humanisation in the countryside. Many have moved to heavy industry, got married, settled down, and have now been added to the dole queue . . .'

But doesn't this seem to be arguing for de-industrialisation of agriculture?

'No. You can't jog backwards and suddenly say, "We'll start doing these jobs by hand." What I'm arguing for is a rural policy — to run parallel with the CAP — that would keep people living in the countryside, but not necessarily in food production.

'With less time at work, and earlier retirement, the great growth industry may be leisure. And a lot of that will take place in the country.'

Would that mean turning the countryside into a fun park, with empty beer cans in every ditch?

'That's the last thing I want to do! I want people to respect the countryside, to realise good land is the workbench of the farmer, that the cheapest lawnmower is sheep on the hills . . . It's a process of education. Much is being done in schools already. But that's only one aspect.

'The second is the development of small industries — timberwork, furniture, toys, paper and so on. Take Bavaria or the Black Forest. You've got these little farms tucked away round the village, farmers working part-time in the paper mills or up in the forest, perhaps with a bit of timber of their own.

'Go into those farms. They're the cleanest and tidiest you've ever seen. The families are healthy, the gardens well-kept. Basically they feed themselves.

'A few years ago we tended to ridicule those part-time farmers. But to work half-time in the factory and half on the farm keeps them sane.

'On the other hand, I know a village in the Hautes Pyrénées where, in 1962, twenty little farms were snuggled away. At five in the morning you could hear the old boy's boots coming over the cobblestones as he went up to church to ring the bell to wake everybody up.

'Today there's only one farm there. The cobbles are covered with tarmac. The houses have gone and there are only 30 inhabitants. And from those acres, where there were twenty farms, they're producing twice as much as before.

'Economically they've done a great job. But they've turned the *people* out. You've got to work with your heart as well as your head in such circumstances.'

But surely economic and business pressures make such a process inevitable?

'I don't think I'm being romantic. I think I'm being hard-headed. I come back to the cost of keeping people unemployed. You and I as taxpayers have to fund that. What are we doing? We're putting people out of work, demoralising them, and doing it at great cost.

It would take a long time, wouldn't it — reversing a trend which began in the industrial revolution?

'Well, just take what's happened. In the early Sixties, 18 per cent of the population of the original Six were in farming. That is now down to seven-and-a-half. Over three years — average it out — one a minute left the land. They're not moving now. But it was mostly the young who left, leaving the old.'

Has the drift from the land been greater in Britain than elsewhere in recent years?

'No. We're down to about two-and-a-half per cent actually involved in farming, a tiny percentage of the working population. Yet it is still our largest industry, and the productivity record is far better than any other. But the drift has stopped. In fact there's a movement back again.

'If you advertise today for extra farm staff you'll be inundated by letters which start off "I was a farm worker and I'm dying to get back." In principle, it almost supports what I'm saying.'

(Source: Europe 82)

in south-west France and Italy. Acknowledging that the situation called for reform,the Commission at the same time drew attention to the fundamental difference in the resource endowment and structural characteristics which underlie these disparities.

Basic geographical factors, it would appear, are operating to produce a result that many people had hoped to circumvent. The force of this important conclusion however, needs qualifying in two respects. Firstly, it must be recognised that in choosing price and marketing policies that have favoured the better endowed northern regions which produce cereals, milk, meat and sugar beet (rather than the fruits, the vegetables and the wines so characteristic of the Mediterranean areas) the policy of the Commission itself has reinforced the effect of that fundamental difference in resource endowment. Secondly, one must not be too deterministic with regard to the effect resource endowment has upon economic results. Capital and technology have been employed to produce a remarkable improvement in the productivity levels of the poorest parts of Italy.

The Commission seems prepared to accept not only the geographical consequences but also the social consequences of its price guarantee schemes. In 1982 it recognised that the general economic crisis in the West imposes strict management of public spending. For those who remain in farming it emphasised that relatively less money will be available for support compared with the past. As a consequence the Commission concluded: 'Farmers will have to act more like businessmen'. Furthermore it continued:

'It is not really surprising that in a market economy farms should become larger and larger. In the long term, there is no valid reason why agricultural production should not follow industry in the trend towards larger and rational economic units with better allocation of resources and economies of scale'.

What went wrong?

As a system of subsidies, the CAP's worst flaw is its open-endedness. For storable products such as milk, cereals and meat, the EEC has until now undertaken to pay farmers a guaranteed price, regardless of quantity. There is a variety of subsidiary help to farmers, but 95% of the farm budget goes on farm support prices to prop up farmers' incomes. Support levels, until this year, were invariably set higher than they need have been, in the name of the poor, inefficient farmer. And, by the perverse logic of the CAP, it was efficient, low cost farmers who benefited most. For them, this misdirected charity was a golden spur to over-production.

As farm production soared, so did the cost of price support, sometimes geometrically (see chart below). In 1983, an 8% average increase in farm prices resulted in a 28% increase in farm spending. Because farm production was growing faster than the EEC's revenues, the CAP share of the budget rapidly grew — to two thirds of the total. The EEC will run out of cash to pay for even its mildly reformed CAP later this year.

But even if there were money to pay for the CAP's extravagance, open-ended price subsidy would be a waste in terms of the CAP's own aims. The original members of the community set out the aims for a common farm policy in the treaty of Rome. Judged against these, the CAP has not been a total failure, despite its cuckooland reputation. The five aims were to:

- increase agricultural productivity
- guarantee a fair standard of living to farmers
- ensure security of supply
- stabilise food markets
- maintain reasonable consumer prices.

The CAP has achieved the first three. In the past decade, it has pushed up production per acre by an annual average of about 2%. It has helped sustain farmers' living standards, although not evenly (big farmers such as cereal growers in the Paris basin or on the prairies of East Anglia profit most) and not during recession. Earnings have sagged less than they might have because the number of those employed in agriculture has dropped from 12m in 1970 to 8.2m in 1982. The CAP also has made the community self-sufficient in practically all farm goods that it can grow itself.

As for stabilising markets, it has achieved this only in one sense: the community has avoided big swings in world food prices. This is not to be scoffed at. In 1973–74, the CAP was a buffer against even higher prices abroad. And world prices could soar again. But in an equally important sense — keeping supply in step with demand — the CAP has seriously unbalanced farm markets. Surpluses are now built-in. The result of all this is that food prices are far higher than they need be. Take two of the community's most important home-grown commodities: butter and wheat. Butter prices are 50% higher than on world markets and wheat prices are 33% higher.

Simple price theory would have predicted this failure. Fixing prices is almost bound to keep supply and demand out of balance. Price is meant to change when supply and demand do not match. And, by changing, the new price should correct the imbalance. By

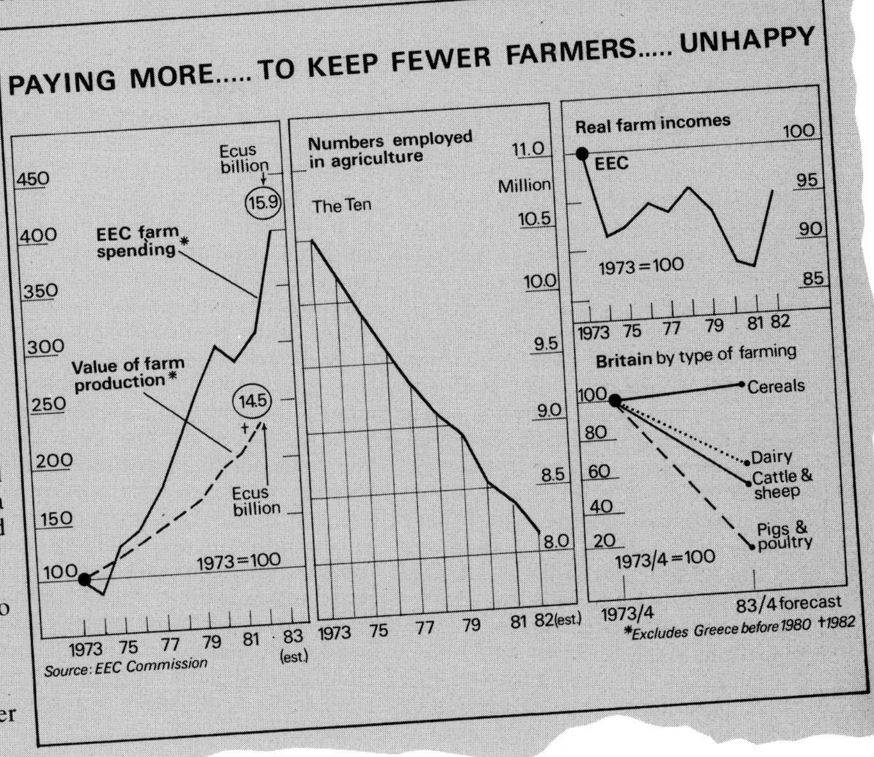

choosing to support farmers by fixing high prices, the community made it virtually impossible to provide consumers with food at reasonable prices.

These distortions might have been more tolerable if they had been less unfair on farmers themselves. But farm price supports have also undercut a sixth, unspoken, aim of the farm policy — supporting peasant farmers, mainly in the Mediterranean and southern Germany. This was partly to provide rural welfare, and partly to control an inevitable surge of people from the land to the cities. Some of that was achieved, but today the CAP gives such farmers little extra help.

Only about a quarter, or 2m, of the community's farmers are on efficient, modern farms. On their own, these could probably provide the EEC with all the food it needs. The other 6m are inefficient smallholders. The less money spent on price subsidies, the more could be freed for direct grant aid to poor farmers who need it. The present inequity is hard to avoid in a price-support system. Smallholders are paid the same per bushel as agribusinessmen. But larger farms are able to achieve economies of scale, making their acres more profitable.

On top of this, the price support system causes haphazard and unfair transfers of real resources from country to country inside the community. Foods in the richer north such as milk, butter, cheese and cereals are more heavily subsidised than Mediterranean products. According to studies by the Institute of Fiscal Studies in London and by the British ministry of agriculture, one farm economy which has done particularly badly out of the CAP is Italy's, despite the fact that, with Greece's and Ireland's, it is the neediest. Why?

Though Italy gets the second biggest share of agricultural aid from the EEC, it is also the biggest importer of north European foods. In 1981, some 13% of Italy's jobs were in farming (as against 8.4% of France's and 2.8% of Britain's); and more than two in three of its farms were five hectares or less. Inequitably, rich Denmark and Holland have benefited most (along with poor Ireland).

The example of Italy is important. Unless subsidies for northern foods are reduced, it will be difficult to say no to subsidising Mediterranean goods. With Spain and Portugal set to join, any widening of price subsidies has to be resisted.

(Source: *The Economist*, April 1984)

?

1 Read through the article above which appeared in *The Economist*. Based on this, and what you have learned while reading through the chapter, how successful do you think the CAP has been in relation to:
a increasing overall production;
b reducing inequalities between rich and poor farmers;
c achieving equality between countries within the community;
d maintaining the infrastructure of European peasant agriculture and society?

2 Read through again 'Snatches from a rural congress' *page 145*. Which speaker would you agree with now? Why?

THE ACCESSION OF SPAIN AND PORTUGAL

Greece is already a member of the EEC. Spain and Portugal have been accepted. From your knowledge of the existing Mediterranean areas within the Community, what problems will their accession bring with it? Plan a further investigation of the issues raised by the new members, taking the problems identified in the article below as a starting point.

● Agriculture. Membership will increase the farmed area of the EEC by 30%, the agricultural workforce by 25% and the number of farm holdings by 32%, but the number of consumers will increase by only 14%. There will be new and costly surpluses of:

1. Olive oil in which the EEC is already 95% self-sufficient. Spanish membership would raise production by 60%. The extra olive oil farmers will cost the common agricultural policy around 800m ecus (£460m) per year, almost doubling the cost to the community of its olive oil regime. Other Mediterranean countries which have traditionally sold the bulk of their crop to the EEC (through preferential trading agreements) will also be affected. The worst hit would be Turkey and Tunisia, each of which sell 75–80% of their olive oil exports to the EEC. And the surplus may be aggravated by a big drop in consumption in Spain as less expensive vegetable oils imported into the EEC at low duty cross the Pyrenees. The Spaniards want to 'solve' this problem by a steep EEC import tax on vegetable oils: something which margarine manufacturers and consumers in Britain, Germany and Holland and trading partners like the United States would strongly resist.

2. Fruit and vegetables. Most EEC production of fruit and vegetables is in Italy, Greece and the Mediterranean area of France. All fear competition from Spain which has a massive surplus of citrus fruit (its output is more than double its needs) and smaller surpluses of peaches, apricots, tomatoes, peas, potatoes, grapes, apples, green beans, pears and onions. As yields improve, and CAP support is introduced, Spanish output could expand even more. Here, too, Spanish supplies will compete with those from non-member Mediterranean states.

3. White wine, which Spain has dealt with by mixing red and white wines. This is forbidden in the EEC, and could spark a row like the Franco-Italian wine war if it is not resolved.

Brighter spot: Spain also has a big deficit in **dairy products** and **cereals**, which should provide new markets for other EEC producers. The commission is anxious to ensure that any new land in Spain farmed through irrigation should be used to produce deficit rather than surplus foods.

Although Portuguese membership is unlikely to provide much of a competitive threat to the existing members, entry will present a hard challenge to Portugal itself. Its agricultural and industrial productivity are both far lower than in either the EEC or in Spain.

(Source: *The Economist*, December 1982)

9 CHINA — CAN SELF-SUFFICIENCY BE ACHIEVED?

Print 1 *Weeding vegetables in southern Guangdong Province. China's very large rural labour force is concentrated on a small fraction of its total area permitting intensive cultivation of the land*

THE IMPORTANCE OF THE AGRICULTURAL SECTOR

The 1982 census figures reveal agriculture's importance through numbers alone. Within the total population of 1008 million (23 per cent of the world's population), farm workers make up around 300 million. Another 400 million people in the countryside work in collective factories, run rural services such as shops and transport systems, as well as involving themselves in some part-time farmwork. Included in this figure are children and old people who are capable of doing light work, tending private plots, looking after animals or working on collective projects. With the exception of the very young, sick or elderly, nearly all of China's 800 million rural population can be regarded as contributing to agricultural work. By the end of the century, even with further industrialisation, mechanisation of farming processes and moderate increases in the urban proportion of the population, the majority of China's people will still be peasants.

In the past, the success and failure of China's dynasties, empires and various power bases depended on the peasant population supplying food, taxes and labour to armies, towns and cities. The success of the Communist revolution rested with the peasantry. Marxist revolutionary ideas had been adapted to fit the peasant majority. The revolution was essentially a rural one, unlike its urban counterpart in the Soviet Union, or the visions of European Marxist political philosophers.

Once established, the Communist government of the People's Republic of China did not ignore its source of mass support. In the early years, the government sponsored the removal of landowners, the redistribution of land and the settling of local injustices and grievances. Mao Zedong (Mao Tse-tung) saw the peasants in 1949 as being 'like a blank sheet of paper' and on such paper 'the most beautiful pictures could be painted'. The rebuilding of agriculture and its peasantry was seen as the basis of New China.

Agricultural development has generally remained a top priority in national economic policy since 1949. Consider the statement by Chairman Hu Yaobang in September 1982 to the 12th National Congress of the Communist Party of China:

'Agriculture is the foundation of the national economy, and provided it grows, we can handle the other problems more easily. At present, both labour productivity and the percentage of marketable products are rather low in agriculture; our capacity for resisting natural calamities is still quite limited; and, in particular, the contradiction between the huge population and the insufficiency of arable land is becoming ever more acute. From now on, while firmly controlling the population growth, protecting all agricultural resources and maintaining the ecological balance, we must do better in agricultural capital construction, improve the conditions for agricultural production, practise scientific farming, wrest greater yields of grain and cash crops from limited acreage, and secure the all-round development of forestry, animal husbandry, sideline occupations and fishery in order to meet the needs of industrial expansion and of higher living standards for people.'

?

1 Analyse Hu Yaobang's statement. What problems does he see for China's agriculture? What means are being proposed to solve these problems? What are the end aims of these efforts?

2 Compare the agricultural problems and proposed solutions with another less developed and/or developed country you have information on e.g. Peru (see Chapter 5), and/or USA (see Chapter 6).

Chinese agricultural development raises a number of issues relevant not only to China's own millions of people but to the world. The major one is whether or not China will be able to feed itself — and to feed itself at the increasingly higher standards demanded. If the answer to this question appears to be 'no' under the present system, what impact will this have on national policies? How relevant is China's model of agricultural and economic development to other developing countries and regions if it doesn't work effectively in China itself? Furthermore, if China were to become self-sufficient what would be the impact on the economies of grain-surplus producing countries? Looked at another way, if China continues to be 'grain deficient', will grain and aid needed by other developing countries be diverted to China?

HOW TO PROVIDE ENOUGH FOOD FOR OVER 1000 MILLION PEOPLE?

How much food is needed?

A useful starting point is the amount of food being used each year in China. In 1982 China produced about 353 million tonnes of grain. Most of this grain was rice and wheat, the main items in Chinese diets. Other grains such as corn, barley, kaoliang, millet and rye are widely grown but in smaller quantities (see Figure 1). (For statistical reasons tubers such as sweet potatoes are included in this figure as 'grain equivalent'.) Another 15 million tonnes of grain were imported, and less than one million tonnes of Chinese rice were exported. Therefore, a total of about 367 million tonnes of grain was available for 1008 million people. This is about 365 kilos for every person for the year. Other foods — vegetables, fruit, meat and eggs are, of course, important additional items to the basic grains.

Most peasant households have grain stores that can be drawn on, mainly in cases of emergency. In successful years these stores should be added to. However, losses due to mildew, rodents, pests and poor transport are widespread, but there are few available figures on such losses to permit valid reductions from the overall figures.

A per capita target

Chinese planners argue that the average grain supply per person should be at least 400 kilogrammes a year, and that this target should be reached as soon as possible. Even so, per person output would be very much below that of the USA and France (see Table 1), but close to the world average.

Even if China reaches its target of 400 kilogrammes of grain per person per year by the end of this century, by then, the nation's population will probably have grown by another 200 million people. In other words, China's grain requirement by the end of the century will be at least 480 million tonnes. This is almost 130 million tonnes more than in 1982. Raising production levels by an average of seven million tonnes a year over 18 years is not a simple task when substantial increases have already taken place — Table 2 shows how China's grain production has varied over the years.

A further dimension can be added to the issue — China's population has been told repeatedly by the party and government that the sacrifices and hard work of the past and present will make for a better tomorrow. People's expectations are that living standards, including food, will be better; that not only will food be more readily available, with no rationing of basic items, but its quality and range will be significantly greater than at present.

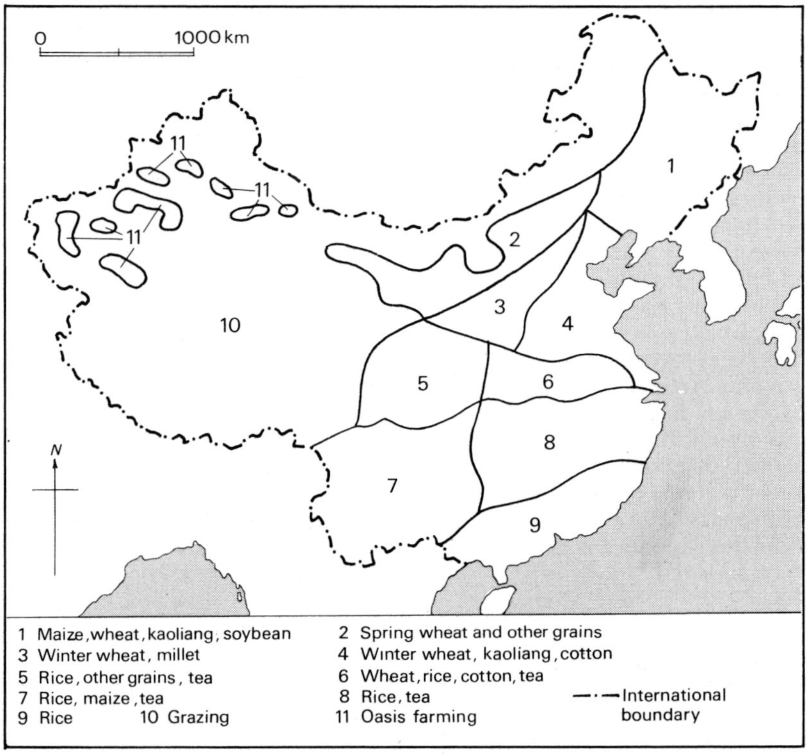

WHY NOT IMPORT GRAIN?

Importing grain is one possible solution to deficiencies, but could create even more problems:

1 For a country as poor as China, grain imports are very costly. The 1981 grain imports totalled a new high of $US 2.9 billion. Preferably imports are paid for with money earned from exporting goods. The alternative is raising money through foreign loans. Such indebtedness was a feature of the old China and seen as an indication of the country's weakness. Further, importing grain which is soon used up is not as cost-effective as importing machinery or fertiliser, plants or frozen cattle semen. Importing grain means there is less money available for importing these more productive goods.

2 Relying on imported grain could be interpreted as a weakening in China's strategy of self-reliance. Government policy has always been directed towards self-sufficiency in grain production, even at the cost of growing more profitable crops. Without becoming self-sufficient, China is more dependent on outside influences such as fluctuations in wheat supply on the international market. The effects of the 1981–82 Australian drought, for instance, temporarily put into question one important source of supply.

3 A dependence on imports could provide an incentive for peasants to discontinue grain growing and switch to more profitable non food crops such as cotton and tobacco. Whatever advantages this may bring to individual peasants and industry, it would further accentuate the problems outlined above.

Figure 1 *Agricultural regions showing major crop combinations and pastoral areas*

Chinese wheat-buying team in Canada

China has been one of Canada's biggest customers of wheat since 1967. Last year the Chinese took more than 3 million tonnes in spite of serious transportation and handling problems which forced some sales to be deferred.

(*Source: Financial Times* 1982)

1 From Table 2, suggest the likelihood of China's grain production continuing to increase by an average of seven million tonnes a year to the end of the century. Why might using past increases to predict future increases be unreliable? 1962 and 1980 were both periods of severe drought: how does this show in the figures?

2 What do you notice about the amounts of grain available on a per person basis for China as recorded in Table 1? What do you notice about France's 1950 figure and China's 1983–4 figure? Why might a comparison with France or the USA not be particularly valid? Which other countries might have made a better comparison?

Table 1 *Grain output per person*

	1950	*1983–84*
China	238	380
USA	1000	1520
France	405	864

Table 2 *China's grain production (millions of tonnes)*

Year	*Production*	*Change from last year shown*
1952	163.9	
1957	195.1	+31.2
1962	182.0	−13.1
1967	215.0	+33
1972	242.0	+27
1977	276.0	+34
1978	304.7	+28.7
1979	332.1	+27.4
1980	320.5	−11.6
1981	325.0	+ 4.5
1982	353.4	+28.4
1983	387.3	+33.9
1984	407.1	+19.8
targets:		
1985	360.0	
2000	460.0	

3 Read the newspaper extract and study Figures 2 and 3. Then answer the following questions.

a Locate Hubei and Hebei Provinces on the map in Figure 3. Why do you think floods in the former province and drought in the latter might cause more serious problems in terms of food supply than if similar natural disasters occurred, say, in Guizhou Province?

b Having looked at the maps, do you think the aid official is right when he says that China is 'a country of abundance in some places and shortages in others'? If so, what particular demands does this inevitably place on the government and the economic organisation of the country?

c 'Subsistence economies are less in need of a complete transport network than commercial economies.' How true do you think this statement is, particularly in relation to China?

4 Look at Figures 1, 2, and 4

a Write a short paragraph drawing out the major correlations between China's natural landscapes, the distribution of the major cropping areas, and the types of agriculture practised.

b What factors, in particular, make the eastern and southern provinces more suitable for wheat and rice production than the northern and western provinces?

c Does Figure 4 go some way to explaining why only 11 per cent (100 million hectares) of China's total land area can be cultivated? How?

5 China's stock of cultivable land represents a low 0.1 hectares per person. On the fertile and highly productive Chang Jiang (Yangtse River) Delta the figure is close to 0.07 hectares per rural person (see Figure 4). The Japanese and UK figures are 0.2 and 0.12 hectares per person respectively. What differences between the Chinese economy and the economies of Japan and the UK, make the Chinese figure particularly critical?

6 'China, with one-third of the population of the developing world, could threaten aid supplies to other countries if food crops failed in any year.' Discuss in class why world aid organisations might be particularly anxious for China to become self-sufficient in food production. Why might a country such as Canada not be so keen for this to happen?

Chinese millions join hungry queue for aid

by Isabel Hilton

A CONFIDENTIAL United Nations report claims that China urgently needs seven months' food for an estimated 20 million people in two provinces. A high-level Chinese mission will fly to Geneva this week to discuss China's 700-million-dollar aid application with representatives of potential donor countries and the United Nations Disaster Relief Organisation (Undro).

Last week an EEC ministers' meeting voted only a token £1 million. Western analysts and aid agencies are desperately trying to find out if reports of millions in China facing starvation are true and to assess the impact of China's huge population joining the queue for food aid.

Undro supports the Chinese application, although an official in Geneva last week insisted there was no question of starvation. But western analysts are puzzled by the application, pointing out that Chinese officials have claimed that the 1980 harvest was good. If millions really are facing severe shortages, it may mean that China has been hiding agricultural chaos more serious than the authorities have admitted.

China's request is a departure for the People's Republic. Until last year China had prided itself on an ability to feed the 1000 million population. But last December, after natural disasters affecting more than 43 million people, an Undro inspection team was invited to flood-damaged areas of Hubei Province and drought-stricken Hebei. During their visit in January, China formally requested $700 million worth of relief.

The scale of the Chinese request reflects the sheer size of the Chinese population — one prefecture in Hubei, Jingzhou, is the size of Holland. China, with one-third of the population of the developing world, could threaten aid supplies to other poor countries.

A World Health Organisation official who accompanied the UN team reported signs of malnutrition in both provinces. But western analysts are puzzled by Chinese claims of food shortages in a year which the same authorities have claimed overall as the second best harvest since 1949. On top of the officially-reported harvest of 321 million tonnes, China has raised wheat purchases this year to an estimated 14 million tonnes. If both these figures are correct, China should have enough grain to meet local shortages.

One possible answer to the puzzle is that official harvest figures were grossly exaggerated. It is also likely that peasants are retaining more grain for themselves and the government is having difficulty buying enough from areas of abundance to relieve shortages elsewhere.

One aid official said: 'China is going to have a shortage somewhere every year — it's a country of abundance in some places and shortages in others. Now she is a member of all the appropriate organisations, she may just be testing out international willingness to help. If it works this year, she could be back every year.'

(*Source: The Sunday Times*, 22 March 1981)

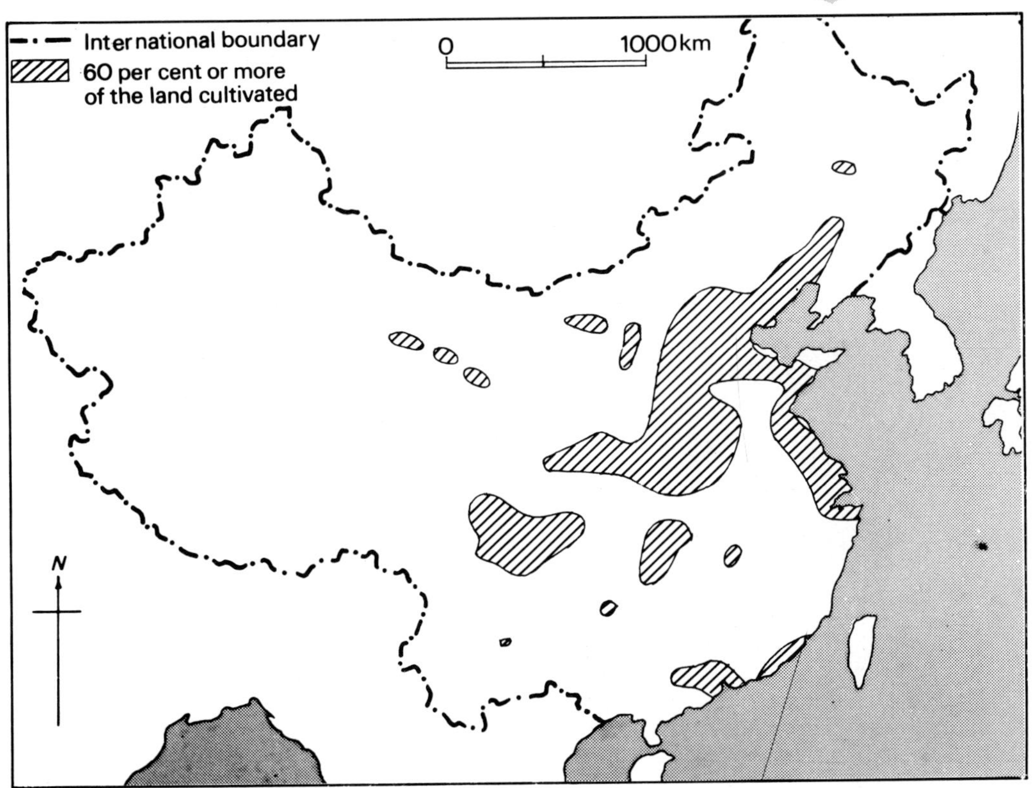

Figure 2 *Major cropping areas*

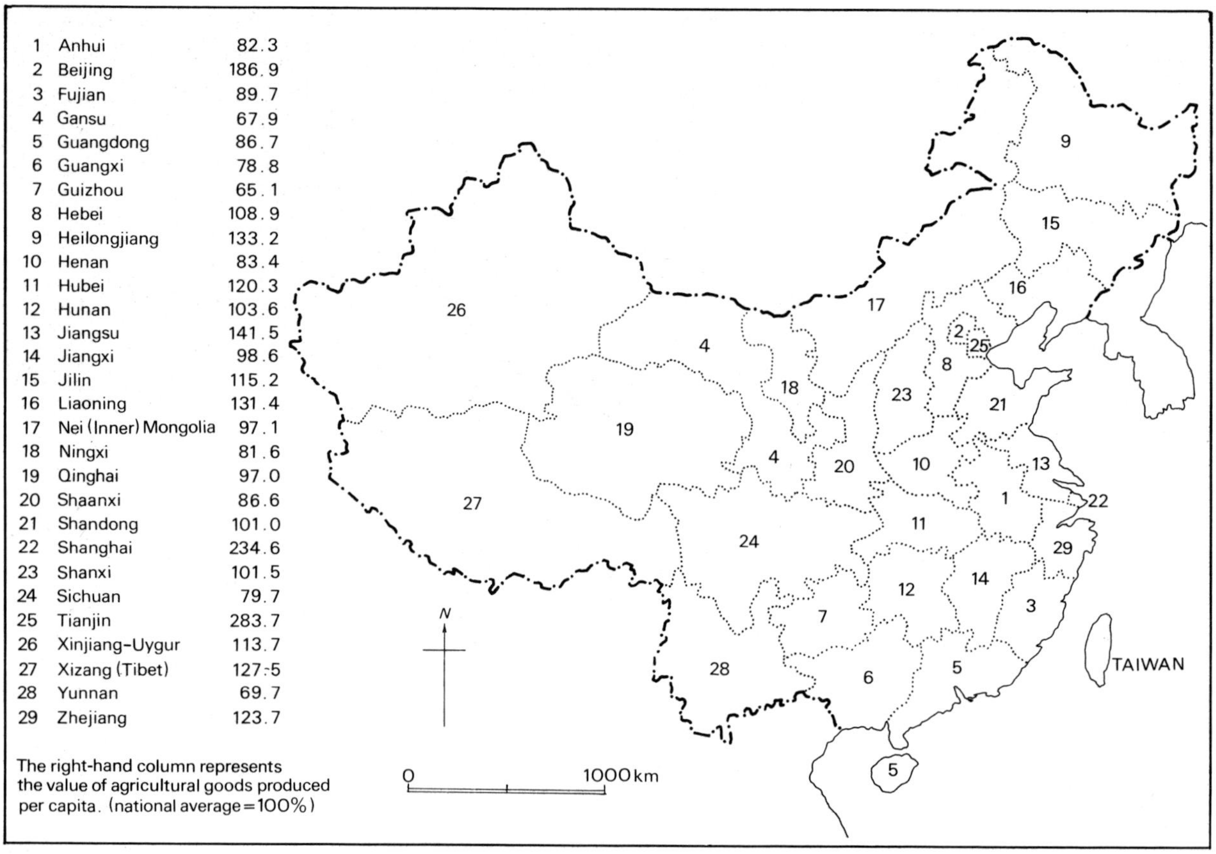

1 Anhui 82.3
2 Beijing 186.9
3 Fujian 89.7
4 Gansu 67.9
5 Guangdong 86.7
6 Guangxi 78.8
7 Guizhou 65.1
8 Hebei 108.9
9 Heilongjiang 133.2
10 Henan 83.4
11 Hubei 120.3
12 Hunan 103.6
13 Jiangsu 141.5
14 Jiangxi 98.6
15 Jilin 115.2
16 Liaoning 131.4
17 Nei (Inner) Mongolia 97.1
18 Ningxi 81.6
19 Qinghai 97.0
20 Shaanxi 86.6
21 Shandong 101.0
22 Shanghai 234.6
23 Shanxi 101.5
24 Sichuan 79.7
25 Tianjin 283.7
26 Xinjiang-Uygur 113.7
27 Xizang (Tibet) 127.5
28 Yunnan 69.7
29 Zhejiang 123.7
The right-hand column represents the value of agricultural goods produced per capita. (national average=100%)
N
0
1000km
TAIWAN

Figure 3 *The agricultural productivity of China's administrative regions, 1979*

Amur
Huang
Chang
Xi
EAST CHINA SEA
BAY OF BENGAL
N
Land over 2000m
1000—2000m
under 1000m
International boundary
Major rivers
1 North China Plains
2 Sichuan
*3 Lower Chang Jiang (Yangtse R.)
4 Quinghai-Tibet Plateau
5 Tarim Basin
6 Loess Plateau
7 Taklimakan Desert
8 Himalayan Mountains
0
500
1000km

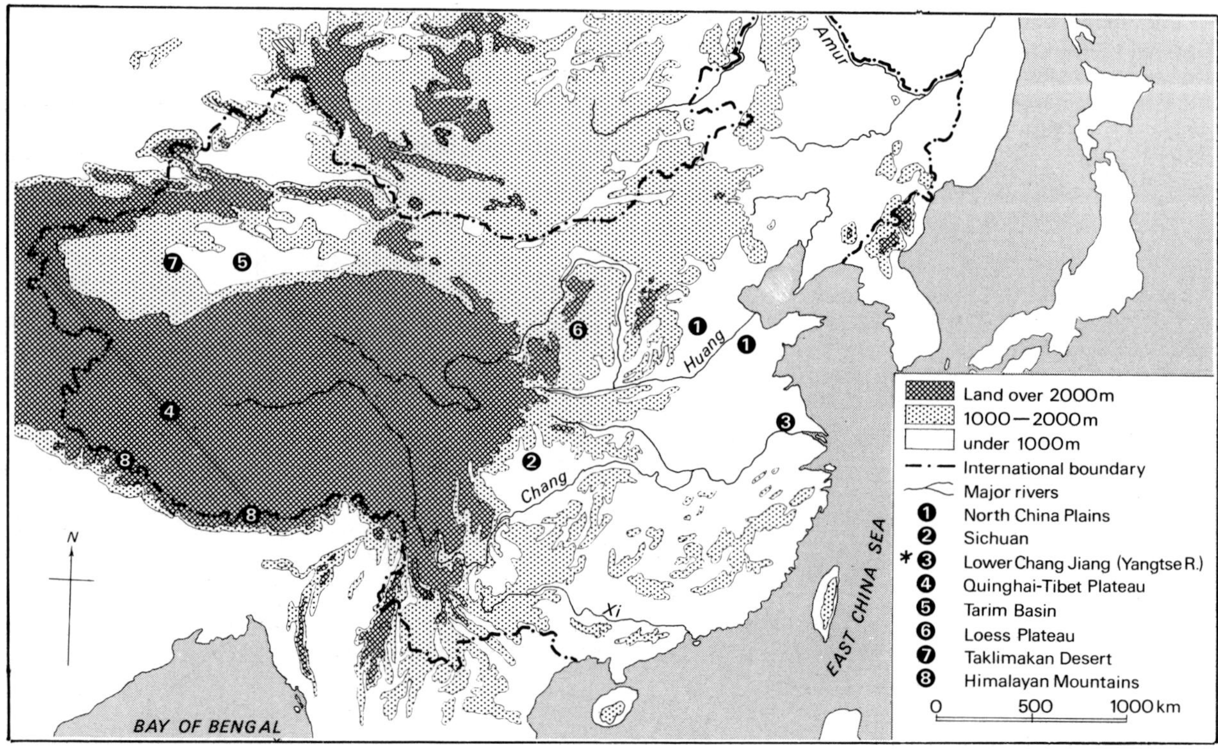

Figure 4 *China's natural landscapes*

? **5** Study Figure 3. The value of agricultural goods produced per person in each province before agricultural reforms were introduced, are listed on the left. 100 per cent is taken as the national average. Below this figure a province is performing as a whole less satisfactorily than other provinces. Above this figure, a province is performing more successfully than the national average.

It is generally believed, by both Chinese and foreign experts, that agricultural productivity is bound to be highest in areas with favourable environmental and marketing conditions, i.e. the well-watered plains where multiple cropping can take place, and areas nearest to cities, where denser transport networks exist to move surplus goods to a market. In other words, the large areas of the provinces of eastern and southern China. Test this hypothesis as follows:

a Rank the figures from highest to lowest.
b Decide on five categories of figures including two below the national average.
c Decide on five shadings from darkest to lightest to correspond with your categories of agricultural performance.
d Copy the map outline from Figure 4.
e Shade in the map according to your categories.
f Analyse what you find. How much of the hypothesis stated above is supported by your map? How much of it is rejected? Can you go any way towards explaining the anomalies? Is the hypothesis over simplistic? Should future planning and investment decisions relating to agricultural improvements be based on it?

Print 3 *Galvanisation of irrigation pipes for local use being carried out in a commune workshop*

THE ORGANISATION OF FARMING LAND AND PEOPLE — WHICH WAY IS BEST?

THE STRUCTURE FROM 1949–79

After 1949, the Communist government organised the peasants first into mutual aid teams, then into agricultural co-operatives.

Communes

By 1958, adjoining co-operatives were merged into *People's Communes*. It was this unit that was regarded, until 1979, as the most effective way of improving agriculture and providing a better, more equal life for all.

There was considerable variation in area, population and activity between communes. Typically, communes were broken down into *production teams* which were usually organised within one village. A team would decide on what, when and how to plant — as long as it met with local party approval.

Peasants were paid according to work points earned for specific jobs each day. The value of a work point was determined by the income earned by the team through the sale of surplus produce, divided by the total number of work points. The greater the value of the surpluses, the greater the value of the work point. This, together with the peasant household's small private plot from which any surplus could be sold privately, provided the material incentives for raising production.

Teams operating together formed *brigades* which ran industrial and service units for the teams. Overseeing all activities was the commune, whose committees planned finances, directed major changes in farming, co-ordinated activities with neighbouring communes and the provincial government on regional schemes, and developed more elaborate services such as high schools and hospitals. Their role was multifunctional, being responsible for county administration, education, industrial and agricultural development.

State farms

In addition to the communes there were over 2000 state farms, mostly where less favourable environmental conditions required large amounts of government money to develop agriculture. The workforce, who were paid fixed wages, frequently included directed labour such as soldiers. Few state farms became profitable because of their high operation costs in agriculturally marginal areas.

Through the communes and state farms the central government directed the people of certain areas to concentrate on particular crops and animals. Agreements between communes, provincial governments and central government authorities were made on the amounts to be purchased by the central authorities. Within a commune, production teams would be directed to sell grain to deficient areas, cities or the armed forces.Surpluses could be sold to the state or at local markets by the production teams.

The priority of grain

Growing grain remained the key to agricultural development, frequently at the cost of not growing crops or raising livestock more suited to particular areas. In Shandong Province for example, production teams had to reduce their cotton area to plant more grain. Low grain yields returned a low income, particularly in the north-western parts of the province, which in turn prevented teams, brigades or communes buying chemical fertiliser or new seed strains that would raise either cotton or grain production.

1979 — A YEAR OF CHANGE

Between 1949 and 1979, grain production increased by almost 300 per cent, and rural living standards in terms of literacy, life expectancy, health, housing and material possessions also increased quite dramatically. However, it is justifiable to ask whether even greater advancements could have been made if a different organisation of the rural sector had been adopted.

In 1979 major policy changes were introduced by the central government in the hope of increasing agricultural production. The newspaper extracts below highlight just how 'revolutionary' these changes were.

Chinese launch farm revolution

From David Bonavia, Peking

Continuing anxiety about crops starved of rain in northern China coincides with moves by the Communist Party to change the basis of the Chinese diet and fundamentally reorganise the rural communes.

A spokesman for the United Nations Food and Agricultural Organisation has said China's demand for imported grain will become a serious world problem if it continues to rise at the present rate. China has been importing between three million and 10 million tonnes of grain annually, mainly wheat and maize, over the past few years, the big fluctuations being explained by vagaries of the weather.

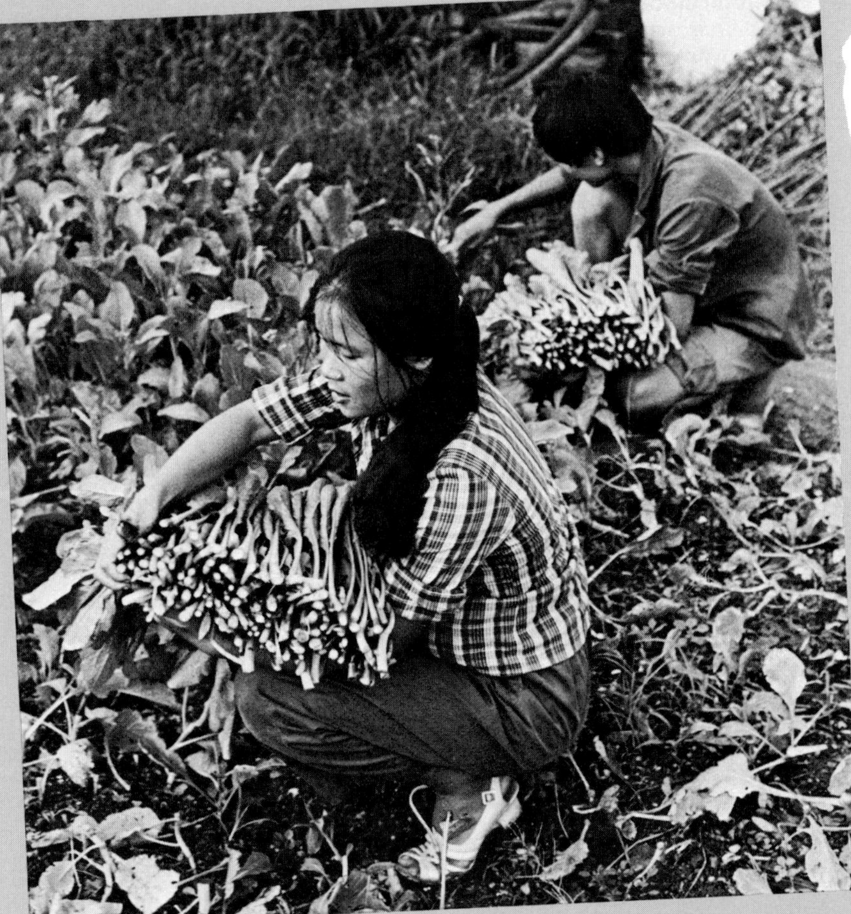

The drought which caused a serious shortfall in the national harvest last year persists in areas of southern and south-eastern Hebei province, surrounding Peking (Beijing).

The province's wheat crops, covering more than six million acres, will be poor if more rain does not fall soon, though in well-irrigated parts the wheat is growing well.

Flooding, which brought havoc in central China last year, is also thought likely to recur, especially in the Yellow River basin in Hunan province. Upwards of 20 million people risk famine again this year if the weather is unfavourable.

The Communist Party's plan to counteract the bad weather and other farming difficulties is not based simply on conserving water more effectively.

The whole organisation of the people's communes — units numbering from a few thousand up to 20 000 or more people — is being changed to permit the peasants greater initiative in the interests both of enriching themselves and improving food supplies to the state, the city dwellers and the armed forces.

More and more communes across the country are encouraging peasant families to sign individual contracts for their deliveries of grain and other products, thus encouraging them to work harder out of self-interest.

The first casualty in this change of policy is the system of work-points, on which Mao Zedong's concept of the people's communes was founded in 1958.

Instead of being graded according to their strength and enthusiasm, then paid according to the number of hours spent working communal lands, peasants who sign delivery contracts with commune organs will now be permitted to sell or retain all surpluses.

Economic planning is being personalised. The peasants will be assigned portions of communally owned land to work with communally owned implements, and their earnings will depend exactly on their work, the first principle of Marxist socialism.

This system, it is hoped, will end time-wasting paperwork and individual squabbles connected with the post-harvest share-out of grain and other basic products practised under Mao's system.

The new policy is a streamlined version of one put into practice after the end of the disastrous Great Leap Forward, which by 1961 had beggared the peasants and brought about widespread starvation.

It was designed by the late Liu Shaoqi, former head of state and Mao's arch-enemy, and by Mr Deng Xiaoping, who is today Vice-Chairman of the party and the leading policy maker.

Huge increases in production have been claimed by rural areas practising the contract system (though observers have learnt to be wary of such boasts), and it will probably become standard throughout China within a year or two.

Under this policy, the communes and subsidiary organs become no more than units of local government, education, health care and so on, with little say in the work or incomes of the peasants.

The other policy change being pushed just now is to remove from grain production those parts of upland, forest, steppe, or water-logged areas laboriously reclaimed at Mao's insistence to grow wheat, maize and other crops which could never flourish there.

The catchphrase of the moment is 'Greater Grain', meaning food supplies as a whole. However, it will take long and patient work to persuade the Chinese, and especially the peasants, that a diet based more on meat, fish, eggs, fruit and other nutritious foods can ever be satisfying without a huge bowl of rice or a pile of steamed bread.

(*Source: The Times*, 12 June 1981)

China's rural communes on the way out

From TONY WALKER in Peking

In the flat, sandy soil of the Yellow Mound Commune, south of Peking, peasants are hard at work in the weak spring sunshine planting vegetables and digging wells. It is a scene as regular as the seasons themselves here on the broad North China Plain.

But behind this peaceful cameo of rural Chinese life, a quiet revolution is under way that will affect the working conditions of hundreds of millions of peasants.

China's new leadership, which is more interested in results than dogma, is dismantling the late Chairman Mao's proudest achievement — the rural commune — and reverting to a pre-1958 system of agricultural management based on the xiang or county town.

In 1958, almost at the stroke of a pen, Mao, in his efforts to stamp out the last vestiges of capitalism, established a rigid system of communisation which was to have the effect of stifling initiative, thereby reducing the flow and variety of produce through State-run markets.

If the object of successful farming is to feed the population adequately then Mao's more than 50 000 communes could not be judged a crowning success. Instead, they became instruments of strict political and economic regimentation which reached its most intense phase during the Cultural Revolution, when almost all private production was ruthlessly stamped out.

Many older-generation Chinese peasants remember with pain the 'bad years' of the so-called Great Leap Forward which followed the establishment of the commune system, when Maoist exhortations to plant more grain even in areas unsuitable for such cultivation produced widespread chaos and famine.

When the winds of economic change started blowing across China after the death of Mao and the rehabilitation of Deng Xiaoping, the pragmatic party vice-chairman, one of the new leadership's first reforms was to allow peasant farmers more freedom in the crops they grew and in the way in which they marketed their produce.

This led to the re-allocation of private plots of land to the peasants and the re-establishment of what the Chinese describe as 'free markets' as an alternative to State-run outlets. Prices in these so-called 'free markets' are, in many cases, higher than in State stores, but the quality of the produce is often better.

The free market system, based as it is on production from peasants private plots of land, has, judging by all reports, helped re-vitalise agricultural production in China. Peasants work harder and are better off.

(*Source*: *The Age*, 1 April 1982)

Back to the family in the fields

By Greg O'Leary and Andrew Watson

Xian: The 1981 harvest in north China brought clear evidence of the impact of land and labour management changes in the communes. Using stalks of harvested maize as markers, peasants divided up the large collective fields of communes into smaller plots for cultivation by households or labour groups. The variations in width of the long strips of farmland — from 2–3 ft up to several yards — underlined the major role that the household unit is again playing in Chinese agriculture.

As the newly-planted winter wheat began to sprout, these changes became more obvious. Some strips remained unplanted, others bore wheat in various stages of development and others retained the stalks of previously-harvested wheat. At harvest time there will thus be wide variations in yield, and those families with skills and abundant labour will generate large, income-boosting surpluses.

For those lacking skills and labour there will be problems in paying the state grain levy, let alone in producing a surplus. Subsequent years will see these problems increase for families unable to afford fertiliser and other inputs — in many cases now the responsibility of the family unit. This potential for polarisation is already apparent in places where the systems of contracting land to the household were introduced in 1980, and while most commune leaders and agricultural economists insist that the overall effect will be to raise average standards of living, they concede that differences in incomes will grow. A few feel that major social differences will eventually emerge.

It is clear, however, that the move away from collectivisation and towards household units has gained momentum. Perhaps 30% of collective units have 'gone household,' according to official estimates; unofficially there is a consensus that the level is 50% and rising quickly.

Many of the changes were banned at the Communist Party's third plenum in December 1978. They have become acceptable only as part of the current 'production responsibility system', encompassing various land and labour management forms, which is intended to relate work done to income. Document 75 of September 1980 identified three main types of production responsibility systems:

- ► In high-income collectives, peasants (individually or in groups) contract with the collective for specific work within a collective economy.
- ► In intermediate forms, land is divided into small plots for farming by individuals, families or groups on a piece-work basis but accounting and distribution of income is collectivised.
- ► In *bao qan dao hu* (contracting land use to households), land is turned over to the family unit based on its size, and the family can farm as it likes provided it guarantees to meet state quotas and any collective levy.

Document 75 envisaged that this third system would be appropriate for poor, remote areas which had progressed little under collectivisation. Those units which had to 'buy grain back from the state, borrow funds for production and rely on relief funds for survival' were permitted to divide land for household use. In those areas, it was stated, the peasants 'have lost faith in the collective'.

(*Source: Far Eastern Economic Review*, 26 February 1982)

China brings back the profit motive

From TONY WALKER in Peking

When a Chinese newspaper reported recently that more than 1000 restaurants in Shanghai were tossing out the 'big pot', it had nothing to do with a change in cooking methods. Rather, the report indicated a new contract system was being implemented to encourage profitability.

Under the new system, Shanghai's restaurants will be responsible for their profits and losses. If service is poor and the quality of the food bad, employees will get less money.

China is now in the midst of a nationwide campaign against 'big pot' practices, a widely used euphemism to describe old systems of management which guaranteed workers wages and benefits whether the enterprise prospered or not.

There was little incentive for individuals to work harder because all employees fed out of the same 'big pot' or, to use a variation on the same theme, were in possession of an 'iron rice bowl' which could not be smashed.

The English-language 'China Daily' said in a recent commentary: 'It is unquestionable that the iron rice bowl and eating from a single pot must go. Everybody in the country, except the lazy, supports the application of the principle, he who works more earns more.'

'China Daily' might be right — up to a point, but there is considerable evidence of significant and widespread opposition to Deng Xiaoping's new policies which are considered anathema by old-line officials brought up on a Maoist diet.

Mr Deng's carrot-and-stick policies are resulting in big disparities in income. 'Peking Review', an official weekly news magazine, attempted to answer criticism against the new responsibility system in agriculture in a commentary last June which said that the differences between rich and poor had nothing to do with 'class polarisation.'

'The responsibility system excludes the possibility of getting rich by exploiting others and therefore class polarisation will not take place with one family dominating the land while a thousand others go bankrupt', the magazine said.

'Peking Review' claimed that differences in income were 'unavoidable' in socialist society and need not be feared. It shows, it said, the way to become prosperous by one's own sweat, and in this it is a prime mover for the growth of production.

This message obviously hadn't got through to officials in Hunan Province, south China, who stymied efforts by local peasants to establish a transport business.

'People's Daily', the Communist Party newspaper, reported late last year that the officials, when they saw the peasants prospering from their newly-established business, declared that these rural entrepreneurs were really capitalists and had locked their vehicles up.

(Source: *The Age*, 21 January 1983)

? Summarise in less than 100 words, the major agricultural changes discussed in the preceding articles.

Table 3 *Changes in Crop Areas (millons of hectares)*

	1969–71	*1979*	*1980*	*1981*	*1982*	*1983*
Wheat	25.4	29.3	28.8	28.3	27.9	28.8
Rice	33.3	34.6	33.9	34.5	33.7	33.9
Barley	2.4	1.4	1.5	1.2	1.1	1.2
Maize	15.9	20.1	20.0	19.4	18.5	19.9
Rye	1.2	0.8	1.0	0.9	—	—
Oats	1.0	0.4	0.5	0.4	0.5	0.5
Millet	6.9	4.1	4.1	4.0	4.0	4.1
Sorghum	5.4	3.2	3.1	2.6	2.8	2.8
Roots and tubers	11.2	11.3	11.1	10.9	10.4	10.5
Soya beans	7.8	7.2	7.5	8.0	8.4	7.6
Seedcotton	4.9	4.5	4.9	5.3	—	—
Sugar-cane	0.5	0.6	0.6	0.7	—	—
Sugar beet	0.2	0.3	0.5	0.6	—	—
Cotton lint	2.2	2.2	2.7	3.0	3.2	3.5

The responsibility system
Collective obligations, such as working on infrastructure projects or harvesting collective grain still remain, and payments are still via work points earned for efforts in these areas. However up to 15 per cent of the collective land can now be assigned to individual households as part of a *responsibility system.* Households or groups of households sign contracts with the collective to be responsible for producing particular quantities of grain, vegetables or livestock. That output is sold to the state through the production team. Sometimes the contract may be to provide services such as ploughing or processing. After fulfilling the contract, the balance of the output or service opportunities can be used by the household or sold for personal profit. This is the scheme's greatest advantage — it doesn't repress individual initiative and effort in producing either quantity, quality or in developing more efficient techniques.

All variations of the responsibility system have been aided by increases in state purchasing prices for agricultural goods, and expansion of local fairs for the sale of surpluses. Land, tools and the infrastructure are still collectively owned. Land cannot be sold or rented out. The commune still plans production for its area in line with state guidelines, and it still initiates developments in agriculture and living standards.

Since producing more brings direct increases in income to individual households, there have been widespread increases in production and diversification into new types of farming wherever the system has been taken up. Nationally, grain production has risen (see Table 2, *page 160*), although since 1979 the total area planted with grains has fallen. Between 1977 and 1981 cotton, sugar, oil seeds and raw silk production rose by 10, 15, 26 and 10 per cent respectively. Yet these years include a period of indecisive central authority directions, a severe drought in north China and widespread flooding in Sichuan Province. Some of these changes can be picked out from Table 3.

Earliest successes with the responsibility system have come from areas where previous agricultural development was slow and peasant incomes frequently below the national average. Many of these areas were ones least suitable for intensive grain production yet, because of previous policies, found themselves concentrating on this. For instance, counties in north-west Shandong (see Figure 3), Anhui, Henan and Gansu Provinces have now been able to show considerable gains in production and incomes.

Critics within China argue, however, that if the responsibility system develops further it may be capable of reproducing the social injustices and economic inefficiencies of the past. The main features of this discussion are taken up in the interviews that follow.

Peasant 1

'In our area we used to concentrate on growing grain because the State and Party cadres said we should. We even terraced nearby hills to produce very small strips of land on which to grow grain. Now some families are replanting them with fruit and eucalyptus trees. We are concentrating on crops that give us a better return for our efforts — such as cotton and fruits. I'd like eventually to use some of my allocated land to build sheds for raising pigs in stalls. The market in pigmeat in nearby villages and the city would make this very profitable. Besides, our family would really like to buy some of those items that are advertised; a bike or television set for example. Under the present system it has taken us over two years to save for just one bike. As for grain, there must be enough households in enough brigades throughout China that produce a surplus for people like us.'

Peasant 2

'There have been many changes in Chinese agriculture — some for the better. Peasants are very enthusiastic about the present system because they have greater freedom to grow what they feel is

? **1** Draw divided rectangles to show the production of crops in millions of hectares, for **a** 1969–71 and **b** 1981.
2 Pick out the major changes in crop areas between these two periods. In what way could these statistics *on their own* give a false picture of China's production changes over the same period?

best, particularly in relation to market prices. I'm all for keeping enthusiasm up if it increases production. However, it should not mean that we peasants move away from what is our national responsibility of growing grain. We should, of course, grow grain and vegetables for our own needs. But we should concentrate on producing a surplus of grain for sale to the cities through the state. The cities need grain, but grain imports are much too expensive. The cities provide us with manufactured goods and educated people. I think the state should continue to set production targets and that we should meet them.'

Peasant 3

'We thought the old commune system wasn't too bad. It did, after all, raise our living standards by guaranteeing food, housing, work and education. We no longer plough with animals since machines are becoming so common. The provincial government's irrigation system allows two crops of grain to be grown in a year. Now we are responsible for producing an agreed amount. I like working my own area as I want, but there is only me. We have two very young children and my wife isn't strong enough to work in the fields. She's been doing light work in a commune factory. I have to rent machinery from the brigade and this year I'll have to hire a labourer. That's a bit like the old system my parents grew up with and fought against in the 1940s. I'll keep at it, but hopefully the commune will be around to stop our living standards falling and to keep other households with more able-bodied family members from getting too rich.'

Peasant 4

'We are very happy. There are five of us who work our allocated piece of land. We are now growing early vegetables under plastic sheeting. We aim to be first into the city market and that means we ask and get a higher price than later in the season. Then we buy a lot of chemical fertiliser for our land and grow grain through the summer and autumn. I used to argue in brigade and team meetings over the years that this is what we should all be doing, but I was always shouted down. So now we are doing it our way, and it works. Our first major purchase was our own tractor to use on our area and move products to the city. Now we are planning an extension to our house. I have three reservations: Firstly, I know some of the older party officials are jealous of our success and call us capitalists — I'm sure they'll do everything they can to stop our success and take away our profits. Secondly, I just hope the supply of raw materials will keep coming — deliveries in the past often didn't happen. Finally, and this really bothers us, what if the government changes their policy in a few years to something else?'

?

1 Identify from the four interviews above the following:
a those who most regret changing the commune system.
b peasants most likely to gain from the responsibility system.
c peasants most likely to lose within the responsibility system.
d peasants who would possibly support one another in changing the system further;
e national needs;
f ideological viewpoints.

2 Divide into small groups. Each assume the identity of one of the above peasants. Plan your strategy of production for the next year and outline plans for the next five years. How do your plans differ from those of others in your group? For success, on what outside factors does your plan depend?

3 Consider the following questions, adopting first the viewpoint of one of the above peasants and then that of an interested outsider. Do the conclusions differ? Reread the newspaper extracts as a further source of ideas and arguments.
a Should national needs be more important than individual aims?
b Should the state dictate which food crops are to be grown or should it be left to peasant judgement?
c Should the state determine prices for all food or allow peasants to get whatever price they can in the markets?
d If the current responsibility system favours families with more able-bodied workers, won't this encourage larger peasant families and therefore be in conflict with the national policy of one-child families?
e Will the responsibility system lead to a new class of rich peasants and a disadvantaged poor class?

4 Are there any alternative ways of organising Chinese agriculture? What are they and how might they function? Which groups of people would benefit most or lose out with these alternatives? A starting point for your answer could be to compare agricultural systems in other parts of the world, such as the US (Chapter 6) or Mexico (Chapter 7).

Print 4 *Mechanisation of ploughing is increasingly common*

IMPROVING FARMING TECHNOLOGY — WHAT ARE THE PRIORITIES?

The organisational changes outlined in the previous section are not solely responsible for China's higher yields. Improved farm technology has had an equal, if not greater, impact. Chinese planners recognise that future improvements do have economic limits and generate controversy as to the best way to spend the available money.

Development of farming technology is often complex. Using more chemical fertiliser, for example, involves a manufacturing process somewhere else, knowledge, techniques and money for purchases as well as a soil type capable of benefiting.

THE CONTROL OF WATER

This has been the most important advance in agriculture since 1949. Water control is needed to regulate and extend growing seasons, to avoid droughts and prevent flooding. Utilised for hydroelectricity it is the most important source of rural electricity. Arable land under irrigation now exceeds 45 per cent compared with 26 per cent in the late 1940s — figures high for anywhere in the world.

In the immediate past many large-scale water control and supply projects were built. These were complemented by commune and county schemes. Table 4 shows the effect that the completion of the comparatively small (66 000 hectares) Shaoshan Irrigation Area Project in Hunan Province had on the local agricultural scene.

More effective use of water is partly responsible for the substantial increases in area — as much as 60 million hectares — that is multiple cropped.

Within the Chang Jiang (Yangtse River) Basin which supports one-third of China's population, over 500 large- and medium-sized reservoirs have been built to control the flow of water. Control involves the age old problem of flood prevention as well as supply for irrigation.

Table 4 *Double cropping and yields, Shaoshan Irrigation Area*

	1965	*1976*
Percentage of crop land with two rice crops	25	95
Yields per hectare a year, in tonnes	3.3	8.25

Investment on the drought prone North China Plain exceeds $US13 billion. The problem remains: North China has 51 per cent of the national cultivated area but only seven per cent of its surface water flow. Diverting Chang Jiang floodwaters northwards via canals is a costly project that in 1983 ensured irrigation water for one million hectares on the North China Plain. The decision to divert twice as much water again has been a controversial one. Certainly more crop land will be guaranteed water, but the effects of such large-scale changes on silting patterns, stream flows and soil ecology are largely unknown. Would it have been better to divert this project's funds to smaller projects where greater returns are more likely?

Certainly the emphasis with current water projects is towards improving conditions in existing irrigation areas to maximise past investments. Future large-scale schemes are likely to be few in number as the most profitable ones are already in operation.

FERTILISERS

Animal manures, human wastes, organic garbage, green manures, composts of weeds and stalks and mud from fish ponds are major sources of organic fertilisers. Of these, China's pigs, numbering over 300 million, are most important and, at one stage, pig production was encouraged with the slogan 'a pig is a walking fertiliser factory'.

Government policy has been to encourage small- and medium-sized chemical fertiliser plants run by communes or counties, together with large-scale plants on the outskirts of cities. Since the mid-1970s, imported plants each producing over 100 000 tonnes a year have added significantly to total output.

Increasing applications of both organic and chemical fertilisers, together with the extension of multiple-cropping and the more effective use of water,have increased China's yield per hectare of major crops. Table 5 summarises the change for rice.

Applications of chemical fertiliser in China are considerably higher than in other developing countries. Within China the variations are considerable. The more successful farming areas are able to afford to buy more chemical fertiliser and further increase their chances of success, while the poorer farming areas have very low application rates. Government subsidies to poorer areas are increasing but under the responsibility system, some families experience difficulty in affording fertilisers as the article 'Back to the Family in the Fields' points out (*page 166*).

Table 5 *Changing rice yields (kg per hectare)*

1952	2300
1969–71	3295
1979	4248
1980	4200
1981	4334
1982	4884
1983	4872

MECHANISATION

In areas of multiple cropping, timing is critical. Land for a second crop can be mechanically ploughed, levelled, watered, fertilised and replanted in a shorter period than by hand. The time saved extends the new crop's growing period.

Successful mechanisation of farming processes has tended to be limited to the richer communes of the plains, to the large state farms in the north-east and to areas closest to the cities. Mechanisation of the lengthy, labour intensive job of rice transplanting, is still far from widespread. In the 1980s animal and human power still dominate many areas but, overall, by 1980, close to 40 per cent of China's 100 million hectares of arable land was worked by machines.

FURTHER AREAS FOR IMPROVEMENT

Training People

More agronomists and agricultural technicians seem an obvious means of improving agricultural performance. Since the 1950s China has trained 860 000 agricultural technicians but, because of low pay and poor working conditions, more than half of these workers have become employed in other activities.

Improved plant and animal breeding

The diffusion of new seed varieties that have greater resistance to cold and drought is paying off. Development of hybrid rice and cotton that are particularly responsive to fertilisers and controlled water supplies have become increasingly profitable when applied widely. The increasing use of frozen semen from better animal herds within China (the imported product is often too expensive) has yet to have a substantial impact on herd quality, but offers considerable potential.

The 1958 Eight Point Charter

Many agricultural techniques don't require advanced farming technology or a great deal of money. The Eight Point Charter for agriculture promoted in 1958 includes many such techniques: correct management of the soil through deep ploughing; drainage and prevention of soil erosion; greater use of fertiliser; water conservation; better seed selection to increase yields on improved soils; close planting and intercropping to give higher yields; protection of plants from winds, insects and rodents; improved farming tools partly through mechanisation; more efficient management of fields.

In the future, the cost of increasing yields from existing farming land will be greater than in the past. Yields in many areas are already high (compared to the highest yields elsewhere in the world) and pushing them higher will cost a great deal in terms of developing new seed types, more effective fertilisers, education of the workforce to handle the associated advanced agricultural technology, as well as providing material incentives for producers to operate at these higher levels.

OPENING UP MARGINAL AREAS — IS IT COST EFFECTIVE?

Marginal lands are those where slope, soil and water supply have made farming activities much less profitable than elsewhere. Marginal lands are not necessarily remote ones — hilly areas adjacent to and within cultivated plains fall into this category. Over centuries these marginal lands have been victim to various attempts to cultivate them or to intensify their use as forests or pastures. In the 1950s and 1960s this was especially so as the policy of growing grain was heavily promoted. In the late 1970s and early 1980s a more pragmatic approach has developed. Hillsides cleared for crop production are being reafforested or laid out for orchards and other tree crops.

Livestock are now seen as a major source of protein and as an efficient user of much marginal farming land. Many peasant households in north

?

1 Which of the processes outlined above could be carried out within a small area without reference to neighbouring communes or regions?
2 Which of the processes seem most dependent on large-scale investment from authorities outside the communes?
3 China is not the only country with water supply problems and sometimes controversial solutions are employed to combat them. Select another country with a similar water problem and compare it with China's. Useful for comparison are Egypt, the Soviet Union, India, Pakistan and south-west USA.
4a In what ways might the various improvements outlined in this section risk increasing inequalities between families, communes and provinces?
b 'Even if improved farming techniques reduce equality, they must be pursued to achieve the ultimate goal of agricultural self-sufficiency for China.' Discuss this statement in a small group or as a class.

China have had their private plot allocations increased with areas of land most suited to livestock or to the raising of fodder crops. The result is the development of small herds managed by the family rather than attempts to grow grain on unsuitable land.

Chinese experts estimate that there are between eight and 13 million hectares of land that can be reclaimed for farming. Much of this land is in China's northern provinces and autonomous regions such as Inner Mongolia, Gansu, Xinjiang-Uygur, Jilin and Heilongjiang. Despite the desire to produce more food, Chinese planners now question whether using scarce national resources in such areas is really worthwhile.

Heilongjiang Province is one of China's frontier farming areas. In the 1950s most of its land was a wilderness with extensive bogs and exposed plains. Five million hectares are being brought into cultivation largely through state farms. Farming is difficult here as there are wide areas with less than 150 frost-free days a year.

The cost of turning wilderness areas into productive farmland is particularly high and such costs must be seen against long-term returns if they are to be considered viable agriculturally. The argument that money and resources spent developing these areas would be better spent in existing farming areas is a powerful one.

Print 5 *Intensive rice cultivation of gentler slopes in Hunan Province. The once terraced, steeper slopes in the background have now been reafforested*

1 Outline the main agricultural arguments for and against developing frontier areas. Should development costs be the major consideration in assessing a project's viability?

2 What other areas of the world can be considered marginal farming areas? What attempts have people made to open them up for farming? Evaluate their success. Are there any lessons to be learnt by the Chinese from these experiences?

HOW MUCH PROGRESS HAS CHINESE AGRICULTURE MADE?

Chinese agriculture has made progress. A comparison with the past shows this. Before 1949 few areas produced regular, large food surpluses. The majority of peasants found it difficult to survive from year to year because they could not build up supplies of food to withstand recurring natural disasters or local wars — today the Chinese population is relatively well fed. This is despite an increase of nearly 500 million people in 33 years, and a continuance of large-scale natural disasters such as the 1980 drought of north China or the summer 1981 flooding of the Chang Jiang (Yangtse River). To some extent the latter have been neutralised with the completion of large-scale projects on the major rivers. While some rural areas are still not self-sufficient in grain, surplus food from other rural areas feeds most of the 200 million urban population.

Grain yields have risen considerably in many areas. In some places, where triple cropping takes place, yields commonly exceed 20 tonnes per hectare per year. (Compare this with the UK average of 4.8 tonnes.)

Nevertheless, grain imports from the USA, Canada, Australia and Argentina are still necessary to supply continuing shortfalls. And there is evidence that Chinese agriculture could have developed further, faster and more efficiently than it has. Critics both within and outside China, point out that the majority of its agriculture is little better than semi-subsistence. They point to continuing rationing of grain, bean curd and oil in the cities. The most damning criticism of all would seem to be the changes in the organisation of agriculture since 1979, which have served to show up the imperfections of the old system. Did the strict commune system, far from answering China's agricultural problems, hold back overall agricultural development?

The exercises below may help you to come to some conclusions about China's agricultural progress. When completing them, consider the following points:

1 How valid or fair is it to make international comparisons?

2 Are comparisons best made with the past or with the current situation in other countries?

3 Should you be drawing any overall conclusions, or is it best to judge each piece of 'evidence' separately?

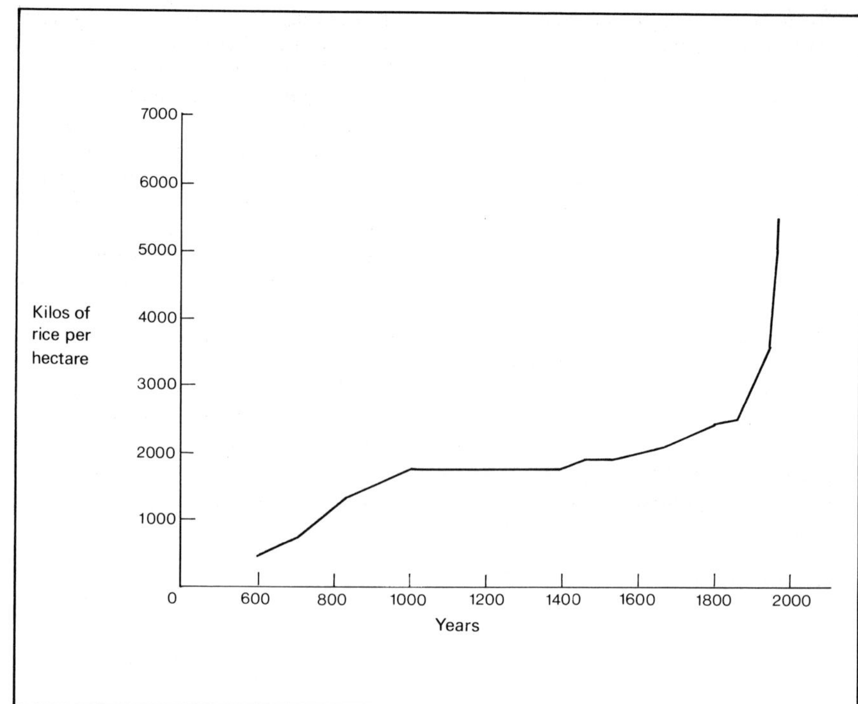

Figure 5 *Japan's rice yields*

? **1** The graph in Figure 5 shows Japan's rice yields over the last 1000 years. Make a copy of it. Then use the data in Table 6 to plot onto the graph the rice yields of other Asian countries. At what stage is China? How does it compare with other countries?

Table 6 *Rice yields in Asia, 1981 (kg per hectare)*

Bangladesh	2047
Burma	3085
China (1952)	2300
China (1983)	4872
India	2122
Indonesia	3637
Japan	5701
Kampuchea	969
Korea, North	6341
Korea, South	6193
Laos	1325
Malaysia	2867
Nepal	2127
Pakistan	2612
Philippines	2470
Sri Lanka	2376
Taiwan	3500
Thailand	1972
Vietnam	2523

Table 7 *Crop yields in China and the rest of the world 1983* (*tonnes per hectare*)

	China	*World*	*Developed countries*	*Developing countries*
All cereals	3.5	2.3	2.5	2.1
Rice	4.9	3.0	5.3	2.9
Wheat	2.8	2.2	2.3	1.9
Maize	3.2	2.8	4.4	1.9
Sorghum	3.6	1.4	2.6	1.2
Soyabeans	1.3	1.6	1.7	1.5

? **2** Table 7 shows average crop yields in China and the rest of the world. How does China compare with **a** other developing countries **b** developed countries for each of the types of crop? For which crop is China above the world average?

3a From the 1960s, grain imports were seen by China's critics as proof of the failure of the Chinese agricultural system. In the 1960s and early 1970s imports averaged 4.6 million tonnes a year or about two per cent of the total grain consumption. Look at Table 8 and say whether, since the mid-1970s, grain imports have increased or decreased in line with increased domestic production.

b For comparison, the total grain imports of all developing countries have been estimated to be from 1976–77 — 58 million tonnes; 1980–81 — 94 million tonnes; and possibly by 1990 — 145 million tonnes. Might these figures provide added impetus in the drive for self-sufficiency in China?

Table 8 *China's grain imports (millions of tonnes)*

1974–75	5.3
1975–76	2.3
1976–77	3.2
1977–78	8.6
1978–79	9.0
1979–80	10.1
1980–81	13.8
1981–82	13.0
1982–83	15.0
1983–84	8.7
1984–85	4.8 (estimate)
1985–86	0.5 (estimate)

BARRIERS IN THE PATH TO SELF-SUFFICIENCY?

As shown in the previous sections, China has greatly improved its agricultural productivity through a variety of means, but are there major reasons why self-sufficiency before the end of the century may prove an impossibility?

POPULATION TRENDS

China's population rose from 541 million in 1949 to 1008 million in 1982. Arguably, a lower rate of increase, producing a lower total population, would allow more food per person. In little over a decade it would also mean less labour to grow food. Increased mechanisation of farming processes should however reduce labour demands in the countryside.

Current population policy promotes one-child families with income and social services penalties for parents who choose to raise more children. Success of this policy is more marked in the cities (where only 20 per cent of the population lives) than in the countryside. Rural prosperity in the not so distant past, combined with lack of social welfare for the aged, depended on many children, particularly sons. Peasant families are still often large. In 1983, 13 million potential couples reached marriageable age. Thus a possible baby boom exists if the policy were to be relaxed.

The current responsibility system may also have a negative influence on reducing family sizes. Officially, the system promotes the principles of hard work and individual enterprise. However, a family with several able-bodied children might be advantaged in cultivating land or building new animal stalls compared to one-child families.

Rural–urban migration is no solution. A ten per cent move from the countryside involves some 80 million people. Cities at present are barely coping with the strain of rehousing and creating employment opportunities. Considerable amounts of scarce resources would have to be redirected into the construction of urban facilities. This is no solution if China is to avoid the problems of other Third World cities.

THE SUPPLY OF LAND — EROSION AND DESERTIFICATION

The supply of land for agriculture is large and very diverse. But it has not remained static in its quantity or quality. Over 400 000 square kilometres of China suffers from some form of erosion. Some of this erosion is particularly severe and much of it is on marginal farming land — land that is currently being promoted as valuable livestock pasture.

China's once vast forest areas cover less than 13 per cent of the nation and in provinces such as Sichuan and Yunnan they have declined particularly rapidly. The Sichuan floods of summer 1981 which inundated over 15 per cent, or one million hectares, of the province's arable land have been partly attributed to excessive removal of forest cover. In many arable areas topsoil was removed to the underlying gravel.

Large-scale forestation is now underway at the rate of about two million hectares each year but it will be years before the expanding desert and semi-desert areas of the north are arrested with the extensive shelter belts.

As much as 6 million hectares of China's agricultural land is saline enough to affect cropping yields severely. Of this land, 3.3 million hectares are on the North China Plain where a high water table and high evaporation rates have led to salt concentration on the surface. At present, a $US 60 million loan from the World Bank is being spent to neutralise 230 000 hectares of this land with network drainage, wells and reshaping of terrain.

Land reclamations have to some extent been offset by losses due to

Print 6 *The rural–urban fringe around Guangzhou*

the growth of city areas, especially by new industrial zones, transport areas and new water storage facilities. As much as 20 million hectares may have been lost to farming this way. For example, between 1971 and 1981 the outskirts of Beijing lost 4100 hectares of cultivated land to other purposes — some 12.3 per cent of the city's total cultivated land. Land adjacent to cities is often intensely farmed and carefully maintained. Its loss is therefore a significant one.

TRANSPORT TECHNOLOGY

Storage and transport of grains are well below the standard they should be. A more complete road and railway system would, for example, be able to move wheat and other products effectively to the industrial population of the cities and into grain deficient areas. Farming areas, at present too far away from transport networks to supply agricultural goods to other regions, would benefit from a denser network of railways and roads.

Such improvements, however, require diverting scarce investment resources with the knowledge that results are not necessarily immediate or equal across the nation.

THE ROLE OF THE AUTHORITIES

Chinese agriculture is subject to a range of authorities: central government directives, provincial and county administration together with the Communist Party which includes the local cadres who interpret policy for immediate conditions and oversee the implementation of party decisions. It could be argued that these diverse entities have handicapped the development of agriculture. In the past, directives which made 'grain the key link' in agricultural development were widely interpreted as meaning 'grow grain to meet targets even at the expense of other more successful activities'.

Organisational changes which have always been promoted as the best solution have created problems and the current responsibility system has this potential as well.

THE LARGE SCALE OF CHINA

In the 1980s there are around 54 000 communes and several thousand state farms populated by several hundred million people. The Chinese landscape and the agricultural situation developed on it are far from uniform. Innovation at village or county level is restricted to relatively minor matters (although there have been some notable exceptions). Central authorities insist on a very active role simply because agriculture is so important and deviations away from currently accepted policies may become threats to political authority.

While central authorities have shown that they can be flexible in changing policies, such as in 1979, flexibility nevertheless has ideological limits. Will the countryside be allowed to become even more capitalist than at present? Would such a policy, in any case, produce solutions to China's food problems and achieve self-sufficiency? And if it did, would another set of problems emerge?

PART III
DEVELOPMENT AND REGIONAL PLANNING
ALTERNATIVE APPROACHES TO DEVELOPMENT

Views of Development

The views quoted on the following page define Development variously as a question of:

- a move to greater social justice, within and between countries;
- economic growth, achieved through investment on the part of the rich and better trading terms;
- industrialisation;
- fair distribution of wealth;
- independence of one country from others;
- limiting exploitation.

The viewpoint shared by all, except that of 'Exploitation', is that Development occurs in the poor countries of the Southern hemisphere, with the implication that the rich countries of the North have already 'got there', and are in a position to aid or hinder the 'Development' of other countries.

1 Discuss the possible implications of the different viewpoints for development policies in terms of who defines the framework in which development takes place; who decides the priorities for development; the role 'experts' play. Try also to include views of Development not included in the statements.

2 Rank the statements on criteria such as:

- which views sound most familiar?
- which views do you identify with most?
- which views discuss issues which you see as being of prime importance?

In deciding on your rankings, consider such points as:

- who holds the different views? What reasons might they have for holding them?
- are there any assumptions common to all the viewpoints?
- what do the different viewpoints say about people?

3 Each member of the class take one of the viewpoints, read it and identify with the attitude it expresses. Then form into groups with those holding similar viewpoints. Spend ten minutes reinforcing your feelings of 'rightness' in holding these attitudes, and deciding how best to justify them to others. Select an 'ambassador' from each group whose role it is to visit other groups and convince them of the rightness of his/her group. At the end of about fifteen minutes, discuss the following points as a class:

- Did you find it easy to identify with your allocated viewpoint?
- How would you summarise the pros and cons of the different viewpoints you came across?
- Which viewpoints were in conflict? What sort of arguments could be used to support them?

(*Source: Priorities for Development, A Teachers' Handbook for Development Education* (1981), Development Education Centre)

Investment

Despite the international uncertainty our experience in Unilever leads us to believe that we should go on investing in those countries where we are welcome, and where the investment climate is such that we can expect a reasonable, long-term return on our investment and remit a dividend to the home country.

(*Source*: *Unilever and Economic Development in the Third World*. Unilever, 1976)

Colonial Links

The Development Policy of the EEC countries is not based on equality, need or justice, but on historical, economic, political or military considerations. The 56 African, Caribbean and Pacific countries which have a special agreement with the EEC are mostly ex-colonies, which, because their level of economic development is so basic, do not challenge EEC industries. By contrast the countries of the Indian sub-continent, where many of the World's poorest people live, have no special relationship with the EEC.

(*Source*: WDM *Fold Out Factsheet* No. 1, 1979)

Industrial Take-Off

The term industrial take-off has become part of our economic vocabulary since Walt Rostow first published his *Stages of Economic Growth* in the 1950s. He believes, 'It is possible to identify all societies . . . as lying within five categories: the traditional society, the preconditions for take-off, the take-off, the drive to maturity, and the age of mass consumption.' The traditional society has limited production capabilities and has a conservative attitude to technology. In the next stage, there are the beginnings of transformation which come from the influence of more advanced society from outside. The third stage marks the great watershed between the old and the new, the poor and the rich, the agricultural and the industrial. This is industrial take-off, when resistance to growth is overcome, and expansion continues exponentially.

(*Source*: *New Internationalist*, February 1978, Summary of Walt Rostow)

The Economic Challenge

The rich countries are on the same planet as the Third and Fourth World Nations: human beings inhabit both. If the rich nations go on getting richer and richer at the expense of the poor, the poor of the world must demand a change, in the same way as the proletariat in the rich countries demanded change in the past. And we do demand change. As far as we are concerned the only question at issue is whether the change comes by dialogue or confrontation.

(*Source*: Nyerere, Arusha Declaration)

Economic Growth

It would be highly misleading to present the Third World as an unchanging picture of widespread poverty. Even among the low-income countries progress is occurring, the beginnings — and in some cases more than the beginnings — of structural transformation. In a number of developing countries, moreover, there have been truly remarkable advances. In terms of sheer economic growth rates, the most striking cases have been the *newly industrialising countries*, which have been thrusting ahead with manufacturing growth. The Latin American ones — Argentina, Brazil, Mexico — have a quite old-established industrial base, which has increased rapidly in the post-war decades. A spectacular example is Brazil, whose economy at current growth rates will, by the year 2000, rival in size that of the Federal Republic of Germany. It is also an important trading partner and thus a stimulus to growth for other countries in the South.

(*Source*: *The Brandt Report*)

Economic Development

The private investor is important not only for bringing about a high growth rate, but also for solving the problems which may accompany it. A high growth rate brings with it rapid change. Aspirations, wants and tastes alter as people get more money and more knowledge. Profit leads the private entrepreneur to search out and satisfy these changing needs in the most efficient way.

(*Source*: *Unilever and Economic Development in the Third World*, Unilever, 1976)

Sharing Wealth

You don't make poor people richer by making their government richer. Often you seem to make more of them poorer. Our experience is that if you induce economic growth (by aid or trade or whatever) in a society that is unequal, you will end up with more poor people. The richer (even the slightly richer) buy out the poor, mechanise agriculture (in Indonesia one tractor destroys over 2 000 man-days of work a year) and such. As the Report says, it's very hard for the rich to share their wealth, hard for the privileged to give up their privileges voluntarily.

(*Source*: *Brandt and the Poor*, Oxfam Public Affairs Unit, May 1980)

Justice and Equity

Development is more than the passage from poor to rich, from traditional rural economy to a sophisticated urban one. It carries with it not only the idea of economic betterment but also of greater human dignity, security, justice and equity.

(*Source*: *The Brandt Report*)

Exploitation

The relationship between the rich and poor in the world is like the structure of the universe. The central planet or metropolis is the United States. It is unique, no-one's satellite. It has internal satellites in the poorer, underdeveloped regions like the Appalachians, and externally in the underdeveloped countries. These countries in turn have their own local metropolis in major cities like São Paulo, Nairobi or Johannesburg and are surrounded by provincial satellites — smaller country towns. The picture is one of a series of ever-smaller constellations down to the farm surrounded by peasants. These metropolises and satellites are linked in a universal exploitative relationship. The central planet, the United States, sucks capital out of the periphery and uses its power to maintain the structure. There is a continual process of underdevelopment.

(*Source*: *New Internationalist*, February 1978. Summary of view of Gunder Frank)

'Views of Development' should have started you thinking about the meaning of development. This is the connecting theme and problem which the four chapters in this section explore in detail.

Lesotho provides a classic case of development difficulties where higher population survival rates have led to increasing pressure on agricultural land and the deterioration of that land. Viewed in a systems framework we have a parallel to the trampling problem of chapter 19. Possibilities for the way ahead for Lesotho are discussed and evaluated in the chapter *Lesotho — Population Pressure and Resource Deterioration*. *Brazil — Regional Planning and Inequalities* presents a very different case study but one, nevertheless, where questions of development policy can be debated and better understood. India, like Brazil, has adopted a western model of development and the successes and difficulties experienced there and outlined in *India — Planning and Priorities* provide us with criticisms of the model. *Nigeria —Choices in Agricultural Development* concentrates on development and lack of development in rural areas, on the priorities which have prevailed and the choices which have been made.

Contradictions abound in all the chapters. 'What is development?' and 'Does development bring development?' are key questions. What alternatives exist? What strategies are most appropriate? These are questions and problems which are likely to be with us for a very long time, for the rest of your lifetime and that of your children's lifetimes and beyond.

HAS THE NORTH UNDERDEVELOPED THE SOUTH?

One of the most easily understood explanations for the great divide in prosperity between North and South is the idea that the North has underdeveloped the South. The North has dictated or managed the pattern of resource exploitation; the North has set the terms of trade (the prices of exports and imports) in favour of the North; the North has taken a very high proportion of the profits from companies set up in the South.

The extract below enlarges on this.

? Read through the extract and decide which view or views of development in the previous exercise it supports. Then read through the next extract taken from an Oxfam pack on the Brandt Report. This complicates the issue a little by moving beyond the broad view of trade and trade barriers to conditions within the society of the developing country, which aren't at all unlike situations which exist in Northern societies also. How does the fable complicate our view of development and underdevelopment?

You may be better able to answer this question after working through the chapters.

EXPORT-LED DEVELOPMENT

Industrialisation has been seen as a prerequisite of 'development'. The theory goes as follows: a developing country should concentrate its productive efforts where it has a natural advantage. So that instead of every country being self-sufficient there should be an international division of labour, one country growing cereals, another coffee, a third cocoa, achieving economies of scale and efficiency. Trade will then be mutually beneficial, and will generate the surplus required for industrialisation. Industrialisation will enable a country to get a higher rate of return from its labour and capital and so launch a 'take-off' into self-sustaining growth. The wealth generated by this process trickles down to all classes in society. Loans and grants made bilaterally and multilaterally from the developed to the underdeveloped world are designed to speed up this process. Such trade-aid development, based on export-led growth, has dominated conventional thinking in the last two decades.

THE RESULTS

Developing countries overall met the 8% target for annual growth in manufacturing output set for the second UN Development Decade. They also met the 6% growth target for gross domestic product, confirming that the generation of wealth has commenced. But who is benefiting from this growth? Not the poorest, who constitute a larger proportion of the population of developing countries now than they did at the start of the sixties. Two apparently incompatible trends are occurring: GDP is increasing, and the poor are becoming poorer. The trickle-down theory fails in practice. Take one example, that of Brazil. The much-publicised 'economic miracle' has left the poor worse off. The growth rate has been substantial: over 9% per annum between 1970 and 1978. But the income of the top 5% is more than 30 times that of the bottom 20% — a statistic unmatched even in South Africa, which has institutionalised inequality. Nor

THE PERSISTENCE OF POVERTY

Why is the number of absolute poor increasing, when governments, international agencies and development experts have spent the last two decades trying furiously to work out a solution? Conventional wisdom points to two factors: the galloping rate of population growth, and the massive increase in the price of oil in the 1970s.

But why should population growth be the major problem if there is enough food to go around? And, as we have seen, even the World Bank admits that this is the case. Furthermore, while the oil price rises have had serious ramifications for the world financial system, and have hit the Third World hardest, most developing countries increased their gross domestic product per capita in real terms during the 1970s. So, although these two factors undoubtedly contribute to the difficulties developing countries face, they cannot account for the spectacular failure of the development policies to help the poor. We must look to the policies themselves — and their inherent shortcomings — for an explanation.

is Brazil unique. All the evidence suggests that at the start of the export-led development process the poorest 40% will lose both absolutely and relatively. This could be seen as the price to pay for later prosperity, but there is no hard evidence to show that in any market economy trying to industrialise in the present global situation the share of the bottom 40% will ever increase. There is positive evidence to the contrary.

Furthermore, not only is inequality within countries increasing, but so is inequality between countries. To quote Shridath Ramphal: 'At the end of three decades of international action devoted to development the result, by 1985, is likely to be an increase of 50 dollars per capita in the annual incomes of the poorest group (of countries), compared with an increase of 3 900 for those of the richest who were already, in 1965, 3 000% better off.' Another way of looking at the same figures is to say that it would take Brazil, at her present fast growth rates, 362 years to close the income gap with the rich world.

EXPORT-LED FAILURE

Part of the problem lies in the whole concept of export-led growth. Primary commodities are for many Third World countries the main source of foreign exchange earnings, and the prices of these commodities are determined in Northern markets. Throughout the seventies the price of commodities fluctuated wildly and failed to keep pace with the increase in cost of manufactured goods from the West needed to industrialise and modernise. Such fluctuations make a mockery of economic planning. As Julius Nyerere of Tanzania has pointed out: 'In 1965 I could buy a tractor by selling 17.25 tons of sisal. During the much-talked about commodity boom of 1974 I still needed 57% more sisal than I did 9 years before. And now the sisal price has fallen again and the tractor price has gone up still further. Take another example — sugar. In 1975 sugar fetched 57.6 US cents per lb. By 1981 this had dropped to 17.2 cents.

It is against the background of such difficulties that negotiations for a New International Economic Order have taken place, with Third World countries arguing in organisations such as the United Nations Conference on Trade and Development (UNCTAD) for commodity agreements to stabilise prices and supplies. More radical action has been attempted. Inspired by the example of OPEC, producer cartels have been formed. In 1977 the Central American banana exporting countries formed Comunbana. Their experience points to some of the reasons for the comparative weakness of such cartels: in the first place they enter markets dominated by multinational companies. Castle and Cook, United Brands, and Del Monte control 70% of the European, and 91% of the American banana market between them. As we shall see later, such monopoly control is reproduced in many other sectors. To avoid open competition Comunbana have turned to the East European market with some success; nevertheless its future will be fraught with difficulty.

Even when commodity prices are high, multinationals (MNCs) can still subvert the advantages to developing countries. During the sugar boom, US-based agribusiness corporations began marketing high fructose corn syrup, which quickly displaced the use of sugar in certain areas of food production, notably jams and confectionery. Susan George comments: 'Sugar is not, to say the least, a very sure bet for "export-led" economies in the Third World.'

In fact, MNC control over food production brings many other problems, and gives rise to some bizarre contrasts. In the Caribbean people starve beside fields growing tomatoes and flowers for export. In Haiti companies own land in the fertile valleys while the peasants have tiny subsistence farms on the barren mountain slopes. Among the produce of the valleys is alfalfa — to be fed to cattle owned by MacDonalds Hamburgers.

Part of the strategy of modernisation adopted by developing countries is to add more to the value of their commodities by processing them before they are exported. In terms of industrial development it makes good sense to set up a plant to turn cocoa beans into cocoa butter. But in spite of their superior position in world trade Western countries still impose higher tariffs on manufactured goods from developing countries than on raw materials. Cocoa beans can enter the UK with a tariff of 3%. The rate for chocolate is 19%. This policy extended over a whole range of goods makes it virtually impossible for a developing country to penetrate the markets it still needs.

INDUSTRIALISATION

Where Third World countries have begun to industrialise they have frequently called on the multinationals for help. In some countries, notably in South-east Asia, there are free trade zones designed to offer the maximum concessions to multinationals in terms of controlled labour (trade union activity being limited or proscribed), tax advantages, import and export controls.

MNCs make dangerous partners. Among the developing countries, only five have a Gross National Product in excess of Exxons's (Esso's) turnover. Ford has subsidiaries in 29 countries. UNCTAD figures indicate that two-fifths of all international trade takes places *within* MNCs. This makes MNCs very much more powerful than the host country, and this power is used for one purpose only — the pursuit of profit. The magazine *Business International* summed up the strategy of these democratically unaccountable institutions in 1967: 'The global company views the world as a single entity . . . Decisions are made not in terms of what is best for the home country or any particular product group, but in terms of what is best for the corporation as a whole.' In some cases a MNC may bring benefits to the host country — higher employment, higher wages for some, provision of goods for the local market, stimulation to the local economy. But if it does bring these benefits they are incidental. The MNC would not be there unless it were making a healthy profit.

MNCs claim that they export and transfer technology and capital to Third World countries. However the question has been posed frequently as to whether the technology exported is appropriate to local needs. With high unemployment,it is argued, labour-intensive, and not the MNC's capital-intensive technology is what is really needed. Equally fundamentally, is the technology transferred in any real sense? Third World governments gain no control over it: such control remains with the company, and its expatriate technical experts. All too often the only 'control' vested in the local people is the operation of increasingly sophisticated machinery on a production line. Some comments from a woman worker on a semi-conductor line in Penang, Malaysia: 'Microscope work affects the eyes, so they used to give us insurance. Now they say that the work does not affect the eyes and they don't give any insurance policy . . . No, we don't have any unions and we are not allowed to join any unions.

The multinationals who control the 'New Technology' are attracted to places like Penang for exactly that last reason. We might question what benefits microprocessor technology will confer on any of us while it is outside democratic control.

The myth that Third World countries receive massive injections of foreign capital through MNC investment can be similarly exploded. Sometimes this may be the case, but often a company will raise money to finance its operations by borrowing on the local market — Ford Philippines was financed entirely in this way. This deprives local industry of much needed capital. Moreover the profits then generated are not ploughed back into the host country, but largely repatriated either directly or in disguised forms such as transfer pricing. From 1950 to 1965 3.9 billion dollars flowed from the USA to Latin America. But it was a worthwhile investment. In the same period 11.3 billion dollars flowed the other way in the forms of repatriated profit, interest on loans, royalties on patents, etc.

(*Source*: *Beyond Brandt, Alternative Strategy for Survival*, Third World First)

CURRENT TRENDS

The trend in the commodity market has been one of instability. Prices set through the market place alone and determined mainly in London and New York have ricocheted up and down during the 1970s, but with an underlying trend downwards.

These fluctuating prices help no-one except stock-market speculators. Unstable prices fuel inflation in the North, as depressed prices lead to slumps in production, resultant shortages and dramatic price rises in a vicious circle. This happened in the sugar trade with shortages leading to price rises in 1974 and 1979. A group at the University of Pennsylvania studied the impact of price fluctuations of eight commodities. They found that they cost the United States approximately $500 000 000 in 1977 in lost output through reducing employment and productivity. More obviously, price fluctuations do not help the South either. Uncertain income makes an integrated development programme almost impossible to sustain and plan for. The temptation for a poor country to overcommit future earnings when prices boom helps to explain the present debt crisis now facing many countries. The cost to the South of fluctuating and basically declining prices is poverty.

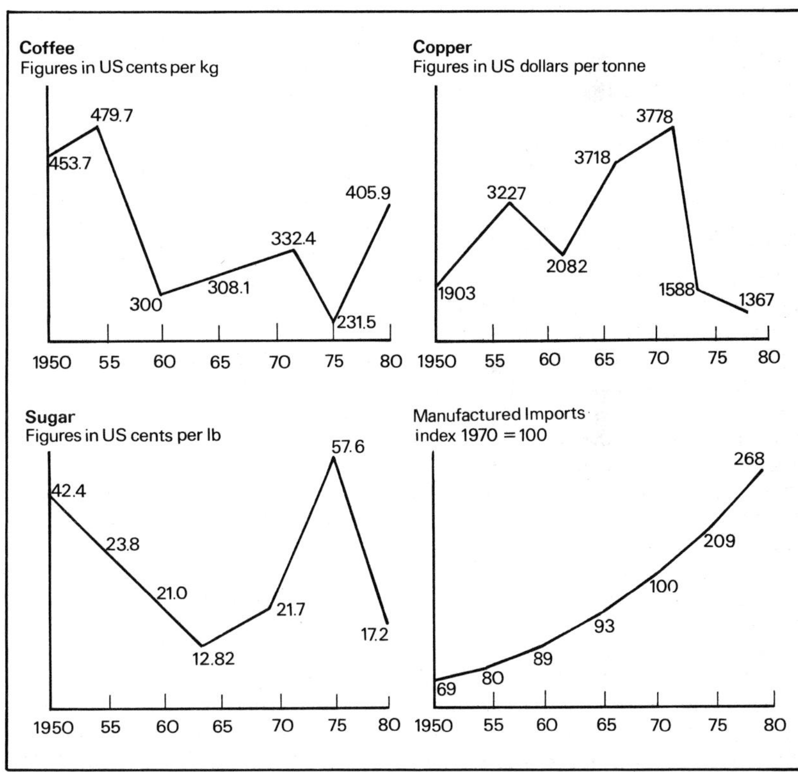

Stable prices would seem to suit everyone; but they must also be fair. Countries of the South must control their own trade so that they realize the profits. The growth in the Third World of marketing and distribution enterprises is an important step towards achieving producer power and a fairer world. There are moves within the world trade system to stabilise prices and get a fair deal for the South, but progress is painfully slow.

(*Source*: Third World First Factsheet No. 8, *Commodities*)

BRANDT, TRADE AND THE PEOPLE: A FABLE

Measures to facilitate the participation of developing countries in processing and marketing should include the removal of tariff and other trade barriers against developing countries' processed products, the establishment of fair and equitable international transport rates, the abolition of restrictive business practices, and improved financial arrangements for facilitating processing and marketing.
(*Source*: *The Brandt Report*)

Manchhubhai unrolled a small mound of cotton from his tattered garment and handed it to the trader, who weighed it. The balance swung from a huge smoke-darkened beam in his shop on the main street of Valod, Western India.

'One rupee 60 paisa' he drawled, not looking at the man.

'Yes, sahib.' Manchhubhai looked sadly at the cotton laboriously picked by his wife from their tiny field a few miles away.

'It seems little,' he muttered shyly.

The trader glared angrily. 'Your cotton is only *deshi*. You know that. The price is less. He waved him away.

Yes, the cotton was the local, short-staple variety which is all the poor can grow. Hybrid cotton, worth two or three times as much, requires expensive seed, fertilisers, insecticides and, above all, irrigation. The poor do not have irrigation any more than they have cars.

Manchhubhai wandered off through the bazaar to buy himself a shirt. It would cost him a lot of money. He sold cotton cheap and bought it back dear. He only possessed three acres of land and he only bought one shirt a year.

As he walks dejectedly past the shops he meets, let us imagine, ex-Chancellor Willy Brandt, who has been taking the trouble, let us also imagine, to learn Manchhubhai's language.

'Ah, Manchhubhai, I was looking for you. So you did not get much for your cotton?'

'No, sahib.'

'What we need, first of all, is a liberalisation of trade. India has surplus capacity in its mills. If they can export there will be a tremendous demand for cotton from your farmers. We must relax the Multi-Fibre Agreement that restricts trade from developing countries.'

'Yes, sahib.'

As is customary in a small Indian town, a crowd gathered to discuss the point. One man pushed forward and said:

'How will that help us? We use most of our land to grow food. We grow a little *deshi* cotton for cash, but we cannot afford to grow much. The rich farmers grow cotton and we work on their fields.'

'Then you will get more work from them.'

'But they pay so little.'

A big man dressed in white elbowed the poorer people out of his way.'

'Last year I grew cotton on my fields. Ten acres of Shankar 4 Hybrid. It cost me 10 000 rupees to grow. Of course I did not sell it to the traders, I went straight to the big dealers. But they would not buy. They were operating a monopoly to bring prices down. My brother took a cartload of cotton to their depot and burnt it in front of them. We can do nothing. We had to sell it in the end, and I made only 1 000 rupees after all my trouble. How can I pay the workers more?'

'But you don't even pay the minimum wage!' shouted an old woman. The big farmer glared at her angrily and was silent.

'It seems that you have a problem here,' said Herr Brandt. 'We are trying to liberalise trade but unless you sort out these problems the poor will not get as much benefit as we want. We have made recommendations about land reform and so on, but I suppose it is really up to your government to do something about it.'

At this point there came a great wail of police sirens and noisy horns. The group scattered to the edge of the street as a huge Mercedes drove along, barely missing the old woman as she limped towards a doorway. The driver in peaked cap looked straight ahead, self-importantly.

'Ah, that's Arvind Vaidya, the mill owner,' someone said and bowed his head. Most of the others did likewise.

'And who was that with him?' asked Herr Brandt.

'That was the Minister of Trade. No doubt they are discussing "liberalisation".'

There was an awkward silence. They all knew that liberalisation was a good thing for which they should strive. But they wondered how they would actually see the benefit of it.

'Willy-bhai,' they said, 'come and have a cup of tea.'

Let us now imagine that Herr Brandt leaves the village in India and lands by magic carpet (after due Customs examination as to its origin) in Lancashire. A number of unemployed men and women gather around as Herr Brandt steps down on to the ground.

'You offering any jobs, mate?' asks a man carrying a placard 'Jobs, Not Bombs'.

'In a roundabout way, yes. You see, if the world is to survive, we must do something to help the poor in countries like India. The first thing we can do is liberalise trade . . . '

'Hold on. You must look around. See that mill? Closed. See the chimneys? Where's the smoke? And you know what the trouble is? It's textiles and clothes from India coming in here and undercutting British-made goods. That's why! That's why we've lost our jobs! And now you want to liberalise trade!'

Herr Brandt straightened himself up as he used to when answering tricky questions in the *Bundestag* in Bonn.

'It just isn't true that textile imports from India and other poor countries have put you out of work. Let's look at the figures. There was a recent OECD study which showed that 80 per cent of the jobs lost in the textile industry of industrialised countries were due to technological changes aimed at increasing productivity. As for imports, they are mainly from other industrialised countries, not from the poor world. During the past three years, textile imports to the UK from poor countries increased by only 19 per cent, compared with a 58 per cent increase from industrialised countries.

Textile imports from the USA have actually gone up by 90 per cent!'

'What, the Yanks?'

'Right. And now let's look at the the potential benefits to you in Britain from greater prosperity in Third World countries. Just imagine what the vast populations of Asia and Africa could buy from Britain, if only they had a little more purchasing power. The world economy would boom (that is your English phrase) as never before.'

'OK, mate, but if we get more trade from poor countries, how are we to know it will mean more jobs for us up here in Lancashire?'

Herr Brandt hesitated. 'Well, that raises some rather difficult questions. You will have to ask your own Government, I suppose. It isn't really for me to say. Well, goodbye for now.'

The magic carpet hovered above the ground for an instant and then disappeared over the dark empty mills and up into the warm sunny air above the clouds.

(*Source*: Oxfam pack on the Brandt Report)

THE DEVELOPMENT PROCESS

We have yet to achieve a full understanding of the processes at work which have led to such contrasts in world development and well-being, and to the self perpetuation of a rich North and a poor South. The extract below may take your understanding a little further.

THE DEVELOPMENT PROCESS

Why did Europe grow rich in the first place? One reaction has been to suggest that in the case of Europe, and more particularly of the kind of industrial capitalism that it produced, the crucial factor was the relatively compact and closely-linked towns. Initially, the multiplicity of small political centres in the Middle Ages, sustained by agricultural surpluses, led to a large number of trade centres of surprising and growing vigour. At the same time their variety prevented any centralised bureaucratic power from growing (like the Chinese Southern Sung empire in the twelfth century, which some people think was sufficiently wealthy to have developed an industrial revolution, but did not). This situation allowed commercial interests to develop policies uniquely convenient to their trading, and later industrial, interests. Significant centres of economic power in Europe shifted a great deal (from 12th–15th century Italy to 15th century Flanders to 17th century Holland, 18th–19th century England and 19th–20th century Germany) but the important thing was that dozens of independent market centres and trading interests fostering an interchange of ideas and techniques grew up and eventually took power in some countries (e.g. Holland, England), which led the way in turn to a new industrial civilisation.

This is only one explanation and underlying it are some complex causes. The fact of the matter is that the process of economic development is one where, in spite of much impressive work, our understanding has not advanced as far as we would like. No automatic formula has been found which can be applied to the problems of any state to raise it to the status of a developed nation. If this were not so, there would be a few less poor people in the world. There is little agreement amongst the specialists about what the necessary or sufficient conditions of development are. And without clear agreement on conditions of development, it is equally difficult to identify correct development strategies. It is, for instance, quite clear that there are *alternative paths* to development: the UK, the USSR and Japan represent three such different ways. The debate is further confused by the two-edged qualities of many of the conditions we can identify; to give a prominent example, foreign investment (and its major vehicle, the multinational corporation) has the potential at once for transmitting technical information from more to less advanced countries and for the strong to control the weak. The extent to which advanced countries help or hinder the less advanced countries is bedevilled by a failure of those taking part to agree on, or even explicitly state their assumptions about, conditions of successful economic development. To give an example of the confusion in the debate: measures taken by advanced countries to support the real level of income that developing countries earn from exporting primary commodites are often criticised, for being on the one hand too ungenerous and on the other sufficiently generous as to discourage developing countries from trying to develop through industrialisation. The debate, especially as it concerns aid, is made even more confusing by the fact that the motivations of advanced countries, in the policies they pursue towards the less developed, are very diverse and almost impossible to disentangle. For instance, it would probably be true to say that European policies toward developing countries are a mixture of humanitarianism, self-advertisement, and enlightened and selfish self-interest.

To illustrate the dilemmas in the debate we will try to note in this section some of the main alternative groups of explanations of the conditions of development, some idea of the types of strategies that can flow from these conditions of development, and a first categorisation of ways in which advanced countries like some of those in Western Europe get involved in these processes.

It is useful to divide possible conditions of development into their internal and external components, with the idea that you may need a combination of these as the basis of successful economic development. Under *internal conditions*, which are often interconnected, we can list:

1 Endowment and physical resources — raw materials, land, climate;

2 The size of the domestic market — i.e. size and income level of the

population; this is particularly important when international conditions mean that foreign trade is difficult.
3 Social and political conditions — this may mean the degree of modernization, or the kind of skills and organization a country has;
4 The policies the government pursues — for instance, the degree of openness of the economy, the relative roles of the state and private capital as owners of the means of production and as entrepreneurs;
5 The pattern of income distribution — some would argue that uneven income distribution (meaning more money in the pockets of the rich) helps economic development because the rich save and invest more; others would question this and argue that more evenly spread income means that more and more people are buying basic goods, i.e. that there is a mass market which encourages mass production;
6 Population growth rates — high rates of population growth, which may be necessary because poor health facilities mean high death rates, nonetheless mean that if the population of productive age (i.e. 15 or above) has to work very hard to feed a very large child population.

Under *external conditions* we mean to emphasise those conditions which affect a given country's ability to specialize within the international division of labour (by importing and exporting) *and* to learn about better ways of doing things from other countries (i.e. new technologies) *and*, in the case of developing countries, to get the help of the advanced countries in the process of development. We can list the following:
1 The growth of the international market. Obviously the greater the demand for a country's exports, the more that country can grow; but this growth in international demand depends on the growth of the individual economies of the rest of the world and/or on the progressive lowering of international trade barriers.
2 The stability of the international market. Inflation in the world prices and periodic fluctuations in commodity prices destabilise world markets; they make the gains from international trade more uncertain and their removal thus may be better for everybody.
3 The terms of trade. If the prices of the goods one country exports do not rise as fast as, or faster than, the goods they need to import, this is not going to help their development, as they have to work harder to earn the same amount.
4 The rate of development of technology. This has obvious positive effects since cheaper ways of producing goods, or the invention of new products, can *potentially* benefit all countries; on the other hand, technical progress *may* also have the effect of producing products that substitute for the exports of poorer countries — for instance, synthetic fibres — or processes that are, initially at least, too complex for poorer countries to operate.
5 Means of passing on this technical progress. This is more difficult than would at first appear; for instance, patent laws give some protection to inventors and stop the early imitation of many new technologies; foreign investment, as already mentioned, is a two-edged weapon which might both transmit technology to, and exploit, poorer countries.
6 The extent of foreign political influence over individual countries. While the empire no longer formally exists to any great extent, there are many subtle and less subtle ways in which stronger countries can so influence the domestic culture and policies of poorer countries as seriously to hinder their development effort, though not all foreign influence is unequivocably harmful — for instance, international aid agencies and research institutions dispense a lot of well-meant advice.
7 The size and nature of international capital transfers (i.e. aid).

These partial lists of internal and external conditions of economic development give a sense of the complexity of the development process and the analysis of the development problems any one country faces gives rise to a specific development strategy. While in reality the actual strategies pursued are complex and difficult to generalize about, it is nonetheless useful to mention some of the major strategy debates that go on about economic development. It will be immediately apparent that they relate to a number of the above conditions. Perhaps the oldest debate is between the proponents of socialism and capitalism. The *socialist model* of development stresses: the ability of planning to change the structure of the economy (through industrialisation) faster than the 'blind' forces of the market; the virtues of centralising resources when these resources are scarce (for instance the collectivisation of agriculture, the creation of large industrial complexes); the application of the surplus (profits) from production to specially beneficial uses (for instance, for more investment, rather than for the consumption by capitalists); and the humanitarian features of a more equal distribution of income. The Eastern European countries, Yugoslavia, China and Cuba provide the mainstream, but often different, examples of this model; some Third World countries, such as Tanzania, provide partial examples. The modern *capitalist system* emphasises: that many individual, competing entrepreneurs end up, through a process of trial and error and through the allocation of resources via market prices, coming to better economic solutions (which may mean faster growth) than the collective decision-making process of the socialist economies; that the market (perhaps somewhat modified by government intervention) stimulates the individual incentive to work; and that finally, because capitalism needs a mass market, it requires a diffusion of wealth throughout society. The 'stars' of this system are the Western advanced economies, Japan, and a variety of developing countries that have recently shown some development promise, such as Korea, Hong Kong, Brazil, Ivory Coast and Kenya (though planning is not absent in all these economies). The socialist/capitalist dichotomy is obviously more complex than this and there are many forms of economic organisation that fall between the models, for instance the mixed economies of Western Europe, the significant elements of private economy in most socialist economies, and the state capitalism (i.e. large public enterprises) of some developing countries.

Another strategy debate, with very close parallels to that of socialism and capitalism, is the open versus the closed economy. At one extreme, countries such as Hong Kong and Singapore have no barriers to imports or exports of goods, services or capital; at the other end of the scale a country like Burma or China has very limited and closely-controlled relations with the outside world. The more an economy is closed, the more the government is able to exercise control over, and plan, the activities of that economy and the more self-sufficient that economy can be. On the other hand, smallness is a barrier to being closed, for unless a country is very big, there is no way that it can be self-sufficient in the thousands of products that are required by a modern industrial society. Moreover, the importance of the openness of the economy is that international trade results in gains from specialisation and exposes producers to competition that is an important stimulus to efficiency. Also, international contacts, whether through trade, investment, or mere personal contact, are an important means of effecting the transmission of technical change. There are, of course, many degrees of openness and closeness. To give an example of the complexity of this, the highly successful Japanese economy has historically been relatively closed to industrial imports and foreign investment, yet very open in the sense of its high rate of exports, and higher rate of imports of technology.

A new strategy debate has come to the fore in the 1970s; this concerns the difference between a 'basic need' and a 'trickle down' development strategy. Conventionally, many Western development economists have believed that investments in large projects that immediately went to help the advanced part of the economy, such as ports, airports, roads, sophisticated industries, etc., produced benefits that 'trickled down' to the poorer, less developed parts of the economy. This is because the incomes created for factory workers (for instance), would be spent on services or goods provided by those who did not have factory jobs. Because of a feeling that this process did not in fact work in practice, there has been an increasing emphasis recently on a strategy that would directly tackle the basic needs of the poorer parts of society, such as shelter, health care, food and basic education, a strategy that, as well as being humanitarian, would stimulate economic development through the stimulation of a mass market for low income goods.

There are many other strategy debates, some of which have gone out of fashion. Twenty years ago, the debate used to be about exporting primary products versus import substitution in manufactures, or about balanced and unbalanced growth (i.e. would the economy best develop through a harmonious development of its component parts or through an unharmonious development in which bottlenecks in one sector created a strong demand for increased production in another). More recently, there has been much debate on the relative merits of small- and large-scale industry, a debate not only about the relative efficiency of different sized firms, but also about the stimulation of entrepreneurship, the quality of society we want to live in ('small is beautiful') and the stimulation of employment. There is also a debate between proponents of the advanced, labour-saving technologies produced by the advanced countries and more appropriate (for instance, more labour intensive) technologies for poorer countries.

We will not delve deeper into these issues here. Instead, one may note that the fact that different nations will want to adopt *different strategies* for development does not mean that outside help is ruled out. Not everyone is in the position of being able to haul themselves up by their own bootstrings.

(*Source: Exploring Europe The Third World*, 1981, Schools Unit, Sussex European Research Centre, University of Sussex)

It is as though, in the development process, a relationship with an undesirable positive correlation (see also *page 368*) has been set up and we do not know how to manage the system to break the positive feedback loop and achieve an equilibrium, an evenness in development and well being. The diagram below illustrates the situation. Human systems indeed seem infinitely more complex than physical systems — there are so many more variables, operating at many more scales.

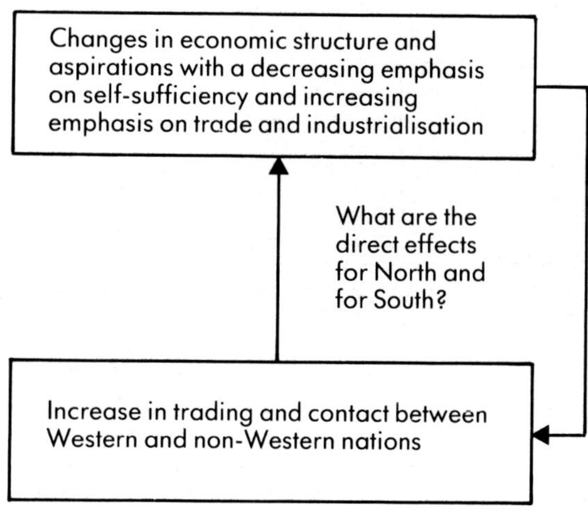
Changes in economic structure and aspirations with a decreasing emphasis on self-sufficiency and increasing emphasis on trade and industrialisation
What are the direct effects for North and for South?
Increase in trading and contact between Western and non-Western nations
What is the nature of the feedback loop? Is it positive or negative? Is it self-regulatory or self-destructive?

10 LESOTHO: POPULATION PRESSURE AND RESOURCE DETERIORATION

POPULATION AND RESOURCES

THE GEOGRAPHICAL BACKGROUND

Lesotho is a small African state which gained independence from Britain in 1966, and is now a member of the Commonwealth. It is entirely surrounded by the Republic of South Africa and consists of an area of about 30 000 square kilometres, lying above 1000 metres in altitude throughout. It can be seen from Figure 1 that less than 25 per cent of the total land area is regarded as 'lowland'. This has important implications for population patterns and national development potential. The remainder of the country is mountainous, deeply dissected by the Senqu and its tributaries, and generally inaccessible. Climatically, Lesotho is characterised by extremes related to its altitude and continentality: summers are hot (January mean temperatures of over 20 °C in the lowlands) and winters cold — many mountain stations record over 170 days of frost per year. A mean annual rainfall in the order of 730 mm is concentrated between the months of October and April; convectional rainfall predominates and is highly unreliable. The region is subject to severe drought, and, even in good years there are parts of the Senqu valley, in the rain shadow of the Drakensberg, which receive less than 500 mm of rain per year. This, then, is a harsh and uncompromising environment, with strictly limited agricultural potential.

In this chapter we shall be examining a problem central to the development process in many developing nations — the crisis created by increasing demands for

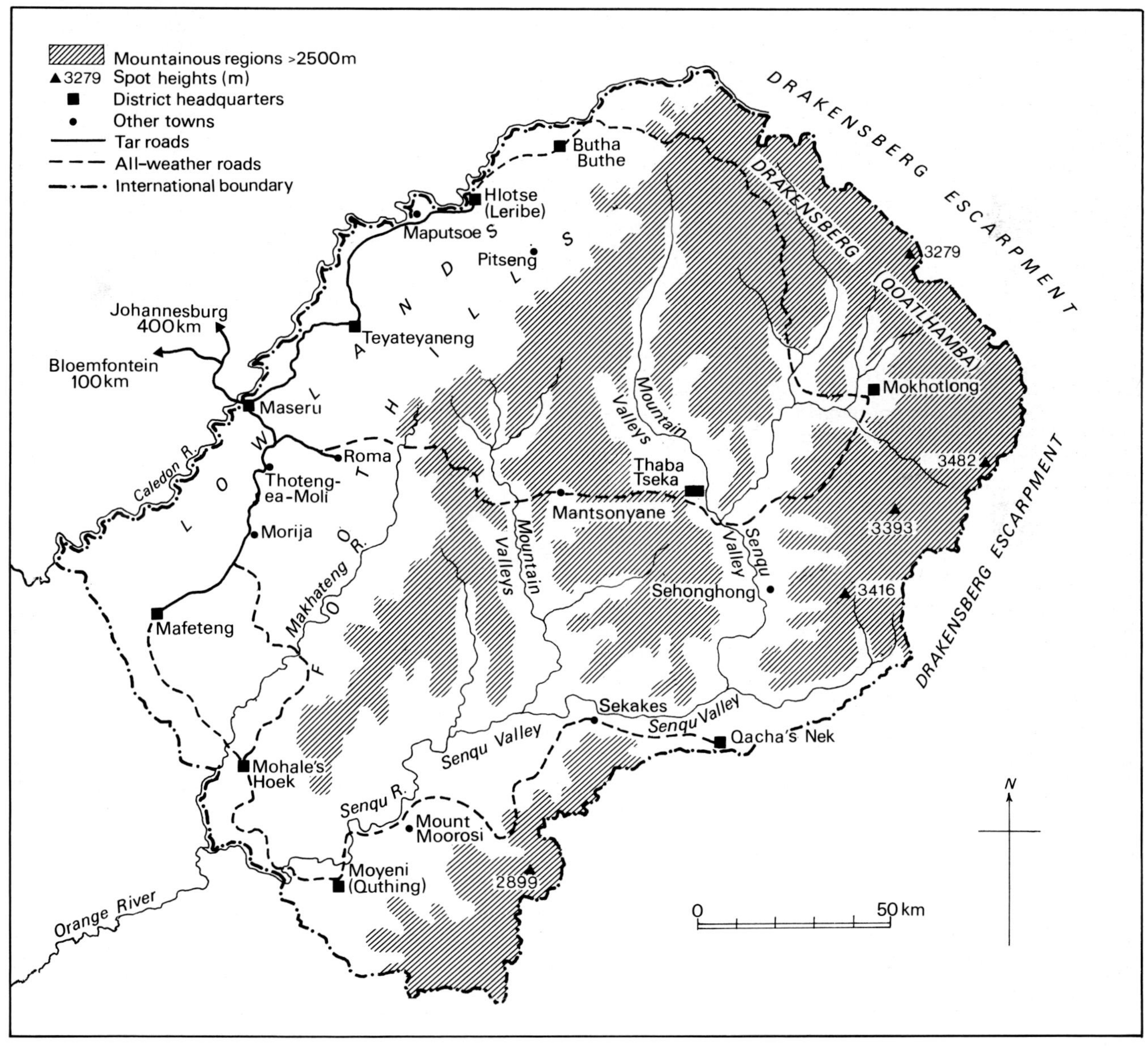

Figure 1 *Lesotho: major features*

land from a growing population in severely limited environments. Population growth in Lesotho is around 2.3 per cent per annum. Peasant agriculture is by far the dominant economic activity within the nation, and, as a result, the expansion of the population is reflected in a growing demand for land. Although the absolute size of the population is small, population pressure is considerable: land unsuitable for agricultural use is being brought into the farming system, resulting in soil erosion and associated environmental breakdown, and the productivity of the good land in the lowland region is jeopardised by over-use and reduced fallow cycles.

The fact that there is little alternative economic activity makes the crisis in agriculture particularly profound. In the absence of a viable farming unit, the rural *Mosotho* (singular of *Basotho* — the people of Lesotho) has little choice but to migrate to the small urban centres (or the national capital) in the remote hope of finding employment, or to join the tide of international labour migration to South Africa.

WHAT ARE THE KEY ISSUES?

The population/resource equation in Lesotho is severely imbalanced, and the development priority, given the absence of industrial resources, must be the improvement of the agricultural sector. The essence of the problem, though, is what sort of agricultural development strategy should be adopted — should the emphasis be on commercial, or subsistence activity? Should the government concentrate on the provision of investment capital or on the restructuring of the agricultural system? Should traditional farming practices and attitudes to the land resource be regarded as sacrosanct? And what influence should the international agencies and expatriate agricultural advisers be allowed?

Such complex planning decisions require urgent action not just in Lesotho but generally in all developing nations if constructive development is to be achieved. In this chapter we shall be exploring some of these issues, looking at how the population/resource crisis has arisen, and evaluating the various solutions being proposed.

POPULATION PATTERNS

POPULATION DISTRIBUTION

The pattern of population distribution in Lesotho is strongly influenced by relief and altitude — a glance at Figure 2 will reveal the concentration of the population in the lowland region where the bulk of the cultivable land occurs. There are pockets of relatively high population density within the mountains along the Senqu and its major tributaries and on mountain flats, but these do not show on a general map such as that shown in Figure 2. Population densities of well over 200 people per square kilometre are recorded in parts of the lowlands; these are largely rural densities, as the urbanisation process is in its early stages in Lesotho. Agriculture is by far the most important sector of the economy and the majority of households are involved, seasonally if not permanently, in some form of agricultural activity. Less than ten per cent of the national population is classified as urban, and of this, 47 per cent lives in the capital, Maseru. Here, population densities of over 2500 persons per square kilometre are recorded. Such data is indicative of the early stages of the development process; a predominantly agricultural economy with a small (but rapidly growing) urban population.

THE DEVELOPMENT OF MIGRATION

Traditionally, each household has had access to sufficient land to support itself, through the cultivation of maize, sorghum and legumes, and through the grazing of stock. But in recent decades, population growth has led to land shortage and landlessness, and rural–urban migration has been the result. Within Lesotho, the scale of rural–urban migration is relatively small, but there is evidence of increasing momentum. For example, the population of the effective built up area of Maseru increased by over

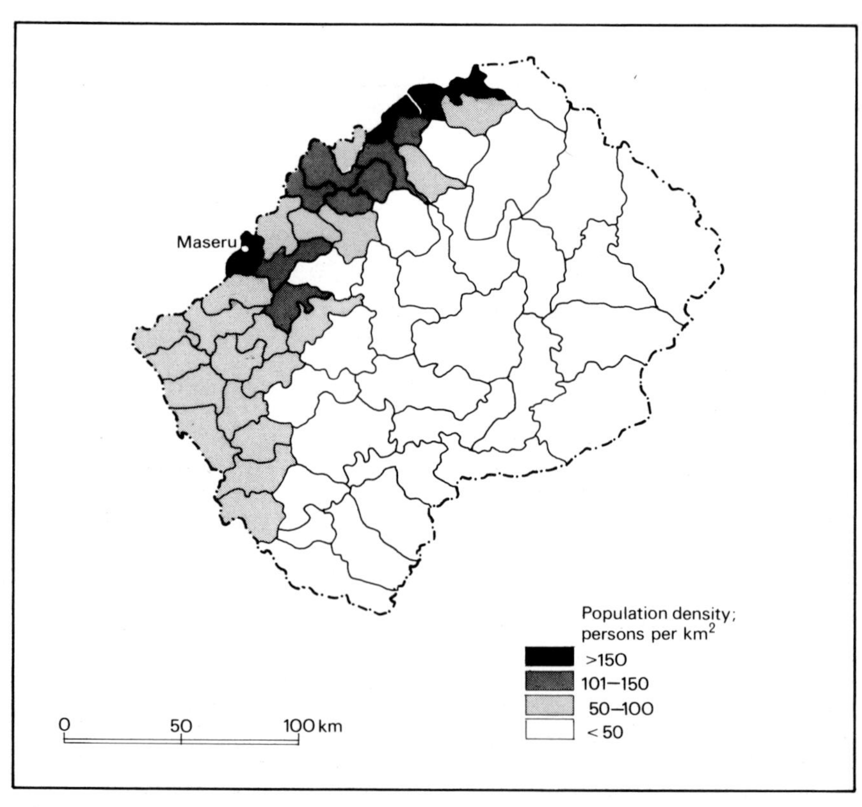

Figure 2 *Population density by constituency*

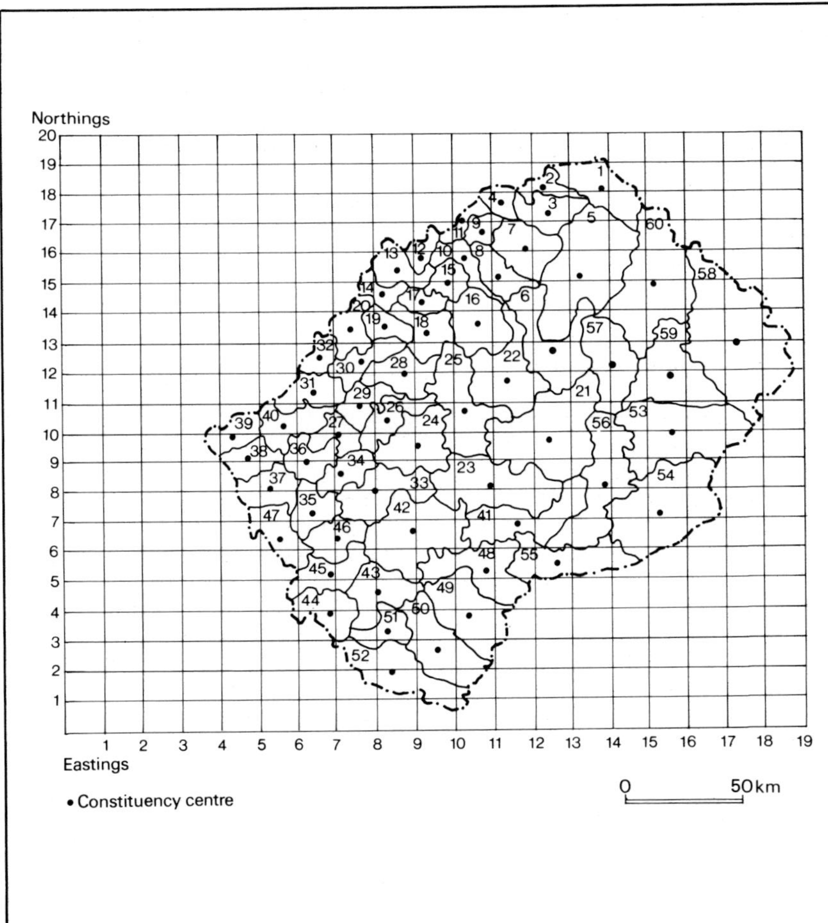

Figure 3 *Lesotho constituencies*

80 per cent in the decade 1966–76. However, a simple model of economic and spatial change is not appropriate to Lesotho, because the economic history of the nation has been governed by the international labour migration system. Lesotho, like many of her sister states in southern Africa, has traditionally been a source of cheap labour for the mines of South Africa, and the absence of over 60 per cent of the male labour force (earning 42 per cent of the GNP) at any one time has been a dominant influence on the development and change process in Lesotho.

The labour migration system has been used by South Africa as a device for maintaining relative poverty in the independent states around her borders. By opting for a labour-intensive industrial sector, South Africa is able to offer considerable employment opportunities to the labour of the surrounding (and relatively impoverished) independent African states. Wages paid to such migrant labourers are low by international standards, but relatively high by those of this sub-continent, and consequently South Africa has considerable political and economic leverage over her neighbour states. The threat of the withdrawal of migrant employment opportunity in South Africa has, until recently at least, been sufficient to maintain an 'authority/dependency' relationship between South Africa and the adjacent nations. At the level of the individual labourer, the economic gain consequent upon migration may well be satisfactory compared with domestic alternatives, if indeed they exist. At the national level, however, the migration of the majority of the domestic labour force to a neigbouring country can be a disastrous loss to the internal development effort.

?

1 What are the implications of the population structures outlined above for a small nation such as Lesotho, with limited land resources and little industrial development potential?

2 Draw a sketch map showing the approximate boundaries of the regions of Lesotho as listed in Table 1. Arrange the data in Table 1 in decreasing rank order. Then devise a suitable shading system to represent the regional variation and shade your map accordingly.

a Describe in your own words the major features of the population density pattern shown on your map.

b What do you think are the major disadvantages of such a distribution for a nation in the early stages of the development process?

c Conversely, can you suggest any development advantages of this pattern?

Table 1

Region	*Mean population density (Persons/Km²)*
Lowlands	103
Foothills	42
Mountain valleys	25
Senqu valley	49
Mountains	14

?

3 Calculate the mean centre of the population distribution in Lesotho. Use Figure 3 as a source for your calculations, and tabulate your results as shown in Table 2. (*Note*: Population figures can be reduced to thousands for ease of calculation, e.g. constituency 1's population of 11 119 may be considered as 11.1 thousand.)

$$\text{Easting of mean centre} = \frac{\Sigma\,(5)}{\Sigma(2)}$$

$$\text{Northing of mean centre} = \frac{\Sigma\,(6)}{\Sigma\,(2)}$$

Table 2

(1) *Constituency*	*(2)* *Population*	*(3)* *Centre easting*	*(4)* *Centre northing*	*(5)* *(2) × (3)*	*(6)* *(2) × (4)*
1	11 119				
2	23 300				
3	21 667				
4	23 286				
5	16 467				
6	13 289				
7	22 565				
8	22 920				
9	19 029				
10	21 511				
11	25 769				
12	26 132				
13	26 549				
14	23 459				
15	21 492				
16	15 939				
17	17 461				
18	21 688				
19	16 421				
20	24 507				
21	23 657				
22	7 779				
23	19 998				
24	11 817				
25	16 318				
26	18 672				
27	19 015				
28	18 630				
29	26 383				
30	29 275				
31	25 907				
32	48 115				
33	17 011				
34	16 449				
35	18 222				
36	20 481				
37	27 818				
38	18 803				
39	18 475				
40	23 318				
41	14 555				
42	20 336				
43	15 827				
44	22 123				
45	21 345				
46	18 107				
47	22 463				
48	16 791				
49	14 941				
50	17 692				
51	17 563				
52	21 949				
53	16 965				
54	20 819				
55	19 820				
56	18 448				
57	17 282				
58	21 012				
59	18 707				
60	16 426				
	$\Sigma(2) = 1\,216\,833$			$\Sigma(5) =$	$\Sigma(6) =$

4 Plot the position of the mean centre with a bold symbol and co- ordinate lines on a copy of the map in Figure 3 .

5 Using the data shown in Table 3, plot a Lorenz curve. Construct the framework for your diagram as shown in Figure 4. Estimate the position of the maximum deviation of your curve from the 45° line, and express this as a percentage read off the *Y* axis. This is called the *index of concentration.*

POPULATION CHANGE

In common with many developing nations, the population growth rate in Lesotho is high. At the present time, it is in the order of 2.3 per cent per annum, and the trend of population growth since 1946 can be summarised as follows:

Although the population is small in absolute terms, the rapid rate of growth gives cause for concern. The capacity of this harsh environment to support a rapidly growing population is strictly limited. There is already evidence of excessive population pressure, especially in the lowlands, where yield deterioration and soil erosion are widespread. Given current technology and finance, the cultivable area is already over-occupied. More than five per cent of the population classified as 'dependent on agriculture' have no land at all, and a much larger proportion have insufficient land for successful subsistence. The total cultivable area is estimated at 405 000 hectares, or only 13 per cent of the national area. This figure

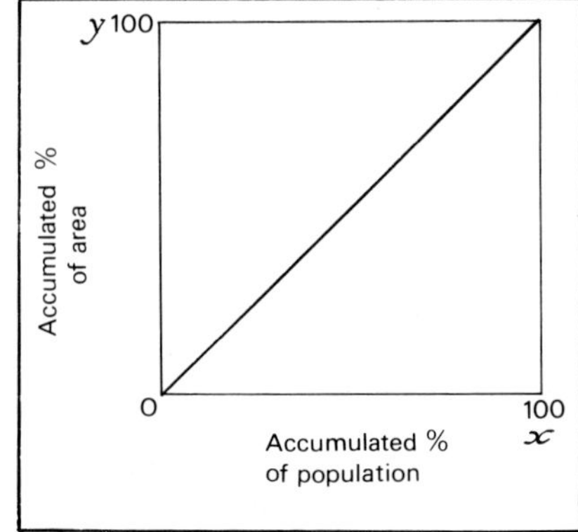

Figure 4

Table 3 *Population distribution, Lesotho*

(1) Region	(2) % of national population	(3) % of national area	(4) Accumulated (2)	(5) Accumulated (3)
Lowlands	59.34	21.26		
Senqu* valley	24.18	37.57		
Foothills	12.23	11.76		
Mountains	4.25	29.41		

* includes mountain valleys

Table 4 *Population, 1946–85*

Lesotho total population	Census date
656 748	1946
793 639	1956
972 025	1966
1 216 815	1976
1 500 000	1984/5*

* unofficial estimate — in the rest of the chapter data is used from the 1976 census as these are the most up-to-date official figures.

probably cannot be significantly increased, regardless of the technology applied. Given such a limited resource base, the rapid growth in population can only lead, in the short term, to a deterioration in the agricultural system and to further out-migration, either to urban centres within Lesotho or as labour migration to South Africa.

? **1** What do you feel will be the major problems created by rapid population growth
a in rural areas, and
b in urban areas?

PHASES OF POPULATION GROWTH

There is no simple answer to the population growth problem. Its explanation lies in a falling death rate in the face of a constant, relatively high, birth rate — this high birth rate would be reduced either by some form of population control or by increasing urbanisation, as proposed by demographic transition theory. The theory suggests three phases of demographic evolution as indicated in Figure 5. Phase 1 is regarded as typical for undeveloped economies where subsistence agriculture predominates. Death rates are high due to poor nutrition, limited hazard control, poor communications and a lack of medical facilities. The birth rate is high in order to maintain the population. Phase 2 shows rapid population growth in response to falling death rates and constant birth rates — many developing nations display such demographic characteristics as nutrition, communications, health care and development technology improve. In Phase 3, the birth rate falls as the population becomes increasingly urbanised and as industrial and tertiary occupations predominate — such conditions prevail in the highly developed regions of North America and Western Europe (see Chapter 1). In these regions, the need for a large family as a source of labour and security has decreased. High levels of mobility and the emancipation of women have also contributed to falling birth rates.

The erratic birth rate and death rate lines at the start and the end of the sequence of events indicate similar but fluctuating values for each measure. In Phase 1 for example, BR may be slightly higher than DR in a given year; the opposite may be true in the following year. In developed countries in Phase 3, BR and DR are generally low, but both fluctuate slightly year by year.

While such models should not be taken as a definitive pattern for a nation's demographic evolution, it is nevertheless generally true that in Western industrialised nations, the birth rate has fallen as urbanisation has increased. There is no case for a direct extrapolation to developing nations today, but it is fair to say that one approach to the population growth problem is to concentrate on the acceleration of the development process so that birth rates will start to fall of their own accord. Attempts at population control as a response to the growth problem have often foundered through a lack of appreciation of the central role of the large family in a subsistence farmer's life. The children are a

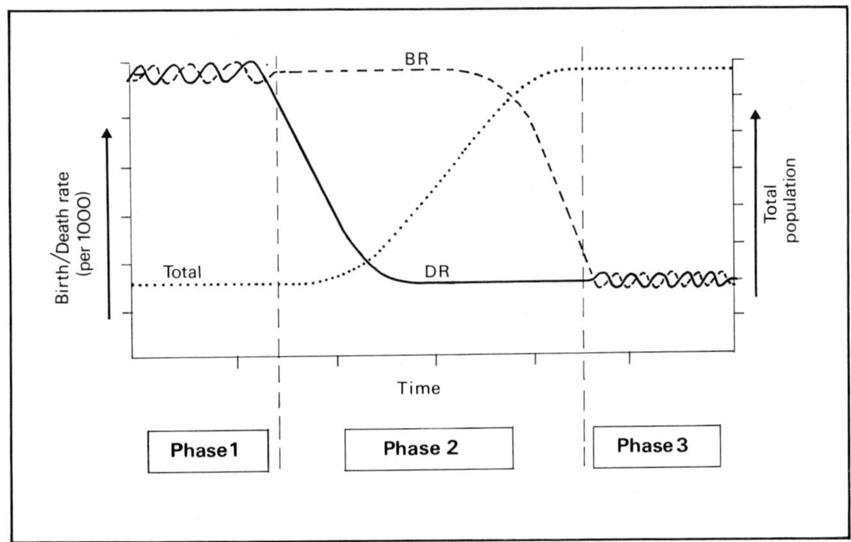

Figure 5 *Demographic transition model (simplified)*

labour force and an insurance for security in old age. Parents will be encouraged to have smaller families only if they are given access to alternative institutions.

In Lesotho, the official response to an average annual population growth rate of 2.3 per cent is to recognise the paramount need for economic development, while at the same time accepting that 'a general lowering of the standard of living seems inevitable unless effective means are found to control population growth' (*Kingdom of Lesotho Third 5-year Development Plan* 1980–85). However, there is little evidence to date of any decisive government action in the area of population growth control. A recent publication on law and population growth in Lesotho states:

> At present it appears that the government would prefer to maintain a low profile on family planning and leave the matter largely in the hands of a voluntary agency, the Lesotho Family Planning Association. No new blueprint for a population policy has been published. This can probably partly be explained by the strong influence of the Roman Catholic Church within government circles.
>
> (*Source*: S. Poulter et al., 1981, *Law and Population Growth in Lesotho.*)

Table 5 *Population change in Lesotho, 1966–76*

Region	*Total population 1966*	*Total population 1976*	*% change in pop. 1966–76*
Lowlands	488 188	651 402	+33.5
Foothills	126 518	149 987	+18.5
Mountains	66 175	76 564	+15.7
Mountain valleys	195 783	237 563	+21.3
Senqu valley	76 092	91 537	+20.3

At the regional scale there has been considerable variation in population change (see Table 5). As shown in Figure 6, the lowland region has experienced a more rapid growth in population between 1966 and 1976 than other regions,and this would suggest a pattern of population movement into the lowlands from other parts of the country. If it is assumed that natural increase (the surfeit of births over deaths) is roughly constant throughout the nation, then any regional variation in population change must be due to migration. Given that the overall growth in the national population over the decade 1966–76 was 25.14 per cent, then the probable migration characteristics of the regions can be considered as shown in Figure 7, *page 193*.

MIGRATION AND POPULATION CHANGE

It is not possible with currently available census data to investigate precisely the nature of the migration pattern in Lesotho, but the trend of movement into the lowlands is clear. Population pressure is felt first in those areas of most limited resources — the mountain regions, where cultivable land is in short supply, are thus losing population to the lowlands. Such general patterns probably obscure a large degree of local variation, however. Parts of the mountain valleys, for example, may also be experiencing a certain amount of in-migration from adjacent uplands, and development projects such as those reviewed later in the chapter may encourage some reverse migration back into the mountain region. And overriding the details of the pattern of internal migration is the fact that all regions are sources for the international labour migration process discussed earlier.

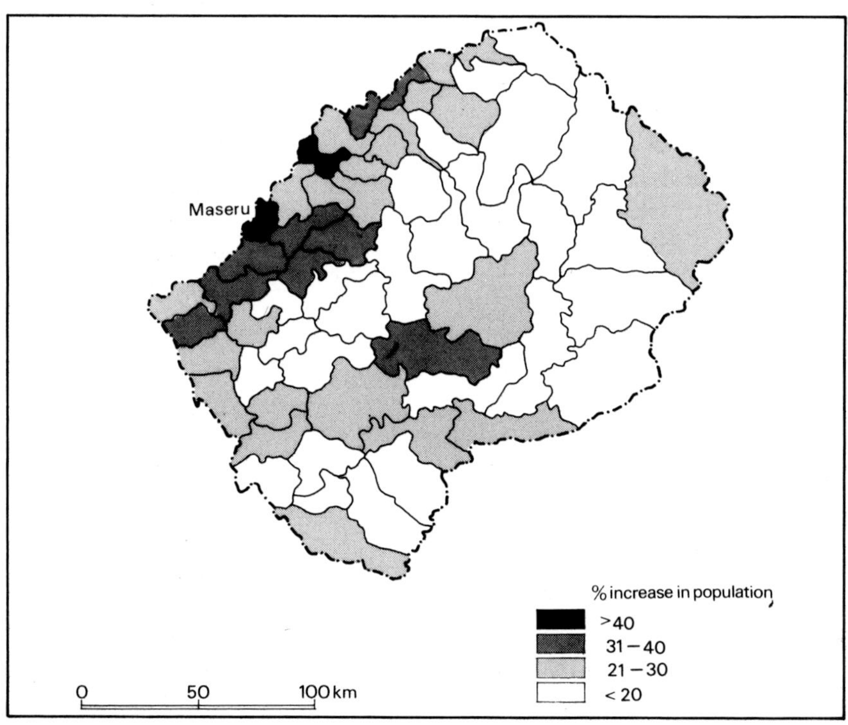

Figure 6 *Population change by constituency 1966–76*

? The regional variation in population change suggests that there is an overall migration of people away from the mountains towards the lowlands. One way of testing this hypothesis is to correlate population density with subsequent population change. If the hypothesis is correct, then areas of high population density (e.g. in the lowland region) should subsequently register a relatively higher percentage increase in population due to in-migration. The converse should apply to areas of low population density.

? **1** From Table 6:

a Calculate the Pearson Product Moment Correlation Coefficient for 1966 population density and population change 1966–76.
b Test the result for significance.
c Plot a scattergraph.
d Plot a regression line.
e Plot standard error lines.
f Identify residuals and suggest possible explanations.
2 What general conclusions would you now draw regarding the hypothesis that areas of high population density tend to register relatively higher percentage increases in population, and that, by implication, there is a general trend of population migration from mountains to lowlands in Lesotho?

(Note: This exercise is particularly suitable for use with micro-computing facilities. If such equipment is not available, it is suggested that a *sample* of data is selected for hand calculation.)

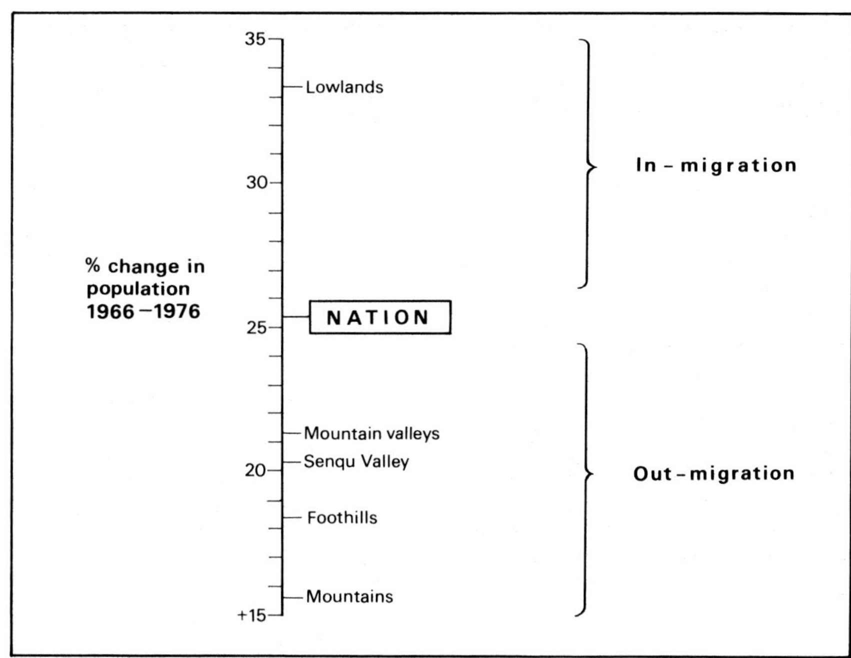

Figure 7

Table 6 *Population density and change in Lesotho, 1966–76*

Constituency	*Population density 1966 (Persons/km²)*	*% Change in population 1966–1976*
1	17	+18.46
2	98	+31.73
3	57	+13.84
4	139	+28.47
5	12	+ 6.58
6	16	+12.89
7	40	+25.36
8	53	+37.58
9	103	+23.92
10	88	+22.08
11	148	+36.62
12	128	+33.52
13	79	+29.32
14	116	+27.05
15	87	+20.85
16	25	+16.21
17	92	+24.81
18	50	+28.42
19	72	+25.58
20	70	+23.07
21	14	+31.30
22	14	−25.70
23	15	+45.28
24	18	+14.81
25	19	+15.97
26	47	+11.00

Table 6 *continued*

27	65	+16.92
28	36	+32.79
29	75	+33.10
30	79	+35.92
31	67	+31.00
32	259	+79.67
33	24	+15.64
34	57	+14.96
35	54	+16.26
36	57	+29.27
37	69	+24.59
38	59	+38.12
39	62	+20.10
40	56	+31.32
41	21	+ 8.77
42	18	+20.23
43	33	+10.33
44	56	+19.55
45	52	+27.68
46	42	+25.43
47	49	+26.23
48	24	+21.06
49	17	+16.52
50	20	+19.27
51	62	+12.71
52	26	+24.75
53	12	+19.31
54	13	+19.43
55	30	+27.03
56	17	+19.98
57	23	+18.78
58	9	+29.27
59	19	+19.26
60	12	+19.98

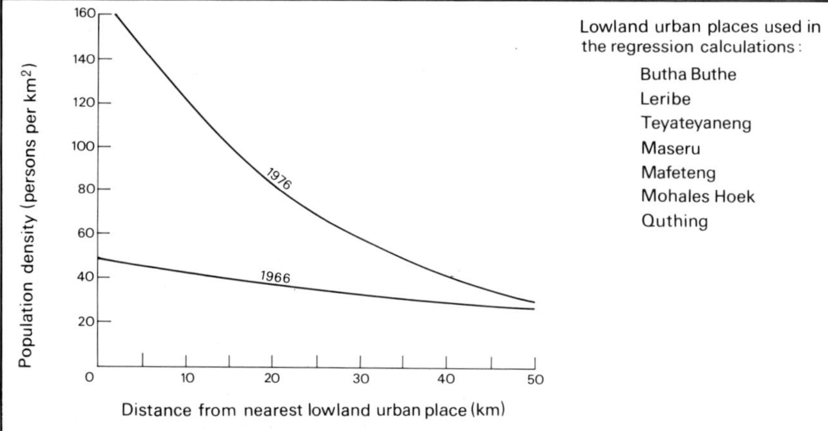

Figure 8 *The relationship between population density and distance from lowland urban centres, 1966 and 1976*

There is evidence that, coupled with the general trend of migration into the lowlands, there is a progressive concentration of population in and near the lowland urban centres. The regression relationship between population density and distance from nearest lowland town is illustrated by the graph in Figure 8.

Curve equation: $y = ae^{bx}$
Where y is population density
x is distance
e is base of natural logarithms
a, b are regression coefficients derived by computation.

These curves are 'lines of best-fit' derived from scatter-graphs showing the actual values of x and y per enumeration area in 1966 and 1976.

? What do you observe on the graph in Figure 8 with respect to
a the shape of the relationship?
b the strength of the relationship in 1976 relative to 1966?

The implication of these results is that not only is population migrating into the lowlands, but that the destination of the majority of the migrants is a location in, or close to, the lowland urban centres. A proportion of the migration towards urban areas will originate from within the lowland region itself. Certainly, Maseru, the national capital, is representative of the trend and indeed accounts for most of the strengthening of the population density/distance relationship noted above. The population of Maseru increased by 82 per cent between 1966 and 1976 — almost double the rate of growth of any other urban place.

The concentration of migration on a single urban destination is a frequently observed phenomenon in developing countries. While the absolute size of Maseru is still small, and while 'primate cities' elsewhere in the developing world reflect migration and urbanisation on a much larger scale, the problems caused by rapid growth in Maseru are by no means insignificant. The scale of in-migration is greater than that of new job creation, and the problems of housing and service shortage are acute. There has been dramatic expansion of the built up area which now extends well beyond the formal city limits. The extension of the periphery comprises largely unplanned and unserviced low quality residential expansion.

POPULATION STRUCTURE

The structure of a population, as defined by age and sex distribution, is a reflection not only of the state of demographic evolution but also of the prevailing social and economic conditions of the nation in question. Developing countries, characterised by falling death rates and relatively constant birth rates, tend to display 'broad-based' population pyramids, as in the case of Lesotho below — the fall in infant mortality in the early stages of a nation's development is clearly of the greatest impact.

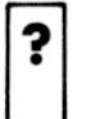

1 Construct a population pyramid for Lesotho on the basis of the data in Table 7.

Table 7 *De Jure population of Lesotho by sex and 5-year age groups*

Age	*Male*	*Female*	*% male*	*% female*
0–4	83 963	84 649		
5–9	76 723	76 543		
10–14	75 861	77 566		
15–19	58 419	66 186		
20–24	48 324	54 807		
25–29	40 477	42 210		
30–34	32 290	34 160		
35–39	30 092	29 453		
40–44	30 150	32 413		
45–49	22 982	23 225		
50–54	18 896	20 464		
55–59	19 877	21 558		
60–64	12 256	15 734		
65–69	8 518	11 272		
70–74	5 978	9 625		
75–79	5 398	10 151		
80 and over	3 695	8 664		

Unspecified: 24326
Total population: 1 216 815 (1976)

RESOURCE PRESSURE

As mentioned earlier, the increase in population in Lesotho, and the redistribution of that population between the regions, is taking place within the context of a severely limited resource base. Flat land is at a premium, soils are of (at best) moderate fertility, and there are no economically viable natural resources other than land and water. Development planners are faced with the following key issues:

- What will be the regional and national consequences of continued population growth?
- What inter-regional stresses will arise from internal migration?
- Should government or other institutions become engaged in attempts at population control?

The evidence of resource pressure as a consequence of population growth and redistribution is presented by a number of key indicators, especially

yield deterioration — the decline in agricultural productivity over time as the land is over-used;

accelerated soil erosion — the increased rate of soil removal by rain and wind due to poor land management and overgrazing;

landlessness — the increasing numbers of rural people without access to the fundamental agricultural resource.

Such are the parameters of resource pressure in Lesotho today — but it would be wrong to look at this problem purely in the context of the population growth issue. We must ask searching questions about the colonial past when seeking explanations for today's problems: during the period of British administration from 1868 to 1966, what base was laid for subsequent national development? What levels of colonial investment were recorded? What nature of administrative, commercial and physical infrastructure was bequeathed on independence? The record, in Lesotho as in many other developing nations, suggests that the net consequence of colonial involvement was a patchy improvement of basic welfare and services (reflected now in population growth), but with an absence of long-term development investment, in either the rural or urban sectors of the local economy. There was investment — but in colonial enterprises designed primarily for the benefit of the colonial power (e.g. plantation agriculture and mining and quarrying). Certainly in the Lesotho case, the colonial involvement was inspired by strategic political motives within the context of the sub-continent rather than by a commitment to the country, and levels of subsequent investment were, at best, modest. In 1966, the year of Lesotho's independence from Britain, the colonial investment per head of the Lesotho population was only about £3.50 per year and ten years previously it had only been about 20 pence.

Lesotho's Third Five Year Development Plan, 1980–85, recognised the pivotal role of the land and water resources in the nation's development, and acknowledged the creation of a sound agricultural base as the priority objective. Effective urbanisation and industrial development will ensue only from sustained improvements in agricultural productivity and from stable rural land-use systems. At present, serious resource–use problems relating largely to the increased demands of a growing population, but exacerbated by poor infrastructure and unskilled management, are barriers to the realisation of national development objectives. In this section, selected resource–use problems will be investigated.

Print 1 *Senqu Valley near Fort Hartley. The foothill landscape is characterised by cultivation on available level ground, village location at break-of-slope, and communal grazing land on upland pastures. Note the degraded pasture in the foreground*

YIELD DETERIORATION

Well over 90 per cent of the Lesotho population lives on the land and derives its livelihood from crop agriculture or the keeping of livestock, or both. Owing to increased demands for land from a rapidly growing population, the average size of a rural household's landholding has now dropped below 2.0 hectares, and nearly 15 per cent of rural households have no land at all. In order even to maintain current standards of living in the rural sector, in the absence of substantial improvements in non agricultural employment opportunities, agricultural productivity would need to increase in step with population growth, which is currently running at an average of 2.3 per cent per annum.

In fact, average yields have shown a persistent decline since 1950 (Table 8a). While climatic hazards have recurrently lowered yields (e.g. drought, late frost), the decline is largely related to farming practices and fertility deterioration. Meanwhile, food imports run at about 50 per cent of the national requirement (Table 8b).

SOIL EROSION

One of the major causes of yield deterioration has been widespread accelerated soil erosion caused by overgrazing and overcultivation. Much of the soil of the lowland region is derived from relatively unconsolidated parent material, and this is very susceptible to erosion. On any degree of slope, rapid runoff removes unprotected soil with frightening speed. Various estimates have been made regarding the overall rate of soil loss from Lesotho — all are probably imprecise, but the statement in the First Five-Year Development Plan 1970–75 that '. . . if soil erosion is not halted by soil conservation measures, and continues at its present rate, there will be no arable land left in Lesotho in two or three generations from now . . . ' is probably close to the truth. The Ministry of Agriculture currently estimates the average soil loss for the worst affected parts of the nation

Table 8a *Average yields of selected crops (kg/ha)*

	1950	*1960*	*1970*	*1977*
Maize	11.9	7.4	5.1	5.7
Sorghum	8.7	7.8	6.9	5.5
Wheat	10.1	8.5	5.4	5.9
Peas	9.8	6.8	5.7	5.6
Beans	3.4	2.4	3.0	2.9

Table 8b *Cereal imports in tonnes*

	1950	*1960*	*1966*	*1970*
Maize	15 621	20 983	30 336	45 318
Wheat	2 111	4 200	23 258	26 789

to be about 300 tonnes per hectare per year: something like 20 per cent of the best soil has already been lost. In a recent study of part of the lowland region, it was calculated that 85 per cent of the cultivated area was subjected either to surface or to gully erosion. In the early 1970s an aerial survey was carried out to identify the percentage of cultivation (by region) occurring on land *unsuitable* for cultivation as defined by the slope criterion — any slope greater than 9° being generally accepted as unsuitable (Table 9).

Table 9 *Agriculture on land unsuitable for cultivation, Lesotho*

	Regions			
	Lowlands	*Foothills*	*Mountains*	*Senqu valley*
% of cultivated area on slopes >9°	8.6	37.9	81.1	49.8

Such data indicates the magnitude of the problem, namely that approximately 80 940 hectares of land are under cultivation on slopes greater than 9°. Figure 9 shows the distribution of the cultivated area, and the extent of cultivation in unsuitable areas can clearly be

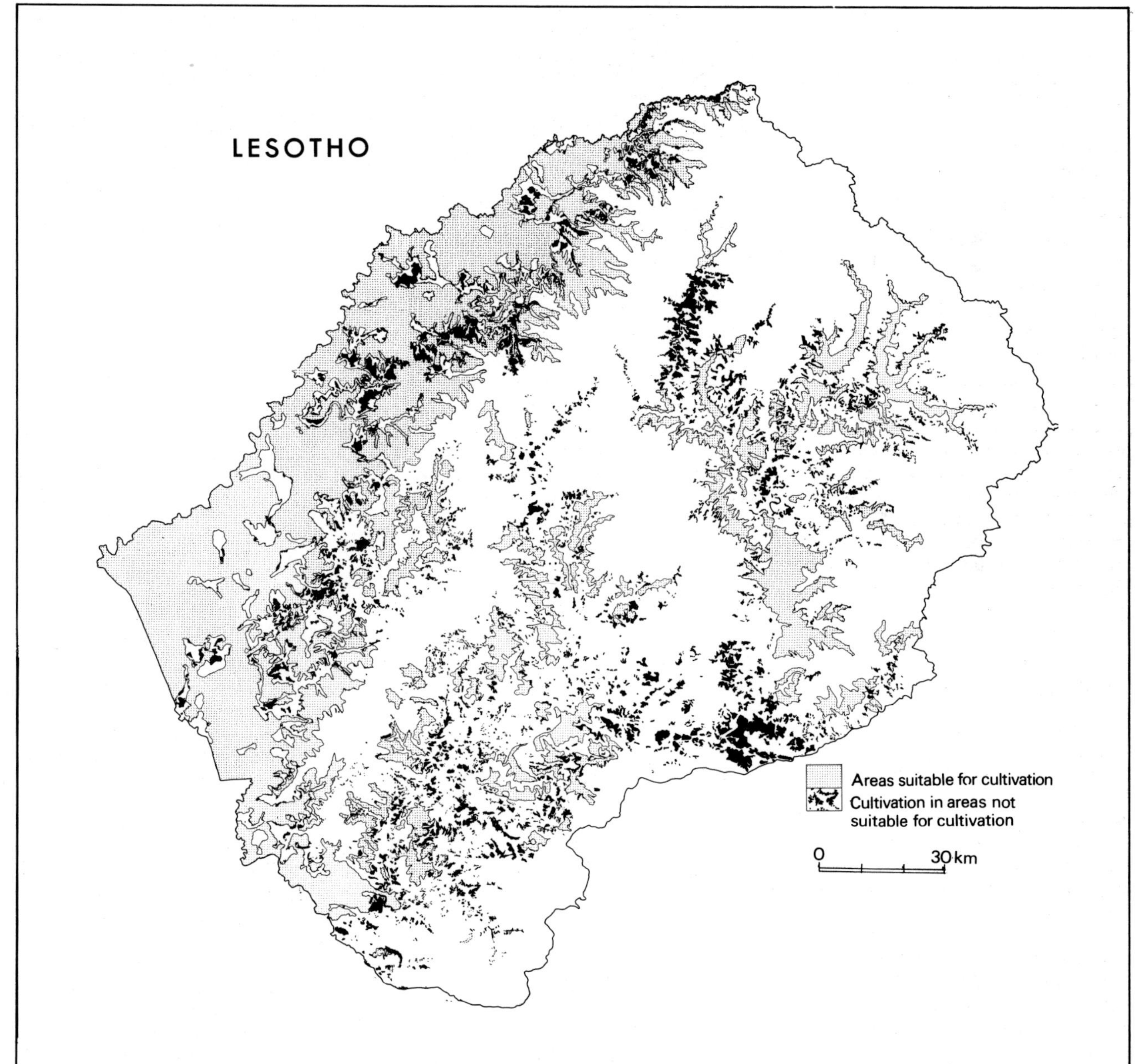

Figure 9 *Lesotho: showing all land under cultivation and highlighting the areas which are unsuitable for the purpose*

gauged. As population has grown and as the demand for land has increased, so cultivation has extended into areas of greater slope and consequently greater erosion hazard. Overgrazing in the mountainous areas has compounded the soil erosion problem. Traditional society in Lesotho stresses quantity rather than quality in livestock ownership, and this coupled with loose management, has led to pasture deterioration and soil erosion. In all cases of soil loss, slope is clearly a dominant influence, but the problem is compounded still further by the nature and seasonal distribution of rainfall. The winter dry season reduces soil stability considerably, and rainfall then occurs in a relatively concentrated period between October and April. The rainfall is of a predominantly convectional nature and short, but heavy afternoon rainstorms are the summer norm. Hail is often associated with such storms and the consequent damage to the soil surface is increased. The elements of the erosion process are summarised in Figure 10.

The consequences of soil erosion are not restricted to soil loss and reduced agricultural potential at the site where such erosion occurs. Downstream, a more extreme pattern of discharge will occur as runoff rates increase, and the resulting cycle of flood and drought hazard compounds the difficulties already faced by lowland agriculture. Increased rates of runoff lead also to the rapid siltation of small earth dams which are numerous in the foothills and lowlands as erosion control devices, and as sources of water for stock and domestic uses (see Figure 11, *page 200*). In a 1973 study of dams in the Roma Valley in the central lowlands, the following observations were documented:

Print 2 *Severe soil erosion near Quthing, southern Lesotho. This has been caused by overgrazing and overcultivation of a fragile resource base. Poorly consolidated soils have been stripped down to bedrock by increased rates of runoff and consequent erosion*

> . . . At construction, the dam wall was 12 metres deep, but it was almost filled up in 1972 when a further three metres were added to the wall. The present maximum depth of the dam is only three metres. Thus in the period 1967 to 1972, 12 metres of deposits were accumulated in the dam. Out of this and the area of the dam the rate of denudation over the whole catchment (27 km^2) could be calculated as 1–5 cm per year . . . a number of dams which were constructed in the late 1950s as conservation measures have been completely silted up and new gullies have developed through most of them.
>
> (*Source*: Qalabane Chakela, *Studies of Soil Erosion and Reservoir Sedimentation in Lesotho*, Uppsala University 1973)

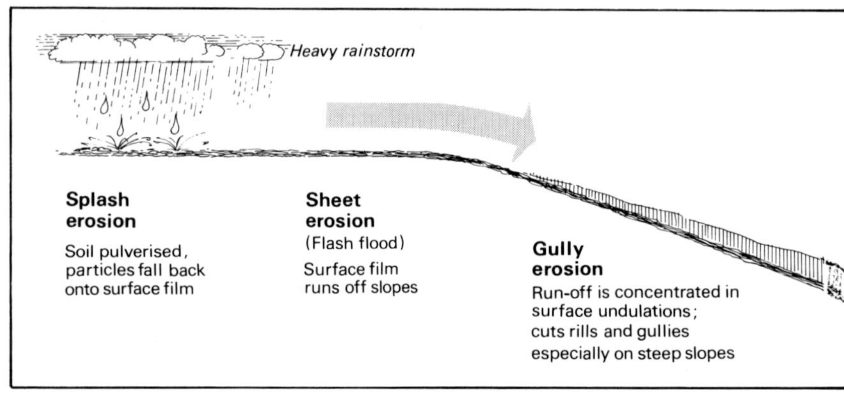
Heavy rainstorm
Splash erosion
Soil pulverised, particles fall back onto surface film
Sheet erosion
(Flash flood)
Surface film runs off slopes
Gully erosion
Run-off is concentrated in surface undulations; cuts rills and gullies especially on steep slopes

Figure 10 *The erosion process*

Such a sequence of events results from the agricultural responses to the demands of a rapidly growing population — reduced fallow, the extension of cultivation upslope, and overgrazing. The Ministry of Agriculture, and externally financed development projects, hold the control of soil erosion as a priority objective, but at present a number of traditional institutional mechanisms retard progress.

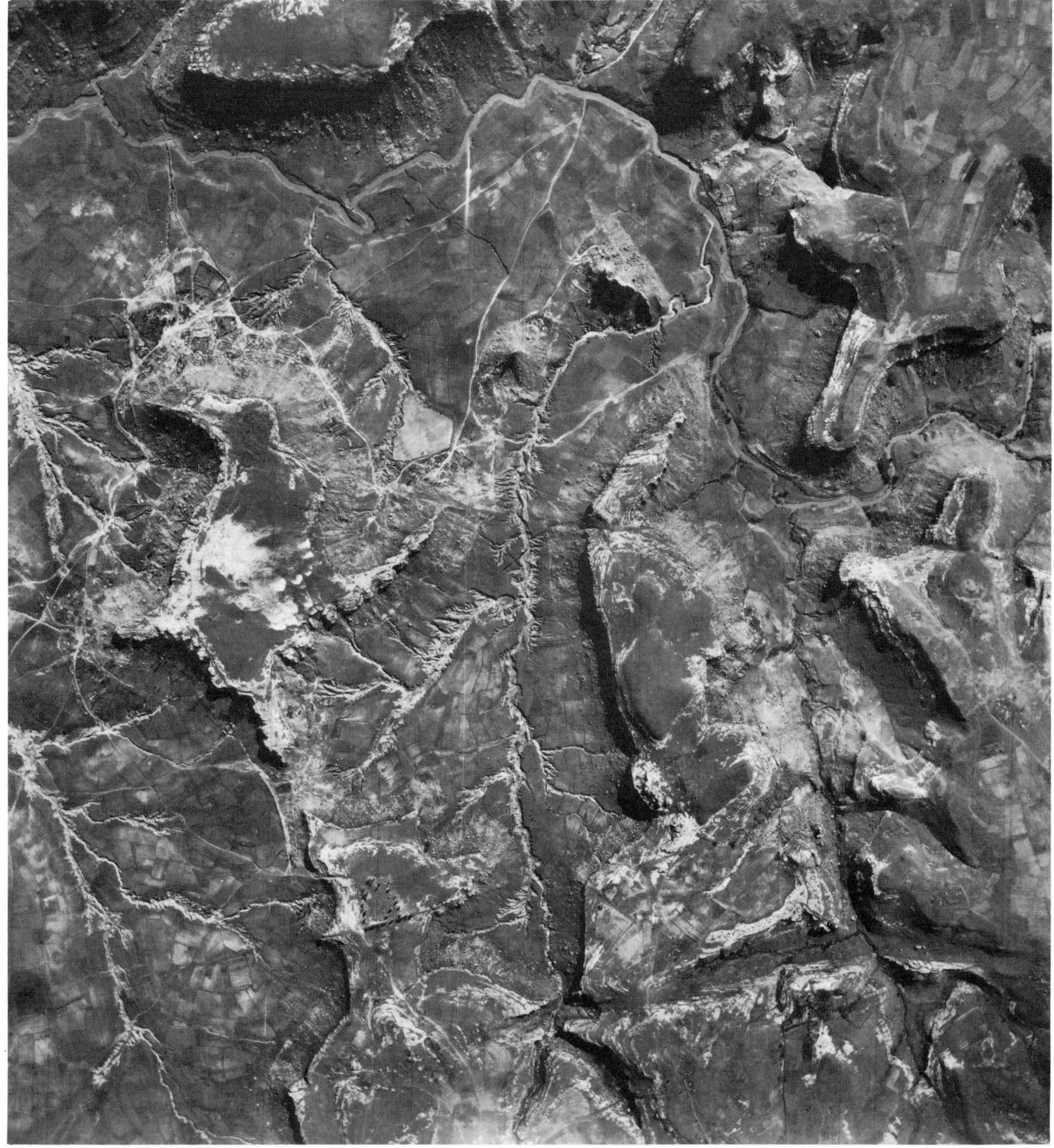

Print 3 *Aerial view of Thaba Bosiu, 30 kilometres from Maseru*

Study Print 3 taken in 1951 from 5 000 metres which shows a number of sandstone plateaux rising from the surrounding lowlands near Thaba Basiu, about 30 km from Maseru. Cultivated areas can be seen, as can villages surrounded by planted enclosures. Place a piece of tracing paper over the photo and mark: major rivers, areas worst affected by erosion, including gully and sheet erosion. Show general patterns rather than accurate detail. Write a short description of the distribution of the erosion evidence you have observed, and consider the implications for agriculture, bearing in mind that this photograph is over 30 years old.

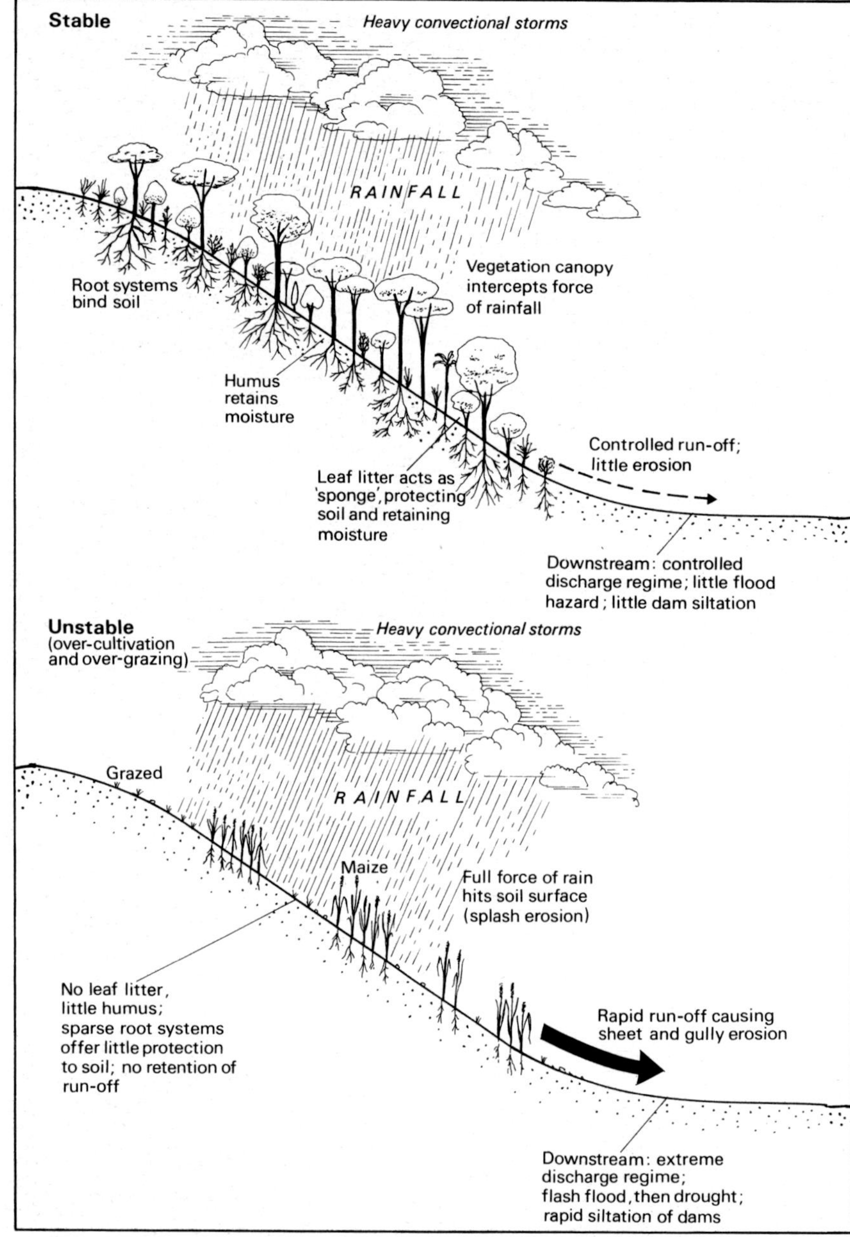

Figure 11 *The environmental breakdown associated with accelerated soil erosion*

SOCIO-CULTURAL PROBLEMS

Paramount among these is the system of land tenure, which has remained largely unchanged since the inception of the Basotho nation in the early nineteenth century. As a general rule there is no individual title for land — in law the king holds all land in trust for the nation, and land is allocated for the use of individual farmers through the hierarchy of chiefs and headmen which makes up the traditional system of government. Each farmer will normally be allocated three parcels of land; one good, one poor and one indifferent, and by this mechanism a fair distribution of lands of varying quality is deemed to have taken place. The total area available to a farmer under this system is usually less than 2.0 hectares and the physical separation of land parcels is a serious barrier to improvements in efficiency and productivity in farming. As the population grows the degree of separation and the further fragmentaion of parcels increases as individual demands for land become more numerous. By the same measure, land allocation is extending into more steeply sloping areas which are ecologically unsuitable for cultivation, and the erosion hazard increases.
Land shortage is exacerbated by the fact that the right to hold land extends even to those former members of the rural community who have migrated into the towns. Such absentee land holding, while it may eventually be penalised by the withdrawal of the allocation, means that some good land is out of use at any one time. The lack of individual title to the land means that farmers are unlikely to invest in the improvement of their plots; without security of tenure they cannot take that risk. Similarly there is little incentive to farmers to involve themselves in conservation and soil protection infrastructure. The lack of security leads to a disinterested attitude towards the long term improvement and conservation of the land resource. As the land is essentially communally owned, the individual's right to use effectively ceases once he has harvested his crop; thereafter the land reverts to communal grazing until the next planting season, and stubble that could be mulched as a soil conservation measure is grazed off by stock, further exposing the soil to the erosion hazard.

Whilst government is committed to overhauling the land tenure system as one element in the programme for agricultural development, progress is slow. The land issue is a fundamental and sensitive one in the fabric of the traditional society, and the power of the chieftainship in preserving the status quo (which is to their general benefit) is considerable. Even given a readiness on the part of the people to accept change, the design and implementation of a new land tenure system would be a complex procedure. Certainly the land tenure problem is central in the search for agricultural improvement, and is generally recognised as such. A comment in a recent publication of the National University of Lesotho is typical:

. . . But the role of the current land tenure system in stimulating improved husbandry practices and a more productive agricultural sector is almost unanimously regarded as negative.
(*Source*: Huisman and Sterkenberg, 1981, *Spatial Organisation for Development in Lesotho* National University of Lesotho)

The traditional role of livestock in the agricultural system is a further constraint on the control of erosion and the search for improving productivity. In Basotho society an important measure of a man's wealth has historically been the number of stock he possesses. The exchange of livestock plays an important part in social transactions — for example, a groom's father will 'pay' a certain price in stock to the bride's father in the *lobolla*, or brideprice, system. In such ways the quantity of cattle, sheep and goats has been the dominant criterion, and this, coupled with loose control over grazing has led to severe pasture deterioration, especially on the more steeply sloping land which is held by the community as common grazing. Many areas in the mountain region, which comprise summer grazing or 'cattle posts' in the traditional transhumance system between lowlands and mountains, are heavily overgrazed and the problem is extending into even the most remote high mountain pastures. Bare and scarred hillsides testify to the rapid soil erosion caused by heavy overstocking.

Print 4 *Herdboys at Koro-Koro, Central Lowlands. During the summer months herdboys drive cattle, sheep and goats to mountain 'cattle posts' and tend them there before returning to the lowlands with the onset of winter*

APPROACHES TOWARDS A SOLUTION

Clearly Lesotho faces a resource crisis as population growth and movement exert profound stress on the national economy. Of prime importance is the design and implementation of an effective development programme, and yet there is considerable controversy regarding the most appropriate course to be taken. Some development planners adopt a 'top-down' approach; others a 'bottom-up' philosophy, the former involving high technology and capital investment as the motive force for development; the latter stressing the

Print 5 *Maluti mountain region. Cultivation at this altitude (over 2500m) and on such a degree of slope is marginal and a major contributory factor in increased soil erosion*

need for harnessing local skill and accumulated wisdom, and the adoption of modest yet relevant technology in the solution of essentially rural problems.

? Read the two scenarios for economic development along with the extracts on *page 203*, illustrating some of the aspects of each approach. Kenya has opted for a free market approach to economic development, while socialist Tanzania under Julius Nyerere has pioneered the principles of self-reliance and appropriate technology.

1 What do you think are the advantages and disadvantages of each of the scenarios outlined? Give your answer in table form, listing advantages and disadvantages in each case.

2 Write your own, third scenario for the development of a small, land locked developing country, explaining carefully what you regard as the major priorities in the early stages of the development process, and using information and data from countries or regions with which you are familiar.

Scenario 1

A small, land locked country opts for a capital-intensive approach to development — external finance and international aid is sought for investment in both the rural and urban sectors. In the countryside, major water management schemes are implemented based on multiple-purpose dams; cash crops are introduced and a network of marketing and input-supply (seed, fertiliser etc.) points is established. Irrigation systems are brought into operation, local communications are improved, and a concerted effort is made to reorient the rural sector away from subsistence farming towards commercial supply to the towns. The pastoral element is rationalised through herd culling and controlled grazing, and the land tenure system is overhauled to provide security of tenure and purchase rights to progressive farmers. Many expatriate advisers and technicians are introduced to service the various development schemes, and attractive conditions prevail for private overseas capital to be invested in the nation. In the urban sector investment takes place in, initially, agricultural processing industries and aid finance is provided for extensive resource prospecting throughout the nation. Multinational enterprises establish branches in the national capital. Capital is borrowed for investment in urban housing and services, and for the development of all-weather communications between the urban centres.

Scenario 2

A small, land locked country opts for an 'appropriate technology' approach to development — external capital and aid finance is treated largely with suspicion and is used sparingly. Through education and training, abroad where necessary, a corps of rural development specialists is established to operate at village level throughout the nation. Labour intensive and low cost innovations are vigorously promoted; contour ploughing, stubble mulching, small earth dam construction, biogas projects and the harnessing of solar energy are examples. Local crafts and skills are directed towards ox-drawn agricultural implement construction and maintenance. Farmers are given intensive programmes of instruction and demonstration in improved husbandry and cultivation. The land tenure system is left largely untouched, save for the reallocation of land held by absentees. The national education system is restructured to provide training in agriculture and rural skills; pottery, weaving and other handicrafts are promoted as supplementary sources of rural income. The improvement of subsistence agriculture is regarded as the national priority, and the limited development finance available is channelled into the provision of improved agricultural inputs (seed, fertiliser, appropriate technology tools) rather than into the urban sector. 'Urban-wards' migration is actively discouraged at this time. Self-sufficiency is the prevailing ethic.

Print 6 *What form should development take in Lesotho? These women are harvesting on a 'food-for-work' scheme that promotes the use of manual labour and local skills.*

Mumieng, Kenya: A Successful Multinational Enterprise

The government of Kenya wished to increase sugar production and also to provide some of its subsistence farmers with a cash income without reducing their own production of food. They approached the large British company, Booker McConnell, and received loans from the World Bank and the East African Development Bank. The company's profits were to be limited to 5 per cent and to be adjusted to the rates received by the participating farmers: that is, if the farmers received no profit, neither did the Company.

Instead of establishing a large plantation, and employing local people as labourers who would consequently grow no food, the company provided seed cane, fertiliser and some mechanical help to farmers on condition that they grew sugar on only half their land. Farms averaged 4 hectares. 5000 farmers entered the scheme, and the company built a mill to employ 2500 more. At the start 60 Europeans were involved, and this was reduced to six by 1976 as Kenyans learnt to take over technical jobs in management and research into plant diseases. The sugar was sold to the Kenyan government.

The three main reasons for the success of this scheme appear to be important clues for other developing countries:

i. Labour intensive technology was used wherever possible.

ii. Food production was maintained on the workers' own farms, and sometimes increased, since their cash income enabled them to buy better seed and fertiliser.

iii. The government of the country laid down limitations to the profit and method of operation of the multinational company. This is underlined by the fact that this same company grows sugar by the plantation method elsewhere.

Tanzania: Small-Scale Development through Increasing Food Supplies

Tanzania is just south of the equator, and is a country of plateaus, rift valleys and high mountains. Much of its area has a low rainfall for the tropics, averaging about 500 mm a year.

The following activities took place on the dry plains just south of Mount Kilimanjaro, an extinct volcano and Africa's highest mountain. An agricultural development project was started there called the Kalicha Production Training Centre. Here an attempt was made to put into effect the Tanzanian government's policy of self-reliance rather than the importation of overseas technology. Local farming customs were used, but efforts were made through minor changes to make them more efficient as food producers.

i. Rabbits. New Zealand white rabbits, which do not dig burrows, have been used in Tanzania for some time. They have up to eight in a litter and breed every two months. Their meat is very high in protein. They are kept in sheds and looked after by the children. Several rabbit houses were built at Kalicha to encourage the use of meat.

ii. Hens for eggs. Simple incubators, operated by kerosene, were built at the centre: the type could operate anywhere in the country. The President of the country paid a visit when the centre had hatched 100 000 chickens in a year, and immediately asked them to hatch a million. The hens were raised in movable A-frame hutches: their scratching cultivated the ground and their droppings fertilised it. After a week or so the hutch was moved to another place to repeat the process.

iii. Pasture was poor and the problem was met in two ways. Zebu cattle — a cross between Jersey and Sahiwel — were bred, since they give good milk from poor fodder. Also, local goats were crossed with Toganberg milking goats, and carefully herded rather than being allowed to run wild.

iv. Agriculture. Tanzania was once one of the main exporters of sisal, used in binder twine and some kinds of rope. Sisal has now largely been replaced by nylon, which does not rot when wet. So the centre was given a sisal site which had become uneconomic. In clearing the ground they could have used a D-8 caterpillar tractor, a long roller, two people and expensive imported fuel. Instead they used local people with bush knives employing 100 to 150 people each year. Part of the land was still used to grow sisal, which the women weave into mats, used locally and also exported. Other crops grown were maize, soya beans and sunflowers, for the oil from their seeds.

v. Oxen. The traditional method of cultivation in Tanzania was to use a hoe. Tractors would be quite impossible for 95 per cent of the people. So the government developed a programme which they called 'Oxenisation'. Using oxen to pull a plough was their intermediate technology. Kalicha centre tried to apply this principle further. After five years experimentation, they devised a scheme in which oxen walked in a circle round the differential of a tractor, which then transferred the power to pumping water or grinding maize.

vi. Recycling waste. The centre used most of its waste materials profitably. Fluid wastes from animals went to the crops as liquid manure. Infertile unhatched eggs from the incubators and residues from the sunflower crops together provided a complete diet for pigs. Manure from chickens and rabbits was used to produce methane gas, which then heated the incubators that produced the chickens.

In all these activities the policy was to keep the work *labour intensive* rather than *capital intensive*, and to modify locally accepted processes rather than replace them by imported mechanisation. Production units were established first: training units were based on these, and have since received overseas grants for further development.

(*Source*: McKay et al, 1981, *The Third World Integrated Development Centre for Human Progress in the Disadvantaged World*, Adelaide College of the Arts and Education)

EXTERNAL HELP

The complex problems of population growth, agricultural decline and resource deterioration are not amenable to simple solutions. Indeed the number and variety of institutions and individuals involved in 'development' in Lesotho is remarkable not least for the fact that the standard of living of the majority of the population has been, at best, stagnant over the past two decades. The need to conserve the resource base and yet to improve the productivity of agriculture is pressing, and only an integrated nationwide approach to rural development is likely to succeed. The recent history of the development effort in Lesotho is characterised by a lack of co-ordination between the various agencies involved. The government, through its own ministries, has attempted to define an overall strategy as documented in the official Five-Year Development Plans — but on the ground a wide variety of largely unsynchronised projects may be operating in any one place at any one time. The big development institutions — the UN agencies (e.g. UNESCO and UNICEF), the World Health Organisation, the Food and Agriculture Organisation, represent one level of involvement. Quasi-governmental institutions represent another; the Anglican, Roman Catholic and Evangelical Churches are active in education and health care, and individual volunteers from a number of overseas nations are involved in teaching and other development activities.

Which of these aspects of the development effort is the most effective is a matter of considerable debate; a reasonable case could probably be argued for any one. Of greatest concern is the danger that the proliferation described above may lead to a self perpetuating 'development business' which may exist largely for its own sake (and own reward) rather than for the people it was originally created to help. The burgeoning of the development machine may be at the expense of the poorest in the land who are least able to adapt to the changing economic, social and environmental context of the development process.

It is essential that the precise nature of rural problems is defined before any formal development strategy is drawn up, and those problems are best defined by the peasant farmers who endure them rather than by the outside 'expert'. Equally, the aspirations of those farmers should mould the objectives of development, and will do so more effectively and realistically than gross productivity targets defined by visitors from London or New York. While some expatriate technical skill is undoubtedly necessary in, for example, the quantification of the soil erosion processes and the identification of land capability, the most potent development resource lies in the hearts and minds of the people themselves, and the development strategy must capture both.

AREA-BASED DEVELOPMENT PROJECTS

During the 1970s, the favoured approach to rural development was the Area-Based Development Project (ABDP), which as the name implies, was something less than an integrated national scheme for rural improvement. Specific areas were singled out for particular development effort, as shown in Figure 12, *page 205*. The Thaba Tseka project is a special case in that the project activities were related to the creation of a new administrative district and regional

Print 7 *Development has been seriously hampered by the inaccessibility of much of the highland areas and road building is now a priority. These women are constructing a road as part of a UN project aimed at improving communications between the lowlands and mountainous areas.*

capital, and as a result, there is a clearer relationship between overall government development objectives and the activities of the project authorities. The other three project areas shown in Figure 12 are more typical examples of the autonomous ABDPs of the 1970s.

Such projects were characterised not only by autonomous decision making regarding the area within their boundaries, but by heavy investment from international donors such as the World Bank. The staffing was largely by expatriates

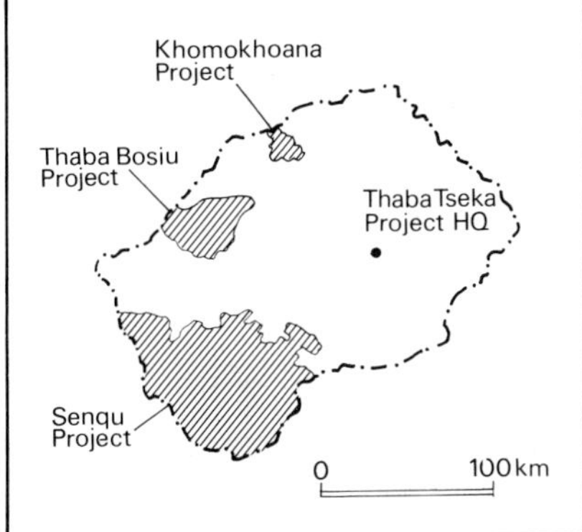

Figure 12 *Major Area-Based Development Projects*

Print 8 *The first all-weather route into the mountainous interior of Lesotho. The road runs from Maseru to Thaba-Tseka*

from Europe and North America, and an increase in agricultural productivity (often based on the introduction of mechanisation and heavy fertilisation) was invariably a priority objective. Government commitment to this approach during the 1970s is revealed in the Second Five-Year Development Plan:

> . . . For the non-mountain area, the first priority of investment will be the three fully operational area-based projects, Khomokhoana in the northern lowlands and foothills, Thaba Bosiu in the central lowlands and foothills, and Senqu in the two southernmost districts . . .

Over 50 per cent of total crop sub-sector expenditure was reserved for the four large ABDPs during the second plan period. Clearly at this time the highly capitalised approach to rural development was very much in favour. It was hoped that rural conditions could be improved by the injection of large amounts of capital, machinery, fertiliser and expatriate expertise. But the design of the large ABDPs was flawed in certain crucial aspects:

1 The areas were largely defined without adequate reference to existing spatial units and decision making institutions, and this eventually led to inefficient communication between government, project and people.

2 There was too much emphasis on the increase of crop productivity rather than on a broad programme of rural development.

3 *'. . . the area based projects were usually characterised by a top-down approach with little attention for the participation of the local population.'* (Huisman and Sterkenberg, 1981)

Table 10 summarises some major characteristics of the ABDPs.

The Thaba Bosiu Project

The following summary of the Thaba Bosiu Project, the best-documented of all the ABDPs, illustrates the general failure of this approach to agricultural development:

Objectives:

a Control soil erosion, mainly through mechanical means.
b Improve subsistence crop productivity over an area of 24 000 hectares with improved seed, fertilisation and mechanisation.
c Introduce cash crops such as asparagus.
d Introduce a dairy farming scheme to the foothill region.
e Demonstrate progressive farming methods at nodal points throughout the project area.
f Improve marketing channels through the construction of local access routes.
g Provide credit facilities for project farmers.

Failures:

Huisman and Sterkenberg (1981) made a detailed study of the results of the Thaba Bosiu Project, and presented the following conclusions:

a The production per hectare decreased during the project period,

Table 10

Project	*Total area (ha)*	*Total population*	*Duration*	*Funding ($ US million)*	*Donor*
Thaba Bosiu	121 000	95 000	1973–77	9.8	World Bank US aid
Senqu	637 000	225 000	1972–76	1.8	UNDP[1]
Khomokhoana	19 000	40 000	1975–79	4.2	SIDA[2]
Thaba Tseka	53 000	Unknown	1975–79	5.8	Various

(Notes: [1] United Nations Development Programme. [2] Swedish International Development Association.)

the proportion of fallow land strongly increased, and production costs went up because of a more intensive use of inputs. In other words, project activities did not result in higher farm incomes.
b The project was too ambitious in terms of the size of the area selected and the scope of the objectives laid down.
c There was insufficient preparatory research on local environmental and socio-economic conditions.
d There was insufficient appreciation of the real space of the farming system. Many farmers had fields within the project area, but grazed their stock on mountain pastures outside the project boundary.
e There was too much emphasis on increasing crop productivity and too little on comprehensive rural development embracing conservation, communications, services, etc. as well as a cultivation-livestock-woodlot element.

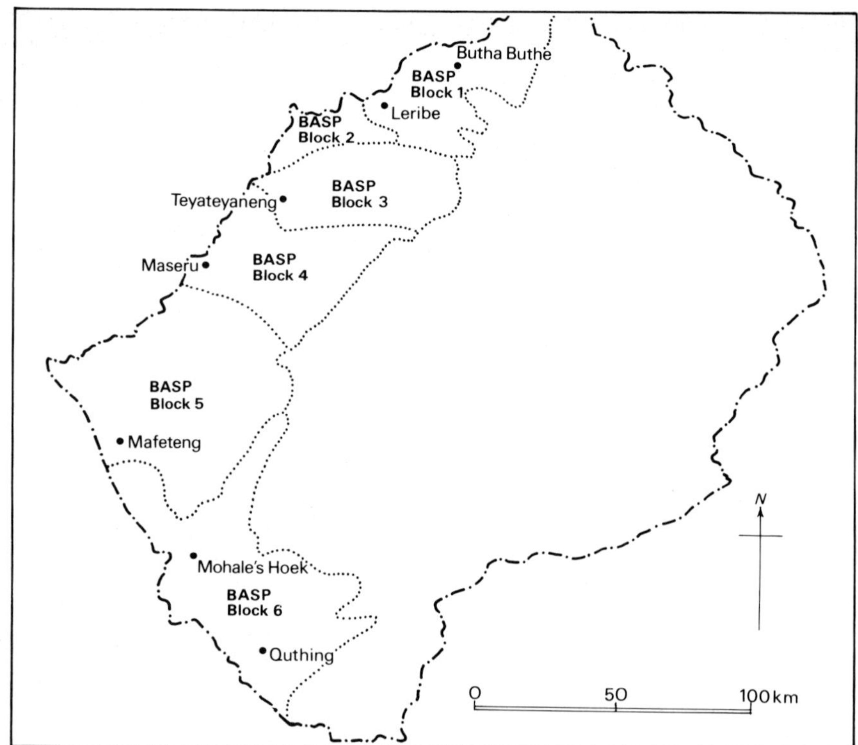

Figure 13 *Basic Agricultural Services Programme Areas*

ALTERNATIVE APPROACHES

By the late 1970s it was clear that the large, highly capitalised ABDP approach was not working, and an alternative strategy was devised. The government turned to a more modest and yet more relevant and cost-effective programme for agricultural development, formalised as the *Basic Agricultural Services Programme*, or BASP. The object is to provide a more limited range of development services than the ABDPs, but to relate those services to the immediate needs of the peasant farmer. The priority tasks are seen as the improvement of local communications, the provision of local input-supply and produce marketing points, and the mobilisation of appropriate technology in farming. The emphasis in this latter area is on low cost, labour intensive and locally-designed machinery such as ploughs, planters, pumps, grinders and storage bins. Crucially, the services of BASP are being made available not by an independent, autonomous and distant project authority but via the existing extension services of the Ministry of Agriculture and within the framework of traditional decision making by chiefs, headmen and village communities. BASP areas relate closely to existing administrative boundaries, especially at ward level, and cover the entire lowland region (see Figure 13).

The government commitment to the more modest package offered by BASP appears to be strong. The recently published 'Blueprint for Action' in agricultural development issued by the Ministry of Agriculture marks the official break with the selective ABDP concept and the concentration of admittedly limited development resources on basic, but lowlands-wide infrastructure. Extracts from the blueprint illustrate the new approach:

'. . . (strategies include) *"bottom-up" planning to ensure that programmes and projects are based upon needs identified by the farmers themselves*'

'. . . (motivation) *will not come by imposition from above. It will come only when the masses of the rural population are convinced that the programmes of action are meaningful to them.*'

It seems likely that BASP will achieve more than the ABDPs, but the obstacles to substantial improvements in agriculture, and thereby in the economy as a whole, are formidable. The context is austere: a small, land locked country with few national resources and a rapidly growing population, a legacy of limited colonial investment and a future clouded by the volatile political issues of the sub-continent as a whole.

? **1** Look back at your responses to the exercises on economic development on *page 202*. Having now read about some of the official responses to development problems in Lesotho, are you in a position to refine your views on the comparative advantages of the various approaches to development? You might choose to rewrite your original third scenario if you feel that substantial modifications are now necessary.
2 Read through the sections on ABDPs and 'Alternative Approaches' again, and write a short report to brief a newly-arrived development officer from the International Monetary Fund, telling him the history of the development effort so far and its present priorities.

11 BRAZIL: REGIONAL PLANNING AND INEQUALITIES

A new colonist town in Rondônia, Brazil

BRAZIL — A WORLD LEADER?

? Examine Table 1 (below) and assess the position of Brazil as a world leader before reading the first section of this chapter.

Brazil is the fifth largest country in the world, with an area of 8 512 000 square kilometres. In 1985, it had an estimated population of 138 million; only China, India, USSR, USA and Indonesia surpassed it. Brazil has an anomalous position in world statistics (Table 1). It appears to be a country in transition, possessing the attributes of both a developing and an underdeveloped country. Of the nine countries shown in Table 1, Brazil has the highest rate of natural population increase, and therefore the shortest doubling time. Nearly 70 per cent of the population is resident in urban areas and just less than 40 per cent of the labour force is engaged in agriculture, whereas underdeveloped nations have more people engaged in agriculture. The level of illiteracy is 25 per cent for males and 28 per cent for females — only India has worse levels than these.

Despite its size, Brazil is not a promiment world power. However, according to Merrick and Graham, in their book, *Population and Economic Development in Brazil — 1800 to the Present*, Brazil possesses many attributes comparable with the USA in the last century, and could therefore become a world power of the 21st century. The Brazilians themselves see their future as that of 'a free, prosperous, strong and independent nation, with a prominent place among the other great nations' — President Medici, as quoted in the *Brazil Herald*, April 1, 1973.

THE CONSTRAINTS

The economic balance of power within the world capitalist system does not aid and assist Brazil to become a world leader. Brazil is a country of the 'periphery' and as such is influenced by the 'core-periphery' relationship that exists between the developed and underdeveloped worlds. Brazilian development has been dependent on foreign investment; capital equipment; technology and knowledge, and further development may be constrained by those foreign powers that assisted development. In addition, international trading practices and the terms of trade constrain Brazilian economic activity; Brazil is dependent on the international institutions controlled by the advanced industrialised world. In addition, Brazil may never become a world leader simply because, at the moment, it lacks any political and military strategic significance.

If not a world leader, Brazil would like to see itself as a leader of Latin America: a modern rapidly developing continent. Brazil is the largest country in Latin America but, as seen in Table 2, does not rank highly in any of the statistics except infant mortality. Isolated from the rest of the continent by language (Portuguese rather than Spanish) and traditions, Brazil is probably not in a position to be the leader of Latin America.

One final possibility is left for satisfying the Brazilian aspiration of becoming a leader of nations, and that is by becoming the 'spokesperson' for other developing nations, and acting as an example of how to plan development. If Brazil is a country in transition, the Brazilian experience in development planning may well be valuable to other

Table 1 *World statistics. The six largest countries in the world according to area, plus an additional three countries with a population over 100 million*

Country and total area (km² thousands)	*Population mid-1985 (millions)*	*Infant mortality rate per 1000 live births (1985)*	*Percentage rate of natural increase per annum (1985)*	*Population doubling time (years)*	*Percentage of population aged 15 or under (1985)*	*Urban population as % of total population (1985)*	*Percentage of labour force engaged in agriculture (1982)*	*Percentage ratio of illiterate adults male/female (1982)*	*GNP per capita US $ (1983)*
USSR (22 402)	278.0	32.0	1.0	71	25	64	17	1/2	6 350*
Canada (9 976)	25.4	9.1	0.4	90	22	76	5	1/1	12 000
China (9 597)	1042.0	38.0	1.1	65	34	21	61	n.a.	290
USA (9 363)	238.9	10.5	0.7	100	22	74	3	1/1	14 090
Brazil (8 512)	138.4	71.0	2.3	30	37	68	39	25/28	1 890
Australia (7 687)	15.8	10.3	0.9	82	25	86	6	0	10 780
India (3 288)	762.2	118.0	2.2	32	39	23	64	44/72	260
Indonesia (2 027)	168.4	87.0	2.2	32	41	22	65	22/42	560
Japan (372)	120.8	6.2	0.6	110	22	76	9	1/1	10 100

(n.a. not available) * 1982 figure
(*Source*: *1985 World Population Data sheet* and *1982 World's Children Data Sheet.*)

Table 2 *Latin American statistics*

Country and total area (km² thousands)	*Population mid-1985 (millions)*	*Infant mortality rate per 1000 live births (1985)*	*Percentage rate of natural increase per annum (1985)*	*Population doubling time (years)*	*Percentage of population aged 15 or under (1985)*	*Urban population as % of total population (1985)*	*Percentage of labour force engaged in agriculture (1982)*	*Percentage ratio of illiterate adults male/ female (1982)*	*GNP per capita US $ (1983)*
Brazil (8 512)	138.4	71.0	2.3	30	37	68	39	25/28	1 890
Argentina (2 777)	30.6	35.3	1.6	44	30	83	13	4/6	2 030
Mexico (1 973)	79.7	53.0	2.6	27	42	70	40	13/19	2 240
Peru (1 285)	19.5	99.0	2.5	28	41	65	40	11/28	1 040
Bolivia (1 099)	6.2	124.0	2.7	26	44	46	46	21/42	510
Venezuela (912)	17.3	39.0	2.7	25	41	76	19	16/21	4 100
Chile (757)	12.0	23.6	1.8	39	32	83	16	6/9	1 870
Paraguay (407)	3.6	45.0	2.8	28	42	39	44	10/17	1 410
Cuba (115)	10.1	16.8	1.1	64	29	70	24	4/5	n.a.
Guatemala (109)	8.0	62.4	3.5	20	46	39	61	41/57	1 210

(n.a. not available)
(*Source*: *1985 World Population Data sheet* and *1982 World's Children Data Sheet*)

developing countries. However, time and changing circumstances mean that the Brazilian development strategy may not be good for other countries. Brazil has many natural resource advantages (except oil) which bodes well for its future development, while other developing countries are not so fortunate.

REGIONAL DIVERSITY

THE FIVE MAJOR REGIONS

The republic of Brazil consists of twenty-six territorial units which can be grouped into five major regions (Figure 1), each one distinctive, separate and homogenous, with its own physical economic and social structure.

The Northern Region This is the area of Amazonia with dense tropical forests cut by many rivers, covering 42.1 per cent of the total area of Brazil and yet housing only four per cent of the population. The area includes the Guiana highlands; and very little of the Amazon basin is, in fact, lowland. The soils are poor and the climate can cause hard pans to develop on, or just below, the surface.

The Centre-west The Amazonian forests merge into this area of savanna grasslands dominated by the undulating Brazilian highlands. The soils here are good but require careful management and the addition of lime to prevent soil erosion to ensure a good crop.

The North-east This is the one area of Brazil without a good rainfall regime supporting lush vegetation. In fact, virtually all of this peneplained plateau suffers from the effects of drought and as a result the principal vegetation is a drought-resistant scrub brushwood known as *caatinga*.

The South-east and **The South** These are similar in physical composition. The Serra (mountain ranges) run parallel to the coast, while the interior is open tabular upland composed of layers of lava that have been weathered to form rich *terra roxa* soils (not to be confused with *terra rossa*, the mainly limestone area soils of the Mediterranean). Between them, the South and the South-east comprise only 17.6 per cent of the total area and yet house 60 per cent of the population.

PATTERNS OF SETTLEMENT AND COMMUNICATIONS

There are regional variations in settlement patterns and communication in Brazil. Most settlements are concentrated along the coastal plain and 75 per cent of Brazilians live within 160 kilometres of the coast. There are nine cities with over a million inhabitants (see Table 3). Only Belo Horizonte and Brasília lie outside the coastal belt. The larger urban areas are in the South-east with São Paulo dominating. Communications within the coastal zone are much better than in the rest of the country. Road transport is the main form of transport in all regions except the North, where river transport is still the major means of communication.

The densest road networks are in the South-east and South (Figure 2). Rail transport has never been very important in Brazil and such lines

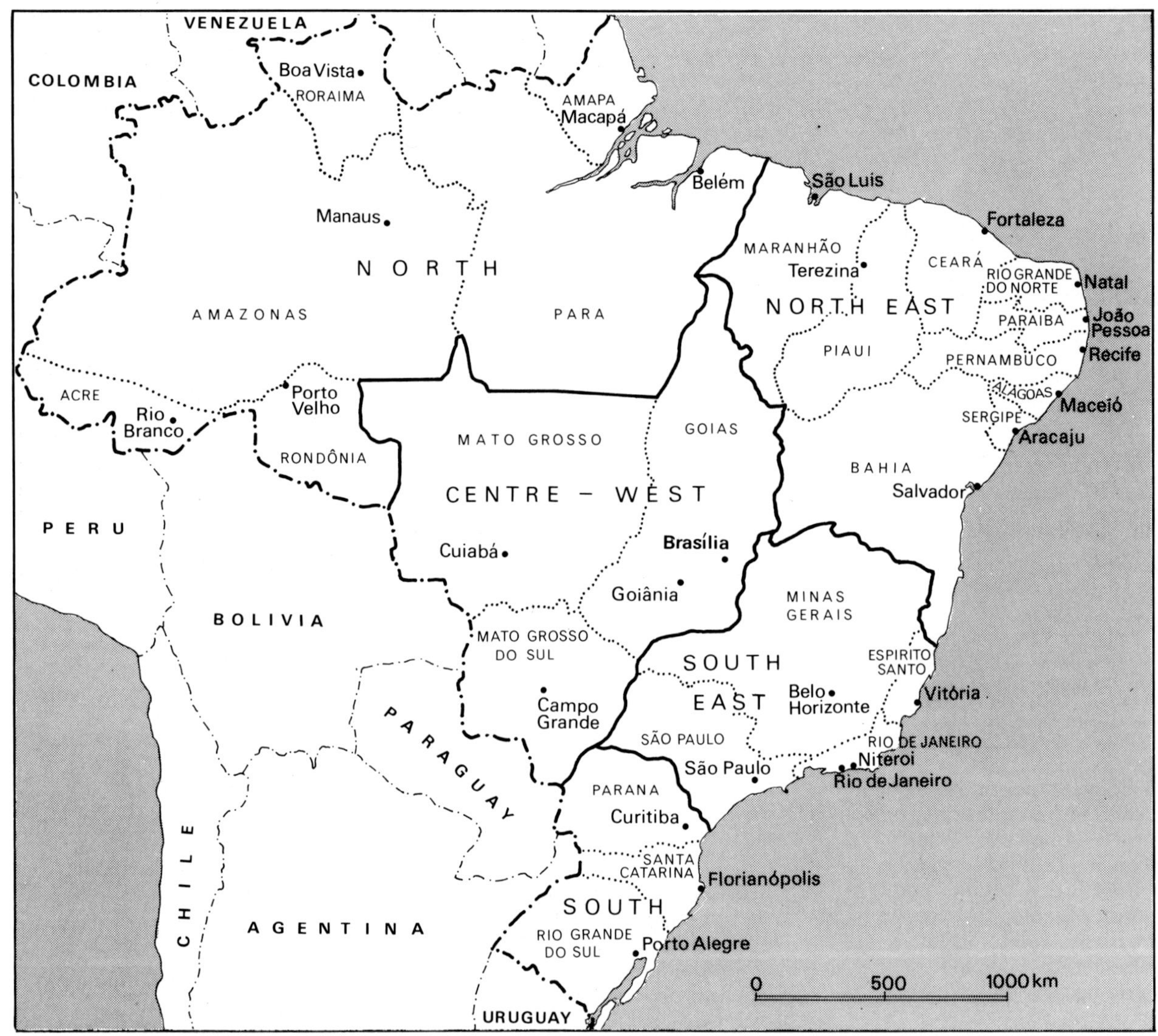

Figure 1 *The major regions and states of Brazil*

that do exist are used to transport heavy bulky commodities along export corridors to the coast. The best inter-urban railway links are between Rio de Janiero and São Paulo and both cities now have their underground railway systems. The size of Brazil makes a network of air transport essential.

? Locate each of the cities in Table 3 on Figure 1 and note their spatial distribution.

Table 3 *Cities with over a million inhabitants in 1980*

City	*Population*	*Metropolitan region*
São Paulo	8 490 763	12 578 045
Rio de Janeiro	5 093 496	9 018 961
Belo Horizonte	1 774 712	2 534 576
Salvador	1 501 219	1 766 075
Fortaleza	1 308 859	1 581 457
Recife	1 204 794	2 346 196
Brasilia	1 176 748	–
Porto Alegre	1 125 901	2 232 370
Curitiba	1 025 979	1 441 743
Belém	934 330	1 000 357

(*Source*: IBGE Anuário Estatistico do Brasil, 1980)

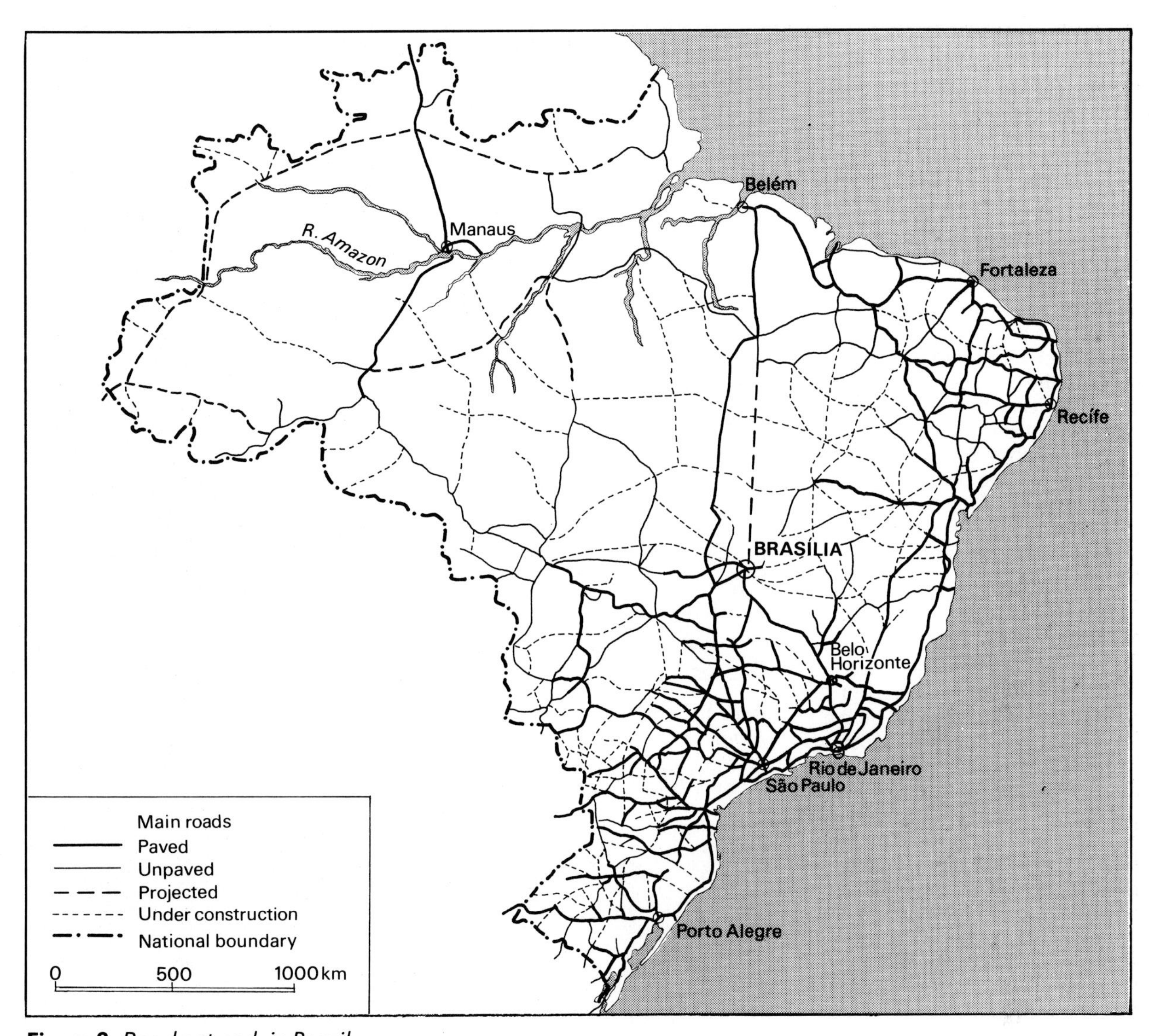

Figure 2 *Road network in Brazil*

A LAND OF INEQUALITIES

The rapid economic growth in Brazil since 1945 as a whole has not been experienced equally by every region, and Brazil is often quoted as one of the best examples of regional inequality within the world. It is estimated that ten per cent of the population hold 50 per cent of the income. Although inequality is often talked of in terms of income, the contrasting regions of Brazil display inequalities in natural resources, infrastructure, levels of industrialisation, standard of mechanisation in agriculture and general socio-economic development.

Perhaps nothing better illustrates the contrasts of Brazil, than that a disastrous drought affecting the North-east for the past 18 months is passing almost unnoticed in the fairly prosperous South-east. Brazil is still a country of extremes, and one result of the fast economic growth of the past decade and a half, when numbers employed in the industry rose from 12 million to 20 million, and when the economy quadrupled in size, is that those who have not shared in the boom have slipped further and further behind. The best paid 10 per cent in Brazil saw their income grow by 160 per cent between 1976 and 1980, the miracle years; the poorest 40 per cent saw theirs grow by a quarter as much.

(*Source*: *'Regional Contrasts Reflect Severe Distortions'* by P Knight in *The Times*, A Special Report, 23 July 1980)

The inequalities in Brazil are largely due to the way in which the government planned and managed the development of the country. Investment capital and managerial power were centralised in the industries of the South-east, particularly in the large urban areas. As a result, the social and economic infrastructure of the South-east is far superior to that of the rest of the country. A pool of wealth and affluence was established here; domestic and international capital flowed into the area and perpetuated its growth. Even the shift of the administrative capital to Brasilia which took place in 1960 has not reduced the political and economic power of the South-east. São Paulo city still acts as the financial capital of Brazil.

Intra-regional inequalities

Social and economic inequalities exist not just between regions, but *within* regions as well. Although the South-east is considered the most developed, there are poor underdeveloped areas within the region. For example, the agricultural areas in the *serras* are some of the poorest in Brazil, and *favela* ('shanty' town) settlements in São Paulo and Rio de Janeiro are centres of urban poverty. Similarly, the generally poor North-east is not totally without wealthy industrial areas, mainly in and around Recife and Salvador.

Print 1 *São Paulo and Rio de Janeiro in the South-east of Brazil are regarded as the two commercial centres of the country. Their combined official population is 34.5 million and they continue to attract migrants from all over Brazil. This shot of Rio de Janeiro shows an affluent area of the city*

Print 2 *Shanty towns are now an everyday feature of developing world cities. São Paulo and Rio de Janeiro are surrounded by sprawling, unplanned urban settlement. These children have lived all their lives on this* favela *in Rio de Janeiro*

Table 4 *Brazilian statistics*

Regions and States	Total area (km^2 thousands)	Area as % of total area	Population in 1980 (thousands)	Population as % of total population	Density of population (persons per km^2)	Population growth rates per annum per 100 persons (1970–80)	Urban population as % of total population (1980)	Economically active population aged over 10 (1970, thousands)	Population in agriculture as % of economically active population (1970)	% of population over 5 that can read and write (1970)	Number of private cars (1978)
North	3 581	42.1	4 923	4.0	1.4	5.04	49.9	2 401	24.4	54.3	82 147
Rondônia	243	2.9	172	0.14	0.7	15.80	56.1	74	21.6	54.9	5 551
Acre	153	1.8	288	0.23	1.9	3.39	32.2	138	31.2	35.6	3 847
Amazonas[1]	1 564	18.4	1 252	1.01	0.8	4.12	49.1	629	25.8	50.2	27 201
Roraima	230	2.7	56	0.05	0.2	6.81	42.9	27	22.2	52.9	1 879
Pará[1]	1 248	14.7	2 981	2.42	2.4	4.67	51.3	1 460	23.8	57.7	40 741
Amapá	140	1.6	174	0.15	1.2	4.35	56.4	73	15.1	58.7	2 928
North-east	1 569	18.2	36 251	29.5	23.4	2.16	47.4	19 053	27.3	39.0	659 450
Maranhão	329	3.9	3 698	3.0	11.2	2.93	31.4	2 023	37.7	34.6	25 074
Piauí[2]	251	2.9	2 307	1.9	0.1	2.44	38.1	1 110	31.3	31.8	23 392
Ceará[2]	148	1.7	5 891	4.8	39.8	1.94	44.9	2 929	25.6	37.8	100 760
Rio Grande do Norte	53	0.6	2 163	1.8	40.8	2.04	52.4	1 046	23.0	40.1	36 536
Paraiba	56	0.7	2 964	2.4	52.9	1.52	47.1	1 625	27.0	35.8	47 345
Pernambuco[3]	98	1.2	6 607	5.4	67.4	1.75	61.8	3 558	21.5	43.0	178 458
Alagoas	28	0.3	2 013	1.6	71.9	2.26	44.8	1 072	30.1	33.2	35 263
Sergipe	22	0.3	1 094	0.9	49.7	2.39	52.4	606	26.7	42.4	26 201
Bahia	561	6.6	9 515	7.7	17.0	2.35	46.5	5 084	28.3	42.0	186 421
South-east	925	10.8	51 575	41.9	55.8	2.64	83.8	29 354	11.9	71.6	4 341 122
Minas Gerais	587	6.9	13 689	11.1	23.3	1.53	64.6	8 080	21.3	58.9	569 056
Espirito Santo	46	0.5	1 860	1.5	40.4	2.38	61.1	1 109	20.7	59.7	86 587
Rio de Janeiro	44	0.5	12 022	9.8	273.2	2.30	93.2	6 831	3.7	77.4	924 069
São Paulo	248	2.9	24 004	19.5	96.8	3.48	91.9	13 334	10.0	77.5	2 761 410
South	578	6.8	22 495	18.3	38.9	1.43	49.0	11 599	25.3	70.1	1 275 958
Paraná	200	2.4	10 274	8.3	51.4	0.96	39.2	4 700	30.6	62.3	445 346
Santa Caterina	96	1.1	3 881	3.2	40.4	2.26	50.5	1 993	22.7	74.5	230 528
Rio Grande do Sul	282	3.3	8 340	6.8	29.6	1.55	60.5	4 906	21.3	76.1	600 084
Centre-West	1 879	22.1	7 788	6.3	4.1	4.04	54.4	3 452	24.0	57.0	308 392
Mato Grosso do Sul[4]	350	4.1	2 489	2.0	2.0	3.19	44.5	1 085	27.5	56.8	71 083
Mato Grosso[4]	881	10.4				6.62					
Goiás	642	7.5				2.77					114 567
Federal			5 299	4.3	8.2		59.1				
District (D.F.)[5]	6	0.1				8.13		2 367	22.4	57.0	122 742
Brazil as a whole	8 512	100.0	123 032	100.0	14.5	2.48	63.5	65 862	19.9	60.3	6 667 069

Notes:

[1]Boundary dispute between Amazonas and Pará, area covered 2 680 km^2.

[2]Boundary dispute between Piauí and Ceará, area covered 2 614 km^2.

[3]Includes the Island of Fernando de Noranha.

[4]The Former state of Mato Grosso was divided into Mato Grosso and Mato Grosso do Sul on 1.1.1979.

[5]The Federal District of Brasilia situated in the state of Goias.

1 Study Table 4 and rank the regions in terms of each of the following: size, population density and socio-economic conditions.
2 Using the data in Table 4 calculate the projected population for each of the five major regions until the year 2035, giving the population totals at five yearly intervals. What would the consequences be if the population continued to grow at the present rate? What do you think Brazil's policy should be towards population growth?
3 Given the data in Table 5 (below), make a population projection for the four states of the South-east until 2010. Are such growth rates realistic?
4 The Lorenz curve is one way of displaying inequality between areas. Draw Lorenz curves using the data in Table 6. What do you deduce from your results?
5 From the data and information so far is it possible to say which region is the more developed, and if so why?
6 Organise the reasons which account for the regional inequalities into a paragraph.

Table 5

State	*Population in thousands in 1980*	*Estimated population growth rate per annum per 100 persons*					
		1980 –85	*1985 –90*	*1990 –95*	*1995 –2000*	*2000 –2005*	*2005 –2010*
Minas Gerais	13 689	1.53	1.50	1.20	1.00	1.00	.80
Espirito Santo	1 860	2.38	2.36	2.30	2.30	2.10	2.00
Rio de Janeiro	12 022	2.30	2.30	2.10	2.00	1.90	1.85
São Paulo	24 004	3.48	3.50	3.00	3.10	3.05	3.00

Table 6

Areas	*Data for 1970 in cumulative percentages*			
	Literate population of 5 years and more	*Total population of 5 years and more*	*Households with electricity*	*Total households*
1	6.9	4.9	10.8	5.4
5	51.2	40.1	64.4	41.6
10	74.7	63.8	85.2	67.9
15	80.7	70.3	89.9	75.4
20	93.4	88.7	96.7	89.1
25	100.0	100.0	100.0	100.0

WHAT IS DEVELOPMENT?

List the attributes of a developed country. What do you consider to be a good indicator or measure of development?

Development is probably a universal goal; however it is not an easy word to define. Over the last 40 years the definition of development has changed radically, and during that time there have been four major definitions, reflecting the progression of thought about what development means to countries, regions and individuals.

1 Economic Growth

After the Second World War, development was synonymous with the term 'economic growth'. Massive injections of capital were given to European countries to recover from the ravages of the war. The 1950s and '60s were a time of industrial and economic growth and expansion for many countries. It was suggested that economic growth was the way forward for all countries; growth having a 'knock-on' effect of promoting social and political change.

One great disadvantage of economic growth as a definition of development is its disregard of how development affects the individual and how economic growth contributes to the creation of a dual economy within many countries. Spatial inequality of individual and regional well-being and development are intrinsically linked.

2 Modernisation

By the 1960s the idea of 'modernisation' was used to explain the development process. A modern society is one that consumes the goods and services manufactured and consumed in advanced industrial countries. This implies that to be a developed country people must emulate Western societies. Such action threatens national cohesion, because people in developing countries are likely to become dissatisfied with traditional society and 'non-modern' goods and services. It has been said that 'modernisation is the process of social change in which development is the economic component'. Within less developed countries the most modern areas, where modern goods and services are available, are the urban centres. These centres act as a force for increasing inequality by attracting migrants, and thus polarising the urban and rural areas. However, the concept of modernisation does relate more to the individual, because it reflects an interest in what he or she consumes.

3 Distributive justice

A just distribution of goods and services throughout a society theoretically establishes social equity, because the population will have equal access to those goods and services. Distributive justice aims to meet the 'basic needs' of all individuals and therefore to reduce or remove poverty. It is, however, difficult to establish an individual's needs.

4 Total socio-economic transformation

More recently there has been a call for the whole issue of development to be seen as the socio-economic transformation of all major institutions in a country, thus establishing the fourth major definition of development. Development is a multidimensional process that affects the whole country and all individuals. In his book, *The Development Process: A Spatial Perspective*, A. Mabogunje states that geographical space intrudes into every concept of development, and therefore one must see development as essentially a socio-spatial process.

According to J. P. Cole in his book *The Development Gap*, six fundamental requirements for integrated development are: population, natural resources, means of production, products that will be consumed, labour including the government, and links between these things (see Figure 3). National and regional development plans that fail to take all six elements into account are probably insufficient and inadequate. A country functions as a system with each element playing a vital role. It is not a closed system unaffected by outside influences. To chart the progress of a country one should monitor all parts of the system.

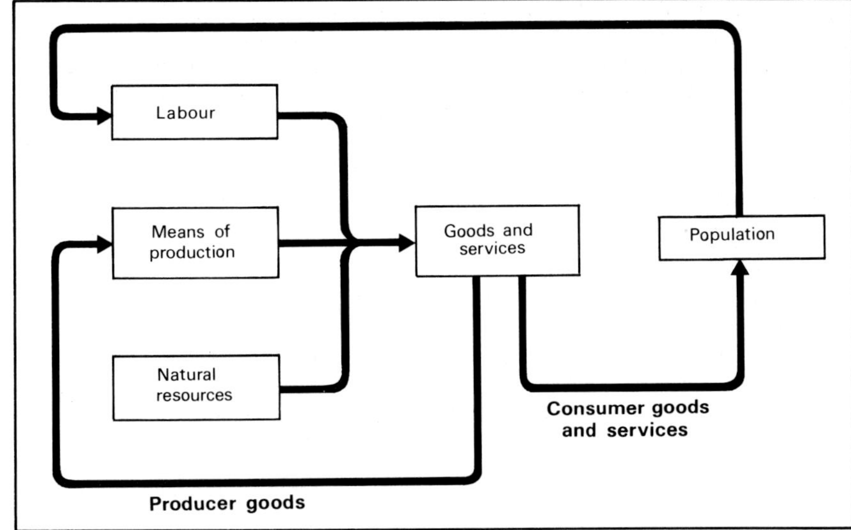

Figure 3 *The fundamental requirements of integrated development*

? **1** How can the model, in Figure 3, be improved to represent reality more closely?

? **2** What other aspects of development need to be incorporated into a 'National System' diagram like Figure 3 to reflect reality more closely?

DEVELOPMENT PLANNING IN BRAZIL

All Brazil's national and regional development plans, outlined in Table 7, have attempted to reduce regional inequalities.

Table 7 *Brazilian Development Plans*

Years	*Plan*	*Aims and Objectives*	*Omissions and Failures*
1957–61	Plano de Metas	Continuation of the policy of import substitution industrialisation. Resolve any problems that hinder industrial progress.	Lack of any real policy for agriculture. No social programme for health and education.
1963–65	Plano Trienal de Desenvolvimento Econômico-Social	Industrial and economic recovery — to reach the growth rate experienced between 1957–61. Control inflation. Reduction of regional income inequalities. Improve scientific and technological education. Reform of various economic bases, e.g. banking, administration etc.	Lack of any real policy for agriculture. No social programme for health and education. No mention of increasing employment. No mention of import-export growth rates. Did not control inflation. Did not reduce regional income inequalities.

Years		*Aims and Objectives*	*Omissions and Failures*
1964–66	Plano de Acão Econômico do Governo	Improve economic growth rate. Combat inflation. Correct the balance of payments situation. Increase employment. Reduce regional and sectoral income inequalities.	No coherent social development programme. No mention of import-export growth rates. Few jobs created. Regional and sectoral income inequalities persisted.
1968–70	Plano Estratégico de Desenvolvimento	Accelerate economic growth. Contain inflation. Control the balance of payments situation. Increase employment. Obtain social progress.	Nothing on regional and sectoral income inequalities. No mention of import growth rate.
1970–72	Programa de Metas e Bases para a Acão do Governo	Establish the country as a developed nation. Decrease rate of inflation. Establish equilibrium in balance of payments. Increase employment. Gradual correction of regional and sectoral inequalities. Redistribute personal income. Land reform and reform of administration.	Lack of a social development programme. No real reduction in regional and sectoral inequalities. Only very gradual progress in other areas.
1972–74	Primeiro Plano Nacional de Desenvolvimento (I PND)	Categorisation of Brazil as a developed nation. Decrease rate of inflation. Reduce regional income inequalities.	Virtually no mention of agriculture.
1975–79	Segundo Plano Nacional de Desenvolvimento (II PND)	Maintain and accelerate economic growth. Correct inflation. Maintain a relative equilibrium in the balance of payments. Improve the regional and personal distribution of income. Improve social and political conditions. Obtain development without a deterioration in the quality of life. Gradual realisation that agriculture was vital to the economy.	Inflation not mentioned. Little said about social development. Employment opportunities not growing rapidly enough. No population policies.
1980–85	Terceiro Plano Nacional de Desenvolvimento (III PND)	Reduce regional and sectoral inequalities. Improve income distribution among the social classes. Increase employment. Improve standard of living for all. Obtain equal access to social facilities. Encourage and improve agriculture. Reduce urban 'hypergrowth'. Control inflation.	No specific mention of a population policy.

(*Source*: Cintra and Haddad and *The Brazilian Gazette*)

? 1 What views of development seem to be reflected in the national policies, particularly the III PND?

There now follow four case studies of Brazilian development. The first is a national policy; the other three are regional ones. As you read each case study make notes on the advantages and disadvantages of each one, both to individual regions and to the country as a whole. Think about the following questions.

?
1 Who benefits from the progress made in each case study and why?
2 Do the case studies indicate integrated national development or concentrated regional development?
3 Are the regions growing further apart?
4 Why might national and regional development policies and plans fail? What hinders development?
5 Are there any problems common to all cases that restrict development; if so, what?
6 What definition of development or view of development, prevails?

CASE STUDY 1 — THE PROALCOOL PROGRAMME

Oil is a major energy base for industry, transport and some domestic consumption in Brazil. Forty-five per cent of the energy requirements of Brazil are met by crude oil. During the years of the 'economic miracle', between the late 1960s and early 1970s when Brazil made dramatic economic progress, and throughout the 1970s, Brazilian oil consumption rose rapidly. In 1980 Brazil imported over 80 per cent of its oil consumption requirements. Heavy oil import bills have forced Brazil to look towards internal sources for other forms of energy. Energy supply was one of the main priority issues in the III PND (see Table 7).

Brazil chooses alcohol to replace petrol

Brazil has initiated a plan to use sugar-based alcohol as an alternative to petrol. Nevertheless, there are severe criticisms of the government's plan to make automobiles a development priority

An ambitious programme is underway in Brazil to free the country from its overwhelming dependence on a petroleum-based economy.

Half of the country's exports go to pay for its imports of crude oil (about 800 000 barrels a day), which provide 45 per cent of its energy needs. Brazil produces a fifth of its oil needs itself.

But since 1973, when oil prices began escalating, it has been looking for other sources of energy. Its search focused on biomass, the idea of growing its own fuel on a renewable basis, with ethyl alcohol, or ethanol, being extracted from plants. Consequently, the Brazilian alcohol programme, *Proálcool*, was officially launched in November 1975.

During the 50s, Brazil adopted the development model of the United States, with the car industry as the centrepiece. There followed an American-style collection of concrete cities and motorways, together with the urban traffic jams, all of which came to be regarded as indelible signs of progress.

The country's transport system became dependent on giant trucks, while the private car became the symbol of the person who had 'arrived'. The importance of petroleum soared. Once a fuel representing only 9 per cent of the total energy needs, it now accounts for 45 per cent.

The Proálcool programme launched in 1975, aims to produce 10.7 billion litres of alcohol in 1985.[1]

One thing has been established beyond doubt: the bio-energy idea works. There are more than 400 000[2] alcohol powered cars in Brazil and the number is expected to double by the end of this year. The increased production will come from an agreement the motor industry signed with the government last year, providing for a quarter of vehicles coming off the assembly lines to be designed specifically to run on alcohol. The cars can fill up at any one of the 6 000 alcohol pumps dotting the country and travel throughout Brazil in an alcohol car poses no hazards.

Alcohol fuel has already given Brazil savings in foreign exchange and created jobs in the country's poorer North-east region. A bonus is that there is no need to add lead to alcohol fuel for cars, because alcohol has a higher octane rating than petrol and does not cause 'knocking' in the engine. Exhaust pollutants are fewer, too, than with petrol. But alcohol cars give off aldehydes in the exhaust, the effects of which on humans are not fully known.

The alcohol programme has led to the creation of mini-distilleries to extend the advantages to the country as a whole. Multiple fuel producing centres, some with alcohol-based chemical industries, are expected to cut costs in many areas of the economy.

Nevertheless, liberation from a petroleum economy is still a glint in the eyes of administrators. In the short term, the aim is to provide a valuable strategic complement to the existing, fairly large petrochemical industry. To this end the government has offered incentives for setting up and improving alcohol plants.

One area where the alcohol energy programme has encountered opposition is in its impact on food prices and availability. Critics say that the programme is jeopardising crops and causing food shortages in the country.

Evidence indicates that areas previously used to grow cotton, rice, beans, corn, manioc and oranges, have been invaded by sugar-cane plantations. And as agricultural progress requires investment in land, materials and technology, one can easily imagine the threat involved in the conversion of agricultural areas into sugar-cane plantations for industrial use.

This train of thought leads to a fundamental question which not many critics ask: to what extent is it justifiable to allow the vehicle industry to be the most important force in the economy? Is what is good for the automobile industry good for Brazil?

Any long range resolution of the energy problem in Brazil must involve a reconsideration of the development model. Halting the expansion of the car industry and redirecting the resources towards goods and foods within the reach of the common people, is seen by some as one solution — one which would benefit people and cut fuel costs.

Whatever the new energy source is used for, there is no doubt now that biomass is the answer to the increasing scarcity of fossil fuels.

(*Source*: *South*, November 1981)

[1] In 1984 8.1 billion litres of alcohol were produced.
[2] In 1983 alone over 58 000 alcohol-powered vehicles were sold.

? 1 '. . . to what extent is it justifiable to allow the vehicle industry to be the most important force in the economy? Is what is good for the automobile industry good for Brazil?' Write down three points for and three against the government's alcohol programme.

2 Form a group with three other people. Look at all your points for and against and decide on the best four points for and against. Discuss the attitudes and values underlying each. Decide what sections of society benefit or not as the case may be.

3 Take a class vote on the proposition that 'Brazil should not have chosen to put so much emphasis on the development of alcohol production'. Two people may speak for and two against the proposition. They have to identify themselves in terms of their occupation and financial well-being.

NATIONAL AND REGIONAL BENEFITS

The benefits of the *Proálcool* programme are twofold: Brazil has found a substitute for oil and can therefore reduce its oil import bill, and the programme has had limited regional benefits.

The traditional sugar-cane growing area of Brazil is the North-east, and the *Proálcool* programme generated wealth and therefore assisted the region's development. In recent years Brazilian sugar-cane producers have suffered from increased competition and a world surplus. Now alcohol production has alleviated these problems and has ensured increased employment, not only on sugar-cane plantations, but in distilleries. This should ease the situation in the depressed North-east, and arrest migration out of the area.

CRITICISMS

The *Proálcool* programme is a good example of how agriculture and industry, or rather agro-industry, can lead to the integrated development of Brazil. However, the programme can be criticised for perpetuating the traditional agricultural plantation system by continuing to grow sugar-cane in certain areas where land reform is needed, for example, the North-east. The programme places a heavy burden on certain areas to maintain and increase their levels of production which may mean an increased use of oil-based fertilisers. Also more land may be given over to sugar production, when in the long-term the land should be used for food crops. The programme has done very little to alleviate the employment problems of the rural producer areas.

Many believe that biomass energy can never totally replace the need for large amounts of imported oil, therefore the balance of payments problems of Brazil will not be eased, since energy consumption continues to rise. It is suggested that Brazil should invest more money in to the research and development of alternative energies, which are not *substitutes* for oil, but real *alternatives*. Substitutes only prolong the problem and possibly defer the crisis point.

Finally, the main criticism of the *Proálcool* programme is its intensification of development in the South-east. While the North-east may be the traditional sugar growing area, São Paulo is now the major producer state and it is here that research and development takes place. Again this means the 'core' in Brazil dominates development in the 'periphery' and ultimately the *Proálcool* programme is unlikely to lead to integrated development in Brazil.

CASE STUDY 2 — AMAZONIA

Amazonia has long been considered the 'pandora's box' area of Brazil. To many Brazilians it is an unknown, underutilised and inaccessible area, but one with great potential. The area has vast reserves of timber, plant and animal life, plus enormous mineral wealth. The tropical climate favours agriculture but the soils are of a poor quality. Therefore, any agriculture in the area must be well planned and managed to prevent soil erosion and the spread of disease. Many Brazilians have been reluctant to move to Amazonia. The climate, environment and isolation have, in the past, inhibited the colonisation and development of the area.

Amazonia has been described as the 'last frontier'; an area to be brought into the national economy 'to change the face of Brazil'. Therefore, since 1945 Brazilian governments have made every attempt to develop the area through the auspices of regional development agencies. The first agency was SPVEA (Superintendency of the Plan for the Economic Development of Amazonia) which was replaced by SUDAM (Superintendency for the Development of the Amazon) in 1966.

SUDAM supported the development of existing settlements — Manaus became a free trade zone and was to act as a counter balance to Belém. Both towns were to encourage and cause development effects to spread throughout the region. SUDAM's main responsibility was to improve the agriculture of the area through:

- Assistance grants to farmers
- Application of modern techniques
- Increased production
- Encouraging ranching
- Promoting legalisation of settled land. (Legalised land ownership enables the settler to borrow money.)

Print 3 *Untouched Amazon forest — a rare shot of a natural habitat which contains a wealth of rare animal and plant species. It is estimated that an area of rainforest the size of Ireland is being cleared from the globe each year*

Only 26.7 per cent of Amazonia lay under the control of SUDAM. INCRA (the National Institute for Colonisation and Land Reform) was responsible for the rest, some 73.3 per cent of Amazonia. INCRA promoted road construction and the colonisation and development of land 100 kilometres either side of the roads. Settlers were encouraged to settle in these designated zones and an infrastructure of essential services was to be provided. Migrants to Amazonia came from all over Brazil, with a large percentage coming from the North-east. It was hoped that Amazonian colonisation would relieve the population pressures and social tensions of the North-east. In actual fact, the government gave very little assistance to settlers, and the infrastructure was hardly improved, except for the roads, which generally assisted the large multinational corporations in the area. There were insufficient supply posts to provide the farmers with the necessary tools, seeds, fertilisers and pesticides. Marketing facilities were virtually non-existent. Many settlers found themselves isolated, without electricity, water supply, health or education services, and on poor infertile land and unable to grow new crops. As a result many settlers left with dreams and high expectations broken and unrealised.

Colonisation programmes in Amazonia suffered from insufficient financial and technical resources to meet the development objectives. These objectives were too grandiose for such a large area, 42 per cent of the country. The planners have been criticised for not obtaining detailed information about the area which would have aided and assisted programme success. Also, the success of SUDAM's objectives was dependent on factors outside its control.

? Read the *Observer* extract and assess the environmental impact of development in Amazonia.

THE RAPE OF THE AMAZON

Most of the clearing was done by foreign enterprises such as Daniel Keith Ludwig's Jari Forestry and Ranching Company, the Italian firm Liquigas, Volkswagen do Brasil, and King Ranch of Texas. These and many more had been encouraged in their attack on the forest by financial incentives offered by the Brazilian Government. Great fires — some of them ignited by napalm bombing — raged all over Amazonia, consuming trees by the hundred million, and for months on end travellers on planes on their way from Belém to Manaus or Brasilia saw little of the landscape beneath them through the smoke.

There were few parts of the world left in this century where uninhibited commercial adventures of this kind were still possible, where land could be picked up for next to nothing, wages were about a tenth of those paid in Europe or the US, and a modest investment in stock and equipment offered the prospect of spectacular profits. The government's early enthusiasm for the giant ranches began to falter when it was found that, like the trees they replaced, they seemed to live on themselves, and produced little surplus to help with the balance of payments. Nor did they relieve unemployment, because when a ranch became a going concern it took only one man to look after 1 000 head of cattle.

. . . Brazil abounds with vigorous and articulate conservation groups, but they are powerless in the presence of one crushing fact: the desperate need of the country's many poor. *O Journal do Brasil* published figures showing that in Rio de Janeiro alone 918 000 people were living in 'absolute poverty', and in the city's total population of nine million, 27 per cent lived in 'relative poverty'. These are the statistics that blunt conservationist scruples. There is a constant pressure to develop more sources of food, joined to an irrepressible belief that sooner or later a way will be found to turn the relatively unproductive five million square kilometres of the Amazon Basin into a bottomless larder.

The Government seeks to put a brake on the excesses of 'developers' by measures that are too often evaded or ignored. Official approval must be obtained for large-scale forest clearances, but nobody seems to bother. Regulations exist prohibiting the burning-off of forest close to river banks where animals tend to congregate. These go unheeded. Slopes are not allowed to be cleared, because to do so is to guarantee immediate erosion, but in our experience landowners give priority to clearing the slopes on their estates. They do so because it is easier to drag or roll the tree trunks down the slopes and leave them to rot at the bottom, than to go to the trouble of extracting them. A promising law forbade the clearing of more than 50 per cent of any concession, but many methods exist by which it is dodged. A common one is to clear half one's land in compliance with the regulation, and then sell the forested remainder, a half of which will be cleared by the buyer in his turn — and so on.

Of course for those who are not there to stay, the short-term yields they derive are paid for by a huge and irreversible loss suffered by Brazilians who have to live with the results. Ninety per cent of the soil of the Amazon Basin is so poor that it cannot be converted into adequate pasture without the addition of costly fertilisers, and as soon as these fertilisers are no longer forthcoming, the coarse African grass it supports will go. For a multitude of small-scale ranchers three acres of land can hardly feed a single cow, and after two or three years, when the rains have washed the nutrients from the soil, over-grazing, over-trampling take effect and the dust bowl process begins . . .

(*Source*: *The Observer Magazine*, 22 April 1979)

POLAMAZONIA — GROWTH AROUND NUCLEI

The sheer size of Amazonia forced a change of policy and the *Polamazonia* (Programme for Agricultural and Agro-Mineral Nuclei) was created in 1974. The programme favoured 'unbalanced' growth in 15 selected growth poles (see Figure 4). These 'poles' were sectors or areas of high economic potential for minerals, agriculture or cattle breeding (although recent studies have, in fact, shown that Amazonia is not ideal for cattle breeding and ranching). Foreign investment was sought for these high potential projects — The World Bank, the EEC and Japan have all invested in the region, often in association with the Brazilian government.

One of the 15 growth poles of *Polamazonia* is the Carajas project. This project forms part of the larger programme of Grande Carajas. The Carajas project involves the extraction of high grade iron ore from the Serra dos Carajas in Para, 550 kilometres south-west of Belém

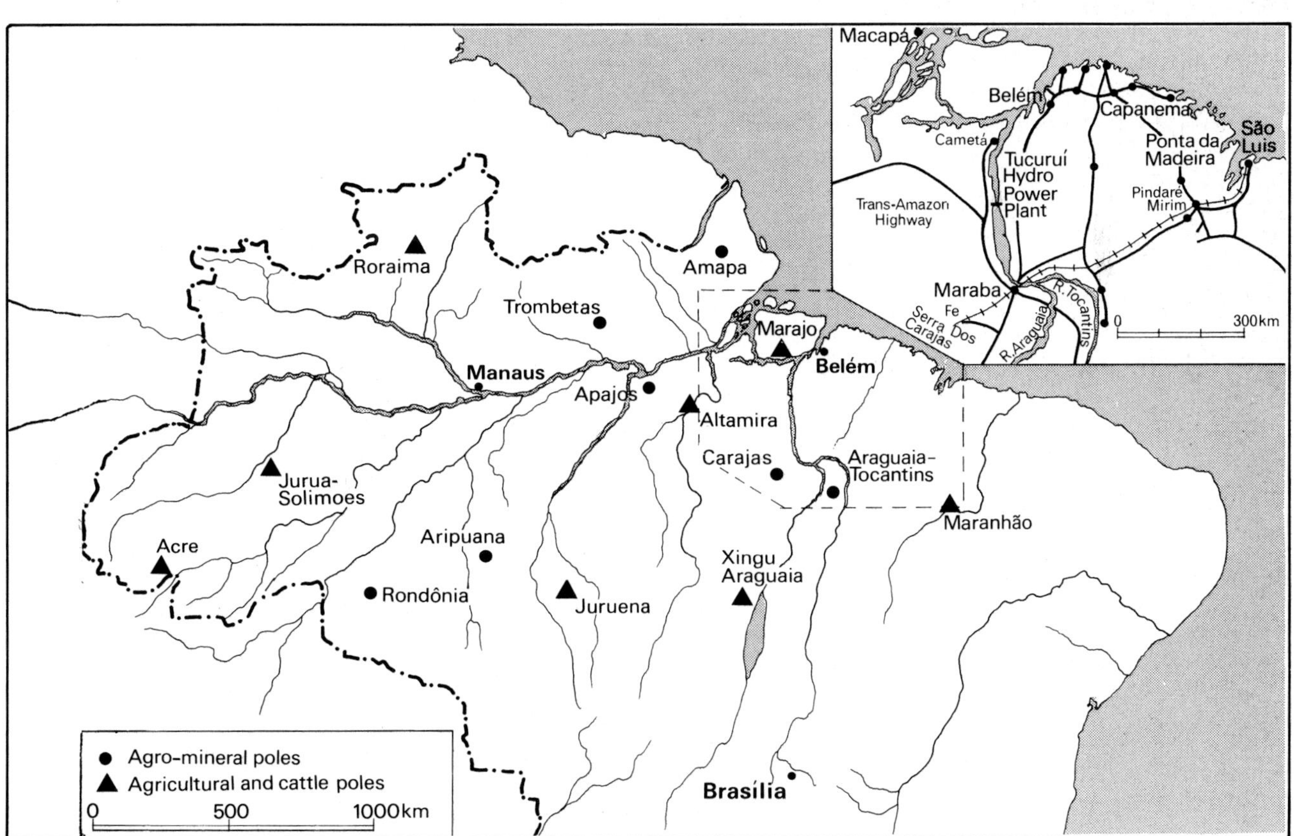

Figure 4 Centres of regional development in Amazonia, inset of the Carajas Iron Ore Project

Greater Carajas scheme 'the project of the century'

'The project of the century' is how Brazilians describe the Greater Carajas scheme. It is no exaggeration or hyperbole. Located south-east of the Amazon, the area under development covers approximately 800 000 square kilometres — the size of France and West Germany put together — of jungle and scrub.

The aim is to turn this neglected corner of Brazil into a prosperous industrial and agricultural region by the end of the decade. It could also be an export base rivalling any other in the country.

Communications are the key to the opening up of these unused tracts shared by Para, Maranhao and Goias states. Hence, projects already underway will provide the basis for an eventual north–south waterway of 2 200 kilometres, from Belem at the mouth of the Amazon to Aruana in Goias, and a transverse railway of 900 kilometres to São Luis on the coast. The Trans-Amazonian highway already runs through the region.

The total cost is $60 billion on present estimates and at today's prices. Much of the fixed investment is expected to come from foreign loans or foreign risk capital.

Discovery

None of this would be possible if it had not been for the chance discovery in 1967 of a low range of hills stuffed full of minerals of a quantity, quality and variety surpassing anywhere else in the world. These hills are the Serra Dos Carajas, 550 kilometres south-west of Belem.

The Carajas iron ore project, to produce 35 million tonnes a year of sinter-feed by 1987, is being handled entirely by CVRD. The latest cost estimate is $3.62 billion. That price includes infrastructure, which can later be used by other minerals as well, but excludes the burdensome financial costs.*

Meanwhile, the world's big mining companies are waiting on the sidelines for the iron project to get off the ground (and for a nationalistic debate in Brazil over the multinationals' place in Carajas to be settled) before making their bid for the non-ferrous ores: copper, bauxite, manganese, cassiterite, nickel — and gold.

The commercial logic of the multinationals has undoubtedly been correct. The preliminary work involved in opening up this remote, unpopulated site could only have been handled by a government agency. Furthermore, Carajas would probably not have been economically feasible if it were not for another giant undertaking in the jungle, the Tucurui dam and hydroelectric plant on the nearby Tocantins River.

Tucurui is second in size in Brazil only to the Itaipu HEP project in the south-west, but is a far more taxing undertaking. In its first stage the power station is designed to produce 4 000 MW of electricity, at a cost today of $2.65 billion, or $1 509 per installed kilowatt. Again, financial costs, adding appreciably to the price, have been excluded.

For the long-term future of the region an almost equally important, if less spectacular, aspect of the development of Carajas is likely to be the agricultural possibilities opened up by the availability of rail, road and water transport.

The Greater Carajas Council, which was set up in Brasilia in 1981 as an inter-ministerial body to co-ordinate the entire programme, talked of establishing an initial 300 cattle farms over a 3 million hectares area. Large-scale plantations of oil palms are envisaged and even the return of the rubber tree to its natural habitat as a commercial proposition is dreamt about.

* Due to crisis in the world steel industry the project was delayed by at least a year in 1983 until signs of improvement in the international market. The postponement of the project, specifically excluded by the Brazilian government from its general spending cuts, on the grounds of its prime national importance, means that virtually all the major development projects in the Amazon have now been set back.

(*Source*: *Financial Times*, 18 November 1981)

?

1 The article above written as recently as 1981, is optimistic and enthusiastic in tone. What view of development does it represent? Name some of the people and organisations who are going to benefit most.

2 Brazil's recent financial crisis threatens the continued development of the Carajas scheme and other Amazonian development projects. What do you consider to be the effect on Brazil if investment in the area is curtailed? Would you concentrate what little money there is on one or two projects? If so, which and why?

Print 4 *A vision of the future? A Nambiquara Indian of western Brazil standing amid the wasteland left by a highway which has been cut through several of her tribe's villages*

?

1 Read the extract 'Amazon Tribe in Danger'. What power do the Amazonian Indians have? Who is likely to consider their rights? Do they have any? Write a report to a Minority Rights group outlining a possible policy. Is there any way your policy could be given 'teeth'?

2 Consider the question: 'Should the Amazon Forest be developed at all?' Gather viewpoints and evidence on such matters as the Amazon as a genetic store, the fragility of equatorial soils, and the needs of the Indian inhabitants. The various issues of *Ambio* listed in the Further Reading would be particularly useful.

AMAZON TRIBE IN DANGER

The largest remaining isolated tribe of Amazonian Indians in Brazil is facing destruction.

Anthropologists from universities in Brasilia, São Paulo and Rio de Janeiro — and missionaries working with Indians are protesting against plans by a government agency to create 221 reserves for the Yanomami tribe which inhabits vast areas of forest near the Venezuelan border.

The reserves, proposed by the much-criticised National Foundation for the Indian (FUNAI) but still to be decreed, would be scattered pockets of land in about 6 million hectares of the tribe's traditional area.

These reserves, the critics say, will inevitably become surrounded by colonists bent on exploiting the land and its resources. The forest will be felled for agriculture and ranching and a 'mining invasion' will take place. Large mineral deposits, including uranium, have been discovered in the region where one of the largest concentrations of the Yanomami live.

Part of one of the great roads the Brazilians are driving through the forest has already been built in the area.

Protestors say that the reserves would be too restricted to sustain the Indians. They are farmers and hunters who move their villages every few years and thus need large areas of land to survive. Their system is much more suitable to the area than the modern methods of colonisers, which usually swiftly exhaust the fragile soil.

By planning reserves around settlements, the critics say Funai is neglecting the Indians' agricultural patterns, and ignoring a tradition of long-range neighbourliness central to the tribe's culture. The Yanomami have a sociable habit of taking long treks through the forest to visit other groups and arrange frequent inter-group rituals.

The critics add that there are groups of the Indians, as yet uncontacted, for whom there is no provision in the plans.

The Yanomami have proved extremely vulnerable to contacts with outsiders. Wave after wave of infectious diseases of 'white' origin, such as 'flu, measles and tuberculosis, have killed hundreds.

When Westerners first reached Brazil, in 1500, there were an estimated six to nine million Indians in the country. Now there are only 200 000.

(Source: *The Observer*, 25 February 1979)

CASE STUDY 3 — THE NORTH-EAST

Once the dynamic area of Brazil, with some of the country's earliest industrial plants, the North-east is now a major problem area. Much of the North-east is dry and prone to drought. The area affected by drought is known as the 'drought polygon', a part of the arid remote interior known as the *Sertao* (see Figure 5). The states most affected are: Ceará, Rio Grande do Norte and Paraiba. The whole region has few natural resources, and suffers from economic stagnation, population pressures, poverty, starvation, rural unemployment and an archaic land tenure system. Just under 30 per cent of the Brazilian population live in the region and over 50 per cent of them live in rural areas. There are some modern industries situated in the coastal plain; most are closely associated with the major cities of Recife, Salvador and Fortaleza.

Successive governments have tried to solve or tackle the numerous problems of this depressed region. The people of the North-east have reacted to the area's problems by emigrating. The rural migrants tend to go to the large regional urban centres and eventually go to the South-east. Some migrants go directly to the South-east.

Basically there are two major problems in the North-east:

1 the climate and the associated droughts, and

2 the land tenure system.

These two problems have caused the North-east to be caught in a 'vicious circle' of poverty. The following two newspaper extracts highlight many of the inadequate solutions which have been tried over the years.

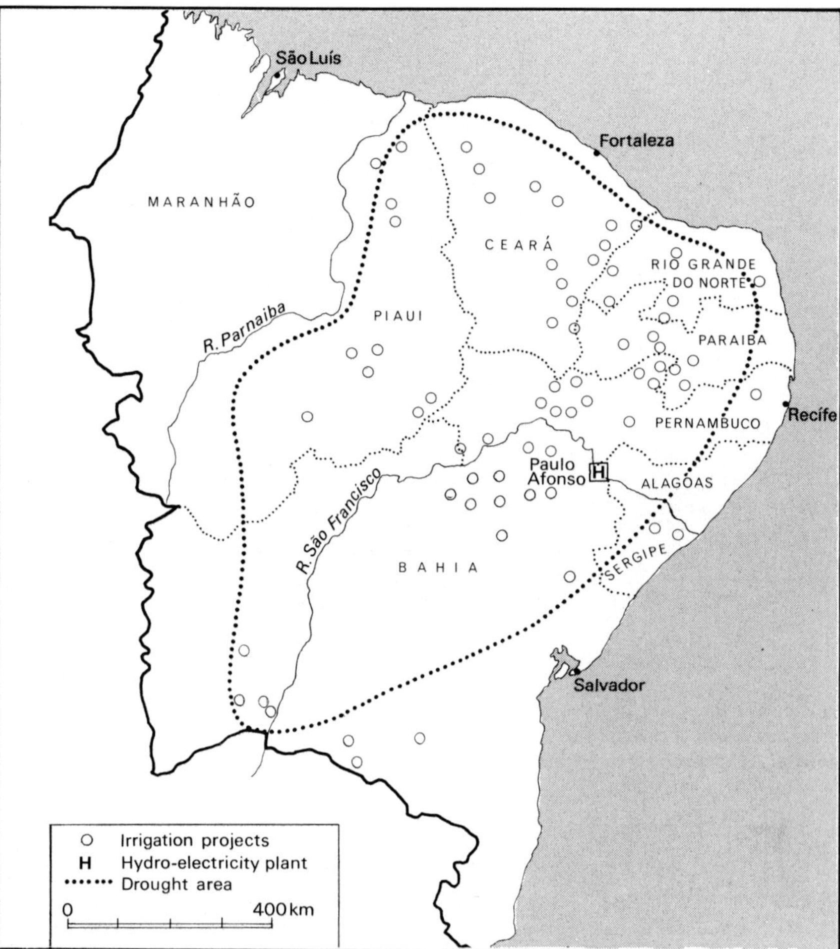

Figure 5 *The 'Drought Polygon' of the North-east and the location of irrigation schemes*

?

1 The first article 'From Nothing to Nowhere' describes a project, the building of the Transamazon Highway, which was undertaken in the early 1970s and which proved unsuccessful in alleviating the problems of the North-east. The second article describes more recent projects. Has the Brazilian government learnt something from its earlier failures? Do political considerations still stand in the way of any far-reaching solutions to the serious social problems of the North-east? Consider particularly the issue of land tenure.

2 Outline the cases for and against past, present and possible future policies in the North-east. Write an essay of examination length. The following may be useful headings: The emergency programme; The aid programme; The irrigation development programme.

FROM NOTHING TO NOWHERE
THE TRANSAMAZONIAN HIGHWAY

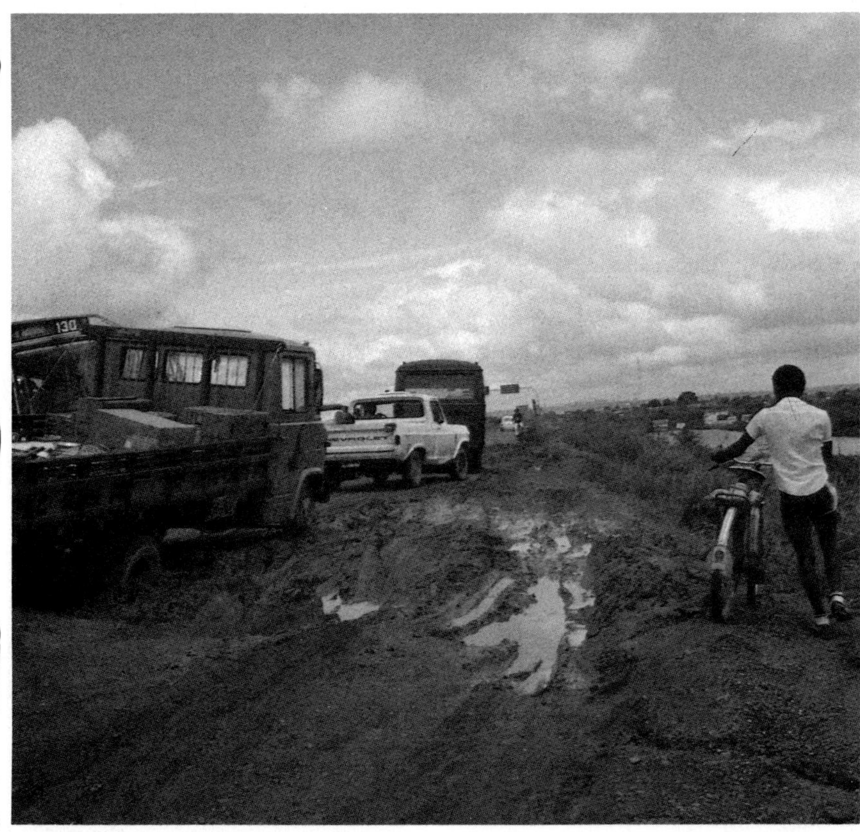

In 1970 the huge, backward, North-east of Brazil was undergoing one of its periodic droughts. Tens of thousands of peasant families were being driven off their tiny plots of land, *minifundio*, into the swollen cities which offered little prospect of employment. Hundreds of children were dying from starvation.

General Emilio Garrastazu Medici, president of the tough authoritarian government then ruling Brazil visited the region. By all accounts, he was profoundly shaken by the suffering he saw. He commented: 'Nothing in my whole life has shocked and upset me so deeply. Never have I faced such a challenge.' The president clearly felt that he must take decisive action.

Rationally, a long-term solution to the human suffering imposed by the droughts should have been sought within the region itself. If the president had pushed through a radical programme of land reform, giving each family an adequate plot of land and providing them with reliable credit facilities and technical advice the peasants would have become much less vulnerable to the droughts.

However such a policy was inconceivable, then. The government would never have declared war on the large landowners — faithful and important supporters of the regime. It would have contradicted the essence of the military government which was busy promoting an elitist, non-populist form of capitalist development.

Instead, the president searched for some kind of *deus ex machina*, an emergency solution outside the region that would end the intolerable suffering without changing the existing social and economic structures. The rapid construction of the Transamazon highway, in itself creating a heavy demand for unskilled labour, was to be followed by a massive colonisation project that would settle millions of landless peasant families on virgin forest land. It seemed to be a heaven sent solution.

After President Medici's visit to the North-east in 1970, the National Integration Plan was unexpectedly announced. It was launched as the master plan that would solve simultaneously the problems of both the North-east and the Amazon. Under the plan, about $400 million was to be spent on road construction, irrigation and colonisation projects. The money was to come from a drastic 30 per cent cut in the resources going to SUDENE, the North-east development agency, which had been grappling ineffectively with the region's huge problems for over a decade.

The most spectacular and costly of the projects was the Transamazon highway and its colonisation scheme. Transport Minister Mario Andreazza explained why the government had decided to regard the construction of the road as one of its urgent priorities: 'On the one hand, the North-east, ravaged by periodic droughts, with a huge sector of its population lacking even the basic conditions for survival, sees many of its inhabitants emigrate to the Centre-south where the large cities are not in a position to absorb this unskilled labour. On the other hand, the population of Amazonia, which is a vast region with fertile valleys and important mineral deposits, is concentrated in tiny hamlets beside the river. The solution was to let the two regions solve each other's problems. The slogan became: 'Land without people for the people without land.' It was predicted that two million people would be settled along the road within two years.

The project was presented as a fearless patriotic undertaking, carried out by a government in a hurry to develop the hinterland and to bring progress to the poorer sectors of the population. All leading government officials dutifully expressed enthusiastic support.

However, a few middle-rank civil servants dared to challenge these facile assumptions. José Sergio de Paz Monteiro, director of the road department for Amazonas, one of the states to be cut by the Transamazon, gave an interview to a leading São Paulo newspaper, *O Estado de São Paulo*, in June 1970. He commented: 'The simple fact of building roads does not mean that we are creating conditions for the occupation of the demographic vacuum. As well as roads, we must provide the settlers with technical and financial assistance so that they can produce and fix themselves on the land.' The engineer estimated that a successful colonisation project would demand an investment twice that calculated on for the road.

The engineer did not have to wait long for the vindication of his predictions. The idea of settling millions of North-easterners was given up within a year. The project struggled on until June 1974, when it was finally abandoned. By then, only 4 969 families had been officially settled. In all about 20 000 families had come into the region.

The families we visited in 1975 were facing serious problems. They were housed in flimsy, prefabricated wooden houses with corrugated iron roofs which looked incongruous in the midst of the tropical forest. And according to the settlers they were less suited to the humid climate than the traditional wattle-and-daub huts. Many complained of failures in the government's back-up programme: little technical assistance, highly expensive farm inputs (pesticides, fertilisers, sprays etc.), inadequate marketing facilities and so on. One settler told us that the road should really have been called the Trans-misery highway.

In parts, the failure of the Transamazon is due to inherent weaknesses in an over ambitious project. However, the main reason for the fiasco was probably political. From the very beginning, one of the main objectives of the project — to solve a serious social problem of poverty in the North-east — was jarringly at odds with the principal aim of the successive military governments in Brazil, which have been to further the interests of a small élite of powerful landed, industrial and banking groups. In January 1985, when the 'Brazilian Democratic Movement' was elected to power, one of their first moves was to pledge increased support and assistance to the ailing North-east.

(*Source*: *The New Internationalist*, October 1980)

THE AID DRAIN

Brazil's North-east, a region ravaged by a five-year drought which has rendered more than 20 million people destitute, is powerless to determine its future. It is dependent on external financial support from a distant and authoritarian central government, or from the World Bank.

At the hub of the vast government bureaucracy that administers this aid is the superintendency for the development of the North-east (SUDENE), which last year swallowed US$500 million, and is a major employer in the region.

At a recent SUDENE council meeting it was announced that the World Bank had agreed to finance a US$3.6 billion public works programme for the region.

The World Bank is no stranger to the *sertao*, where it is already involved in a series of smaller projects, and SUDENE officials admit that this latest World Bank project is designed to consolidate investments already scattered throughout their inefficient empire. It is also an attempt to prevent funds being used for trafficking in political favours.

Interior Minister Mario Andreazza, who is chairman of SUDENE, rejects the view that funds are used inefficiently. But one West German economist working for his government on a technical aid project says that up to 20 per cent of funds disappear into the wrong pockets, or are used to consolidate local political fiefdoms.

A recent critical study by the federal planning ministry discovered that funds granted to more than 400 projects were in many cases being misused. The beneficiaries are often big ranchers who in some cases cannot prove legal title to their land; not all the companies who receive development grants are engaged in productive agriculture.

Land of the tormented ones

Though a third of Brazil's population live in the North-east, the region lags badly behind in terms of wealth. Growth of agriculture, the region's most important activity, has been half that in the south. Per capita income in the state of Ceara, for instance, is just 16 per cent of that in the southern industrial state of São Paulo. 'These states have no financial independence, and therefore no political autonomy — we depend on the central authority for everything,' say Roberto Magalhaes, governor of Pernambuco.

The Catholic Church, which for centuries helped to maintain the rural landowning structure, is now critical of the government's record, especially its enthusiasm for large-scale projects. The church wants land reform. 'An authentic land reform is a much more serious and urgent problem than even the drought,' argues Dom Helder Camera, archbishop of Recife. 'It's ridiculous for the government to describe as land reform the simple handing over of land titles to peasants, without accompanying measures to allow them to make use of that land.'

The trouble is that the government has no more funds to expand the handful of successful projects that have been launched on the back of land reforms.

FAMINES AND FEASTS

'When it doesn't rain at all I'm happiest — rain just brings me problems,' said agronomist Jorge Roberto Garzeira at his farm near Juazeiro on the São Francisco river. In an area where it has scarcely rained for five years, he represents a tiny minority.

Behind him lush vineyards slope gently down to a broad river, forming a landscape that might have been transported from central France into the semi-desert of Brazil's North-east. In the same municipality, farmers with crops decimated by drought curse the lack of rain and depend on government water carts for survival.

The secret of Garzeira's success is irrigation, which has brought a new breed of agronomists and investors to the São Francisco river valley, a region many believe has the potential to rival some of the world's richest fruit and vegetable producing areas. 'This vineyard is a machine that never needs to rest,' Garzeira says. 'When we sign an export contract we just bud the vines, turn on the water and then send the grapes to Miami.

Satellite surveys show that in the centre of the arid North-east lies a potential bread basket, comprising a fertile crescent of one million hectares of irrigable land beside the São Francisco, and another 1.5 million in the parched upland valleys of the *sertao*.

A handful of capital intensive irrigation projects in the lower valley of this 2 600 kilometre river already produce record yields of melons, tropical fruits, sugar-cane, tomatoes and onions. Studies show the region could easily yield US$2.5 billion of produce a year. A new airport near Juazeiro has an international runway; a weekly cargo service to Europe and the United States is planned for high-value tropical fruits. This, the agronomists say, is the region's future.

But for the great majority, agriculture in the North-east is centuries away from such projects, and is dominated by a structure that has changed little since the abolition of slavery. Of the region's 30 million people, only 20 per cent have sufficient food. In the semi-arid area which comprises 70 per cent of the region, farming is sharply divided between the few who have land, subsidised bank credit and local political weight, and the millions who do not.

'Agriculture could rapidly generate millions of jobs here; we could solve the problem of total hunger without relying on palliative measures,' argues Joao Alves, governor of the state of Sergipe. But though the North-east is not self-sufficient in staple food and prices are almost double those in the south, little of the area's farmland is devoted to primary food production.

The vast bulk of the land is divided into *fazendas*, vast private holdings farming cattle and cotton. The big ranchers have little interest in food crops. Subsistence agriculture is carried out on their land by share-croppers who harvest maize and black beans, but in times of drought the share-croppers are forced to join small farmers in the ranks of government work gangs.

Of Ceara state's 323 000 rural producers, 300 000 farm less than 100 hectares; only half this number own the land, which represents about 10 per cent of the total area. Although such small farmers are responsible for 70 per cent of the region's food production, the system of government credit has been largely dedicated to helping the big ranchers produce meat for the southern cities.

The traditional wisdom that large-scale cattle raising is the only viable activity in the *sertao* is being challenged at an agriculture research station near Juazeiro, run by a government agency. Irrigation will remain a costly option serving a small part of the region, and must be balanced with new techniques for farming the dry uplands and its *caatinga*, or scrub, according to the station's assistant director, Edson Lustosa de Possidio.

As well as experimenting with such non-traditional crops as sorghum and millet, researchers are testing traditional irrigation techniques borrowed from such countries as Mexico and India. Lustosa believes that the region's native vegetation could sustain large numbers of people, producing foodstuffs, charcoal and medicinal herbs.

José Matias Filho of Ceara state university's agriculture department also believes that farming based on a combination of irrigation and semi-arid culture could render the region self-supporting. 'With our present water resources and technology we could produce five times as much food', he says. But at present 'all the agricultural emphasis is on supporting industry instead of food products.'

But an ambitious project does exist to realise the agricultural potential of the region. At present the waters of the São Francisco, after passing a string of hydro dams, flow directly into the sea. Of the one million hectares of irrigable land in its valley, only 32 000 have been developed. Now it is proposed that a third of the São Francisco's waters could be diverted northwards into the dried-up river beds in the *sertao*, to allow irrigation of another 1.5 million hectares of fertile land pinpointed by satellite survey.

World Bank aid is being sought for the US$2.8 billion project, which would employ 75 000 people in constructing canals, dams and electric pumping stations to raise the water 170 metres up to the *sertao*.

Some officials remain to be convinced, arguing that land by the river's edge must be irrigated before debt-laden Brazil embarks on another major project.

But some foreign bankers have been captivated by the dream. Donald Pearson, president of the investment subsidiary of the UK's Midland Bank, says US$200 million has already been invested, and he is keen to back the diversion instead of industrial and energy projects in the south.

Such a project, though, would not be completed for two decades — perhaps in time for the region's next major drought. In the meantime it will take years to make good the damage to agriculture caused by the present drought.

(*Source*: *South*, May 1984)

CASE STUDY 4 — THE CENTRE-WEST

THE POLOCENTRO PROGRAMME — A SUCCESS STORY

In 1974 the Polocentro programme was founded for the agricultural development of the area. One of the major areas of agricultural expansion is the Central Plateau of Brazil (see Figure 6 *page 229*). It is an area of *cerrado* (savannah grassland with some woodland) covering 50 million hectares. The Brazilian agricultural policy was outlined in the III PND (see Table 7 *page 215*). Agriculture was to be improved by means of a land policy plus developing and improving agricultural techniques via mechanisation, soil quality and water control. Improvement in the sector would lead to an improved rural standard of living. The plan recognised the need for change in the size of holdings — small and medium holdings were to be encouraged, and intensive use of land would be rewarded. Basic foodstuffs and export crops were preferred; the quality and quantity of cattle herds should be controlled. Marketing, transport and storage should be greatly improved to minimise losses. Modern mechanised techniques should be used where appropriate and there would be increased research into soil quality, fertiliser use and seed types.

Print 5 *Conditions on the land are hard in Brazil's North-east region where the climate is hot, soil poor, droughts frequent, poverty desperate and development schemes have gone over the heads of the poor, producing a huge rural population of dispossessed — pushed out by large-scale agriculture using machines instead of people. Small holders, nevertheless, grow most of Brazil's staple food. The Brazilian church has been advocating a radical change in land ownership for many years but it has only been since the election of the civilian government under José Sarney that any significant reforms have been discussed*

Figure 6 *'Polocentro' development poles in the Centre-west and South-east regions*

1 Read 'The plough wheels under wider skies'. Set out your evaluation of the agricultural development taking place in the savannah lands under the following headings; *Statements for* and *Statements against*. List three points for each side pretending that you are a UN observer in this area of Brazil.

2 Discuss your statements with others and then draw up a common list of observations and recommendations as if you had been asked to submit them to another developing country.

THE PLOUGH WHEELS UNDER WIDER SKIES

The savannah lands are a vast area of 2 500 000 square miles, with rolling plains under generous skies. Hitherto it has been largely uncultivated rough grassland studded with bent arthritic trees. The Brazilians used to regard it as an area good only for grazing.

Even now the savannah land has not benefited from the kind of exhaustive surveys devoted to Amazonia when that region was such a fashionable area for would-be developers. Yet the research into the savannah's potential has revealed that it should become the second great agricultural power-house of Brazil; so much so, indeed, that Brasilia, which is now situated right in the middle of the region as an administrative capital with mainly symbolic logic, may, within 25 years, find that it has also become a huge agricultural centre rivalling Chicago in scale and importance.

The savannah is one of Brazil's most strategic assets; and by no means just the dull plainland of its reputation. It has sunlight, constant temperatures, adequate average rainfall and large amounts of surface water. Its soil has a good texture and is of an adequate depth; the topography of gently rolling plainland is conducive to mechanised agriculture; land is in abundance and is still fairly cheap for a population expanding at about five per cent a year as immigrants arrive either from the depressed areas of the North-east or the overcrowded and highly priced South-east.

If the savannah can be harnessed fully to agriculture Brazil will have doubled its area of cultivation far more effectively than through any of the more far-fetched schemes which were half applied to Amazonia. It is true that the chemical quality of the soil on the plains is poorer than that found in the Amazon basin but the combination of climate and rainfall help to make it easily responsive to fertilisers.

A meticulous research programme has been planned to overcome the built-in drawbacks of distance, lack of basic services and untried agricultural conditions. The Government does not intend to let the area be overwhelmed with pioneer farmers lacking both expert guidance about the best crops on which to concentrate, and facilities for getting their produce to market.

So the research programme proceeds in gear with the colonisation of the area. Thus, since 1975 the population has grown to 10 million; 2 400 000 hectares have been incorporated; 54 per cent of Brazil's coffee is now grown there, 40 per cent of its rice, 16 per cent of its maize, as well as 36 per cent of its beef and 24 per cent of its pig population.

The types of soil and the climate of this area can also be found in some of Brazil's neighbours such as Venezuela, Colombia, Bolivia and Paraguay, all of which hope to benefit from one of the world's most comprehensive research programmes into seasonal acid soils. In Africa, Mali, the Ivory Coast, Kenya and Tanzania they may learn something to their advantage from Brazil's efforts to cultivate these plains.

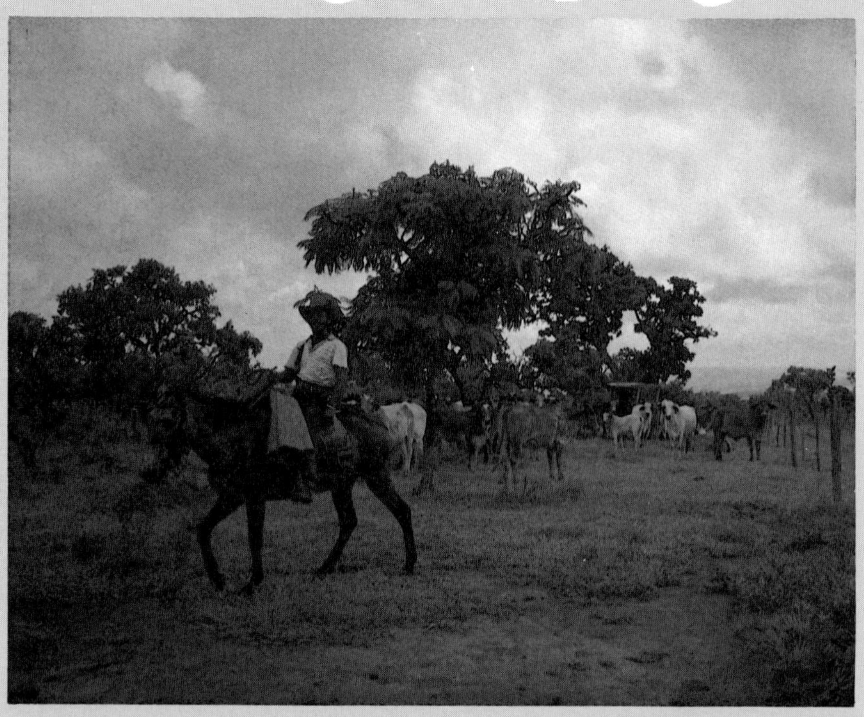

Cattle farming on the Cerrados *Experimental station in Brazil's savannah*

The conclusions of the first few years of agricultural experiment are: crop rotation is essential; there should be no question of monoculture such as exists elsewhere, particularly in the sugarcane areas of Brazil; it is an inherently risky environment so that farmers should diversify their output in at least three main areas: (a) annuals, such as soya, maize, rice and wheat: (b) perennials, such as cassava, coffee, fruits and forestry; and (c) pasture for making hay or grazing beef cattle.

Every new holding, by law, must have 20 per cent of its land set aside for afforestation. In research, the Brazilians have found that eucalyptus trees can grow as much as nine metres high by their second year, and can be cropped with profit at least every four years.

The other crops which have shown real promise so far are soya, where there is already a million hectares of growth, and wheat which promises higher yields than the patchy performance achieved in Brazil's southern states.

There are limiting factors. Fertilisers and minerals have to be added to compensate for chemical deficiencies in the soil. The threat of dry spells during the rainy seaon could also endanger crops when at their most critical period of growth. Equally, the long dry season of Brazil's winter, from May to September, will presumably require some measure of irrigation for certain types of crop.

However, the great significance of this strategic agricultural asset at the heart of Brazil is in its effect on the future prosperity of the country. Recently rising population has not been fully met by comparable increases in food production. Thus a bad harvest too often turns Brazil into a net food importer —

something it can ill afford and should never be, with its space and climate.

It is not just that the savannah will create food surpluses to earn Brazil foreign exchange. It is that the creation of more agricultural production gives Brazil's decision makers more power and influence, since in an overpopulated, undernourished world food is power; they now realise that the rich red, almost edible, soil of Brazil is as important a strategic weapon as all its many minerals, hydropower or any nuclear programme.

Finally, the savannah should go to the heart of Brazil's chronic economic imbalance. Hitherto, in the developing world, it has been fashionable to imagine that only industrial development can lift a country's economy out of the rut. It is now accepted that the foundation for most countries' wealth, stability and self-sufficiency is a broad-based agricultural economy.

This helps prosperity to percolate out from the industrial classes to those millions who are now on subsistence and who are bound to be casualties of, or at least left out of, the progressive industrialisation of parts of Brazil. The hope for their future lies in the savannah.

(Source: *The Times*, 23 July 1980)

The future for the Centre-west looks promising, probably because the government wants to maintain control over the area to ensure success, as it is vital to national development. The philosophy behind regional development is virtually unchanged; what is different is the organisation and management of such regional development schemes. Past experiences and mistakes have taught the regional planners many valuable lessons — it has been established that regional planning depends on clear, coherent, efficient and effective action backed up with sufficient capital finance. Critical to the success of this scheme is the relationship of this area to the South-east and Brasília. The links and lines of communication to the capital and the South-east are very good; produce can go quickly to the markets of the South-east or for export.

However, the development of agriculture in the Centre-west can be viewed simply as an extension of the South-east, particularly when a large amount of investment capital has come from the South-east. POLOCENTRO is a case of highly capitalised modern agriculture rather than a case of rural development to ease the population pressures in other areas. The Centre-west agricultural development scheme still has the potential to increase inequalities within the country by having very successful agriculture in the South, South-east and southern Centre-west and not the North and North-east.

BRAZILIAN ACHIEVEMENT AND THE FUTURE

Since 1950, national and regional planning programmes have become increasingly important in the development of Brazil. Development programmes face many problems, which differ from region to region and time to time. It is obvious that certain short-term policy directives have to be reassessed and reconstructed to meet the needs of changing circumstances. Some would argue for changes in long-term policy commitments, too. However, Brazil still maintains its overall objective of a developed integrated country, with the hope of becoming a world leader.

THE REASONS FOR BRAZIL'S LIMITED SUCCESS

In the past, regional development plans have had limited success because of:

- The size of the country
- The size of the task
- Insufficient finance and capital
- Insufficient back-up services
- A lack of regard for the social and economic development of the region

As a result many sectors and areas have experienced polarised development and few 'trickle down' effects. There is still within Brazil, a newly industrialised country, a strong dual economy and a large gulf between the regions. There are large regional inequalities in income distribution, purchasing power and social service provision. In the III PND, 'Particular emphasis has been placed on social development, the better distribution of wealth through taxation and large investments in the social sphere' (Lloyds Bank, 1980).

The election of the PMDP (Brazilian Democratic Movement) in January 1985 was the culmination of a gradual political change within the country — trade unions had become stronger, the *abertura* (transition towards democracy) policy had been introduced and in 1979 the political party system had altered dramatically. It is too early to judge the effectiveness of the new government in tackling Brazil's acute financial and social problems but it has pledged to pay special attention to the rural poor and regional disparities as well as recognising the need for austerity and careful management if Brazil's crippling debts are to be paid off. Many projects are being considered by the World Bank and other aid organisations for possible funding and if the new government can maintain a stable programme of development, with a close eye on its foreign debt there is every hope that the late 1980s will be a period of expansion and improvement for Brazil.

The international scene and ideas of development can also account for the limited success of certain Brazilian development projects. For example: the wrong notion of development; foreign ownership of companies; an imbalance in the overall structure of world trade; and too heavy reliance on foreign aid with consequent debts can all influence the direction, value and success of a development project.

FOREIGN DEBT

Since the oil crisis, Brazil has achieved world renown by virtue of possessing one of the largest national debts (Table 8). Therefore, other countries are forced to recognise Brazil's importance in the world economic system simply because if Brazil defaulted on its debt many other countries would be affected.

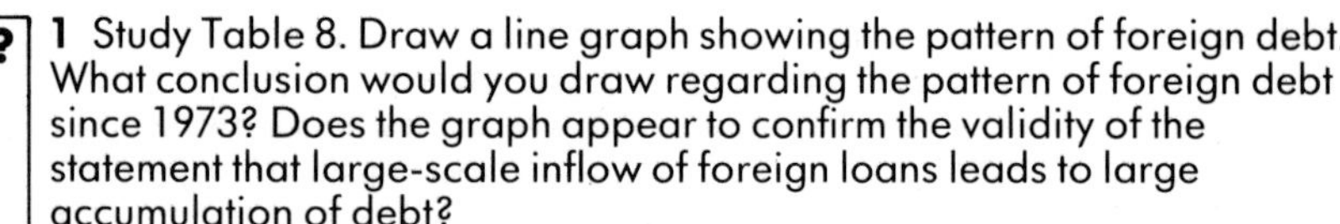

? **1** Study Table 8. Draw a line graph showing the pattern of foreign debt. What conclusion would you draw regarding the pattern of foreign debt since 1973? Does the graph appear to confirm the validity of the statement that large-scale inflow of foreign loans leads to large accumulation of debt?
2 Look at Figure 7. What does the debt repayment schedule assume about the performance of Brazil's economy from 1986 onwards?
3 Does Figure 8 give support to the argument that Brazil's economy, as reflected in the balance of trade, is becoming more healthy? Does the debt repayments schedule seem realistic to you?
4 Read the folk tale below and look at the diagram which summarises the story. Do you think that there is any relationship between the diagram and the Brazilian experiences in connection with foreign debt and its general impact?

Table 8 *Brazilian foreign debt and foreign investment ($US billion)*

Year	*Debt*	*Year*	*Debt*
1973	12.6	1979	49.9
1974	17.2	1980	53.8
1975	21.2	1981	61.4
1976	26.0	1982	69.6
1977	32.0	1983	81.3
1978	43.5	1984	99.5

(*Source*: Latin American Bureau Special Report, and Banco Central do Brazil, 1978–1983)

FOLK TALE: WHY HENS SCRATCH THE GROUND

Once upon a time, there was a hawk who owned an axe that he used to earn a living. One day the hens asked him to lend them the axe. He let them have it. The hens didn't know that they were getting themselves in a jam; they were very happy and took it to the mealie fields. They cut lots of mealies and sat down to a fine feast. It wasn't long before the hawk came back and asked for his axe. The hens searched and searched, but the axe was nowhere to be seen. The hawk waited until his patience wore out and he spurned them saying, 'I do not want the axe any more. From this day onwards, I am going to live on your young ones.' It wasn't long before the hawk was seen stealing chickens. Up to this date, the hens still scratch the soil, searching for the axe. We believe that the old hens left word among the generations that succeeded them that the only way of obtaining the hawk's forgiveness was to find his axe.

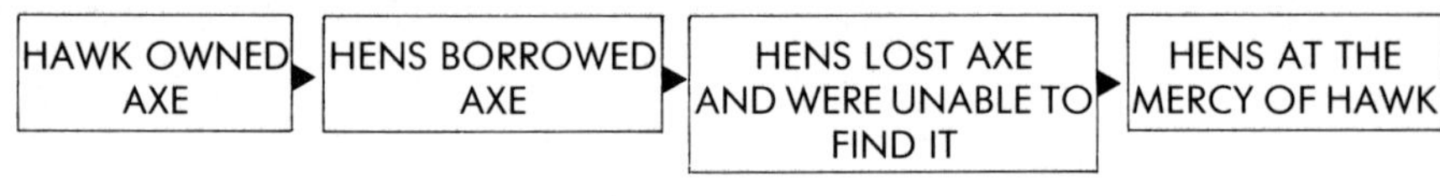

(*Source:* Fashion Phiri, Department of Education, University of Zambia)

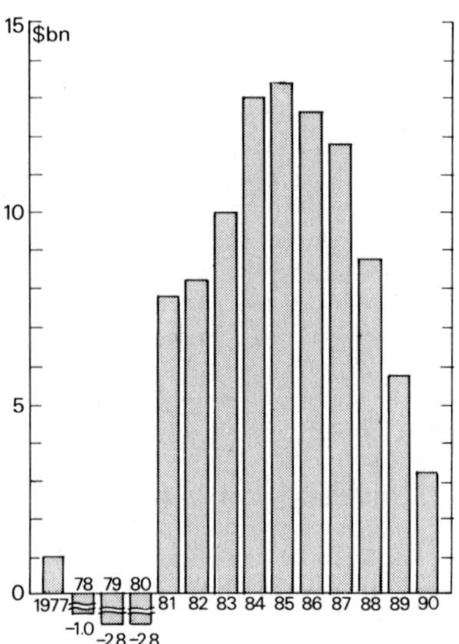

Figure 7 *Brazil's projected debt repayments schedule — medium- and long-term loans*
(*Source*: *The Financial Times*)

? **5** From your general reading about Brazil, or any other developing country, state other probable social and economic consequences due to accumulation of foreign debt.

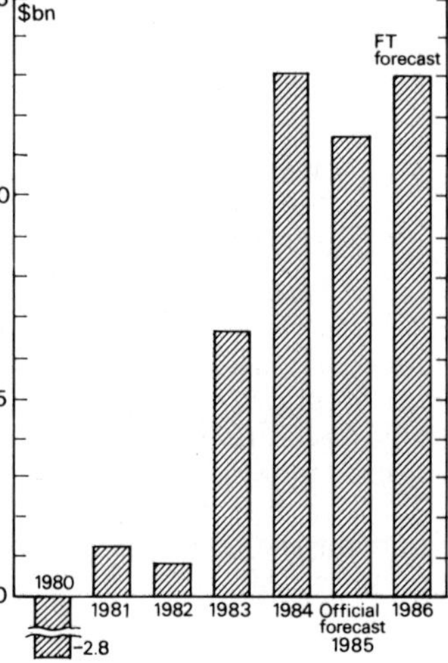

Figure 8 *Brazil's trade balance, 1977–1984*
(*Source*: *The Financial Times*)

AN ALTERNATIVE FUTURE

Brazil has experienced rapid economic development in the last 30 years but at a cost; socio-economic inequalities have grown and development has been spatially unequal. The majority of Brazilians, the urban and rural poor, have not benefited from regional planning policies which were designed to achieve economic development and social justice. Increasing numbers of Brazilians are marginal to society; they have no power to influence government policies and their basic needs and wants remain unmet.

There are Brazilians working within the country such as Paulo Freire (an educationalist) and Milton Santos (a geographer) who call for reforms and for social action by the people to obtain freedom, liberation and self-determination. Challengers to the government's policy on development, which is one of 'development from above', favour the policy of 'development from below'. Reforms that are considered necessary to achieve 'development from below' are:

- True political electoral freedom
- Equal rights for all
- The reorganisation of land ownership and an integral rural development programme for all areas
- The state ownership of all means of production
- Social assistance for the poor and unemployed
- A public housing programme
- A higher statutory minimum wage linked to the rate of inflation
- Controls on inflation
- Production of essentials rather than sophisticated consumer goods
- Use of appropriate technologies
- Environmental controls
- Inter-regional co-operation
- The possible removal of all foreign investors in Brazil

Such reforms question the present socio-economic and political system of Brazil. For radical Brazilians the only way forward is when all Brazilians participate in organising and orienting development to meet their needs.

Which way forward?

What are the real priorities of Brazilian development? What style of regional development should Brazil adopt in the future and who will benefit from the chosen style of development.

? Read the newspaper extracts below which give up-to-date insights into the problems and choices facing Brazil. Then carry out the role play.

'REALISM': THE NEW WATCHWORD

Brazil and the International Monetary Fund have settled down this week to the third, and final, set of their gruelling and long-running negotiations. After 27 months of an up-and-down contest, the stage has been set for a grand finale.

The novel feature this time is that during the recent three-month break, Brazil has brought on a fresh player, who promises new tactics and is committed to a much tougher game altogether.

A clear favourite with the crowd, the new contestant—the civilian government of President Jose Sarney—has already proposed that 'extended fund facility' should be abandoned in favour of 'stand-by loan'—a change in the ground rules which appears to be agreeable to an opponent who, as usual, is giving little away about his own strategy.

The conundrums facing both teams as they prepare to face each other once more are depressingly the same. How can a developing country with a moderate-to-high population growth balance austerity against economic growth?

Realism is the watchword of the day: realism in dealing with the economic situation in Brazil as it is, rather than what one would like it to be, and realism in setting targets that the country can achieve 'without excessive strain.'

The starting point for any debate is the balance of payments and, within it, the envisaged trade surplus.

Last year's spectacular performance, when the trade surplus doubled to \$13.1bn, is highly unlikely to be repeated. Officials agree that 1984 was an atypical year, for all sorts of reasons, such as the strength of the US recovery, an orange juice export boom and the coming on stream of new offshore oil wells.

The Government is thus forecasting an \$11.5bn or \$12bn trade surplus in 1985, based on imports of \$14.5bn—marginally up on 1984—and exports of \$26bn to \$26.5bn, slightly down on last year's record. Most private forecasters are less sanguine, believing that a surplus of \$10.5bn or \$11bn is more likely.

In the first quarter, industrial output was up by 9.5 per cent compared with the same period last year. Measured over a 12-month period to March, it showed a healthy growth of 8 per cent, the strongest since 1979, prior to the severe recession of the early 1980s.

Industry, however, takes second place to agriculture in the calculations of the New Republic. Farm exports are pounding along strongly—output grew by 4.3 per cent in 1984—but the Government has said it intends to restore the tilted balance between domestic food crops and export commodities.

Whether this rate of growth can be maintained is an open question. Much is due to import substitution and recent improvements in disposable income levels.

Where there is no doubt is about the dynamism of the economy: nor about the fact that a once-and-for-all shift has taken place on the trade balance, giving the country the theoretical ability to service the foreign debt—if it should choose to apply the resources to that end. The rest is politics.

(*Source*: *The Financial Times*, 29 May 1985)

'Robin Hood' share-out angers Brazil landowners

Brazil's fledgling civilian government is planning to settle 7.1m landless families by the year 2000. For this, it will need an area of at least 490m acres, eight times the size of the United Kingdom.

The newly created Ministry of Land Reform and Development has identified an area of 920m acres as being legally suitable. But 840m acres of it is in private hands. With a month to go before the plan, announced last week, becomes law landowners all over the country are preparing their counter-attack.

On the desk of Nelson Ribeiro, the land minister, sits an aluminium bowl perforated with 32 bullet holes to remind him of the task that lies ahead. Ribeiro is fully aware of the obstacles to land reform on this scale. But without it, he believes rural Brazil could explode into civil war.

He was given the bowl by Otilia Maria Nogueira da Silva, the widow of a peasant farmer killed, with 18 others, by hired gunmen two weeks ago. After a land dispute the 33 gunmen surrounded Otilia's house, where the peasants were holding a meeting, and opened fire. They had been sent by a local 'grileiro', or robber baron. Before leaving, they ransacked Otilia's house, but left the bowl punctured by bullets.

Last year, there were 180 deaths in 916 land disputes. Most of the conflicts rage in the recently cleared lands of the southern Amazon, and along the southwestern agricultural frontier, where the jungle to the east of Bolivia is giving way to large agri-business. But there are disputes on a lesser scale in the rich southeast.

They present a chilling problem for Brazil's new president, José Sarney, whose traditional family base of political power has been in the backward northeast, and who took over six weeks ago.

As available land shrinks, the onslaught of the big landowners has forced smallholders and squatters to retreat inland, often into Indian reservations. Smallholder becomes pitched against Indian. But in recent years, both have been fighting back.

Ribeiro's ministry believes that an equitable land distribution will reinforce capitalism. Settled on smallholdings, the dispossessed should become supporters of property-owning democracy. Food production, which has declined over the past two decades as agri-business has invested in export crops, should increase.

The government will also collect more taxes. It is the small and medium farmer who traditionally pays his taxes on time. The treasury is currently owed £380m in land taxes, nearly all from the big farmers.

In 1980 the 0.9% of landholdings classified as large properties represented 46% of land in Brazil. The 9.4% of medium-sized properties accounted for 34% of total area, and the 89.1% of smallholdings for 18%.

The land ministry has identified 10.6m landless rural workers. Settlement is planned to start fairly modestly, with 100,000 families in 1985/86, and 300,000 in 198/87.

The necessary legal instruments to effect the reform already exist in a 1964 land statute which was created by the first military president, Hugo Castelo Branco, but never enforced. Ironically, one pretext for the military takeover that year was an attempt to promote timid land reform by the leftwing Joao Goulart government.

The critical question now is who will lose land and how. The statute stipulates that the first to lose land will be the speculator, the big landowner who is not farming his land. Next will be the landowner whose property, even if worked, is more than 600 times the minimum area needed to support a family, which varies from region to region. And this is where the controversy lies.

(*Source*: *The Sunday Times*, 1985)

Role Play

?

Introduction In response to Brazilian requests for development assistance, the World Bank, with guaranteed legal and administrative help from the Brazilian government, is willing to give financial and technical support to one regional development project. The cost of each project may vary considerably but total cost is not a limiting factor.

Aim Each development project must be designed to further the aims of the Brazilian national development plan 1980–85 (see Table 7 *page 215*). Each project will be assessed on its merits to act as an agency to encourage regional development. The World Bank–Brazilian government consortium will make the final decision as to which is the best project to receive financial and technical support.

Players Six groups, one per region and one to act as the World Bank–Brazilian consortium. If possible, each group should consist of four people. One member of the group should act as the organiser and co-ordinator, one as the presenter, one as the fact finder and one as the problem solver. The consortium group will have two World Bank and two Brazilian government representatives. If there are insufficient players to meet these requirements, then each regional development group can be represented by one person and the teacher can act as the World Bank–Brazilian consortium.

Some extra regional data is also given in Table 10.

? **Planning and Presentation** Each Regional planning group must present a detailed plan of the project, as well as a map of the proposed location indicating infrastructure development and the project's possible sphere of influence. Each region is allowed one consultation with the World Bank Brazilian government consortium for guidance. While constructing the project plans, the Regional planning group should consider the following:

1 What is a suitable project for the area?
2 Where is the best location for the project?
3 What are the advantages of this location, or are there several suitable locations?
4 What are the requirements of the project? e.g. infrastructure, communications and services, etc.
5 Does the project require specific financial and specialised technical support?
6 What is the time-scale of the project?
7 What are the likely effects of the project on the region and nation in terms of employment, infrastructure improvement and services?
8 Are all the effects of the project beneficial or are some detrimental?
9 What might be obstacles to the project?
10 Is the project part of a larger scheme?
11 Do you require any additional information?
12 Is the project possible and desirable? What are its *major* selling points?
13 Will debt be incurred? Should the project be implemented by foreign finance?
14 What role will peasants have in the project?
(*Note* This is not an exhaustive list; the planners may well think of other considerations.)

At the end of each presentation, the project team will be questioned and challenged by other groups and by the World Bank–Brazilian government consortium. (They might want some reassurance that the project will 'take-off' or request detailed plans and information.) The final decision of the World Bank–Brazilian government consortium is binding.

Suggested projects

- A motor car assembly plant, for alcohol cars, in the North-east.
- A centre of scientific research excellence in the South.
- A new national exhibition/conference centre, or sports complex in the South-east.
- A national park (for tourists, outdoor pursuits and recreation) in the North.
- A tropical agricultural research centre of an agricultural implement factory and distribution centre, in the Centre-west.

Table 10

Region	*% of total energy production (gigawatt hours, 1978)*	*% of total energy consumption (gigawatt) hours, 1978)*	*% of country's exports from points of embarkation*	*% of bank accounts held in each region, 1979*	*No. of universities in each region*
North	1.67	1.56	2.39	3.37	12
North-east	12.69	12.60	12.17	25.07	89
South-east	72.83	70.93	58.86	26.00	216
South	9.76	12.02	25.53	39.16	72
Centre-west	3.05	2.89	1.05	6.40	20

(*Source*: IBGE)

12 INDIA: PLANNING AND PRIORITIES

The central objective of planning in India is to raise the standard of living of the people and to open to them the opportunities for a richer and more varied life.
(*Source*: *First Five Year Plan*, 1952)

Print 1 *Ploughing with bullocks on a plain near Barvanitola village, Bihar*

There are as many opinions on the performance of Indian planning as there are commentators. It is nevertheless abundantly clear that planning has played an important part in the path followed by the Indian economy since independence from Britain on 15 August 1947, and that, despite relapses and various changes in direction, the government remains firmly dedicated to planning.

The First Five Year Plan was drawn up in the heady days following independence and ran from 1951 to 1956; currently the Sixth Five Year Plan (1980–85) is at the end of its term and the Seventh Five Year Plan (1985–1990) is underway. This chapter undertakes an appraisal of the impact of the various plans on people–environment relationships, and in particular on agricultural and industrial changes, together with an assessment of the effect of demographic change on the physical environment.

THE PHYSICAL ENVIRONMENT

Geology and soils

India is a land of great diversity, and this in itself has been the cause of fundamental problems in attempting to provide overall planning proposals for the country as a whole — the spatial dimension has too often been ignored. India can be divided into three great physical zones: the mountain wall centred on the Himalayas in the north, the alluvial trough of the Indus and Ganga valleys, and then the old peninsular block in the south formed from ancient gneisses and granites. While a great variety of soils exist throughout the subcontinent, they too follow this basic physical divide. In the north are found skeletal mountain soils; in the plains of the Ganga and Brahmaputra rivers are found alluvial soils of varying depth and age; and in the peninsular in the south there is a threefold division of latosols on the coasts, black soils mainly of a clay character in the central western states of Maharashtra and Karnataka, and red soils of sandy to loamy texture in the east, south and north of the peninsula.

Print 2 *Cauvery Falls in Southern Karnataka — India has a great diversity of landscapes*

Climate

The great seasonal rhythm of the monsoon dominates. December to February is the cool dry season. Except in the south, mean temperatures are normally below 21 °C and night temperatures often fall to 0 °C. In the north a westerly airstream brings some rain. From March the temperature begins to rise and a hot dry season continues until mid-June. Daytime temperatures in the plains of the north frequently reach 40 °C. Humidity, though, remains low. From mid-June the country waits for the arrival of the south-west monsoon. Once the rains have come, in June or July, humidity increases rapidly, and, although temperatures fall somewhat, the air feels hot and sticky. Gradually through September, the monsoon withdraws and the temperature begins to fall. Eventually in the north the return of a cool wind in November heralds the beginning of the cool dry season.

While this is a very simplistic picture of the climate, the overall seasonal rhythm it describes is fundamental to the agrarian economy of India which is divided into two main harvests, the *kharif* and the *rabi*. The *kharif* crops — rice, jowar and bajra — are sown just after the onset of the monsoon in June and July and are harvested in the autumn. The *rabi* crops — such as wheat and barley — are sown after the rains of the summer are over and are harvested in the spring.

Vegetation

The combination of geology and climate, and people's interaction with this environment, has given rise to a complex pattern of vegetation. Most of India was once forest; today the Forest Department classifies about one-fifth of the country's total area as forest, and less than half of this is under adequate arboreal cover. It is now estimated that the current rate of deforestation is well over 1 million hectares each year, and this loss of forest cover is a subject to be discussed at greater length later in this chapter. The greatest amount of forest remains in the northern mountain regions, in the eastern states of Assam, Manipur, Tripura, Meghalaya and the Union Territory of Mizoram, in the south-west in the state of Kerala, and in a wide band of scattered forest across Madhya Pradesh, Orissa and southern Bihar. Much of the western part of the western state of Rajasthan is desert, and the vegetation of the higher mountains of the north grades from forest to pasture and then alpine waste. Arable land totally dominates the Ganga valley as it does much of Maharashtra, Andhra Pradesh,

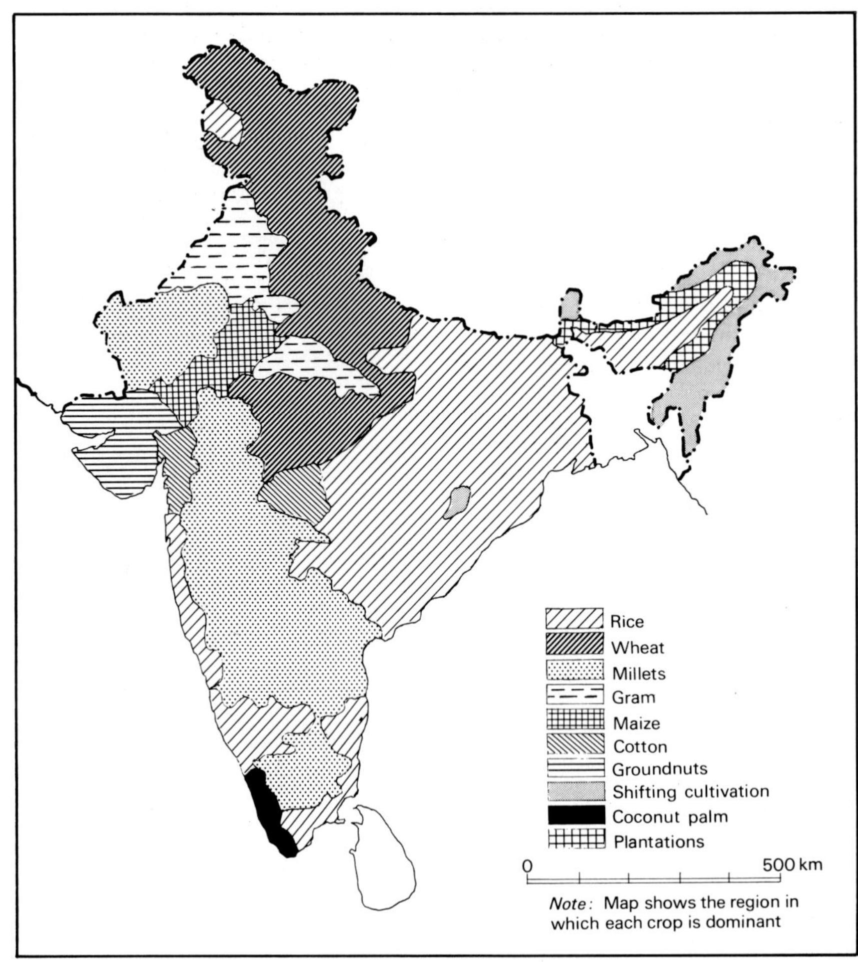

Figure 1 *Main crop regions*

Karnataka and Tamil Nadu (see location map).

It is against this physical background, with a population of 361 088 000 in 1951, that Indian planning began. Upon independence in 1947, the Indian subcontinent had been split into the two countries of India and Pakistan. This caused considerable social and economic disruption, and it was with the hope of creating a new economic and social structure that Indian planning was set underway with the avowed objective of raising the standard of living and increasing economic opportunities.

? Having read this introduction, write down in note form what you consider to be the ten most important facts or ideas about India. Draw from what you have read so far in this chapter and also from any studies of India you made lower down the school. When you have finished the chapter look back at your notes and decide how you would change them.

PLANNING IN THEORY AND PRACTICE

Since its inception Indian planning has been based on five year plans, each providing for a sectoral allocation of financial resources associated with projected increases in economic indicators. There was nevertheless a period of three annual plans between the Third and Fourth Five Year Plans, and a further gap between the Fifth and Sixth. Tables 1 and 2 provide a broad breakdown of the proposed and actual levels of expenditure,and they illustrate clearly the changes of emphasis that have taken place in planning over the last 30 years.

? **1** Divide into six groups, representing the following sectors of the economy: agriculture, irrigation, power, industry and mining, transport and communications, and social services. Within each group/sector half of the group must act as suppliers of the services and commodities and half as consumers. As both suppliers and consumers of a particular sector, each group must then, by itself, analyse Tables 1 and 2 to see how they have fared in terms of government support and emphasis over the duration of the plans. Each group should discuss the question 'What are the constraints placed by the suppliers on the consumer demand?'

2 After this initial discussion, the groups/sectors should come together and discuss each plan in turn as a class to see which sectors have received support and which have not at various times over the last 30 years. How has the emphasis placed on different sectors changed through time? Why do you think this is?

The previous exercise should have highlighted several important features of the planning process. It is now necessary to place these within the broader spectrum of overall changes in planning policy. A recent summary of national planning in India has stated that:

According to the objectives of the national plans, economic growth was to be achieved through full employment, higher national output and improvement in the general standard of living. In the First Plan (1951–56) the vehicle for achieving economic growth was overall rural development; in the

Table 1 *Proposed financial outlay in Indian five year plans (rupees crores*)*

	1st Plan 1951/2–1955/6		*2nd Plan 1956/7–1960/61*		*3rd Plan 1960/61–1965/6*		*4th Plan 1969/70–1973/4*		*5th Plan 1974/5–1978/9*		*6th Plan 1980–1985*	
Agriculture and community development	361	17.5%	568	11.8%	1 068	14.2%	1 600	7.8%	4 935	13.3%	5 695	5.8%
Rural development											5 364	5.5%
Irrigation and Power	561	27.1%	913	19.0%								
Irrigation/Flood control	168	8.1%	381	7.9%	650	8.7%	1 073	5.2%	2 681	7.2%	12 160	12.5%
Irrigation and power multi-purpose	266	12.9%										
Power/Energy	127	6.1%	427	8.9%	1 012	13.5%	2 523	12.3%	6 190	16.6%	26 535	27.2%
Industry and mining	173	8.4%	890	18.5%	1 784	23.8%	6 044	29.5%	9 029	24.3%	15 018	15.4%
Large/medium industry			617	12.9%	1 520	20.3%	5 298	25.8%				
Village/small industry					264	3.5%	746	3.7%				
Transport and Communications	497	24.0%	1 385	28.9%	1 486	19.8%	4 117	20.1%	7 115	19.1%	15 546	15.9%
Transport											12 412	12.7%
Communications and information											3 134	3.2%
Social services	425	20.5%	945	19.7%	1 300	17.3%	3 560	17.3%	6 800	18.2%	14 035	14.4%
Social Services	340	16.4%							5 074	13.6%		
Rehabilitation	85	4.1%										
Education									1 726	4.6%	2 544	2.6%
Miscellaneous/other	52	2.5%	99	2.1%	†	†	1 600	7.8%	500	1.3%	3 147	3.2%
Total	2 069		4 800		7 500		20 517		37 250		87 500	

* Indian statistical documents dealing with very large figures frequently use words rather than numbers. A crore is equal to 10 000 000. As a rough guide £1 equalled 15.33 rupees (Rs) in May 1984

† figure for miscellaneous is included in Social Services

(*Sources*: *Sixth Five Year Plan*, 1980–85, Government of India Planning Commission, 1981, Bombay. *Statistical Outline of India 1982*, Tata Services Ltd., Bombay)

Second Plan (1956–61) it was heavy industry; in the Third Plan (1961–66) it was the application of science and technology to agriculture with emphasis on the development of medium-sized industries. The Fourth Plan (1969–74) and Fifth Plan (1974–79) used these vehicles but with different emphases. But the early plans rarely contained specific suggestions or policies for spatial development.

(*Source: Service Provision and Rural Development in India: A Study of Miryalguda Taluka*, Sudhir Wanmali, International Food Policy Research Institute, 1983)

This statement emphasises two crucial issues: firstly that the seeds of planning policy were sown in the first three plans, with no major change of direction since then, and secondly that spatial aspects of development were generally ignored. Until the Sixth Plan, very little attention was paid specifically to the environment in India's Five Year Plans, other than regular, but short, sections on soil and forest conservation.

THE FIRST PLAN

As Table 1 indicates, this concentrated in particular on the provision of irrigation and power, and the development of transport and communication links. It was aimed primarily at building up the rural economy, and it is pertinent to note that the percentage of the total proposed outlay allocated to the agricultural sector in the First Plan was higher than in any other of the plans. Actual expenditure levels (Table 2) were indeed approximately similar to those proposed, and the First Plan did achieve considerable success with national income increasing by 18 per cent over the five years, and

Table 2 *Actual expenditure in Indian five year plans (rupees crores)*

	1st Plan 1951/2–1955/6		*2nd Plan 1956/7–1960/61*		*3rd Plan 1960/61–1965/6*		*4th Plan 1969/70–1973/4*		*5th Plan 1974/5–1078/9*	
Agriculture and community development	357	15.2%	530	11.5%	1 460	14.3%	3 466†	20.7%†	4 865	12.3%
Rural development									5 364	5.5%
Irrigation and power	661	28.1%								
Irrigation/Flood control	384	16.3%	420	9.1%	650	6.4%			3 876	9.8%
Irrigation and power multi-purpose										
Power/Energy	260	11.1%	445	9.6%	1 062	10.4%	2 448	14.6	7 399	18.8%
Industry and mining	179	7.6%	1 075	23.4%	2 995	29.4%	3 729	22.2%	9 581	24.3%
Large/medium industry	148	6.3%	900	19.6%	2 570	25.2%				
Village/small industry			175	3.8%	425	4.2%				
Transport and communications–	557	23.6%	1 300	28.3%	1 736	17.0%	3 887	23.2%	6 870	17.4%
Transport										
Communications and information										
Social services	533	22.6%	830	18.1%	1 697	16.6%	2 324	13.9%	6 839	16.2%
Social services										
Rehabilitation										
Education										
Miscellaneous/other	69	2.9%			600	5.9%	920	5.4%	446	1.1%
Total	2 356		4 600		10 200		16 774		39 876	

† figure includes both agriculture and irrigation

(*Sources*: same as Table 1 and *Sixth Five Year Plan Mid-Term Appraisal*, Government of India Planning Commission, New Delhi, 1983)

food grain production rising by 20 per cent.

THE SECOND PLAN

By the mid-1950s the planners had turned for their inspiration to the Soviet model of development, which emphasised the construction of heavy industry. As Table 1 indicates, the percentage of proposed outlay devoted to agriculture and irrigation was to fall, whereas that devoted to industry and mining was to rise from its actual level of 7.6 per cent in the First Plan to a level of 18.5 per cent. In reality, as Table 2 shows, the actual expenditure on industry and mining in the Second Plan represented 23.4 per cent of total expenditure. It is of particular significance that 84 per cent of all of this industrial expenditure was on large- and medium-scale industry. The four overall aims of the Second Plan laid the foundation for the direction in which future plans would go. The aims, as stated, were as follows:

- A large increase in national income so as to raise the standard of living in the country.
- Rapid industrialisation with particular emphasis on the development of basic and heavy industries.
- A large expansion of employment opportunities.
- Reduction of inequalities in income and wealth and a more even distribution of economic power.

THE THIRD PLAN

This followed along similar lines and continued to emphasise heavy industrial development, with large- and medium-scale industry accounting for over 25 per cent of total expenditure over the plan period 1960/61–1965/66. As the Third Plan stated:

> While agriculture and industry must be regarded as closely linked parts of the same process of development, there is no doubt that industry has a leading role in securing rapid economic advance.

This continued emphasis on heavy industry to achieve *rapid* economic growth coincided with two years of drought in 1966 and 1967. The resultant food shortages precipitated a major crisis, with the importation of substantial food aid and the devaluation of the rupee in 1966. Planning had failed to provide for a situation where the environment presented a constraint on 'development'. Consequently, for the next three years, *ad hoc* annual plans, which attempted to balance short-term income and expenditure, were promulgated, providing the opportunity for a rethinking of the overall planning framework. In the agrarian sector during this period the High Yielding Varieties Programme, which had been initiated in 1966 to introduce new varieties of grain with the aim of increasing food production by 5 per cent per annum, first began to have an impact.

THE FOURTH PLAN

Despite much debate over the direction in which future plans should go, by the time the Fourth Plan (1969/70–1973/74) was announced it was clear that no major change of direction was in sight. As Table 1 indicates, heavy industry, transport and communications continued to take the lion's share of the government's proposed financial outlay.

THE FIFTH PLAN

The Fifth Plan (1974/5–1978/79) continued to follow a similar pattern of allocations, although the amount devoted to agriculture and irrigation was marginally increased. The political crisis consequent on the Prime Minister, Mrs Gandhi's, announcement of the Emergency and her 25-Point Economic Programme proclaimed on 1 July 1975, eventually led to her downfall and a

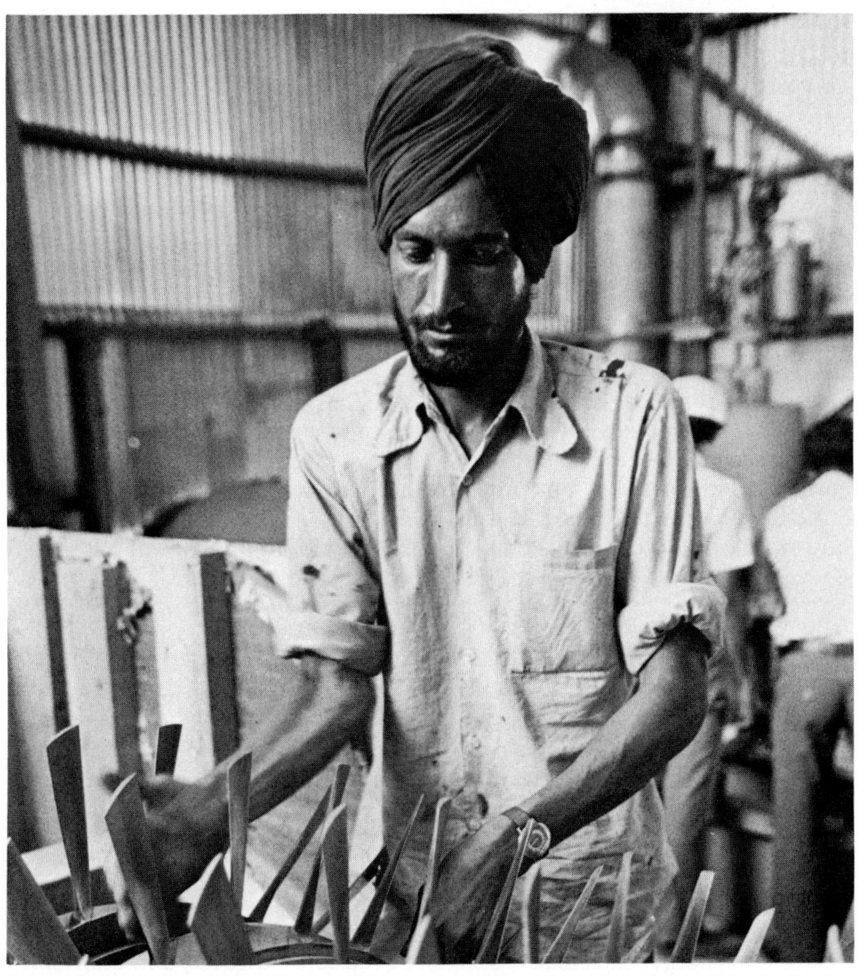

Print 3 *A worker polishing blades at a fertiliser plant near Bombay*

Print 4 *Large-scale irrigation projects in India — work being carried out on the Narmada dam, Gujurat in 1985*

new Draft Five Year Plan 1978–83 was put forward. In its proposals this plan paid specific attention to the problems of the poor. In an honest appraisal of the situation it advocated a new development strategy, stating that:

> What matters is not the precise rate of increase in the national product that is achieved in five or ten years, but whether we can ensure within a specified time frame a measurable increase in the welfare of the millions of the poor.
>
> (*Source: Draft Five Year Plan 1978–83*, 1978)

THE SIXTH PLAN

The fifth plan was never properly implemented, and with the return to power of Mrs Gandhi a new Five Year Plan, the Sixth (1980–1985), was introduced. As Table 1 illustrates, this again represented an important departure from the previous plans, and possibly a new sense of realism. A major emphasis was placed on the provision of power and energy, and industry's allocation was down to just over 15 per cent of the budget. The provision for irrigation was also much higher than in previous plans.

The Sixth Plan, though, got off to a bad start, and by 1982 it was clear that changes would have to be made. An article in the *Eastern Economist* argued that:

> The Sixth Plan is dead, killed by a resources failure. After suffering a series of coronary attacks on this account — through inflation, inefficiency, rising defence bills and falling aid — the patient has given up the ghost.
>
> (*Source: Eastern Economist*, vol. 78, no. 1, p. 4, 1982)

Annual plans were introduced within the Five Year Plan framework, and on 14 January 1982 Mrs Gandhi announced the introduction of a Revised 20-Point Programme. This programme paid particular attention to increasing agricultural output, providing basic needs to the rural poor, streamlining industrial production, and tightening up on corruption. However, only one item, Point 12, was specifically concerned with environmental issues. It called for the vigorous pursuit of afforestation, social and farm forestry together with the development of biogas and other alternative sources of energy.

THE SEVENTH PLAN

By 1984 some of the priorities of the Seventh Plan (1985–1990) had become clear, although its economic aspects were given something of a political gloss by Mrs Gandhi in election year. The Seventh Plan will concentrate on the further improvement of agricultural yields, more efficient use of existing capital investment, and increasing employment. It is anticipated that the plan will aim for just over a 5 per cent annual growth rate in GDP. Particular emphasis will be given to electricity, power, irrigation, fertilisers, tea, oil exploration and refining, transport and telecommunications, and general poverty alleviation programmes.

As the above summary has illustrated negligible interest was paid in the first five plans to environmental factors:

> Very little attention has been paid to the proper management of our land and soil resources with the result that they have suffered very serious degradation.
>
> (*Source: Sixth Five Year Plan*, 1980–1985)

This is not, however, the same as saying that the policies followed within the first five plans had a negligible impact on people–environment relations. As the remainder of this chapter will illustrate, people–environment relations have indeed changed appreciably over the last 30 years. While it is difficult to assess how much they would have changed without the implementation of the particular planning policies chosen, it is nevertheless possible to identify at least some changes which can be attributed to the Five Year Plans.

Table 3, *page 243*, presents an overall summary of the achievements of India's Five Year Plans. This clearly illustrates agriculture's relatively poor performance, particularly during the Third Plan and also the very low overall percentage increase in per capita consumption.

Table 3 *Percentage growth rates in Indian Five Year Plans*

% Annual Growth Rates	*1st Plan 1951/52–1955/56*	*2nd Plan 1956/57–1960/61*	*3rd Plan 1961/62–1965/66*	*Annual Plans 1966/67–1968/69*	*4th Plan 1969/70–1973/74*	*5th Plan 1974/75–1978/79*	*Overall 1950/51–1978/79*
National Income	3.6	4.0	2.2	4.0	3.3	5.4	3.5
Agricultural Production	4.1	4.0	−1.4	6.2	2.9	4.2	2.7
Industrial Production	7.3	6.6	9.0	2.0	4.7	5.9	6.1
Per capita consumption	1.7	1.8	0.1	2.0	0.4	2.3	1.1
Gross fixed investment	3.0	5.8	8.7	1.5	3.1	6.6	5.5

(*Source: Sixth Five Year Plan 1980–85*, Government of India Planning Commission, 1981)

? The class should divide in half. Using the data in Table 3 and the following quotations, one half should argue for the overall success of planning policies to date and for their continuation into the future, while the other half should argue against them. This debate could be imagined to be taking place in Parliament.

It is now widely accepted that planning in this country has been a charade almost since the start of the Annual Plans in 1966–67. Since then, planning has come to mean largely the setting out of budgetary allocations under various heads. The imperatives of genuine planning are such that existing socio-economic institutions and vested interests would have to be shaken up.

(*Source:* Charvaka in *Economic and Political Weekly*, November 1977)

The launching of the Sixth Five Year Plan synchronises with the beginning of a new decade during which our major goal should be the realisation of an economic and social order based on principles of socialism, secularism and self-reliance. During the 30 years of planned development which we have just completed, we have made impressive progress in developing agriculture and industry, science and technology, health and education and the infrastructure for a wide range of services. Technological change in agriculture has led to improved production and productivity, and a great stability in the output of several crops. The steps taken during the Fifth Five Year Plan period for building a national food security system enabled us to face the widespread drought of 1979 without food imports. Wide ranging advances in industry have enabled us to be self-reliant in many critical areas. We now manufacture an impressive array of both capital and consumer goods. At the same time, progress in the development of new skills and technology in the small and village industries sector has also been considerable. The export base of the economy has been widened and strengthened. In science and technology, our scientists have shown that they can achieve well-defined goals as has been demonstrated in agricultural, nuclear and space research. We can thus enter the new decade with confidence in our capability to face and solve successfully the complex problems confronting the different sectors of our economy.

(*Source: Sixth Five Year Plan 1980–85: A Framework*, 1980)

The Government of India institutionalised its alliance with big business and launched 'planned economic development', the two applying their combined mighty organisational strength reinforced by foreign governments and foreign big business.

The result has been a largely self-contained expansion characterised by a 'suspension pyramid', controlled by the magnetic pull of the power elite at the top, drawing resources from the base, and principally serving the top layers.

(*Source:* Krishna Murti in *Economic and Political Weekly*, May 1967)

Because of her natural resources, India has considerable potential for industrial growth. She has extensive known reserves of iron ore, manganese, bauxite, coal and atomic materials such as thorium ores. Surveys and exploration have already indicated the prospects of oil reserves. There is a large potential for hydro-electric power.

(*Source: Third Five Year Plan*, 1961)

Having provided an overall summary of the main thrusts of Indian planning over the last 30 years, it is now possible in the next section to look more closely at how people–environment relations have changed, firstly in the agricultural sector and then as a result of industrial change.

AGRICULTURE

Agriculture has been India's success story since the mid-1960s . . . The symbols of agriculture's new dawn are the thousands of inappropriately drab, windowless buildings that are dotted around the Indian countryside. They are storing surplus crops in a country that for 20 years and more had to import on average 5 per cent of the grain it needed. In 1966 and 1967, both years of disastrous drought, imports went up to 14 per cent and 12 per cent of the total. Yet in 1979–80, when the monsoon failure was almost as bad, India did not import a gram of grain and still ended the year with stocks of 14m tonnes – 10½ per cent of 1978–79's record 131 million tonne harvest.

(*Source*: 'India: Treadmill or Take-off? A Survey', *The Economist*, 28 March 1981)

This is an optimistic view of India's agricultural 'progress' over the last decade. The Sixth Five Year Plan, which aimed to increase agricultural production by 4 per cent per annum, presents a more sober picture:

Agricultural production during 1967/68–1978/79 has grown at an annual compound rate of 2.8 per cent. The task is one of moving on to a new trend line and this will require a determined effort considering that it will involve programmes extending to difficult crop segments and also difficult agro-climatic regions and locations. Moreover the increase has to accrue mainly from a substantial step-up in the yield levels and from multiple cropping in view of the virtual exhaustion of the scope for increases in the net cropped area. This is apart from the problem of inter-year fluctuations and the need to stabilise yield levels and production performance.

(*Source: Sixth Five Year Plan 1980–85: a framework*, 1980)

THE 'GREEN REVOLUTION'

As Tables 3 and 4 indicate, agricultural production has picked up appreciably since the devastations of the mid-1960s. As was mentioned in the previous section, this was largely through the application of new technology, known as the 'Green Revolution'. In essence, the 'natural' agricultural environment, with all its associated risks of rainfall variability and pest attack, was to be changed to a man–controlled environment, where special high-yielding varieties of crop were to be irrigated with water from new wells, fertilised with new inorganic chemical compounds, and protected from pest attack by a variety of pesticides. The achievements of the past decade in terms of output levels have indeed been impressive.

Since the end of the Third Five Year Plan, fertiliser consumption, irrigation and the use of high yielding varieties of seed have increased dramatically, and average yields of rice and wheat have doubled. Increased control of the environment has enabled people to increase food production, and as the quotation at the beginning of this section noted, India now almost has a food surplus.

However, access to these new agricultural inputs, at least in the

Table 4 *The impact of the green revolution*

	1950/51	*1955/56 End of 1st Plan*	*1960/61 End of 2nd Plan*	*1965/66 End of 3rd Plan*	*1966/67*	*1967/68*	*1968/69*	*1973/74 End of 4th Plan*	*1978/79 End of 5th Plan*
% gross irrigated area to gross sown area	17.1	17.4	18.3	19.9	20.8	20.3	22.2	23.7	28.0
Fertiliser consumption (kg/ha)	0.5	0.9	1.6	5.1	7.0	9.4	11.0	16.7	29.5
% rice area under high yielding varieties					2.5	4.9	7.3	26.1	41.8
% wheat area under high yielding varieties					4.2	19.6	30.0	59.2	71.1
Average yields:									
Rice (quintals/ha)	6.7	8.7	10.1	8.6	8.6	10.3	10.8	11.5	13.3
Wheat (quintals/ha)	6.6	7.1	8.5	8.3	8.9	11.0	11.7	11.7	15.7

(*Source: Sixth Five Year Plan 1980–85*, p. 11, Government of India Planning Commission, 1981)

1960s and 1970s, was limited. The High Yielding Varieties Programme was first instigated in the parts of the country, such as the Punjab and Haryana, where yield increases were expected to be greatest. This immediately led to greater spatial inequality at the national scale than had previously existed. More importantly, though, countless studies have illustrated how it was the richer, better educated and younger farmers, who already had large farms, who benefited most from the new technology. It was they who could afford to adopt these agricultural innovations; it was they who had the education to be able to use the new methods required. The 'Green Revolution' can be seen as having polarised society. The increased domination of the environment by some people has enabled them to achieve greater domination over many of their fellows. Before moving to the next section carry out 'The Green Revolution' exercise.

THE EFFECTS OF THE GREEN REVOLUTION

While there are numerous indications that the Green Revolution has exacerbated social inequalities, it is also clear that the introduction of the new technology has had an important impact on the environment. This can be assessed by studying together the four main inputs: crop varieties, water, fertilisers and pesticides. Table 4 (*page 244*) well illustrates the point that the crop which received most attention in government policy was wheat, and at the end of the Fifth Plan, just over 71 per cent of all the wheat area in India was under the new high yielding varieties. These new varieties of crop have the following main characteristics:

- They respond well to increased plant nutrients.
- They are normally dwarf varieties so that more nutrients go into the grain rather than the remainder of the plant, and so that they are more resistant to rough handling by the weather and by reapers.
- They include more grain bearing heads.
- They can be made to produce a crop within a specific number of days.

THE GREEN REVOLUTION

1 Basic scenario
Assume the class to be a village in rural India. Divide into four categories as follows:

A 10 per cent of class: large farmers; 75 per cent have access to GRP; all require two labourers.
B 20 per cent of class: medium-sized farmers; 50 per cent have access to GRP and require one labourer each.
C 40 per cent of class: small farmers; 25 per cent have access to GRP.
D 30 per cent of class: landless labourers.
(If you are in a small class or group, these percentages can be rounded off.)
GRP = Green Revolution Package: high yielding varieties of seed, fertilisers, pesticides and wells for irrigation.
Trad = Traditional varieties of grain and no new inputs.

Unless otherwise stated, all farmers are assumed to grow traditional grains and to have no other inputs. Labourers must sell their labour in return for food; if they are unable to they starve.

Once the class has divided into farming categories, a short while (5–10 minutes) is allowed for labourers to find employment. The Organiser then decides on the type of farming year: good, drought, pest attacks, or drought and pests. The crop yield for that year for each type of farm is then found from the Table below:

		Good year	*Drought or pest*	*Drought and pest*
A	GRP	+	+	0
	Trad	+	0	–
B	GRP	+	+	0
	Trad	0	–	–
C	GRP	0	0	0
	Trad	0	–	–

\+ = surplus enabling labourers and farmer's family to be fed.
0 = sufficient grain to feed farmers and their families (but *not* labourers) and to sow grain next year.
– = deficit: must buy, borrow, beg, steal food and seed grain for next year.

The outcome for each farm is then determined, and the village decides what to do if certain farms are deficit.

2 Development of exercise
The above scenario illustrates one year's cycle in a 'village' with some new technology. Moving on from here it is suggested that the following procedure is followed:
a For the first year assume that no farmers have GRP, but that the social division of the village, in terms of farm sizes, is the same as in the basic scenario. Run the exercise for both good and bad years. Discuss who has survived and what remedies might be implemented to overcome crises in bad years.
b Then run the exercise as in the basic scenario with some farmers having GRP, for either good years, droughts, or pest attacks. After each year a number of alternative options could be introduced such as:

i If a farmer is in deficit one year, he could become a landless labourer the next year
ii if three small farmers (category **C**) decide to form a co-operative then they can be treated as a medium farmer with access to GRP.

c After each year, discuss who has benefited and why.

It should be noted that this exercise assumes that if farmers have any new technology they have the complete package. In reality farmers frequently do not have access to all of the requirements. If pesticides are not available, or water and fertiliser are not applied in the correct amounts at the right time, the yield of the new seed varieties can in fact be *lower* than for the traditional varieties.
(The ideas in this game are derived from Chapman, G., Douler, E., and Payne, P., *The Green Revolution Game*. This game is available from Marginal Context, Cambridge.)

New varieties of wheat and rice by themselves would not have led to the improvements that occurred, and a major characteristic of the implementation of the High Yielding Varieties Programme was that the new inputs had to be introduced together in a package. Thus the new varieties of grain had to be given water and fertiliser at the correct times and in the correct amounts to achieve maximum results. Unfortunately in the early stages this was not always achieved, and production was often well below the possible maximum. In addition, the new varieties were also more susceptible to pest attack, giving rise to the need for an increased use of pesticides. In order to provide sufficient and regular water, major irrigation projects were established and tube wells, often powered by electricity, were also introduced to replace the old traditional hand-dug wells. By 1979–80, the government estimated that it had created the potential for 56.7 million hectares to be irrigated. Table 4 also indicates the rapid increases that occurred in fertiliser application, which consisted mainly of nitrogen, phosphorous and potassium nutrients. Much of this was derived from hydrocarbon feedstocks, and the oil price rises that took place in 1973–74 and 1978–79 placed an added burden on India's agriculture, which by the late 1970s had come to rely heavily on the fertiliser and pesticide part of the package to sustain yields.

The impact of these changes on people-environment relations has been appreciable. One of the reasons for the concentration on intensification of agricultural production and thus the introduction in the mid-1960s of the High Yielding Varieties Programme, rather than on extensification, was that it was argued that little new land was readily available for agricultural production. The new varieties, as well as producing more grain through their good response to fertilisers, also enabled double cropping to take place, thus further increasing production. This has therefore required a greater input of labour throughout the year, and has as a result somewhat reduced the problem of the seasonality of labour demand. The quotation from the Sixth Five Year Plan framework, given at the beginning of this agricultural section, nevertheless emphasises the extent to which further developments in agriculture have to concentrate on extending agriculture to even more difficult environments through the expansion of irrigation, and also to concentrate on multiple cropping and the better management of farms. As the framework says:

> There has to be greater emphasis on improved farm management with as much attention being paid to non-cash inputs as to cash inputs. In particular group/community endeavour in the areas of water management, pest control and post-harvest technology will have to be promoted.
>
> (*Source*: *Sixth Five Year Plan 1980–85: A Framework*, 1980)

The Sixth Plan period also suffered from a number of relatively poor harvests, particularly in 1982–83, due largely to low and erratic rainfall. By the end of the Plan agricultural production prospects revived as the rainfall in January 1984 was well above average, and food grain production nearly reached 150 million tonnes for 1983–84, compared with levels of around 130 million tonnes in each of the preceding three years. This gave rise to headlines in the press such as:

> **RAIN GODS SMILE ON FARMERS AND PLANNERS**
>
> **The first really good monsoon for five years has put the elusive production targets of the Sixth Plan well within reach**
>
> (*Source*: *The Financial Times*, 11 June 1984)

This illustrates how Indian agriculture is still subject to the vagaries of the weather and the environment. Nevertheless, 48 of the 65 major irrigation projects of the Sixth Plan were behind schedule in 1984, and 1983–1984 saw almost a trebling of fertiliser imports compared with the previous year.

UNDESIRABLE EFFECTS

There have also been other, more immediately dramatic, impacts of this increased agricultural activity on the environment. Two of the most significant of these have been the increase in soil erosion and waterlogging.

Soil Erosion

Ever since the First Five Year Plan there has been a section within each plan on soil conservation, and until the Sixth Plan this was the only direct environmental consideration within government planning. The First Plan saw soil erosion as being caused by the destruction of the forests and faulty farming practices, and in the Second Plan 200 000 000 rupees were set aside to redevelop 3 million acres where there was severe soil erosion. The Third Plan stated:

One of the principal reasons for low productivity in agriculture in certain parts of the country is the progressive deterioration of soil due to erosion.

The Third Plan also estimated that 200 million acres, approximately one-quarter of the country's surface, was suffering from the effects of soil erosion. Despite the injection of large sums of money into areas suffering from soil erosion, the pressure of population on the land has meant that the farmers have had to try to squeeze more and more production out of that land, and consequently soil erosion has, in many places, increased. Recent reports from the Ministry of Agriculture estimate that there are now 150 million hectares of land that have become unusable as a result of wind and water erosion — this is reported to be more than the entire cultivated area of India.

Waterlogging
This is a result of the vast increases that have taken place in irrigation, and is now also seen as a growing environmental problem. Thus a report in 1983 stated:

Irrigation can, however, be a mixed blessing. Waterlogging is a big problem if the system is badly planned, and is not easily remedied. Reclamation through drainage is an expensive business; newly waterlogged lands are often left to rot. Also by raising the water table, irrigation can cause salination. Water pulls with it salts from lower layers of soil and then evaporates, leaving the salts concentrated on the surface. Plants do not take to salty soils. There are no precise calculations of the area of land in India affected by salination and waterlogging but the problem is widespread. Mr B. Vohra, chairman of India's National Committee on Environmental Planning, reckons that, of the 40 million hectares under irrigation, 10 million hectares are in danger of becoming infertile.
(*Source: The Economist*, 5 March 1983)

Print 5 *Thirty years ago, Akbarpur-Barota was surrounded by forest. Today the scene is one of deforestation, erosion, soil loss and flooding*

Overall, agricultural change in the Indian context has seen the attempted replacement of a 'natural' agrarian environment, subject to the vagaries of the weather, pest and variable soils, by a man–made environment in which, theoretically, a higher and more constant food output may be achieved. The intensity of use of the environment has undoubtedly increased, as has food output, but, because access to the necessary environmental control elements is limited, this has also exacerbated the potential for increased social inequality.

?

1 Figure 1 (*page 238*) indicates the main agricultural crop regions of India. On the basis of the above section, remembering that the High Yielding Varieties Programme initially concentrated only on rice and wheat, write a short account describing the parts of India where the programme would have been expected to have had greatest impact.

2 Table 4 (*page 244*) indicates the percentages of the wheat and rice area that were under High Yielding Varieties at different dates, and also the rice and wheat yields measured in quintals per hectare (1 quintal = 100 kg). Tables 5 and 6 illustrate the overall production of India's main crops and the area sown with them. Within the areas of India already noted in question 1 of this exercise as being subject to the High Yielding Varieties Programme, write a short account stating which areas are likely to have had greater increases in yields. To emphasise the differences between rice and wheat you may find it useful to draw up the information in Tables 4, 5 and 6 on graphs.

3 Figure 5 (*page 254*) provides a generalised map of population density. Write a short essay comparing the above information on the likely spatial impact of the High Yielding Varieties Programme with the distribution of population. Identify areas which have high population densities and yet which were not likely to receive the benefits of the High Yielding Varieties Programme. Likewise, identify areas of low population density which can be expected to have had high increases in yields. What are the implications of this for the requirements of the transport network to move food from some parts of India to others?

Table 5 *Production of principal crops ('000 tonnes)*

	1950–51	*1960–61*	*1970–71*	*1978–79*
Rice (cleaned)	22 400	35 153	42 900	54 630
Jowar	6 350	10 057	8 235	11 623
Bajra	2 722	3 338	8 157	5 659
Maize	2 394	4 181	7 606	6 299
Ragi	1 375	1 906	2 189	3 251
Small millets	1 804	2 042	2 020	1 920
Wheat	6 931	11 171	24 213	36 078
Barley	2 558	2 856	2 828	2 174

Table 6 *Area under principal crops ('000 hectares)*

	1950–51	*1960–61*	*1970–71*	*1978–79*
Rice (cleaned)	30 810	34 128	37 592	40 480
Jowar	15 571	18 412	17 374	16 140
Bajra	9 023	11 469	12 193	11 390
Maize	3 159	4 407	5 852	5 760
Ragi	2 203	2 515	2 472	2 710
Small millets	4 605	4 955	4 783	4 400
Wheat	9 746	12 927	18 241	22 640
Barley	3 113	3 205	2 555	1 830

(*Source* for both tables: *India: a Reference Annual 1981*, Delhi, Ministry of Information and Broadcasting, 1981)

INDUSTRY

Since agriculture accounts for about half of India's GNP, its continuing success is felt throughout the economy. A good harvest brings a surge in industrial demand some six to nine months later, because farmers are big customers — for inputs like fertilisers and irrigation pipes as well as finished products like clothes and radios.

The trouble is that industry does not return the compliment . . . The impetus that agriculture's success has provided is not being carried through.

(*Source*: 'India: treadmill or take-off?: A survey', *The Economist*, 28 March 1981)

India's main problem of air pollution is due to the fact that 80 per cent of her industrial production is concentrated in eight or ten large industrial centres forming isolated pockets. Industrial chimneys, power houses, burning of fuels and auto exhausts emit such pollutants into the air as suspended matters like smoke, dust and sprays, SO_2, CO, nitrogen oxides, fluorides . . .

(*Source*: 'Environmental Problems — a Bird's Eye View', Ravi Sharma, *Current Trends in Indian Environment*, 1977, New Delhi)

A short section on India's industry can do no more than highlight some of the major impacts of development planning on this area of the economy. Nevertheless, here too, the relationships between people and the environment have changed appreciably, with the last 30 years seeing a far greater concentration placed on heavy industry and mineral extraction, with little regard to the impact of this on the environment. An example of the creation of an industrial wasteland can be seen in the fact that in the late 1970s roads across parts of the coalfield in the Damodar valley in Bihar carried signs stating 'road under fire' — the underground coal seams were steadily burning away.

INDUSTRY AND THE FIVE YEAR PLANS

The Second Five Year Plan set the scene for the future of India's industry with its fundamental concentration on the heavy sector. Heavy industries were going to be run by the state, and would receive the lion's share of government investment. Manufacturing industry was to concentrate mainly on the domestic market in an attempt to reduce imports to a bare minimum. The industries specifically to be the responsibility of the National Government were categorised as Schedule A industries, and included coal mining, iron and non-ferrous mineral extraction, petroleum, the iron and steel industries, heavy engineering, machine tools, power, atomic energy, and air and rail transport. Schedule B industries, which were to be shared by the government and private enterprise, included chemicals, aluminium, fertilisers, and road and sea transport.

As can be seen from Table 2 (*page 240*), industrial production did increase appreciably during the first three Five Year Plans, averaging a growth rate of 7.6 per cent per annum overall. However, since 1966 the annual growth in industrial production has only averaged 6.1 per cent. At independence, India's industrial structure was dominated by textiles, but since the 1950s heavy industry, in the form of iron and steel, engineering and chemicals, has grown greatly in importance. Table 7 illustrates the pattern of growth in a number of Indian industries, and from this the relative stagnation of the textile sector is readily apparent.

Table 7 *Progress of selected industrial production*

	1950–51	*1960–61*	*1965–66*	*1970–71*	*1976–77*	*1979–80*	*1979–80 / 1950–51*
Coal (100 000 tonnes)	328.0	557.0	703.0	743.0	1 048.0	1 063.0	3.24
Steel ingots (100 000 tonnes)	14.7	34.2	65.3	61.4	86.6	80.3	5.46
Aluminium (000 tonnes)	4.0	18.3	62.1	166.8	208.7	191.9	47.97
Bicycles (000)	99.0	1 071.0	1 574.0	2 042.0	2 677.0	3 837.0	38.75
Refined petroleum products (100 000 tonnes)	2.0	58.0	94.0	171.0	216.0	255.0	127.5
Nitrogenous fertilisers (000 tonnes of nitrogen)	9.0	98.0	233.0	830.0	1 900.0	2 226.0	247.33
Cement (100 000 tonnes)	27.3	79.7	108.2	144.0	188.0	175.0	6.41
Jute textiles (000 tonnes)	837.0	1 097.0	1 302.0	958.0	1 186.0	1 355.0	1.62
Cotton cloth (10 000 000 metres)	421.5	637.8	744.0	759.6	810.5	837.0	1.99
Electricity generated (Public utilities only; 10 000 000 KWH)	530.0	1 690.0	3 300.0	5 580.0	8 920.0	10 541.0	19.89

(*Source: India: a Reference Annual, 1981*, Delhi: Publications Division, Ministry of Information and Broadcasting, Government of India, 1981)

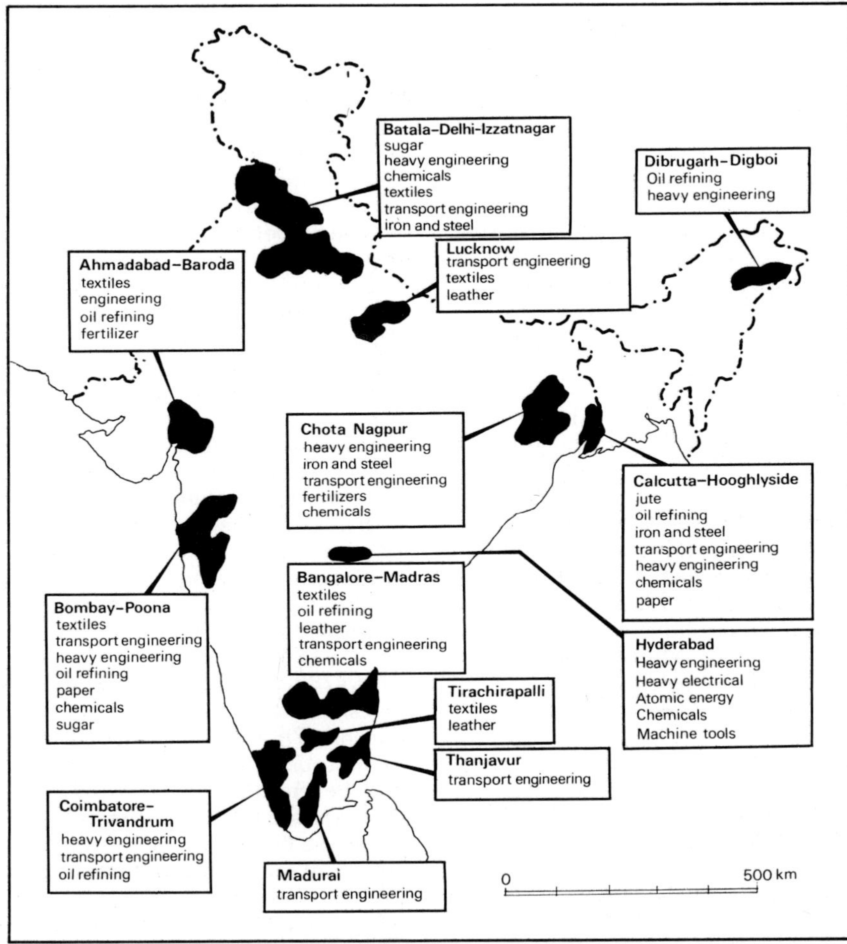

Figure 2 *India: main industrial regions*

? Depict the information shown in Table 7 graphically. Of the ten industries, which three grew most rapidly in the 1950s, the 1960s and the 1970s? Which three grew least rapidly over each decade?

THE INDUSTRIAL REGIONS

In spatial terms there are five main industrial regions in India:

Ahmadabad–Baroda in the north-west, dominated by textiles;

Bombay–Poona on the west coast, with an emphasis on cotton and engineering;

Madurai–Coimbatore–Bangalore in the south, also concentrating on cotton and engineering;

Chota Nagpur astride the Bihar–Orissa border, dominated by mining and metals;

Calcutta–Hooghlyside in the north-east, the traditional home of the Indian jute industry and also an important engineering centre.

Figure 2 indicates the location of these major zones and also a number of other less important industrial centres. In studying this map it must be remembered, though, that smaller clusters of industry are to be found elsewhere in India.

THE PROBLEMS OF INDIA'S INDUSTRY

Despite the government's concentration on public sector heavy industry it is this sector that has failed to produce the requisite levels of growth.

> India's nationalised industries today employ 4 million people, hold three-quarters of the country's industrial assets, contribute only one-third of industrial output and make a thumping loss.
>
> (*Source*: 'India: Treadmill or Take-off?: A Survey', *The Economist*, 28 March 1981)

Part of the explanation for this recent failure of industry lies in the constraints imposed upon it by three main factors: a shortage of power, massive trade deficits, and transport problems (particularly on the railways).

Print 6 *Woman working at a brick works in Poona*

Power and Energy

Tables 1 and 2 (*pages 239 and 240*) illustrate that until the Sixth Plan relatively little attention was paid to power and energy in the planning process. Certainly the installed capacity of electricity did increase appreciably over the period, as indicated in Table 7 (*page 249*), but the demands for power increased even more rapidly. India's concentration on capital intensive industries, which were by nature energy intensive, greatly increased the demand for power without increasing the income available to pay for it.

Table 7 also illustrates that while the decades 1950–60 and 1960–70 saw public utility electricity generation rising by 319 per cent and 330 per cent respectively, this rate of increase fell dramatically to 189 per cent in the decade up to 1980. To make matters worse, power stations, and in particular the thermal coal-fired ones which supply over 60 per cent of India's electricity, are currently running well below capacity. This is partly due to problems encountered in increasing coal supply, but it is also due to poor maintenance and low prices. While these problems of underutilised capacity are not so serious in the states where hydroelectric power is the dominant source of electricity, as in Himachal Pradesh, Jammu and Kashmir, Karnataka, Kerala and Meghalaya (see Opener), in the remainder of the country problems of power supply have severely hampered industrial expansion. As a result the Sixth Plan paid considerable attention to increasing power capacity, which is planned to rise over its five-year period by 19 666 MW. This represents a considerable task when it is noted that installed capacity at the end of September 1982 was 32 938 MW. The increases in power supply are now somewhat below planned levels.

Trade Deficits

Trade deficits have also severely hampered industrial growth, and during the late 1970s and early 1980s the position worsened considerably. While exports increased 4.4-fold over the decade 1970–71 to 1980–81, imports increased 7.5-fold. This situation was undoubtedly made more serious by the rapid increases in world oil prices that took place in the 1970s — 35 per cent of India's imports are currently accounted for by oil. Nevertheless, India's concentration on import substitution in the early stages of planning has meant that it has not been easy to step up exports. In an attempt to overcome the constraint imposed by oil prices, hydrocarbon exploration is being increased, and the oil sector appears to be one part of the economy where production levels will be higher than those proposed in the current plan.

The railways

Although India's transport network has indeed been undergoing modernisation, this has not kept up with the increased demands being placed on it. In 1960–1965 the railways received 9.6 per cent of India's total investment, but since then there has been a decline in their relative position as far as investment is concerned and there has been a similar decline in the percentage of the freight and passenger traffic carried by the railways. Thus in 1975–78 only 2.3 per cent of total investment went to the railways. At the end of the 1970s, only 10 per cent of the country's track was electrified, and the overall system consists of three separate gauges. The railways have also been plagued by poor industrial relations and disruptions.

SMALLER INDUSTRIES

While the bulk of government

attention has thus been paid to heavy industries (see Tables 1 and 2) small-scale industry has nevertheless also continued to expand. Government figures record that between 1973–74 and 1979–80 the value of production in the small-scale sector rose by 265 per cent and employment in this sector grew by 163 per cent. The increasing inability of heavy large-scale industry to provide sufficient employment has, though, meant that more attention has had to be given to the village and small industry sector in recent plans. The framework for the Sixth Five Year Plan argues as follows:

> The plan will give a high priority to the speedy development of small, tiny and village industries with a view to enhance employment opportunities in these industries on a large scale. Existing traditional industries will need to be revitalised and their productivity raised by upgradation of skills and techniques. A positive effort will be made to disperse these industries over a wider area, particularly in the rural and semi-urban areas.
>
> (*Source*: *Sixth Five Year Plan 1980–85: A Framework*, 1980)

THE ENVIRONMENT AND INDUSTRIAL CHANGE

What then has been the impact of this industrial change on people–environment relations in India? In attempting to answer this question it must be remembered that in 1980 agriculture still contributed 35 per cent of GDP and employed about 70 per cent of the labour force. Industrialisation has therefore only directly influenced a relatively small percentage of the population and a small percentage of the land area of India. Nevertheless, where industries have developed, as in the Damodar valley in the Chota Nagpur industrial region of the north-east, or in the major cities such as Bombay, the local impact has been appreciable. It is difficult to obtain accurate figures on industrial pollution, but it seems that the Prevention of Water Pollution Act of 1974, which was introduced to attempt to restrict the dumping of industrial effluent, has been ineffective. A report on Bombay's river, the Kalu, well illustrates the current situation:

> Industrial wastes pour into it from 150 factories, depositing, among other things, high levels of mercury. The city's outskirts are being compared to Minamata, the village in Japan where 300 people were crippled, paralysed, or killed by mercury poisoning in the sea — Bombay's Institute of Science says mercury levels in the Kalu are similar to those recorded in Minamata Bay, and people are suffering from some of the symptoms of the Japanese victims.
>
> (*Source*: *The Economist*, 5 March 1983)

Elsewhere, too, industrial pollution is significant, although mainly local in its impact. In contrast to the widespread deterioration of the environment as a result of changes in agricultural practice, the direct environmental impact of industrial change has been less extensive. Nevertheless, mineral extraction has led to scarred landscapes; the hillsides and streams around the town of Noamundi in South Bihar are stained red with iron ore. Likewise, around most of the cement factories constructed in the last 30 years there is a swathe of white dust covering fields, trees and buildings.

Many of India's new industrial works have been constructed away from traditional urban concentrations of population, in the hope that they would act as growth poles. The expected positive feedback effects, though, have frequently not taken place, and as a result, particularly in parts of northern India, it is quite common to find factories situated in the middle of wide expanses of agricultural land. The impact of air pollution from these factories is thus made especially severe, in that it can seriously reduce agricultural yields and damage people's health.

On a larger scale, industrial towns, such as the major steel town of Jamshedpur in Bihar, have an

Print 7 *December 1984 — Uncontaminated water is handed out to survivors of the Bhopal disaster in which 40 tonnes of poisonous gas was released into the air*

important, but again mainly local, impact on the environment in which they are located; ten miles away from the centre of Jamshedpur one can be in as rural, agricultural, a setting as any part of the vast agrarian landscape of India. The government is now beginning to take legislative action in an attempt to stop industrial air pollution. In 1981 the Air (Prevention and Control of Pollution) Act was passed, providing for state boards to be set up to investigate pollution and for fines to be imposed where standards are not met. Many feel though, that, as with the Water Pollution Act, it will prove to be ineffective. Dramatic evidence of the potential for environmental crisis resulting from India's industrialisation was revealed in the methyl isocyanate gas leak in the town of Bhopal at the end of 1984 which killed more than 2 500 people and injured over ten times as many. The disaster hit the international newspapers:

BHOPAL NUMB WITH SHOCK AS BODIES PILE UP

(*Source*: *The Guardian*, 12 December 1984)

BRITISH SPECIALISTS HELP TREAT BHOPAL GAS VICTIMS

(*Source*: *The Times*, 11 November 1984)

The disaster revealed the extent of the involvement of multinational companies in countries such as India, for the pesticide plant from which the gas leaked was owned by the US-based Union Carbide Company.

The pressure for *rapid* industrial and agricultural expansion has largely been fuelled by India's ever increasing population, and it is therefore to the demographic aspect of the people–environment equation that attention will be turned in the next section.

Imagine that World Bank officials are meeting in the USA to consider a recent batch of applications for loans. They sometimes find it hard to reconcile their thinking as bankers with the actual needs of developing countries. Should they consider only the need for aid and risk the disapproval of the countries who supply their money for lending, or should they look for maximum profit and the least risk of losing their capital?

You know from reading the previous section what India's needs are. Adopt one of the roles below and then, after a class debate, take a class vote. Write a paragraph explaining the class decision and the reasons for it to the chief of the Bank of Chicago and another paragraph to the head of Oxfam.

Chairperson
On the agenda are two requests for loans from India. Only one can be granted:
1 To build a new hydroelectric power station.
2 To build small factories around the sub-continent manufacturing small hand water pumps for village water supply.
Official A You are a banker who has been to India and has shares in a huge plant which manufactures goods for hydroelectric plants. Ask why they want another plant and why they want their own people to run it when plenty of American 'know how' exists.
Offical B You are in favour of lending for the power plant, but insist on high interest rates and a prompt return of the money. You must consider the welfare of the banks who lend to the World Bank.
Official C You are in favour of the hydroelectric plant and consider training Indians to run it to be of great importance. Eventually they may build the next plant without outside aid.
Official D You are accustomed to big business deals and find the request for small factories to make hand pumps trivial and humorous. Suggest this is beneath the notice of the World Bank.
Official E You have visited similar factories in some Latin American countries. Explain that the policy is to encourage the development of local technology. You are enthusiastic about making the loan.
Official F You are sympathetic to the proposal on hand pumps but suggest quicker results would be obtained by importing from the States or Europe.
Official G You have shares in a factory making the machinery for setting up hand pump factories. You are in favour of making a very low cost loan.
Official H You are pessimistic about the local population across the continent learning the skills needed for the hand pump factory and suggest one large factory in Calcutta.

? Write an essay on one of the following:

1 What have been the main sectoral and spatial implications of planning in India since 1947?

2 Compare and contrast the impact of Indian planning on industry and agriculture.

3 Illustrate the main problems facing those planning Indian industry in the late 1980s.

POPULATION AND THE ENVIRONMENT

The need for human 'population control' is no longer a seriously debated question among the scientific community that is bothered about the quality of our environment. There is barely any environmental or conservation problem that is not made worse by the problem of rapidly increasing population.

(*Source*: Shanti Balakrishnan, 'Population pollution?', *Current Trends in Indian environment*, New Delhi, 1977)

Print 8 *A novel way of getting across the message of population control — this family planning elephant in northern India travels from village to village helping to spread the message that two children per family is more than enough for a country that had an estimated population of 700 million in 1985*

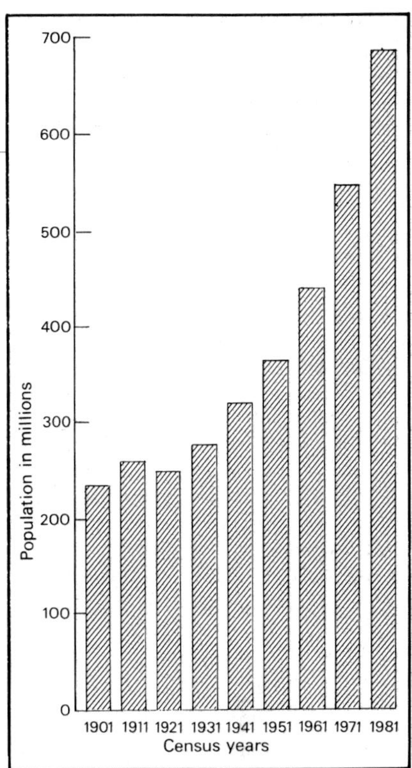

Figure 3 *Population growth, 1901–1981*

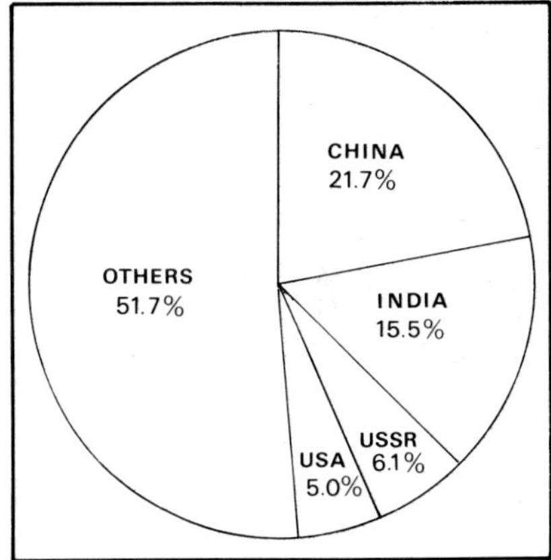

Figure 4 *India's share of the world's population, 1980–1981*

INDIA'S LARGE AND EXPANDING POPULATION

In 1911 India's population was estimated to be 252 million; at Independence in 1947 it was approximately 344 million; and in 1985 it was estimated to be between 700 and 740 million (Figure 3). It is against this background of a very large and ever growing population that all of India's environmental problems and development planning must be evaluated. India has approximately 15 per cent of the world's population on 2.4 per cent of the land (Figure 4). Between 1971 and 1981 its population grew by 136 million — a figure larger than the total population of Brazil, and more than twice the population of the United Kingdom. India's population has continued to grow at this fast rate despite government attempts at population control in the various plans since 1950. The growth rate over the decade 1971–1981 at 24.75 per cent was virtually identical to that of 1961–1971 at 24.8 per cent.

The areas of India with the densest population are the vast plain of the Ganga in the north and the coastal regions both in the west and the east. In addition, the major towns and cities act as dense focuses of population. The percentage of the population that is urban has risen from 17 per cent in 1951 to 24 per cent in 1981. This urban population increment has in itself given rise to major environmental problems. The drawing together of concentrations of population in cities such as Calcutta with a population of over 9 million in 1981 and Bombay with over 8 million has meant that there is an increasingly concentrated demand for, on the one hand, resources such as water, land and food, and on the other for the efficient removal of waste. An idea of the magnitude of the pollution problems faced by Indian towns can be gained from the observation that in the mid-1970s only 188 out of 2921 towns in India had sewerage facilities. Shanti Balakrishnan has even estimated that the accumulation of one year's sewage from Bombay would form another Mount Everest! With an increase in the urban population alone of 46 million between 1971 and 1981 the environmental problems of congestion have become much more severe, both in terms of resource use and the effluent from that use in the form of pollution.

Taking India as a whole, though, it is apparent that, in general, it is the states with the least dense populations, such as Sikkim, Nagaland and Assam, that have the highest population growth rates. Compare and contrast Figures 5 and 6. By contrast, Kerala in the south, the state with the highest population density, has one of the lowest growth rates. This situation also gives rise to environmental problems, since, to some extent, areas which have always had dense populations have evolved ways of managing an approximate balance between people and environment. It is in the less densely populated areas, where shifting cultivation has traditionally been practised and where there is now rapid population growth, that the increasingly intense use of the environment is leading to adverse consequences, particularly in the form of soil erosion. In the major areas of shifting cultivation (which are mainly in the states of Assam, Manipur, Tripura, Arunachal Pradesh, Nagaland and Orissa) the increasing population density has meant that the tribal cultivators have been unable to give the land a sufficient period of rest to enable it to regain its fertility. The destruction of forest vegetation and increasingly permanent agriculture has led to increased soil erosion on the hill slopes, and it has also increased the deposition of silt on lower crop fields gradually decreasing their fertility.

Population pressure has therefore led to growing environmental problems in both the urban and rural context. A survey, entitled *The State of India's Environment — 1982: A Citizens' Report*, produced by the Centre for Science and Environment in Delhi, highlighted a number of subjects of concern relating to the impact of planning on relationships between people and the environment. We can now look at two examples from this — the destruction of forests and the construction of major dams.

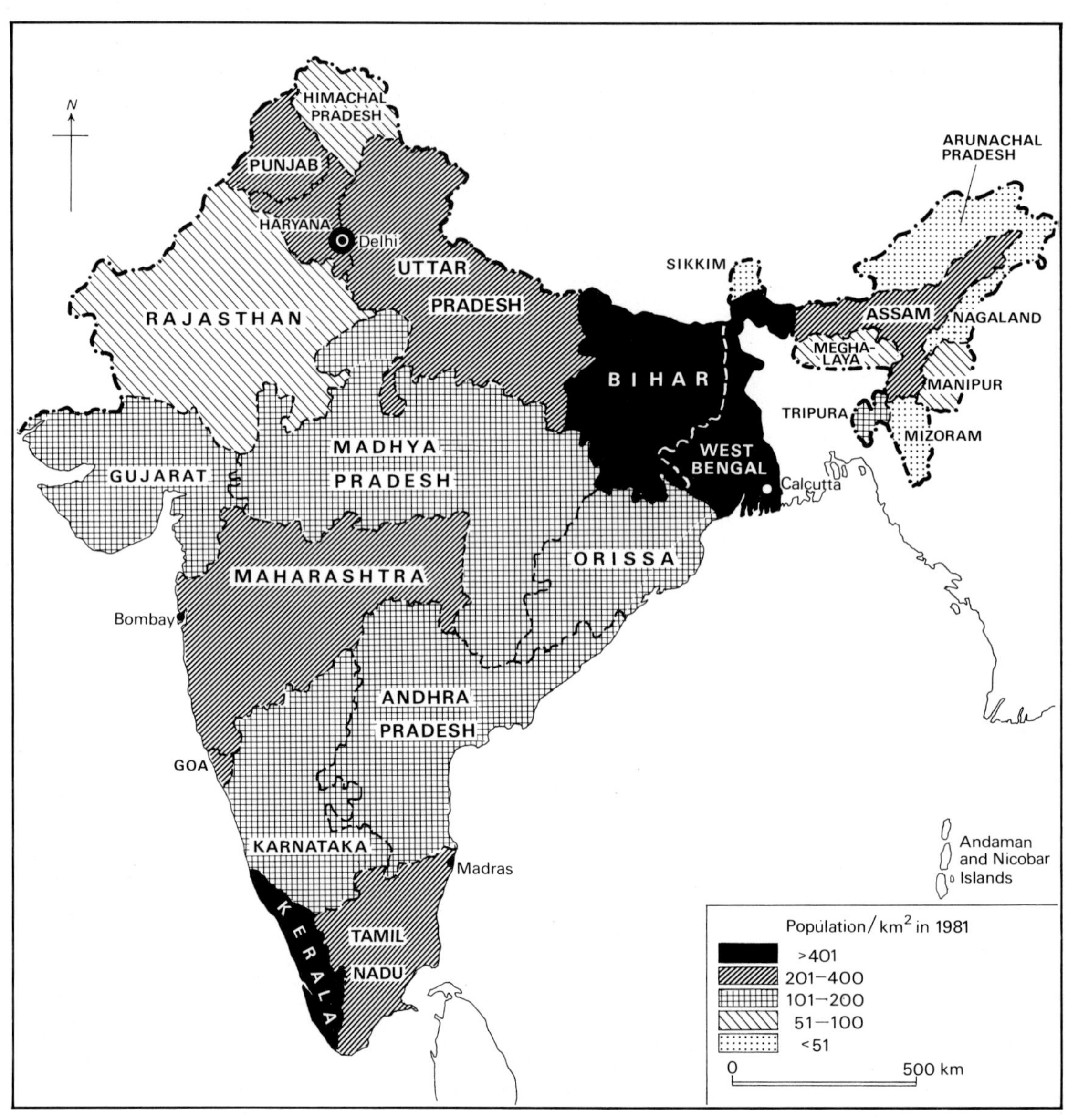

Figure 5 *Population density by state, 1981 (Source: Census of India, New Delhi, 1981)*

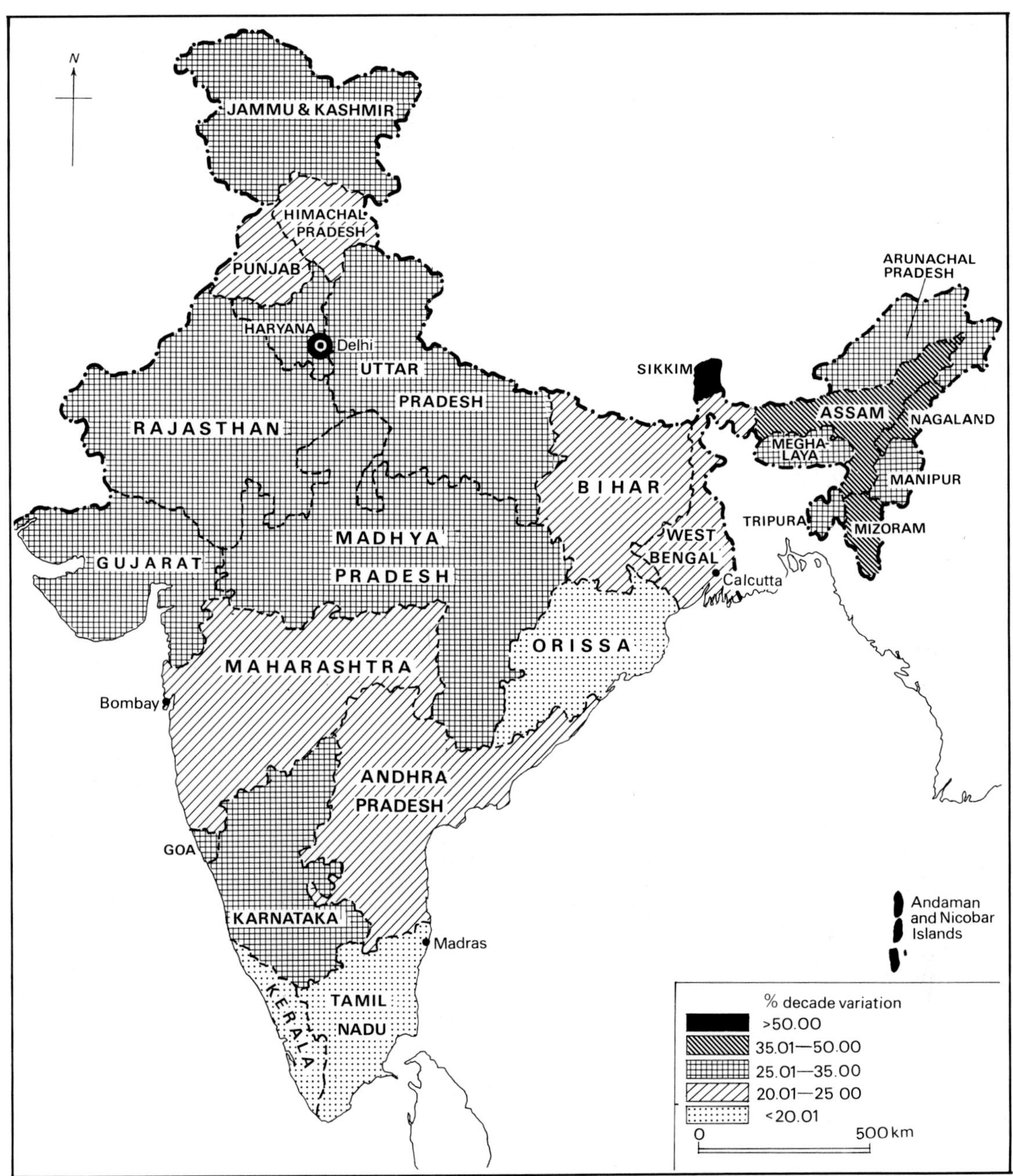

Figure 6 *Decennial population growth by states, 1971–1981*

THE DESTRUCTION OF INDIA'S FORESTS

This has already briefly been mentioned in an earlier section of this chapter (*page 338*). It remains a crucially important issue, with the Forest Department estimating that 3.4 million hectares of forest land were lost to dams, new crop lands and industry between 1951 and 1972. This appears to be an underestimate of the actual extent of the forest loss, but even as these figures stand they represent a physical and human impact of great magnitude. In physical terms the destruction of forests leads to soil erosion, increasing siltation of dams (and thus a reduction in their effective lives), increasing floods, and a depletion of the country's ecological bank.

In human terms, the loss of forest is at least as severe: the main source of fuel for India's poor is wood, and the destruction of the forests means an ever diminishing supply of firewood. In addition forests supply much needed grazing ground for buffaloes, sheep and goats, and this too is disappearing. Impending legislation seems likely to weaken further the position of the poor — the collection of forest produce such as wood and fruit is to be severely penalised, yet wealthy private contractors felling wood for industry will still be able to continue their devastation of the forests.

THE IMPACT OF DAM CONSTRUCTION

By 1979, the government had invested 105 000 million rupees (14 per cent of total planned expenditure) in dams and canals. Their use for irrigation and hydroelectric power has been one of the keystones of India's development planning. While the beneficial increases in irrigated area and electricity output that they have generated have already been noted, it is necessary to mention their adverse effects. In this context there have been four major impacts of dam construction:

1 By raising the water table, dams have led to waterlogging and an increase in soil salinity.

2 In a number of cases the government's rehabilitation programmes for people displaced as a result of dam construction have been inadequate. Notable examples of refugee problems arose as a result of the building of the Sabarnarekha dam in Chota Nagpur and the Srisailam dam in Andhra Pradesh.

3 The provision of new expanses of water has given rise to an increase in diseases connected with water. Reservoirs provide ideal breeding grounds for the malaria-carrying mosquito; elephantiasis (Filariasis) may become endemic; and the risk of the bone-disease fluorosis is often increased.

4 Dams can destroy vast tracts of landscape, and with it the biological diversity of species found therein.

These adverse environmental impacts have all resulted from planning aimed at increasing the standards of living of India's population. The rate of increase of that population and its sheer magnitude have meant, however, that until very recently, consideration of the environment has been omitted from the planning process.

Any attempt at evaluating the actual impact of planning on people–environment relations in India is problematic. If, as has been suggested, it is the immense scale of the population increase that has occurred in India that has largely been responsible for the degradation of the environment, then it could be argued that planning has actually had very little impact on the environment. Against this, though, two alternative arguments must be cited:

1 By failing to affect the country's demography in any significant way, Indian planning has itself also failed to tackle the environmental crisis to which it has given rise.

2 It is not just the population increase that has given rise to environmental problems, but rather, it is the economic and social strategy followed by successive Indian governments, in conjunction with the demographic changes that have taken place, that has caused the problems outlined in this chapter. Population increase by itself need not necessarily lead to environmental degradation, but the course followed by Indian planners over the last 30 years has done little to prevent increasing exploitation of the environment by people, and with it the domination of some people by others.

? In small groups, discuss what the term 'environmental crisis' means in the above paragraph. Identify the dimensions of the crisis and, thinking back over the chapter as a whole, come up with a definition.

THE SIXTH FIVE YEAR PLAN AND THE ENVIRONMENT

The Sixth Plan made a specific attempt to tackle environmental issues. The twentieth chapter of this Plan begins by stating:

> The environment must not be considered as just another sector of national development. It should form a crucial guiding dimension for plans and programmes in each sector. This becomes clear only if the concern for environmental protection is understood in its proper context.

Here, then, is a conscious attempt to locate 'development' within the environment, and this concern is expressed in a depressing catalogue of the adverse environmental changes that have taken place in India since Independence. The section of the Plan on natural resources and the environment summarises changes in land and water, forests, other natural living resources, and in marine ecosystems. It notes in passing that loss of nitrogen, phosphorous and potassium nutrients through water erosion represents an annual loss of 7000 000 000 rupees. This is followed by a discussion on water, air, land and noise pollution, and quotations from this section bring into sharp focus some of the problems already outlined in this chapter:

> Only 8 cities in India are provided with complete sewerage and sewage treatment facilities. They are Ahmedabad, Bangalore, Bijapur, Sangli, Nanded, Nasik, Thana and Durgapur.

> Air pollution is usually associated with industrial growth and urbanisation. However, in many towns and cities of India domestic sources which burn coal, cow dung, firewood and trash can be a significant source of pollution particularly under conditions of stagnant air in winter months.

> Pollution of land results largely from the insanitary disposal of solid wastes. In India open dumping of such wastes (of municipal and industrial origin) on low-lying land is a common phenomenon. This serves as a breeding ground for pests and disease carrying vectors.

The various and different problems faced by the rural areas, where some 160 million people have not yet been provided with potable water, and the urban areas, which suffer from a high population density leading to a breakdown in services, are also surveyed. The

Sixth Plan tackles problems caused by earlier planning practice face on:

The unintended environmental impacts resulting from the execution of development projects like those involving thermal or hydropower generation, mining industry, agriculture, human settlements etc. manifest themselves through one or more of the environmental problems discussed in the preceding sections. For instance, mining operations in India have often led to serious problems of water and air pollution, land subsidence and scarring of huge tracts of land. Indiscriminate discharge of wastes by industries has caused a whole variety of pollution problems including those due to heavy metals and other 'exotic' chemicals that are inimical to all life forms. The unplanned, intensive use of agricultural chemicals has led to cases of water pollution and appearance of pesticide residues in food products.

The government in drafting the Sixth Plan was therefore clearly aware of the environmental problems facing India, and in an attempt to solve them advocated the introduction of four main new proposals.

1 A strong programme of environmental research and development was to be set up.

2 Environmental Impact Analysis was to be made an integral part of the entire planning process.

3 Various Eco-Development programmes were to be initiated, including the creation of an Eco-Development Force consisting of ex-servicemen who would be involved in a massive afforestation and soil conservation programme.

4 A new government bureaucracy was to be established, based on a new Department of the Environment.

The prospects for the success of such a policy depended very much, though, on how much power was given to such environmental bodies to enforce legislation. A bureaucracy with a bark but no bite would be incapable of slowing down, let alone stopping, the deterioration of the environment brought about by the implementation of previous planning policies.

The Mid-Term Appraisal of the Sixth Plan, though, paid scant attention to environmental issues, giving them only two pages. Its three main conclusions were that some infrastructural sectors would have to improve their performance, that investment in irrigation, coal, power and the railways must be protected against short-falls, and that more efficient use should be made of infrastructural and productive investment. The political unrest of 1984 also meant that relatively little attention has recently been paid to environmental issues.

? Write a short essay on one of the following:
1 What were the main environmental problems identified by the Indian government in 1980, and how has it attempted to overcome them?
2 What have been the main impacts of industry and agriculture on the Indian environment since 1947?
3 Assess the differential spatial impact of demographic change on the Indian environment.

THE PROSPECTS FOR INDIA'S FUTURE

The last two sections of this chapter have emphasised that the environment has received relatively little attention in Indian development planning over the last three decades. Nevertheless, over this time period India's population has almost doubled in size, and it is a remarkable achievement that people–environment relationships have not been more seriously disrupted. Today the Indian environment is still just about capable of sustaining its population.

Perhaps the main changes that have taken place as a result of specific government planning have been the development of a capital intensive heavy industrial base and the introduction of a form of agriculture also requiring large capital inputs. In a country where the population is its major resource, the Indian emphasis on a capital-intensive economic base seems to be surprising. It can possibly be partly explained as a result of a desire by the government to achieve rapid 'progress', but it might be argued that the time has now come for India to concentrate on developing its human resources and realising the vast potential of its population, both as a form of labour and as a form of inspiration.

A concentration on capital has also meant that those without access to the necessary 'development' inputs, be they fertilisers, seeds or iron ore, in other words the poor and those without political power, have become further disadvantaged. Although in theory Indian development planning has been avowedly socialist, in practice it has provided the framework in which the rich have got richer and the poor poorer. As human pressure on the environment has increased, those best able to dominate that environment have been able to do so at the expense of their fellow citizens.

The assassination of India's Prime Minister, Mrs Gandhi, in 1984 illustrated the delicate nature of the political balance which must be maintained by any government. Mrs Gandhi's assassination was a direct result of the political moves for independence by elements of the Sikh population, but it also reflected the potential vulnerability of India's leaders when they have to make any 'unpopular' decisions. The surprisingly smooth transition of her son, Rajiv Gandhi, who as the new Prime Minister won an overwhelming victory in the December 1984 elections, suggests that economic policies will not be greatly changed, and the future may well hold out some real hope for the country's poor.

While the rhetoric of the Sixth Plan was full of concern for the alleviation of poverty, it seems that in practice, it fell behind its targets in this field. Thus in the rural development programmes, the minimum needs programmes, and the special area programmes, commentators noted that expenditure figures in the Plan's Mid-Term Appraisal were somewhat lower than the target figures:

> The performance of our rural and poverty orientated programmes does not offer any cause for the kind of optimism expressed in the Mid-Term Appraisal about the economy as a whole.
>
> (*Source*: Paul S. 'Mid-Term Appraisal of the Sixth Plan: Poverty Alleviation Lags Behind', *Economic and Political Weekly*, vol. 19, no. 18, 5 May 1984, p. 760)

> India's real development problem is the inability to make sure that those who are very poor have access to the income that is generated.
>
> (*Source*: *The Financial Times*, 11 June 1984, p. XI)

The Seventh Plan is therefore making a considerable effort to come to grips with this problem, and when its full details are published it is likely that expenditure on poverty alleviation programmes will be appreciably increased.

The Sixth Plan's acknowledgement of the adverse impact previous development planning has had on the environment, and its rhetoric advocating a more sensitive approach to environmental issues, are steps in the right direction. It nevertheless remains to be seen whether any major change will result in practice, during the Seventh Plan period up to 1990. Much is likely to depend on the future political stability of the country, and the power of the new government under Rajiv Gandhi to maintain order and control within which economic and social change can be implemented. India's development planning since 1956 has continued to emphasise the heavy industrial sector. The vested interests within this sector seem unlikely to relinquish their power base without a struggle.

On the basis of the information provided in this chapter divide, as a class, into particular interest groups. Possible groups might include landless labourers, wealthy farm owners, private industrialists, conservationists, poets, family planning officers and tribal forest dwellers. The following lists of concerns may help you formulate your ideas:

Landless labourer Major concerns: need to find employment; difficulty in getting food; children dying through lack of water and medicine; wish to have some land to grow crops.

Wealthy farm owner Major concerns: prices obtained by different crops; cost of electricity for pumps; cost and availability of fertilisers and pesticides; availability of labour; education and marriage of children; the weather.

Private industrialist Major concerns: impact of government legislation on exports/imports; distribution of population as labour and market; international commodity prices, especially oil; possibility of labour unrest and strikes.

Conservationist Major concerns: industrial pollution; soil erosion; destruction of forests; disappearance of wildlife; nuclear power.

Poet Major concerns: social justice; inequalities; ties to family and homeland; images of natural and industrial landscapes; religion and cultural heritage.

Family planning officer Major concerns: population growth; provision of health clinics; literacy, particularly of women; availability and expense of contraceptives.

Tribal forest dweller Major concerns: reduction of forest area available for shifting cultivation; loss of cultural identity; presence of informal markets; availability of plants for traditional medicine; impact of conservationist policies on animals hunted for food.

Within each group, produce a summary of the impact of development planning on your specific 'interests' since 1947, present this to the class as a whole.

Evaluate, within each group, what you consider to be:

a The most pressing needs facing your members over the next decade.
b The best means of achieving your desired aims.

In reality, there is no forum in which all of these interest groups can meet, but to illustrate the complexity of the problems facing Indian planners and the incompatibility of many 'solutions' put forward to solve these problems, each group, in turn, present your 'needs' and 'solutions' to the other groups in the class as a whole. They can then debate the suitability of your 'solutions' with respect to their own, very different, needs and interests.

13 NIGERIA: CHOICES IN AGRICULTURAL DEVELOPMENT

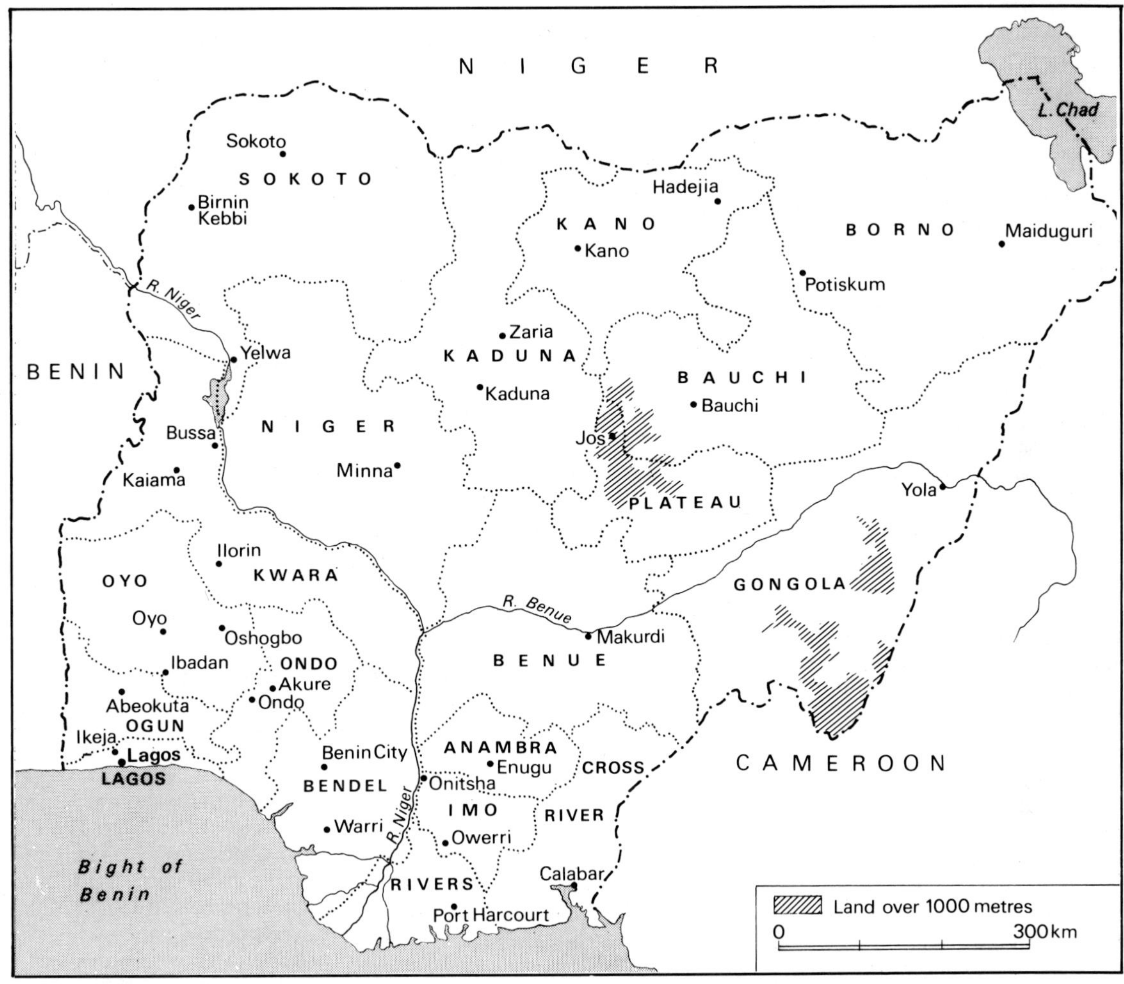

THE DEVELOPMENT OF AGRICULTURE IN GONGOLA

GONGOLA GOVERNMENT GETS N 3.8M LOAN FOR FARMING

The Gongola state government has secured a loan of 3.775 million naira from the Union Bank of Nigeria for the development of agriculture.

About 3400 farmers in all parts of the state are expected to benefit from the loan during the next rainy season.

Speaking at the annual agricultural show in Song, the Commissioner for Land and Survey, Alhaji Ayuba Musa Mamawa, stressed that the government would not relent its efforts in providing the necessary inputs like fertilisers, insecticides, tractors, equipment-hiring schemes and land clearance services to farmers.

Alhaji Ayuba declared: 'The state government will continue to appeal to the private sector to come in full force and supplement the government's efforts in the development of the vast agricultural potentials of the state.'

The commissioner noted with delight the role of the Nigerian Agriculture and Co-operative Bank, agricultural loan guarantee schemes and commercial banks in the development of agriculture in the state.

He commended the role of extension services workers and other professional personnel in the field of agriculture.

On the inherent dangers of bush burning, Alhahi Ayuba stressed that all existing by-laws and regulations would be enforced to curtail the increasing rate of indiscriminate bush burning in the state.

(*Source*: *New Nigeria*, 16 February 1982)

? The article above appeared in one of Nigeria's daily papers, the *New Nigerian*. Gongola State lies in the middle belt of Nigeria on the east of the country (see introductory map). Make a list of the main methods to be used to improve farming in the state. Which of the following sentences best describe the view of agricultural development put forward by the Commissioner for Land and Survey?

- Local scale development using simple tools and local resources.
- Development by providing credit and appropriate technologies to all farmers in the state.
- Improvement of agriculture by direct government intervention and investment using high technology methods.
- Expansion of agriculture by small numbers of large farmers in the private sector using high technology methods with government help.

The Gongola state government's approach to the development of agriculture is in fact the last using government resources to assist a small, and hopefully efficient, number of farmers with large farms. The 3400 farmers who will receive loans represent less than one per cent of the farmers in the state. If they are assisted in this way, how would you expect them to benefit compared to the other 99 per cent of farmers? Assuming that the large farmers *are* efficient, then this policy will increase their food production, but if the small farmers are not helped, what might happen to overall production?

NIGERIA: OIL RICH AND FARMING POOR?

Nigeria is, with the exception of South Africa, the largest and richest country in Africa south of the Sahara. Its population was estimated to be 85–90 million in 1984, although because of political influences and accusations of false counts there has been no acceptable census since 1963. Nigeria's gross national product (GNP) in 1979 was $US 55.3 billion, far greater than its immediate neighbours of Cameroun (4.6 billion), Mali (0.9 billion) and Benin (0.8 billion).

So Nigeria is rich: or is it? The GNP figures hide as much about a country's economy as they reveal. Compare Nigeria's GNP per capita with that of other African countries (Table 1), and also compare other indices such as life expectancy at birth and the number of people per doctor. Using these measures of quality of life, it is clear that Nigeria does not do anything like so well.

THE DEVELOPMENT OF OIL

A lot of Nigeria's apparent wealth comes from oil. Commercial production of oil began in 1957 in the south-east of the country. In 1965 petrol refining was begun in Port Harcourt, and oil output began to rise very fast. There was a short interruption during the Nigerian Civil War in 1967 and 1968, but by 1969 production was 200 million barrels in the year. It doubled to 823 million barrels in 1977 (2.3 million barrels a day), since which time production has remained at a high but fluctuating level (Figure 1). The revenue to the government from this oil production has been enormous, and has played an important role in the form and scale of Nigeria's development planning.

However, although oil and the subsequent public spending that has been made possible by it have been important to the country, 70 per cent of Nigeria's population still lives in the rural areas and has been little affected by the oil wealth. It is estimated that about 60 million people depend on agriculture in Nigeria. In the last decade population growth has outstripped

Table 1 *Indices of standards of living in Nigeria and other African countries*

	Population (millions) mid1979	Area (thousands of square kilometres)	GNP per capita: Dollars 1979	GNP per capita: Average annual growth rate (per cent) 1960–79	Life expectancy at birth (years) 1979	Population Per physician 1960	Population Per physician 1977
Low-income countries	**187.1 *t***	**15 718 *t***	**239 *w***	**0.9 *w***	**46 *w***	**50 788 *w***	**32 241 *w***
Low-income semi-arid	***28.0 t***	***5 745 t***	**187 *w***	**0.0 *w***	**43 *w***	**67 302 *w***	**36 781 *w***
Chad	4.4	1 284	110	−1.4	41	72 190	41 940
Somalia	3.8	638	. .	−0.5	44	36 570	. .
Mali	6.8	1 240	140	1.1	43	67 050	28 150
Upper Volta	5.6	274	180	0.3	43	81 650	48 810*
Gambia	0.6	11	250	2.6**	42	21 800	13 171*
Niger	5.2	1 267	270	−1.3	43	82 170*	42 720
Mauritania	1.6	1 031	320	1.9	43	40 400	15 160
Low-income other	**159.1 *t***	**9 973 *t***	**247 *w***	**1.0 *w***	**47 *w***	**47 756 *w***	**31 539 *w***
Ethiopia	30.9	1 222	130	1.3	40	100 470	75 320
Guinea-Bissau	0.8	36	170	. .	42	. .	100 094
Burundi	4.0	28	180	2.1	42	96 570*	45 020*
Malawi	5.8	118	200	2.9	47	35 250	40 680*
Rwanda	4.9	26	200	1.5	47	138 100*	38 920
Benin	3.4	113	250	0.6	47	20 030	26 880
Mozambique	10.2	783	250	0.1	47	20 390	33 980
Sierra Leone	3.4	72	250	0.4	47	20 420	n.a.
Tanzania	18.0	945	260	2.3	52	18 220	17 550
Zaïre	27.5	2 345	260	0.7	47	37 620*	15 530
Guinea	5.3	246	280	0.3	44	48 000*	16 630
Central African Rep.	2.0	623	290	0.7	44	41 580	17 610
Madagascar	8.5	587	290	−0.4	47	8 900	10 240
Uganda	12.8	236	290	−0.2	54	14 060	27 600
Lesotho	1.3	30	340	6.0	51	23 510	18 640
Togo	2.4	57	350	3.6	47	37 760*	17 980
Sudan	17.9	2 506	370	0.6	47	33 500	8 690
Middle-income oil importers	**65.2 *t***	**3 690 *t***	**532 *w***	**1.5 *w***	**50 *w***	**20 971 *w***	**11 877 *w***
Kenya	15.3	583	380	2.7	55	10 690	11 630*
Ghana	11.3	239	400	−0.8	49	21 600	9 920
Senegal	5.5	197	430	−0.2	43	24 540	15 710
Zimbabwe	7.1	391	470	0.8	55	4 790	7 030*
Liberia	1.8	111	500	1.6	54	12 600*	9 260*
Zambia	5.6	753	500	0.8	49	9 540	10 190*
Cameroon	8.2	475	560	2.5	47	48 110*	16 500
Swaziland	0.5	17	650	7.2**	47	10 134	9 185
Botswana	0.8	600	720	9.1**	49	26 200	9 597
Mauritius	0.9	2	1 030	2.3**	65	4 662	2 410
Ivory Coast	8.2	322	1 040	2.4	47	29 190	15 220
Middle-income oil exporters	**91.6 *t***	**2 781 *t***	**669 *w***	**3.2 *w***	**48 *w***	**67 250 *w***	**15 494 *w***
Angola	6.9	1 247	440	−2.1	42	14 910	. .
Congo	1.5	342	630	0.9	47	16 430	7 290*
Nigeria	82.6	924	670	3.7	49	73 710	15 740*
Gabon	0.6	268	3 280	6.1**	45	9 722	3 029
Sub-Saharan Africa	**343.9 *t***	**22 189 *t***	**411 *w***	**1.6 *w***	**47 *w***	**50 096 *w***	**23 904 *w***

Notes: Figures marked with an * are for years other than specified   ***t*** = total   ***w*** = weighted mean   (*Source*: World Bank, 1981)

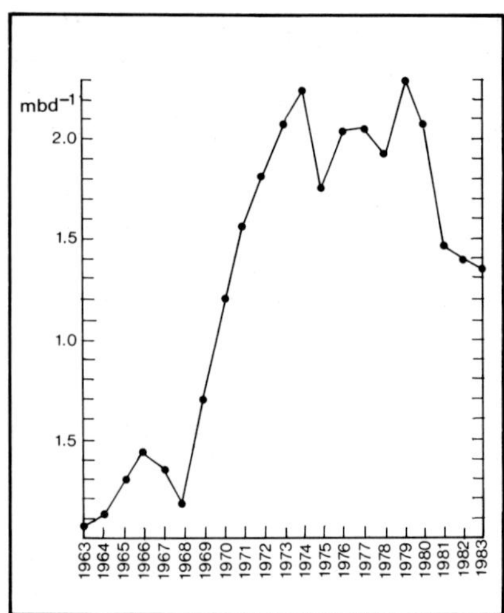

Figure 1 *Nigerian crude oil production 1963–1983*

Table 2 *The importance of agriculture to Nigeria's Gross Domestic Product*

	Percentage of GDP				
	1958–9	*1966–7*	*1970–1*	*1975–6*	*1982*
Agriculture	66	54	42	24	22
Industry (including oil)	9	20	32	49	39

(*Source*: World Bank, 1984)

Table 3 *Agriculture and other exports of Nigeria*

	Percentage of exports by value				
	1962	*1967*	*1971*	*1975*	*1978*
Cocoa	20	23	11	4	7
Palm oil (including kernels)	15	4	2	<1	<1
Groundnuts (oil and cake)	24	19	3	0	0
Oil	10	30	73	93	89

? Examine Tables 2 and 3 and what you consider to be the significant points and compare your list with the commentary which follows.

agricultural production. In 1979 population growth was estimated at 2.5 per cent per year, whereas over the previous ten years the average annual value of agriculture production fell by 0.8 per cent.

Tables 2 and 3 show the place of agriculture and agricultural products in the Nigerian economy. In 1958, before Nigeria got its independence, agriculture was contributing 66 per cent of Nigeria's GDP, and agricultural exports were the mainstay of the country's economy. With the rise of oil production, agriculture has become progressively less important. Its proportion of GDP fell to 24 per cent in 1975–6 (Table 2). Exports of the major agricultural products have also declined (Table 3). In part, this decline simply reflects the overwhelming importance of oil in the modern Nigerian economy, but it also shows the stagnation of agriculture. Agriculture contributed 1.5 billion naira to Nigeria's GDP in 1973–4, very little more than the 1.4 billion a decade before.

So while Nigeria has been becoming a major oil producing country, and becoming very wealthy on paper, in the rural areas the agricultural economy has stagnated. The new oil wealth has not penetrated to the rural areas. A lot has been invested in costly capital projects, such as irrigation schemes which have, in general, not worked well, and prestige projects such as the new Federal Capital at Abuja. Enormous heavy industrial plants, such as steel mills, have been built, yet light- and middle-range manufacturing industry has remained dominated by the assembly of imported components (e.g. Land Rover and Volkswagen), and the largest proportion of the oil money has been spent on the importation of consumer items such as luxury foods, radios, cassette players and manufactured cars.

The oil money has allowed Nigerians to develop the West's desire for consumer goods without the economy to make them themselves. The increasingly educated young men are leaving the land and moving to the towns to swell the ranks of the unemployed and underemployed. In the new rural backwaters food production has fallen behind, and the growing population's increasing demand for food is not being met.

Following the large-scale exploitation of oil after the Civil War, spending power increased rapidly in Nigeria and both food prices and the level of imports rose rapidly. It was not just luxury food that was imported. World Bank figures show a rise in grain imports from none in 1970 to 406 000 tonnes in 1975 and a further rise to 1.52 million tonnes in 1978. Rice imports also grew, as it became a fashionable and popular urban food. When the government tried to curb imports in 1980 there were angry outbursts throughout the country. The Chairman of the Presidential Task Force on Rice said in 1980:

'We are still battling, because the more rice we bring in, the more people want to eat rice. The only real solution lies in growing rice ourselves in this country and then setting up rice mills so rice can be milled and processed to the same degree and quality as the rice we import.'

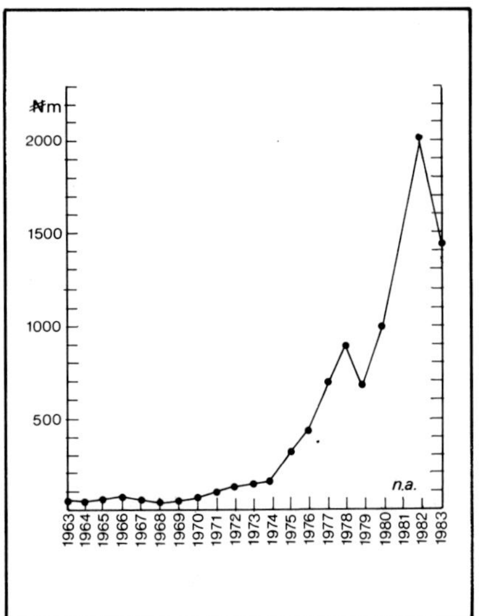

Figure 2 *Nigerian food imports (including live animals) by value 1963–1983*

?

1 Examine the graph showing food imports to Nigeria between 1972 and 1980 (Figure 2). Compare it with the graph of rising oil production (Figure 1). Why is there a lag in the rise of the two graphs? Look at Figure 3. What linkages does it suggest between oil production and the import of food?

2 The Chairman of the Presidential Task Force on Rice commented that rice was becoming more popular. If a further breakdown of the data were available, what information would you want to be able to decide what kind of food is being imported and who is consuming it? Think about the following categories:

- Importance of basic foods: grains (rice, wheat, others), meat
- Importance of processed food, e.g. flour
- Importance of luxury goods, e.g. tinned, dried and fancy foods
- Source of imports of different kinds
- Seasonality of imports
- Number of importers (monopolies?)
- Food consumption in rich and poor families in rural and urban areas
- How food is marketed (hoarding?)

3 Imagine that you are an official of the Nigerian government. Your political superior in the Ministry of Agriculture has instructed you to set out a memorandum of points for and points against banning the import of luxury items. Write up your memorandum under the following headings:

For the ban

1 Waste of foreign currency
2 Adverse effects on home production and marketing
3 Policy unworkable on economic grounds in long run
4 Imports encourage elitism and do not help the poor

Against the ban

1 Imports encourage healthy competition with local produce
2 Imports prevent food shortages
3 Imports provide a breathing space to increase home-grown food
4 Importers are rich and powerful men.

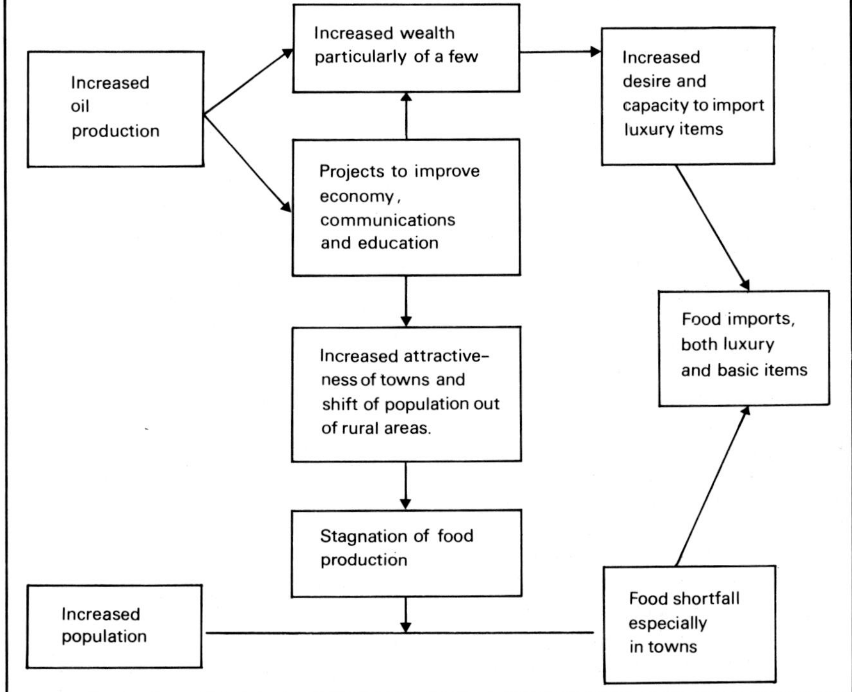

Figure 3 *Possible linkages between increased oil production and food imports in Nigeria*

?

4 This chapter has been written from a certain standpoint — what attitude do you think the author shows towards the nature of Nigerian development? Contrast his values with those that the Nigerian government has clearly held in relation to the opportunities given by wealth.

ENVIRONMENT AND AGRICULTURE

NIGERIA'S RAINFALL

It is difficult to generalise about Nigerian agriculture because the country has such a diverse range of environments. Rainfall is seasonal throughout the country, but varies greatly in amount from south to north. In the low lying Niger Delta, the annual rainfall is over 400 centimetres, while far north in Sokoto and Borno it is less than 50 centimetres (Figure 4). The length of the rainy season also varies, from barely three months in the north to ten or 11 months in the south (Figure 5.

In the south the rainfall is not only high, but is well distributed throughout the year. As a result, crops can grow all year. It is a zone of staple root crops, such as yams, cooyams and cassava, and tree crops such as oil palm, plantain (a kind of large banana eaten as a vegetable) and citrus fruits. Except in the Niger Delta where flooding is a problem, agriculture in the south is generally productive.

In the middle belt of Nigeria, further north, the rainy season is shorter. Root crops like cassava are still grown, but they are mixed with cereals better adapted to dry conditions such as sorghum and millet. In the middle belt rice has become important in a number of places.

In the north, dry land and grain crops predominate, and conditions for agriculture are much more difficult. The area is in the climatic zone known as the Sudan Savanna. Its short rainy season and low rainfall give a very restricted growing season for crops (Figure 6). At the end of the dry season the ground is hard baked and everything is dessicated. The growth of crops depends entirely on the arrival of the rains. The dry season ends with a series of short heavy storms, often several weeks before the proper rains begin. If farmers wait too long before planting they may miss part of the short growing season. But if they plant too soon, they risk a dry spell killing their crop in the ground.

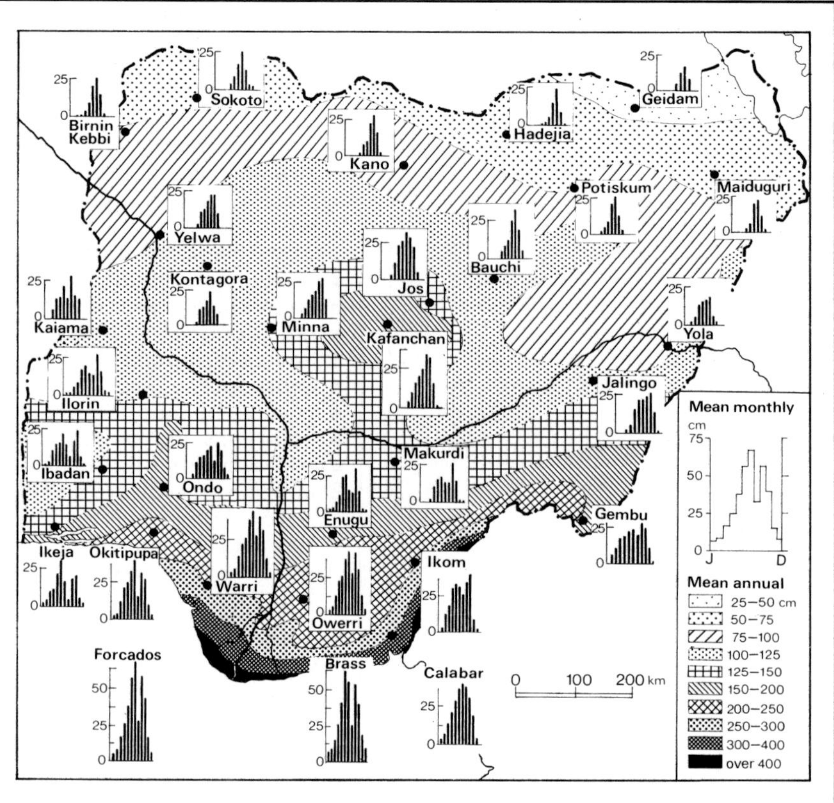

Figure 4 *Mean annual rainfall in Nigeria*

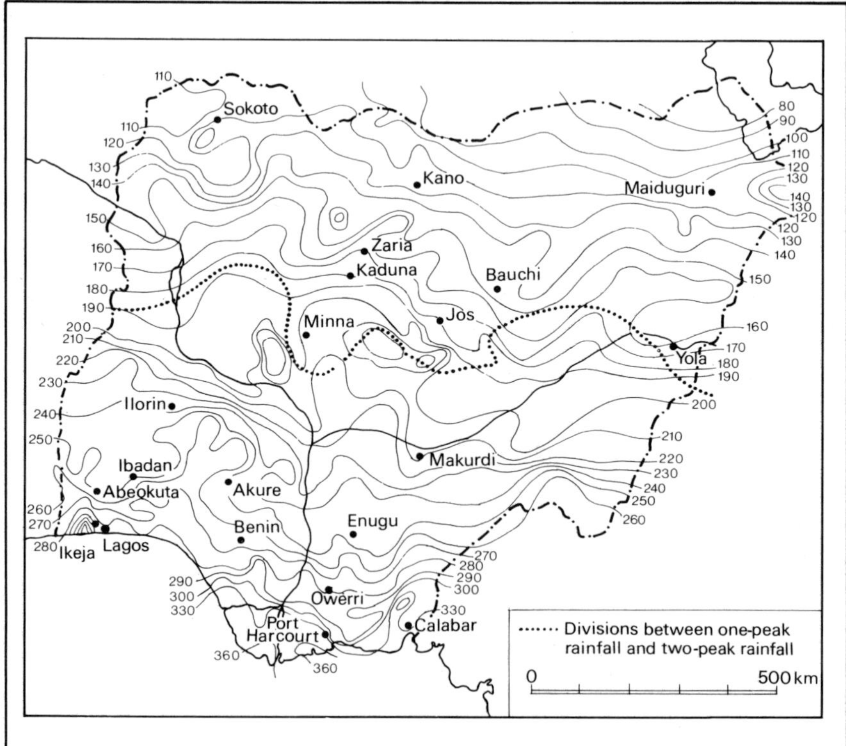

Figure 5 *Duration of the wet season*

? **1** Look at Table 4, which shows how the yields of millet vary with the date of planting at two sites in northern Nigeria. What is the best time to plant millet at Samaru? What is the best time at Kano? Does Figure 7 help to explain the difference?

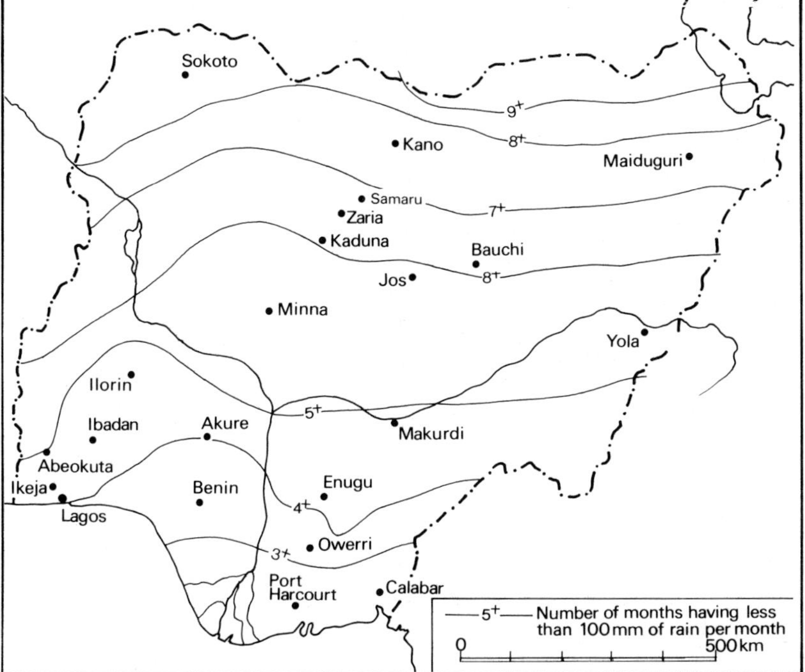

Figure 6 *Months with less than 100 mm rainfall*

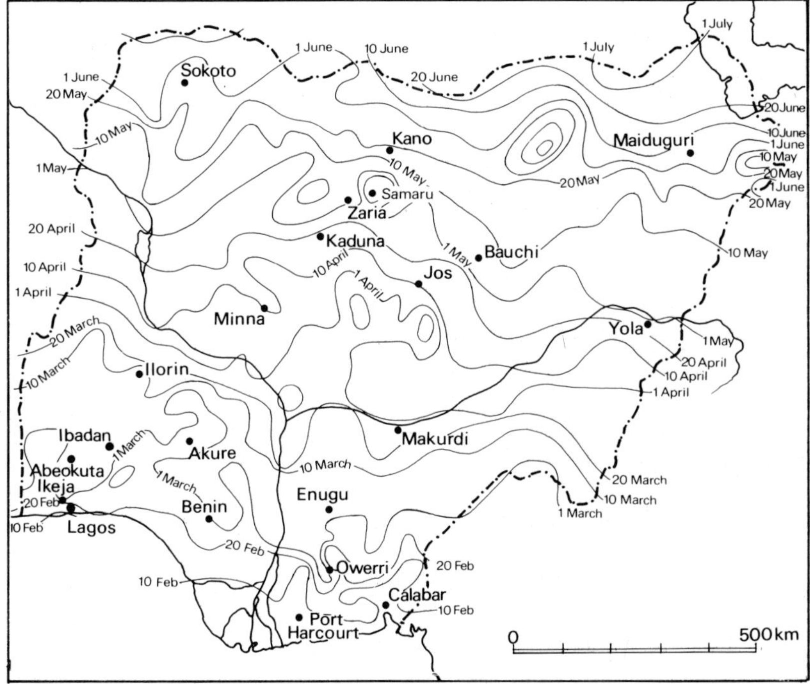

Figure 7 *Date of the onset of rains*

? **2** Using the data in Figures 5 and 6, expand Table 4 to show the approximate duration of the wet season, and the number of months with over 100 millimetres rainfall at Samaru and Kano. Average yields of millet recorded in one survey were between 2500 kilos per hectare at Kano and 2720 kilos per hectare at Samaru. Does this difference surprise you? If a farmer at Kano and one at Samaru have equal skill and work equally hard, which will get the higher yield? What economic implications might this have?

Table 4 *The effect of time of sowing on grain yield of millet — (after Egharevba, 1978)*

Planting date	*Yield kg/ha*	
	Samaru	*Kano*
Mid-April	350	*
Early May	1072	*
Mid-May	2239	208
End of May	2460	1026
Mid-June	1300	2280
End of June	278	1942
Mid-July	**	847
End of July	**	194

* Not planted because it was too dry
** Grain yield not worth reporting

The date the rains arrive is controlled by the movement of air masses in the upper atmosphere, and in particular the position of the Intertropical Discontinuity (ITD) which marks the boundary between the moist maritime air over the Atlantic and the dry air of the desert. The ITD reaches its furthest south (7 °S) in February and then moves north over Nigeria and into the Sahelian countries like Niger and Mali to its nothernmost point (20 °N) in August (Figure 8).

However, in some years the ITD moves less far north, and the rains in each place are later and shorter. In the north of Nigeria crops fail and famine is never far away. The most notorious drought of recent years was the so-called Sahel Drought of the early 1970s. It struck most

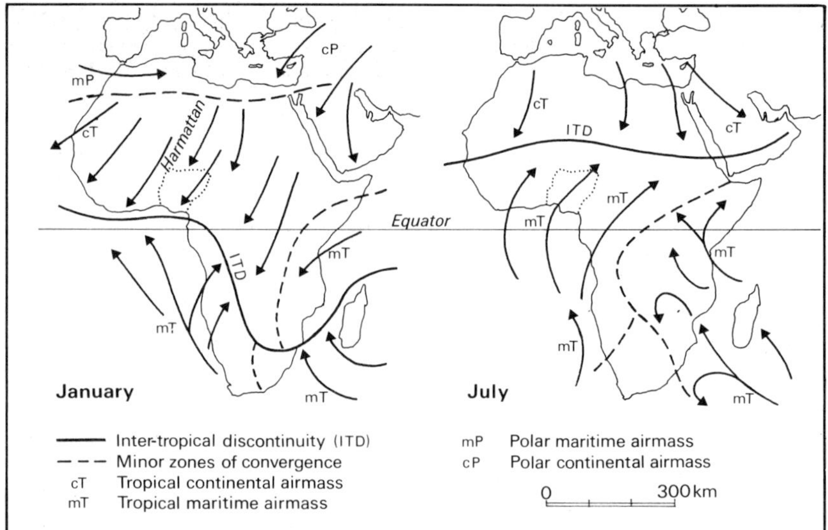

Figure 8 *Movement of the intertropical discontinuity (ITD) in West Africa. In January, air movements over Nigeria are dominated by the dry continental air mass from the north, and the dry Harmattan wind blows. In July, the ITD has moved north and the maritime air mass brings rain to Nigeria.*

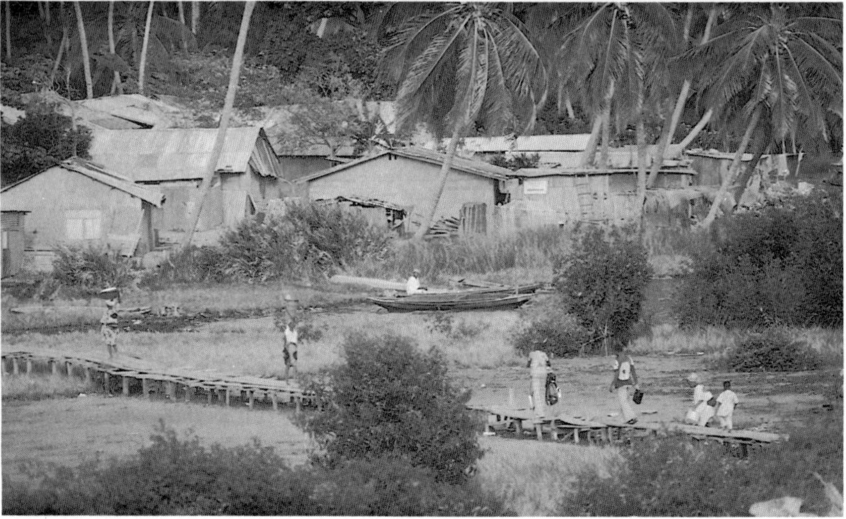

severely in the drier countries to the north of Nigeria, but even in Nigeria itself there was widespread failure of crops and farmers had to leave their homes to seek work in the towns in order to survive.

The 1984 wet season was again bad, but these droughts are not only recent occurrences. There was a severe drought in 1913 for example, with rainfall two-thirds of the mean of previous years, and again in 1919. In that year the serious situation in Sokoto in the north-west of the country was described like this:

'Crops along the River Rima and River Sokoto were practically complete failures. The stalks grew well, but there was no grain in the ear. The same is true of the rice crop. The river flood remained low and lasted so short a time that the grain had no time to mature.'

Farmers in the north of Nigeria have had to learn to live with periodic droughts in the past. Population increases have left the farmers with less room to manoeuvre when faced with drought, and the problems are becoming increasingly severe. Table 5 shows the way in which a series of years of poor rainfall can leave a male farmer in debt and struggling for several years after better conditions return. Women also farm, and can be heads of households. They can be even more vulnerable to adverse conditions.

Prints 1 and 2—*The type of agriculture practised in Nigeria is very much controlled by the climate and physical environment. Prints 1 and 2 show the contrasting landscapes that Nigerian farmers have to deal with; the North has a long dry season and is prone to drought whilst the South is lush with plenty of water although in recent years it has also experienced periods of drought*

? Look at Table 5 and discuss, in groups, the options open to the small farmer. What kind of government policy might you try to formulate for the small farmer?

Table 5 *The effect of poor harvests on a farmer's livelihood in northern Nigeria*

Year	*Rains*	*Harvest*	*Farmer's situation*
1	Early, good	Good	Plenty of grain in store. Decides to marry a second wife and sells some grain to meet expenses.
2	Late, short	Poor	Sells animals (cattle, goats) to buy extra food in the dry season.
3	Very late, very short	Almost total failure	Has to borrow money to buy grain in the dry season.
4	Late, short	Bad	Has to sell some of his crop at harvest to pay debts, but gets a low price. Has to borrow even more to buy food again later in the year. Sells the rest of his animals.
5	On time, good	Good	Despite good harvest he still has to sell and buy grain again, so he goes back into debt.
6	On time, good	Good	Despite a second good harvest, he is still in debt. Another bad season next year and he will have to sell land or flee south to seek wage employment

THE CROPS

The farmers of northern Nigeria grow crops which are well adapted to the dry and uncertain conditions. The basic grain crops are millet and sorghum. Both are resistant to drought — sorghum for example has waxy leaves which curl up when the plant is short of water, to cut down water loss. Both millet and sorghum are ground into flour and made into a kind of porridge. Rice is also grown, in the wetter areas like flood plains. It is now being realised that the flood plains of the major rivers, where the soils stay slightly damp long after the rains end, are very important both for wet season agriculture and for vegetable crops in the dry season.

The other main crops of the north are groundnuts (peanuts) which are usually crushed for oil, cowpeas which are small beans rich in protein, and cotton, which has long been an important cash crop. It is unusual for a farmer to grow only one of these crops on its own, as is normal in Europe. Instead, several are grown together in the same field, being planted at different times. This practice is known as *intercropping*, and means that as one crop is harvested the next will be starting to ripen. Thus, perhaps millet, which grows quickly, will be planted first in June; by the time it is harvested in early September, a crop of groundnuts will be nearly mature and can grow out to fill the space left. Intercropping has several advantages:

1 It keeps the soil covered throughout the wet season so that sudden intense storms do not cause disastrous erosion.

2 It gives the farmer several chances to get a good crop, since there is a chance that whatever happens to the rains one will grow well.

3 It ensures that none of the short growing season is wasted.

SMALL FARMERS AND BIG PROBLEMS

Farmers in Nigeria, like farmers everywhere, face other problems, in addition to those connected with the environment. Farms tend to be small (Table 6, *page 268*). In the southern states of the east and west over 90 per cent of farms are smaller than two hectares — they are more like allotments than farms in the European sense. Even in the north where, as we have seen farming is less productive, and where it might be expected that larger areas would be needed to grow sufficient food for a family, 69 per cent of farms are smaller than two hectares.

Table 6 tells us that most farmers are growing food for themselves and their families rather than for sale — in other words most are primarily subsistence cultivators. In many cases they grow enough in a good year to be able to sell some of their crop, and often they do grow crops like cotton or onions specifically for sale, but their involvement in the cash economy is cautious and small-scale. The majority of farmers have no capital to spend on new agricultural inputs like improved seed varieties, fertiliser and pesticides, let alone money to spend on new implements like an ox plough or a tractor.

VICIOUS CIRCLES AND POVERTY TRAPS

A series of poor harvests, or heavy losses of stored grain due to insects, or even expensive events in the family like the marriage of a daughter, can easily push farmers into debt. If they try to borrow money from a richer man in their village they may have to pay high rates of interest, and they rarely have access to banks. Anyway, banks will not lend money to people already in debt, because there is no guarantee that it will be repaid. Often a farmer in debt will sell his crops in the field or soon after harvest, to raise some money. But at that time of year there is a lot of grain on the market and prices are low. If he sells too much and runs out of food before the next harvest he will have to buy grain in the dry season when it is scarce and prices are high. To do so he might have to borrow some more money, or sell some of his livestock, thus becoming bound more tightly into the cycle of debt. Richer men can buy grain when it is cheap and sell when it is expensive, and often benefit from the poverty of their neighbours. There can be great unevenness of wealth even in a small village.

So most farmers in Nigeria are skilled in their work, but are poor and — schooled by circumstances — cautious about new ideas. When a development 'expert' arrives with a theoretical solution to their problems they are likely to want to see the idea tested very carefully before they commit themselves. That caution is not, as it has been popularly described in the past, a kind of 'peasant stupidity'. In fact it is a vital strategy for survival in an uncertain world full of natural and man-made hazards.

Tables 7 and 8 show why farmers will be cautious about innovations, and why some farmers are more likely to benefit from them than others. Table 7 shows hypothetical data for five years with harvests ranging from very good to very bad. The farmer has to find the seed to plant and labour to cultivate the fields, and the net cash income he could get per hectare of land by selling his produce is shown in the second row. The income varies from 40 to 100 naira per hectare in different years (a naira was worth about 88 pence in June 1985).

It is suggested to the farmer that he might start to use fertiliser (at a fixed cost of 20 naira per hectare) to increase his yields by 25 per cent (fourth row). The cost of the fertiliser has to be met out of the increased yields, so the net income per hectare is shown in the fifth row. Notice that in good years (year 1 and year 5 especially) the fertiliser gives a good return on the investment, but in bad years (years 3 and 4) the extra cost of the fertiliser actually decreases the farmer's income from the land, even though the yields themselves are greater. The visiting 'expert' knows that the fertiliser will increase yields, but he will not have the experience to know what happens in a year of bad rainfall. The farmer has learned how bad the rains and the harvest can be, and therefore — quite rightly — he is more cautious.

Table 6 *Farm sizes in Nigeria*

Farm size	*Western states* %		*Eastern states* %		*Northern states* %	
less than 0.1 ha	6		21		1	
0.1–0.2 ha	14		21		3	
0.2–0.4 ha	24	94	21	95	9	69
0.4–1.0 ha	33		24		28	
1.0–2.0 ha	17		8		28	
2.0–4.0 ha	5		4		22	
over 4 ha	< 1		< 1		9	

(*Source*: J. T. Uyanga *A Geography of Rural Development in Nigeria*, University Press of America)

Table 7 *Hypothetical returns per hectare of farmland with and without fertiliser in a sequence of 5 years*

	Year 1	2	3	4	5
1 Harvest	very good	poor	very bad	bad	fair
2 Income per ha (Naira)	100	75	40	50	80
3 Cost of fertiliser (Naira)	20	20	20	20	20
4 Income using fertiliser (Naira)	140	100	50	70	110
5 Net income once fertiliser is paid for (Naira)	120	80	30	50	90

(*Note*: In the year of very bad harvest (year 3), it is disadvantageous to use fertiliser)

Table 8 *Hypothetical income from farms of three sizes with and without fertiliser*

	Year 1	2	3	4	5
Harvest	very good	poor	very bad	bad	fair
Farm size:					
Large (10 ha)					
Income with fertiliser	1200	800	300	500	900
Income with no fertiliser	1000	750	400	500	800
Medium (6 ha)					
Income with fertiliser	720	480	180*	300	540
Income with no fertiliser	600	450	240	300	480
Small (2 ha)					
Income with fertiliser	240	160*	60*	100*	180*
Income with no fertiliser	200	150*	80*	100*	160*

* Years with shortfall, if it is assumed that 200 Naira is needed to feed each family

Now look at Table 8. This shows the situation of three farmers, one with ten hectares of land, one with six and one with two. The income each farmer can make from his land in each of the five years, both with and without fertiliser, is shown. Assume that each has the same size family, and needs a minimum of 200 naira a year to feed them. See how the large farmer has more than enough in all years — he can afford to weigh up the relative merits of using fertiliser or not, to give him the best returns. But see how the small farmer cannot produce enough food in any but the best years, whether he uses fertiliser or not. He must look for income outside farming, and certainly will be very unlikely to risk more outlay on fertiliser. The medium farmer has enough income every year if he does *not* use fertiliser, but has too little in the third year if he *does* use it.

Obviously, it seems on the surface that using fertiliser is always going to be a good idea. However, looking at year to year variations in harvest, and at the individual economic situation of different farmers, it is clear that there are risks as well as benefits with its use. Some farmers would be better off without it, some would profit. In this way, the apparent paradox of the farmer's refusal to adopt a seemingly beneficial innovation can be explained.

WEALTHY FARMERS — A NEW BREED

Not all farmers in Nigeria are poor. Read the following account of changes in the nature of farming in the country:

> A new class of farmers is emerging in this country. This class comprises civil servants, military men, businessmen and landowners who now see agriculture as a new frontier for becoming wealthy or for storing their windfall wealth. This class of absentee landlords and farmers are capable of using their position of power in national policy making institutions to frustrate meaningful efforts at establishing self-reliance and agricultural changes in rural areas.
>
> (*Source*: *New Nigerian*, 1982)

This article suggests that successful people from diverse occupations are investing in agriculture to make money, and by doing so are standing in the way of the majority of small farmers. In what ways might the new big farmers and the traditional small farmers be in competition? Which would be more likely to get government loans or materials and new equipment? The large farmers need to obtain land for farming: if they bought out the smaller farmers instead of clearing new land, where would the small farmers go? How likely would they be to get jobs on the new farms? What would their prospects be in the towns?

Look back to the article at the beginning of the chapter about the Gongola State government's programme of agricultural development loans. The government is trying to encourage investment from 'private enterprise' — in other words it wants just this kind of interest in agriculture from big landowners and businessmen. The reason the government gives is that the large farmers can use modern and mechanised methods, and produce a lot of food, but in encouraging the larger farmers is the government doing anything to help the vast majority of small farmers?

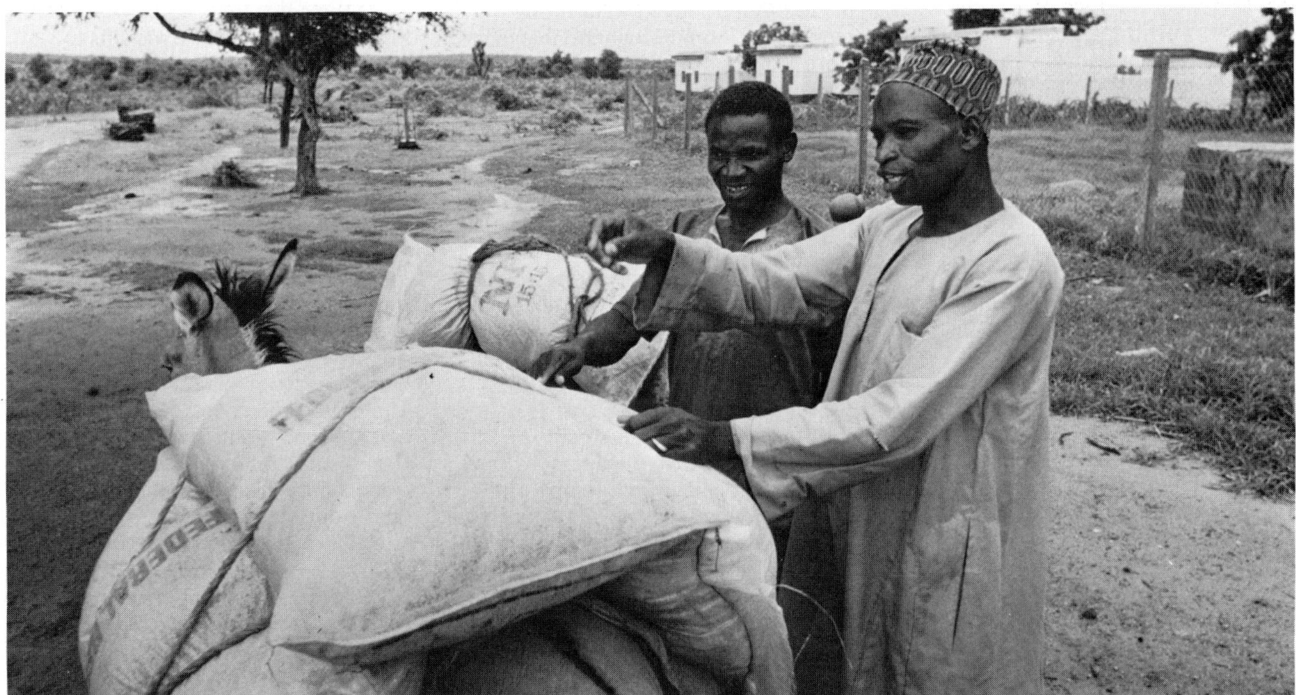

Print 3 *Farmers buying fertiliser on the Gusau Agricultural Project; funded by the World Bank—the decision to invest in fertiliser is a gamble for many small farmers*

> **?** Imagine you are a banker, and you have to choose between the following two applications for an agricultural loan:
>
> *Businessman farmer* I have a 100 hectare farm on which I grow sorghum and cotton, both of which I sell for good profits. I have records of my harvests and my sales for the last five years, and can show that my farm makes a profit. I now want a loan of 10 000 naira to purchase a tractor to start mechanised cultivation which will be more efficient. I expect to be able to pay back the loan within ten years through increased profits.
>
> *Traditional farmer* I have a three hectare farm on which I grow millet, sorghum and beans, most of which I need to feed my family. There have been some bad seasons, and I am getting into debt — I have to sell grain after harvest to repay loans but I get a bad price, I have to borrow again to buy again later. I need a bank loan to get me out of the trap I am in, because I can pay it back over a long period and not pay much interest. Even 100 naira would do. If I don't get help my family will really suffer.
>
> Which farmer needs the money most? Which of them is most likely to repay the loan? If your bank demands that you lend on strictly economic criteria, to which farmer will you loan money? These are difficult decisions.

A FARMING DEBATE

One might imagine the following arguments at a meeting to decide on the future of agriculture in a state in Nigeria.

Minister of Agriculture:
Oil has made us rich in Nigeria, and the population of our towns is growing very fast. These people need to be fed. The policy of my government is to increase agricultural production and my Ministry intends to do this by the most efficient means possible. The peasant farmers are backward and unprogressive. In time they will adopt new methods and become efficient farmers, but at the moment money spent on them will not increase the amount of food available very much at all. We are therefore concentrating on the modern agricultural sector, and are making *loans* available to the largest and most progressive farmers for machinery, seeds and fertilisers. Their production will feed the people in the cities, and their example will inspire the traditional farmers to greater achievement in the long run.

Large modern farmer and businessman, Alhaji Musa:
I am a businessman. I am investing my money for the benefit of the nation, and growing food by modern methods. I bought my 500 hectare farm from local people, and paid a reasonable price for it. Since then I have spent 50 000 naira on machinery for it, and have a full-time manager. I employ a lot of labourers locally, and without the wages I pay, a number of villages would be very poor. Mine is the largest rain-fed farm in the district, and I have plans to introduce irrigation in the future. I was at school with the Minister of Agriculture, and he is a great friend. He brings visitors to my farm, and loans from the Ministry have helped me to develop my land and buy machinery. I am convinced this is the right way for Nigeria to advance.

Traditional farmer:
When I look out from the village at my old farm all the land I see now belongs to Alhaji Musa. He says he paid me for my land, but I only sold it to him because I owed money after the great drought. The money he paid seems nothing now that I have no land to farm. I still have a few fields left, and share others with my brother, but I can't grow enough food now so I have to work as a labourer on Alhaji Musa's farm. I work for five naira a day on my old land, but he uses tractors now and doesn't need much labour. Anyway he makes most of his money on business deals, and his manager doesn't know how to farm well. The land is not looked after. His children won't have any farm left.

Of course, my neighbour's son has learned to drive a tractor and he gets good wages. His family is rich now, and the second son is at school. I suppose he will get a job with the government, and one day he will come back like Alhaji Musa and buy land and grow even more rich. My family will not be rich. My sons have gone away to the city to try to find work because I have no land to give them. A man pays them to sell combs and sweets to the rich people in the cars, but they have no training and no money — they don't even know how to farm. They think I am a fool to have sold the farm, but a greater fool for staying here and trying to live in the village. They have no respect. They are angry and say their problems are all the government's fault, but I know it is because I sold the farm.

Critic:
The government says it is concerned with welfare in the rural areas, but really they don't care. They are not so much concerned with growing more food as making sure all their friends and political supporters make money out of farming. By encouraging big agricultural businesses and large landowners they are contributing to the very problem they are trying to solve — the small farmers get pushed more and more towards the poverty line and eventually they give up and go to the cities to look for work. So that makes even more mouths to feed, even more angry and unemployed people demanding cheap food and jobs.

What the government ought to do is to help the small farmer, not the big man. Even if each farmer only has a small increase in production, the overall effect would be enormous, and above all it would make the country areas centres of prosperity and not derelict backwaters which the young people flee away from as fast as they can. The government should forget tractors and mechanisation and big farms, and concentrate on giving guaranteed credit to small farmers, and inputs like fertilisers and new seeds that they can really use.

Minister of Agriculture:
All right, I hear you. The small farmers are losing out at the present time, it is true. But look at it from my point of view — the problem is too big to be tackled slowly. Our population is growing at two to three per cent per

year. We need big increases in food production and we need them now — that is why I must support the bigger and more efficient farmers like Alhaji Musa. However, I have listened to your complaints, and in the future our policies will also include more help for the small farmer. Let us all progress together, rich and poor, to greater prosperity for Nigeria.

? You are a rich man. Originally you came from a village where your father was a small farmer. Your brother now farms his land. You see your rich friends investing in land, and you know it is a good investment. You have heard from your brother that several farmers are badly in debt and would probably be willing to sell their land. You could buy it and you do want a secure investment for your capital but you know it would mean even poorer lives for those who sell to you. Your brother urges you on as he wants to be the farm manager, but how can you make other people landless? What do you decide to do?

GOVERNMENT DEVELOPMENT SCHEMES

Before the Second World War, the government of Nigeria (then the colonial government) left agriculture very much to itself. Its main activities were to set up research stations and to introduce new machinery and ideas on a very small scale. After the war the attitude changed, and the government began to concern itself with economic development, and especially agricultural development. A series of national marketing boards was set up in the 1940s to try to guarantee prices to farmers. These covered the main export crops of cocoa, groundnuts, cotton and palm-oil. In practice the boards were inefficient and expensive to run.

The other main post-war initiative of the government was the establishment of agricultural development schemes. The 1947 Shendam Scheme, for example, attempted to settle hill people from the Jos Plateau on the plains. Another idea was the attempt in 1959 to establish co-operative farms on the model of the Israeli *moshav* (similar to the kibbutz but a collective in which people have their own strips of land from which they keep the profit) in south-west Nigeria. Neither of these ideas, nor others like them, were particularly successful, and they made little impact on the majority of farmers. Nonetheless the attempts persisted.

The idea of persuading or coercing traditional farmers to move to a new 'model' settlement and take up 'modern' agriculture is an old one. It has rarely been popular with those chosen to participate, and the schemes have seldom been judged successful from any point of view. Nomadic communities in particular, and shifting cultivators like those on the Jos Plateau, have always been key targets for this kind of development. In 1977 someone writing about development in West Africa gave a high priority to 'the stabilisation of currently nomadic communities, giving them a sense of civic consciousness and making it easier for the benefits of social progress to reach them'.

From the point of view of the government, it is almost impossible to bring facilities like schools or health clinics or the provision to vote to isolated communities far away from roads. Also, as we have seen, traditional agriculture and ways of life tend to look 'backwards' to the person in a government office, and to call for drastic transformation. Equally, from the point of view of the traditional farmer, although he might want to be able to use a school or a clinic, he is likely to value his own freedom and way of life, and he certainly knows the value of his own proven system of agriculture. Also it is likely that he will not trust the government very much — he knows that once he is resettled he can be more easily taxed, and he will be much more vulnerable to any other changes the government might wish to introduce.

THE NIGER AGRICULTURAL PROJECT

Perhaps the best example of one of the problems of a post-war settlement project is the Niger Agricultural Project at Mokwa in the

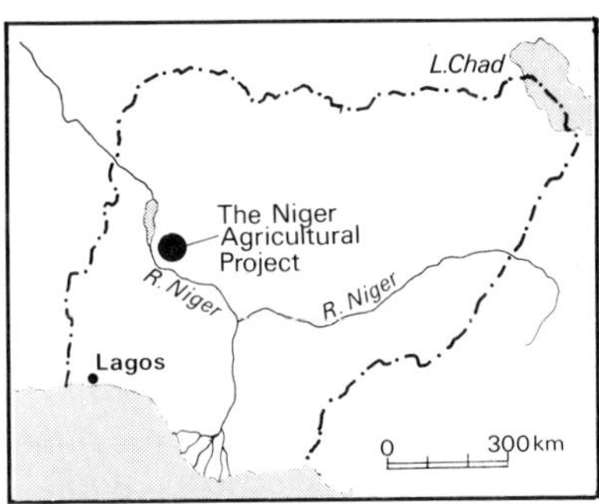

Figure 9 *Location of the Niger Agricultural Project*

middle belt (Figure 9). After the war there was a powerful feeling that the same meticulous and large-scale central planning that had gone into the military machine could be harnessed for peaceful purposes. A British mission was sent to West Africa to find possible places for large-scale cultivation of crops from which vegetable oils could be extracted. They suggested that an extensive area near Mokwa could be cleared and developed for growing groundnuts using mechanised methods. In 1948 a project covering 50 square miles was fixed upon.

It was felt that the 'plantation' approach where the land was taken over by the government and run as one unit was not suitable in northern Nigeria, so instead the project was established as a settlement area. The Niger Agricultural Project Company was to clear land, and plan and control agricultural operations. The farming, however, was to be done by people from more densely settled localities in the country. The plan was to combine the benefits of large-scale mechanised cultivation methods with those of individual care by farmers of their own holding. The scheme would grow sorghum for local consumption and groundnuts for export as oil to Europe. It was a good plan, but unfortunately it did not work.

The first task was land clearance. Initially progress was rapid, with 740 hectares cleared in 1949 and 1300 hectares the next year. However, mechanical clearance was expensive, and heavy crawler tractors were needed. Even then, the bush grew again, and hand clearance was found to be necessary to prevent regrowth. Next there were problems persuading settlers to come to the area. It proved impossible for political reasons to bring in settlers from outside the local area, yet those living nearby had enough land already and had no desire to join the scheme. They knew the project area, and disliked the scarcity of water and the poor soils. Those settlers who did come objected to the small two-room houses provided for them to live in — most men had more than one wife, and the houses were small, and widely-spaced, which was unpopular. There were also no funds for public buildings.

On top of this, the farming itself did not go well. It was hoped to demonstrate the benefits of mechanised production at an early stage so that the farmers would adopt it enthusiastically. Unfortunately, in the first two years the average income being earned by the settlers was probably only a half to two-thirds of that which the local farmers were getting with smaller farms and without mechanisation. The new methods looked so unattractive to the farmers that it was said that 'when local farmers were asked to take part in the scheme, they must have thought they were being called to the rescue'.

By 1952 it was clear that the scheme would never make money, but it was still thought that it could contribute to rural welfare. By 1954 it was realised that even this aim was not being achieved, and the scheme was abandoned. None of its primary aims were fulfilled. There was no increase in cereal and oil seed production, the village communities did not develop, and the project failed to establish much new land, or improve its fertility and productivity as had been hoped. Looking back it is clear that too much was attempted too fast in the Niger Agricultural Project. There was too little information on the area and its existing agriculture, and the plan for developing the area with settlers was unproven. The project had been summed up like this:

'Changes in the social habits of the local farmers were being superimposed upon drastic innovations in their agricultural technique. However desirable such changes may have seemed, it was a most unfortunate decision to thrust them on the farmers before the profitability of the new technique had been clearly proved. Since this profitability was by no means apparent, it is not surprising that the settlers felt their existing way of life was being disturbed to no purpose.'

WAYO SETTLEMENT PROJECT

Imagine that the Wayo Settlement Project has been planned by a state government on an extensive stretch of well-drained soils close to a river in the middle belt of Nigeria. Part of it is being developed with machinery as a state farm, and part is to be cultivated by smallholder farmers brought to settle on the project. Each farmer will receive a four hectare block of land and a new house made of mud with a tin roof. However, he will have to sign papers as a tenant for ten years, and pay back the cost of the house out of the promised profits from farming over this period. The new houses may also get piped water, but this is expensive and could make the scheme too costly to build, so the government has not made any decision on this yet. The farmer will get advice with his farming, and credit to buy seeds, fertiliser and pesticides. However, he will have to repay any loans that he receives, grow maize instead of his traditional crops of millet and sorghum, and market his crop through the scheme only.

An official from the project is visiting a village in the hills nearby to try to find farmers to come to settle and farm. The village is small and isolated, and the people are traditionally suspicious of all outsiders. Most of their farming is for subsistence and not for trade, and they are not familiar with the maize and other crops they will have to grow on the scheme. They also have many goats and sheep as well as some cattle which they fear may not be allowed on to the scheme if they move. They are used to moving their farm plots every few years, rather than concentrating on one place and maintaining its fertility. Also they like the hills with their cool winds and good water supplies — they have traditionally seen the valley as a place of sickness. At the same time they realise that they are missing out on things like schools which villages on the plain have, and they know that they need to have money to buy things like bicycles, radios and colourful imported cloth which they see in the markets they visit.

Divide into three groups.

The first group will prepare a speech for the official from the Wayo Project who has been sent to find recruits. He will try to persuade the farmers that they want to join in, and will have to be ready to answer their questions. You will have to decide as a group what you can offer the farmers to attract them, but remember that the government is making decisions about the scheme still, which will affect what you can offer.

The second group will prepare a speech for the village headman. You will need to decide what his advice should be to his people, particularly to the younger men who might be attracted to the project. He will surely have many questions he will want answered, and will probably want to move slowly, whereas the scheme official needs recruits quickly.

The third group represents a visiting agricultural consultancy firm. Your job is to write an appraisal of the prospects of the project. You should analyse how successful the official was in attracting settlers, and assess how likely it is that his promises would be fulfilled. You should try to predict what might happen to the farmers over the next ten years.

IRRIGATION: THE ANSWER TO DROUGHT?

Print 4 *Ojirami irrigation dam in Bendel State, Southern Nigeria*

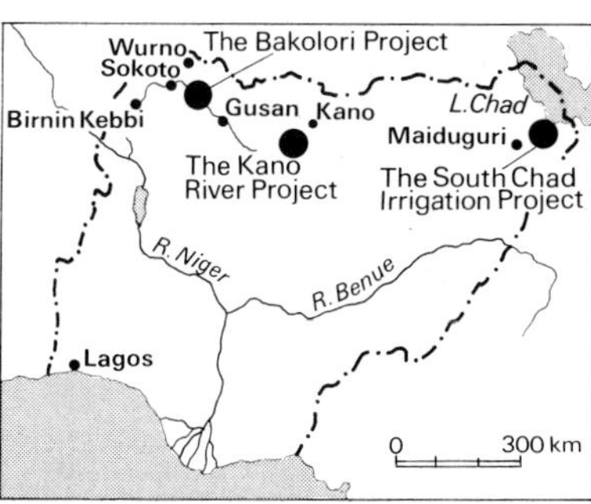

Figure 10 *Location of major irrigation schemes in northern Nigeria*

The limitation which the low rainfall and short growing season in the north of Nigeria places on agriculture can be offset to some extent by irrigation. In theory, if water can be supplied throughout the year, crops can be grown one after the other. There are, of course, other factors affecting crop growth, for example, high temperatures in the dry season which can prevent proper flowering and hence grain development in some crops. Nonetheless, it is lack of water which is the main constraint, and there have been suggestions for irrigation in Nigeria as far back as the beginning of this century. However, apart from a few small pilot projects, nothing was done about irrigation until the 1960s.

In 1966 the Food and Agriculture Organisation of the United Nations (FAO) wrote:

> *'Full exploitation of the agricultural potential of Nigeria requires development of the water resources of the country for irrigation.'*

At the same time, it was realised that irrigation could not be developed instantly by any means:

> *'It must be recognised that the accomplishment of this objective will require very heavy capital investment and the indoctrination of farmers in radically different agricultural techniques; it must therefore be accomplished under a very long programme.'*

So, irrigation was seen to be very important, but not at all easy to get right.

Despite these cautious words, irrigation development went ahead rapidly in Nigeria. There was a big FAO study of the Sokoto River Basin in the north-west of the country published in 1969 which suggested a series of dams and irrigation schemes. The first stage of this development, the *Bakolori Agricultural Project*, was begun in 1974 (Figure 10). In 1978 the dam on the River Sokoto which would hold wet season floodwater for use in the dry season was finished, and work began on the irrigation scheme itself. The whole project covers 30 000 hectares of flood plain and terrace land downstream of the dam is supplied with water from a 30 kilometre long concrete-lined canal.

At the same time that work began on the Bakolori Project, two other irrigation schemes of over 10 000 hectares each were being built in the north of Nigeria. The *Kano River Project* (18 000 hectares) is supplied by the Tiga Dam near Kano, and the *South Chad Irrigation Project* (27 000 hectares) pumps water out of Lake Chad. Across the whole of Nigeria, eleven River Basin Development Authorities have now been established to plan and develop water resources. In the south, flood control is a major concern, although this too can be a big contribution to agricultural development. In the north, the irrigation schemes and associated dams are the most important projects. Although the area of irrigated land fully developed was estimated in 1982 to be only 30 000 hectares, there are plans to develop a full 2 million hectares in the near future. Irrigation has come to stay in Nigeria.

HOW WELL ARE THE NEW SCHEMES WORKING?

The answer to this question is, unfortunately, not very well. There are a number of problems. First, they are expensive to build, and take vast amounts of the Nigerian government's income from oil. Bakolori is likely to cost over 300 million naira to build, which for only 30 000 hectares of irrigated land is very costly. Secondly, the costs of maintaining and running the schemes are very high, and it is proving difficult to find crops which have a high enough value to cover those costs. Crops like tomatoes are theoretically valuable enough, but only if processing plants are built, and their construction is often delayed or not planned at all. Farmers on the schemes also often refuse to grow the special high value crops, preferring to ignore instructions and grow their familiar food crops like sorghum, rather than risk a new crop and an unproven marketing system. In addition to the simple running costs of the schemes, interest has to be paid on the money borrowed to build the dam and other structures, and this is an extra financial burden faced by the new irrigation schemes.

There are also social problems on the new irrigation schemes. In order to understand how these come about it is useful to look in detail at one scheme in northern Nigeria. The problems in this case were very great, so it is easy to see what happened, but of course not all schemes are as bad as this. Indeed one hopes that the lessons of this scheme can be learned, and the same kind of problems minimised in the future.

THE BAKOLORI SCHEME — A CASE STUDY

The scheme lies in an area of low rainfall, confined to 3–4 months of the year. Traditionally all cultivation has been done during this short period. River discharge is great during the rains, but the rivers are virtually dry for 8 months of the year. It was realised at an early stage that if some of the plentiful wet season flow could be stored for use in the dry season it could be used for irrigation. Detailed studies of one valley located a suitable site for a dam, and aerial photographs and soil surveys identified an extensive area of land nearby which was adequately drained and suitable for irrigation. Development began.

The project involved a dam, a large reservoir 120 square kilometres in area, and a canal 30 kilometres long to supply the irrigation area. Both the reservoir site and the irrigation area were quite densely inhabited. There were several thousand people in the reservoir area, and they all needed to be resettled. Plans were made to build proper resettlement villages, and to survey all the property to be flooded so that compensation could be paid. But time ran out and at the last moment the evacuees were moved to three sprawling settlements without proper water supplies, following a perfunctory property survey. The new houses, with square mud walls and tin roofs in place of the traditional thatched huts, were all laid out in rows — they looked, as one commentator put it, like a prison camp. The houses were not well made and started to collapse in the rains the next year, while the land around was unsuitable for farming. To add to the misery of the evacuees compensation payments were delayed.

In the irrigation area itself, a network of canals down to the level of a 30 hectare field had to be constructed, and land levelled so that it could be irrigated. Part of the area was to be irrigated by gravity, water being syphoned out of field canals by plastic pipes and allowed to flow slowly across the surface of the carefully levelled field, sinking in as it went. Any surplus would be removed in another network of field drains. In areas where the land was too steep for this surface irrigation, sprinklers fed from buried pipelines were used, powered by electric pumping stations. Both kinds of irrigation required a large amount of engineering work on the farmland. The old farm boundaries and most trees had to be destroyed and a completely new irrigated landscape was created.

The idea was that all the work could be done in one dry season, when the farmers would not be using their land anyway. In theory the engineers could take over the land, develop it, and hand it back in time for the next wet season, all ready for irrigation to begin in the following dry season. In practice this programme proved far too complicated, the land development fell behind schedule and the process of reallocating farmland was too slow. Instead of losing their land for a short period, some farmers were without land for several years. Without land on which to grow food, they had nothing to live on. Eventually they were compensated for trees destroyed during land development, and for crops they were prevented from growing, but there were delays, and a lot of dissatisfaction.

Both the farmers whose land was flooded by the reservoir and those on the scheme felt that they had been cheated. Eventually they rioted, blockading the project several times. They were promised full compensation, but still there were delays. An account of one of the blockades is given below.

ANGRY FARMERS HALT WORK ON N200M. DAM

— For non-payment of compensation

Angry farmers have halted work at the 200 million naira Bakolori Dam near Talata Mafara in Sokoto State.

Reports reaching us from Sokoto said the farmers have blocked all entrances to the dam site with huge stones and logs of wood and formed an impregnable picket line. They numbered over 5000.

In an attempt to disperse the farmers and remove the barricades the police were reportedly given a tough time in which one of their Landrovers was completely burnt.

Our correspondent in Sokoto, Ibrahim N. Salibu, reports: When we arrived at the scene, we were given a hostile reception.

Many of the irate farmers converged on us throwing out daggers, home-made revolvers, dane guns, clubs, cutlasses and other lethal weapons.

They immediately engaged in threatening dialogue with us asking 'who sent you here, what do you intend to do, why should you come here at all, you must not take any photographs here'.

Later we were allowed to interview two of their spokesmen who asked me to swear by Allah that I have nothing to do with the Sokoto state government and that I was from Kaduna. It was then the interview started. According to their spokesmen, the farmers were poised for a showdown and were even ready to damn the consequences, in the event of confrontation with the government.

The spokesmen said about 800 000 farmers were affected and complained that both their houses and farms have been taken over for the dam on the pretext that they would be fully compensated in two weeks time. 'The two weeks have extended to three years and we are now fed up with the promises,' they said.

They added that they had not intended the issue to drag on for so long, without any action, but the Sultan of Sokoto had promised them full payment, but the authorities turned round to say that they would not be compensated for the houses and farmlands but only for their economic trees.

They recalled that the initial agreement was that cement block houses would be built for them and new farmlands to be provided in place of those taken over. But until now only a handful of them have received even a token fee for the 'economic' trees, they lamented,

'Look at our number here, many of us have parted ways with our families owing to poverty and hunger, a situation imposed on us by our fellow human beings.'

They said they were there not to beg for compensation any more or for anybody to plead with them again but to die so that their sufferings could be over.

Meanwhile all efforts to get the authorities to talk to me on the issue proved abortive.

When I called at the State Police Headquarters to see the Commissioner of Police, Alhaji Salisu, Zakari Daura, his secretary, asked me to fill a form and wait, which I did for more than one hour without reply.

When I enquired again, I was told that the commissioner was busy, but has, however, sent for the Police Public Relations Officer.

After signifying my intention to leave, the secretary asked me to leave any phone number so that he could ring me whenever the PPRO reported to the commissioner but nobody rang.

The Chairman of the Caretaker Committee, River Basin Development Authority, and his acting general manager were also said to be at Talata Mafara apparently trying to resolve the issue.

The Bakolori Dam project was formally commissioned by the retired Chief of Staff in December 1978.

(*Source: New Nigerian*, 28 February 1980)

The numbers of farmers said to be involved in the riots is undoubtedly exaggerated. But note the sense of total frustration felt by the farmers. They obviously believed that there was no other way to get their message across to the project management.

In this project a theoretical scheme was worked out on paper and put into practice without properly consulting the farmers who would be involved. This is in fact a very common pattern of project development — it is exactly how the Niger Agricultural Project described in the last section was begun.

? **1** Do you think farmers should be consulted before beginning a project? If this had been done in this case, what would the implications have been for the size of the project and the speed with which it could be built? Even if slower development might produce less adverse aspects in a project, why might the government feel it had to develop it quickly?

? **2** You are a geography graduate, working now as a reporter for *The Guardian*. Explain to your English readers the reasons behind the farmers' actions.

3 Take the role of a government official in the Department of Agriculture in another African country which is considering an irrigation scheme. Write a report on the Nigerian schemes. Your report will contain recommendations for your country.

SMALLER SCHEMES — A RECIPE FOR THE FUTURE?

Not all irrigation projects have to be big. There is an increasing interest within Nigeria in small-scale schemes, involving only a few farmers. It is widely hoped that a lot of the problems of large-scale schemes can be avoided on smaller ones. Small diesel pumps can irrigate up to 100 hectares, and there have been experiments with solar-powered pumps to irrigate one or two hectares. It is also being realised now that the traditional ways of lifting water out of wells or pools, the shadoof and even the basic bucket-lift can be very efficient. Recently, development 'experts' in Nigeria have been discovering how suitable and efficient some of the farmers' own methods are, and there is a reaction against the expensive and not very successful large schemes in favour of small-scale irrigation developments of all sorts.

INTEGRATED DEVELOPMENT?

Since the 1970s the Nigerian Government has put forward a series of national programmes to boost agriculture. Neither of the first two, the *National Accelerated Food Programme* (launched in 1973) and *Operation Feed the Nation* (launched in 1976) were particularly successful. Neither was aimed at the small farmer, who as we have seen is the backbone of Nigeria's agriculture. The latest programme, *Operation Green Revolution*, was launched in 1980, with the aim of reaching food self-sufficiency by 1985. This target was judged by many experts to be quite unrealistic, although the programme was in some ways quite encouraging. For the first time agriculture won more than ten per cent of government funding: over a tenth of the 4th National Plan budget (1981–1985) of 82 million naira was allocated to agriculture. In addition a massive 75 per cent subsidy on fertiliser was set by the government, and the contract signed for a fertiliser factory at Port Harcourt which will cost 450 million naira.

However, these grand plans have met a series of damaging setbacks, amid a period of considerable turmoil in Nigeria with military coups in 1984 and 1985. Economic problems, particularly a fall in oil prices, and drought in 1983, 1984 and 1985 have caused problems, and in the 1985 budget agriculture's share fell to six per cent. The problems are described in the following article.

Print 5 *'Operation Feed the Nation' poster in Sokoto Village*

Michael Holman sets the scene for the vital farming industry

Drive to boost food output

The so-called 'Green Revolution', launched as a national priority by former President Shagarl in April 1980, set out to transform Nigeria's dismal agriculture sector which has seen stagnating food production and declining output of cash crops over the past decade or more.

The importance of the sector can hardly be overemphasised: directly or indirectly it provides a livelihood for over three-quarters of the population.

Inadequate extension and research services, neglect of the rural infrastructure, a low level of investment prior to the current development plan and perhaps above all the lure of better-paid jobs in the towns and cities during the oil boom had severe consequences.

Once a major exporter of cocoa, groundnuts, rubber and oil palm, and almost self-sufficient in food, today only cocoa is a significant foreign exchange earner (and, even here, Nigeria is steadily falling in the ranks of world producers) while food imports have soared from barely $100 million a year to around $2 billion in little more than 10 years.

'By international standards, and at existing exchange rates,' comments one agriculturalist, 'Nigerian agriculture has become a high-cost, inefficient and highly subsidised producer.'

The new military government must rapidly decide its attitude to some of the key reforms which are needed. The most contentious issue is pricing which, in turn, is inextricably linked to the devaluation of the naira, the central issue confronting Nigeria's economic planners today.

On the face of it, points out one report, there should be sufficient producer incentives. Internal prices for crops are, on average, about 50 per cent higher than obtained internationally, when converted at the official exchange rate.

The explanation for the failure of most growers to respond, some observers believe, is three bottlenecks: lack of adequate technology, erratic supplies of inputs, and transport problems including poor feeder roads.

Yet the report goes on to attach as much weight to the fact that while prices may be high compared to international levels, in reality the picture is different. The naira is overvalued, it maintains, a factor in the rapid climb in food imports, which in turn limit the rise in producer prices.

The report calls for a rationalisation of trade policies in agriculture. Policies in the past, it says, 'have been based largely on quantitative restrictions and have introduced a large element of uncertainty into the incentive structure.'

Thus, large imports of wheat and rice have at various times pushed prices down. The alternative is a tariff based system set at levels which will not erode local incentives.

This is one of the areas under discussion with the World Bank. Meanwhile government has to cope with two major setbacks to the Green Revolution target of self sufficiency.

Austerity measures

The first is the impact of the fall in oil prices and production, and agriculture has not been shielded from the impact of austerity measures.

Federal and state contributions to agricultural development projects (ADPs), which form the backbone of the Green Revolution, have fallen substantially.

In addition, the severe strain on foreign currency allocations have led to cuts or delays in the imports of crucial inputs — notably fertiliser.

Shortages of fungicides and insecticides will damage tree crops such as cocoa and coffee in particular, while the poultry industry — one of the few success stories in the food programme — is already seriously affected by the unavailability of essential feedstock.

The second setback is the current drought affecting at least nine northern and central Nigerian states — Gongola, Borno, Kano, Sokoto, Niger, Plateau, Bauchi, Kwara and Benue.

The extent of the damage on 1983 production was still being assessed at the time of writing, but preliminary reports all point to considerable animal and food crop losses.

Kano state — which normally 'exports' food to neighbouring areas — is particularly hard hit. Rainfall is the lowest this century, at between half and a third of the normal level.

And the drought comes on top of an earlier calamity — the outbreak of rinderpest at the beginning of 1983.

Official estimates of the damage are not available, but one report suggests that half a million head of livestock died, and overall losses are put at 'hundreds of millions' of dollars.

The setbacks followed some encouraging signs in 1982, marked by good weather. Food production that year, according to the Federal Ministry of Agriculture, rose 'at least three per cent . . . a significant increase compared to the one per cent annual average recorded before the inception of the Green Revolution programme.'

Despite the increase, however, there is a long way to go. Bulk agricultural commodity imports in 1982 came to nearly $1 billion. Although this was 14 per cent down on the 1981 level, the reason was the government's need to cut imports rather than a reflection of higher local output.

Total food and agricultural imports (including smuggled items) were put at $2.25 billion in 1982, down 20 per cent on the preceding year for the same reason. But both the main staples, wheat and rice, show an exception to the trend, with the former up 19 per cent in 1982 over 1981, and rice imports 11 per cent higher in 1982 despite an increase in domestic production from 1.2 million tonnes to 1.4 million tonnes.

Clearly the objectives of the Green Revolution are a long way off.

(*Source*: *The Financial Times*, 23 January 1984)

INTEGRATED AGRICULTURAL DEVELOPMENT PROJECTS

One new departure in the Operation Green Revolution, however, has survived. The old emphasis on large development projects was replaced by a focus on the small farmer. In large part this change in emphasis is due to the prompting of the World Bank. They have been involved since 1974 in a series of *Integrated Agricultural Development Projects* (IADPs) in Nigeria. These are jointly funded by the Bank and the Federal and State Governments. The World Bank has contributed $1.2 billion since 1975. The projects are deliberately aimed at the smaller farmer rather than the large landowner, and attempt to increase agricultural production by overcoming the problems met by the small farmer, and not by imposing some wholly new large-scale farming pattern.

Print 6 *Agricultural trainees on a model farm in Nigeria*

Table 10 *Indicators of increased wealth on one Nigerian IADP*

	*Increase in number of goods owned 1976–79**	
	Index 1976	*Index 1979*
Watches	100	349
Motor cycles	100	516
Sewing machines	100	288
Radios	100	277
Bicycles	100	252
Clocks	100	179
Hurricane lamps	100	140
Donkeys	100	125

% increase — value of possession 26%

The IADPs involve the provision of services to farmers. These include extension advice, marketing services, agricultural credit, new inputs such as seeds and fertilisers, as well as rural facilities such as all-weather laterite roads and improved water supplies. The idea is that by offering such an integrated package of services and necessary infrastructure in the project area a sufficient number of the constraints on the small farmer will be removed and he will be able to expand, intensify his farming and grow more food.

Between 1975 and 1980 the original three projects boasted the construction of 163 farm service centres, 132 dams for water supply, and 1700 kilometres of unsurfaced rural roads. Almost 152 400 tonnes of fertilisers were distributed. Impressive results in terms of increased agricultural production were claimed. These included dramatic increases in the yields of some crops, and as a result increased value of overall agricultural output (Table 9). It was suggested that as a result of this extra production, incomes in the project areas had increased. One measure of this was the recorded rise in the number of consumer goods (Table 10).

Table 9 *Evidence of increased production on Nigerian integrated agricultural development projects*

Crop	*% Increase in yield 1976–79**	*Increase in value of production (million naira)*
Millet	+ 130%	+ 16
Maize	+ 11%	+ 0.9
Sorghum	+ 57%	+ 15
Groundnuts	+ 56%	+ 6
Cotton	+ 18%	− 7
Cowpeas	− 15%	− 4

* For both these tables this is the most recent date available.

The statistics look very impressive, but not all observers are happy about them. It is difficult to see, for example, how millet production could increase by 130 per cent, since crop breeding has not yet produced a high-yielding variety, and fertiliser adds little to the yield of the usual varieties. Another problem is that although there were evaluation programmes on each project, they did not take into account the pace of rural change outside the projects, and therefore gave an inaccurate picture of how well the projects were doing.

There has also been some criticism of the way in which the IADPs work by selecting certain progressive farmers, called master farmers, and concentrating efforts on these men. It is hoped that as the master farmers adopt the new ideas and show them to be successful, other farmers will copy them. However, it is argued that as a result of this policy, the master farmers are becoming wealthy in contrast to their neighbours, and instead of investing in further development of their land they are giving up farming and going into trading, money lending or other lucrative activities. In other words, critics argue that even the IADPs which claim to focus on the needs of the small farmer are really helping the slightly richer person. They seem to be helping those most able to help themselves, and instead of helping the whole community equally, they may be promoting division between richer and poorer farmers.

Despite this criticism, it is clear that the early IADPs stand out as success stories in the history of Nigerian agriculture. The World Bank itself attributes this to a number of factors.

1 The areas chosen for the projects were favourable, with good soils and adequate rainfall.

2 The projects focused on the priorities for development which were those the farmers themselves thought most important (in order: rural roads, water, farm inputs and improved extension).

3 The projects involved no compulsion on farmers to become involved: attractive services were offered to farmers to accept freely if they chose.

4 The projects were large and ambitious and involved a lot of skilled and enthusiastic staff from overseas who worked hard to make them a success. As a result they attracted attention and interest within Nigeria at all levels from government downwards.

However, the big question is how easily can the successes of the early projects be reproduced in other areas where the environment is less suitable? The answer seems to be that the formula is far from a guarantee of success. Nonetheless, Nigeria has gone ahead to push the core elements of the IADPs nationwide. The *Accelerated Development Area Programme* is intended to create a project involving the core elements of the IADPs (extension advice, agricultural inputs and rural roads) in all 19 states, at a cost of 2000 million naira. By 1982 and 1983 a number of projects had been announced, but the provision of funds from the state and federal governments had become increasingly erratic and the World Bank was beginning to demand a tightening up of procedures before they committed more money.

By 1984 state wide projects with World Bank funding had begun in a number of states, including Bauchi, Kano, and Sokoto. The loans for these three states alone were US $421 million, and there are signs of exciting new developments such as experiments with small-scale irrigation of the extensive floodplain areas.

RURAL DEVELOPMENT PLANNING EXERCISE

An investigation of an area of 500 square kilometres of land in and around a river valley in northern Nigeria has come up with a set of alternative ways in which the area could be developed. You will represent one of a series of groups concerned about the form that the development should take.

The area consists of 20 000 hectares of good flood plain land on both sides of the river, and a further 30 000 hectares of higher land of mixed quality away from the river. The rains in the area are short, lasting only four months. The river flows strongly for five months, but is virtually dry for the other seven. The main crops in the area are millet, sorghum, groundnuts, cotton and cowpeas, but a wide variety of vegetable crops are grown in wet corners of the flood plain after the end of the wet season. Crop yields vary greatly from year to year, and although usually good, they can be very poor because of drought.

The mapped land lies on both the east and west banks of the river (Figure 11), but only one side is served by a metalled road. There is no bridge across the river for 100 kilometres upstream or downstream. A sizeable town lies on the road in one corner of the area (it contains 20 000 people), and there are two large villages and a number of small villages in the area on both sides of the river. Access to them is by dry season dirt roads and paths. The river is crossed by canoe in the wet season and on the dry river bed in the dry season.

The area is chiefly inhabited by Hausa farmers who practise permanent cultivation, and also have some livestock. The land is intensively farmed in the flood plain, but there is some fallow land away from the river. The main town is an important regional centre with a secondary school and a small hospital as well as a large market and government offices. The two large villages each have a small primary school and a clinic, but most of the villages have no facilities. Although there are no government schools many of the farmers read and write in the arabic script of the Koran, taught by the *malams* in each village.

The villages get their water supplies from shallow wells, and when they dry up they use the river itself or residual pools and springs. The disease hazard from these water sources is great, and malaria, bilharzia and various gut infections are widespread. Throughout the area a number of young men go away in the dry season to seek work in other areas further south, and in towns especially. They began to do so in the drought a decade ago and have continued, partly because they value the contact with the changes that are going on outside the villages and partly because their families increasingly need the cash income they obtain.

There are three alternative proposals for the development of the area:

1 The large-scale irrigation scheme

A dam on the river and a canal to carry water to irrigate part of the flood plain, and a small area of higher land on both sides of the river. The dam will flood 100 square kilometres of land, and the area to be irrigated will be 20 000 hectares (15 000 hectares of flood plain and 5000 hectares of other land). Under irrigation, it is claimed that the yield of crops will be doubled in five years. The proposal is to acquire all the land which can be irrigated, compensating all the farmers, and to manage it efficiently for irrigation. The rest of the area will be left undeveloped, except that some of the proposed benefits of development may percolate through to surrounding areas. Apart from the irrigation works, new roads into the area and a bridge over the river will be required. The cost is expected to be about 200 million naira.

2 Small-scale irrigation development

A series of low barrages on the river to retain just some of the wet season flow for use in the dry season, and a

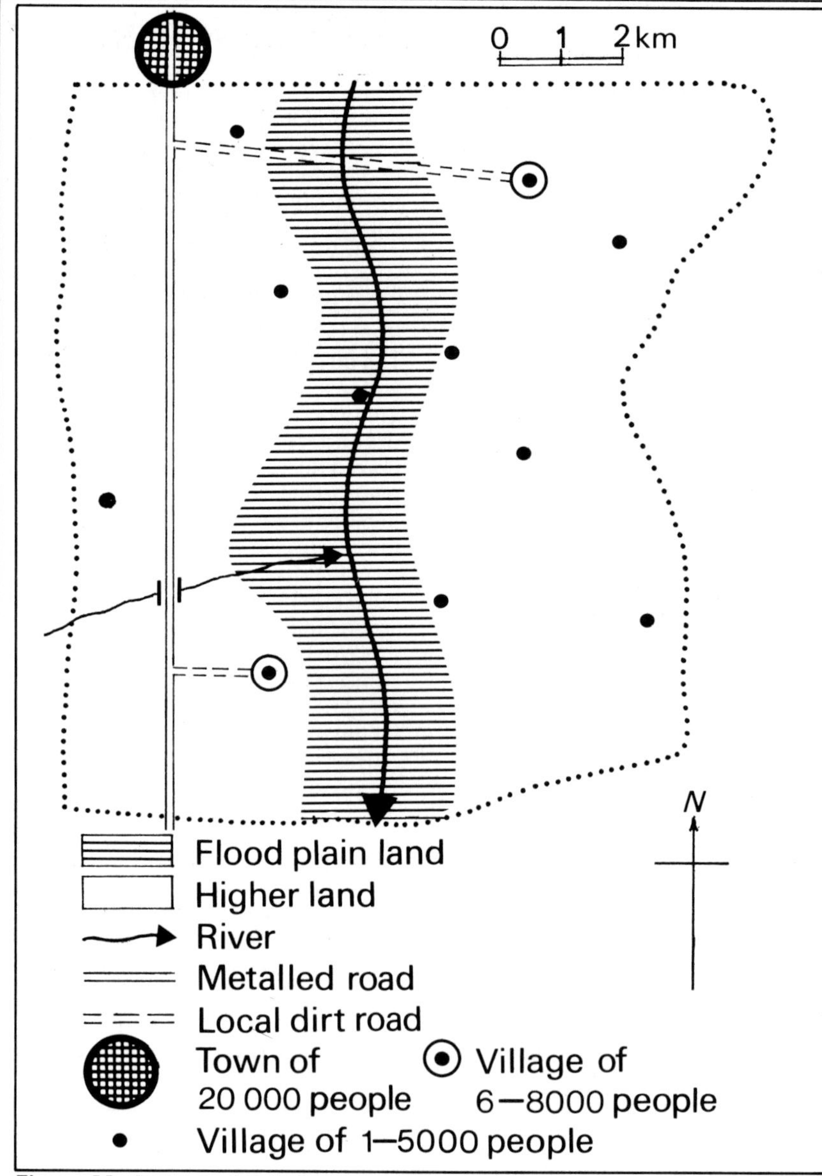

Figure 11 *Proposed development project area*

series of small pumps to irrigate riverside farmland. The barrages will create small lakes totalling 1000 hectares. The total area to be irrigated will be 8000 hectares on both sides of the river. Yields are expected to rise by 50 per cent in ten years. The proposal is to set up an extension and marketing service in the riverside villages and to develop co-operatives which can purchase diesel pumps at subsidised cost to irrigate the land. Farmers will keep their own land, and build their own canals to take the water from the pumps to their fields. Existing tracks will be improved, but no major new roads or bridges will be built. At a later date the agricultural services could be extended through the whole area. The cost is expected to be about 70 million naira.

3 Dryland development
A system of credit, extension, agricultural inputs and roads and water supply over the whole area of 50 000 hectares. No major engineering works will be built, but an extensive network of new all-weather roads, service centres and wells will be created. By concentrating efforts on master farmers and by improving rural infrastructure it is expected that crop production over the whole area will increase by 25 per cent over ten years. The cost is expected to be 100 million naira in the same period.

Divide into five groups. Each group represents an interested party in the decision over which form development in the valley should take. The groups are as follows:

a An engineering planning company, Greengrow Irrigation Limited
You are most interested in the large-scale irrigation project. Most of your staff have worked on such schemes before, and you believe that if the scheme is built well it can be a success. You are also interested in further work after the planning is finished so you are keen to see an important scheme proposed. However, you are aware that in the past it has been difficult simply to take over the land belonging to farmers, and that the problems of survey and administration of compensation are great. You also know that the scheme will be expensive, and perhaps, therefore, it will not be attractive to the government who will have to find the money for it. You are therefore also interested in the possibility of the small-scale irrigation. You will need to know what the farmers think about the different proposals, and about the attitude of the River Basin Authority (one of the other groups) whose recommendation to the Federal Government will carry great weight. You do not believe that development of the area without irrigation can succeed because of the long history of drought.

b An agricultural development company, Dryland Futures Inc.
You are experts in planning and carrying out agricultural schemes involving extension advice, credit and rural services of the kind used on the Nigerian World Bank projects. You therefore support that approach to development. However, you are also interested in the possibilities of small-scale irrigation. Your experience of working with farmers leads you to doubt that large-scale schemes will

work, but you are aware that the relatively high cost of the dryland development option and the low increase in production it promises may not make it attractive to government. Nonetheless, you are convinced that it will help the area even if it is not a success financially.

c Rich farmers, represented by Alhaji Ibrahim
You are a group of the richer farmers in the area. You already have contacts outside the area for trading and other activities, and are aware of the rapid changes going on in other parts of the country. You can see the opportunities both within and outside agriculture if a major development project is begun. Obviously the opportunities will be different with different kinds of project, and you will need to find out what each type of development might bring you. Although you are concerned about your own position, you have considerable responsibilities for the whole community who tend to look to you for leadership, and you cannot ignore their needs or interests. The wider significance of the different kinds of development will thus also be important.

d Poor farmers, represented by Hassan Bakudi
You are a group of the smaller and poorer majority of farmers in the area. You have been forced to leave your villages to seek work in recent years, and have seen and heard something of the changes going on in Nigeria. However, you are deeply suspicious of change, especially radical change. You can see the possible benefits of a major development project, but you can also see many difficulties and dangers. You will want to find out exactly what each type of development means for you. Unlike Alhaji Ibrahim, you are worried, not about future benefits from development, but about the lives of your family.

e The Gulbi River Basin Authority
You are a Federal agency established to develop the land and water resources of the whole river basin. At first your main interest was irrigation, but early schemes were not a success and were expensive. Social problems on the schemes were embarrassing to you when they reached the national press. Recently you have turned your attention to small-scale irrigation and to rain-fed dryland farming, but you are still uncertain about the effectiveness of projects of this sort. You will be responsible for the project that is built in this area, and must present a good case to government for the money you will have to spend. You want, above all else, for the project to be a cost-effective and peaceful contribution to national welfare.

? In your group, spend some time discussing which proposal, if any, you think should go ahead. You will have to defend your proposal in detail, so your arguments need to be well worked out. Each group then has ten minutes in which to put its case, saying which scheme they support and why, or else what their main concerns or reservations are. They can also present modifications of the set proposals if they wish.

Then you will have further time in which you can strengthen your case in the light of the arguments of other groups. If you wish you can have discussions with other groups — you can try to persuade them to do what you want, or to persuade them to change their minds on their own plans. At the end of this session, each group will make a short statement of their revised view on the proposals.

- If the proposals of each group were now sent to the Federal Government, which do you think would gain their approval?
- Which groups (if any) changed their minds? Why?
- How effective were the farmers' views in influencing the other groups? In practice, how could the farmers' opinions be included in planning?
- In practice which organisations (or which people) would have the biggest influence on development project planning?

WHAT KIND OF AGRICULTURE FOR NIGERIA?

This chapter has explored some of the problems of agricultural development in Nigeria, and some of the possibilities for the future. Go back to the first newspaper article and your paragraph written for the first exercise. Make a list of the questions you would now ask the Commissioner during a 20-minute interview he has granted you.

PART IV INDUSTRIAL CHANGE AND RESOURCE DEVELOPMENT

INDUSTRY AND RESOURCES — CHANGING PATTERNS

INDUSTRIALISATION AND DEINDUSTRIALISATION

One of the main contrasts in this section is between the deindustrialisation occurring in Britain (along with the less severe decline of some manufacturing industries in Western Europe and the USA) and the industrialisation of some newly developing countries. As you work through the section, look out for newspaper stories about industrial and manufacturing changes and difficulties being experienced by the Western world, and of the successful development of industry in some countries in the South. Be aware of possible bias.

REASONS FOR DEINDUSTRIALISATION

?

1 The table gives an immediate insight into Britain's industrial decline. Examine it and rank 'redundancies occurring' and 'employees in employment, net change'.
2 Before studying the first two chapters in this section, make as long a class list as you can of the reasons you think account for Britain's industrial decline. Then as you study the section keep a list of reasons given for the changing location of industry and at the end of the section rank the reasons from what seem to be very significant reasons to less significant.

Rates of job losses by industry, Great Britain, 1976–81; Redundancies occurring and net changes in employees (percentages of employees in employment, 1976, expressed at annual rates)

	Redundancies occurring			*Employees in employment Net change*[a]		
	1976–79	*1980*	*1981*	*1976–79*	*1980*	*1981*
Agriculture etc	−0.2	− 0.3	− 0.2	−1.3	− 0.7	− 1.7
Mining and quarrying	−0.5	− 0.8	− 1.6	−0.2	− 1.6	− 4.3
Food, drink, etc	−1.9	− 3.1	− 4.2	−0.5	− 5.5	− 5.8
Petroleum products etc	−0.7	− 0.6	− 0.5	+0.1	− 3.1	− 6.6
Chemicals etc	−0.7	− 2.7	− 4.4	+1.3	− 6.8	− 5.9
Metal manufacture	−1.9	−13.0	−10.1	−2.1	−18.1	−12.8
Mechanical engineering	−2.0	− 5.7	− 6.2	−0.9	−10.3	−10.1
Instrument engineering	−1.2	− 3.2	− 4.0	+0.0	−11.0	− 7.8
Electrical engineering	−1.8	− 4.5	− 6.6	+0.5	− 8.1	− 7.9
Shipbuilding etc	−2.9	− 4.7	− 2.8	−2.8	− 8.5	− 3.5
Vehicles	−1.3	− 5.5	− 6.7	+0.2	− 8.7	−11.5
Other metal goods	−1.1	− 6.4	− 5.9	+0.3	−11.8	− 9.2
Textiles	−2.7	−11.7	− 5.4	−3.0	−16.2	− 6.3
Leather etc	−2.7	− 4.8	− 3.5	−2.5	− 9.4	− 5.8
Clothing and footwear	−2.1	− 7.9	− 5.4	−0.5	−12.5	− 6.6
Bricks, pottery, etc	−1.3	− 6.4	− 5.1	−0.6	−12.0	− 8.9
Timber, furniture, etc	−1.6	− 4.0	− 4.0	−0.9	− 8.8	− 5.5
Paper, printing, etc	−0.9	− 3.0	− 3.6	+0.1	− 6.1	− 4.5
Other manufacturing	−1.7	− 6.4	− 4.7	−0.9	−15.2	− 4.4
Total manufacturing	−1.7	− 5.7	− 5.6	−0.5	−10.1	− 7.8
Construction	−1.8	− 3.0	− 3.9	−0.6	− 5.5	− 7.2
Services	−0.2	− 0.4	− 0.6	+1.1	− 1.9	− 2.4
Total	−0.8	− 2.2	− 2.4	+0.4	− 4.6	− 4.5

(*Source*: Unpublished data of Manpower Services Commission, Sheffield; *Employment Gazette*, Table 1.2)
Note: [a] Years ending December. All data are based on official estimates subsequent to June 1978.

Changing production costs and levels of profit have affected the location of manufacturing at a regional scale in Britain. We should not, then, be surprised at changing locations of manufacturing on a world-wide scale as labour costs, transport costs and raw material costs change. In *Changing Trade Patterns — Newly Industrialising Countries in SE Asia pages 312–327*, the new international division of labour and the international product cycle are described. These two ideas help to explain why location of manufactured goods is changing. Read the section through now and add to your list of reasons for deindustrialisation in the Western world.

FROM INDUSTRIAL SOCIETY TO INFORMATION SOCIETY

The years 1956 and 1957 were a turning point, the end of the industrial era. Confused, unwilling to give up the past, even our best thinkers were at a loss to describe the coming epoch. Harvard sociologist Daniel Bell termed it the *post-industrial society* and the name stuck. We always name eras and movements 'post' or 'neo' when we don't know what to call them.

It is now clear that the post-industrial society is the information society, and that is what I call it throughout this book. (In any case, Daniel Bell was one of the earliest, and perhaps the best, thinker on the subject, and much of what I have to say builds on his work.)

Bell's post-industrial society was misunderstood. Again and again, scholars told us the post-industrial economy would be based on services. At first glance, the notion seems logical, since we traditionally think in economic terms of either goods or services. With most of us no longer manufacturing goods, the assumption is that we are providing services.

But a careful look at the so-called service occupations tells a different story. The overwhelming majority of service workers are actually engaged in the creation, processing, and distribution of information. The so-called service sector minus the information or knowledge workers has remained a fairly steady 11 or 12 per cent since 1950. The character of service jobs has changed — virtually no domestics today, and thousands of fast-food workers, for example — but the numbers have remained quite steady; about one-tenth of the US workforce can usually be found in the traditional service sector.

The real increase has been in information occupations. In 1950, only about 17 per cent of us worked in information jobs. Now more than 60 per cent of us work with information as programmers, teachers, clerks, secretaries, accountants, stock brokers, managers, insurance people, bureaucrats, lawyers, bankers and technicians. And many more workers hold information jobs within manufacturing companies. Most Americans spend their time creating processing, or distributing information. For example, workers in banking, the stock market, and insurance all hold information jobs. David L. Birch of MIT reports that only 13 per cent of our labour force is engaged in manufacturing operations today.

It is important to acknowledge the kind of work we do because we are what we do, and what we do shapes society.

Farmer, labourer, clerk: That's a brief history of the United States.

The occupational history of the United States tells a lot about us. For example, in 1979, the number-one occupation in the United States, numerically, became clerk, succeeding labourer, succeeding farmer. Farmer, labourer, clerk — that is a brief history of the United States. Farmers, who as recently as the turn of the century constituted more than one-third of the total labour force, now are about 3 per cent of the workforce. In fact, today there are more people employed full-time in our universities than in agriculture.

The second largest classification after clerk is professional, completely in tune with the new information society, where knowledge is the critical ingredient. The demand for professional workers has gained substantially since 1960, even more dramatically than the rising need for clerical workers.

Professional workers are almost all information workers — lawyers, teachers, engineers, computer programmers, systems analysts, doctors, architects, accountants, librarians, newspaper reporters, social workers, nurses, and clergy. Of course, everyone needs some kind of knowledge to do a job. Industrial workers, machinists, welders, jig makers, for example, are very knowledgeable about the tasks they perform. The difference is that for professional and clerical workers, the creation, processing, and distribution of information *is* the job.

In 1960 the approximately 7.5 million professional workers were the fifth largest job category and employed about 11 per cent of the workforce. By 1979 that group had doubled to 15 million workers and made up some 16 per cent of the overall workforce.

The new wealth — know-how

In an industrial society, the strategic resource is capital; a hundred years ago, a lot of people may have known how to build a steel plant, but not very many could get the money to build one. Consequently, access to the system was limited. But in our new society, as Daniel Bell first pointed out, the *strategic* resource is information. Not the only resource, but the most important. With information as the strategic resource, access to the economic system is much easier.

The creation of the now well-known Intel Corporation is a good example of this. Intel was formed in 1968 when Robert Noyce and Gordon Moore split off from their former employer, Fairchild Semiconductor. Intel was started with $2.5 million in venture capital, but it was the brainpower behind the financial resource that led to the technological breakthroughs that brought the firm annual sales of $850 million by 1980. Noyce is credited with being the co-inventor of the integrated circuit and Intel with developing the microprocessor.

Noyce, the firm's founder, is well aware of which resource is the strategic one:

'Unlike steel, autos, and some others, this industry has never been an oligopoly,' he said about the field of semi-conductors. 'It has always been a *brain-intensive* industry, rather than a capital-intensive one.'

(*Source*: J. Naisbitt, 1984, *Megatrends*)

A word which is used less frequently perhaps than deindustrialisation is *post-industrial*. In the United States and in Western Europe, but less so in Britain, references are made to the post-industrial society, a society in a phase or stage beyond the production of the old manufactured goods. Both words, 'deindustrialisation' and 'post-industrial' describe, but do not really explain, what is happening. Perhaps a phrase such as ' from an industrial society to an information society' explains the changes a little better. Such a phrase at least points to the new raw material of the new age — information. Coal, iron, steel, fibre, chemicals provided the raw materials on which the old industrial society rested. Many people believe that the raw material of the post-industrial world is information — the collecting, buying and selling of information. John Naisbitt in his book, *Megatrends: Ten New Directions Transforming Our Lives* (MacDonald, 1984) describes the transition from 'industrial society' to 'information society' in the US. He considers that the trends are also evident in Western Europe.

The information society is billed as a high technology, *brain-intensive* industry. Computers lie at the heart of storing, processing, retrieving and using information which is, itself, the human-generated raw material of the post-industrial society. That it is a brain intensive industry is illustrated in the extracts below. They give some insight into the type of industry and the reasons for its development in Cambridgeshire. This development is being called 'The Cambridge Phenomenon'. This new, high techonology industry exists because we have generated an enormous market for information and we are using the computer as the key machine in storing the information to be bought and sold. Some possible side-effects could concern us. Is information likely to get more expensive and therefore less accessible in some senses? Is it likely to become more influenced by business decisions? Like the issue of whether we are wise or not to develop nuclear power — the subject of the last chapter in this section — these are deep questions which you will see being debated throughout the rest of your life.

THE CAMBRIDGE PHENOMENON — THE GROWTH OF HIGH TECHNOLOGY

? Read through the following account of the growth of industry in the Cambridge area. From the general text and the case studies sort out what the 'high tech' industries are dependent upon.

Regional context

Cambridgeshire, of which Cambridge located in the south of the county is the principal city, forms the western part of the region of East Anglia. Until some 25 years ago this region was long regarded as something of a rural backwater, albeit with fine historic towns such as Cambridge and Norwich and a prosperous agricultural sector, as well as a few towns such as Peterborough and Huntingdon with a successful record of manufacturing industry. The population was small and dispersed and the general sense of remoteness and slow change was underlined by the poor transport links with the rest of the country.

Around 1960, however, change started taking place rapidly, principally as a result of people and industry previously based in London and the South-east relocating to the region. Movement from these congested areas had of course been happening already, partly facilitated by the new and expanded towns programmes of the early post-war period. But initially the concentration of growth was mostly in towns closest to London and the South-east, and it was not until the 1960s that the scale of change in East Anglia became appreciable.

The single most important component of East Anglia's manufacturing growth, until the mid-to late-1970s, was the establishment of factories by firms located in London in particular, and also elsewhere in the South-east. More often than not, the move involved a complete relocation of the entire business, not just the establishment of a subsidiary operation, a reversal of the typical pattern for industrial movement into the more distant assisted regions. Another distinctive feature of industrial migration into East Anglia has been the relatively small size of the establishments involved, in keeping with the size of the local labour markets throughout the region. Several factors underlay the industrial movement: amongst the 'push' factors was the drive to escape the congestion of the London area and limited unionisation in East Anglia, which was a consequence chiefly of the historic dominance of agriculture as an employer.

During the 1970s an additional component of growth started becoming evident: the indigenous formation of entirely new enterprises. A survey of manufacturing firms in this category in the 1971–81 period has shown that East Anglia and Cambridgeshire in particular, fared particularly well by UK standards in terms of new firm formation.

Time profile and age structure

The graph below which charts the time profile of establishment of 261 firms, clearly shows a significant increase in births in the 1960s followed by an 'explosion' starting in the early to middle 1970s.

If the sheer number of companies is striking, then their youthfulness is even more so. Fully 60% of the

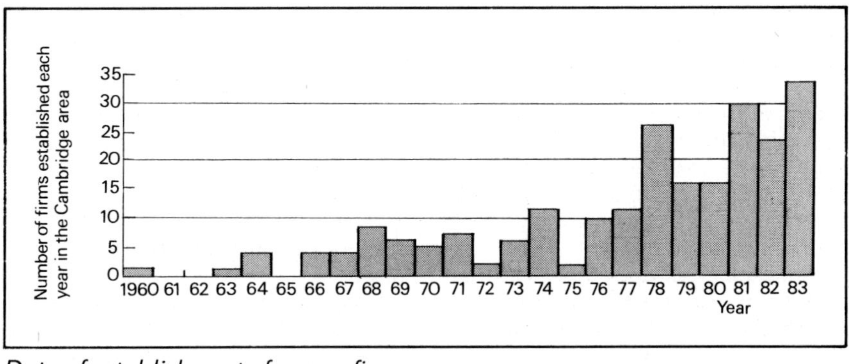

Date of establishment of survey firms

population was established after 1978; on another measure, about 190 of the enterprises set up in the area in the decade 1974–1984. These data may be contrasted with an estimate of 250 new high technology firms started in the 1960s on the San Francisco peninsula.

In the 20 years since 1964 the number of companies establishing operations in the Cambridge area has averaged nearly one per month; in the past 10 years the monthly rate has been just over 1½. The peak years have been 1981 (2½ per month) and 1983 (just under 3 per month). The average annual birth rate (new operations established in a year as a percentage of the number of companies, surviving to 1984, in operation at the start of the year) has been 11.4% over the 20-year period from 1964 and 11.8% over the past 10 years. The peak birth rate so far has been a little over 20% (in 1978).

CAMBRIDGE INTERACTIVE SYSTEMS LIMITED

CIS is one of the first and so far the most successful spin-out company from the Computer Aided Design Centre in Cambridge. The company was started in 1977 by four of the leading people in the CADCentre. They felt that their expertise in CAD, developed at the University and the Computer Aided Design Centre, had outstanding business potential and that the Centre, as a government research institution, was not the right vehicle for realising it. When in 1976 the Centre's future status became uncertain they decided to set up on their own.

The founders recognised that although (and because) they were at the leading edge of CAD-CAM technology, it would take time to develop significant clients among their target group of large engineering and vehicles companies. Consequently, in the start-up period, they undertook a miscellany of consulting assignments (including graphics for TV commercials) to generate cash flow while simultaneously developing a variety of new products including one particular drafting system which became the core of their future products and services.

Growth was rapid: turnover had risen fifty-fold by 1981 (and a further six-fold since then, to reach some £30m in 1984). The company was split among several sites in Cambridge and there was a growing need for consolidation not only in physical but also in management terms especially as their operations were increasingly becoming international. The need for capital to finance long-term expansion also became pressing. After considering a number of purely financing packages, the company decided to sell out entirely (through a share swap) to Computervision based in Bedford, Mass, which is one of the world's largest CAD companies.

CIS currently employs 70 people in Cambridge (14 with higher degrees) and 15 are graduates of Cambridge. There are few formal and operational links with the University but the company benefits in marketing and recruitment terms from the prestige of Cambridge, and there is continuing informal interaction with the academics and other CAD experts in the city. CIS has recently renovated and extended an old mill in Harston, a village about 6 miles south of Cambridge, on which all its UK operations are now based, and it sees this as providing an excellent long-term base for remaining closely involved with the high technology scene in Cambridge.

EICON RESEARCH LIMITED

Eicon Research is a young company, founded in 1979. The founders had met while students at Cambridge University and had consciously kept in touch since they were each keen on starting a company.

Their start-up strategy was to undertake consultancy and contract R & D in electronics and computer software to finance product developments. They also became sales agents for Apple and NEC. Wherever possible they sought to do only those R & D contracts on which they could keep the intellectual property rights and hence could potentially be converted into their own products.

Eicon is committed to continuing development of new products and its business thinking concentrates on product planning over at least a 2–3 year time horizon.

Eicon has taken advantage of several government aid programmes. These have included the small firms loan guarantee scheme and the support for innovation scheme, the latter being crucial to one of their R & D projects. In 1984 the company, confident of its business strategy, negotiated external finance from a City of London institution in return for a minority equity stake.

Eicon finds Cambridge an outstanding business location. Apart from the founders' keeping up their personal links with the University, the company benefits by being able easily to recruit first-rate people who are willing to relocate to Cambridge if necessary because they know that there are a significant number of other interesting job opportunities in the area. A further advantage of Cambridge is the presence of a network of sub-contract companies who are able to produce very high quality, low volume work with minimal delays.

(*Source*: Segal, Quince and Partners, 1985, *The Cambridge Phenomenon*)

CONTEMPLATING RESOURCES AND THE ROLE OF CAPITAL

Each chapter in this section is to a greater or lesser extent about the exploitation of natural resources and as you work through this section you should be able to compile a long list of what counts as a resource. Another outcome of studying the chapters in this section should be that you will become aware of and be able to discuss profound issues raised by resource development. Should scarce capital which has alternative uses be used to develop nuclear energy or fossil fuels? Should a developing country use foreign capital, technology and skilled labour to develop a rich natural resource knowing that most of the income and wealth generated will flow out of the country to the foreign owners of those factors, at least initially?

Difficulties *are* faced by different regions and countries as companies move their capital to new areas and businesses, and as governments make decisions based on political ideologies, to shift government support from one industry to another. What role does capital play in the development and consolidation of nuclear power in Britain? What are the effects on families and whole communities of steel works and coal mines closing down? What choices do we, as nations, make about developing one resource rather than another? Who benefits and who loses as a result of these decisions?

? After you have studied the chapters in this section, imagine you were born in a poor developing country with apparently no significant natural resources. Your government provides finance for you to study at the University of California. You qualify as a geologist and return home after several years experience with Amax Inc., a US based transnational mining corporation. Your government employs you in the Department of Geology and Mapping. Soon you discover a mineral orefield of world significance, similar to the central African copperfields discovered in the 1920s.

Your country is labour abundant but has virtually no capital, skilled labour or modern technology. In effect, it can only supply unskilled labour. If the rich natural resource is to be developed it will be necessary to work in partnership with a foreign transnational mining corporation. The foreign corporation will supply the capital, mining technology and foreign labour to exploit the resource.

1 Draw up a brief report for your departmental manager identifying the probable effects of resource development on the regions of your country; and the probable share of income and wealth between your country and the mining corporation and expatriate labour.

2 Debate possible policies your government could introduce to accelerate the development of your country and to alleviate regional inequalities. (Remember that mining companies are not charity organisations and expect to make an above average rate of profit on risky development projects and in high risk parts of the world.)

14 THE IMPACT OF INDUSTRIAL CHANGE: SOME EUROPEAN EXAMPLES

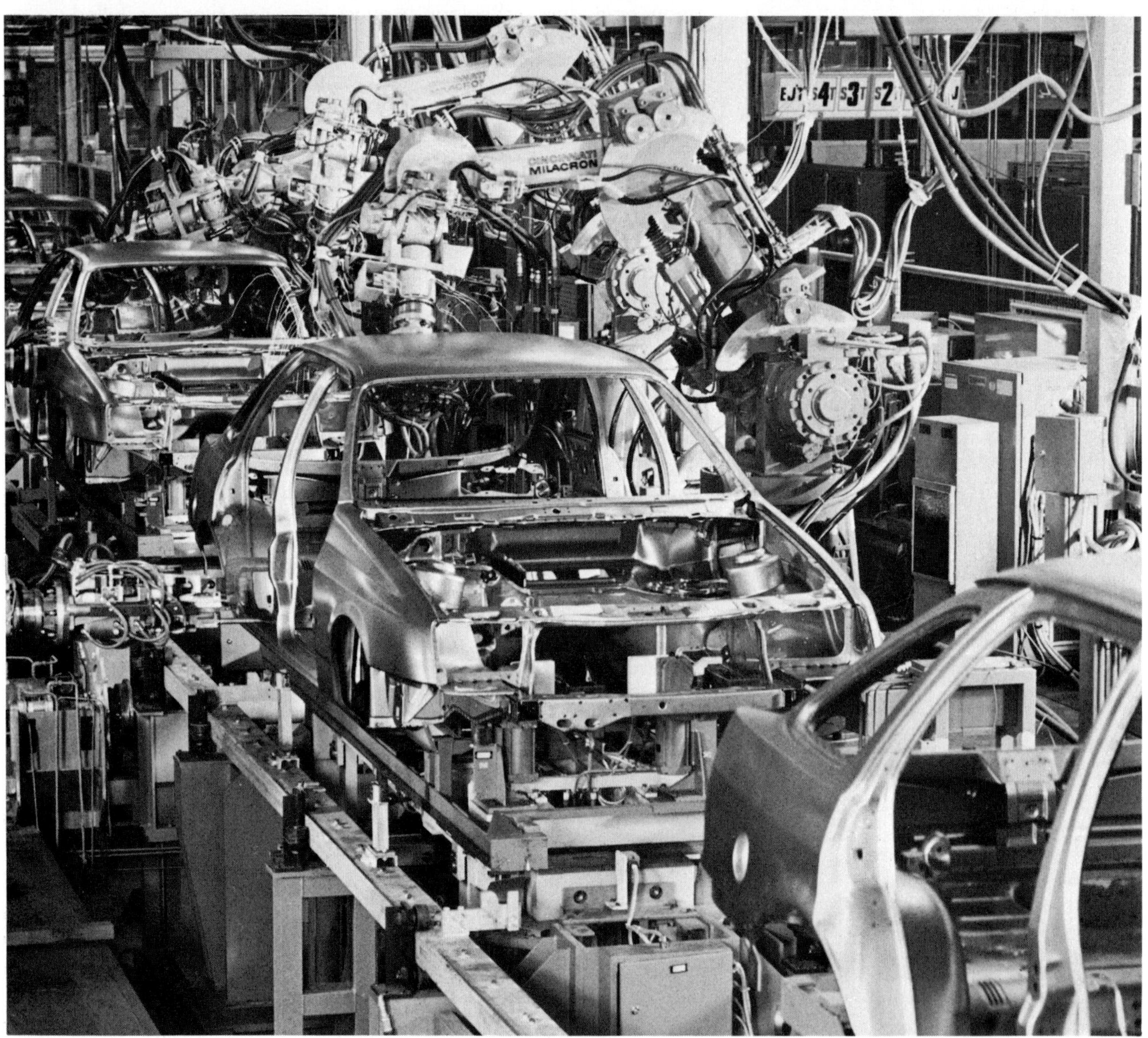

Automated body assembly of Ford Sierras at Dagenham in Kent

WHAT IS INDUSTRY? WHAT IS INDUSTRIALISATION?

? Write down your own definitions of industry and industrialisation before reading through the next section. Then, when you have read the section, decide whether you want to amend or expand your original definitions. The distinctions are important in understanding our world today.

Usually the word *industry* is used to refer to a subset of those human activities that result in the production of useful goods and services. It is a subset for two main reasons:

1 Industrial production is normally considered to include only those goods and services which can be exchanged for money and which are bought and sold on the market. As a result the domestic work performed mainly by women, and the products of a peasant or artisan family produced for its own consumption, are not counted as part of socially useful economic activity.

2 The word *industry* is normally limited to the activities of manufacturing, energy production, mining, and construction.

? Study Figure 1. Compare employment trends within the industrial sector, the agricultural sector and the service sector for the seven countries illustrated.

The word *industrialisation*, on the other hand, is used to refer to the processes of economic, social and political change which were set in motion in the late 18th century with the advent of the Industrial Revolution. Since then the pace of technological change has been very fast – new methods of making industrial and other goods have been introduced and new products have appeared. In addition, the introduction of new techniques and products has been associated with a sharp increase in the volume of goods produced and exchanged.

As a result of the Industrial Revolution, the level of material wealth increased. Industrial employment expanded. What had been predominantly rural and agrarian societies were transformed into urban and industrial ones. An industrial working class, subjected to new work practices and new forms of industrial discipline, was created and trained. The patterns of consumption and leisure and the ways of life of the overwhelming majority of people were transformed. But not all of these developments were unambiguously for the good.

? Discuss what is meant by, and expand on the phrase 'not all of these developments were unambiguously for the good'. Examine the phrase from the viewpoint of management and workers, using Figure 2 as a source of ideas.

INDUSTRIALISATION AND ECONOMIC ORGANISATION

Industrialisation always occurs in the context of particular forms of economic organisation which themselves play an important part in determining its path. In other words,

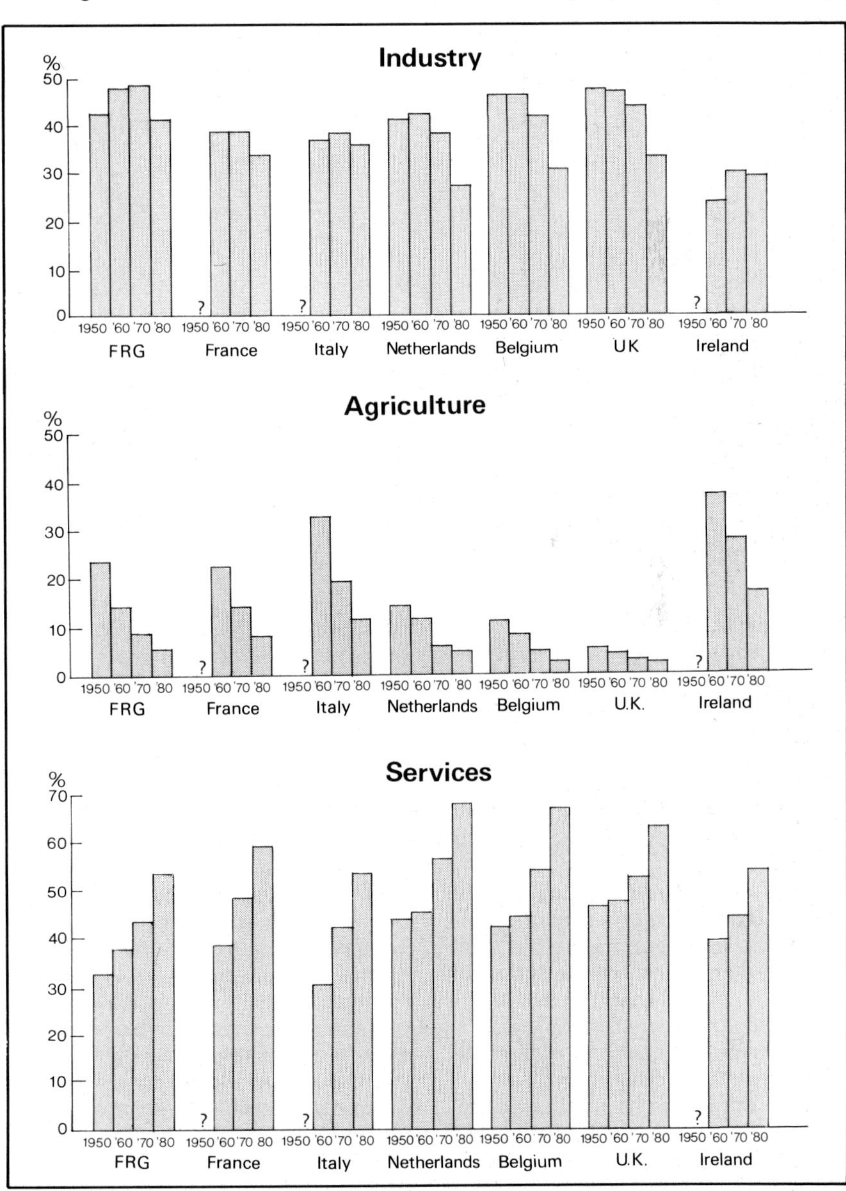

Figure 1 *Changes in structure of employment by sector of activity, 1950–80*

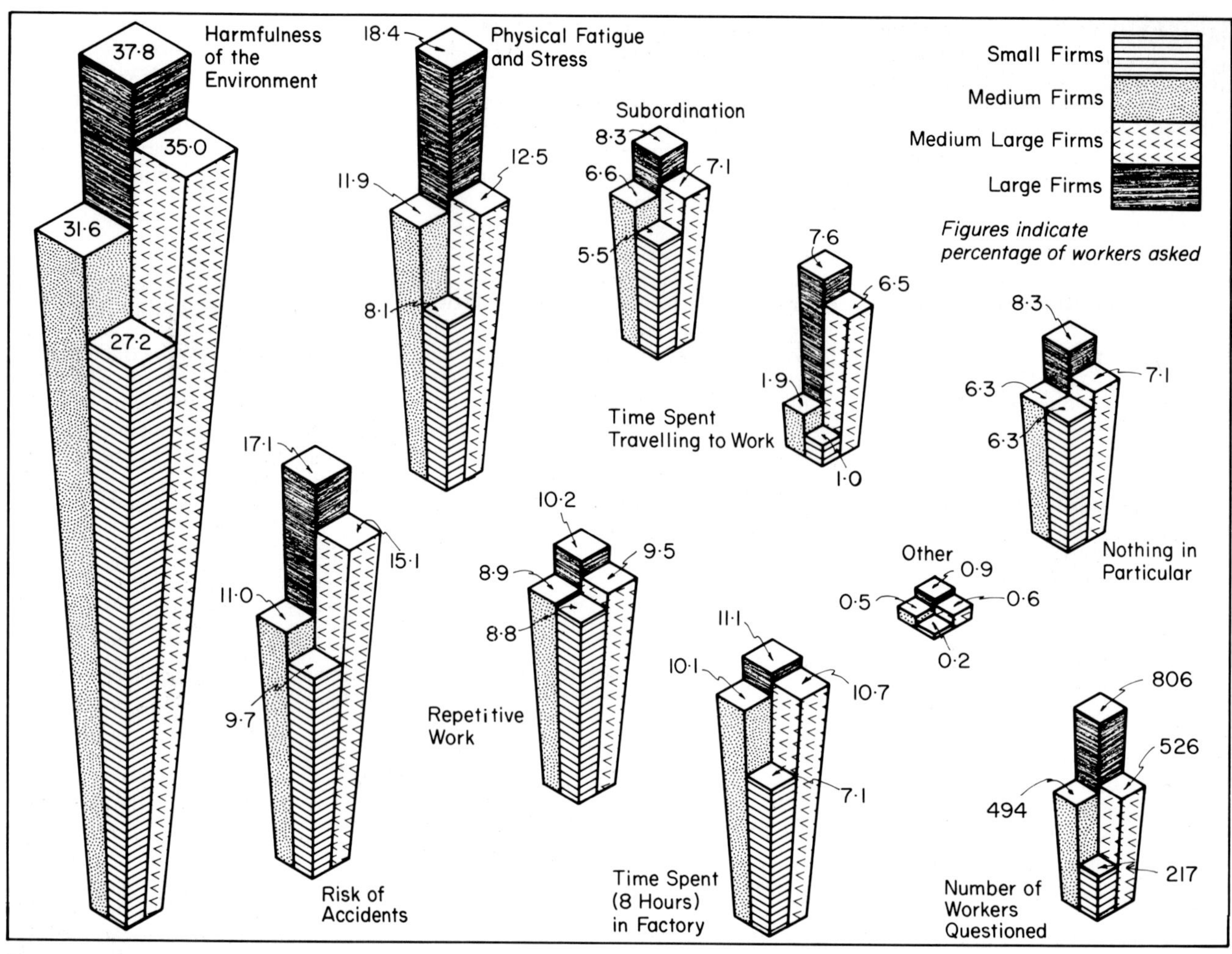

Figure 2 *The most unsatisfactory aspects of work*
(*Source*: Osservatorio Sociale, Rinascita No. 22, 1982)

questions about what is to be produced, what sectors are to grow and which are to decline, what kinds of technology are to be developed, how work is to be organised, who is to be employed and on what terms, are determined by decisions people make and by the economic and social environment in which they live and work.

In the last 200 years, economic life in Western Europe has been shaped by the fact that people were and are living in a market economy and in an economy characterised by private ownership of the means of production. In a market economy almost all the means of production, including raw materials, semi-finished goods, industrial buildings, and machinery as well as labour are bought and sold on the market. In addition, they are allocated between different lines of economic activity mainly on the basis of calculations of profit and loss. One of the main functions of market signals is to indicate what lines are likely to yield a satisfactory profit. As a result, the market determines what should be produced and where people are to be hired or workers are to be laid off.

Within industry, the organisation of production itself is shaped by the need for a manufacturer to produce goods of an adequate quality and to raise productivity or keep down costs so that a profit can be earned when they are sold at the prices ruling on the market. Accordingly, an employer will try to ensure that the people he or she employs are working continuously and rapidly and that there are no delays in the arrival of parts and materials.

Productivity can also be increased by introducing new technology and investing in new equipment. In many cases such new methods of production enable employers to reduce the job control enjoyed by workers with traditional craft skills, to increase managerial authority in the workshop, to employ fewer skilled workers, or perhaps to replace qualified workers by less skilled workers or machine operators. Change in the organisation of work is one of the factors underlying industrial conflicts. These conflicts, themselves, play an important role in determining the direction of industrial change, even if it is employers who have most say over what is to be produced and how work is to be organised.

? Read the following extract on technological change in the newspaper industry. Does it substantiate what has been said about the reduction of craft skills and individual job control?

It is important to think about other ways in which industrial and trade union conflict is significant in shaping the process of industrialisation (or, indeed, deindustrialisation). Think about the kinds of compromises that are reached in industrial disputes and how this determines the deployment of resources, capital and labour

Technological Change in the Newspaper Industry

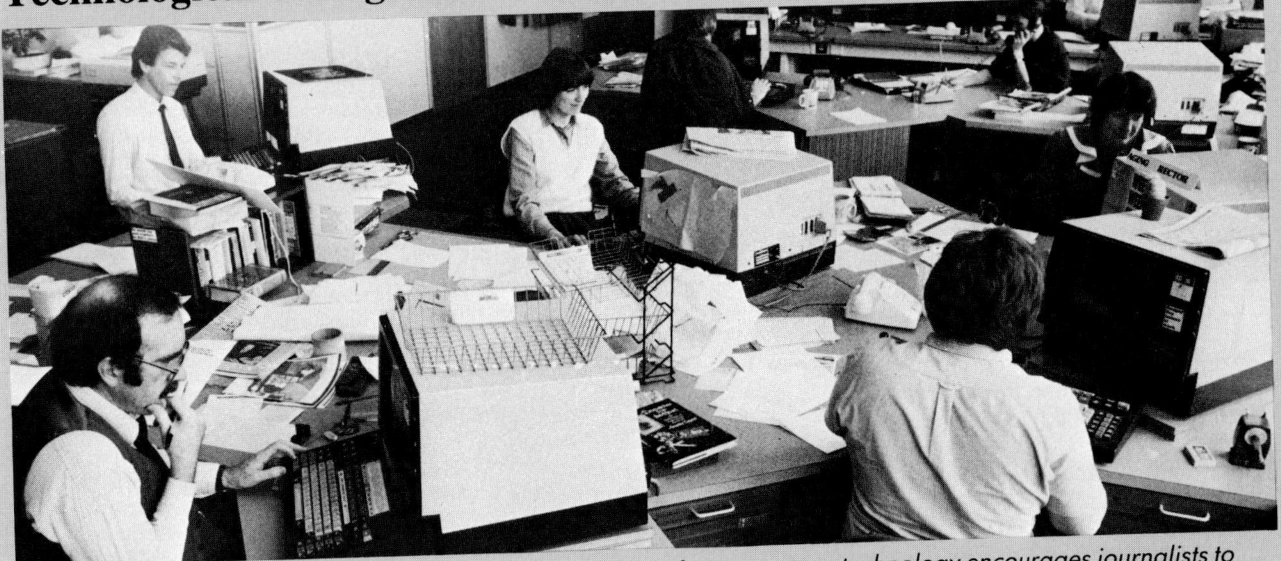

Is the traditional craft skill of the compositor about to be lost forever as new technology encourages journalists to input their material directly?

One group of workers at present being confronted with radical technical change are the compositors in the newspaper industry in London.

Hot metal and craft control

Until recently the preparation of type for letterpress printing was based on hot-metal processes. In the newspaper composing room, the men who were apprenticed to the inclusive craft of composition for print used to specialise to some extent as *linotype operators* (setting the type), as *piece case hands* (handling the large characters for headlines), as *stone hands* (organising the type into page form) or as *readers* (checking the proofs for accuracy).

As a group, the compositors organised by the National Graphical Association were skilled, well-paid, and secure in their employment. The tradition in which they were apprenticed as lads differed little from that of their fathers, uncles and grandfathers, many of whom were also in the print. It was a patriarchal craft culture, with a strong trade union identification.

At work, the chapel of the compositors' union was intricately involved in the actual organisation of production. In Fleet Street this tendency was developed further than elsewhere. Even in 1980 it was accepted that the Fathers of Chapel should organise and co-ordinate the production job.They drew up rotas, placed men on various tasks to cover the work, and were responsible to some extent for discipline and welfare.

Crisis in the industry

In the 1960s and early 1970s the national newspaper industry was experiencing recurrent financial crisis. Customarily most of the blame was accorded to 'the gremlin in the works': the Fleet Street printers. The high earnings and restrictive practices of craft labour and the degree of chapel control had, it was held, inhibited technological innovation. Various official investigations have also, however, attributed to management its share of responsibility for the Fleet Street 'disaster'. (To this list should also be added the growth of competing channels of information and communication and competition for advertising.)

New Technology: Cold Composition

The 'weapon' with which to smash the compositors – the new technology the British print employers needed – was forged in the USA. What is sometimes called 'cold composition' was an innovation with two distinct phases. The first was the shift from molten metal to a photographic principle, which was well underway in the USA by 1955. The second was a further conversion to a totally electronic process.

The first change in technology opened up the possibility of using semi-skilled labour in place of linotype workers. With the second step, production could be integrated. Journalists, editors, accountants, advertising personnel, administrators and proof readers all have keyboards with associated video screens, all on line to the computer. Any one of these may be used to tap in material, which may be corrected simply by overtyping errors as they are noticed. Stories may be called up and edited, classified adverts added to the computer's store, invoiced, recalled, updated or 'killed' as required. Work can be supervised and measured. The input methods are designed for 'direct entry'. The composing room is bypassed, and its craft personnel eliminated.

Every Fleet Street proprietor dreamed of an American style, direct entry, computerised photocomposition system. In the 1970s the corporations that own the newspapers began a sweep of modernisation which the trade unions are regarding as a technological offensive. Often a new labour process implies a new labour force or a new location perhaps in another region. In the case of London's newspapers, however, the need for speed and the need for proximity to the main railway stations holds the firms to central London and here, of course, the compositors and machine men have constructed their fortress. As a result the industry has up to now been obliged to convert its existing craftsmen.

Occupational and Employment Change

With computerised photocomposition many composing jobs will be lost and opportunities for entry into the occupation will become scarce. Some compositors have been shed from the labour process altogether because they failed to adapt or because they were too costly. Young people who might have become compositors will not now get access to this work because of narrowing job opportunities due to the greater productive capability of the new technology. The redundant compositor and the unemployed school-leaver are ghosts that hover round the new composing area and they are never far from the minds of the men who work there. They are the 'industrial reserve army' that weakens the worker's bargaining power and urges him to accept the new discipline. On the other side of the coin, however, and equally absent from the composing room, are those compositors and others who have trained for, and found different and better work, as the new corps of higher-paid, technical and managerial workers who minister to the computerised system.

What the unions sought was compositors on the keyboards and retraining of the hot-metal compositors. Up until now the unions have proved strong enough to retain the right of many existing compositors to the transformed job.

The Craftsman's Changing Position

Once retrained and working on the new technology, some of the men became *keyboard operators*, some *paste-up hands*, some *computer* or *photosetter operators* and others did integrated work involving several processes.

The status and role of the surviving craftsmen have changed very sharply. Their position with regard to the employer is weakened and their relationship with others in the industry – with editors, reporters, technicians as well as with the semi-skilled and unskilled men they have been classing as their social inferiors – all have to be reassessed. New technology is also disrupting gender relations. The men find the new work less manly. Women are entering typesetting – once the unique preserve of the craftsman.

What is more, the men have been taught the new keyboard by typing instructors, unconnected with the printing industry. The new boards are Q-W-E-R-T-Y boards named after the characters of the left-hand side top row of a conventional typewriter, or, in the words of the men, 'idiot boards'; the linotype's 90-key board had an entirely different layout. In some cases, the teacher has been a young woman. 'They weren't even operators, they didn't understand things,' the men said. They were typists. 'She gave us tests, which I hate, to be quite honest.'

The outcome of such a change is a sharp drop in input efficiency in the early months following changeover. At Mirror Group News, where new technology had been in operation for two years, the men's average typing speed was around 30 words per minute. A good typist is at least twice that fast. The workforce as a whole has, however, gained in productivity.

The men are caught in a contradiction: they must either acknowledge themselves totally deskilled, or acknowledge that many women are as skilled as men. The dilemma accounts for much of the bitterness they feel.

The Next Step: Direct Entry and Beyond?

The upheaval in printing has only just begun. As well as the typesetting side, the page make-up side of the photocompositor's job is likely to be obliterated. With the use of more complex screens, known as graphics display terminals or page-view terminals, display adverts and even whole newspaper pages can be made up on the screen and emitted as a whole sheet from the typesetter, so eliminating paste-up work. Other technological innovations on the way will make the photographic department irrelevant, which at present reproduces the composed image and transfers it to the printing plate. The printing plate will be made by laser direct from the information on the page-size video screen. Even the massive rotary presses which churn out the finished print may one day be redundant as ink-jet printing enables a high-speed printed image to be produced, one copy at a time, direct from the computer memory without the use of a printing surface at all. Beyond that again, newspapers themselves may be made irrelevant in the future by a fully electronic media. Electronics will make printing itself increasingly marginal to our information needs.

Already newpspaper proprietors in the provinces are introducing direct entry by journalists. In London the large national proprietors are investing in new printing equipment in the Docklands where rents and rates are lower, congestion is less, and the job control of the Fleet Street unions may be circumvented. Outside London, Eddie Shah who achieved notoriety with the Stockport Messenger dispute in 1983, is speaking of a national tabloid whose directly entered and fully made up pages will be sent electronically to five regional printing centres near motorways for distribution by road. In the struggle for survival and in view of the dramatic economic advantages of new methods of production change is on the agenda and many jobs will be eliminated.

(*Source*: based on C. Cockburn, *Brothers, Male Dominance and Technological Change*, London, Pluto Press, 1983)

Is the market self-adjusting?
In a capitalist market economy an income is distributed to those participating in economic activity, or to those who own necessary resources, in the form of wages and salaries to workers, profits to capitalists, and rents to the owners of land and property.

The incomes formed in this way underlie market demand. In a situation where unemployment is high, and wages and incomes are low, demand is correspondingly low, with the result that products cannot be sold and workers are laid off in other parts of the economy. It follows that a market economy will function effectively not only when individual employers can all earn a profit but also when adequate effective demand exists.

Often these two conditions are not satisfied. As is indicated by the historical record the process of industrialisation has been very uneven temporally (and also geographically). Since the Industrial Revolution, phases of expansion have alternated with phases of slackening growth, of stagnation, or of contraction. In the 1930s, for example, there was a deep depression. After 1945 the economies of the West grew rapidly. But since the early 1970s the capitalist world has found itself in the midst of a new crisis.

As a result of these ups and downs, some people have concluded that the market is not self-adjusting. They believe that markets are not capable of the following:

1 securing an efficient allocation of resources via the price mechanism;

2 ensuring that resources are fully used and not left idle;

3 promoting economic adjustment and growth

In the West, people who hold this view often advocate various forms of state intervention in the economy. Indeed, the extent and character of state intervention in economic life is one of the major factors differentiating the development paths of different regions and countries, with some economies reflecting the impact of a free market strategy and others reflecting much greater state protection and the fostering of industrial development via state subsidies, state entrepreneurship, economic planning and so on.

We shall now go on to explore the development of one specific area in the UK, the North-east, and one industry in France, the steel industry. While reading about them, think of the ways in which the case studies highlight the issues outlined in this section.

? Draw out the relationships outlined in the last section in the form of a systems diagram.

?

THE RESTRUCTURING OF THE FRENCH STEEL INDUSTRY

You are the economics reporter for your local newspaper and over the years you have had a number of assignments to report on the crisis in the steel industry.

You have a background knowledge of the steel industry and a few years ago you did an in-depth study of the French industry and its situation and difficulties. You have updated your knowledge and you have copious notes and an account very similar to the section following. Read it all through thoroughly once or twice and then complete the following tasks:

You have been asked to give a talk to a local geography and economics sixth form on the French steel industry. Prepare overhead projector transparencies for the talk. Specify what maps and tables you would use. Make the overhead transparencies as visually attractive and clear as possible. Your introduction could deal with:

1 changes in world steel output;

2 weaknesses in the European steel industries;

3 strengths of new steel industries; and

4 the role of the EEC.

You could then go on to look in more detail at the French steel industry – its structure, changes in technology, the *Plan Acier* and social unrest.

In the discussion at the end of your talk, you may be asked what alternatives you can see to closing plants and making workers redundant. Set out notes which would enable you to answer this question.

THE EUROPEAN CRISIS

In the years between the end of the Second World War and 1974, steel production increased rapidly, if irregularly, in Europe and in the rest of the world.

The expansion of the industry and the growth in productivity were accompanied and reinforced by the development of a new geographical distribution of production. New plants were established outside traditional steel producing areas, and older and less efficient installations were closed or run down.

At a world level, the market shares of the less developed economies and of the centrally planned economies increased, while those of the developed market economies fell. In the EEC, the shares of the Italian and Dutch steel industries increased, while that of the UK in particular declined. As Table 1 shows, the share of the European Coal and Steel Community in world production fell from 28.4 per cent in 1960 to 22 per cent in 1974 (and to 16.6 per cent in 1983). Within the ECSC itself, the Italian industry increased its share

Table 1 *Crude Steel Production in the ECSC and the Rest of World, 1950–83 (in millions of tonnes and percentages)*

	1950		*1960*		*1974*		*1983*	
	Million tonnes	*%*	*Million tonnes*	*%*	*Million tonnes*	*%*	*Million tonnes*	*%*
Belgium	3.8	(7.8)	7.2	(7.3)	16.2	(10.4)	10.1	(9.2)
F R Germany	14.0	(29.0)	34.1	(34.9)	53.2	(34.2)	35.7	(32.6)
France	8.7	(17.9)	17.3	(17.7)	27.0	(17.4)	17.6	(16.1)
Italy	2.4	(4.9)	8.2	(8.4)	23.8	(15.3)	21.8	(19.9)
UK	16.6	(34.2)	24.7	(25.2)	22.4	(14.4)	15.0	(13.7)
ECSC of 9	48.4	(100)	97.8	(100)	155.6	(100)	109.5	(100)
ECSC of 9			98	(28.4)	156	(22.0)	110	(16.6)
DME[1]			231	(67.0)	461	(65.0)	332[4]	(50.1)[4]
CPE[2]			105	(30.5)	216	(30.5)		
LDC[3]			9	(2.5)	31	(4.4)		
World			345	(100)	596	(100)	663	(100)

[1] Developed market economies [2] Centrally planned economies [3] Less developed countries [4] OECD Countries
(*Sources*: Chambre Syndicale de la Sidérurgie Française, *Bulletin de la CSSF*, Paris: CSSF, Various Years, and OECD, *The Iron and Steel Industry in 1983*, Paris: OECD, 1985)

from 4.9 per cent in 1950 to 15.3 per cent in 1974. The French share remained fairly steady, while the share of the UK industry, which had the lowest levels of productivity fell from 34.2 per cent in 1950 to 14.4 per cent in 1974, (and to 13.7 per cent in 1983).

After 1974 the steel industries of the more developed countries entered a deep and protracted crisis, and competition between European, US and Japanese producers was translated into open conflict. One of the reasons for the crisis lay in the fact that the demand for steel fell at the same time as a large amount of new capacity was becoming available. As a result a considerable amount of excess capacity was found to exist, steel prices fell sharply, and so did profits, with many companies in the more developed countries finding themselves unable to meet interest charges, let alone to repay capital debts.

The impact of the steel crisis was in fact much more severe in the advanced countries because, by international standards, much of their existing capacity was inefficient. In the EEC, only the Italian and Dutch industries had blast furnace installations that were fully realising potential scale economies and that had attained internationally competitive levels of productivity, while methods of continuous casting were not used as extensively as in Japan. At the same time, many plants were not well located in relation to sources of raw materials such as high grade iron ore, which was being imported on an increasing scale in bulk carriers from West Africa and Brazil, and coal, which similarly was beginning to be imported from abroad.

These events accelerated the fall in employment in the European steel industry. Many of the resulting job losses occurred in largely non-industrial regions and at times when little alternative employment was available. (In the past, steel owners have traditionally opposed industrial diversification in steel-producing districts, fearing competition on the local labour market.)

The role of the state

Attempts by governments to attract new jobs to the most hard-hit areas were not very successful. As a result an attempt was sometimes made to control the speed of contraction of employment. Relatively high redundancy payments along with a variety of other social measures were introduced in order to defuse the discontent of groups of workers, to redeploy some of them, and to alleviate some of the distress caused by rising unemployment.

In addition, the state was often forced to intervene to help firms on the verge of bankruptcy, by writing off or reconstructing their debts and providing funds for investment, sometimes to save jobs but often in return for a restructuring of the industry in order to restore its international competitiveness.

EEC Policy

The action of national governments was supplemented by that of the EEC. The Social and Regional Funds were used in order to help in areas in which economic activity had contracted, and the EEC *Iron and Steel Employees Readaptation Benefits Scheme* was aimed at the alleviation of social distress. At the same time, funds for investment were provided by the European Coal and Steel Community, and investment and reconversion assistance was also given by the European Investment Bank.

In addition, two anti-crisis plans were introduced with a view to protecting the European steel industry while its restructuring was proceeding. In January 1977 the *Simmonet Plan*, an 'anti-crisis' plan for the European steel industry was launched. Six months later it was succeeded by the *Davignon Plan*. One of the aims was to secure increases in steel prices and in company profits by introducing production and delivery quotas and anti-dumping measures. A second aim was to provide a breathing space in which capacity could be reduced and the industry restructured. At first it proved to be relatively successful, but as the underlying crisis persisted, it started to crumble, and in October 1980 a situation of manifest crisis was declared, and voluntary price and production targets were replaced by obligatory controls on European steel producers.

The French steel industry

At the end of the Second World War, the French steel industry was made up of about 30 family-based firms and their subsidiaries, and 177 plants, of which 66 were integrated. These plants had an annual steel-making capacity of 11 to 12 million tonnes. The industry's main centre was near the ore deposits in Lorraine (see Figure 3) where, in 1938, 78 per cent of the country's pig iron and 67 per cent of its steel were produced.

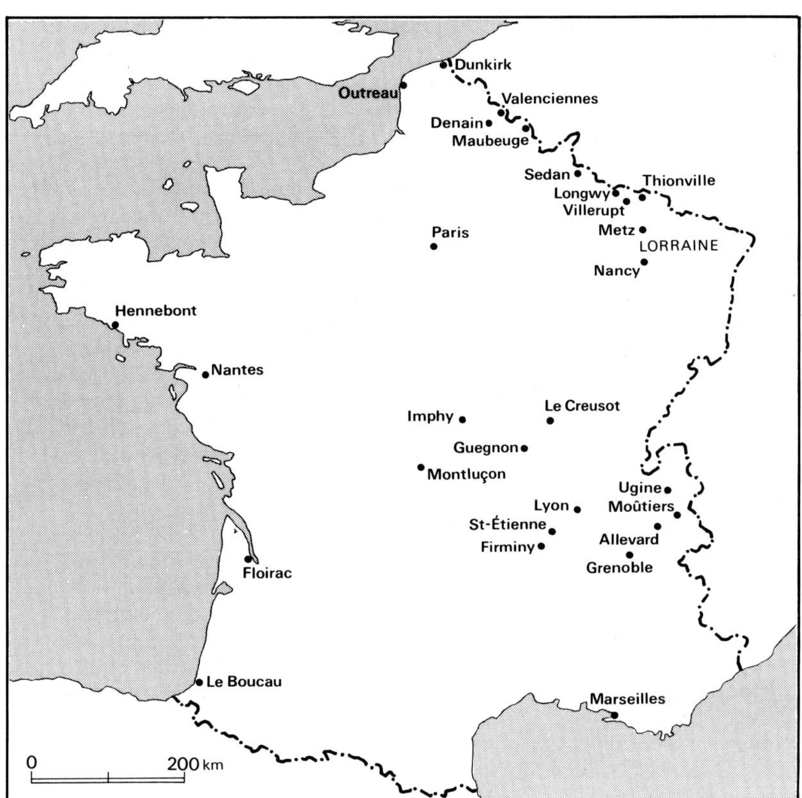

Figure 3 *The main centres of French steel production, 1945–50*

Too many plants?

In the view of the members of the Steel Modernisation Commission of 1948, only 24 plants were necessary, but the modernisation of the industry was impeded by the fragmentation of ownership and by the priority given to the production of semi-finished *long* products. (In general, ingots are reduced to blooms, billets and slabs in a primary mill. The resulting pieces are then reheated and passed through mills which roll them either into flat products such as plate, sheet, strip and coil or into long products including bars, angles, shapes, rails and wire rod.) In many cases the investments carried out were of a piecemeal kind, and the industry was tardy in mechanising.

Changes in technology

By the mid-1960s important changes in steel-making technology were occurring. In particular, the growing use of oxygen steel-making processes was making it possible to produce a wide range of good quality steels in less time and with three times less labour than the open hearth process. Also reductions in the cost of transporting materials by sea, and the mechanisation of mining were beginning to make it profitable to exploit high grade ore deposits in West Africa, Brazil, and elsewhere. As a result, the development of new coastal complexes, where imported ores and coal could be used most advantageously, was considered a profitable step. So in the wake of the Japanese, Italian, Dutch, and German industries, the French company Usinor proceeded to construct a small integrated shore-based plant at Dunkirk.

Economic change and mergers

In the early 1960s demand stagnated, and competition increased. In addition, some of the early stages in the automation of steel production were occurring. All of the French companies except Usinor were faced with problems of profitability. In the mid-1960s the companies had to announce that they did not have the resources necessary to meet the objectives of the Fifth National Economic Plan. The result was an agreement with the government in which financial concessions and assistance were given in return for a geographical regrouping and consolidation of the industry. Soon afterwards Usinor merged with Lorraine-Escaut, and a new company, Wendel Sidélor, was formed. Between 1964 and 1968, some 12 000 jobs were lost in the industry, but in the late 1960s employment increased once again.

In the Sixth Plan (1971–1975), the strategy of industrialising the country was translated into a projected increase in steel-making capacity from 25.8 to 35.6 million tonnes per year. The capacity of the Dunkirk plant was to be increased from three to eight million tonnes per annum, and a new coastal complex with an initial capacity of some three million tonnes per year was to be constructed at Fos-sur-Mer near Marseilles.

However, the conditions on which the financing of the Fos complex was based were not realised, and Wendel Sidélor announced that it was not in a position to meet its share of the projected expenditure. The outcome was the *Plan de Conversion de Wendel Sidélor*, and the inclusion of Usinor in the Fos project. In all, 10 650 jobs were to be lost. The effects of the fall in employment were most severe in the steel-producing districts in Lorraine, where in five years only 2 200 new government-aided jobs were created.

In 1975 the demand for steel fell just as significant amounts of new capacity were coming on stream in Europe and in countries developing steel industries of their own. Output and prices fell sharply, and the main French companies, which were heavily in debt, made large losses. It was at this juncture, as we saw, that the EEC used the powers in the ECSC Treaty to protect the steel industry, while plans to restructure it were prepared and implemented.

The Plan Acier (Steel Plan)

This plan was introduced in France in 1976 and adopted a strategy of closing older installations, of fully loading the more modern ones, and of increasing the capacity of the more efficient plants. Through it, about 16 100 jobs were to be lost between April 1977 and April 1979 – 13 400 were to be lost in Lorraine alone.

The assumptions on which the *Plan Acier* was based were not realised, although employment in just three companies fell by more than 20 000 in three years. The main French firms were losing money and were on the verge of bankruptcy. In 1978 the government decided to step in to save the industry, and a new plan was announced, *Le Plan de Sauvetage de la Sidérurgie*. Instead of nationalising the industry, a couple of holding companies were set up, with the state assuming majority control of Usinor and Sacilor-Sollac (formerly Wendel-Sidélor). In addition, more plants were to be closed or rationalised. The plans for restructuring the industry implied a loss of 21 750 jobs between April 1979 and the end of 1980. Of this total, some 14 000 were to be lost in the steel-producing districts in Lorraine, with Longwy in particular being badly affected. Major losses were also to occur in the internal parts of the Nord region. At the same time new jobs already promised by the government were not being created on a significant scale, and were often fewer than had originally been announced. In Longwy only 850 jobs had been created since 1970, even though steel employment had already fallen by 5075 jobs.

The size of the job losses, the ending of the possibility of early retirement, and the concentration of redundancies in the towns of Longwy, and of Denain in the Nord, led to violent working class opposition. Some factories were occupied, roads and railway lines were blocked, police stations and the offices of employers' organisations were attacked, a free radio station was set up at Longwy, and wide-ranging demonstrations were organised. All activity was brought to an end, the town was closed off from the outside world in order to highlight the likely impact of the collapse of its economic base. Ultimately, in the short term, all that was achieved was a number of measures aimed at buying out some of the workers and at limiting the social effects of the plan.

Print 1 *French steel workers in Paris protesting against job losses, 1979*

A change of government

In May/June 1981 a socialist president and government were elected. Very soon afterwards the loss-making steel firms, Usinor and Sacilor, were nationalised and, via takeovers and the acquisition of majority shareholdings, they gained control of almost the whole of the French steel industry. In 1982 a new steel plan was approved, but with large losses continuing to be recorded, it was revised in 1984 (see Table 2). Underlying these decisions were three major considerations:

1 The European Commission had set the goal of getting the steel industry out of the red by 1986.

2 The European Commission required that the French close one wire rod mill.

3 The French decided that greater emphasis should be placed on the recovery of scrap metal in electric furnaces and less on the conversion of iron ore into pig iron and steel in blast furnaces and traditional steelworks.

As a result some 20–25 000 jobs were to be lost by 1987. In the view of the government, which was divided over the choice of plan for the industry, the job losses would occur without redundancies, but in Lorraine, where the job losses were to have most impact, the government's decisions were met with great anger and despair.

These events have had two major consequences. One has been a sharp drop in capacity, output, and employment. The other has been a striking regional transfer of steel production from old steel-producing districts in the internal parts of the Nord and in Lorraine to the coast at Dunkirk and Fos-sur-Mer.

Why was industrial adjustment so problematic?

Owing to the relative backwardness of the French steel industry, which was second only to the UK in the lowness of its productivity levels, the process of adjustment has involved severe dislocation, an enormous waste of resources, and has been fraught with tensions and conflicts.

The roots of these problems lie in the following factors:

- the delay which ocurred in mechanising and automating the French steel industry;
- the domination of the industry by small family-based firms;
- the slowness and shortcomings in the processes of concentration and centralisation of capital;
- the continuing lack of adequate links between the steel industry and metal using industries.

Table 2 *The French steel industry measures of March 1984*

Lorraine	*Outside Lorraine*
Sacilor Serémange: *confirmation* of construction of a cold rolling mill producing flat products	**Usinor Valenciennes:** *modernisation* of the steel girders mill
Usinor Longwy: *retention of* the blast furnace supplying the wire rod mill; *switch* to electric steel making for the steel girders mill	**Usinor Dunkirk:** *project approved for* heavy (33 mm or more) sheet and plate mill
Usinor Neuves-Maisons: *switch* to electric steel making for the wire rod mill	**Sacilor Fos:** *closing* of Ugine Aciers
Sacilor Rombas-Gandrange: *closing* of the wire rod mill and decision *not* to build a universal rolling mill	**Sacilor Caen:** blast furnace to be retained and reconsidered in 1986
Sacilor Hayange: *retention and modernisation* of rail rolling mill	These measures are in addition to those decided on in the 1982 Steel Plan involving the closing in whole or in part of installations at Vireux-Molhain in the Ardennes (Usinor), at Denain in the Nord (Usinor) at Joeuf Lorraine (Sacilor) at Villerupt in Lorraine (Sacilor) and at Pompey in Lorraine (Sacilor)

Government responses to the problems

On several occasions the French state has attempted to overcome these problems, but, as has been indicated, with little success. Accordingly, before it came to power, some of the supporters of the current socialist government, spoke of the need for a different model of development. What they advocated was a high level of investment aimed at 'reconquering the domestic market'. Investment would have to be in activities with a potential for long-term expansion which would enable the French economy to meet many of the social needs of the whole population.

It was claimed that such a strategy would enable the economy to return to full employment, providing alternative jobs in growing sectors for those which would simultaneously be lost in contracting ones. As a result redeployment would occur, it was argued, without the levels of unemployment experienced in countries pursuing different types of strategy. In practice, a strategy of reflating the economy was abandoned after one year, and the government's industrial policy will only have effects in the medium term.

In a situation in which the steel industry is required to return to financial viability in a relatively short period of time and in which technology, sources of material inputs, methods of work organisation, and other determinants of location are changing, a strong case can be made for closing plants and making workers redundant. In the conditions prevailing at present it seems as if very little can be done by way of providing alternative jobs to replace those that are being lost.

Wider responses

The seemingly inexorable logic of the last argument can, however, be challenged in a number of ways. One is by noting the way in which the collapse of the steel industry is linked with what is happening in related parts of the economy. Similarly the lack of alternative jobs is closely bound up not only with the fall in investment and in industrial production, but also with the length of the working week and the performance of overtime by those in work.

The question that now needs to be asked is whether or not a different strategy of development at a national and at a European and international level, would alter the situation. For example, would the expansion of industries producing machinery that is at present imported, and a reequipping of other sectors of the economy, lead to an increased demand for steel? Is a programme of new construction likely to raise the demand for steel and for other construction materials significantly? Can the length of the working week be reduced and job-sharing schemes be introduced? Would a process of co-ordinated expansion at a European level generate a sufficient number of satisfactory new jobs to ensure that a rundown of the steel industry does not mean a significant lengthening of the dole queues?

All these questions imply that the discussion should be extended beyond the question of financial viability in a narrow sense. Of course, a different pattern of development would alter the prevailing financial situation. But social and political questions, to which we shall return at the end of the chapter, are also involved. Before we do this, however, we shall look at the development of one region of Great Britain in which there have been great changes in the pattern of industrialisation.

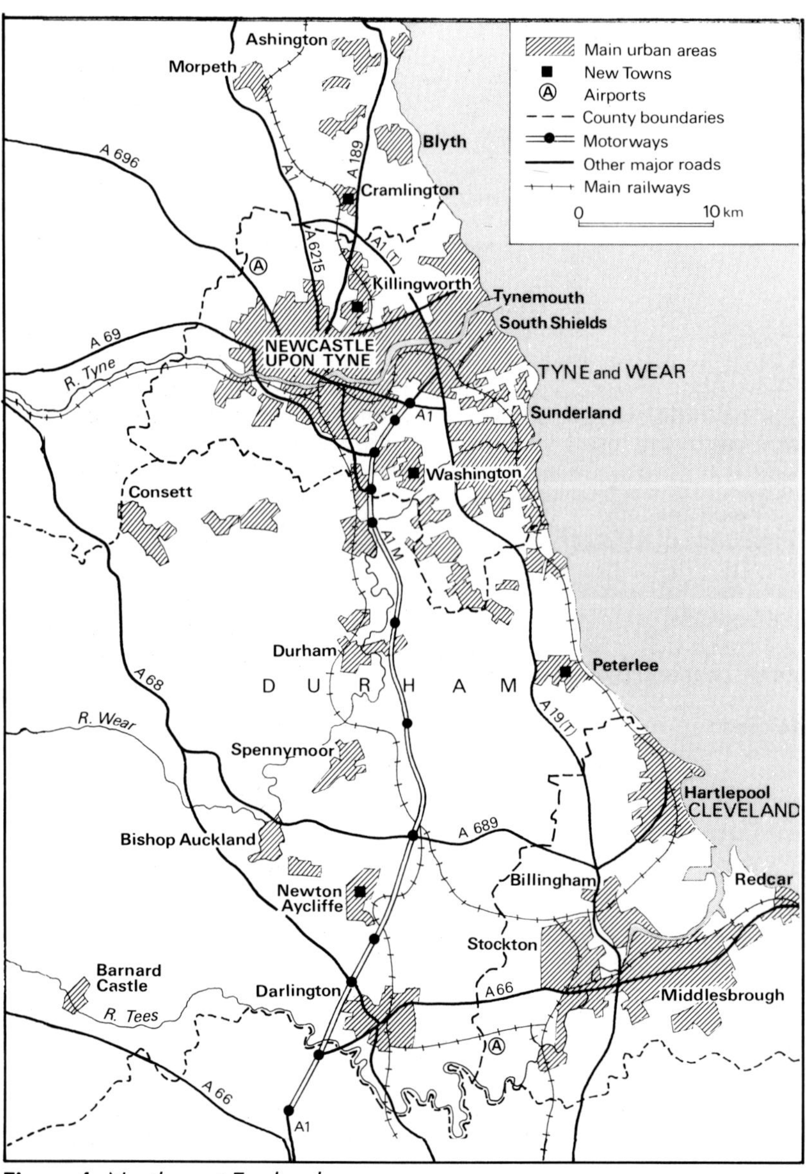

Figure 4 *North-east England*

INDUSTRIAL CHANGE AND REGIONAL INEQUALITY IN NORTH-EAST ENGLAND

In 1950 a sharp dichotomy existed in Britain between the old industrial regions in the North and West of the country and the more prosperous Midlands and South-east. A large gap also existed between declining rural areas in Scotland and Wales and areas of modern industry, but it was of less importance than the contrast between rural and urban areas in other parts of Europe because of the earlier rationalisation of farming in Britain.

During the 1950s the main trend in the location of manufacturing employment was a concentration in central areas with some dispersion of industry to overspill and new towns, and a decline in peripheral regions. Nevertheless until 1958 many of the industrial sectors that had caused such problems in the 1930s seemed to be doing reasonably well. They were enjoying protected markets or faced little competition. As a result, existing capacity was being fully used, and some modernising investment was being carried out.

JOB LOSS IN THE TRADITIONAL INDUSTRIES AND THE HAILSHAM PLAN OF 1963

In 1958, however, a recession was accompanied by massive reductions in capacity and in employment, especially in the coal and shipbuilding sectors. In the case of coal mining a major switch to oil as a source of energy coincided with a fall in demand caused by the adoption of increasingly successful techniques of fuel saving and the beginnings of the development of nuclear energy. In the late 1950s and 1960s a large number of pits were closed with a high proportion of the closures occurring in the North-east, South Wales and Scotland where they prompted the cartoon depicted in Figure 5. At the same time the level of capital investment in the coal industry was raised. As a result productivity increased sharply, and the loss of jobs was accelerated, with overall employment dropping from 711 000 jobs in 1952 to 419 000 in 1967.

The consequent rise in unemployment in the depressed areas led to an amendment and intensification of regional policy. In 1963 plans for modernising and diversifying the economies of Scotland and the North-east which were most hard hit, were prepared. The 'Hailsham Plan' (a growth strategy for the North-east) and the plan for West-central Scotland were both introduced by a Conservative Government, but the plans themselves were subsequently to be closely linked with the short-lived experiment in indicative planning that was conducted at a national level mainly by the Wilson Government of 1964.

Figure 5 *A message from the minister of power was sent to all miners yesterday*

(*Source*: based on cartoon reproduced in Austrin, T. and Beynon, H., *After Blackhall*, University of Durham, Department of Sociology, 1982)

THE DEVELOPMENT OF A BRANCH PLANT ECONOMY

The economies of many of the old industrial regions were transformed in the 1960s as a result of the continuing decline of employment in many of the traditional industries and of the implantation of new manufacturing firms and the growth of the service sector. For example, in the county of Durham, employment in the coal mining industry fell from 98 000 jobs in 1951 to 83 000 in 1961 and to 33 000 in 1971. Ninety-nine collieries were closed in the years between 1958 and 1970, with 42 being closed in 1964–68 (Table 3). In addition, deep mining was moved from the west to the east of the county. At the same time employment declined in other traditional sectors in the region such as the metal manufacturing, shipbuilding, and rail transport industries, although remaining more or less stable in the chemical industry, as Table 4 shows.

In the manufacturing sector the new establishments implanted in the old industrial regions were, in a high proportion of cases, branch plants

Table 3 *Trends in coal production and employment in the North-east, 1947–1982*

	Number of collieries			*Employment (thousands)*			*Output (thousand tonnes)*		
	Total	*Durham*	*Northumberland*	*Total*	*Durham*	*Northumberland*	*Total*	*Durham*	*Northumberland*
1947	201	134	67	148.7	108.3	40.4	35.4	24.5	10.9
1959	163	113	50	130.7	93.0	37.7	35.7	23.7	12.0
1965	112	78	34	93.5	65.9	27.6	31.7	20.7	11.0
1970	50	34	16	48.6	34.5	14.1	20.8	13.8	7.0
1982	22	16	6	29.0	21.3	7.7	13.4	9.1	4.3

Note: The data for 1947 and 1959 refer to the relevant calendar year, while that for 1965, 1970 and 1982 refer to the 12 months ending in March.
(*Source*: NCB, *North East Coal Digest*, 1981–2, London: NCB, 1982)

Table 4 *Employees in Employment by SIC Order in the Northern Region, 1952–81*

	Employees in employment in 1952		*Change in employment (in 000s)*			
	(in percentages)	*(in location quotients)*	*1952–60*	*1960–66*	*1966–74*	*1974–81*
Agriculture, forestry & fishing	3.2	0.9	−5.0	−9.2	−6	−2
Mining and quarrying	14.9	3.5	−23.7	−51.3	−55	−11
Food, drink & tobacco	2.8	0.8	0.8	0.2	3	−4
Chemicals and allied industries[1]	3.7	1.6	15.7	−2.5	−6	−6
Metal manufacture	4.9	1.7	3.4	−4.2	−13	−19
Mechanical engineering	4.5	0.9	4.5	8.1	4	−18
Instrument engineering	0.1	0.2	0.2	0.6	2	−1
Electrical engineering	2.5	0.9	9.3	13.9	7	−20
Shipbuilding & marine engineering	5.1	3.3	−1.2	−17.9	−10	−10
Vehicles	1.2	0.3	1.0	−3.9	2	−2
Metal goods not elsewhere spec.	1.1	0.4	−0.6	2.3	2	−4
Textiles	1.1	0.2	5.2	1.8	5	−16
Leather, leather goods & fur	0.2	0.7	−0.6	−0.6	0	−1
Clothing & footwear	2.3	0.8	3.6	3.0	1	−15
Bricks, pottery, glass, cement etc	1.6	0.9	−0.5	0.7	−1	−5
Timber, furniture etc	1.2	0.8	−2.2	1.9	0	−4
Paper, printing & publishing	1.1	0.4	1.7	2.8	6	−1
Other manufacturing industries	0.6	0.5	4.8	3.2	2	−2
Construction	6.2	0.9	12.1	21.6	−9	−26
Gas, electricity & water	1.6	0.9	0.5	3.3	−4	0
Transport & communication	8.4	1.0	−9.1	−5.5	−19	−2
Distributive trades	10.9	0.9	27.0	7.7	−11	−2
Insurance, banking,finance & business services	1.0	0.5	4.7	3.6	8	8
Professional & scientific services	6.9	0.9	21.4	28.6	33	17
Miscellaneous services	6.9	0.8	4.0	16.4	12	26
Public administration & defence	6.0	1.0	−2.4	4.8	11	4
Totals			74.6	29.4	−36	−117

[1] Including coal and petroleum products
(*Source*: based on statistics in S. Fothergill and G. Gudgin, 'Regional Employment Statistics on a Comparable Basis 1952–75'. Centre for Environmental Studies Occasional Paper No. 5 (August 1980) 1–72, and Census of Employment Statistics, published in Department of Employment Gazette)

? Use Table 4 to write a paragraph on changes in employment in the Northern Region between 1952 and 1981. Revise the meaning of location quotients.

of multinational firms. At this time many of these firms were expanding by setting up new plants in the UK, and they were attracted to the Assisted Areas by the availability of workers, by the quantity and quality of the infrastructural facilities provided by the state, and by the availability of grants (see Figure 6). But this decentralisation consisted, in most cases, of production operations and assembly work, requiring unskilled or semi-skilled workers, and routine managerial jobs. The research and development and the decision-making functions of such firms were located or remained outside the region. The plants themselves were completely integrated into the operations of the companies to which they belonged, and established few links with other parts of the economies of the regions in which they were located. As a result the decisions to buy up, to expand, to rundown, or to close a plant, on which the livelihoods of the people in an area depended, were increasingly determined in board-rooms located outside the region. In addition, the only factors taken into consideration when making such decisions were:

1 how the plant fitted in with the company's overall production and market strategy; and

2 whether or not there was sufficient return on the company's investment.

? Look at Figure 6 and write a short paragraph describing how the map of Assisted Areas has changed in the period 1966–1984. Does the North-east appear to have been more or less affected than other parts of the country?

JOBS FOR WOMEN IN LIGHT INDUSTRY AND SERVICES

The type of employment provided in the new establishments was radically different from the kind of work provided by the traditional sectors. For example, in the North-east, the work in the traditional sectors was in general skilled or heavy manual work done entirely by men and often organised on a shift system. By contrast the new light manufacturing firms attracted to the area tended to employ semi-skilled and unskilled workers to carry out routine assembly work. These firms gave employment for the first time to large numbers of married women, organising production in such a way as to enable them to combine part-time work or twilight shifts with domestic work. The increase in the number of women in the workforce was strengthened by the growth of the service sector and the transfer of central government offices to the periphery. This growth of female employment was one of the most important consequences of the new pattern of industrialisation. Another was an increase in the importance of the *mass worker*, graded bureaucratically within each of the corporations, not according to occupational groups. (A mass worker is a worker lacking specific craft skills who can easily be employed to perform a variety of routine tasks requiring little initial training.)

THE CHANGING GEOGRAPHY OF INDUSTRIALISATION IN THE NORTH-EAST

The new firms attracted to the North-east were not normally located in the places in which employment was contracting – in coal mining and other traditional industrial districts. Instead they moved to areas where pools of labour, unused for the most part to the conditions of factory life, had been assembled and organised by the area's local authorities whose investments in economic and social infrastructural facilities were concentrated in certain growth points.

In the Durham area, employment and population were allowed to fall in the west. On the east coast, where the pits remained open, a new town had been built at Peterlee. The firms attracted to it, however, were mainly

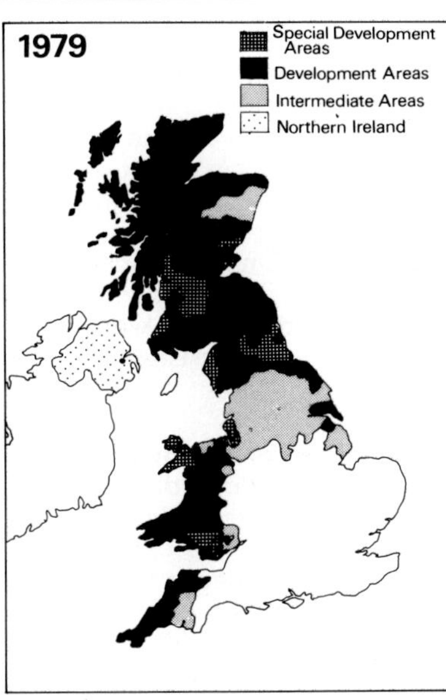

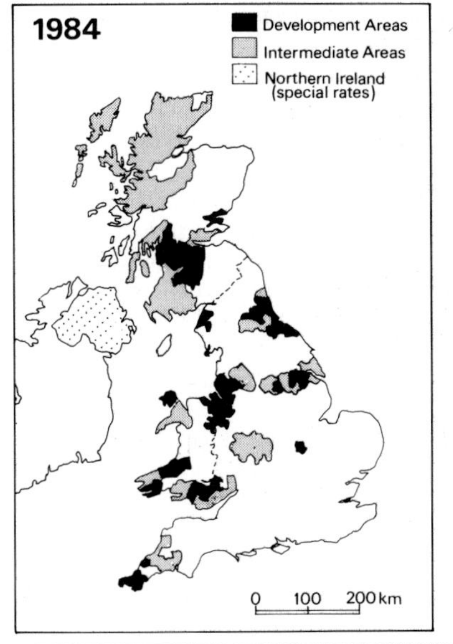

Figure 6 *The changing map of the assisted areas, 1966–84*

Print 2 *Women working at the Fisher Price Toys factory, Peterlee*

employers of women workers who, because of the constraints on their mobility (posed by the continuing need to perform domestic tasks and by the pattern of public transport provision), formed a captive labour market. The industrial development of Peterlee actually dates from the mid-1950s when three firms opened factories in the North-eastern Industrial Estate. Two of these employed mainly female workers. In the early to mid-1960s five other large firms also employing mostly female workers were established, and in 1978 Fisher Price Toys built a plant in the town. Four of the eight firms with a predominantly female workforce produced clothing.

A number of other new towns were built in a central corridor of growth extending from the estuary of the Tees in the south to those of the Tyne and Wear in the north where the chemicals, steel, and shipbuilding sectors were concentrated (see Figure 4). The old coal and steel town of Spennymoor slightly to the west of this central belt was converted into what two sociologists have called a 'global outpost' of Thorn Electrical Industries (which expanded its existing plant as the pits in and around the town were closed), and of Courtaulds and the Black and Decker Manufacturing Company, which moved to the area in the middle of the 1960s.

In order to understand the movement of firms to the North-east it is necessary to consider not only locational factors but also the overall economic strategy of the groups involved. After 1962, for example, Courtaulds was pursuing a dual strategy. On the one hand existing capacity in fibre-using sectors located mainly in the North-west was being acquired and rationalised. On the other hand, ultra-modern fibre producing plants of high capital intensity were being developed in the Assisted Areas.

The Anatomy of Deindustrialisation

In the 1970s Courtaulds' strategy proved unsuccessful. The rush of investment in the late 1960s and early 1970s resulted in the provision of excess capacity, while the costs of production of fibre plants in low wage developing countries were much lower than those of British plants. In 1979 the worsted spinning mill at Spennymoor was closed, as were many other branch plants which had been located in the Assisted Areas in the preceding decade. Montague Burton closed several tailoring factories in Sunderland and Gateshead, Plessey closed a large telecommunications plant in Sunderland, GEC reduced employment heavily in Hartlepool, and the region's last effort at building oil rigs was also terminated.

In the 1970s and 1980s, the traditional industries in the region were once again severely hit adding to the lengthening dole queues. In 1976–9 large job losses were recorded in engineering and shipbuilding. On Tyneside the rationalisation of several important old firms each resulted in the loss of over 1000 jobs. Included were Parsons (turbine generators), Vickers (heavy engineering and armaments) and the nationalised firms in merchant shipbuilding, marine engine building and ship repairing. In Tyne and Wear 3800 workers were made redundant in 1978, 1979 and 1981 by British Shipbuilders. In 1980 and 1981 the iron and steel industry was the leading source of job loss. In the largely mono-industrial town of Consett a steel works was closed at a cost of 4600 jobs, while 11 300 jobs were lost in Cleveland between 1976 and 1981. Automation and reduced demand along with high energy costs were associated with job loss in the chemical industry on Teeside.

Print 3 *Consett steel works — an ex steel worker looks on*

Table 5 *Numbers employed and percentage change in employment for the largest firms in the UK, 1973–82*

	Ranking	*Numbers employed in UK*				*% change in employment*			
	82	*1982*	*1981*	*1977*	*1973*	*73–77*	*77–82*	*73–82*	*81–82*
GEC	1	145 000	157 000	156 000	170 000	−8.2	−7.1	−14.7	−7.6
British Steel	2	103 700	120 900	209 000	229 000	−8.7	−50.4	−54.7	−14.2
BL	3	80 600	104 000	171 943	171 296	0.4	−53.1	−53.0	−22.5
British Aerospace	4	77 475	79 000	66 000	71 000	−7.0	17.4	9.1	−1.9
Thorn-EMI	5	74 000	79 000	74 661	78 568	−5.0	−0.9	−5.8	−6.3
Unilever	6	73 252*	79 148	91 923	88 544	3.8	−20.3	−17.3	−7.4
ICI	7	68 000	74 000	95 000	104 000	−8.7	−28.4	−34.6	−8.1
Ford	8	65 200	70 000	73 333	70 143	4.5	−11.1	−7.0	−6.9
British Shipbuilders	9	64 600	67 500	n/a	n/a	–	22.1	–	−4.3
Courtaulds	10	62 636	77 405	112 009	125 000	−10.4	−44.1	−49.9	−19.1
Rolls-Royce	11	50 000	55 000	57 164	61 446	−6.2	−12.5	−18.6	−9.1
Lucas	12	49 440	53 728	68 778	71 330	−3.6	−28.1	−30.7	−8.0
BAT Industries	13	44 860*	47 450	36 388	36 782	−1.1	23.3	22.0	−5.5
GKN	14	40 000	51 000	73 196	78 351	−6.6	−45.4	−48.5	−21.6
Hawker Siddeley	15	35 800*	39 500	39 200	66 900	−41.4	−8.7	−45.6	−9.4
BP	16	35 700	39 000	33 708	26 072	29.3	5.9	36.9	−8.5
Tube Investments	17	34 000	42 600	61 777	64 728	−4.6	−45.0	−47.5	−20.2
S Pearson & Son	18	29 648*	31 222	28 946	27 282	6.1	2.4	8.7	−5.0
BICC	19	29 600†	32 100	32 200	36 200	−11.0	−8.1	−18.2	−7.9
Northern Eng Inds	20	26 560	26 822	33 678	–	–	−21.1	–	−1.0
Philips	21	25 000‡	30 000	45 698☆	61 339	−25.9	−45.3	−59.2	−16.7
Dunlop	22	24 492	32 000	48 000	52 000	−7.7	−49.0	−52.9	−23.5
STC	23	20 644	25 009	29 000	35 500	−18.3	−28.8	−41.8	−17.5
Vauxhall	24	20 123	21 000	30 180	34 141	−11.6	−33.3	−41.1	−4.2
Michelin	25	18 629*	19 229	18 658	17 274	8.0	−0.2	7.8	−3.1
Vickers	26	18 500‡	20 879	17 706	29 724	−40.4	4.5	−37.8	−11.4
IMI	27	18 200	22 255	26 664	28 173	−5.4	−31.7	−35.4	−18.2
Distillers	28	18 125†	18 943	19 156	19 300	−0.7	−5.4	−6.1	−4.3
ICL	29	16 000	21 114	23 000	23 000	0.0	−30.4	−30.4	−23.8
IBM	30	15 590*	15 362	13 814	12 428	11.2	12.9	25.4	1.5
Burmah	31	15 200‡	16 500	17 300	19 300	−10.4	−12.1	−21.2	−7.9
Babcock International	32	15 098‡	16 248	22 521	20 716	8.7	−33.0	−27.1	−7.7
Glaxo	33	13 188	13 725	15 924	16 568	−3.7	−17.2	−20.4	−3.9
Dowty	34	13 041	14 474	12 395	11 656	6.3	5.2	11.9	−9.9
Racal	35	12 800	14 135	5 373	3 533	52.1	138.2	262.3	−9.4
Rank Xerox	36	10 900‡	12 520	11 615	10 762	7.9	−6.2	1.3	−12.9
English China Clays	37	10 600	10 860	10 977	10 900	0.7	−3.4	−2.8	−2.4
John Brown	38	10 040	11 697	14 719	1 387	961.2	−31.8	623.9	−14.2
Kodak	39	9 789*	10 496	11 364	12 768	−11.0	−13.9	−23.3	−6.7
Davy	40	8 358‡	9 113	5 929	4 938	20.1	41.0	69.3	−8.3
CIBA-Geigy	41	8 000	9 234	10 933	7 465	46.5	−26.8	7.2	−13.4
Wellcome Foundation	42	6 750	6 688	6 996	6 610	5.8	−3.5	2.1	0.9
Rothmans International	43	6 679	8 057	6 349	5 596	6.6	5.2	12.1	−17.1
Massey Ferguson	44	6 500‡	15 870	21 486	18 907	13.6	−69.7	−65.6	−59.7
Talbot	45	6 300	10 000	22 800	30 883	−26.2	−72.4	−79.6	−37.0
Johnson Matthey	46	5 761	5 888	5 635	5 748	−2.0	2.2	0.2	−2.2
Cummins Engine	47	5 220	6 109	3 760	3 389	10.9	38.8	54.0	−14.6
INCO	48	5 000‡	6 400	7 552	4 187	80.4	−33.8	19.4	−21.9

Table 5 (*continued*)

Texaco	49	4 839	5 177	4 422	n/a	n/a	9.4	–	−6.5
Caterpillar	50	2 961	4 812	5 114	4 655	9.9	−42.1	−36.4	−38.5
Monsanto	51	2 657	3 053	5 380	4 682	14.9	−50.6	−43.3	−13.0
Assoc. Octel	52	2 753	2 847	2 810	2 574	9.2	−2.0	7.0	−3.3
Mobil Oil	53	2 535	2 477	2 343	n/a	n/a	8.2	–	2.3
International Harvester	54	2 500	3 500	6 430	5 829	-10.3	−61.1	−57.1	−28.6
Conoco	55	2 299*	2 238	2 259	1 724	31.0	1.8	33.4	2.7

Notes by 1982 ranking: 1 – 1982 figure at Jan 31 3 and 23 – 1982 figure at Sept 14 – 1982 figure mid-Aug 9 – % change 1978–81 15 – Excludes subsidiaries transferred to BAe 16 – Dec 1982 estimate. Includes redundancies announced but not necessarily implemented. Selection Trust acquired Sept 1980 20 – New group 1977 21 – Floated Cambridge Instruments in 1981 taking 5 000 jobs 26 – Rolls-Royce Motor merger 1980 32 – Excludes construction equipment group sold Oct 1982 35 – Merger with Decca 1980 38 – Totals at year-end March 31 40 – Acquired Herbert Morris and British Testing in 1978 41 – Takeover of Ilford 1975 49 – Includes Station Supreme and shore-based personnel of Texaco Overseas Tankships
* 1980 and 1981 figures used for 1981 and 1982. † Source: London Stock Exchange. ‡ 1982 average figure. ☆ 1976 used for 1977.
(*Source*: *Financial Times*, 9 December 1982)

? In Table 5 the job losses for 1973–82 in the largest firms in Britain are recorded. Consider the North-east's experience in the 1970s and early '80s in the light of national trends in employment in large firms.

New attempts are now being made to run down coal mining and to shed labour in the traditional coalfields.

The amount of coal that is being traded internationally is increasing but, at the same time, the demand for coal from major users like the British Steel Corporation and the Central Electricity Generating Board (CEGB) has slumped. In the case of the CEGB, of course, it is likely to contract as a result of the switch from coal to oil and to nuclear reactors for electricity generation (see *The Nuclear Power Debate, pages 341–364*). Of course, a change in government policy with respect to nuclear power would alter the situation.

According to a recent issue of *The New Statesman*, the reduction in demand for coal is not, however, the main reason for the threat to employment in the industry. Job losses will come from the massive leaps in productivity expected from the new pits and from new technology. It is clear that the Coal Board intends to concentrate production in new high productivity pits at Selby, Asfordby in the Vale of Belvoir, and in the planned new South Warwickshire coal field.

A team of university academics has been working independently for some years on the effect of computerisation and automation on miners' jobs. They point out that the super pit at Selby will produce with 4000 men the amount of coal now produced on average by 16 000. Their report on new technology in mining to a subcommittee of the NUM executive remains confidential and is by no means official NUM policy, but it is understood that their report says that, assuming demand for coal remains constant and new technologies are fully installed, anything between 55 and 74 per cent of the 1981 workforce of 200 000 will be made redundant.

A standardised computer system for central control at collieries – MINOS or Mine Operating System – would wipe out jobs throughout the industry. It is already partially installed in many pits. In particular, two sub-systems of MINOS will have serious implications for the NUM. The first, called FIDO, for coal face monitoring will reduce 'avoidable delays' in coal cutting, said by the NCB to represent a third of all shift time presently worked. The second is IMPACT – In-built Machine Performance and Condition Testing – which monitors machine faults and potential breakdowns. Along with the computerisation of coal clearance and coal preparation, these technical changes will lead to major job losses.
(*Source*: *New Statesman*, 5 November 1982)

Print 4 *The Wistow Pit at Selby, Yorkshire. Nearly 4000 miners will be employed in the coalfield when it reaches peak production in 1988*

?

As no new industry is likely to be attracted to the North-east and other areas affected by the decline of traditional industries, any reduction in the number of jobs is likely to lead to large-scale and long-term unemployment. In this context, the National Union of Mineworkers took a firm stand against pit closures based on the profitability of particular collieries. They suggest that the introduction of new technology should be linked instead with a reduction in the length of the working week.

1 Compare the introduction of new technology in the printing industry with the introduction of new technology in the coal industry. Are the aims the same? Are the effects the same? Write a paragraph (or more) on industrialisation as a process and the different effects it can have on those caught up in it.

2 The NUM is arguing that the introduction of new technology in coal mining should be linked to a shorter working week. Is this a desirable solution? What would be the effects on the towns in coal mining areas – on the people and on the other industries located there? Would there be other more wide-reaching effects?

3 Consider the arguments concerning the restructuring of the steel or coal industry in Britain. Since 1979 the demand for steel and coal has fallen very sharply. As a result the BSC and the NCB have lost revenue and made substantial losses. The leaders of those industries and the present government wish, at the very least, to get rid of these deficits by closing high-cost installations. (Some members of the government wish to go further by closing any plant or colliery that is not making a profit. In that situation, in the coal industry, the surpluses of low-cost pits would no longer be offset against the deficits of high cost pits.) Against them are ranged many of the workers in these industries and the trade unions. What they tend to argue is that:
a in a recession deficits should be allowed, especially as the recession is in part a result of the actions of the government;
b the losses are insignificant compared with the costs to the nation in foregone taxation, increased social security and other unemployment-related spending, and redundancy payments on the one hand, and the social costs of rising unemployment and devastated communities on the other;
c the accounting methods of the NCB and the criterion of profitability are not appropriate in making decisions about the exploitation (and conservation) of non-renewable coal resources; domestic production should be encouraged to ensure security of supplies and to improve the balance of payments position.

In addition to the major protagonists, many other social and political groups are involved including suppliers to the industries concerned, steel and coal users, and, in the case of the coal industry, groups such as the Campaign for Nuclear Disarmament that prefer the use of coal to nuclear-based sources of energy.

In the light of these brief remarks write roles for some of the following people: the chairpersons of the steel and coal industries, the prime minister, a politician opposed to the policy of the current government, a number of steel workers or miners including ones whose jobs are safe and ones whose jobs are under threat, a wife or husband of a worker whose job is threatened, a shopkeeper in a community where the major source of employment is likely to disappear, a police chief required to control any industrial conflict and to enforce the laws passed by the government, a leading member of the anti-nuclear movement, and a trade union leader.

4 Decide what three points you would select as the most significant if you were describing the changes in the North-east to, say, a Canadian not acquainted with the area. Compare your three points with others in the class and try to agree a common list. Discuss the implications of these changes.

INDUSTRIAL CHANGE AND ADJUSTMENT

Out of consideration of the development of the French steel industry and of the North-east of England two more general questions emerge. The first concerns the way in which resources are and should be allocated. The second concerns the overall capacity of the French and British economies to supply new jobs to replace those that are lost and to secure economic growth.

WHAT DETERMINES RESOURCE ALLOCATION?

In the societies we have studied resources are allocated primarily through the market. What is more many people argue that the destiny of any industry and any region should depend only on criteria of commercial viability (see, for example, the extracts below from an article on the coal industry by C. Huhne).

On the other hand, it can be argued that the rundown of an industry when there are no alternative jobs and no increase in output in other sectors, generates large social costs such as growing imports unmatched by exports, rising social security spending, and, in particular the psychological, physiological, and social costs imposed on individual human beings and their families. It may be that the *social* benefits of modernising and keeping open existing plants outweigh the costs.

Of course, if the losses incurred in a particular sector exceed the net social benefits of keeping marginal plants open, and if other sectors are deprived of necessary resources, capacity should be reduced. However, in this case, it could be argued that closures and reductions of capacity should be linked with the provision of other jobs. In short, the question of resource use and resource allocation is a social and political as well as a financial question.

?

After reading the following extracts on the problem in the coal industry, summarise and comment on the competing arguments.

IS MRS THATCHER IN REALITY THE MINERS' BEST FRIEND?

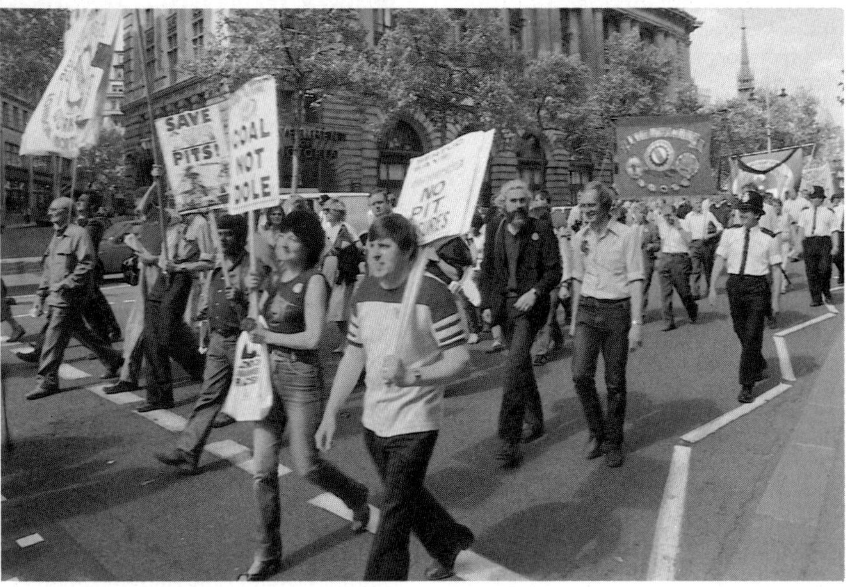

Christopher Huhne

Mr Scargill and the National Union of Mineworkers argue that the NCB's losses are artificially inflated by the depressed state of the British economy, and that subsidies per tonne of coal in Britain are lower than anywhere else in Western Europe. Changes in geology and markets make it foolish to talk about 'uneconomic' pits, since they can rapidly become economic again. Moreover, debt charges distort the real trading position, which would look about £2 million a pit better if there was a full debt write-off.

The NUM also argues that coal, as an indigenous fuel which still provides 36 per cent of all Britain's inland energy consumption, provides a security of supply in a world which could be hit by another oil-shock.

Most interesting, because the point touches on the very nature of profit and loss as a reasonable guide to whether an activity should continue, is the NUM's argument that it would actually cost the Government more to close pits than it would to keep subsidising them even at a higher rate.

Let's take a point at a time. Clearly, the NCB's losses are inflated by the downturn, but the fact remains that the Board wants to expand coal output after the current cutbacks but at lower-cost, more efficient pits like Selby in Yorkshire, where the cost per tonne could be around a third of the present market price. The argument, in short, is not about how much coal should be produced but where; economics say go for the cheapest source.

Certainly, Britain's production subsidies were only £3.19 per tonne in 1982, compared to £17.72 in France, £16.47 in Belgium and £8.65 in West Germany. But all those countries are making efforts to cut their subsidies – by closing faces and pits. The real market-setters are anyway the Americans and the Australians, who seem to be wholly unsubsidised and whose coal is competitive with Britain's even after high transport costs. Nor, of course, do four European wrongs necessarily make a right.

It is hard to see why miners, earning even on the NUM's figures for face workers last September some 114 per cent of male national average earnings, should be more worthy recipients of subsidy than kidney patients, the under-5s or the three million people in Britain who live on less than 120 per cent of the supplementary benefit level. Yet that is at bottom what this argument about comparative subsidy amounts to. With the grants budgeted for the mines at £680 million this year, the opportunity cost is at least five new district hospitals.

Debt charges certainly account for a large part of the NCB's losses, but they represent merely a part of past losses and investment. The NUM has a short memory if it cannot remember the debt write-offs under the Coal Industry Act 1965 or the further capital reconstruction in 1973. The argument about security of supply, after the events of the last few weeks, surely cuts both ways. Those City commentators who have tried to assess the possible effects of the dispute have been surprised to find that Britain is more dependent on coal than it was in 1974.

Mr Scargill's pièce de resistance, though, is the claim that it would cost the country more to shut pits than to keep them going. Ergo, his plan to stop pit closures is a heaven-sent gift to the Treasury in its campaign to cut public spending. So why aren't ministers jumping for joy?

NUM literature claims that in June 1982, the Government estimated the average cost to public funds for each unemployed mineworker at £7188 in the first year and £6326 thereafter. Thus the cost of 70 000 job losses (which, by the way, is nearly four times what the NCB has suggested) over 10 years would be £4.5 billion, while the cost of keeping open the same capacity would be some £2.36 billion (no more detailed calculations are given).

In principle, the NUM is right to say that profit and loss to the NCB is not profit and loss to the Government. The average cost of an unemployed person is more than £5000 a year in benefits and lost taxes. It is therefore in principle possible for the Government to subsidise each job, provided it is assured that the person would otherwise be out of work, by £5000 without a deterioration in public finances. The snag is: which people are to be chosen? And what if we all asked for £5000 more? The argument only works if no one else follows suit, or we would rapidly end up in an orgy of cross-subsidisation. Or, alternatively, paramilitary ways of controlling wage demands, Polish-style.

What is more, the NUM's point does not even work in practice as a clever piece of special pleading, because so much of the NCB's losses are caused by relatively few pits. Last year, the Board estimates that the worst 12 per cent of output cost £275 million, which would have been enough to put it in the black at the trading level (though not after interest payments).

Individual colliery figures are not normally published, but the MMC report last year gave a snapshot for 1981–2. The worst loss-maker that year was the Treforgan colliery in South Wales, which is still open. It's net proceeds covered only 31 per cent of the operating cost per tonne, resulting in a loss of £104.80 per tonne, or £6.5 million for the pit, or £12,969 for each of its employees.

In other words, it would have been cheaper for the taxpayer to pay each Treforgan employee more than double the then male manual average wage not to mine coal. Even on the NUM's calculations, it would have been a lot less expensive to lay them off – at least for that year. At the margin, Britain's uneconomic pits really are uneconomic.

Yet this year, the Government is budgeting cash grants for the NCB worth £4000 per employee. This represents more than 42 per cent of the predicted national average wage, and is an increase after allowing for inflation of 200 per cent since 1978–9.

This, moreover, is what the Government's targets imply after the planned pit closures and 20 000 fewer jobs. Last year, the cash grant to the industry was 137 per cent higher in real terms than in 1978–9, while the external financing limit (including allowance for some genuine investment) was up 37 per cent in real terms.

(*Source*: *The Guardian*, 3 May 1984)

When coal not dole is the best choice

Andrew Glyn

MANY people who sympathise with the miners feel nevertheless that their struggle against the further rundown of their industry flies against the laws of economics. If the pits which the NCB wants to close really are 'uneconomic,' isn't the call to keep them open special pleading for huge state handouts, as Christopher Huhne alleged recently in The Guardian?

Such a view is quite wrong. Of course, the miners concerned do benefit from keeping their jobs. But far from imposing an economic burden on the rest of society, the extra coal the miners produce is worth much more than the additional costs of keeping them at work. 'Unprofitable' the higher cost pits may be, according to the NCB's narrow accounting; 'uneconomic' they certainly are not.

The explanation for this is simple. The net gain to the miners from keeping their jobs is far less than the cost to the NCB of employing them. On the one hand a substantial proportion of their gross pay goes to the government in income taxes and national insurance contributions. On the other hand, if they lost their jobs they would be receiving redundancy pay, unemployment or other benefits.

So the miners receive far less *extra* when they are employed (as compared to being on the dole) than the value of the coal they produce, leaving a surplus which benefits the rest of us. This is reflected in the fact that the government pays out in subsidies to keep the high cost pits open much less than it gains from the income tax etc., which the miners pay and from not having to pay them dole.

Last autumn the NCB estimated that the highest cost 12 per cent of its colliery output incurred £275 million losses (about half its total losses in 1982–83). These collieries employed some 40 000 miners.

The 12 per cent of colliery production coming from the 'appallingly high-cost of uneconomic pits' (Guardian leader, May 12, 1984) was worth around £475 million in 1982–3. With losses at these pits running at £275 million, however, total costs were £750 million. Surely then they really were 'uneconomic' if the costs of production were around 1½ times the value of the coal produced. But this would only be the case if the resources used to produce the coal (the labour of the miners and the inputs of power, steel, etc) would otherwise have been used productively elsewhere. In the present context of mass unemployment and government refusal to expand the economy, this is quite inconceivable.

Wage costs of miners account for only around half the operating costs of the collieries. The rest comprises salaries of other NCB staff (at pit, area or HQ level) and inputs brought in from other industries (notably electricity, steel, engineering goods and transport). If all these costs really were to be saved by pit closure (which would imply that NCB overheads were pruned down in line with the colliery shutdowns) then roughly 35 000 more workers (about one quarter in the NCB and the rest elsewhere) would lose their jobs.

How much then do the 75 000 or so workers employed directly or indirectly (in the NCB or supplying industries) benefit financially from retaining their jobs?

Receiving an average of £50 per week when out of work would mean that on average the 75 000 workers affected are nearly £70 per week better off working – amounting to £245 million in total (after deducting additional indirect taxes like VAT which they would pay). This is only half the value of the coal which they collectively produce. So far from requiring a subsidy to maintain them at work, their labour provides both sufficient resources to meet their extra pay and contributes some £230 million of resources for other uses. Once there is mass unemployment, financial costs do not measure real costs, even to the rest of society, let alone to society as a whole (which does, after all, include the miners!)

The difference between the £750 million financial costs and £245 million real costs is mainly accounted for by payments to and from the Government. Savings in dole and receipts of direct and indirect taxes, rates and extra nationalised industry profits work out at roughly £475 million (well over half as much again as the famous subsidy).

On this basis, with real costs around one-third of financial costs, even the highest cost of the Coal Board's pits, Treforgan in South Wales, produced enough more or less to earn its real costs (according to the 1981–82 figures published by the Monopolies Commission). So Christopher Huhne's suggestion that 'it would have been cheaper for the tax payer to pay each Treforgan colliery employee more than double the then manual wage not to mine coal' is as inaccurate as his notion that subsidising mines deprives us of hospitals is cheap and misleading.

But the NUM's case is most definitely not based on a plea for subsidies from the rest of us. They have an irrefutable economic case which the NCB has never answered. Success for the miners would preserve their jobs, to all our benefit.

(*Source: The Guardian*, 28 May 1984)

WHAT DETERMINES THE LEVEL OF EMPLOYMENT AND GROWTH?

One of the reasons why the process of industrial change has been so problematic is that in the 1970s and 1980s the contraction of old industries has not been matched by a growth of jobs in new sectors and that the potential of technology to meet human needs and aspirations has not been fully realised.

? Write down what you think are the reasons for the recession. Compare them with the points made below.

THE ANATOMY OF THE PRESENT CRISIS

In the years up to 1974 increases in output and productivity were recorded in all the main West European countries. Employment also increased but only by a small amount. The manufacturing sector in general, and industries producing equipment goods (i.e. producers' and consumers' durable goods) in particular, acted as the motor of economic growth. But with the quadrupling of oil prices in 1973 and the ensuing recession, the manufacturing sector lost its dynamism. Output and productivity growth slowed down, stagnated or declined while employment fell. The trends in output and unemployment in the UK and other developed market economies are represented in the graphs shown in Figures 7 and 8.

A fall in the rate of economic growth and inflation

It would, however, be easy but misleading to attribute what has happened since 1973 to the rise in oil prices alone. The events of these years were in fact underlaid by a secular (i.e. long-term) downturn in the rate of growth which had started in the late 1960s and early 1970s, and which was accentuated after 1975. The long-term fall in the rate of economic growth was accompanied by an acceleration in the rate of inflation, with the creeping inflation of the 1950s and 1960s starting, in some cases, to turn into a gallop. At the same time unemployment increased, especially amongst women and young people. In 1979, for example, unemployment in the EEC was 4.7per cent for women, 2.6 per cent for men, and 11 per cent for people under 25 years of age.

The fall in profitability and a new technological revolution

In addition and most critically, there has been a secular downturn in the rate of profit and in the rate of investment. As a result the events of the 1970s and early 1980s are seen by some as representing an ending of the developments of the Second Industrial Revolution and the culmination of the growth potential of the economic mechanisms and institutions on which post-war expansion was based. The question that now has to be answered is whether or not existing types of society are capable of adjusting to automation, the microprocessor and the Third Technological Revolution, and of laying the bases for a new phase of development.

A new international division of labour

Several other components of what is a crisis rather than a recession can also be identified. A number of newly industrialising countries have been emerging. As a result the international division of labour has been transformed (see the chapter on Newly Industrialising Countries *pages 312–327*). In addition to competition in the form of cheap textiles and clothing, cheap radios and black and white television sets from low-wage countries, the old industrial countries are faced with new competitors in the fields of chemicals, steel, and vehicle production.

Unequal development in the developed world

Another major point that needs to be made is that the dynamic of growth has varied quite markedly within the more developed countries as well as

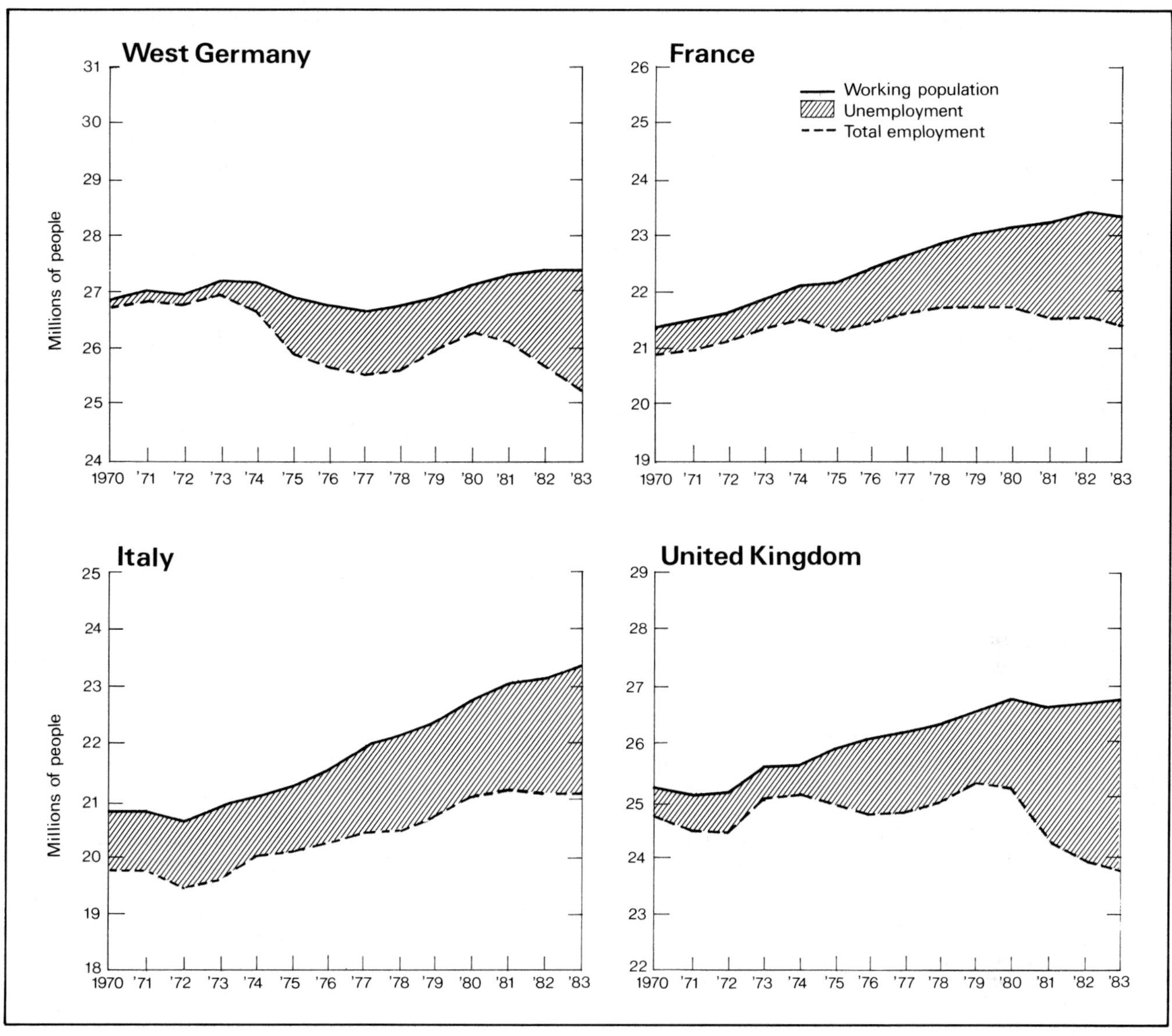

Figure 7 *Working population and employment in four West European countries, 1970–83*

between them and the rest of the world (see Figure 9). In particular the economic supremacy of the United States has been eroded by the rise of the West German and, especially, the Japanese economies. Also, important divergences in response to the crisis can be identified. The economy that has performed worst is that of the UK.

The Decline of the UK

The slow rate of growth of the UK economy and of its manufacturing sector and its lack of competitiveness are of course major elements of current political debates. Not only are there sharply diverging views about the causes, but also radically different ways of coping with these problems have been proposed.

? Write down the factors which in your view lie at the root of the problem. Afterwards read the rest of this section. At the end reformulate your answer explaining why you agree or disagree with the arguments which follow.

In part, the decline of British manufacturing has stemmed from its early industrialisation, from the exploitation of protected markets, and from the fact that industrialists have failed to invest on a sufficient scale in new plant and machinery or in new sectors of production. In the face of the continental weapon of state-protected, bank-financed industrialisation the free market strategy of Britain has been quite inadequate, as described by this economist:

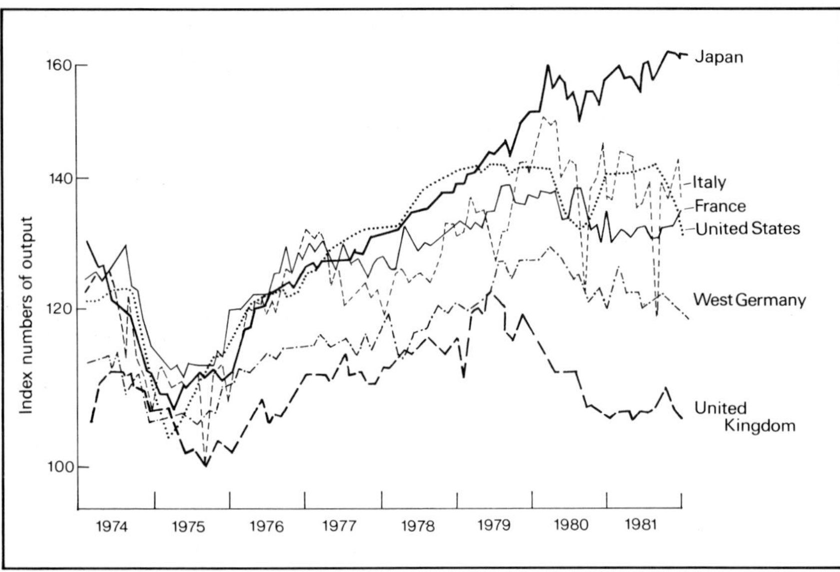

Figure 8 *Trends in industrial production in selected industrial countries, 1974–81*

> The market system creates a vital contradiction between the interests of individual firms and the interests of the economy as a whole. If firms have sunk their capital in equipment and factories which produce goods which can still be sold somewhere, the incentive to scrap perfectly good equipment in the interests of modernisation is rather weak. If all firms could be organised into a modernisation drive, they would provide the demand for each other's products and would all be better off. But in a competitive economy organised only by the market, the uncertainty of demand and the cost involved in a large-scale re-equipment may well outweigh potential profits, even if the finance can be found from meagre current profits and an uninterested banking system.
> (*Source*: Eatwell, 1982)

As a result, a culture of growth has been lacking. The banking system did not develop close links with industry or get involved in the financing of long-term industrial investment. Investment funds were diverted away from industry in Britain into land and property or into assets in other countries. In the field of education insufficient emphasis was perhaps placed on training in the applied sciences and engineering, while the absence of real equality of opportunity in British society meant that much human potential was wasted. More generally, government policy was not harnessed to the achievement of rapid industrial growth.

After 1945, British society also came to be characterised by the existence of a large, strong, well-organised, and economically class-conscious working class, itself divided for historical reasons along craft lines. With the post-war recognition of the right of workers to join independent trade unions, and in a context of relatively full employment, the working class made some striking gains. On the one hand, a high degree of job control was secured. On the other, wages came to form a high proportion of net output squeezing the profitability of the manufacturing sector.

In addition, priority was given at government level, to an international financial function, an imperial role – and substantial military spending. The competitiveness of the manufacturing sector was damaged, while resources that the economy could ill afford were diverted away from the reconstruction of the domestic economy.

Several other factors could be mentioned, but what is important is that neither employers, nor the state, nor trade unions can be held individually responsible, as none of them is fully in command of the situation. Nor are they equally responsible as the distribution of power is very unequal; in particular, an employer determined to introduce new methods of production will usually succeed. However, neither employers nor the government can altogether avoid responsibility by referring to the international recession or some other 'impersonal' factor. A recession itself stems from the action of individuals

Print 5 *Unemployed people looking for work at a Job Centre in Bradford*

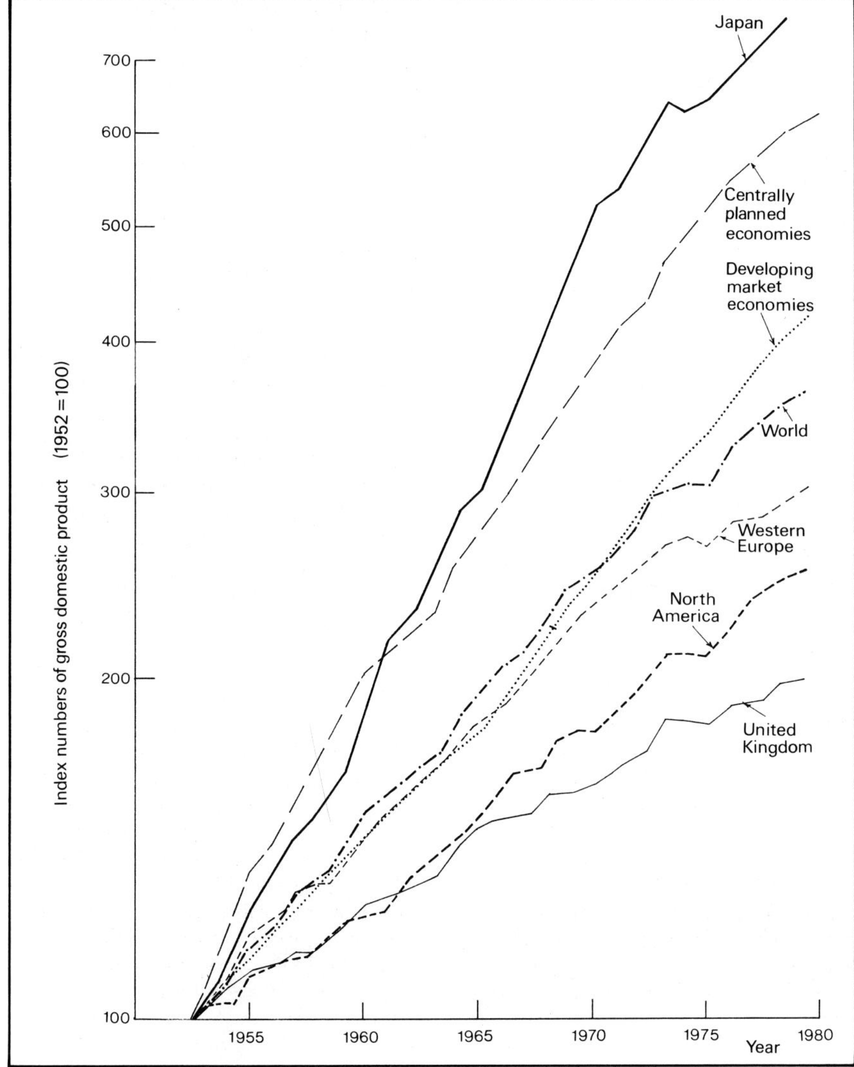

Figure 9 *Gross domestic product in the UK and in the rest of the world (Index numbers, 1952 = 100)*

(*Source*: based on statistics in United Nations, *Yearbook of National Accounts Statistics* New York: UN, Annually)

and firms who are not investing and from the pursuit by governments of deflationary strategies. A decision by a foreign government to reduce overall demand means a loss of export markets for British firms, while a cut in domestic demand weakens the home market. In short, what is happening is a product both of the actions of the main agents in economic life and of the conditions in which they are operating.

The tragedy of the situation is that there are not enough new jobs or sufficient reductions in the length of the working week to compensate for those that are being lost. What is more, the British economy, in particular, is increasingly unable to produce necessary material goods and, without oil, will have difficulty in paying for net imports of manufactured goods.

? In this section, it has been suggested that whether or not industrial change is a problem depends in part on the context in which it occurs. To what extent do you think that a solution to regional problems depends on a solution to the national problem?

15 CHANGING TRADE PATTERNS: NEWLY INDUSTRIALISING COUNTRIES IN SOUTH-EAST ASIA

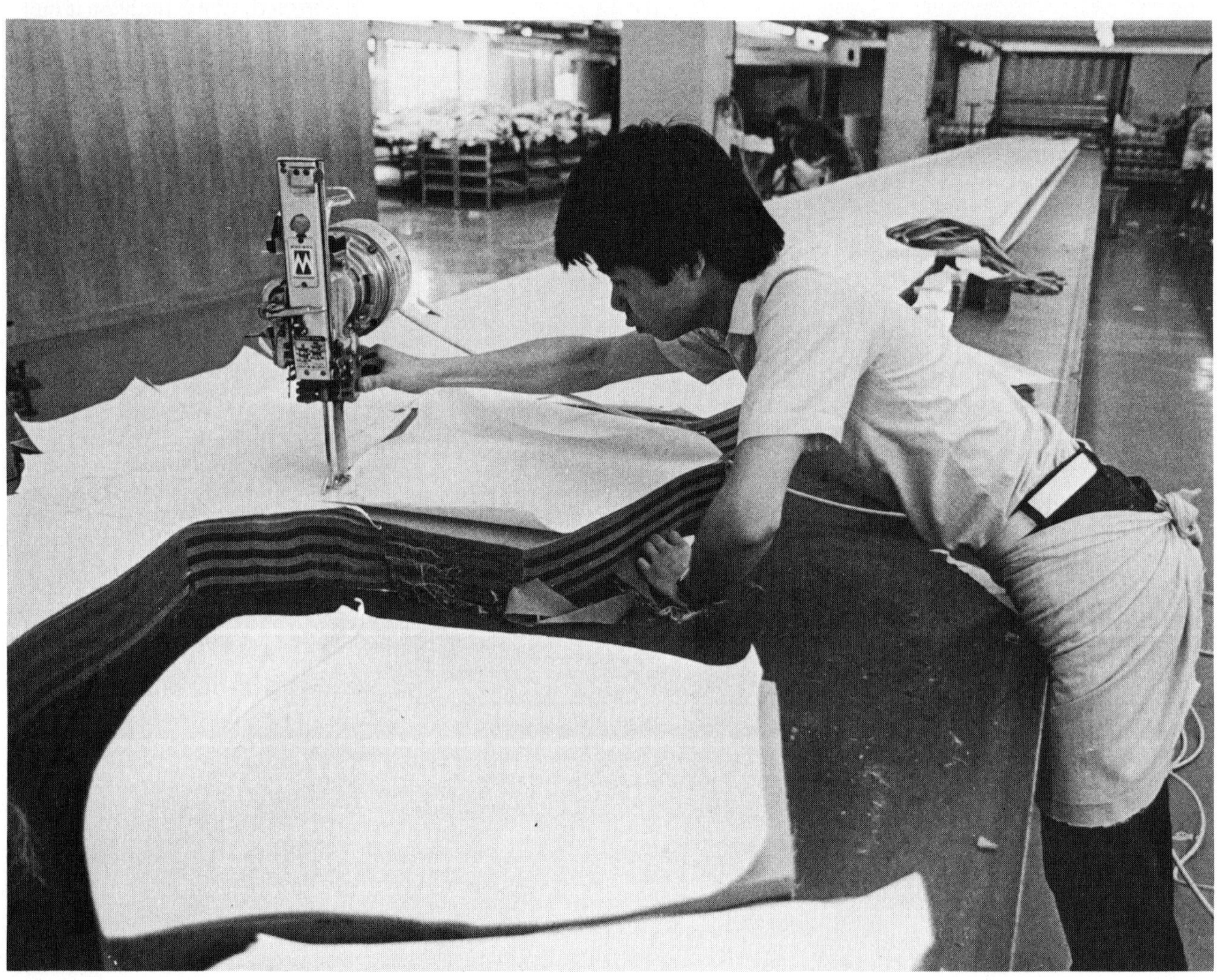

WORLD TRADE PATTERNS

People raise their living standards by specialising in a particular occupation and exchanging their labour services through the money system for the goods and services that they require. Certainly they would be much poorer if they attempted to return to the 'good life' and became self-sufficient. It is the same with geographical areas and nations.

The pattern of world trade reflects the specialisations of different nations. Each nation produces those products and services for which it has a comparative or relative advantage (Figure 1).

Comparative advantage, in turn, reflects a country's factor endowments.Some nations enjoy abundant capital resources and will produce and export machinery and equipment, for example, the UK and West Germany. Other countries have rich land resources and will specialise in producing petroleum, minerals, raw materials, temperate or tropical foods and tourist services. Highly inventive nations may concentrate on high technology products while labour-abundant countries, such as Taiwan, may produce simpler, labour-intensive manufactured products. If all nations and geographical regions specialised according to their comparative advantage and traded freely, world output and living standards would be maximised.

Figure 1 *Specialisation and the law of comparative advantage*

The Law of Comparative Advantage states that 'trade will be advantageous if each of two countries specialises in the production of those goods in which it has a comparative advantage or in which it has the least comparative disadvantage'.

Example

Countries *X* and *Y* can both produce machines and copper. Country *X*, using 10 labour units, can produce 10 machines or 200 tonnes of copper. Country *Y*, using 10 labour units, can produce 20 machines or 100 tonnes of copper. Assume each country has 20 labour units and uses half of them to produce each product. Output will be as follows:

	Machines	*Copper (tonnes)*	*Opportunity cost ratios*
Country X	10	200	1 : 20
Country Y	20	100	1 : 5
Total produced	30	300	

The theory of comparative advantage is based on differences in opportunity costs (i.e. the amount of a product that has to be foregone or sacrificed in order to produce another product). In our example, this means that the cost of a machine is measured in terms of the amount of copper which has to be sacrificed in order to produce that machine. Similarly, the cost of a unit of copper is measured in terms of the machines foregone. Country *X* has a comparative advantage in the production of copper. One tonne of copper 'costs' only 0.05 machines in Country *X*. Whereas in Country *Y* the 'cost' of a tonne of copper is 0.2 machines. Country *Y*, on the other hand, has a comparative advantage in the production of machines. (One machine in Country *Y* 'costs' 5 tonnes of copper, while it 'costs' 20 tonnes in Country *X*.)

Now assume the countries specialise according to comparative advantage. Production will be as follows:

	Machines	*Copper (tonnes)*
Country X	0	400
Country Y	40	0
Total Produced	40	400

More of both goods have been produced. Providing the terms of trade lie between the opportunity cost ratios (swap ratios), say 1 machine for 10 tonnes of copper, both countries will benefit from specialisation and trade.

TRADITIONAL TRADING PATTERNS

A traditional trade pattern emerged in the nineteenth and early twentieth centuries. Western Europe and North America developed a comparative advantage in manufactured products. Less developed countries (LDCs) and Australia and New Zealand, with their rich land resources, specialised in primary product production. Gradually a North-South trade pattern evolved, with Western Europe and the USA exchanging manufactured products for primary products from LDCs. Africa, the Middle East and Australasia supplied Western Europe with petroleum, minerals and raw foodstuffs. Latin America supplied similar products, largely to the USA. This trade pattern was reinforced later in the twentieth century with the emergence of Japan as a new industrial nation. Japan began to exchange manufactured goods for primary products from South-east Asia and Australasia (Figure 2).

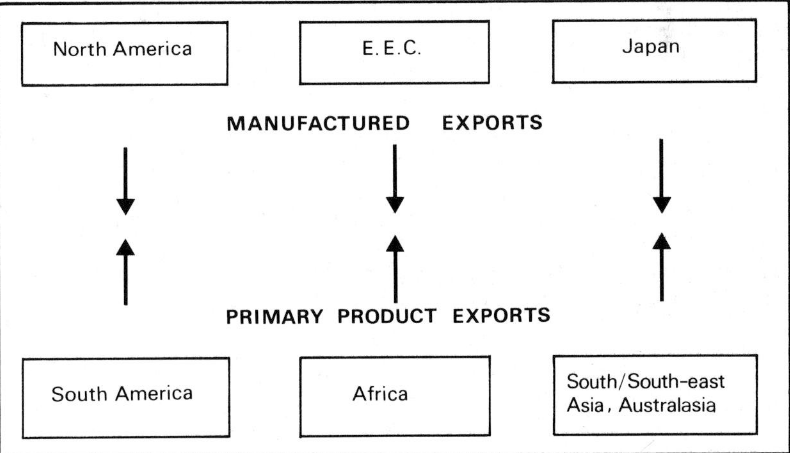

Figure 2 *Traditional trade patterns*

The traditional trade pattern tended to become frozen and was reinforced by colonial and US political and economic policies. By the Second World War it had become clear that the lion's share of the gains from trade were being enjoyed by the rich industrial countries. World trade expanded during the twentieth century much more rapidly in manufactured products than in primary products. Even as late as the 1960s primary products comprised over 80 per cent of LDC's exports while manufactured goods made up over two-thirds of advanced country exports.

Discontent began to grow in LDCs in the post-war period. Countries became aware of the dangers of excessive specialisation in one or two products. The world price of commodities such as cocoa or copper, over which they had no control, would suddenly collapse and hit their economies hard. Demand for most primary products was growing slowly and retarding the development of their economies. This was largely because there is a low-income elasticity of demand for most primary products and a high-income elasticity of demand for manufactured goods and services . That is, as incomes increase in the rich countries, which are the main markets for primary exports, people spend a lower proportion of their incomes on food and other primary products. Conversely, they spend an increasing proportion of their earnings on manufactured goods and services.

Demand for primary products has also been affected by research and development programmes in the rich countries. Over 90 per cent of all research and development takes place in the advanced countries. Fruits from this include new methods which economise on raw materials and synthetic materials, both of which reduce demand for primary exports from LDCs. Only rarely do research programmes create new demands for primary products. As the twentieth century has progressed many LDC governments have attempted to escape from this trap of stagnant or decreasing demand by developing new areas of comparative advantage through a policy of industrialisation.

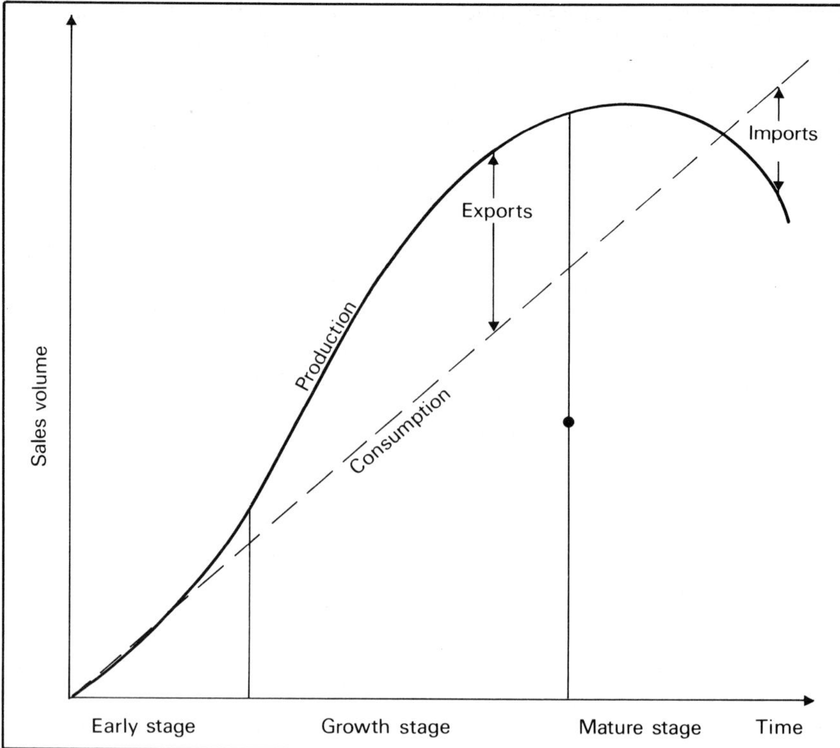

Figure 3 *The Product Life Cycle. A country may initially export a new product but it MAY lose its comparative advantage when the product becomes mature, finally becoming an importer of the product*

THE NEW INTERNATIONAL DIVISION OF LABOUR

A new international division of labour has now emerged. The traditional specialisation of developing countries in primary production, and the advanced countries in manufactured goods is changing rapidly as the industrialisation of the LDCs progresses. The days are over when a small group of industrialised countries, principally in North America and the EEC, supplied manufactured goods in exchange for raw materials or relatively simple industrial products from overseas.

There seems to be an international product cycle which helps to explain trade in manufactured goods (see Figure 3). New products are usually developed in the long-established industrialised nations. At a later stage, when the process is 'matured', production 'migrates' to developing countries in order to benefit from their comparative advantage. A classic case of this process are the US/Mexican border industries. The more labour-intensive processes are performed on the Mexican side of the frontier. In general, we may expect labour-intensive 'mature' industries and products to migrate to the developing world.

In the early phase a product such as a motor vehicle is research- and skill-intensive; sales are low. During the growth phase mass production and distribution are possible and exports of cars build up. (The USA is a good example.) Then the industry becomes mature, the product becomes standardised and the manufacturing process is routine. Overseas producers now gain a comparative advantage in the 'old product' because they can easily acquire the technology and can man plants with cheap unskilled and semi-skilled labour. Japan, and perhaps eventually Brazil and others, export vehicles back to the US, which may become a net importer (see Figure 3). The product cycle suggests that mature industries, such as textiles, iron and steel, ship-building and motor vehicles will gradually migrate to the developing world and change trade patterns. As the recent article below suggests, however, it is also possible that in future the introduction of new labour-saving technology, such as robots, will gradualy eat away at the comparative advantage enjoyed by the developing countries, as regards labour costs – in respect to certain industries.

The conventional idea of the multinational corporation is that it buys raw materials in the cheapest parts of the world, produces them in places like the Far East with low labour costs, and then sells them in the most attractive market place.

That, according to a new book, Triad Power (Macmillan), is old hat. The benefits of chasing the lowest labour costs around the world are short-lived. Such is the increasing capital intensity of industry that labour costs in these industries are only 10 per cent of total manufacturing costs.The advantage of cheap labour is quickly outweighed by the cost of transporting critical components, especially as newly trained labour soon becomes more expensive.

Most competitive Japanese firms, the author (Kenichi Ohmae, managing director of the McKinsey office in Tokyo) says, are now pulling out of south-east Asia and investing in robots and machines. Tomorrow's successful companies, he adds, will have to be triads, with a base in each of the three key regions of the 600 million population OECD area – Europe, Japan and the United States, which represent 54 per cent of the world economy.

(*Source*: *The Guardian*, 28 May 1985)

THE RAPID INDUSTRIALISATION OF LDCs

Rapid industrial development has occurred in many developing countries during the last 20 years (see Table 1). Manufacturing output, although starting from a low base, has grown at around 7 per cent per year, whereas that in the advanced countries has grown at 4.5 per cent. Employment in manufacturing industry has grown at 5 per cent in the less developed countries over the period, whereas it has either stagnated or declined in the advanced nations.

In the early post-war period,

Table 1 *Third World Industrial Production (per cent of total)*

	1963	*1970*	*1975*	*1980*
Brazil	14.0	14.3	16.7	18.8
South Korea	1.4	3.5	7.1	12.1
India	18.8	14.9	12.1	12.0
Mexico	9.1	10.7	10.2	10.7
Argentina	11.1	11.9	10.6	7.8
Total	54.4	55.3	56.7	61.4

(*Source*: based on UN *Yearbook of Industrial Statistics*, 1979 and 1980 editions)

developing countries concentrated on *import substitution* industries. That is, factories were created behind tariff and quota barriers to supply the domestic market. Tariffs of up to 300 per cent were levied on some imported products to protect the sale of similar products produced at home. An important change was discernible from about 1960 when countries, especially those in East Asia and South America, began to concentrate increasingly on building up export-oriented industries.

The industrialisation of Latin America was particularly dramatic in the 1950–75 period. Production of machinery and equipment showed a nine-fold increase and met three-quarters of the region's needs. Steel production increased 13 times, cement seven times and electricity production eight times. Automobile production, which was zero in 1950, reached 2 million units by 1975. Projections for 1990 indicate that Latin America's manufacturing output in that year will be only 20 per cent below the EEC's manufacturing output of 1970.

Industrial development in LDCs is constantly becoming more varied and it is not just restricted to light industry. Heavy industry, starting from a very low point, has been expanding more rapidly than light industry during the last 20 years. Developing countries now have a significant share of the world output of textiles and clothing, and of certain products made by the electrical engineering industry. Moreover, their share of steel and shipbuilding is by no means negligible.

Examples of very modern industrial production include drilling rigs in Singapore, and helicopters and small computers in Brazil. During the last two decades modern industrial sites have been established in virtually all LDCs, often by multinational corporations. Opinions on these developments vary.

'We shouldn't import manufactured goods from the developing world. Cheap imports are knocking out one great industry after another. Textile mills and shoe factories have closed, and now it's our shipyards, steel mills and colour TV plants. It's impossible to find a suitable job in this town. The UK is becoming an industrial wasteland.' (Unemployed steelworker, Consett, North-east England)

'Sweat shops in East Asia are ruining our industries and causing more and more unemployment. Multinational corporations like Dunlop's, with their 'green flash' tennis shoes and rackets, and what-not produced in Taiwan and such places, are to blame. They employ labour for a few pence an hour, women and children as well, in their sweat shops for 16 hours a day. It's exploitation. It should be banned.' (Trade Union Official from the footwear industry, East Midlands)

'We must industrialise to wipe out poverty and raise the living standards of the masses. Exchanging primary products for manufactured goods with the rich North is a disaster. Demand grows too slowly for coffee, sugar, cotton and our other products. The rich countries invent substitutes for our farming and mineral products. The prices of manufactured products go up and up while primary product prices go down and down. Every year we've got to export more and more tonnes of sugar for a tractor. The sooner we transform ourselves into a fully industrialised nation the better.' (Brazilian journalist)

'The emerging new international division of labour should not be seen as a threat to jobs and living standards in the rich countries. If all countries specialise according to comparative advantage and trade, the peoples of the world will enjoy cheaper products and higher living standards. The concerns and worries of some people in the rich countries are totally irrational.' (World Bank official, Washington)

? Analyse why these people express such views. Discuss the implications of the 'new international division of labour' for the advanced industrial countries and the less developed countries. Take into account also the views expressed in *The Guardian* extract, and the fact that the present product cycle may change again.

WHO ARE THE NICs?

Any classification of newly industrialising countries is somewhat arbitrary, but the *Organisation for Economic Co-operation and Development* (OECD) has identified ten countries.

Table 2 *The ten newly industrialising countries*

East Asia	*Latin America*	*South Europe*
Hong Kong	Brazil	Greece
Singapore	Mexico	Yugoslavia
South Korea		Spain
Taiwan		Portugal

(*Note*: Spain was reclassified as an industrial market economy by the World Bank in 1983)

They are a disparate group. Some, such as Singapore and Hong Kong, are city states with few natural resources and small populations, while others, for example, Mexico and Brazil, are very rich in natural resources and have large populations (see Table 3), but they have several features in common. Their economies, and especially manufacturing output and exports, have grown much faster than the advanced industrial countries (OECD countries) in recent decades. Since 1970 their real GDP has grown at more than twice the rate of the OECD countries (see Table 4). They have followed outward-looking or export-oriented growth policies. In the 1970s exports of manufactured products grew in real terms at a massive 13 per cent per year against annual increases of five per cent or so for the UK and six per cent for the US and West Germany. The ten NICs account for around 80 per cent of total developing world manufactured exports. Of this figure the 'gang of four' in East Asia account for over half.

The newly industrialising countries are gradually exporting a wider range of products. Initially they specialised in traditional, labour-intensive industries, such as textiles, clothing, footwear and leather products. More recently, they have developed more capital-intensive mature industries. Exports of machinery, transport equipment, electrical and electronic products are growing in importance, although more traditional areas of textiles and clothing are still very significant (Figure 4).

Table 4 *Growth of real GDP (per cent per year)*

	1960–70	*1970–80*
South Korea	8.6	9.5
Hong Kong	10.0	9.3
Singapore	8.8	8.5
UK	2.9	1.9
USA	4.3	3.0
Japan	10.9	5.0

(*Source*: The World Bank, 1982)
(*Note*: GDP = *Gross domestic product* – a major measure of a country's national output of goods and services. It is equivalent to gross national product minus net investment incomes from foreign nations. Real GDP indicates that any inflation or price effects have been removed from the statistics)

There is no reason to regard the ten newly industrialised countries as special cases. The industrialisation process will probably spread to most LDCs. In 1955 only 10 per cent of total exports from the developing

Table 3 *Indicators: newly industrialising countries 1982*

	Populations (millions) (1982)	*Population growth rate (1970–1982)*	*GDP (billions US $) (1982)*	*GDP real growth rate (1970–1982)*	*GNP per capita (US $) (1982)*	*Manufacturing as % of GDP**	*Agriculture as % of GDP**	*Exports as % of GDP**	*Petroleum imports as % of total imports**
Brazil	126.8	2.4	248.5	7.6	2 240	28	11	8.3	41
Greece	9.8	1.0	34.0	4.1	4 290	19	16	13.0	18†
Hong Kong	5.2	2.4	24.4	9.9	5 340	19	1	88.0	7
Korea, South	39.3	1.7	68.4	8.6	1 910	27	20	29.6	25
Mexico	73.1	3.0	171.3	6.4	2 270	29	10	7.4	n.a.
Portugal	10.1	0.8	21.3	4.5	2 450	37	13	19.3	19
Singapore	2.5	1.5	14.7	8.5	5 910	28	2	158.0	29
Spain	37.9	1.0	181.3	3.1	5 430	25	9	9.2	33
Taiwan	18.4	1.9	49.7††	8.7†††	2 670	35	11	52.0†	14†
Yugoslavia	22.6	0.9	68.0	5.5	2 800	31	12	11.9†	22

* 1979 † 1978 †† GNP ††† GNP 1951–83

(*Sources*: The World Bank, 1984, *Barclays Bank Review*, 1982, *Lloyds Bank Economic Reports*, Taiwan)

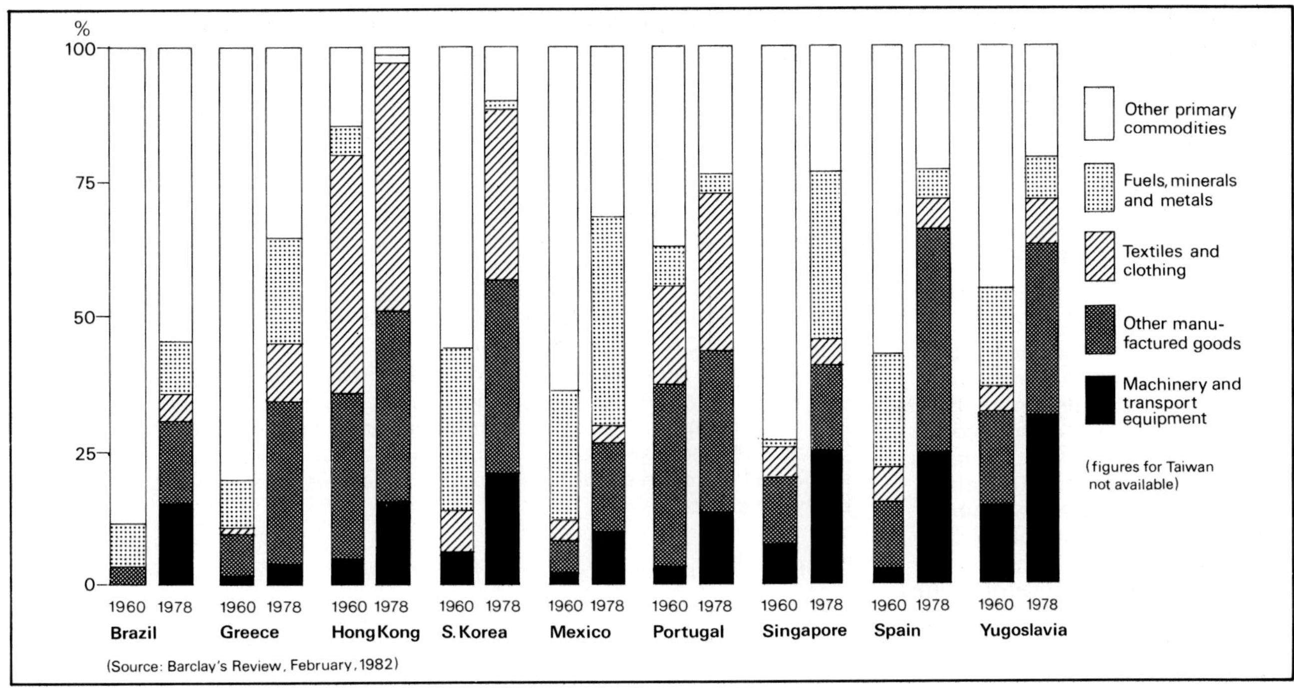

Figure 4 *Percentage share of merchandise exports* (*Source*: Barclay's Review, February 1982)

countries were manufactured goods (excluding petroleum). By 1977 manufactured products comprised 41 per cent, and are projected to reach around 65 per cent by 1990. There are a whole host of aspiring NICs including Indonesia, Malaysia, Sri Lanka, India, Pakistan, the Philippines, Nigeria and Argentina.

THE NEWLY INDUSTRIALISING COUNTRIES OF EAST ASIA

The 'gang of four' are in a class by themselves. Hong Kong, Singapore, South Korea and Taiwan have pursued vigorous export-oriented development policies. These policies have been very successful to date. Real gross domestic product (GDP) and industrial production have grown during the last 20 years at annual rates of over 9 per cent and 13 per cent respectively, far above the rates achieved in the advanced industrial countries. However, there has been some variation from country to country as Table 5 indicates. The 'gang of four's' share of World industrial production rose from a tiny fraction in 1960 to over 2 per cent by the early 1980s. But it is their exports of manufactured

Table 5 *Key indicators: the Asian NICs, 1960–82*

	Hong Kong	*South Korea*	*Singapore*	*Taiwan*
GDP (% real growth average annual)				
1960–67	9.9	7.2	7.5	8.9[1]
1967–73	10.2	10.9	13.0	10.6[1]
1970–82	9.9	8.6	8.5	8.7[1]
Manufacturing production (% real growth average annual)				
1960–67	n.a.	12.9	11.1	14.5
1967–73	n.a.	21.0	17.9	22.1
1973–80	n.a.	18.0	8.9	12.8
Manufactured exports (% growth average annual)				
1960–67	13.5	76.4	2.8	21.9
1967–73	22.7	52.4	33.4	37.8
1973–80	21.3	28.1	28.1	23.7
Manufactured exports per capita (US$)				
1960	183	0	133	15[2]
1967	349	7	129	49[2]
1973	1076	78	660	289[2]
1980	3489	396	3422	1126[2]

(*Note*: [1] GNP [2] Total exports)
(*Source*: *Barclay's Review*, February 1982 and the World Bank, 1984)

products which have grown at a spectacular rate. During the last two decades manufactured exports have grown at an average rate of 28 per cent per year, around four times faster than OECD industrial exports.

AREAS OF COMPARATIVE ADVANTAGE IN EAST ASIA

Exports from East Asia are now penetrating the markets of North America, the EEC and Japan, and making significant inroads into certain sectors of these markets. The Asian NICs' exports comprised less than one per cent of OECD imports in 1960. By the early 1980s their share of OECD imports was about five per cent. The Asian NICs' inroads into markets in the advanced industrial countries have been heavily concentrated in certain products.

Textiles and clothing

It is relatively easy for developing countries to enter the textile and clothing industries, especially the latter, because of its labour intensity. (Compare textiles and iron and steel in Table 6.) This sector tends to be of central importance in the early stages of industrialisation. But there are some problems. Firstly, trade in textiles tends to cause friction with the OECD nations, as the textile industry tends to be concentrated in particular stagnating regions in the advanced countries, as is, for example, the case in North-west England. Secondly, product cycle theory would predict a movement of such mature textile industries first from the OECD countries to the NICs and then from the established NICs to other aspiring NICs, such as Indonesia and Bangladesh. Given that, at present, in the case of the Asian NIC's economies, the clothing and textile industries account for up to 30 per cent of industrial output, 30–40 per cent of manufactured exports and 50 per cent of total industrial employment, this future scenario could cause severe structural problems for them.

Growing textile and clothing exports from the NICs have led to structural changes in the world pattern of industrial production. There has been relative stagnation of production in the advanced countries, especially in the EEC in recent years.Since about 1970 employment has declined in the textile and clothing industries at 4.5 per cent per year in the EEC, two per cent in the USA and six per cent in the Japanese textile industry. Over 1 million jobs were lost from textile industries in the OECD countries in the 1970s. In the early 1980s the 'gang of four' supplied around one-third of OECD imports of clothing.

? You have recently been employed by the North-west England Regional Planning Council. Concern has been expressed in your region with the growing unemployment in the textile industry. Recently, a journalist writing for the *Blackburn Evening News* suggested that the area should diversify into electrical goods, such as TV sets and radio cassettes. Your planning Council has similar views.

1 Identify recent trends in world specialisation and trade.
2 Review, with examples, major labour-intensive, skill-intensive, mature capital-intensive, sophisticated and highly capital-intensive, and high technology-intensive industries.
3 Recommend suitable new industries for this region, giving reasons for your choice.

Table 6 *Physical and human capital per worker in selected US industries*

Industry	*Physical Capital (US$)*	*Human Capital (US$)*
Clothing	2 370	11 000
Textiles	10 000	16 620
Road motor vehicles	12 890	25 400
Chemicals	21 370	25 510
Household appliances (including televisions)	8 290	39 090
Iron and steel	27 710	28 150

(*Source*: B. Balassa, 1980, *Structural Change in Trade in Manufactured Goods between Industrial and Developing Countries*, World Bank, Washington)
(*Note*: **1** Physical capital includes machinery, equipment etc. **2** Human capital is a measure of the workforce's education and skills, calculated from the difference between average and unskilled wages. **3** The table confirms that clothing is at one end of the spectrum and steel and chemicals at the other.)

Print 1 *A 'sweatshop' jean factory in Hong Kong*

Electronics and light electrical goods

A variety of industrial activities are included in the light electrical and electronics sectors. The Asian NICs are particularly strong in this varied sector. Hong Kong is the world's largest exporter of toys, and to a considerable extent, this is due to electronic developments.

It has also developed a strong comparative cost advantage at the lower end of the digital-watch market and is now the world's third largest producer of watches. New and growing electrical and electronic product areas being rapidly developed by the Asian NICs include colour televisions, calculators, home computers, microwave ovens, citizens' band (CB) radios, home TV cameras, burglar alarms, electronic watches, electronic games and audio and video cassettes.

Print 2 *A street vendor in Hong Kong displays a wide range of watches produced in Hong Kong and Taiwan*

The leading Asian NICs are trying to break into an important segment of the consumer electronics industry – colour televisions. This is a highly skilled, but, for the moment, labour-intensive industry. Colour televisions are also a case of an apparently technologically mature product as new sales now meet replacement demand largely.

Between three and five per cent of the OECD total industrial working population is employed in the electronics industry as a whole, and less than one per cent in consumer electronics (which includes television production). In the UK five per cent or so are employed in the electronics sector, and less than 0.5 per cent in consumer electronics. There is therefore not an excessive number of jobs at risk from increased competition from the Asian NICs.

The advanced countries absorbed pocket calculators, electronic watches and radios from Hong Kong without too many problems, but their reaction to television sets has been more sensitive. At first the Asian NICs concentrated on black and white sets (a declining market in the OECD countries), but now they have switched to colour television in a big way. South Korea and Taiwan, in particular, appear to be developing a new comparative advantage in this area. South Korea has recently developed a capacity to export over 1 million colour sets per year, their initial target being an ambitious four or five per cent of the world market. Initially Japanese corporations established subsidiary companies in the Asian NICs (partly to export cheap TV components back to Japan) but the NICs are rapidly developing their own expertise.

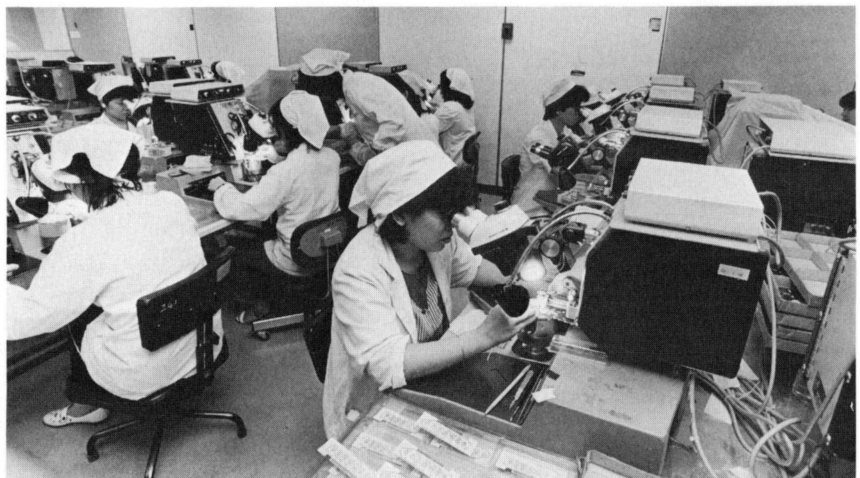

Print 3 *High technology in the New Territories, Hong Kong—a specialist factory making microchips for computer games which are sold worldwide*

Multinationals

Multinational corporations have come to regard the Asian NICs as very attractive bases for new labour-intensive plants. Japanese corporations, such as the Sony Corporation, are actively involved in this area although as the recent *Guardian* extract (*page 315*) points out, this involvement may decline in certain areas. The multinationals may own producing plants, allow their products to be produced under licence or engage in sub-contracting activities. Japanese corporations built up textile plants in East Asia during the 1960s and electronics plants during the 1970s and early 1980s. US corporations have also created electronics production facilities. Workers in the United States have complained that the multinationals have transferred jobs from their own country to East Asia. The multinationals have invested particularly heavily in Singapore and Hong Kong. They account for around 80 per cent of total investment in Singapore and for more than two-thirds of the country's manufactured exports.

HOW THE ASIAN NICs BECAME INDUSTRIALISED

The Asian NICs lack both a raw material base and large domestic markets. How then have they become new industrial nations? The reasons are complex and vary somewhat from country to country. A fundamental reason seems to have been the adoption of an export-oriented policy based on plentiful supplies of cheap labour. This policy enabled them to develop a strong advantage in the production of labour-intensive products, such as clothing and leather goods. But why the 'gang of four' and not other cheap labour countries, such as Zaire, Kenya and Sri Lanka? A critical factor appears to have been active and successful state intervention to promote the establishment of export industries in apparently free market economies.

State intervention

Governments have used a whole battery of methods to develop export industries. They have protected or subsidised carefully chosen infant industries until they became established. Certain governments have shown imagination and foresight. Compare the policies of the UK and South Korean governments concerning the steel industry since the late 1960s. South Korea, which produced negligible quantities of steel in the 1960s, is already exporting steel and is planning to overtake UK steel production by 1991.

The UK government chose medium-sized, relatively high-cost, new steel plants. The Korean government decided to build giant, low-cost steel plants. The *Sunday Times* article describes the Korean success in steel production and ship building.

Korea rules the waves

THE GROWTH of South Korea's shipbuilding industry has been spectacular. In 1974, it ranked 70th in the world; just seven years later it was second only to Japan.

Its premier company, Hyundai, now boasts one of the world's largest shipyards, spreading over 5m square metres and with 45 ships in varying stages of completion. Although it has not escaped the effects of the recession, like many other companies Hyundai and its nearest rival, Daewoo Shipbuilding, are jammed to the gills with work.

As Korea's industry has grown, so too has criticism from western critics. They accuse the Seoul government of granting excessive subsidies and export credit to infant industries, thereby enabling them to bankrupt European yards by quoting prices up to 35% lower.

Yet both Daewoo and Hyundai emphasise that apart from an initial five year tax break, they receive no direct government subsidies. The Export-Import bank of Korea sometimes helps with domestic and with yard export credit, but according to a foreign consultant, "Money is short, and the Koreans are seldom as competitive as the Japanese in credit terms."

They maintain that Korea's attractive low prices and early delivery dates are due to the efficiency of its modern equipment, the comparatively low costs of initial shipyard construction, cheap raw materials – such as locally produced steel – and perhaps most important, its cheap, highly disciplined, hard-working labour force.

Most shipyard workers put in 10 hours a day, six days a week and are not averse to additional early morning, late night or Sunday overtime. Wages are still only about one-third of their Japanese counterparts, strikes are illegal and unions are mostly formed by management. Foreigners visiting or working at the shipyards are nearly always impressed by what one consultant described as the high "willingness to work hard and do it right quotient" of Korean workers

Korean managements are well aware of the value of their cheap, diligent labour force and look after their workers by providing medical, educational, housing and other facilities.

Husbands and wives are encouraged to join the company, and free meals are given to those who volunteer for early or late overtime shifts. Top executives wear the same uniform and work the same hours as shop floor workers. Looking after his workers is "quite frankly, a way of avoiding unproductive labour unions," says Hong In-Kie, president of Daewoo Shipbuilding. "We all work hard and we share the sense of accomplishment."

Ironically, Austin & Pickersgill at Sunderland was one of the international companies to which the Koreans turned for expertise when they launched their shipbuilding industry.

Jacqueline Reditt

(*Source*: *The Sunday Times*, 13 February 1983)

Certain Latin American countries 'featherbedded' their infant industries in the post-war period. Protective tariffs of up to 300 per cent were introduced, virtually guaranteeing that their new industries would never be able to compete internationally and earn export income. In contrast to these inward-looking policies which ignored comparative advantage, the Asian NICs have tended to adopt outward-looking policies. They have selected industries with future comparative advantages perceived.

The Asian NICs have provided a

range of stable tax incentives, cheap loans and other subsidies for their export industries. Their exchange rates have been reduced when necessary, which makes their exports more competitive, as devaluation has the effect of reducing export prices in world markets. (Compare the overvalued pound sterling in the 1979–82 period which had the effect of making UK exports 20–50 per cent overpriced in world markets.)

The Asian NICs have tended to prohibit or restrict unions, so wage pressures have been much less than in the advanced industrial countries. Taxes have been reduced on imported materials and components for the new export industries. Foreign corporations have been encouraged to set up manufacturing plants. They often receive several years of tax 'holidays' (that is, they don't pay for the first few years) and are allowed to remit profits and dividends freely to their own countries. In general, the Asian NICs have created a very favourable business climate for the production of manufacturing exports.

Hong Kong and Singapore

In the case of Hong Kong and Singapore, physical characteristics, location and history all seem to have combined to make them centres of manufactured exports. They are almost totally lacking in natural resources and can provide only a small portion of their food supply. They do not have domestic markets large enough to serve as an initial base for industrialisation. Their very existence depends on their ability to import, which in turn rests on their export capacity. Both developed historically as entrepôts,that is, ports where goods are imported and then re-exported without incurring liability for duty. They were alert to worldwide developments, and this helped the development of financial, communication and distribution skills associated with the entrepôt trade. This, in turn, helped their transition to industrial societies.

South Korea

South Korea decided on the import-substitution route to development (see *page 322*) in the late 1950s, but constraints, such as the absence of natural resources and the small size of the home market, soon led to a rethink. There was a fundamental switch in the early 1960s to an export-oriented growth policy. A whole series of measures, such as devaluation of the currency to make exports more competitive, lavish tax incentives, and subsidies were introduced to stimulate exports. Wholehearted commitment by the government to export expansion created a favourable business climate.

Taiwan

In Taiwan, Japanese occupation and investment (1895–1945), followed by an influx of talented Chinese from the mainland after World War II, laid the basis for modern development. From the 1950s there was considerable expenditure to build up the education and skills of the population (this also happened in South Korea). In the 1950s and 60s US economic aid averaging $100 million a year helped to stimulate further development. Fundamental policy reforms were introduced in 1960, when the country adopted an export-oriented development policy. The Taiwanese government became fully committed to promoting manufacturing exports. The import-substitution policies of the 1950s were abandoned. Tariffs on imports were reduced. Again, manufactured exports were stimulated by tax incentives, subsidies and reducing the exchange rate to make export prices competitive in world markets.

Educating the workforce

Export-oriented industrialisation policies have been reinforced by heavy investment in education and training at all levels. Everybody in the Asian NICs is provided with primary school facilities and over two-thirds are enrolled in secondary schools. Literacy rates compare favourably with the OECD countries (Table 7). While gaps still exist in higher education, they are rapidly catching up. Just over ten per cent of the age group are now receiving higher education, compared with 20 per cent in the UK and 30 per cent in Japan.

The newly industrialising countries of Asia are therefore continually improving the education, skills and productivity of their working populations. Their store of human capital is being increased. This enables them to develop new areas of comparative advantage, and diversify out of textiles and clothing as markets in the OECD countries for these products are being restricted. They are gradually developing an advantage in more skill-intensive and more technologically advanced industries in which they did not have such an inherent natural advantage. Examples include electronics, machinery, chemicals and steel. As skills, including management skills, continue to improve, they can be expected to mass-produce several

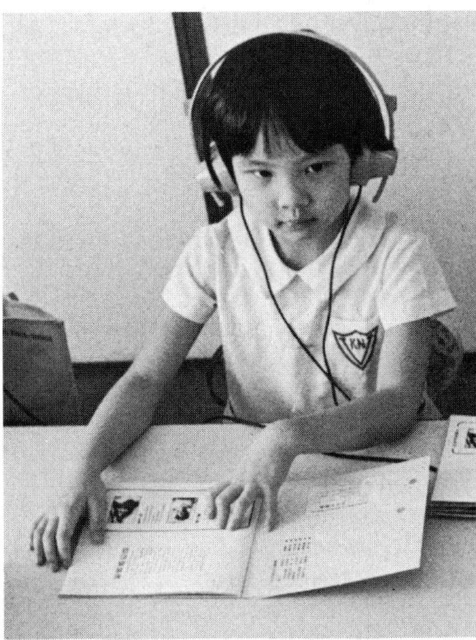

Print 4 *Education is a priority for the Asian NICs—all children now attend primary school*

Table 7 *Asian NICs: Adult literacy rates (1980)*

Country	*Per cent*
South Korea	93
Hong Kong	90
Taiwan[1]	82
Singapore	83
Industrial countries (including UK)	99

(*Source*: World Bank, 1983, and *Barclays Review*, February 1982)
(*Note*: [1] 1979)

technologically mature products. These products may meet lower trade barriers than textiles in export markets.

Labour practices

The East Asian NICs have been able to compete on the basis of cheap labour. A lot of their factories are sweat shops where men, women and children work excessively long hours in poor working conditions for extremely low wages. Trade unions are illegal or severely harassed. Welfare spending is low. In larger companies, management policies tend to be paternalistic, providing welfare services for their non-unionised workforce.

Investment

A final key factor which helps to explain the Asian NICs' rapid growth and industrialisation is the high level of savings and investment. Investment adds to a country's capital stock enabling it to produce more goods and services. For example, it enables Korea to produce more colour televisions, and Singapore is able to produce more banking and financial services. Savings rates have increased dramatically during the last two decades, from very low levels in 1960 to over 25 per cent of GDP by 1980. Investment rates have been even higher as the NICs have also borrowed funds from overseas. (We shall return to the potentially negative impact of this borrowing later.) One important reason for a relatively slow growth rate in the output of goods and services in the USA and the UK has been a lower investment rate.

Table 8 *Gross Domestic Investment (per cent of GDP)*

	1960	*1980*
Singapore	11	43
South Korea	11	31
Hong Kong	18	29
UK	19	16
USA	18	18

(*Source*: World Bank, 1982)

?

1 You are employed as an expatriate expert by the Ministry of Planning and Development in Bangladesh's capital city. Your Minister does not want Bangladesh to remain one of the world's poorest countries for ever. He has been casting covetous eyes on the spectacular growth performance of the Asian NICs. He challenges you to investigate and suggest how Bangladesh can become a 'copy-cat'.

a What key factors help explain successful and explosive development in the Asian NICs?

b Make policy recommendations to your Minister which may help to improve Bangladesh's growth rate and help alleviate her poverty problems.

2 Examine the view that specialisation according to comparative advantage is the major factor underlying the pattern of trade between the NICs and the OECD countries.

Print 5 *Workers exercise during a break at Hyundai shipyard in Ulsan, South Korea*

THE SOUTH KOREAN EXAMPLE

In the mid-1950s, South Korea was an extremely poor, largely rural LDC. Its population was growing rapidly. The modern sector of the economy was extremely small, with manufacturing comprising only six per cent of GDP, and manufacturing exports were almost unheard of. Even as late as 1961, South Korea appeared to be poor in resources and overpopulated, with its population growth rate approaching three per cent. Per capita income was a meagre US $82. There was extensive underemployment and unemployment. The nation had no significant exports, and it had relied heavily on US foreign aid in the 1950s, some of which had been used to provide more extensive education services, to improve the infrastructure and to create some import-substitution industries.

Then, in the period after 1960, the Korean economy entered the take-off stage. High growth rates were achieved and the country experienced one of the most dramatic and rapid reversals of fortune that any country has experienced in modern times. Its growth rate (real GDP) averaged almost 10 per cent per year in the 1960–80 period, exceeding that of the Japanese superstate. In the same period, manufacturing output grew at a massive annual rate of just under 20 per cent. The whole economy was propelled forward by

Table 9 *Key national indicators for Korea*

	1962		*1980*	
Population (millions)	27		38	
Urban population (per cent of total)	28	(1960)	55	
Population growth rate (per cent)	2.5	(1960–70)	1.7	(1970–80)
Infant mortality (per thousand)	60		31	
Life expectancy (years)	53		66	
Per capita income (US$)	82		1520	
Per cent of population below minimum subsistence level	41	(1965)	10	
Growth of GDP (per cent)	8.6	(1960–70)	9.5	(1970–80)
Exports (annual growth rate per cent)	34.1	(1960–70)	23.0	(1970–80)
Manufacturing output (annual growth rate per cent)	17.6	(1960–70)	16.6	(1970–80)

(*Source*: The World Bank, 1982, and *South*, 1982)
(*Note*: Growth rates in real terms)

a 'supersonic' rate of growth of exports. Exports grew in real terms at 34 per cent per year in the 1960s, and at 23 per cent in the 1970s (see Table 9). The UK did not remotely approach these growth rates in its heyday and clearly Korean rates are not sustainable in the long run. Nevertheless, between the 1950s and early 1980s, Korea transformed itself from being a poor LDC to a middle-income country with a per capita income of over US $1500. While there is considerable inequality in income distribution, it does not have the extremes of countries like Brazil.

? What other key national indicator would you like to see in this table which would help you to judge the success of Korea's export-led growth?

Korea's remarkable progress in recent years was the result of an outward-looking development policy in which a phenomenal expansion of exports played a key role. How did Korea manage to achieve such an explosive growth rate when most LDCs grew relatively slowly? Is the international environment of the 1980s now unfavourable to a country like Korea? How should Korea adapt to the more protectionist environment of the 1980s as some countries move away from free trade policies? What part has extensive borrowing had to play in South Korea's 'economic miracle'? We will now try to explore some of these questions.

LABOUR-INTENSIVE MANUFACTURED EXPORTS – THE 1960s

Korean government policy in the 1950s emphasised import-substituting industrialisation. This is an inward-looking policy. It involves protecting new or infant industries by tariffs and quotas. Products previously imported from overseas are produced by new domestic industries for the home market.

In the 1960s the government published two five-year plans. Their primary goal was to achieve a high rate of growth through the expansion of labour-intensive manufactured exports for which Korea enjoyed a comparative advantage. So the government consciously chose an export-led growth strategy in this period. Exports such as textiles, clothing, footwear and leather goods were expanded. Policies to stimulate exports were introduced. There was a massive devaluation of an overvalued currency by nearly 100 per cent. This made exports more competitive by reducing their prices in world markets. Businesses were given tax incentives and subsidised loans. Foreign enterprises were encouraged to locate their activities in Korea by the creation of free trade zones and lavish tax incentives, and they could freely remit profits.

The international environment of the 1960s was also favourable for Korean exports. World trade was expanding, and trade restrictions were being reduced under the GATT Agreements.

DEVELOPMENT OF HEAVY INDUSTRIES – THE 1970s

Rising protectionism, affecting traditional industries, and doubts over whether the USA would continue to defend Korea stimulated a dramatic change of policy during the 1970s. The five-year plans now emphasised the build up of strategic heavy industries in which Korea had no obvious comparative advantage. Conscious policy decisions were taken to create a heavy industrial base, including iron and steel, shipbuilding, machinery, petrochemicals and electronics. Cheap loans and subsidies were provided. Often, huge plants enjoying massive economies of large-scale production were chosen, especially in shipbuilding and iron and steel. Exports rose from US$ 88 million to US$ 2.2 billion between 1971 and 1981.

New exports have been developed in electronics, machinery, steel and shipbuilding. Steel exports totalled 3.6 million tonnes in 1981 and were expected to rise by 20 per cent per year. There has been a spectacular rise in electronics production and exports in the last decade.

Table 10 *Main markets for Korean exports, 1983*

Country	*Per cent*
USA	33.7
Japan	13.9
Saudi Arabia	5.9
UK	4.1
Hong Kong	3.3
West Germany	3.2
Others	35.9

(*Source*: *Lloyds Bank Economic Report*, Korea, 1984)

Print 6 *A woman working on an assembly line at Hyundai car factory in Ulsan, South Korea*

THE 1980s

Korea's export-oriented growth depended heavily on exporting to the West in the early 1980s (Table 10). The recession experienced by the Western market economies has therefore hit Korean exports. These have also been hit by the protectionist policies introduced in recent years by the advanced countries. Quota restrictions on Korean textile and clothing products are one important example. It has been estimated that more than 40 per cent of the country's exports face some kind of restriction in world markets.

New policies were adopted in the early 1980s in an attempt to combat the more hostile international environment of a slower growth of world trade, and protectionist policies. Heavy industry, which received around 75 per cent of total investment expenditure in the 1970s, is being developed less rapidly in the 1980s. The emphasis has switched to light industry. Once more the government is playing a major role. The policy is to move up-market, to produce more skill-intensive, high-value products in order to escape protectionist policies in the OECD countries.

The government is now concentrating its injection of finance into the electronics and textile industries. The quality of textile materials and clothing produced is being improved. Similarly, Korea will move up-market in electronics. In 1980 the bulk of output consisted of consumer electronics (televisions, audios etc), but Korea is now diversifying into more advanced areas, with products such as advanced semi-conductors, including large integrated circuits, electronic industrial plant, computers and telecommunication equipment. Between 1980 and 1990 exports of electronic goods are expected to double.

Certainly the country's strongly interventionist policies to promote industrialisation and exports contrast strongly with the free market approach of the Reagan administration in the USA and the Thatcher administration in the UK during the 1980s. The Korean economy has pulled out of the recession strongly – growing nine per cent in both 1983 and 1984.

Some critics would say that Korea's 'economic miracle' owes much – perhaps too much – to overseas borrowing. They would argue that South Korea's growth in the 1970s was debt-sustained rather than export-led. But it is normal for economies to borrow heavily at an early stage of development and Korea's debt service ratio is below average for the entire group of middle-income developing countries. Korea's debt service rate as a percentage of exports improved significantly between 1970 and 1982 whereas in the same period, a country like Brazil experienced a worsening situation (Table 11). It is Korea's current policy to reduce its debt still further by 1986–87.

Table 11 *Korea's and Brazil's debt service rate as a percentage of exports, 1970–82*

	1970 %	*1982* %
Korea	19.4	13.1
Brazil	12.5	42.1

(*Source*: The World Bank, 1984)

? **1** Which important factors help explain Korea's rapid transformation from a traditional rural society in the 1950s to a modern, middle-income, industrial exporting country in the 1980s?
2 Which Korean policies, if any, would you recommend for adoption by the UK Government? (Policy suggestions should be practical, bearing in mind our more sophisticated workforce and technology, and our membership of the European Economic Community.)

HOW ARE THE OLD INDUSTRIAL COUNTRIES RESPONDING TO NEW TRADING PATTERNS?

THE NICs' SHARE OF THE WORLD MARKET

Undoubtedly, during the last 20 years, the newly industrialising countries have made an impact on the major markets of North America, Western Europe and Japan. Well over half their total manufacturing exports go to OECD markets as Figure 5 indicates. Of manufacturing exports destined for rich country markets, and excluding the European NICs, about three-fifths go to North America, one-third to Western Europe and one-tenth to Japan. Their overall share of these markets is indicated in Table 12.

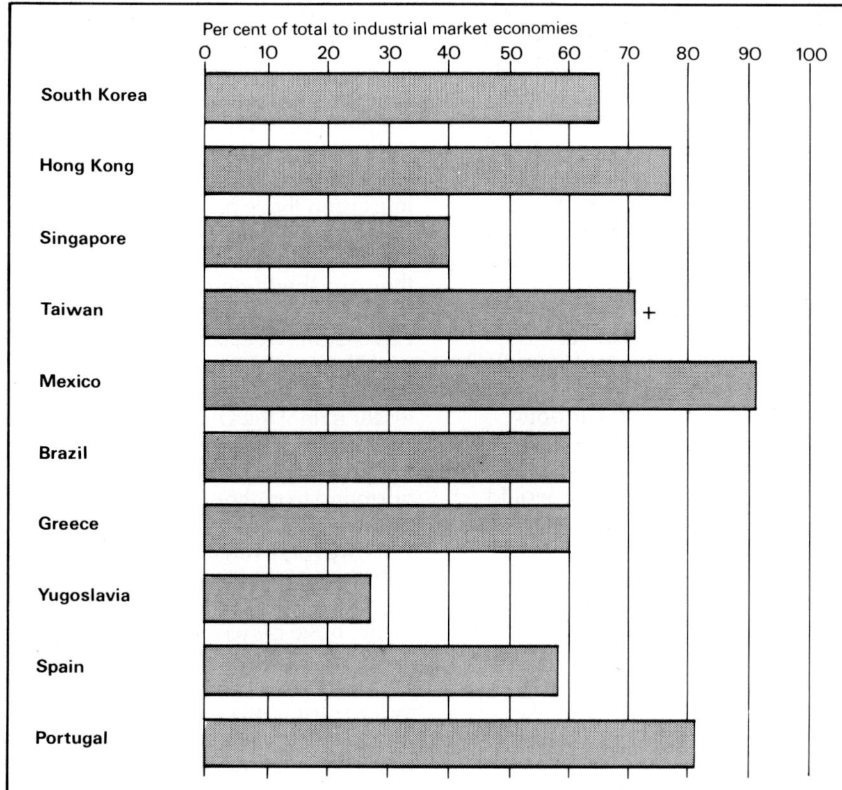

Figure 5 *The ten NICs—the percentage of total exports going to OECD markets, 1982**

(*Source*: The World Bank, 1984)

* Spain was reclassified as an industrial market by the World Bank in late 1983

† Estimate

Table 12 *The ten newly industrialising countries – share of industrial imports in major OECD markets, 1977*

Area	*Per cent*
Japan	18.4
North America	13.6
Western Europe	5.1

(*Source*: C. Saunders ed., 1981, *The Political Economy of New and Old Industrial Countries*, Butterworths, London)

THE IMPACT OF NIC EXPORTS ON ADVANCED INDUSTRIAL COUNTRIES

The NICs main exports, although they are developing many new product areas, are still predominantly labour-intensive manufactured products. One serious problem, already touched upon, is that the affected industries in the advanced industrial countries are often geographically concentrated, such as the clothing industry of London's East End, or the textile areas of North-east France, Central Belgium and North-west England. Various pressure groups in the threatened geographical areas demand protection against 'market disruption' or ask for 'adjustment assistance'. (Market disruption occurs when there is a sudden influx of imports which badly hit a local industry. Adjustment assistance is government grants and other financial assistance to help local industries in difficulty.)

While manufacturing jobs in recent years have been expanding at three or four per cent per year in the newly industrialising countries, parts of the OECD countries have been experiencing deindustrialisation. Traditional industries have become uncompetitive and decline has set in (see Table 13 for the UK).

Table 13 *The decline in employment in selected UK manufacturing industries (percentage decline from peak year to June 1981)*

Textiles	67 (since 1951)
Leather goods	60 (since 1951)
Clothing and footwear	55 (since 1951)
Shipbuilding	54 (since 1956)
Metal manufacture	49 (since 1961)
Vehicles	31 (since 1960)

(*Source*: Department of Employment Gazettes, HMSO, London)

However, compared with industrial jobs lost through technological change, world recession and faulty government policies, job losses through workers being displaced by cheap imports from the developing countries is probably relatively small.

Trade is a two-edged sword. A large part of the export earnings of the newly industrialising countries are spent in the OECD countries. They purchase large quantities of machinery, equipment and components which create jobs in the developed countries. It has been estimated that the net effect on employment of trade in manufactures will be around zero for the OECD as a whole.That is, jobs destroyed by cheap imports from the newly industrialising countries will be roughly offset by expenditure on machinery and other products by the newly industrialising countries which will create jobs. However, more competitive OECD regions and countries will fare better than others.

? 1 Research, identify and list major areas of advanced industrial countries which are experiencing deindustrialisation.

2 Study Table 13. Draw a suitable graph to illustrate the data. Which areas of the UK have suffered most from the decline of these major industries?
(Use library resources for information in order to answer both these questions fully.)

WHAT POLICIES SHOULD OECD COUNTRIES ADOPT?

The OECD nations can choose either positive or negative policy responses. Traditional trade theory predicts that free trade is the best policy, providing all countries obey the rules. If all geographical regions and countries specialise according to their comparative advantage and trade freely, world production and welfare will be maximised. Any restrictions on trade, for example, if countries try to become more self-reliant, will reduce world output and living standards below what they could be otherwise. Trade provides a richer variety of products to satisfy people's wants.

A positive response by the advanced industrial countries would be to shed certain products and industries to the developing world and to develop new product areas and industries. The LDCs, in the main, have a comparative advantage in labour-intensive and natural resource-intensive industries. So the rich countries should shed their traditional labour-intensive and mature industries to the NICs. These could include textiles and clothing, and possibly iron and steel, shipbuilding and the motor industry.

The advanced industrial countries have a comparative advantage in skill-intensive and capital-intensive products. This is especially so in the case of technological and knowledge skills. Specialisation in the OECD countries should reflect their abundant factors, in this case highly skilled labour and capital resources. They need to move out of the labour-intensive, unskilled, low value-added industries into skill and capital-intensive industries. They could specialise more in a whole range of high technology and information technology industries. Product areas could include electronics, the large computer area, video, cable television, office technology and telecommunications. An interesting, if controversial, perspective is, of course, offered by *The Guardian* extract at the beginning of this chapter (*page 315*) which highlights the fact that in certain industries, where labour costs represent only a small proportion of manufaturing costs, multi-national corporations would do better, in future, to invest in robots and to locate their manufacturing companies in OECD countries. (However, if the non-unionised NICs purchase or produce robots, they may protect their comparative advantage.)

A negative response by the OECD countries would be to protect high-cost, inefficient, traditional industries. Protected industries tend to become more and more inefficient, and, if they are old, declining industries. There is also the danger of a 'knock-on' effect, with large areas of the economy becoming inefficient. For example, if an uncompetitive steel industry is protected then the engineering, motor vehicle, household appliances (white goods) and shipbuilding sectors may become high-cost, and decline.

Tariffs and quotas could be used to protect traditional industries and restrict world trade, but they are against international agreements. Since 1973, during a prolonged period of high unemployment, the OECD countries have introduced various informal devices to protect domestic industries. They include negotiated 'voluntary quotas', 'orderly marketing arrangements' (increasing exports to a country in an orderly way to avoid a sudden disruptive surge) and policies of nationalising and subsidising certain industries. Examples in Europe include protection of 'sensitive industries' such as textiles, clothing, steel, motor vehicles and shipbuilding.

Some industrial countries have adapted better than others. Japan has rapidly created new 'sunrise' high technology and information technology industries and has run down its textile, clothing and simpler sectors of the electrical industries (for example black and white televisions and radios). It is importing these products from the East Asian NICs. Similarly, as incomes rise in the 'gang of four', they are shedding parts of their textile industry to third generation East Asian NICs, such as Indonesia and Thailand. It could be argued that the UK and the EEC, and to a lesser extent the United States, have adopted much more protectionist, inward-looking policies. They are adapting to a changing world more slowly.

However, even firm supporters of world specialisation, according to comparative advantage and free trade, have criticised some of the NICs trade practices. Three Korean firms laid down large, export-oriented colour television plants at the same time. At what point do large subsidies to private firms and state corporations go beyond the protection of infant industries and constitute unfair competition? How can OECD private corporations fairly compete with heavily subsidised developing world private companies and nationalised industries? Are subsidies to infant, export industries as opposed to import-substituting ones under hand. Should advanced industrial countries worry about placing tariffs on cheap manufactured goods when some NICs place very high tariffs, often over 100 per cent, on a range of imports from the advanced industrial countries? Should even more stringent conditions be imposed by organisations such as the IMF and the World Bank, when negotiating loans with countries such as South Korea, Thailand, the Philippines and Indonesia to avoid debt-sustained growth?

You will now realise that international trade is a politically sensitive area. Vested interests exist in both the advanced countries and the developing world. The trade issue is complex. However, the principle of comparative advantage may dispel the political confusion and provide a firm and secure base for government policy.

?

1 Below are presented differing views on how the UK should respond to the new trading patterns with the NICs. Analyse why each person holds his or her own particular view and then discuss in groups what you think the UK government should do, and whether it should respond positively or negatively to the challenge.

2 You are working for a rapidly growing UK electronics company. Write a report for your company comparing the pros and cons of locating their proposed new, labour-intensive electronic burglar alarm factory either near Heathrow Airport or in East Asia. Recommend a specific country and town. (You, of course, accept your company's strategy of minimising costs and maximising profits.)

3 Debate, as a class, the motion that 'Geographical factors have played only a minor part in the changing patterns of production and trade in East Asia, and it is enterprising governments which have been the key factor in the transformation of agrarian societies into modern industrial states?'

'The introduction of more protectionist policies by the EEC are ridiculous. We import more from the OECD countries (i.e. the advanced industrial countries) than they import from us.' (Bank Manager, Hong Kong and Shanghai Bank, Hong Kong)

'We should run down our sunset industries and develop high technology products.' (Employee, R and D Dept., electronics firm, Reading area)

'These cheap imports from East Asia are destroying our jobs.' (Motor vehicle assembly worker, Longbridge, Birmingham)

'What we need is not protectionist policies. That way disaster lies. We need to negotiate a system of free and fair trade.' (Lecturer in International Economics, University of Sheffield)

'These cheap products from Korea are destroying our industries. We'll soon be a Third World country.' (Employee, Sunderland Shipbuilders)

'I don't see what all the fuss is about. It's all the unions' fault. I prefer inexpensive, reliable foreign products and I'm going to continue buying them.' (Pensioner, Great Missenden, Buckinghamshire)

16 ZAMBIA: A ONE-MINERAL ECONOMY?

View of Nchanga opencast copper mine, Chingola

GLOBAL IMBALANCE IN RESOURCE PRODUCTION AND CONSUMPTION

The developing world or less developed world includes most of Latin America, Africa and Asia (except Japan). It supports three-quarters of the world's population. Under the present system of international specialisation, more than three-quarters of the exports from these countries are products based on natural resources. Most of their mineral and plantation production is organised by multinational corporations. Their major railways and roads still lead from the mines and plantations down to coastal ports. These less developed countries produce around one-third of the world's minerals and yet consume only five per cent or so themselves (Figure 1). An obvious danger is that the rich natural resources will become exhausted before these countries have developed their economies.

This chapter analyses resource development in a developing country – Zambia. Situated high on the southern plateau in the heart of the African interior, thousands of kilometres from the markets of the rich North, Zambia is a major copper producer. It is endowed with rich copper resources, but is resource development transforming it from a poor, rural-based society into a rich developed country, or is its copper being shipped out by the multinationals so that the rich developed world receives most of the benefits from development? Is the revenue derived from the export of copper serving to increase or decrease regional inequalities within Zambia itself?

ZAMBIA – THE COUNTRY

Location

Zambia is a relatively large, landlocked country in the heart of southern Africa. It is 2000–2500 kilometres from any major ports (see Figure 2). It is served by tenuous rail routes through Tanzania to the port of Dar-es-Salaam on the east coast and south through Zimbabwe to the ports of Beira and Maputo on the Mozambique coast and East London on the South African coast. From time to time these railways and ports have been unable to handle all Zambia's imports and exports – such delays adding to Zambia's importing and exporting costs. The port of Lobito on the Angolan coast has not been used for the last 15 years because of the Liberation wars.

Population

Zambia has a small population. In 1980 its population was six million giving an average density of eight per square kilometre. Over 98 per cent are African while Europeans and Asians comprise most of the remainder. Around 40 per cent of the population live in the urban areas and about 50 per cent in the core region known locally as the Line of Rail (see Chapter 3, *page 48*). The urban population is heavily concentrated in the towns and cities of the central core region, especially

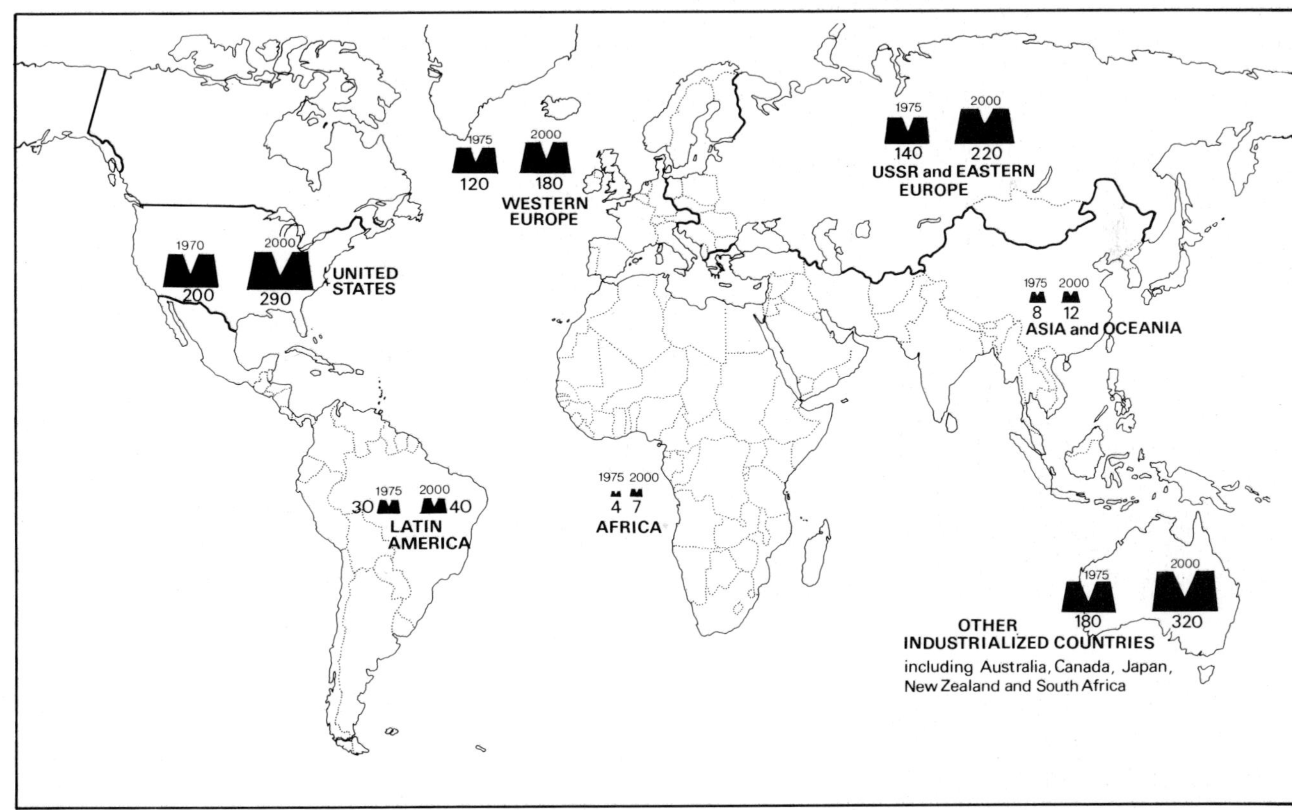

Figure 1 *World metal consumption per capita. Actual consumption in 1975 and projected consumption for 2000, of steel, aluminium and copper ores (given in US dollars at 1975 value)*

in Copperbelt Province in the north and Central Province, which includes the capital city, Lusaka (Figure 3). The rural areas have a low population density, with some concentrations in the Barotse Plain area of the upper Zambezi, the Lake Bangweulu area, Eastern Province around Chipata, and parts of Southern Province. Other areas, often malarial and subject to the tsetse fly, such as the great expanses of the Kafue and Luangwa Game Reserves, drained by tributaries of the Zambezi River, are virtually empty. In recent years the population has been growing rapidly at three per cent per year, which compares with a growth rate of less than one per cent in the advanced industrial countries. The projected population in the year 2000 is 11 million.

Political background

Between 1889 and 1924 Zambia was governed by Cecil Rhodes' British South African Company (BSA), and later, directly by the colonial office. It was a reluctant member of the Federation of Rhodesia and Nyasaland from 1953 to 1963, before achieving independence in 1964, and supported the anti-colonial liberation struggles in neighbouring countries during the 1964 to 1980 period. Today Zambia is a one-party participatory democracy.

Print 1 *Rail maintenance in Zambia*

? You are working for Zambia Consolidated Copper Mines Limited. In the light of ever-rising transport costs, you have been requested to examine alternative transport routes. Copper can be transported by rail to the following ports:

Dar-es-Salaam (Tanzania)	East London (South Africa)
Beira (Mozambique)	Lobito (Angola)*
Durban (South Africa)	

* Assume, for the purposes of this activity, that Lobito may, at some point, become usable again.

1 Calculate the distance from Kitwe to each port in kilometres, using Figure 2 or an atlas map of Southern Africa.

2 Assume the cost of rail transport is four pence per tonne per kilometre. Calculate the cost of transporting a tonne of copper to each port and illustrate with an appropriate graph.

3 The country exports around 600 000 tonnes of copper and at the moment 70 per cent goes out through Dar-es-Salaam with the remainder going through East London. Sea transport costs are approximately half rail costs. Exports are financed by bank overdraft which adds £10 per tonne per week to costs. Sea transport, and delays at ports therefore add to costs. Recommend economic routes for exports to **a**) the EEC, **b**) Japan, and **c**) Brazil.

4 News has just come in of civil unrest in Angola and the railroad is blocked. Assume some copper was going out through Lobito to the EEC. How would you reroute it?

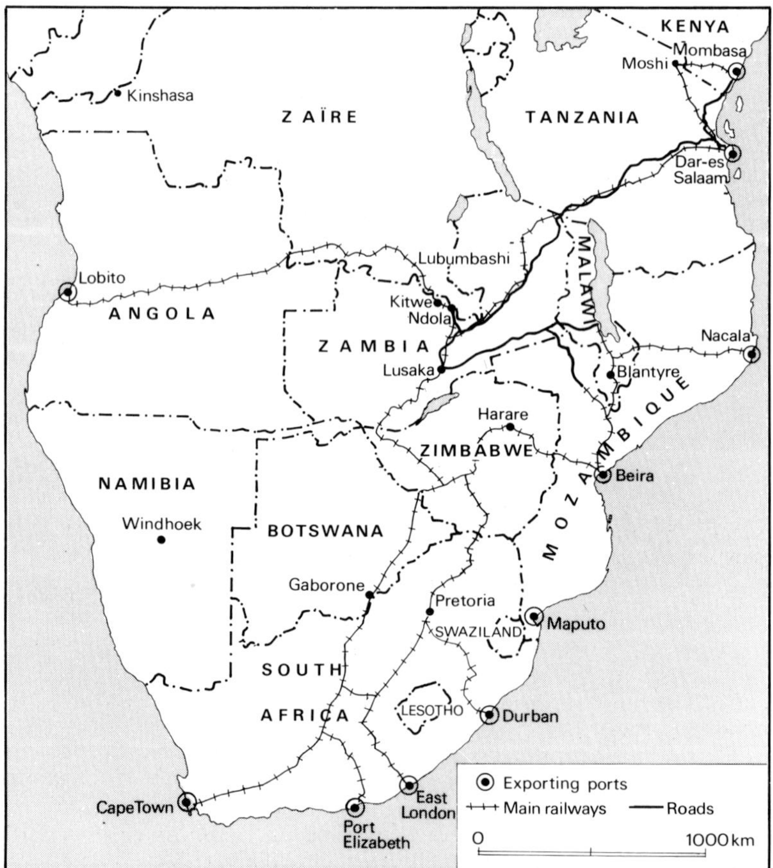

Figure 2 *Zambia – a landlocked country, reliant on routes through neighbouring countries for access to ports*

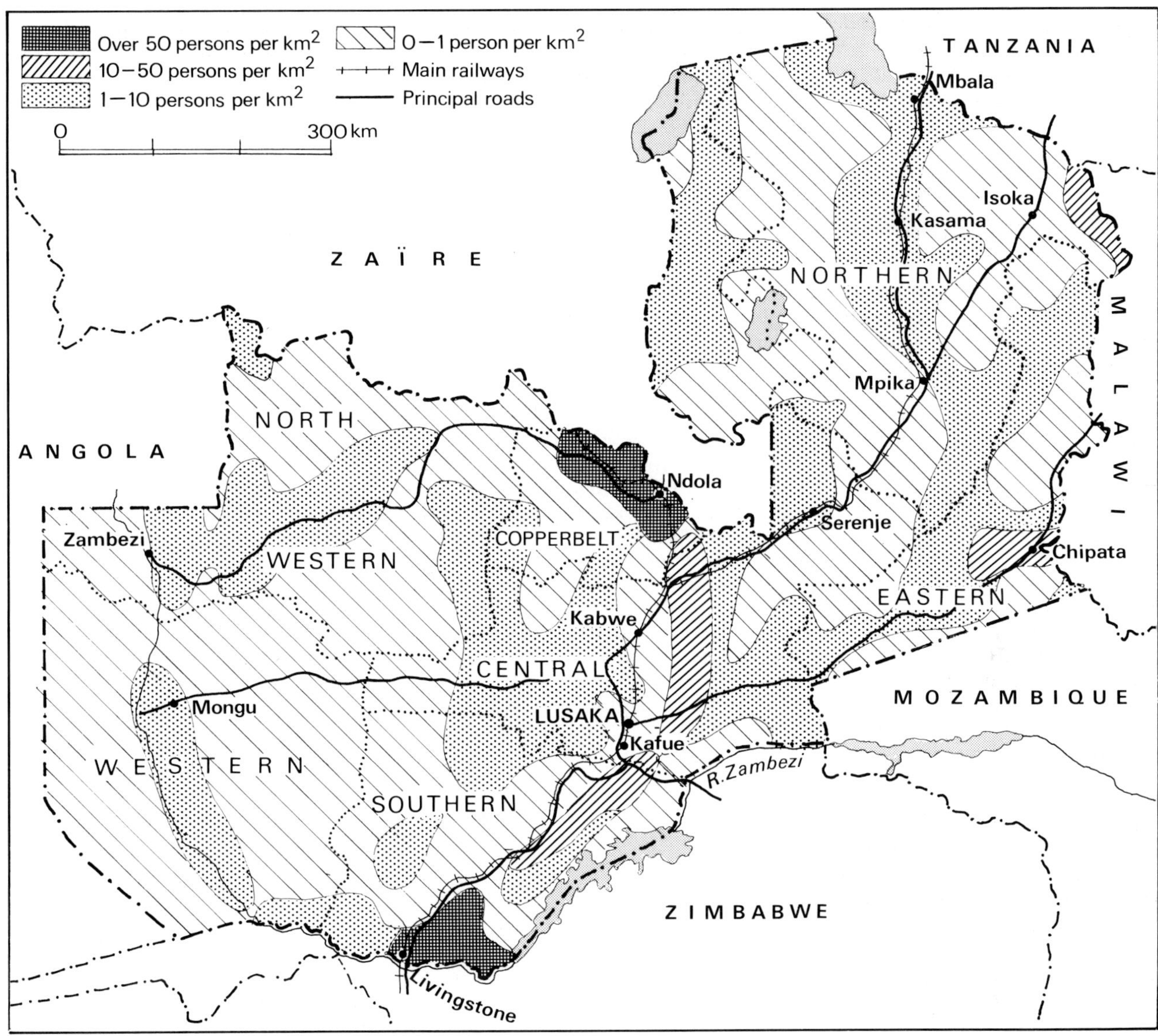

Figure 3 *Population density map of Zambia*

THE DEVELOPMENT OF THE COPPERBELT

Modern development in Zambia began in the period after 1889, with the injection of foreign capital into Zambia's traditional economy by Cecil Rhodes' British South Africa Company. Central Africa was thought to be rich in minerals and the BSA Company acquired land and mineral rights by force and by making small payments to African chiefs. The company extended the rail network from South African ports through Rhodesia (now Zimbabwe) and Zambia to the rich copper ores of southern Zaire. The railroad reached what is now the Zambian Copperbelt in 1909.

But it was in the 1920s that copper production in the northern part of Zambia increased dramatically, stimulated partly by the discovery in the USA of new methods of processing copper which had a high sulphur content, as was the case with Zambia's northern ore. Significant export-led development followed.

In the decade after 1925, two large multinational mining corporations, the Anglo-American Corporation of South Africa, and Amax Inc. of the USA brought in a large amount of capital, advanced technology, and skilled and semi-skilled labour and managerial knowhow. Rich copper ores, foreign technology and cheap local unskilled labour combined to turn Zambia into one of the lowest-cost copper producing areas in the world by the 1930s. Large mines were opened at Kitwe, Luanshya, Mufulira and Chingola in the 1930s (Figure 4) and despite the Great

Depression, production increased dramatically – by 1937 Zambia was the third largest copper producer in the world.

? **1** Construct a line graph from Table 2. Does the graph show that the value of copper per tonne fluctuates and that there is a relationship between this and the total value of copper in a given year?
2 More than 90% of Zambia's total exports consist of copper. Discuss, in a group, the possible effects of copper price fluctuations on the Zambian economy and population.

EXPORT-LED GROWTH THEORY

As Zambia's economy was and still is dominated by its leading sector, the copper export industry, the export-led growth theory is particularly relevant in trying to understand the development of the Zambian economy. This theory is concerned with the idea that development is transmitted through trade from the heartland economies – the USA, the EEC and Japan – to the developing, peripheral countries. It regards trade as an 'engine of growth' and a good thing. In the 19th century, it is argued, trade was a significant stimulator of growth; grain exports from the USA, wool and lamb from Australia and New Zealand and gold and diamonds from South Africa are a few examples of trade that greatly boosted the economies of the countries concerned.

Definition of a dual economy
The theory states that growing demand for minerals and agricultural products in the heartland countries transmits export-led growth to the under-developed, hinterland countries in the following way: expansion in the advanced industrial countries of the North is transmitted to the peripheral countries of the South through strong demand for minerals and farm products. As demand for primary products grows, output and incomes grow in the export sector of the hinterland countries. The economic surplus generated in the modern export sector, mainly company profits and personal savings, is reinvested year after year. The modern core region continually expands as incomes and exports rise, though other parts of the country remain rural. The growth of the core region is reinforced by further 'spread' or 'multiplier'

Table 1 *Copper production, selected years (tonnes)*

Year	*Output*
1925	90
1930	6 000
1940	267 000
1950	281 000
1960	567 000
1970	683 000
1980	596 000
1985[1]	525 000

(*Sources*: 1925–50, *Annual Blue Books*, Northern Rhodesia; 1955–80, *Monthly Digest of Statistics*, CSO, Lusaka)
(*Notes*: [1]estimate)

Table 2 *Volume, value and unit value of Zambia's copper exports*

	Volume (000 tonnes)	*Value (K mn)*	*Unit value (K/tonne)*
1970	683	681.4	996.2
1971	635	450.2	709.0
1972	711	490.9	690.4
1973	671	699.0	1 041.7
1974	673.4	838.9	1 245.2
1975	641.2	471.1	734.7
1976	745.7	688.6	923.4
1977	666.6	645.9	968.5
1978	537.6	604.4	947.9
1979	653.6	905.3	1 385.1

(*Source*: *Bank of Zambia*, Vol 13, March 1983)

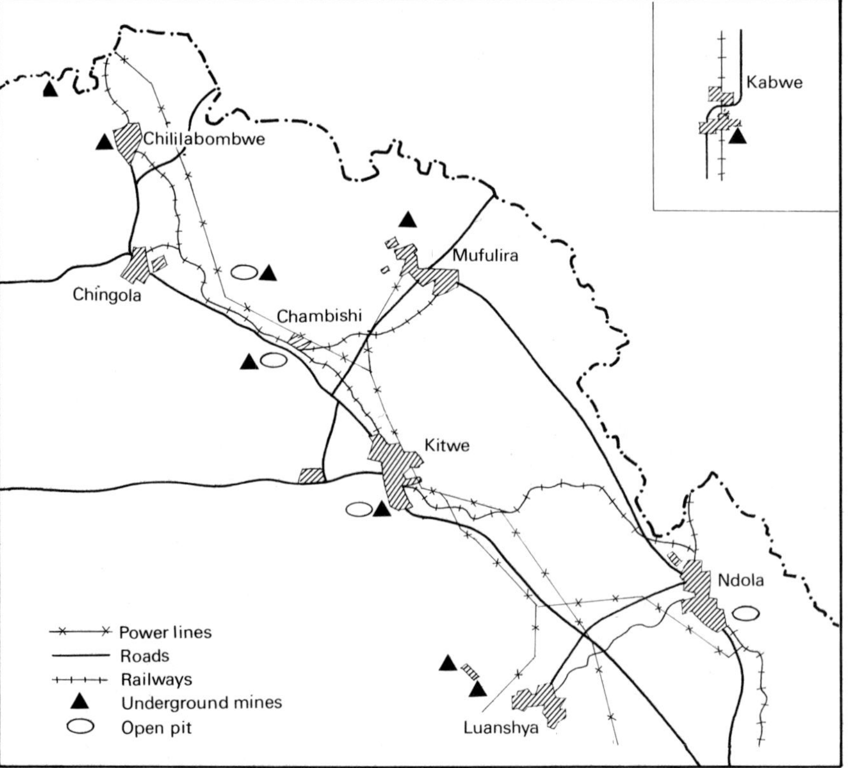

Figure 4 *The Zambian Copperbelt today*

effects, i.e. new manufacturing industries are created to supply the export industry, to process products and to provide goods and services to meet the needs of the expanding wage labour force. The 'take-off' stage is reached, characterised by a high level of investment, and cumulative and general development occurs. The modern market sector expands and absorbs the rural hinterland subsistence areas so that there is an even development throughout the country and not a 'modern' economy in one region and backward economies in others. Where two economies exist in the one country economists and geographers describe it as a *dual economy*. Eventually the developing world becomes industrialised and highly developed throughout. The concept of the dual economy and how it develops is more fully described below. Read through the information carefully and discuss it in small groups so that you understand the assumptions and the process.

THE DUALISTIC SPACE ECONOMY

Assume a traditional, under-developed country with everyone at the subsistence level. The traditional society is then disturbed by an inflow of foreign investment which creates a modern export industry, such as the copper industry in Zambia, and an urban population. The site of production becomes what *The Economist* calls a growth pole or core region. Wages in the modern sector are fixed at a margin above subsistence, say 30 per cent. People in the labour-abundant rural areas find these wages attractive, and migration and urbanisation occur. At this stage of development – the labour surplus phase – there is an unlimited supply of labour to the towns at the going wage. Jobs are created by reinvesting the surplus (profits and savings) in the modern sector. So the modern sector expands.

Assume the export sector is foreign-owned and run by foreign, skilled labour. If profits and savings are largely exported and not reinvested, the modern sector remains limited and fails to absorb labour from the subsistence sector. Development becomes blocked or retarded and the dualistic space economy continues to exist. Dual economies have existed in some parts of the world for over 100 years. Similarly, if population growth offsets job creation in the modern sector dualism will persist.

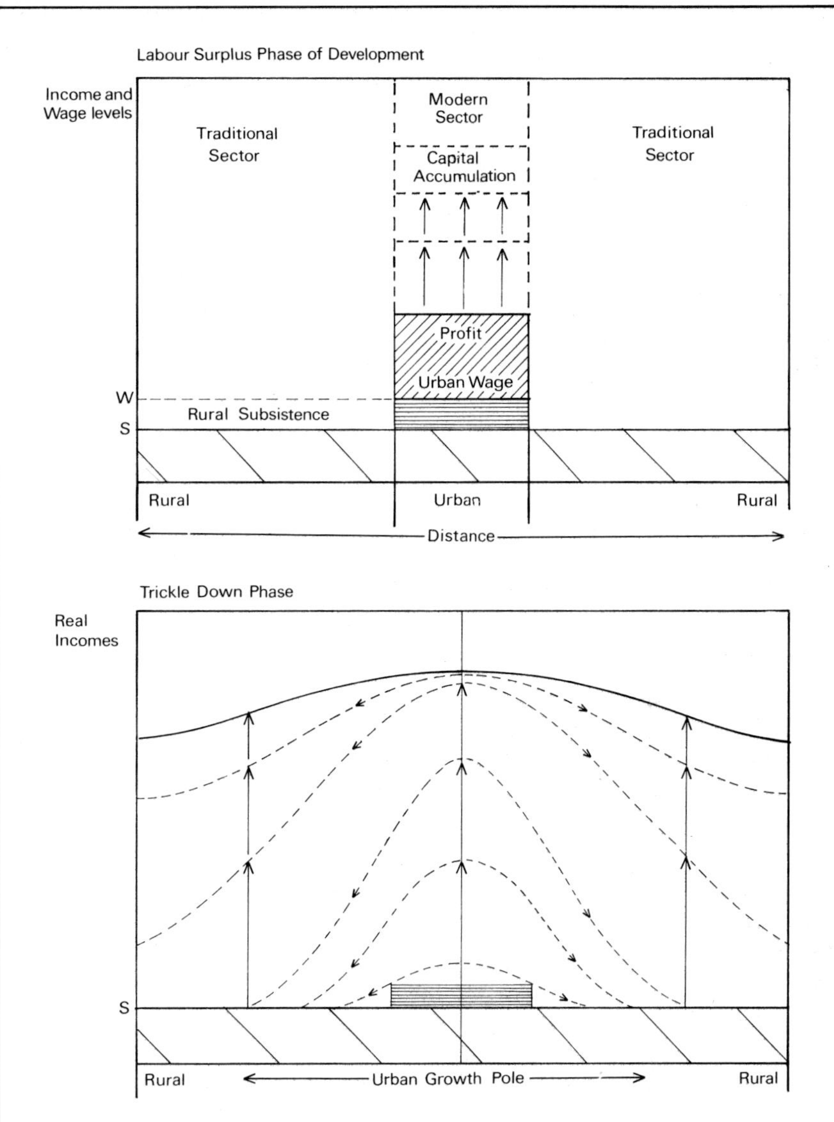

Figure 5 *Stages of development in a dualistic space economy*

Now assume the surplus is reinvested and/or population control methods are introduced. The trickle-down phase of development begins. The modern sector expands and eventually exhausts the labour surplus in the traditional sector. There is now a shortage of labour and market forces begin to drive up wages. The share of wages in the country's national income rises and the huge share of profits and skilled labour falls. This is because unskilled workers are now scarce. Their wages rise and they receive a bigger share of the national pie. The massive inequality associated with dualism is reduced (see Figure 5).

During the last 20 years the rapidly growing new industrial countries of East Asia, such as Taiwan and South Korea, have exhausted the labour surplus and entered the 'trickle-down' phase. On the other hand many countries in Latin America, Africa and Asia are still labour-abundant dual economies exhibiting massive income inequalities. Examples include Mexico, Kenya, and India.

We must now look at the extent to which the theory reaches its final stage of overall and even development by examining the case of Zambia. Do things work out like this in practice? If not, why not?

ZAMBIA'S DUAL ECONOMY: DOES THE DUALISTIC ECONOMY ALWAYS DISAPPEAR? IS FOREIGN TRADE AN ENGINE OF GROWTH?

Zambia's experience in this century suggests that foreign trade is not an automatic engine of growth. By Independence in 1964, foreign investment and trade had, in Zambia's case, produced a highly dualistic space economy. A substantial inflow of overseas investment had been injected into the central core region of the Copperbelt and Line of Rail to expand primary production for export, while most of the traditional hinterland had remained underdeveloped.

The core region, with its advanced technology and highly capital-intensive methods, was a market sector, organised, financed and administered by Europeans, and extremely dependent on overseas trade and investment. The periphery, the undeveloped traditional economy, stretched for hundreds of kilometres west and east of it. Most of the country's population lived and worked on their small, unmechanised, subsistence farms in this great, sparsely peopled outback (see Figure 6).

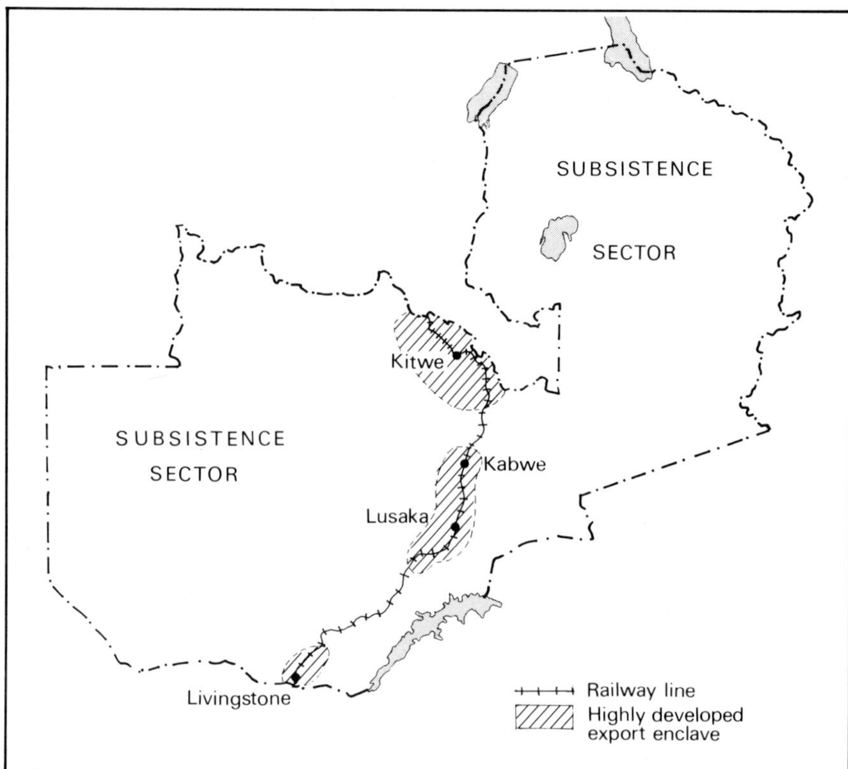

Figure 6 *Zambia's dual economy created by the development of the copperbelt in the middle, with the traditional economy to the east and west*

Between the early 1930s and the mid-1960s Zambia was a classic case of growth without general development. The export sector expanded unevenly but rapidly in the sense that copper production and export income increased. Some minimum health services such as malaria and smallpox control were provided for the African population. Yet at Independence in 1964 the new Zambian Government inherited a dualistic economy in a most extreme form. Approximately 80 per cent of the population were living in the underdeveloped peripheral areas. The remainder, together with almost all the 74 000 Europeans, who provided skilled, professional and managerial labour or owned and managed large, scientific mechanised commercial farms, lived in and around the towns and cities of the central core region. Most of the 11 000 Asians, who were mainly traders and small businessmen, also resided in the developed central area. (In 1981, Kitwe and Ndola, where the Copperbelt's two electrolytic refineries are situated, each had a population approaching 400 000, compared with Lusaka's population of around 650 000.)

Print 2 *Workers planting potatoes on a small commercial farm, Zambia*

The economy was dominated by its export sector, which provided over 90 per cent of export income and around half the country's national output and government revenue (Table 3). However, the country's strategic industry, completely owned and controlled by two foreign transnational corporations, was capital intensive, and so provided relatively few wage jobs. Capital-intensive methods were necessary because of the nature of the industry and in order to contain costs and operate profitably in a competitive world market. Agriculture, despite being the main activity of the mass of population contributed only around four per cent of domestic output, as it was labour intensive and productivity was low.

Table 3 *Economic importance of the copper industry, 1964*

Net domestic product (£m)	*Contribution to net domestic product (£m)*	*(%)*	*Contribution to government revenue (£m)*	*(%)*	*Contribution to exports (£m)*	*(%)*
222.8	107.5	47	28.5	53	151	92

(*Source*: *Copperbelt of Zambia Mining Industry Year Book*, 1967)

What happened to the profits from copper mining?

Various factors help to explain why general and cumulative development did not occur in the period up to independence in 1964, despite rapid expansion in export income (Table 4). A large part of the profits and savings generated in the core region were exported and not reinvested to develop the country (Table 5). Between 1920 and 1968, approximately K60 million of risk capital flowed into the territory to develop the copper industry while over K800 million was exported in dividends. (One year's remittances in the 1950s were equal to Zambia's share of the cost of the massive Kariba hydroelectric project.) Cecil Rhodes' BSA Company (the landlord, in effect) received and exported royalties on every tonne of copper produced. Part of the taxes raised by the colonial, and later the federal government, were remitted to the UK or spent in what was then Rhodesia (now Zimbabwe). Expatriate skilled workers tended, on average, to save and export at least 25 per cent of their high incomes for the day when they returned home. A large part of the surplus generated in the core region disappeared overseas and was not reinvested in the space economy to stimulate development in the hinterland. The rich copper revenues were not sown to diffuse development and to integrate the sparsely peopled hinterland into the modern economy.

Table 4 *Zambian exports by value, 1930–80, million kwacha*

Year	*Exports*
1930	2
1935	10
1940	26
1945	24
1950	98
1955	236
1960	258
1969	380
1970	715
1975	521
1980	1004

(*Source*: 1930–50, *Annual Statement of the Trade of Northern Rhodesia and The National Income and Social Accounts of Northern Rhodesia*, CSO, Lusaka; 1955–80, *Monthly Digest of Statistics*, CSO, Lusaka)
(*Note*: 1 kwacha = approx 50p at the time of the figures)

Table 5 *Total income from the copper industry received by various groups, 1964*

	£ million
Dividends exported to shareholders	29
Royalties exported by BSA company	9
7455 European mineworkers, total earnings	22
38097 African mineworkers, total earnings	16

(*Source*: *Monthly Digest of Statistics*, CSO, Lusaka; *African Development*, Sept 1969, London; R. Hall, *Zambia*, Pall Mall Press, London)

Zambia's trade pattern before Independence consisted of exporting copper (which comprised over 90 per cent of total exports) to industrial heartland economies, and importing manufactured capital and consumer goods. There were few manufacturing plants producing consumer goods and virtually no capital goods industries. Given the trade pattern and foreign ownership of the leading sector – the copper industry – there were few linkages between the copper industry and the rest of the economy. Spread or multiplier effects generated by an expansion of exports (as anticipated in the export-led growth theory) were extremely weak. It was estimated that around 65 per cent of export income earned by the copper industry leaked overseas to heartland economies in one form or another. This process created further employment and income in the advanced industrial countries, but not in Zambia (Figure 7).

Another problem was the industrial, social and educational colour bar which existed from the 1920s until Independence. Every facet of life was segregated – employment, education, health services, housing, sport, entertainment, church services and shopping facilities. Skilled work was reserved for the Europeans. The colour bar prevented the spread of education and skills to the African population and retarded development.

> ? Pretend a small country has just been discovered in Africa. It has rich copper reserves and a geologist employed by a large UK mining corporation is calling for the development of this resource. He has written a paper setting out the eventual benefits to the whole country of such development. Write an answer to him based on Zambia's experience as described so far. Identify benefits to the UK mining corporation and to Zambia, if any.

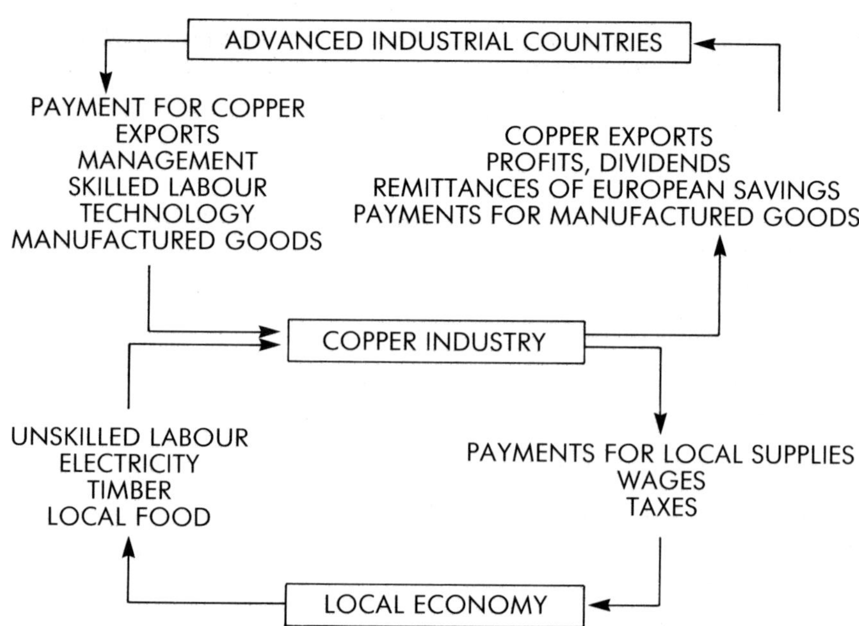

Figure 7 *Linkages between the copper industry and the local and world economies*

THE ROLE OF THE GOVERNMENT AFTER INDEPENDENCE

SOWING THE COPPER REVENUES

After 1964 the new independent government embarked on a series of national development plans designed to capture the economic surplus generated in the economy and to invest it in order to achieve general development. The objectives of the development plans were similar to those of many governments in developing countries:

- Raise the level of investment to achieve a more rapid rate of growth.
- Raise real incomes and living standards.
- Diversify the economy away from dependence on copper.
- Improve the infrastructure and social services.
- Create employment.
- Develop a skilled and educated workforce.
- Reduce inequalities, especially rural/urban disparities.

The government embarked on a policy of sowing the copper revenues to develop the country. In the decade after 1964 it increased taxation on copper industry profits from 20 per cent to just over 70 per cent. In this period the BSA Company was expropriated, so the government became the landlord and received mineral royalties, and virtually all the large enterprises, including the copper corporations, were nationalised in 1969 and the early 1970s. Taxes previously being spent in Zimbabwe were now available for Zambian development. In the period 1964–76 the level of investment was raised from well below 20 per cent of GDP to between 25 and 30 per cent.

Government tax revenues increased six times in value during the 1964–70 period, because the tax increases coincided with rising copper prices during the boom of the 1960s. The copper industry contributed between 50 and 70 per cent of total government revenue, providing the government with unprecedented financial resources for development. GDP grew at a Japanese-type rate of 10.6 per cent per year in the copper boom of 1963–70, and real per capita incomes more than doubled.

The copper revenues were largely used to build up the physical and social infrastructure of the country. By the mid-1970s the independent government had achieved more in a decade than the colonial government had in 75 years. At Independence, there was only one main paved road and one main railway in the country, running from the Copperbelt to Livingstone through the core region. A decade later all provincial capitals were linked by a network of paved roads, and new international road and rail routes had been opened up to Dar- es-Salaam in Tanzania.

In 1964 not one African had completed an apprenticeship, around 1000 had obtained the equivalent of GCE O-level standard and there were only about 100 graduates. The government and business firms embarked on a massive education and training programme. By the early 1980s, around 90 per cent of children were receiving primary education and the percentage receiving secondary education had risen from two to over 16 per cent. The new University of Zambia had produced several thousand graduates.

After 1964, the government also embarked on an industrialisation programme in an attempt to diversify the economy. New industries, mainly of the import-substitution type were developed. High tariffs were placed on imported goods to make them more expensive than home-produced goods. Food processing, beverages, textiles and clothing, and motor car assembly industries were built up. These manufacturing sectors are relatively labour-intensive and typical of the early phase of industrialisation. Some capital goods plants have been established, including fertilisers, explosives, oil refining (there is an oil pipeline from Dar-es-Salaam to the Copperbelt), and a small copper fabricating plant. The country has benefited from transforming raw materials into manufactured products. This process has increased value-added, government tax income for development and has provided additional jobs.

The persistence of regional inequalities

Development has, however, been concentrated in the core central region. Most of the new factories have been sited in the Copperbelt and in Central Province, especially in the Lusaka area. The core region is the lowest-cost site for most new factories, especially those producing consumer goods. Most incomes are generated there, so the pull of the market is important. Business enterprises have also benefited from the external economies created by the highly developed infrastructure of the central area, such as roads, railways and energy supplies.

The fluctuating price of copper

Zambia suffered greatly in the world recession of 1975–82. By 1982 copper prices had fallen in real terms to the lowest levels for over 30 years. National output and living standards fell, the surplus produced by the copper companies evaporated, and government income fell accordingly.

Zambia's engine of growth, which still provides over 90 per cent of export income, has temporarily broken down. But copper booms follow copper recessions as night follows day in the uneven growth of the world space economy. Development will continue in the next upswing. Almost certainly the resource base will expand as new minerals are discovered in the potentially rich southern African plateau. However, a more favourable political and financial climate may be necessary to stimulate exploration. Further industrialisation is constrained by the small size of the home market, so export industries need to be built up to reap economies of size. They will need to be chosen carefully due to the country's landlocked position far from major markets and its relatively high unit labour costs.

Table 6 *Rural-urban inequalities*

Manufacturing establishments, 1969		
		Number
Line of Rail provinces		524
Rural provinces		8
Zambian Total		532
% of total manufacturing employment, 1969		
		%
Line of Rail provinces		97.3
Rural provinces		2.7
% of bitumen road network by province, 1969		
		%
Line of Rail provinces		70.1
Rural provinces		29.9
Distribution of capital expenditure (First National Development Plan, 1966–70)		
	Population 1969	*Total (K million)*
Line of Rail provinces	2 073	539.4
Rural provinces	2 087	119.5

(*Source*: *Second National Development Plan*, 1972–76, Lusaka)

Print 3 *Factories producing consumer goods such as bicycles have sprung up around the Core Region, the Copperbelt and Central Province in Zambia*

Table 7 *Rural-urban distribution of population by province (all races, in thousands)*

	1963			*1969*			*1976*		
Province	*Urban*	*Rural*	*Total*	*Urban*	*Rural*	*Total*	*Urban*	*Rural*	*Total*
Luapula	6	361	367	10	336	346	20	364	384
Northern	8	571	579	10	551	561	20	603	623
Eastern	10	484	494	15	507	522	30	550	580
Southern	52	433	485	68	440	508	116	452	568
Central	179	341	520	330	397	727	572	414	986
Copperbelt	487	76	563	744	94	838	1 100	62	1 162
North-western	–	216	216	–	237	237	–	267	267
Western	5	368	373	11	410	421	20	469	489
Total	747	2 850	3 597	1 188	2 972	4 160	1 878	3 181	5 059

(*Note*: Urban areas include Lusaka, Livingstone, Kabwe and the whole of Copperbelt (excluding Ndola Rural District) and also Chipata, Choma, Kasama, Mansa, Mazabuka, Mongu and Monze. Other townships are included in rural areas.)
(*Source*: Planning Division, Ministry of Development Planning and National Guidance)

Table 8 *Average wages in Zambia, 1973–1980 (Kwacha per month)*

Year	*Mining sector*	*Transport*	*Agriculture*
1973	140	108	35
1974	142	116	37
1975	123	153	38
1976	209	152	50
1977	219	175	53
1978	210	172	60
1979	–	–	–
1980	282	221	88

(*Source*: International Labour Office, *Year Book of Labour Statistics*, 1983)

Table 9 *Balance of payment for Zambia, 1977–1981*

	(Kwacha million)				
	1977	*1978*	*1979*	*1980*	*1981*
Exports	700.8	665.1	1117.6	1048.1	972.2
	645.1*	597.7*	900.7*	885.8*	807.8*
Imports	532	494.8	599.8	884.5	911.3
Trade balance	168.8	170.3	517.8	163.6	15.9

* Amount that copper accounted for
(*Source*: *Bank of Zambia*, Vol. 13)

PLANNING EXERCISE

?

You have recently been appointed as Regional Adviser in the Ministry of Development and Planning, Lusaka. You know that the government inherited a dual economy in a most extreme form in 1964 and you are aware that there is still a need to develop the whole economy and to achieve the objective of reducing regional inequalities.

There are two main views in your Ministry. One group argues strongly that resources should be concentrated in the Line of Rail to expand the core region. Gradually the development of the core region will absorb the surplus labour in the rural areas and development will 'trickle down' to the whole population.

A second group, whose fundamental concern is rural poverty, argues that many more resources should be diverted to the rural areas. Indeed, they argue for positive discrimination in favour of the rural poor. Only in this way, they argue, can the vicious circle of poverty in the rural areas be broken and rural development initiated. They point out that the simultaneous development of modern industry and agriculture complement each other, for example, by providing markets. After all, they say, the UK, the USSR and Japan followed this path.

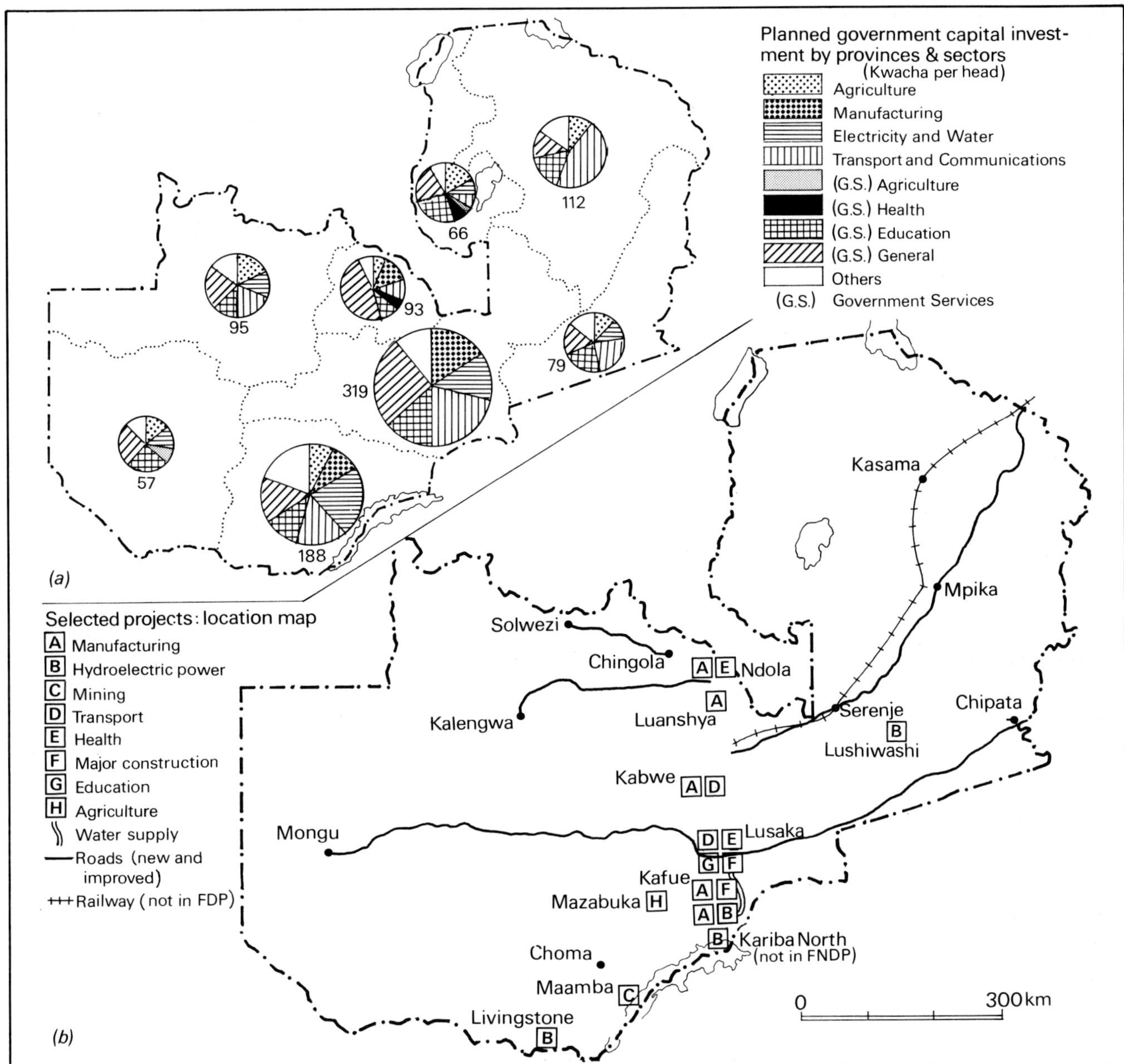

Figure 8 *First National Development Plan, 1966–70*
(*Source*: Davies, D.H. (1971); *Zambia in Maps*, University of London Press, London)

? 1 You decide, as a first step, to get some sort of historical perspective. You therefore look at Tables 7 and 8 and Figure 8 which provide insight into the First National Development Plan, 1966–70.
a Calculate capital expenditure per capita (1966–70) in both the Line of Rail and the Rural Provinces, using 1969 population statistics.
b What evidence is there that the government concentrated largely on expanding the core region during the First National Development Plan, 1966–70?
c How did this affect the pattern of urbanisation in the country? (The three core region provinces are Copperbelt, Central and Southern.)

? 2 You're also in possession of Tables 8 and 9 showing, respectively, average wages in Zambia from 1973–80 and the balance of payments for Zambia, 1977–81. How successful does Zambia appear to have been during the 1970s at achieving her objectives of **a**) raising incomes, and **b**) diversifying the economy away from dependence on copper?

3 Recommend a development policy for the late 1980s and 1990s. Argue your case lucidly as your views will be considered by your minister. The following are some of the possibilities:

- Import substitution industrialisation (expanding core region).
- Rural development strategy (developing agriculture, rural industries and reducing rural poverty).
- Balanced development policy (developing industry and agriculture together, spreading resources thinly over the country).
- Basic needs approach (concentrating on eliminating severe poverty in rural areas and shanty towns, providing basic primary education, clinics, clean water and pit latrines). (Look at Table 10.) Assume the recession has ended and that a copper boom occurs in the late 1980s and early 1990s.

Table 10 *Regional Disparities: Zambia*

Basic Needs	*Year*	*Rural (%)*	*Urban (%)*
Households with incomes below basic needs level (K100 per month)	1980	80	25
Malnutrition (children under 5 years)	1970–2	40	–
Population without hygienic water supply	1976	40[1]	0
Houses with sun-dried brick walling or better	1969	36	87
Occupancy per dwelling	1978	4.0[2]	6.3[2]
Dwellings with flush toilet	1974	5	52
Dwellings with pit latrines	1974	34	39
Dwellings with no sewage disposal	1974	61	9
Population within 12 km radius of health clinics	1974	68	92
Population within 30 km radius of hospitals	1977	2	71
7–14 year olds in primary school	1978	74	100
Children in secondary school (Form I)	1978	17	24
Population within 8 km of all-weather roads	1977	11	94

(*Source*: International Labour Office, *Zambia*, Addis Ababa, 1981)
Notes: [1] Peripheral rural areas. [2] Number of persons.
It is difficult to define a basic needs income. The ILO figure of K100 per month (1980) would be sufficient to meet a family's needs for food, shelter, clothing and other basic needs. (The K100 includes income in cash or kind.)

Write a short essay on any of the following:
1 Explain how natural resource development initially transformed traditional rural societies into lopsided, dualistic space economies, using Zambia as an example and drawing on your knowledge of resource development in other less developed countries.
2 Explain why general development was blocked or retarded in Zambia despite rapid growth of exports in the period up to the early 1960s.
3 Explain government policy changes introduced to achieve general development in Zambia during the last 20 years. What were the spatial effects of these policies?

Debate the following topics:
1 Are mineral resources a more powerful source of development than agricultural resources? Why has Zambia a much higher per capita income than her geographical neighbours (except Zimbabwe)?
2 Do most of the gains from world specialisation and trade accrue to the advanced industrial countries?

17 THE NUCLEAR POWER DEBATE IN THE UK

Could pressurised water reactors become the basis for an expanded nuclear power programme in the UK, or will mounting opposition to nuclear power mean that the programme is run down?

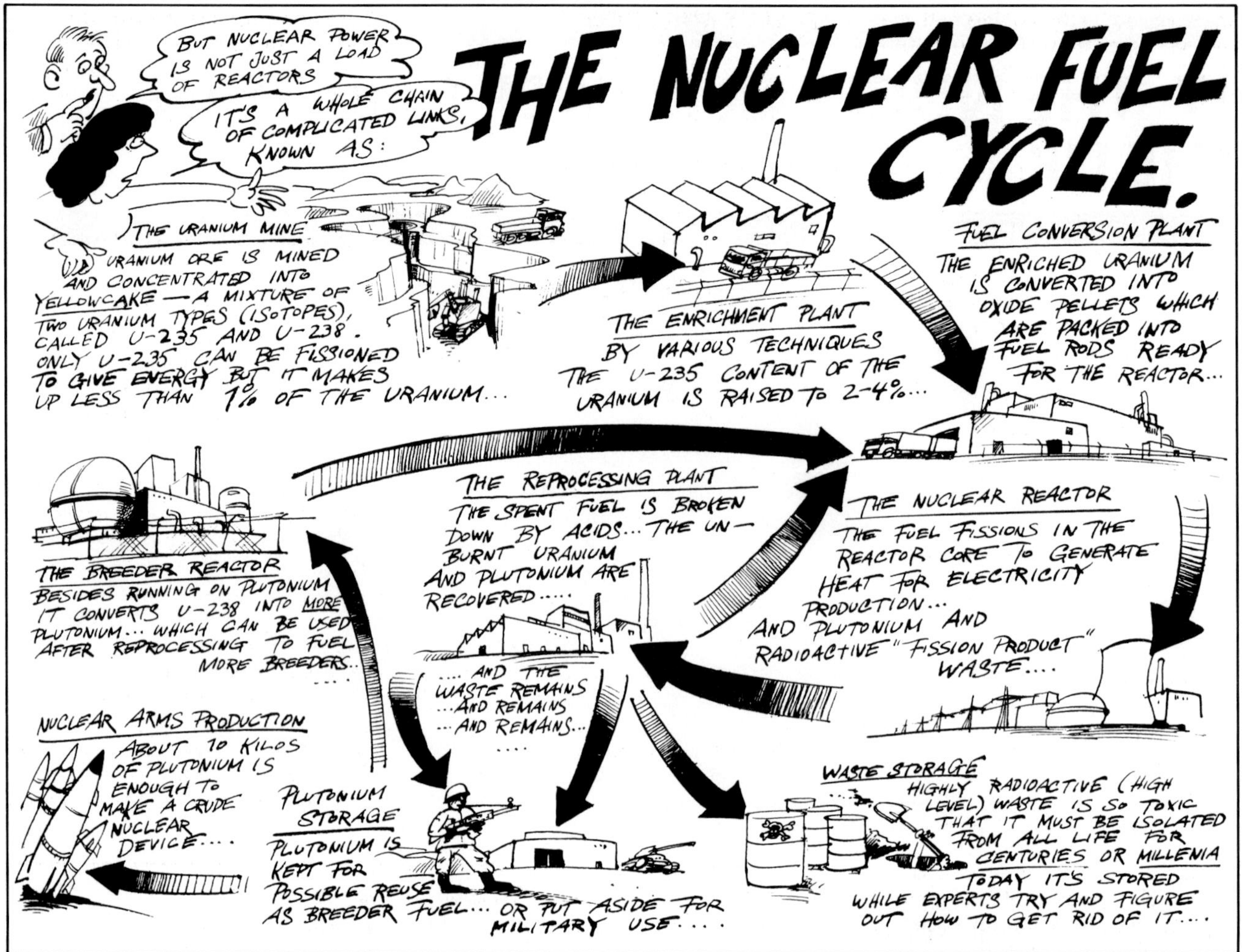

Figure 1 *Outline of the nuclear fuel cycle*
(*Source:* adapted from *Nuclear Power for Beginners* (1978) Writers and Readers)

Nuclear protest couple have electricity cut off

A couple have had their electricity cut off because of a protest about nuclear power. Mr Jim Brewer and his wife Ann have been refusing to pay 11 per cent of their bills—the amount they say is spent on developing nuclear power.

The couple live at Luxulyan, Cornwall, the scene of a protest last year against survey work carried out for the Central Electricity Generating Board, which was seeking a site for a nuclear power station.

The board is no longer considering a site near Luxulyan, but the Brewers have continued an individual protest.

They have withheld amounts totalling £24 from their electricity bills and are expecting the South Western Electricity Board to sue them for this.

Mr Brewer, aged 38 and a carpenter, said yesterday: 'Cutting us off makes no difference to our attitude. Nor will it make much difference to our way of life.'

The couple, who have a daughter, aged 14, and a son of 12, have installed a windmill generator sufficient for lighting and television.

For cooking they had a wood-burning stove and they also had a three kilowatt alternator, driven from a diesel engine in the garden, said Mr Brewer.

(*Source: The Guardian*, 8 October 1982)

SOCIETY AND NUCLEAR POWER

Ann and Jim Brewer obviously have very strong views about nuclear power. They are prepared to live without electricity from the national grid rather than help finance the further development of nuclear generated electricity. This chapter sets out to present both sides of the nuclear debate: that which views nuclear power as a cheap, safe and clean energy source for the future and that which sees it as a threat to our health, environment and society. Right from the start, it has been the allocation of financial, scientific and technical resources for the development of nuclear power which has resulted in intense disagreements and political debate between many individuals and groups within British society. You are encouraged, in the course of this chapter, to examine the arguments of the main parties in the dispute — also their beliefs and values and the power which they have to put forward their points of view — and to clarify your own attitudes to the debate about the further development of nuclear power.

Having read the chapter and completed the exercises, you should be more able both to express and justify your own viewpoint. On finishing the chapter you may not agree with Ann and Jim Brewer, but you should at least better understand what caused them to install a windmill.

?

1 As a class, discuss what you think of the Brewer's action before reading the chapter. Keep a record of your discussion. You could have another discussion after finishing the chapter to see if your views have changed in any way.

2 As you read the chapter, keep two lists. The first should contain the viewpoints and reasons which suggest that the Brewer's action was not merely eccentric and that their protest against nuclear power is justified, and the second should contain the viewpoints and reasons supporting the continued development of nuclear power. It is important that you consider the arguments for both sides.

3 Show the newspaper extract to several adults and ask them for their views. You could ask the following questions:

a Do you have any sympathy for the Brewer's actions?

b Why do you support or oppose the development of nuclear power?

c Do you think the government and the opposition parties have energy policies? If so, what are they?
Report your results back to the class.

ENERGY AND SOCIETY

Energy is a crucial resource. As this is the case, it is important to consider such questions as: 'Who controls world energy resources?' . . . 'How are resources developed?' . . . 'At what cost?' . . . 'Who controls the supply and price of energy?' . . . 'Who creates ideas about energy shortages and surpluses?'.

The answers are generally complex but the choice of which energy resources to develop and which to ignore is a matter for debate. Different ways of producing, distributing and selling energy have differing effects on the economic, political and social relationships between people. A society fuelled entirely by large nuclear reactors, for example, would be fundamentally different from one fuelled by domestic windmills. If it were feasible — and the anti-nuclear lobby argues that it is — a society fuelled by sun, waves and wind, might provide more jobs, cheaper electricity, less dependence on imported minerals, and might also allow people greater control over their own lives.

Let us now look at the overall development of nuclear power in Britain as a prelude to examining both sides in the debate about which sources of energy it would be best to continue to develop.

HOW MANY NUCLEAR POWER STATIONS IN BRITAIN?

Figure 2 shows the location of nuclear power stations and associated nuclear sites in Britain. Thirty-two reactors at these stations provide the power to generate electricity which is fed into the national grid. In 1979, nuclear stations represented eight per cent of Britain's generating capacity and supplied 13 per cent of the nation's electricity. In the same year, a newly elected Conservative government announced plans to build ten large pressurised water reactors, each of 1500 MW capacity, by the year 2000. These would have increased nuclear capacity almost five-fold but there are now signs that these plans will not be realized and have probably been revised downwards.

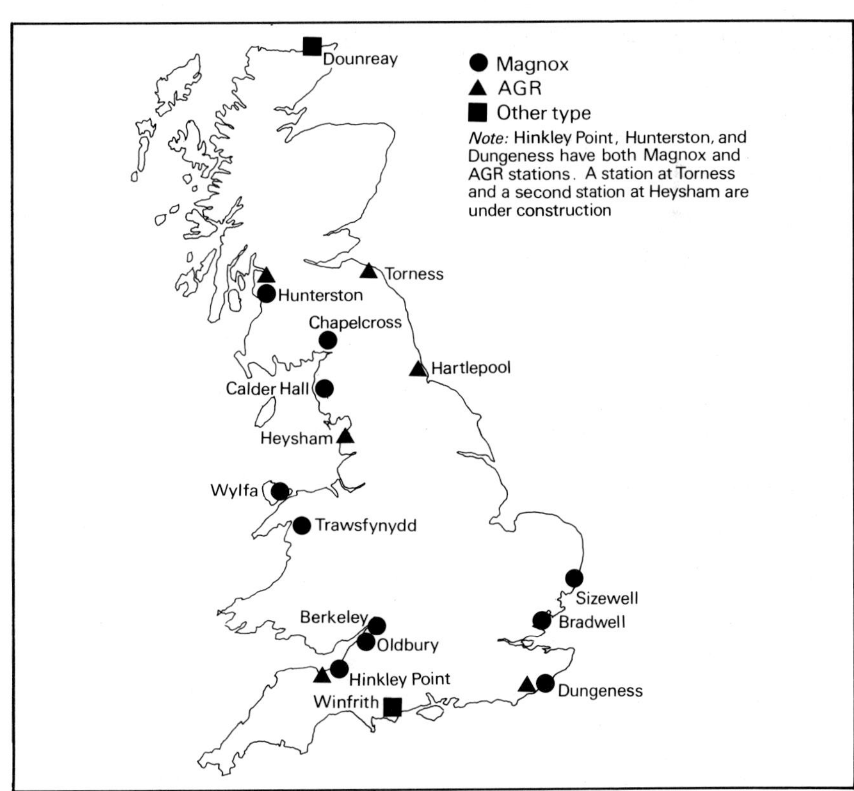

Figure 2 *The site of UK nuclear power stations* (*Source:* TCPA 1978)

The arguments for and against expansion can be viewed from three perspectives — economic, environmental and social.

ECONOMIC ISSUES

THE PRO-NUCLEAR VIEW

The economic case for nuclear power is based on two assertions: firstly, that it will be increasingly needed to fill an energy 'gap' created by rising demand and the depletion of fossil fuels, and secondly, that it will produce electricity more cheaply than competing sources of power.

The energy gap
The 1975 World Energy conference published a report in 1978 which is typical of the sources quoted by the pro-nuclear camp. Its authors concluded that the share of nuclear power in electrical power generation must expand to 45 per cent of the total in 20 years (the year 2000) and rise to 65 per cent by the year 2020.

Typical forecasts of total UK energy resources required by the year 2000 are between 500 and 600 mtce (millions of tonnes of coal equivalent) as shown in Table 1. If the lower forecast figure is taken (column 2), the deficit is still 120 mtce, which is equal to the whole of our present day coal output.

Table 1 *Typical forecasts of total UK energy resources required by the year 2000*

Year	*1976*	*2000*	
		Col. 1 Typical Forecast	*Col. 2 Typical Forecast*
Total UK energy resources required	343	575	490
UK fossil fuel contribution	328	300	300
Possible savings	–	30	30
Renewable energy resources	2	25	25
Nuclear (at present level)	13	13	13
Deficit		207	122

? To fix these forecasts in your mind, draw a graph of Table 1 and make brief notes on the arguments outlined below.

The pro-nuclear lobby argue that with the world population expanding so rapidly over the next 50 years, the energy shortage will appear in the next 15 years. Unless we are able to prevent this, the energy gap will give rise to competition for the dwindling supplies of oil and perhaps cause a major world war. The present conflicts in the Middle East, they

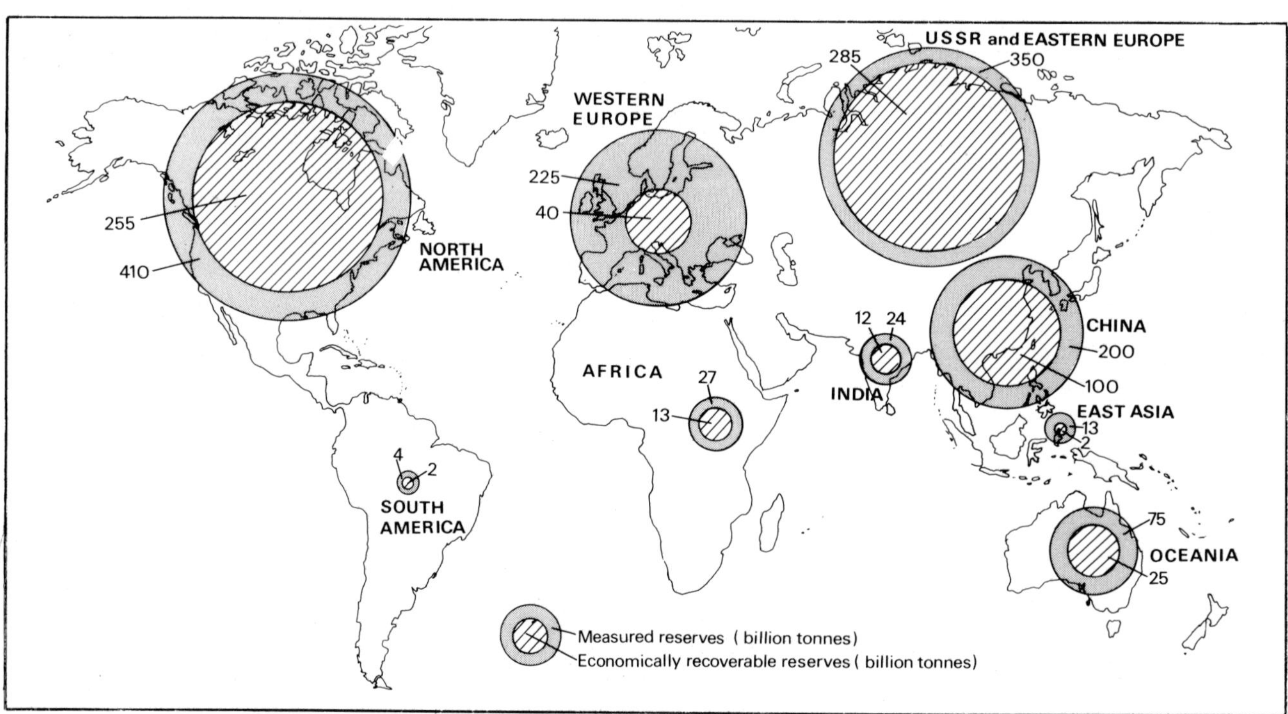

Figure 3 *The uneven distribution of the world's coal reserves*
(*Source: Commonsense in Nuclear Energy*)

argue, are potentially dangerous because they are taking place in the region that supplies so much of the world's oil.

The pro-nuclear lobby argue that traditional sources of energy cannot fill the energy gap.

There is an overwhelming argument against a long-term reliance on coal. Coal is not distributed uniformly over the world (Figure 3). While both the USSR and the USA might contemplate such a long-term reliance, the rest of the industrialised world cannot do so — at any rate without becoming energy dependent on the USSR and the USA. Such a sensitive dependence would be a certain recipe for political unrest. It is true that world resources of nuclear fuel are not uniformly distributed either, but they are better distributed than coal.

(*Source*: *Commonsense in Nuclear Energy*, Fred and Geoffrey Hoyle, 1981, Heinemann Educational Books)

What about traditional sources of energy?

Half the energy used in the world today comes from oil. The discovery of new reserves is not keeping pace with consumption and production rates are bound to fall eventually. The same is true for natural gas. Coal is plentiful in some parts of the world but there would be problems in extracting and transporting it to population centres from such areas of major reserves as China and Siberia. Britain is comparatively well provided with indigenous energy but North Sea oil and gas reserves will be declining by the 1990s, while coal extraction can be increased only gradually. The world has already entered a period of heavy investment in obtaining less accessible fossil fuels. Inevitably this trend is accompanied by rises in the real prices paid by consumers.

(*Source*: *The Case for Nuclear Power*, Nuclear Power Information Group)

The pro-nuclear lobby argue that new sources of energy — from waves, wind or sun – are hardly developed and could not be ready in time to meet the projected energy gap. Most new energy supplies need 40–50 years to develop and manufacture in sufficient quantity to make a substantial impact on world energy requirements. For example, windmills look an attractive option, *but* big 2–3 megawatt machines could not be made environmentally acceptable. They would be incredibly noisy and the dangers of flying ice in winter or broken blades, not to mention television interference, would mean that they would have to be sited at sea. The cost of the windmill plus the loss in power transmission would make them extremely expensive. The pro-nuclear lobby insist that so-called 'alternative sources' fall down in this way when they are examined for their commercial possibilities by experts.

A great deal of money and effort is being devoted to investigating and, where practicable, developing 're-newable' energy sources. For example, tidal power is being examined, in particular the construction of a tidal barrage across the River Severn estuary. Wind power looks promising; prototype wind-powered generators are operating at Carmathen Bay and in Orkney, where a larger machine is also being built. However, several hundred wind-powered generators would be required to equal one large power station.

Nuclear energy is the best way of making a substantial contribution to our energy needs in the foreseeable future.

(*Source*: *The Facts about Nuclear Energy*, CEGB pamphlet)

Nuclear energy — a 'cheap' option

The pro-nuclear lobby argues that nuclear power is a 'cheap' option.

The electricity generating boards have shown that, taking into account the cost of constructing the stations, of fuel reprocessing and waste disposal, and decommissioning stations at the end of their useful lives, electricity generated by nuclear power is comparable in cost to that generated by using coal and less than that generated by using oil.

A recent international study* of generation costs has shown that new nuclear generating plant has a clear competitive advantage over coal-fired plant for base load generation in the UK, West Germany, France, Italy and Belgium.

Given Government approval, the Central Electricity Generating Board propose that the next new nuclear power station to be ordered should be based on the Pressurised Water Reactor (PWR) system. Subject to the necessary consents and the outcome of the public inquiry, this station will be built at Sizewell, Suffolk.

A UK PWR, commissioned around 1990, would be expected to produce electricity over its lifetime at about two-thirds the cost of a contemporary coal-fired power station.

Uranium has no significant commercial use other than for electricity production. As nuclear energy increases its contribution to electricity production it will allow coal to substitute for oil where possible. The use of oil can then be concentrated on those applications for which there is no economical substitute, for example some forms of transport.

One tonne of uranium used in a nuclear reactor can produce as much electricity as 25 000 tonnes of coal or 15 000 tonnes of oil.

*The study was made by the International Union of Producers and Distributors of Electrical Energy.

(*Source*: *The Facts about Nuclear Energy*, CEGB pamphlet)

Print 1 *Oldbury nuclear power station on the river Severn in Gloucestershire. Water for the cooling processes is taken at low tide from the Severn and stored in a 380 acre reservoir*

THE ANTI-NUCLEAR VIEW

The arguments of *need* and *cost* which the nuclear lobby uses to justify the considerable investment in nuclear power are strongly contested by the anti-nuclear lobby who claim that the forecast 'energy gap' will not materialise and that the real costs of nuclear power are far greater than official sources admit.

What energy gap?

Those opposed to the development of nuclear power argue that the energy gap will not, in fact, materialise becase it is a myth, based on an exaggerated projection of the demand for energy in the future. Colin Sweet, in his book *The Price of Nuclear Power* (Heinemann Educational, 1983), argues as follows:

> Of course, if the world is to have greater prosperity, as well as a greater number of people, there will be a need for more energy. But energy is a demand derived from economic activity, not a cause of such activity. Many writers present the future demand as an extrapolation of the energy past. The boom years from the 1950s to the mid-1970s are not a reliable guide to the future. Indeed they can be highly misleading. All we can safely say is that the pattern and the size of the demand will change by the end of the century. Straight line linear projections which show a constantly rising need for energy, both in absolute terms and in per capita terms, produce highly coloured and inaccurate presentations of future energy requirements. In one sense it is not surprising that their predictions are misleading. Energy forecasts are not objective. Perhaps consciously, but more often than not unconsciously, the forecasters build their own subjective bias into their figures, and, as most forecasts originate from the supply side industries, they reflect their view of the world. The high energy scenarios which proliferated in every OECD country during the growth years are necessary, indeed indispensable, to the case for nuclear power.

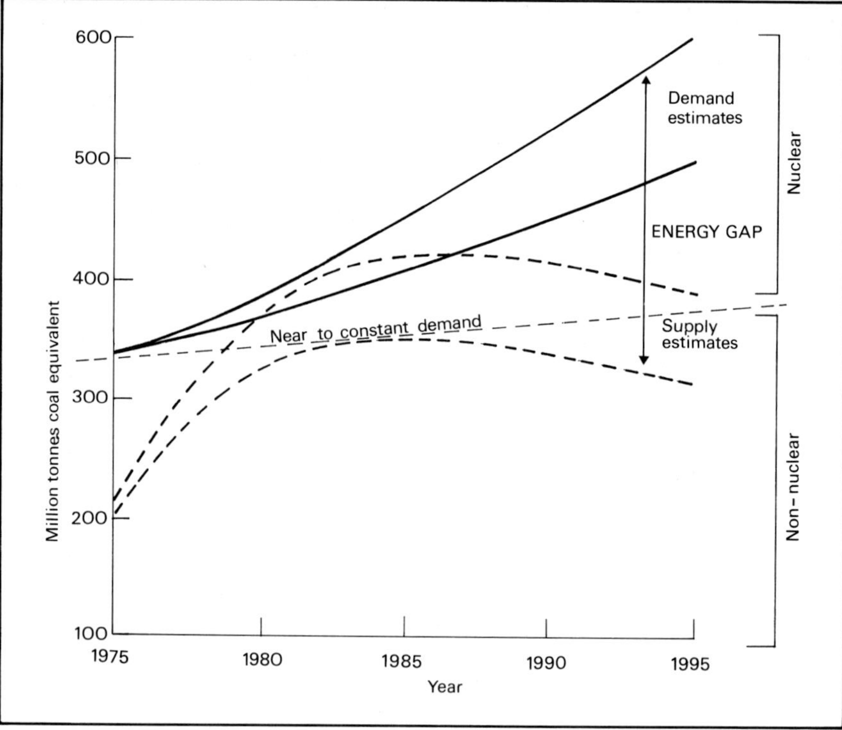

Figure 4 *The energy gap*
(*Source*: SSEB (National Conference, Department of Energy, 1976))

Figure 4 represents the type of energy forecast used by the pro-nuclear lobby to justify heavy investment in nuclear power. It shows that in 1976 the Department of Energy was predicting a future shortage in energy supply relative to demand and was suggesting that this 'gap' should be filled by nuclear power. The anti-nuclear lobby points out that behind such a forecast lies the assumption of continuing economic growth resulting in continuing, increased demand for energy. They would highlight the fact that if one assumes constant or near to constant demand, the forecast energy 'gap' virtually disappears.

The anti-nuclear lobby maintains that the CEGB has consistently over-estimated the demand for electricity in the past (Figure 5) and that this points to the need to radically review forecasts for the future.

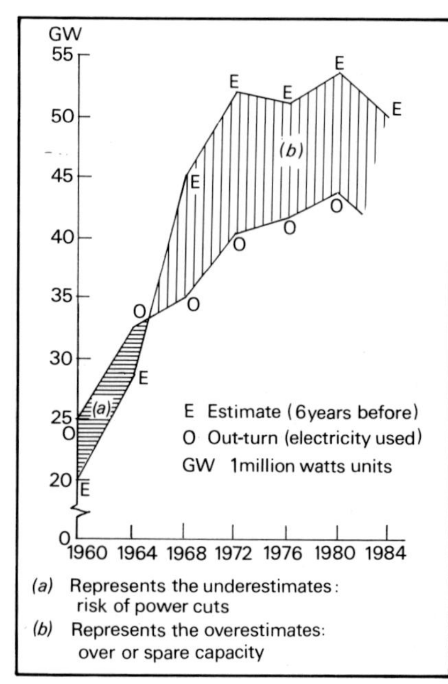

Figure 5 *Load estimates adopted by the Electricity Council*

Do we need nuclear power?

Nuclear power now provides only 2% of Britain's delivered energy that you, the consumer, uses. Even with the proposed 10 Pressurised Water Reactors (PWRs), plus the two Advanced Gas Reactors (AGRs) being built, it will account for no more than 7% of delivered energy by AD 2000. Meanwhile our primary energy consumption has declined steadily to below the 1970 level. Electricity consumption has also levelled off and further growth is unlikely. Nevertheless by grossly inflating electricity consumption forecasts, the Central Electricity Generating Board (CEGB), has built a vast surplus of unnecessary new plant; during 1980/81, despite premature closing of non-nuclear plant (putting thousands out of work), it had 33% more generating capacity than necessary to meet its peak demand. Yet between 1981 and 1988, the CEGB intends to commission another 14GW – *the equivalent of one-third of its present generating requirements.* Meanwhile, the South of Scotland Electricity Board has a surplus of more than 70%: this will rise to well over 100% when the Torness AGR is operating. The escalating cost of nuclear plant being built, or planned, will raise electricity prices still higher, further depressing electricity consumption.

If the money spent on nuclear plant were spent instead on energy conservation, double the energy would be provided for the same cost. Failure to do so has so far cost every household in Britain around £500 since the nuclear power programme started.

(*Source*: ECOROPA Information sheet)

? In pairs, discuss the meaning of the concept 'energy gap' and briefly rewrite, as if for a 14 year-old, the points made in this section.

The alternatives to nuclear power

Reassessment of recoverable fossil fuel reserves

The anti-nuclear lobby query many of the official figures relating to when supplies of fossil fuels will be exhausted. They argue that the UK's recoverable stock of coal is 47 billion tonnes – about 300 years' supply at present rates of extraction. They point out that reserves of fossil fuel are understated by the operating bodies, because maximising reserves means raising costs. Present reserve figures, it is therefore argued, usually relate to what is profitable rather than what is actual.

If the reserves of fossil fuel that are economically viable to recover *are* in fact greater than officially stated, then the argument about the great urgency to fill the 'energy gap' by quickly building more nuclear power stations no longer holds, and, it is argued, there is then time to spend the next decades developing the potential of recoverable energy resources such as wind or wave power.

Conservation of energy

The anti-nuclear lobby argues that much of the energy used in Britain is wasted. It maintains that in our homes some 35 per cent of the heat loss could be prevented by adequate insulation and domestic appliances, including the common light bulb, could be redesigned for far greater energy efficiency. Building standards in the UK are, it is said, less demanding as regards energy conservation than those in Denmark and Sweden, and builders have been slow to build houses with such energy-saving devices as heat pumps to recover waste heat. An environmental group in North-east England calculated that insulating half a million homes would employ two thousand people for ten years and save energy valued at £24 million a year.

Those opposed to the development of nuclear power say there is similar scope for energy conservation in industry where 30 per cent of the energy used could be saved by introducing known and feasible technologies. These would involve changes to production processes, improved insulation, better maintenance of equipment, the introduction of a range of fairly simple heat recovery systems, and more sensitive control systems. There is also, it is argued, much scope for energy saving in the development of public transport.

The anti-nuclear lobby points out that at present only one-third of the energy used to raise steam in our power stations is converted to electricity. Two-thirds of the energy ends up as waste heat which is discharged into the air, rivers, and the sea. Combined heat and power stations operate in such a way as to increase the amount of usable energy obtained in power generation. At the cost of producing rather less electricity, they increase the temperature of the waste cooling water from 25°C to 80°C. It can then be piped to local homes and factories where it provides space heating and hot water. In a CHP (Combined Heat and Power) station as much as 85 per cent of the fuel burnt is converted into usable energy; a great improvement, it is argued, on conventional stations especially as 40 per cent of our energy needs are for space heating and hot water. CHP has been used in Germany, Sweden, Denmark, France, and Finland for many years but Britain only has plans for pilot schemes in Hereford, Tyneside and Sheffield.

The opponents of nuclear power are angry that, despite the huge amounts of energy being wasted, governments in Britain have only relied on market forces and exhortation to encourage energy conservation. While they have put their faith in rising prices and 'Save It' campaigns to bring about change, other European governments have adopted far more positive policies. After comparing Britain's performance with that of other EEC members, the Association for the Conservation of Energy recommended the following:

- an energy certification scheme to encourage all existing houses to be brought up to building regulations standards (Denmark);
- a comprehensive homes insulation plan, backed by government cash aid and covering homes with some existing insulation (Holland);
- tax deductions for energy saving measures (France);

- mandatory labelling of domestic appliances such as dishwashing machines and cookers to detail their energy use (Italy, Holland);
- low-interest loans for industrial energy saving investments (France, Germany, Italy).

Is conservation of energy any help?

Certainly. *At least twice as much energy can be saved by conservation as can be produced with the same money.* About half the energy we now use is for water and space heating *for which electricity is quite uneconomic.* Two-thirds of the fuel which goes into UK power stations emerges as waste heat – *enough to heat all our buildings!* Such heat is already harnessed abroad but such schemes are not viable with remote nuclear stations.

(*Source*: ECOROPA Information sheet)

A further factor which the anti-nuclear lobby puts forward against the large-scale development of nuclear power stations is that uranium is not, itself, a *renewable* resource. Uranium supplies will limit the possible contribution of nuclear power to world energy needs to a maximum of ten per cent and this contribution could not be realised until the year 2005 even with the most favourable assumptions. The pro-nuclear lobby see the development of fast breeder reactors as a means of overcoming the limits of uranium supplies. As these reactors produce more nuclear material than they consume (in the form of plutonium), this development creates a major problem in terms of how this radioactive waste is to be disposed of. The safety of fast breeder reactors is also a matter for debate.

How cheap is nuclear power?
Again, the anti-nuclear lobby disputes the official argument that nuclear energy is a cheap option (see *page 345*). They argue that nuclear power stations are extremely costly to design and build, and to repair when they develop faults. There are also fuel cycle costs and decommissioning costs.

If we abandon nuclear power and, however unlikely it may seem, need more energy, are there alternative sources?

Yes. Coal production could be expanded dramatically and we can use infinite energy from the sun, wind and waves. Wood, waste and plant matter can be transformed into petrol substitutes. Official Department of Energy reports *admit that a significant slice of Britain's energy could come from renewable sources.* Britain is well-placed to develop the 'renewables' some of which are already being used on a large scale abroad. We shall inevitably have to rely on the 'renewables' – indeed many nuclear protagonists believe this – *yet the Government has just decided to cut from £14 millions to £11 millions the budget for their development and to raise to £221 millions the money spent on nuclear development.* Some 'renewables' are already cost-competitive with nuclear power, e.g. wind electricity costs 2p/kwh for large and small systems. The US believes it can produce *more than twice the UK electricity demand from wind alone.* Japan already has several million solar installations and even in northern latitudes, Sweden is developing large solar district heating schemes. Besides being inexhaustible (unlike uranium) and inherently safe, 'renewables' are widely dispersed and would employ far more people. They are also suited to the needs of the Third World and offer considerable export potential. Other countries are far ahead – the US exported £30 millions of solar equipment to Europe alone in 1981. The nuclear industry argues that renewables are dangerous! A windmill blade might fly off and kill someone, or poisonous snakes might be hidden in a woodpile! But such risks are limited. We are not dealing with the kind of delayed, all pervading effects of radiation.

(*Source*: ECOROPA Information sheet)

Print 2 *Solar windmill and solar panels in Wales*

Is nuclear power the cheapest energy available?

The Generating Boards insist that nuclear power provides the cheapest electricity, excepting hydro-electric plants. The only valid way to cost power stations is to add up the cost of construction, operation and final decommission, in comparable inflation-adjusted money and include likely inefficiencies or breakdowns of the plant as well as the safe disposal of radioactive wastes. These costings show unequivocally that *nuclear power has always been more expensive than 'coal-power' by as much as 50%.*

(*Source*: ECOROPA Information sheet)

The anti-nuclear lobby maintains that the true costs are far greater than the industry admits. They base their case on the huge escalation in the costs of building nuclear stations, their failure to generate as much electricity as forecast, and the accounting methods of the CEGB which lead to an underestimate of real costs. If the cost of nuclear electricity is to be compared with that generated from other fuels, it is essential that in each case *all* the necessary costs are included in the accounting. CEGB data on nuclear power fails to fulfil this requirement as the following points illustrate:

1 They have not always included the full cost of capital including the interest charges during construction, or interest charges on stored fuel.

2 They never include the major costs of research and development.

3 They are based on *notional* operating capacities which overstate the *actual performance of stations.*

4 They underestimate nuclear fuel costs relative to coal in forecasting future costs.

5 They underestimate or exclude the costs of decommissioning reactors – nuclear power stations have a useful life of around 25 years after which they must be decommissioned or dismantled.

Concern about the accounting methods of the CEGB and the forecast costs of nuclear power is not confined to its radical opponents. In 1981 the reports of both the House of Commons Select Committee on Energy and the Monopolies Commission were critical of the industry's economic case. The Select Committee found the CEGB's figures to be:

'highly misleading as a guide to past investment decisions and entirely useless for appraising future ones'

and refused to endorse the Government's nuclear programme:

'We remain unconvinced that the CEGB and the Government have satisfactorily made out the economic and industrial case for a programme of the size referred to by the Secretary of State in his statement to the House in December 1979.'

The Monopolies Commission was highly critical of the industry's demand forecasts and found its investment appraisals defective:

'In particular, a large programme of investment in nuclear power stations which would greatly increase new capital employed for a given level of output, is proposed on the basis of investment appraisals which are seriously defective and liable to mislead.'

Perhaps the final word on cost calculations should rest with Colin Sweet:

The reason why nuclear-fired stations are more expensive is not difficult to understand. It is agreed that nuclear power is capital-intensive. Nuclear stations are more costly to build than either coal- or oil-fired plants. In 1975 Sir John Hill – then Chairman of the AEA – gave a 'typical' breakdown in costs, shown in the table below:

	Coal	*Nuclear*
Capital charges and operation	0.34	0.41
Fuel costs	0.51	0.12
Total	0.85	0.53

Note: Costs are in p/KWhr.
(*Source*: *Atom* 219, January 1975, p. 4.)

This cost structure suggests that because nuclear power is more capital-intensive than coal, nuclear fuel costs have to remain low if nuclear power overall is going to be competitive. The implications of this have not always been understood, especially when capital costs have been rising rapidly. As I have shown elsewhere the capital cost differential is such that the financial feasibility of nuclear power must assume that fuel costs are falling. In the period up to 1974 this assumption was usually made. It was but one more of the heroic assumptions upon which the case for nuclear power has been built. The 'logic', of course, is not difficult to understand. Because the capital costs of nuclear power were rising in comparison with those of coal-fired stations, relatively as well as absolutely, then the only variable that could be shifted to keep nuclear power in sight of being competitive was a falling fuel cost. So fuel costs were 'required' to fall. Unfortunately, the reality is that not only did they move upward, but they moved very rapidly upward. This too was concealed for a time by writing down the cost of fuel reprocessing, and by using uranium prices for stocks which had been bought when prices were low. But reality began to catch up very quickly in the late 1970s.

(*Source*: *The Price of Nuclear Power*, Colin Sweet, 1983, Heinemann Educational Books)

?

1 What is the kernel of Colin Sweet's argument?

2 Look at the table only and explain to, let us say, a member of your family, why the figures do not *prove* the economic case for nuclear energy.

3 Look back at the chapter so far and make a list of how figures and statistics have been presented by each side to support their case. Draw up a list of questions which you are beginning to learn to ask yourself before accepting the information in graphs and tables at face value.

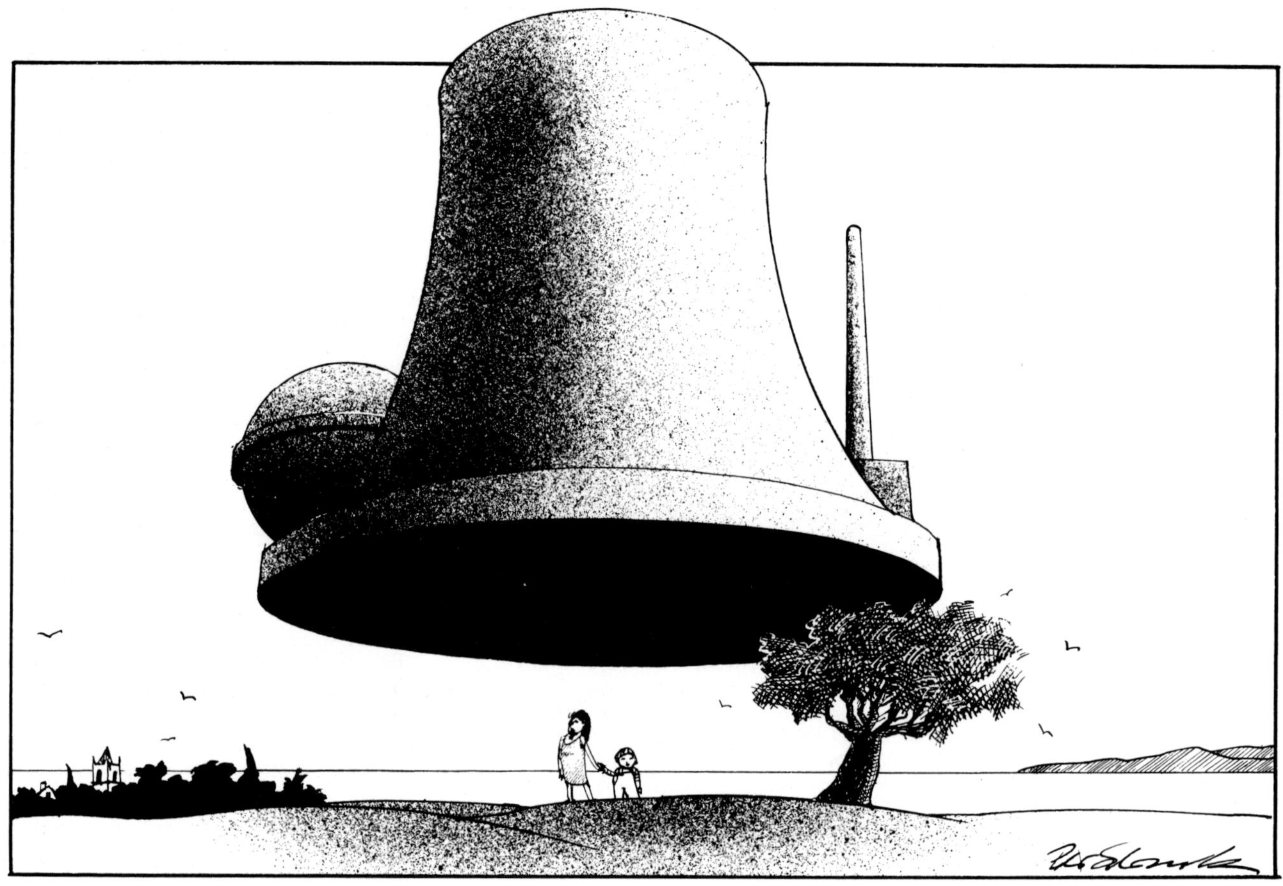

ENVIRONMENTAL ISSUES

The debate concerning the environmental impact of nuclear power revolves around fears of environmental contamination and the resulting hazards to health. There is disagreement about the likely effects of the routine radiation caused by the nuclear industry, the probability of catastrophic accidents during power generation or waste disposal, and the siting of power stations which poses a continuing threat to high quality landscapes. While the nuclear industry claims stringent safety standards and a good record as far as the well-being of workers and the environment is concerned, the opposition cites evidence of the industry's inadequate safeguards, its long record of accidents, and its persistent failure to admit these to the general public. Proponents of nuclear power regard fears concerning health and the environment as essentially exaggerated and alarmist, while opponents stress inevitable human fallibility in handling high technology and the irresponsibility of exposing present and future generations to unknown risk.

Liquid effluents | Gaseous effluents

INGESTION BY THE SEA → Shoreline irradiation → Swimming, fishing, etc → Water consumption → Seafood consumption

INGESTION ON THE LAND → Air inhalation → Air submersion → Ingestion through crops → Ingestion through dairy products

Ingestion by humans

Figure 6 *Generalised exposure pathways for humans*

RADIATION

Does the radiation resulting from nuclear power generation represent a risk to human health and the environment?

?

1 Look carefully at Figure 6 and identify the pathways.
2 Study Information Box A and Information Box B. Based on the two sets of extracts here, do you think the hazard of radiation should be a major factor in the nuclear debate? If so, which side appears to have the stronger case?

INFORMATION BOX A

Sellafield assurance

By James Lewis

Sir Douglas Black, past president of the Royal College of Physicians, and chairman of a team which carried out an inquiry into the incidence of leukaemia in west Cumbria, said he thought that people had been 'over alarmed to an astonishing degree' by allegations of links between the disease and the level of radioactivity around the Sellafield nuclear reprocessing plant.

He thought that parents should be convinced that the chances of dying from leukaemia in west Cumbria were no greater than the prospects of being killed in a road accident, and that a child in the area has as good a chance of surviving to the age of 30 as a child in any other part of England.

Sir Douglas, speaking after meeting members of the Sellafield local liaison committee formed to liaise with the medical inquiry team, said he thought that the committee accepted the qualified assurances given in his report. But he added that there were people whose minds would be closed for ever.

(*Source*: *The Guardian*, 8 September 1984)

Radiation

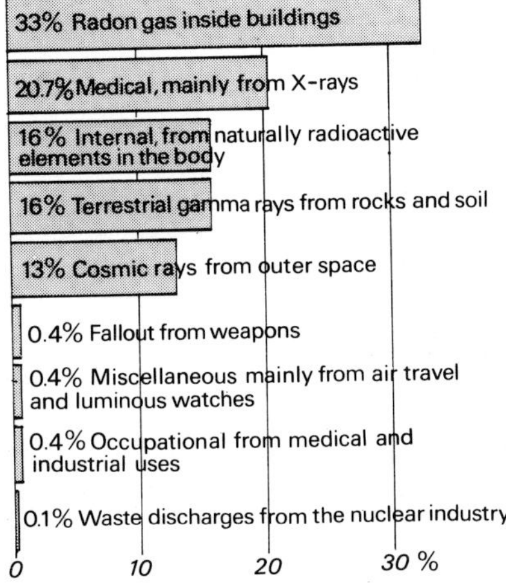

Exposure to very high levels of radiation can cause death through the destruction of body tissues. Radiation can also cause cancer and produce genetic defects in future generations. However, mankind has evolved in an environment exposed to cosmic rays from space and radioactivity from rocks, soil and even from the human body itself. Today, members of the public are also exposed to radiation from medical diagnosis and treatment, from TV sets and, in considerable amounts, from the building materials used in our houses. Nuclear energy is also a source of radiation.

On average, 78 per cent of the radiation to which we are exposed comes from natural sources, with nearly 21 per cent from medical sources and one-eighth of one per cent from the nuclear industry. The maximum annual dose from the nuclear industry is less than the variations in natural background levels between different parts of the country. The average dose is one hundred times smaller than these variations, which are themselves of no practical significance to health. Discharges of radioactivity to the environment from nuclear installations are strictly controlled by Government departments and based on exposure limits. The discharges present no hazard to members of the public who live and work near nuclear plants.

(*Source*: *The Facts About Nuclear Energy*, UKAEA pamphlet)

Print 3 *Sellafield nuclear power and reprocessing plant, Cumbria*

*Inhaber's table of risk in man-days lost per unit energy output**

	Upper limit of estimates	*Lower limit of estimates*
Coal	2500	110
Oil	1800	12
Nuclear	10	1.7
Natural gas	6	–
Ocean thermal	30	25
Wind	700	125
Solar		
Space heating	125	100
Thermal	600	80
Photovoltaic	700	160
Methanol	350	225

*man-days lost are through deaths, injuries and disease and the energy unit is the megawatt-year
(*Source*: *Commonsense in Nuclear Energy*, Fred and Geoffrey Hoyle, Heinemann Educational Books [a pro-nuclear book])

INFORMATION BOX B

Why is radiation dangerous?

You can't feel, see, taste or smell it – but it kills. Radiation can kill cells or alter their behaviour. Damaged cells may reproduce to be diagnosed later as cancer or leukaemia: in sex cells as miscarriages or mutated children. It weakens resistance to disease especially in children. Large 'doses' kill outright; *the effects of small doses may take 30 years to become evident.* This is why most of the routine releases of radiation from the nuclear industry are still to be paid for in unknown numbers of lives. Radiation which can be ingested is infinitely more dangerous than natural background radiation. No exposure is so low that it is safe. 'Permissible limits' are constantly being revised downwards: If UK limits were reduced to those permitted in the US (20 times lower) the UK nuclear industry would have to close.

(*Source*: ECOROPA Information sheet)

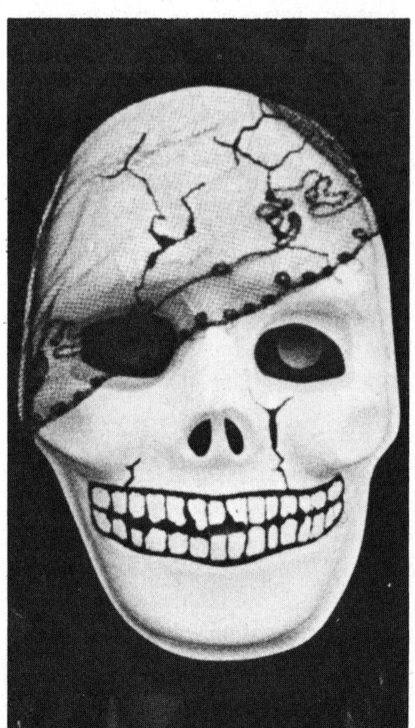

Print 4 *A demonstrator on the last day of the Sizewell Inquiry*

Key safety issues over reactor still unresolved

SIZEWELL INQUIRY

David Fairhall examines the questions still to be answered over the plan to build Britain's first PWR

THE Sizewell nuclear power inquiry, which has now been running for almost a year, resumed yesterday with at least four of the main safety issues identified by the Government's Nuclear Installations Inspectorate still unresolved.

The inquiry is into the Central Electricity Generating Board's proposal to build Britain's first American-style pressurised water reactor (PWR) on the Suffolk coast.

On two safety issues — making the PWR less vulnerable to fire, and providing four boric acid tanks instead of one to shut the reactor down in an emergency–the board reckons it has already satisfied the NII.

On a third issue — whether to install a £5 million electromagnetic filtration plant in the reactor cooling circuit to reduce the radiation doses received by maintenance engineers — the generating board has so far rejected the NII's suggestion as ineffective and financially unjustified.

(*Source*: *The Guardian*, May 1984)

Sellafield leak will slow privatisation

By David Simpson,
Business Correspondent

The leakage of radioactive waste into the sea from the Sellafield reprocessing plant could lead to the Government postponing privatisation plans for British Nuclear Fuels Ltd, the group's chairman said yesterday.

'I can't see a flotation at the same time as we were appearing in court,' Mr Con Allday said yesterday. The BNFL chairman refused to say how the corporation would plead to the charges which are to be brought against it as a result of the leak last November.

Introducing the nuclear fuel group's profit figures for the year to March 31, Mr Allday said that the leakage which resulted in radioactive contamination of a 15-mile stretch of the Cumbria coast, should not have happened.

But, he added, 'no-one, whether an employee or a member of the public, was potentially exposed to other than an extremely small extra level of radiation above natural background levels as a result of the Sellafield incident, and we can confidently claim that no harm has or will be caused to anyone.'

Tests had shown that only a tenth of a curie of radioactive material had been washed up on local beaches, Mr Allday said, compared with the total of 67 000 curies of radioactivity discharged annually into the Irish Sea from the Sellafield plant.

BNFL is allocating £500 million of its spending programme for waste management at Sellafield, Mr Allday reported. Of this, £190 million will be devoted to reducing the discharge of radioactive material into the sea.

(*Source*: *The Guardian*, September 1984)

WASTE DISPOSAL

Do safe methods of dealing with radioactive wastes resulting from nuclear power generation exist? How can the damage to the environment caused by leakage of nuclear waste be evaluated against the pollution caused by other forms of energy generation, e.g. acid rain from coal-fired power stations?

? Here is yet another set of questions regarding the development of nuclear energy. Who has the advantage of the argument regarding waste disposal in your opinion? On what do you base your opinion?

What happens to the wastes?

All industries create some wastes. The amounts of nuclear waste are small by comparison but because they are radioactive they are subject to strict controls. Intermediate waste such as fuel cans and sludges are stored pending the building of engineered disposal facilities. Low-level waste such as contaminated clothing and scrap is disposed of at sea. All disposal is carried out under strict governmental control.

The high-level waste from the Sellafield reprocessing plant is stored as a liquid in special stainless steel tanks which have proved entirely safe. The total high-activity waste so far from the British nuclear energy programme is the equivalent in volume to two average-sized houses. It is planned that the liquid waste will be turned into blocks of glass which will be encased in steel. Storage in this form will be more convenient and the glass blocks will be further cooled in stores for at least 50 years, after which they could be disposed of by burying them deep in stable rock formations or depositing them on or under the ocean bed.

(*Source: The Case for Nuclear Power*, Nuclear Power Information Group)

Why can't radioactive wastes be disposed of safely?

Nuclear power generates radioactive wastes, some–like plutonium–being extremely long-lived (250 000 years) are lethally poisonous. The nuclear industry has allowed a trickle of such substances to escape into the environment believing that they would soon be diluted to inconsequential levels. This has not happened and over the last few years radioactive waste has built up (for example in the coastal areas around Windscale – now named Sellafield). 'High-activity waste' is kept in specially cooled stainless steel tanks, with a limited life, although the wastes remain lethal for thousands of years. The industry intends to vitrify its high-activity wastes and then to store them, either on the surface or in underground vaults, for 50 years or more – and then possibly to dump them in the ocean, hoping that the radioactive impregnated glass blocks will not leach their deadly contents. They will be irrecoverable, *and all the oceans on Earth are insufficient to dilute to a safe level these radioactive wastes.* Independent research into vitrification has not shown that it is a long-term solution. Our own and foreign untreated waste accumulates. Meanwhile overwhelming public hostility has halted attempts to drill experimental holes in Britain for subsequent burial of nuclear wastes. 'Low-level' waste is buried in 55 tips around Britain and, despite informed protest, is being dumped at sea. *Britain is responsible for 90% of the man-made radioactivity known to enter the world's oceans.*

(*Source: Nuclear Power: The Facts They Don't Want you to Know*, ECOROPA)

A British ship dumps nuclear waste watched by members of Greenpeace (1981). Following actions by Greenpeace and the National Union of Seamen in 1983, the London Dumping Convention suspended such dumping in 1985. This put increased pressure on the UK government to find sites for dumping waste on land

Disposal vessels designed, despite government assurances of 50-year delay

UK plans new A-waste dump ships

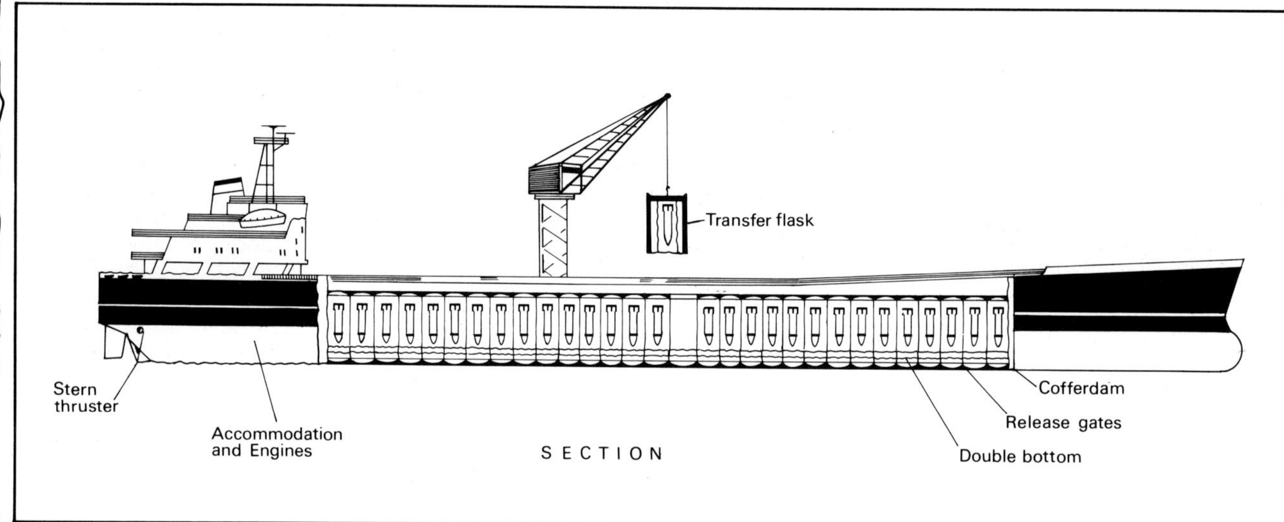

One of the designs, based on existing vessels, for a ship to dispose of high-level nuclear waste at sea

By Paul Brown

Studies into ways of disposing of high level nuclear waste in the sea have been commissioned by the Department of the Environment despite government assurances that it does not intend to put the methods into effect for 50 years.

Plans of ships carrying 150 torpedo-like objects containing high level nuclear waste have been drawn up by Ove Arup and Partners, the Swedish-owned engineers and design consultants.

The ships have the same specifications as those owned by British Nuclear Fuels Ltd and used to transport nuclear waste from all over the world to its reprocessing plant at Sellafield, Cumbria.

The Government is considering manufacturing the torpedo-like objects, known as propellents, packing them with vitrified high level nuclear waste and firing or dropping them into the sea bed from the purpose-built ships.

British Nuclear Fuels Ltd plans to build a nuclear waste vitrification plant at Sellafield and the ships Ove Arup have been asked to design could use the existing BNFL terminus at Barrow-in-Furness.

Greenpeace, who were given the plans yesterday said they were further proof that the Government planned to dump high level nuclear waste despite its assurances.

Mr Peter Wilkinson, a Greenpeace director, said the ships and equipment would be obsolete in 50 years. 'It is too much to ask us to believe they are doing all this for something that is not going to happen until 2034.'

In September, when the Guardian published details of contracts for design of propellents, Mr William Waldergrave, the Environment Minister, said the Government had published the fact it was doing research into the waste disposal.

Part of the research involved waste containers, he said. He did not mention the ships from which to launch the propellents.

In a letter he said there were no plans to dump high level nuclear waste in the sea. 'No such plans exist. What we are doing and doing quite openly, is to continue the research recommended by the Royal Commission on Environmental Pollution (in 1977) into what might be done with the waste in 50 years time when it has cooled down sufficiently to simplify disposal.'

The Government has repeated this assurance several times, and repeated it yesterday. In February the issue was debated by the London Dumping Convention, the international body which controls dumping of waste at sea. The Convention has suspended dumping of low level nuclear waste at sea while scientific investigations are carried out.

Britain and other nuclear nations have claimed that firing or dropping propellents into the ocean bed so they bury themselves in sediment is not dumping but 'sea bed emplacement' and therefore outside the convention ambit.

OVE Arup has drawn up three designs based on existing BNFL ships. Each design includes trap doors for dropping or firing propellents at the ocean floor.

Although it is always said that high level nuclear waste has to be stored until cool for up to 50 years before it can be disposed of, the French have already developed a technique to vitrify it only five years after coming out of a reactor.

BNFL has a plan to build a similar plant at Sellafield, where some high level waste has been stored for 35 years.

Mr Wilkinson said: 'Britain and America are spending $16 million a year on plans to dispose of high level nuclear waste. We cannot be expected to believe that it is all in aid of research they intend to do nothing about.'

(*Source*: *The Guardian*, 10 April 1984)

SITING NUCLEAR POWER STATIONS

A nuclear power station requires a level area of 200 acres with good foundations, an abundant supply of cooling water, easy access to road, rail or sea transport, and a short connection to existing power lines. While stations are best sited in areas of high demand for electricity in order to reduce transmission costs, early nuclear stations in Britain were sited in remote areas for safety reasons. While the Health and Safety Executive still requires new reactor types, such as the PWR, to be built in 'remote areas', Stan Openshaw, a geographer at Newcastle University, has produced maps showing that relaxed siting criteria for 'proven' reactor types now allow their location on 'virtually any site not in or near a large urban area'. Figure 7 allows you to compare the distribution of invalid reactor sites, based on 1955 criteria, with a distribution based on a set of relaxed criteria proposed in 1974. Both sets of criteria relate to the population distribution in the surrounding area, but the more recent criteria are based on more optimistic assumptions about the scale of a possible accident, the efficiency of evacuation procedures, and the radiation dose the public is likely to 'tolerate'. Such siting criteria inevitably involve political/economic judgements in which public safety has been balanced against the additional costs of siting reactors in remote locations.

Despite Openshaw's work showing a large number of potential sites which satisfy relaxed siting criteria, most existing and proposed reactor sites in Britain (see Figure 2, *page 343* and Figure 8) are in rural locations. Their existing or future presence in landscapes of high amenity value arouses the protest of conservationists and amenity groups and much of the opposition to nuclear power has taken the form of campaigns to prevent a reactor being built in a particular place.

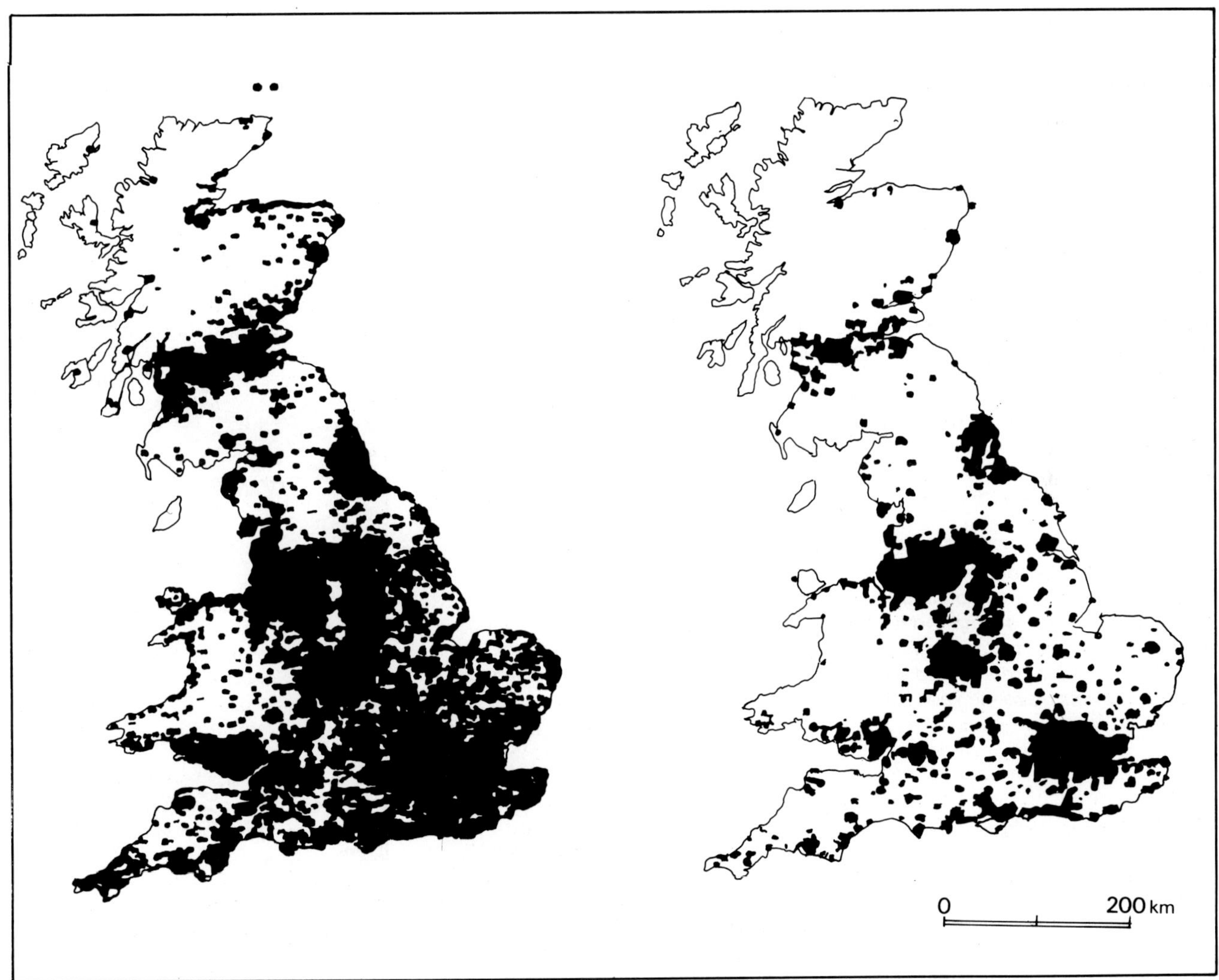

Figure 7a *Invalid nuclear reactor sites, 1955 (according to Roy)*

Figure 7b *Invalid nuclear reactor sites, 1974 (according to Openshaw)*

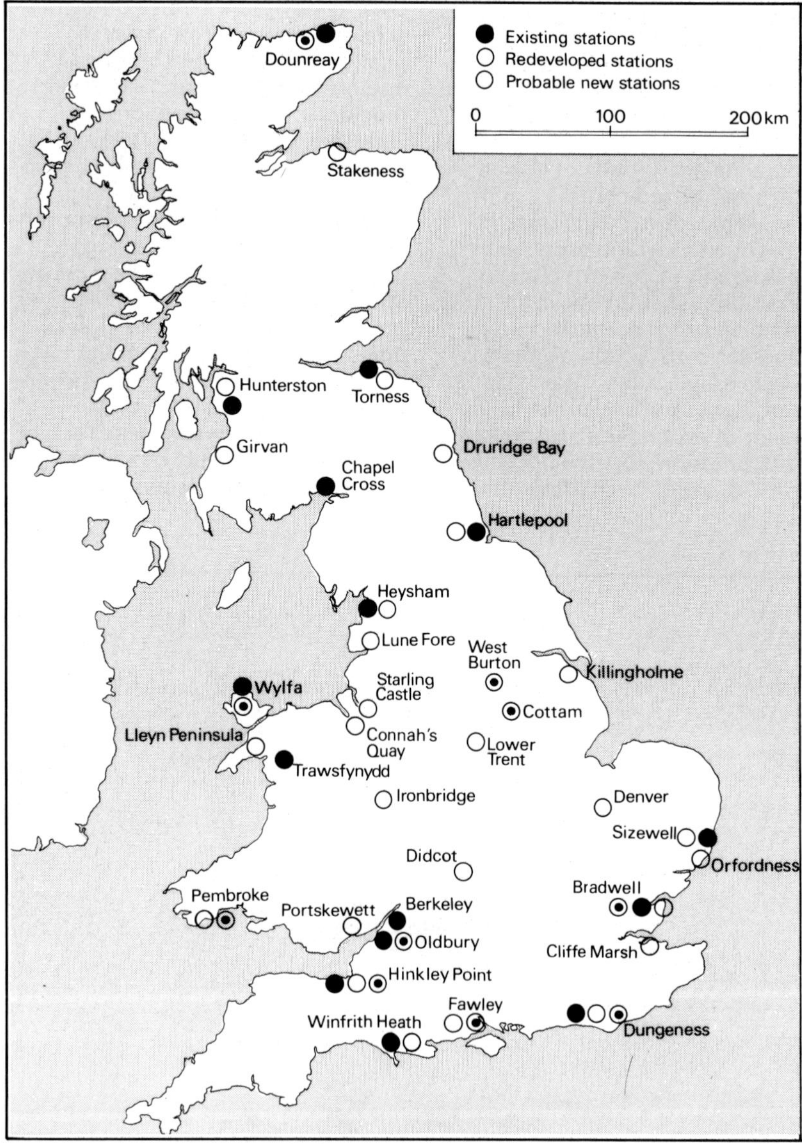

Figure 8 *Possible long-term nuclear power development. This map, which appeared in* The Guardian *in Spring 1982, reflected the Council for the Preservation of Rural England list but eliminated some of the more 'speculative' sites (for example, the site on reclaimed land in the Wash). It included anywhere that had been publicly discussed or mentioned in local authority plans up to the year 2025*

In September 1982, the Council for the Protection of Rural England issued a list of more than 50 sites in Britain which it believed to be earmarked for possible nuclear power stations (see Figure 8). It suggested that plans to expand nuclear power would bring much upheaval to the countryside and coastline, particularly when associated pumped storage schemes and power lines were considered. At that time, the CEGB had announced firm intentions regarding only two new sites, Duridge Bay in Northumberland, and Winfrith Heath in Dorset, although they also proposed multiple stations, or nuclear parks, at Sizewell in Suffolk, Hinkley Point in Somerset, and Dungeness in Kent. The CEGB disputed the authenticity of the CPRE's list, saying that it had investigated far more possible sites for nuclear power stations than it could possibly use, and that the list was out of date.

The proposed site for the building of a fast breeder reactor at Winfrith Heath in Dorset is close to the existing UKAEA establishment. Distant from the sea, the station would require massive cooling towers visible for 30–40 miles, and giant pipelines carrying sea water across the Isle of Purbeck, an area of Outstanding Natural Beauty. The construction of a station at Winfrith could further destroy one of the largest surviving fragments of lowland heath in Britain. Dorset's heathland was described by Thomas Hardy in *Return of the Native* but since his death, 50 years ago, its area has been halved, largely as a result of reclamation for agriculture. Studies by the Nature Conservancy council show the area of heath to have declined from 75 000 acres in

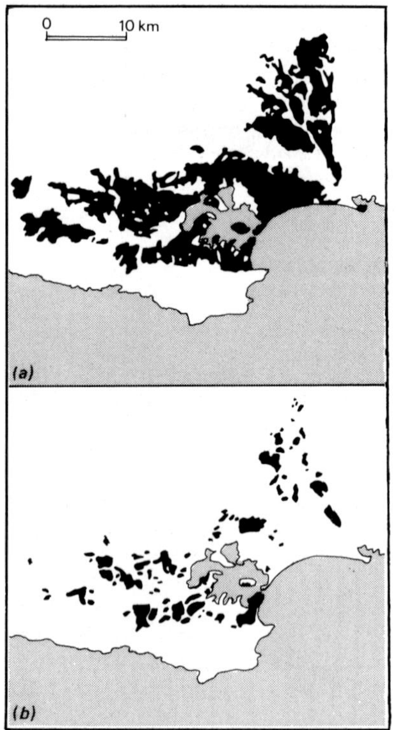

Figure 9a *The heathlands of east Dorset in 1811*
Figure 9b *The heathlands of east Dorset in 1973*
(*Source*: Furzebrook Research Station)

1811 to 25 000 acres in 1960, and 15 000 acres in 1973 (see Figure 9).

?

1 Write a suitable caption for the cartoon in Figure 10.
2 The Dartford Warbler is a rare bird which breeds exclusively on lowland heaths. Its British population is threatened by further development of the Dorset heathlands. Hold a debate on the motion: This house believes that the Dartford Warbler's right to survival outweighs people's right to increased electricity by the establishment of a fast breeder reactor at Winfrith Heath.

Figure 10
(*Source*: *The Radio Times*)

SOCIAL ISSUES

Because energy is such a crucial resource in modern society, the manner in which it is produced, distributed and sold has far reaching social implications. While proponents of nuclear power believe that nuclear electricity can sustain a consumer society with ever rising material expectations, its opponents argue that it could lead to a more dangerous and unequal society with fewer jobs, greater poverty, and a significant erosion of individual and democratic rights. We can only examine some of these aspects briefly here, but you should be alerted to their role in the nuclear debate.

Print 6 *A Dartford Warbler on its nest*

The Weapons Connection

The fundamental overlap of the paths to nuclear explosives and to civilian uses of nuclear energy has been recognised since the mid-1940s. Governments have said from the start that the military and civilian atoms were virtually identical, yet paradoxically, that they wanted to stop the one and promote the other.

The main purpose of Britain's first nuclear power reactor at Calder Hall was to produce plutonium for use as an explosive in atom bombs: the electricity was just a bonus. The main purpose of Britain's latest power reactors is to produce electricity – with plutonium as a by-product. At Sellafield the plutonium is separated from the other radio-active wastes in the used nuclear fuel, thereby making it more accessible to governments, or conceivably terrorists, who might want to acquire nuclear weapons.

In 1974, India exploded her first nuclear device, using plutonium which she had reprocessed from fuel used in a 'peaceful' reactor; and using expertise and equipment which she had acquired from three other countries to assist with her civil nuclear programme.

The possible spread of nuclear weapons in this way has led to stricter and more selective controls on the use of nuclear fuels and technologies, but experience has shown how inadequate these controls are.
(*Source*: Friends of the Earth brochure)

SHOULD DEVELOPING COUNTRIES BE DEPRIVED OF NUCLEAR POWER?

Figure 11 shows the developing countries where, in 1984, nuclear power reactors were operating, under construction, or planned. Of the 18 countries actively engaged (Iran's programme came to a halt after the overthrow of the Shah), only India has built reactors without foreign aid. Other countries have been aided and supplied by the USA, Canada, West Germany, France and the USSR.

While the demand for electricity in Western Countries has decreased in the 1980s, newly industrialising countries are not only consuming increasing amounts of electricity, but their growing populations are also

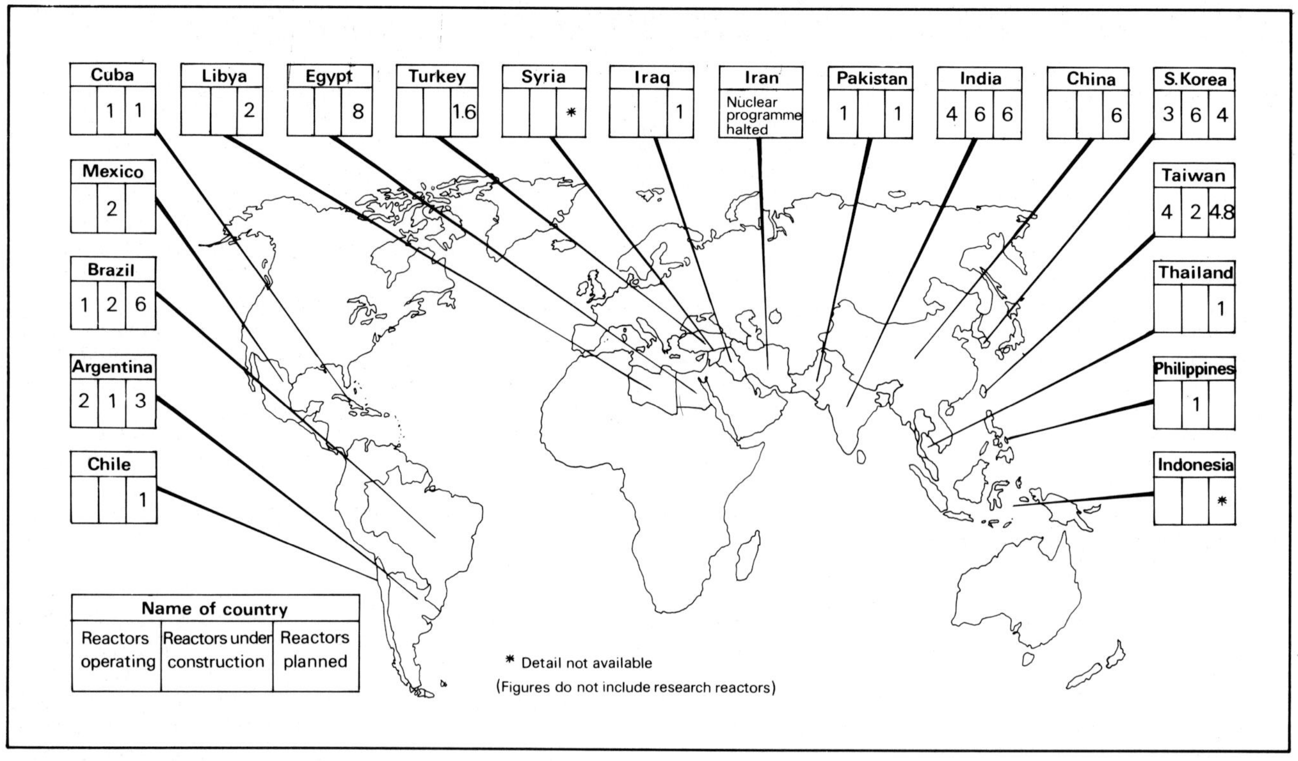

Figure 11 *Nuclear power reactors in the developing world, 1984* (*based on IAEA data*)
(*Source*: based on *Geofile*, September 1984)

expected to sustain an increasing domestic demand for electricity for years to come. But there is a lobby, mainly in the West, which argues that the answer to increased electricity production does not lie with nuclear power. It argues that a capital intensive, high technology industry which is the focus of major opposition in the North is even less appropriate in the South. The poor will not benefit from the development of nuclear power – high costs will mean high prices.

Desire to control weapons proliferations clashes with a commercial desire to export nuclear technology. President Carter became so concerned about proliferation that he got 40 nations to agree to a two-year pause in plutonium recovery and recycling; a moratorium which was lifted when the Reagan administration took over. The Peace Movement is aware of the danger and has linked its campaign against nuclear weapons to the campaign against nuclear power.

> Atomic power is completely unsuitable for meeting the energy needs of Third World nations. It both increases their dependency and undercuts self-reliance and equity.
>
> Stymied at home by widespread public outcry, a slackening of energy consumption and inflated construction costs, the Western nuclear industry is placing more emphasis on Third World sales. The US, France, West Germany and Canada are stumbling over each other in haste to bail out their domestic nuclear industries.
>
> (*Source*: *The New Internationalist*, Issue 102)

On the other hand, some lobbyists in the developing world, argue that developing countries need nuclear power, in the absence of other abundant sources of energy and that the rich world is being deliberately obstructive. The following extract is taken from a long article which appeared in the *New Scientist* in February 1981.

Print 7 *A nuclear power station in Rio de Janeiro, Brazil. Many people are alarmed by the spread of nuclear technology in the developing world*

Why the Third World needs nuclear power

When Munir Ahmad Khan left the International Atomic Energy Agency (IAEA) in 1972 to head Pakistan's Atomic Energy Commission, he promised himself to make Pakistan self-sufficient in nuclear technology.

A quiet, deeply religious man, Khan has earned a tremendous respect amongst his colleagues in Pakistan. He has also acquired a reputation for championing the cause of nuclear energy for the Third World.

But how does he justify nuclear power for developing countries when it is coming under increasing attack in the industrialised states? 'For many developing countries nuclear power is simply a matter of survival,' he says. 'We are pursuing a nuclear power programme not because of prestige or for political reasons. The developed nations have no conception of how the energy crisis has hit us – what it means for our future. In the US, an average citizen consumes some 10 000 units of electricity per year. In Europe, this consumption is between 6000 and 9000 units per capita. The world average is about 1600 units per capita per year and even in the poorer nations of Asia the average is more than 300 units per year. In Pakistan the average consumption of electricity is barely 160 units per person. Because of the energy crisis we cannot meet even this amount without going nuclear. It is easy to be against nuclear power if you are sitting in a fully lit, fully heated house in London or New York. The view is considerably different from where I am sitting,' he says.

Pakistan has few other energy sources. 'The US can consume Pakistan's entire reserve of coal, oil and gas in just over one year,' says Khan.

That leaves hydro-power, which, says Khan, 'is a major resource for Pakistan and many developing countries but is constrained by the total amount available, by its location, and by the rate at which it can be exploited. In Pakistan, most of the hydro resources are in the north and it would be very expensive to transmit their power to where it is needed. As the most favourable sites have been exploited the cost of constructing dams, most of which are in seismic regions, is becoming very high. If Pakistan used all its rivers to the full we would still meet only 30 per cent of our energy requirements.

'So, either we continue to import oil or we must rely to some degree on nuclear power. Last year we spent over $500 million on importing oil for industry and transport. The industrialised nations are threatening to invade the oil fields of the Middle East because they have to spend 10 per cent of their export earnings on oil imports. By 1985 we would have to spend 100 per cent of our export earnings if we did not have nuclear power. Can you imagine the catastrophe it will be for us?'

Khan argues that Pakistan is typical of most non-oil producing developing countries. Most of them are facing an energy crisis that could have serious consequences for their future; and for most of them the options are limited to either developing their own programmes of nuclear self-sufficiency or sinking further into dependency and bankruptcy. At almost every international gathering on nuclear power, Khan has pleaded that 'developing countries must be free to choose the course suitable for their needs and conditions for developing their peaceful nuclear energy programmes.'

But Khan realises that developing countries have to fight for this freedom. 'The developed countries operate a double morality on the issue of nuclear power,' he says. 'Nothing is being done to check vertical proliferation – the continuous research and development of more and more destructive nuclear weapons. While the industrialised countries have the right to develop nuclear technology and consider themselves sophisticated and responsible enough to continue stockpiling nuclear warheads, the developing countries have neither the right to develop nuclear technology nor are they considered responsible enough to safeguard their nuclear power plants adequately and use nuclear energy even for peaceful purposes.'

The oppressive hand of the rich countries

Limiting access to nuclear technology has become one of the most overt tools of domination, Khan argues. The developed countries are particularly anxious to preserve this domination and have become hypersensitive about nuclear co-operation between developing countries: 'the whole proliferation argument must be seen in the broader perspective of the North-South dialogue.'

Khan admits that a nuclear technology developed for peaceful purposes can easily be used for military purposes. 'We have been singled out for criticism because we have chosen to take the plutonium route to nuclear technology. As far as we are concerned, without plutonium there can be no efficient breeder reactor and without breeders the full potential of nuclear power cannot be realised. We need a reprocessing plant to enable us to recover unburnt uranium which can be reused in new power reactors. We do not accept the thesis that for economic reasons reprocessing, enrichment and fast breeder reactors should be confined to a few countries which either have large nuclear programmes or extensive uranium resources. Those who criticise reprocessing plants want us to depend upon them for the essential supplies of the reactor fuel. This is not acceptable. A power reactor is almost a billion dollar project and cannot be left to the mercy of others.'

The developing countries do not possess the trained manpower and the economic and technological backup to produce significant nuclear weapons. Such a programme would pre-empt all available resources badly neded for essential sectors. 'Since we cannot attain a credible nuclear deterrent, possessing a few nuclear weapons will not add to our security; it may in fact endanger it,' says Khan. He nevertheless admits that proliferation is a serious problem – 'but you cannot solve the problem by limiting nuclear technology to the privileged few,' he says. 'Nuclear technology has already spread and cannot be retrieved. Technical safeguards deserve our fullest support but they themselves cannot ensure non-proliferation because the incentive towards proliferation springs from insecurity and the political climate in which we live. To strengthen the non-proliferation regime we must also control the vertical proliferation, the stockpiling of more and more destructive nuclear weapons, which poses the most awesome threat to human survival.'

Khan's arguments made a great deal of sense to many decision makers in the Third World. He has found ready listeners among the member states of the Organisation of Islamic Conference (OIC). 'So far nuclear power has met with little resistance in the developing countries,' he says. He admits that 'as far as the safety and environmental impact of nuclear power is concerned, the developing countries do not demonstrate as deep an anxiety as is evident in the West. This may be partly because they do not fully comprehend the potential hazards and partly because their programmes are not large. However, after Harrisburg the educated elite and policy makers in Third World countries have become much more aware of what can go wrong and this is bound to influence their thinking.'

(*Source*: *New Scientist*, February 1981)

EMPLOYMENT

Much of the support for nuclear power in the UK, particularly among trade unions is based on the premiss that growth in energy consumption means economic growth and more jobs. Against this it could be argued that the UK economy is being restructured in such a way that machines replace workers and that greater energy consumption could equally well mean fewer jobs. While orders for large new nuclear stations may well keep employed a workforce which numbers around 30 000, there are those who argue that the modernisation of smaller coal fuel plants would preserve more jobs in the long term. The premature closing of older plant to make way for nuclear power represents a considerable threat to those who work in the coal industry and it is no accident that the miners' leader, Arthur Scargill, has been a prominent figure in the anti-nuclear campaign.

The electricity supply industry itself is increasingly capital intensive. The CEGB represents the largest single accumulation of capital in Britain with capital assets of £400 000 per employee in 1982. As these have increased, so employment in the industry has fallen. It fell 24 per cent between 1970 and 1982, while sales of electricity rose 18 per cent between 1970 and 1979 and then began to decline. New, highly automated generating plant together with productivity deals with workers, ensure that the CEGB can generate more electricity with a reducing workforce.

While the National Nuclear Corporation suggests that the building of the Sizewell PWR means work for 6 000 people in the construction and engineering industries, others claim that these jobs will be short term, and that the same investment in other energy sources and conservation projects would create far more employment.

So, again, we have opposing views in the nuclear power debate. Will its development provide more jobs or fewer jobs? We have now completed a survey of the views of the pro- and anti-nuclear lobbies in the economic, environmental and social fields. Finally, we need to look more carefully at who makes up these lobbies and what motivates them to hold these viewpoints.

PARTIES IN THE NUCLEAR DEBATE IN THE UK

THE PRO-NUCLEAR LOBBY

Supporters of nuclear power are to be found in the major political parties, in state-owned agencies such as the CEGB, in private corporations and amongst trade unionists. Although the strength of their commitment varies, they generally display a strong faith in science and technology and are inclined to equate human progress with economic growth.

While it is the Conservative Party in Britain which is strongest in its support for nuclear power, the Liberal Party has consistently opposed its development. The Party's 1980 assembly called for nuclear power to be phased out, but the subsequent Liberal/SDP Alliance found it difficult to agree a common policy. The Labour Party's policy has been strongly linked to shifting opinion within the trade unions.

It is as well to be aware of the alliance between big business and state capital when trying to understand the composition of the pro-nuclear lobby. The UK nuclear industry is an amalgam of public and private interest and its structure is shown in Figure 12.

The *UK Atomic Energy Authority* (UKAEA) was set up by the government in 1954 to undertake both military and civil nuclear

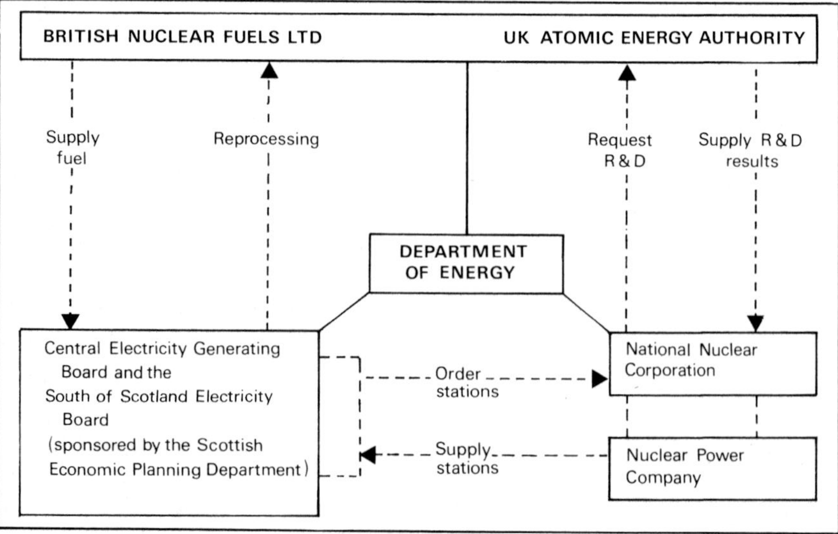

Figure 12 *The structure of the UK nuclear industry as laid down by a UKAEA document in the 1970s. The structure has changed since then.*

projects. It carries out research and development tasks for the industry, for example, problems such as waste storage and radiation damage are studied at Harwell. *British Nuclear Fuels Limited* (BNFL) is a state-owned company financed largely through the UKAEA which is now being prepared for privatisation. It is responsible for the manufacture and supply of uranium and plutonium fuels and the provision of such related fuel-cycle services as reprocessing. BNFL is involved in the European nuclear consortium URENCO. It has investments in many associated and subsidiary companies, and connections with such multinationals as Gulf Oil and Rio Tinto Zinc.

The two Electricity Generating Boards shown in Figure 12 are responsible for generating electricity and commissioning new nuclear power stations with the government's approval. These stations are designed and constructed by the Nuclear Power Company which is a wholly owned subsidiary of the National Nuclear Corporation. The NNC is comprised of the Government (35 per cent shareholding exercised by the UKAEA), the General Electric Company (30 per cent), and British Nuclear Associates, BNA (35 per cent). BNA is made up of many private, civil and electrical engineering interests including MacAlpine & Sons, Taylor Woodrow Construction, Babcock and Wilcox, and Clarke Chapman. A web of interconnecting financial and other business interests runs through the UK nuclear industry and there are joint directorships between its component companies. International links are already well developed for the supply and reprocessing of fuel and growing co-operation in the field of reactor design and construction seems likely. The UK industry has contracts to reprocess waste from Italy and Japan, and wishes to build pressurized water reactors of American design. While the industry argues that such co-operation is vital to provide economies of scale, its opponents point to the mounting problems of maintaining a democratic control over an industry which is owned by multinational capital which is so widely dispersed.

?

1 The essence of the radical left's view of nuclear power is that it is 'a continuing source of profit to private corporations, largely due to state subsidy'. They argue that this is where the power behind the pro-nuclear lobby lies – with multinational capital propped up by sympathetic governments. How far do you feel that the structure of the UK gives credence to this viewpoint?
2 The chart below shows the attitude some major trade unions adopted towards nuclear power in 1981. Try to explain each union's attitude to nuclear power.

For	Ambivalent	Against
Amalgamated Union of Engineering Workers	National and Local Government Officers' Assoc.	National Union of Mineworkers
General and Municipal Workers' Union	Assoc. of Scientific, Technical, and Managerial Staffs	National Union of Public Employees
Electrical, Electronic, Telecommunication, and Plumbing Union	Transport and General Workers' Union	Union of Shop, Distributive and Allied Workers
		Union of Construction Allied Trades, and Technicians
		Confederation of Health Service Employees

?

3 To what extent do you think Frank Chapple's and Arthur Scargill's views reflect the interests of their union's membership?

Life has always, and will always, carry risks. High technology has reduced the risks and improved the quality of life. Nuclear power, far from being a threat, is our only guarantee of survival beyond the year 2000, unless a greatly reduced population chooses to return to subsistence farming. If emotion and hysteria are allowed to delay our existing nuclear programme, we not only limit our choice but reduce the possibility of ensuring a safer nuclear industry.

(*Source*: Frank Chapple (EETPU), *The Observer*, 28 September 1979)

My opposition is based on the three main arguments.

(i) Production of nuclear power puts at risk tens of thousands of people through the possibility of an accident or as a result of nuclear waste dumping.
(ii) Cost – In 1978 electricity produced by AGR nuclear power stations cost 1.52p per unit compared with coal at only 1.23p per unit.
(iii) We have sufficient reserves of coal for one thousand years – enough time to provide viable alternatives – such as solar, wind, wave and tide power. We can afford to wait.

(*Source*: Arthur Scargill (NUM), *The Observer* 28 July 1979)

THE ANTI-NUCLEAR LOBBY

The anti-nuclear power lobby in the UK consists of people of all political persuasions and many who suggest the issue is 'above politics'. Many of those who are opposed to nuclear power are, however, of the Left and Centre and they are attracted by the activities of such organisations as Friends of the Earth, Greenpeace and the Ecology Party. While early opposition focused largely on probable environmental damage, it now includes a wide range of economic and social arguments and stresses the benefits of alternative energy policies.

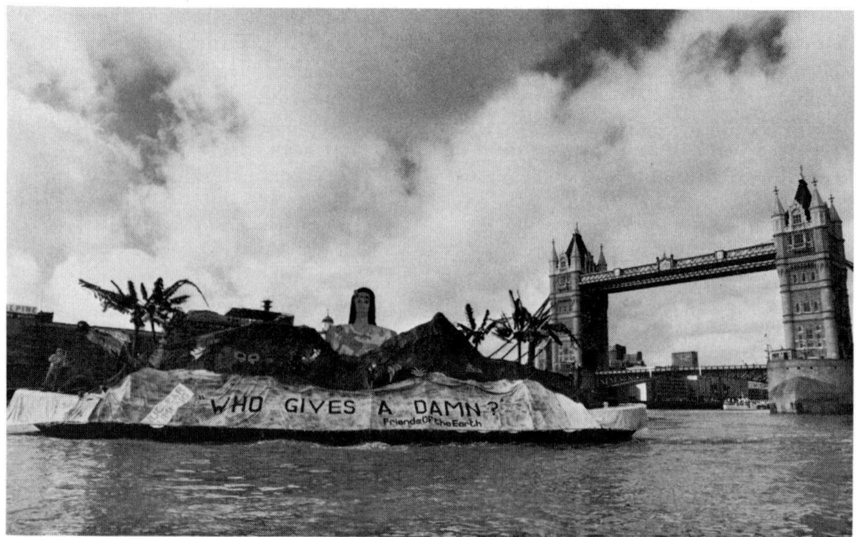

Print 8 *Anti-nuclear pressure groups have been more and more active in recent years — a Friends of the Earth barge in the River Thames protests against the dumping of nuclear waste by European countries in the South Pacific*

? **1** Before the general election of June 1983, Harford Thomas reviewed the manifestos of the major political parties for his environmental column 'Alternatives Notebook' in *The Guardian*. Below is printed an extract from his article. Based on information in this article, write very short speeches for a Tory, an SDP/Alliance and a Labour candidate. What are the basic differences in viewpoints? Using knowledge gained from the rest of the chapter, can you explain these differences?

The Tory manifesto is frighteningly complacent. Its energy section starts by boasting that 'Britain has come from nowhere to be the world's fifth largest oil producer'. Back to nowhere how soon? For a Government which has refused to impose a depletion policy to extend the life of the North Sea fields there is naturally no answer. It just crosses its fingers and says it will encourage development in the North Sea.

And what else? The coal industry must 'return to economic viability' – how is not explained; one is left to wonder whether this is a disguised promise to privatise. The government will 'press ahead' with nuclear power (that means a bit more a lot later, not before 1995 at the earliest). There will be a new Energy Efficiency Office 'to coordinate the conservation effort' (an effort currently treated with indifference and starved of funds).

And that's it. It takes an amazingly skimpy 16 lines to state the Tory energy policy in a manifesto of 47 pages. It does not get a mention in Mrs Thatcher's foreword.

Roy Jenkins and David Steel in their foreword to the Alliance manifesto emphasise that energy conservation will create jobs. The manifesto gives first priority in the energy section to conservation and efficient use of energy, and says that it will give 'urgent priority' to investing in new energy technologies, naming sun, wind, waves, geothermal sources, and combined heat and power systems.

In the Labour manifesto, energy has no place in Michael Foot's foreword, and it features only as energy conservation in Labour's emergency plan of action. In the main section on energy it promises 'a massive conservation programme,' combined heat and power schemes, and 'greatly increased spending on the development of renewable sources.' It proposes an annual review of a comprehensive energy plan.

Both Labour and the Alliance say no to Sizewell and the PWR; the Alliance would continue nuclear power research, and Labour would 'reassess' nuclear power.

The litmus test issue in these programmes is the attitude to renewable energy. It is ignored, without mention, in the Conservative manifesto. Both Labour and the Alliance are encouragingly positive.

(*Source*: *The Guardian*, June 1983)

? **2** Look at the extract from the Ecology Party's manifesto also printed below. Contrast the Ecology Party policies with the viewpoints of the main political parties. What values do they seem to emphasise which other parties seem to gloss over?

Ecology policies:

1 NORTH SEA OIL

Considerably reduce the present rate of extraction. Such an asset is primarily important for its use in derivative industries, rather than as fuel. It must not be squandered merely to conceal the chronic imbalance in our trading position.

2 NUCLEAR ENERGY

No further nuclear power stations will be built. They are prohibitively expensive, unsafe and unnecessary. Development of the reprocessing plant at Windscale should be stopped immediately. As recent reports have indicated, the so-called 'energy gap' simply doesn't exist, and this makes plans to develop the fast-breeder reactor all the more absurd. The security risks are enormous and will jeopardise basic civil liberties. Above all, we still don't know what to do with the waste products. We are not prepared to commit future generations to solve a problem which we ourselves cannot.

3 ALTERNATIVE ENERGY SOURCES
Research into energy from sun, wind and wave must be dramatically stepped up. Choose technologies which encourage local autonomy and initiative, which are non-polluting and renewable, using the earth's 'income' and not its 'capital'.

4 CONSERVATION
Treat this as the best 'energy source' of all. Industry could manage a 40 per cent reduction in its energy requirements without any loss of output. A national insulation programme would provide many jobs. We simply must achieve a steady decline in our overall consumption of energy.

(*Source*: the Ecology Party Manifesto, June 1983)

AN EQUAL DEBATE ?

The capital intensive nature of nuclear power inevitably means that it finds strong support in some branches of industry and the financial world. Such support is able to exert great pressure on government and some suggest that this, more than any other factor, explains the pro-nuclear policies of some governments, for example, the present Conservative government. The State's role in the development of nuclear power, and the huge sums of money already committed, would however make it very difficult for any government to rapidly abandon Britain's programme.

A further concern of many in the anti-nuclear lobby is that parliament has never had a full debate on overall energy policy. Current sources of energy supply have developed in a piecemeal way, and while the Department of Energy produces plans and forecasts for the future; these are rarely subjected to full public scrutiny. Many important announcements of energy policy are given as written answers to MP's questions and cannot be fully debated. Some opponents of nuclear power have become frustrated in trying to use supposedly democratic channels to seek information and challenge policy. They compare Britain with countries like Sweden and Austria where nuclear power has been subjected to wider debate and referenda, and its development eventually slowed or halted.

Since Britain has no national energy policy and there is little discussion of energy policy in parliament, public enquiries, such as those at Windscale (now named Sellafield) in 1977 and Sizewell in 1983–85, have become the major forums for nuclear debate. Set up by the Minister of State to advise on whether a development of major significance should be given planning permission, a planning inquiry has terms of reference and procedures which may inhibit full debate. However strong the local opposition to the proposal, there is always the possibility that planning permission will be granted by reference to national need (see the extract from *The Times* below). A planning inquiry cannot determine this need and some therefore consider it to be simply a form of token consultation. The anti-nuclear lobby feel that the failure of the government to fund objectors with public money adds support to this view.

Fairness of Sizewell inquiry studied

When the Sizewell B inquiry resumed on its 258th day yesterday, it became the longest-running public inquiry in British history.

In addition the cost of the hearing had risen to more than £25m, Dr Raymond Kemp reported in a paper to the Sociology section of the association.

Dr Kemp, of the Centre for East Anglian Studies, at the University of East Anglia, Norwich, is investigating the effectiveness of the inquiry; especially its acceptability for resolving local issues.

He said that most inquiries routinely took place expeditiously, with a minimum of controversy, and placed relatively few demands upon those participating. That was not true of the Sizewell B inquiry.

The hearing took place because of the Three Mile Island incident in the United States which raised public fears about the safety of the pressurised water reactor type of nuclear power station. The strength of concern expressed by local councils close to the Sizewell site was demonstrated by the fact that Mr Stephen Reed, the mayor of Harrisburg, would give evidence on their behalf.

Dr Kemp said calling Mr Reed from the United States encapsulated a central feature of the Sizewell controversy. How could the fears and legitimate interests of local residents opposed to such projects properly be reconciled with national policy proposals?

Was it possible, first, for the inquiry to be 'full, fair and thorough' as the Secretary of State for Energy promised, and second, in what manner were local interests and concerns being addressed by the inquiry? Dr Kemp asked.

He said the Sizewell inquiry was not a normal hearing. The issues were extremely complex and the disparity of resources between opposing sides was very marked.

Consequently, in the relative absence of the more usual adversarial conflict of views, the Sizewell inquiry had actively begun to pursue issues on its own behalf, by actively investigating key topics.

It had become increasingly necessary for the inquiry to take the initiative in testing the Central Electricity Generating Board's case for the PWR. Dr Kemp said the Sizewell proposal was an important watershed in the development of the British nuclear power programme and construction industry. If the PWR proposal were accepted, then that reactor type was most likely to be the established design for future British nuclear power stations.

If it were rejected, then the future of the PWR and possibly the advanced gas-cooled reactor might be thrown into considerable doubt.

Dr Kemp said that those issues were of prime concern to the inquiry, but in attempting to be 'full, fair and thorough', it was possible that the interests of the local population might be overlooked.

(*Source*: *The Times*, 12 September 1984)

SIZEWELL INQUIRY

Many of the opponents of the PWR believe that, apart from delaying the £12 000 million project and therefore increasing its cost, the public inquiry has done almost nothing to influence the outcome. Given the Government's determination to proceed with a substantial nuclear power programme as an insurance against future miners' strikes, and the enthusiasm of the CEGB chairman, Sir Walter Marshall, for the American PWR, they argue cynically that the result has always been a foregone conclusion.

It is true that by the likely date of the next general election, when a government more sceptical of the PWR's attractions might be voted in, the Suffolk power station could be beyond the point of no return.

One of the notable events of the inquiry session just finished was the generating board's announcement that £12 million of key reactor components were about to be ordered, with the Government's approval. By the autumn of 1987, more than £500 million will have been spent.

(*Source: The Guardian*, 4 August 1984)

Should objectors at the Sizewell Inquiry into Britain's first PWR be funded from the public purse? This was the question facing Energy Secretary, Nigel Lawson, prior to the launching of the Sizewell Inquiry in the autumn of 1982. Groups such as Friends of the Earth and the Council for the Protection of Rural England estimated that they needed hundreds of thousands of pounds to fight the proposal and they resented the millions of pounds which the government and CEGB would be able to spend on what was likely to be a long inquiry. They drew attention to countries like Canada where objectors are granted financial assistance.

When the government refused help, several of the objectors, such as Greenpeace and the Political Ecology Research Group, withdrew. Others, like Friends of the Earth, were left to fund their opposition from appeals and jumble sales. The following article is taken from an FoE newspaper of spring 1982:

Lawson told: 'There can be no satisfactory outcome'

Nigel Lawson, Secretary of State for Energy, has been told by FoE that his refusal to publicly fund objectors at the Sizewell PWR Inquiry means that there can be 'no satisfactory outcome for you'.

In December Des Wilson wrote in his capacity as FoE Chairman to Lawson to say that in the event of a refusal by the Minister to enable a genuine Inquiry to take place 'it will become increasingly difficult for responsible leadership to argue with those forces who say that they had a valid input to make but were denied the opportunity to make it; that the opportunity for participation in decision-making on issues of this sort is theoretical but doesn't work in practice; that in our society today only money can buy influence and the opportunity to participate in major Inquiries.'

On the eve of the Inquiry, Lawson replied to say:

'Objectors, in common with others, can reasonably be expected to meet their own costs. I do not accept that 'the dice are loaded' deliberately and overwhelmingly in support of a Government position, or that there is no longer a genuine opportunity for public participation in this Inquiry. A great deal of information has been made available, and the Inquiry is aimed at ensuring that every aspect of the CEGB's application is examined and that there will be every opportunity to question it. I am pleased to know that FoE believes it should work to achieve its objectives through reasoned argument. I am sure you will join with me in expressing the hope that the Sizewell Inquiry will be conducted in just such a spirit and in deprecating any action to the contrary.'

Des Wilson replied:

'Of course we would like the Inquiry into the Sizewell PWR to be conducted with reasoned argument – we would also like it to be fair, and peaceful. It is because of this that we urged public funding of the objectors in the same way as your position and that of the CEGB is publicly funded.

'I have made it clear to you, and to the Inspector,that while FoE will be participating in the Inquiry as best it can with the financial resources at its disposal, as are a number of other environmental organisations, we do so under protest at the circumstances of the Inquiry and at serious disadvantage, and therefore we cannot accept the Inquiry as a fully genuine attempt to reach the truth. We therefore reserve our right to reject its predictable findings on the grounds that it was financially rigged to achieve your Ministerial objective of forcing upon the British people a form of energy provision that is unnecessary, economically insupportable, and unsafe.

You have undermined the integrity of your Inquiry before it even starts, despite every attempt to warn you of this, and you must know that there can be no satisfactory outcome for you, no matter what the Inspector decides.'

(*Source*: FoE Newsletter, Spring 1982)

?

1 Why does FoE suggest that there can be 'no satisfactory outcome' to the Sizewell Inquiry as far as the government is concerned?

2 What grounds does Nigel Lawson give for rejecting FoE's request for public funding?

3 Why might FoE and the Secretary of State hold different views on what is likely to take place at the Inquiry?

Remember Mr and Mrs Brewer and their windmill? Can you now understand better why they acted in the way they did? Were they right to do so? Was it likely that their action would have any direct effect in halting the expansion of nuclear power?

You have now considered some of the many environmental, economic, and social issues related to nuclear power and are therefore better able to make your own decision as to its desirability. It has, of course, not been possible to present all the evidence in this chapter and you may feel you want to find out more before making up your mind. Having weighed up the evidence, in the light of your own values and priorities for society, you will find yourself either supporting or opposing the case for nuclear power.

PART V MANAGING AND MISMANAGING THE NATURAL ENVIRONMENT

NATURAL ENVIRONMENTS — A CHALLENGE FOR PEOPLE

DEVELOPMENT AND DESTRUCTION OF RESOURCES

ASSAULT ON THE EARTH

The world environment is caught in a tightening vice. On the one hand the Earth is squeezed by massive overconsumption and waste; and on the other by the Third World's poor who destroy their resource base just to stay alive.

Cubatão is not a typical Brazilian city. For one thing the Mayor refuses to live there. And for another the small community has the highest per capita income in the country. That is because the city is one of Latin America's largest petrochemical centres with more than 24 major industrial plants.

Many are Brazilian owned but there are also international giants like Dow, Du Pont and Union Carbide. Each spews out its own particular noxious and deadly effluent. The smoke is many colours, but the sky is a uniform grey. Environmentalists in Brazil call Cubatão the 'valley of death'. Pollution has poisoned the birds and most of the insects. The four rivers that dissect the city are lifeless, chemical sloughs — fish from the nearby ocean outlet are deformed.

The residents of Cubatão are just beginning to see the effects of their nation's wink-and-nod approach to industrial polluters. Miscarriages, still-births and birth deformities have suddenly shot upwards causing panic and fear. Respiratory diseases like tuberculosis, pneumonia, bronchitis and emphysema are the norm. Unlike the Mayor (and despite the city's reputation of prosperity) most of Cubatão's citizens can't leave. More than one-third of them live in the tumbledown slums which are cheek-by-jowl with the offending industries. For them the Brazilian miracle is a mirage destroying their community, their health and the world around them.

As the Third World strains to pull itself out of poverty by a quick injection of industrialisation, there are other 'Cubatãos' waiting in the wings. But myopic governments and polluting industries are not confined to the Third World. Nor are they especially new.

Writers from Engels to Dickens have described the 'dark, Satanic mills' of the industrial revolution in Britain. And James Nasmyth, inventor of the steam hammer, wrote this description of the town of Coalbrookdale in the 1890s:

> *'The grass had been parched and killed by the vapours of sulphureous acid thrown out by the chimneys; and every herbacious object was of a ghastly grey — the emblem of vegetable death in its saddest aspect'*

Despite sporadic warning signals of ecological collapse, until the mid-1950s it was assumed that economic growth geared to rapid industrialisation was the *only* way in which human welfare could be improved.

The natural environment was assumed large enough and flexible enough to absorb all the depradations of pell mell growth. Not so. In fact, about the time the post-war consumer society was coming into its own the Earth was showing definite signs of fatigue.

Despite obvious material wealth many in the developed world began to look at the other side of the ledger. A century and a half of economic growth had increased material prosperity. No question of that. But even in the richest countries in North America and Europe poverty was not abolished. And the focus on disembodied economic statistics camouflaged more important concerns about the quality of life.

Although people were in general richer they weren't necessarily much happier. Then, too, there were the mounting costs of sullying the natural environment and attempting to live amidst the poisonous wastes of our own prosperity.

(*Source*: *New Internationalist*, August 1982)

The extract above contains ideas with which you are probably familiar. The destruction and poisoning of our environment often makes headlines, of which the latest concern, acid rain is another example. The development of resources and their destruction go hand-in-hand. The first chapter in this section describes resource use and environmental conflicts in Japan. It is an introduction to the very real idea that bringing about a change in our lived-in environment will also make its impact felt on other parts of our environment — impacts which are often difficult to trace as the investigation of the itai-itai disease shows.

? After you have worked through the chapter, *Environmental Conflicts in Japan — Production versus People*, come back to the extract from 'Assault on the Earth' and reread it. Then add a paragraph, written in similar tone and style, which uses the conflicts and connections between production and people in the Japanese environment to illustrate and reinforce the author's point that there is another side to the ledger of economic growth geared to rapid industrialisation.

ECOSYSTEMS IN DANGER

RELATIONSHIPS AMONG VARIABLES

The world over, a wide range of costs and benefits result from people living, working and playing in their various environments. Consequently, it is very important that we try to understand fully how all the parts of these environments fit and work together. In other words, how each part, or variable, impacts on the others either to maintain a smoothly functioning system or to throw the system out of normal working and thus to change the relationships between variables, possibly for the worse.

Below are extracts from a short article 'Highlands and Islands: ecosystems in danger', which describes how changes in one variable — population movements, in and out of an area — bring in their train other environmental changes. Keep a list of these as you read through.

HIGHLANDS AND ISLANDS: ECOSYSTEMS IN DANGER

In the Alps, population movement has been in two directions. The traditional mountain populations are abandoning some of the higher altitude areas, while adjacent lower areas are becoming comparatively crowded. But the main cause for concern is the transformation of what was formerly a relatively stable agricultural and pastoral system due to seasonal inflows of people from outside the area on such a vast scale that the carrying capacity of these areas may well be exceeded.

These inflows are made up of summer and winter tourists and weekend visitors from the densely populated neighbouring regions. To these must be added owners of second homes who often become permanent residents upon retirement. Many traditional villages, Alpine pasturelands and, indeed, whole valleys can no longer meet the demands of this additional population for avalanche-safe building land, water, roads, energy supplies, ski-lifts and ski-runs. There is also the risk that excessive population concentration may lead to water, air and land pollution.

Certain areas may risk losing their attraction for summer tourists if landscapes are more marked by ski-lifts and ski-run erosion than by the beauty of the traditional Alpine landscapes, and even winter tourism will be discouraged if the areas become too crowded, too built-up and if traffic on the mountain roads rivals that in the towns.

Agriculture, too, is very much affected — not only in tourist areas but throughout the Alpine region. New attitudes towards agricultural work and rural life are as important as the profitability of agricultural and pastoral use of steep and isolated mountain lands in provoking the abandonment of certain areas, or the appearance of phenomena such as undergrazing. As fewer and fewer pathways for the herds remain available, the old practice of transhumance (the moving of cattle from winter to summer pastures) is disappearing in the western Alps. One immediate side effect of this abandonment of grazing land is that changes in the vegetation cover create surfaces more conducive to avalanches.

The direct effects of the massive increase in the numbers of skiers over the past decade must also be taken into account. Aerial photographic surveys have shown how much the impacting of snow by thousands of ski descents every winter hinders the run-off of water in the spring. Vegetation is destroyed by the cutting edges of skis and it has been found that along the ski-runs the insect population is decimated and the number of earthworms per square metre along some trails, scientists have discovered, has been reduced from 130 to ten.

The village of Obergurgl, at the heart of the Oetz valley in the Austrian Tyrolean Alps faces many of the problems affecting Alpine areas described above. Situated at an altitude of about 2 000 metres, the village draws an ever-increasing number of summer and winter tourists. The first signs of the environmental consequences of this development expressed largely in terms of hotel and ski-lift construction and a sharp increase in the presence of people (and cars) in the village and its surroundings, are beginning to appear. The people of Obergurgl (300 inhabitants) perceive these as negative, and fear that eventually tourists might stay away from their village because they no longer find it attractive. Another potential limit to this type of economic growth is simply the availability of avalanche-safe land.

The MAB* National Committee of Austria, following proposals put forward by researchers of the University of Innsbruck and working in collaboration with the International Institute of Applied Systems Analysis in Laxenburg, Austria, chose Obergurgl as a site for intensive study.

The disciplines working together in this project included meteorology, botany, zoology, microbiology and pedology, as well as economic geography, anthropology, sociology, regional planning sciences and economics. The great success of the project lies, however, in the fact that it was possible to involve in its work not only the research workers, but also those actually concerned by the research and in need of the information generated by it.

* *Unesco's Man and the Biosphere Programme*

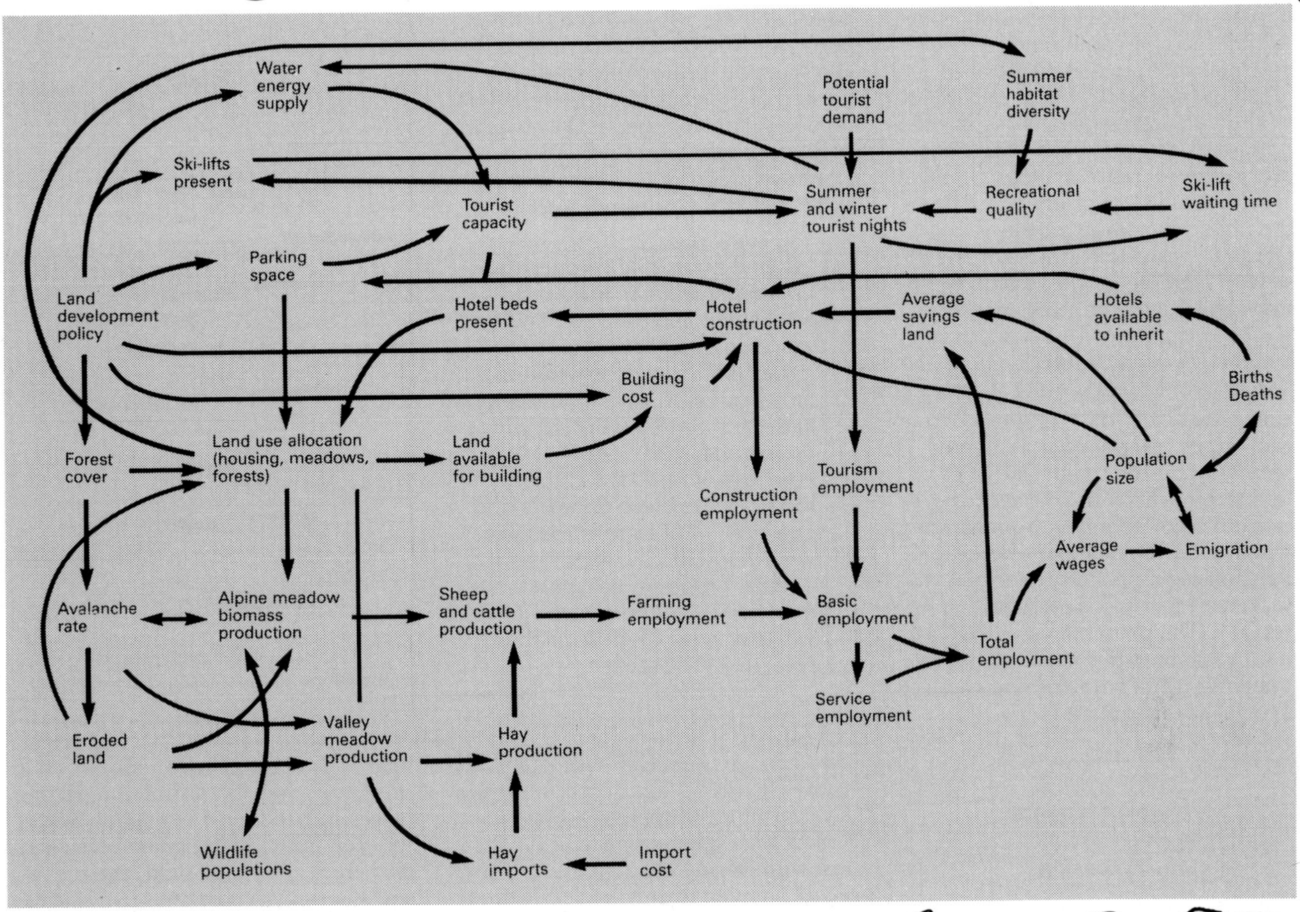

The diagram, above, shows the many interlinked factors which have to be taken into account in planning for the balanced development of even a small, well-defined community like the village of Obergurgl in the Austrian Alps. Alteration of any one of these factors may have complex and sometimes unexpected repercussions on the others. To take the simple example of ski-lift waiting time — if winter visitors have to queue up for more than a certain length of time, they will stay away in future years; if more ski-lifts are built, more winter tourists will come, and more hotels will have to be built, thus reducing the land available for cattle grazing and the production of forage; summer tourists will stay away as the scenic attractions of the area are increasingly marred by unsightly ski-lifts and additional building; summer employment in tourism will fall and emigration will increase.

(*Source*: *UNESCO Courier*, April 1980)

Fortunately, most natural systems are self-regulating. That is, they are able to respond to changing circumstances by adjusting the way in which they function. The result of this adjustment is usually the return to 'normal working', although some systems will take much longer than others to achieve it. If self-regulation is to occur, a system must be able to check or correct itself. For this to happen, at least one part of the system must be able to compensate for the behaviour of another part. This compensation is known technically as 'negative feedback'. Without it there can never be a return to normal, and self-destruction is virtually inevitable.

A simple example of a self-destructive relationship between three parts of the soil-slope system is shown here:

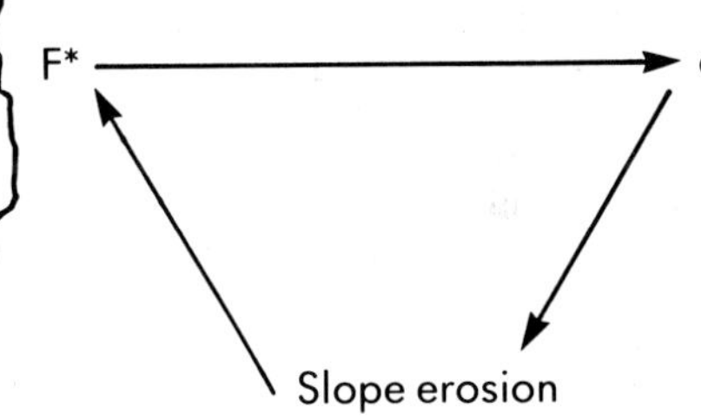

Suppose that a decrease in the infiltration capacity of the soil (F*) causes an increase in the amount of surface water running-off the soil (q). In its turn, this increased run-off causes an increase in the erosion of the slope leading to the removal of its soil in the process. Removal of the soil creates a further decrease in infiltration capacity (less soil, hence less capacity) resulting in even more run-off and more slope and soil erosion. In other words, the cycle begins again. Such behaviour exhibits no self-regulating mechanism. In fact, it displays 'positive feedback' where one initial change (the decrease in F*) sets off a pattern of activity which becomes progressively more acute.

Where physical or natural systems are linked to human activity systems — as, for example, in the case of trampling and erosion — the risk of positive feedback occurring is much increased. This is because human activity may interfere with the normal working of a natural system and prevent negative feedback from playing its compensating role. A very long period of time may be needed for recovery, if it is at all possible. For example, poor farming techniques can devastate an area if they initiate a positive feedback loop which causes severe soil erosion. The problem is how to manage linked physical and human systems so that a destructive chain of events is not set in motion.

Change in environmental quality — decrease in vegetation, increasing erosion.

Feedback loop

Change in population density leading to trampling

INVESTIGATING ECOSYSTEMS

In *Managing Recreational Areas*, Carolyn Harrison defines and describes the concept of an ecosystem and explains how the effects of trampling, very simply illustrated in the diagram, can be investigated scientifically (i.e. in an ordered, logical way) by path studies and field experiments. It is important to become aware of the aims of path studies and the methods used and also the aims of the field experiments. The way such studies and experiments are set up have been described carefully here particularly to give you an insight into scientific design and study. At the end of the chapter, you should be able to set out in a general way how you would go about studying the effects of grazing or burning, for example, on a moorland area.

? When a class of fifth years were presented with the following worksheet and data relating to land use and traffic management in Borrowdale, they produced varying answers to the problem, several of which are reproduced. After you have studied the chapter look back to these answers and consider what suggestions you would add, if you were answering a similar type of question on management. You will also have ideas from the role play to consider.

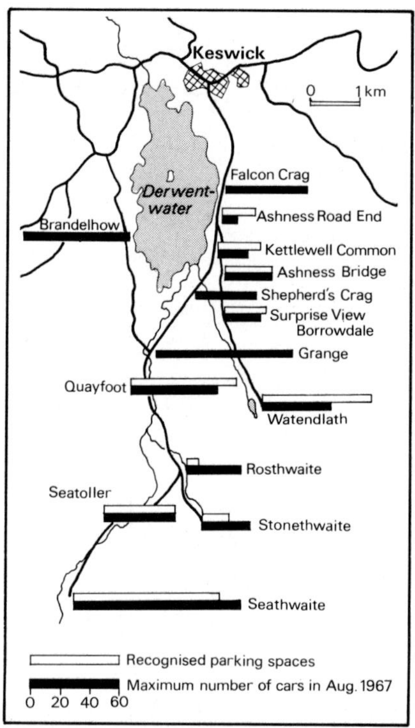

The average weekend visitor to the Lake District — and especially the day or half-day tripper — is less concerned with open-air recreation than he is with driving about in his car. This seems to emerge pretty clearly from the painstaking report just published by the Countryside Commission which indicates, among a mass of other details, that there can be more than 50 000 tourists moving around in 16 000 motorcars on peak days in the summer time. Almost half the day or half-day trippers admitted in answering the questionnaire that they were 'just driving around', but it was in detailing their leisure activities during their short stay that the visitors, departing by car, proved most revealing. Forty-six per cent had spent their day picnicking in or around their cars, whereas only 15 per cent seem to have used their legs for more than one mile. (This was the count for day visitors; only four per cent of the half-day tourists went walking or climbing but perhaps they hadn't the time.) But only seven per cent of the day visitors went boating or sailing, five per cent fishing and two per cent swimming, although 28 per cent at least got some fresh air in 'sitting by the lakeside'.

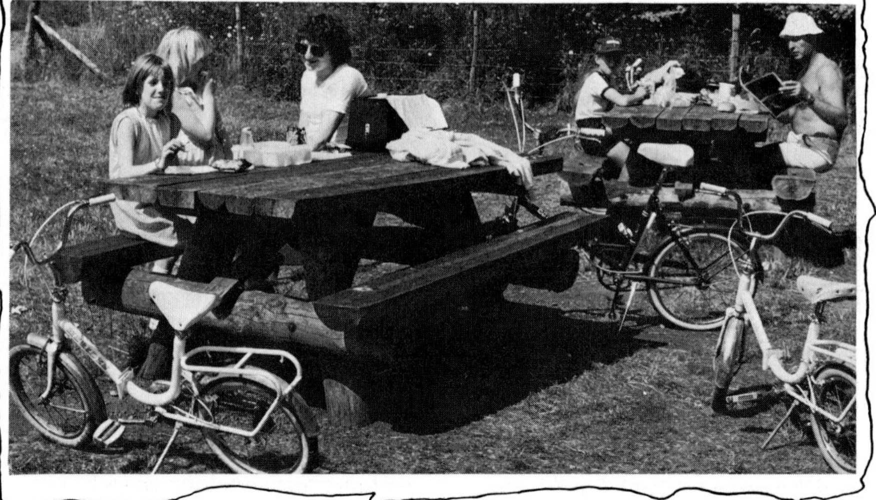

WORKSHEET

LAND-USE AND TRAFFIC MANAGEMENT IN BORROWDALE

1. Why exactly is there a problem in Borrowdale? (e.g. Is it only an occasional problem? Whose problem is it?)
2. What are some of the different ways of dealing with the problem? (Give at least three alternative suggestions).
3. Which of these suggestions do you think is the best way to dealing with the problem? Consider:

   a) What is practical, e.g. cost;
   b) The views of people using Borrowdale for different purposes;
   c) What you think really matters most.

Three answers to Question 2

1. Pupil A

   Three ways of dealing with the problem:

   a) Build up fences, causing a little inconvenience to the tourists.

   b) Try and get a toll road, to make the tourists pay to drive on particular roads.

   c) Get rid of all the old people who complain about the inconvenience of it all.

2. Pupil B

   a) I think perhaps if there were car parks going off the motorway at different places there would be no parking problems at the different villages.

   b) A toll gate might be a good idea to pay for damage the tourists do. The people who lived in the village could appoint someone who knew the villages to take the money so that the villagers did'nt have to pay to get in and out of the village.

   c) If you leave it to work out on its own maybe it will. Otherwise you could have litter bins for rubbish and fines for any offenders.

3. Pupil C

   a) They could build more car parks, but there might not be anywhere to put them without spoiling the landscape and interfering with the farmer.

   b) An underground car park could also be built. It would solve the problem about spoiling the scenery etc., but the cost might be too great.

   c) A toll gate could be set up to get money to repair the damage the tourists do. This could cause a problem for the people living there: they wouldn't want to pay to get home to their villages and even if passes were given they could lose them.

UNDERSTANDING A RIVER SYSTEM

Understanding how all the variables in a river system work and how the variables mesh together is obviously very important if we are to manage drainage basins and water resources in a constructive way. The need to manage water resources is bound up with our need to appreciate that the activities of people in urbanising and farming more and more intensively in drainage basins — often changing infiltration rates and other variables — has upset the balance of nature and very often introduced a destructive relationship into the system.

For example, when soil and vegetation are covered with more and more concrete and tarmac, what effect does this have on the permeability and infiltration capacities of the ground? And what link is there between infiltration and run-off, between storage capacity of the soil, interception by vegetation and incidence of flooding? The graphs below should help you in starting to sort out these relationships. Look at the graphs closely to see what they suggest. What factors other than rainfall control flooding — intensity, rate of rainfall, water already in storage ? Describing and explaining the linkages between variables in a drainage basin is the main purpose of the third chapter in this section, *Managing Drainage Basins*. Once the linkages are understood, drainage basin management can be appreciated as a complex undertaking.

? Add another paragraph to 'Assault on the Earth' after studying the chapter. Keep in mind the idea of a ledger of credits and debits in the conflicts between economic growth and our natural environment ?

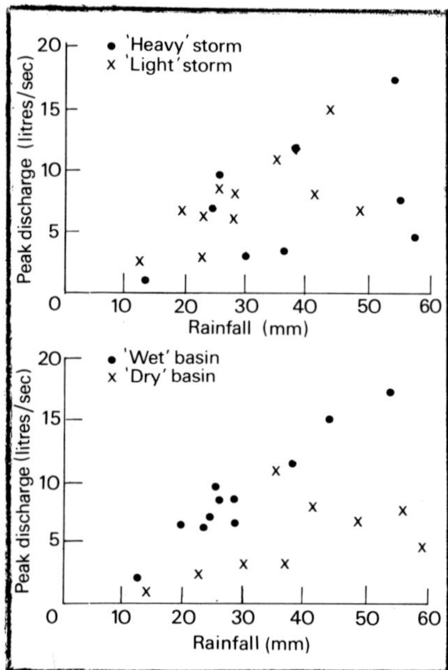

CONFLICTS OVER WATER

Water is certainly a resource we have long taken for granted. However, it is now becoming a significant export commodity. The demand for water is outstripping its supply in many parts of the world (see article on *page 424*). It is not surprising that conflicts can occur over water and rights to its use — who has the right to use and benefit from available water? *The US Midwest — The Problem of Water Allocation* examines water management at three levels, the individual, regional and international. The water in the Ogallalla aquifer in the American West is being used up at ever increasing rates. How is the water supply to be replenished or augmented? Different decision-makers have different goals and different ideas about water management as the decision-making exercises in the chapter are designed to show.

? After you have worked through these chapters, ask yourself, 'Are human systems less complex than physical systems?' If you constructed diagrams of the relationships detailed in these chapters you could begin to give an answer. Consider the variables which enter into human systems which are not present in physical systems. Is it more helpful to think about the linkages between physical and human systems so that we think in terms of physico–human systems and do not separate the one from the other?

PHYSICAL SYSTEMS AND HUMAN SYSTEMS

Compare the focus of *Managing Drainage Basins* and *The US Midwest — The Problem of Water Allocation* — the focus of the first is largely on understanding the interlocking of variables in a physical system; the focus of the other is on the debate and exchange which characterises human systems held together or held in conflict by competing economic, social and political variables reacting to a situation of resource scarcity. We most often think about these reactions, and the solutions and compromises reached as ways of *managing* the resource and the physical and human systems it is linked into. Are we good managers? Who benefits from our management? Again go back to 'Assault on the Earth'. We need to remember that human systems have human variables operating in them. The most significant and often the most difficult of these to deal with are the state of knowledge and the attitudes and values which people hold. These attitudes and values — the drive for economic growth valued by Japanese society; the profit-making attitudes and values of government and farmers in the Pennine area of the Yorkshire Ouse Basin; and the attitudes and values, the perceived self-interest of farmers, regional and national interest groups in North America, for example, give rise to some very sticky problems.

WEATHER AS A RESOURCE TO BE MANAGED

We are used to thinking of the air as a resource, water resources, food resources and people as resources. Do we often think about our *knowledge* of how drainage basins function or how trampling affects plants as a resource? Our knowledge is in fact a resource which we can draw upon and use. In the last chapter in this section, *Weather Impacts and Weather Management*, weather is viewed as a resource and information and knowledge about the weather is equally viewed as a useful resource for people. The author, John Maunder, a pioneer in this field, considers that, given an understanding of physical-biological-sociological interactions with the atmosphere, and given sufficient information about atmospheric events, people can, at times, use their management abilities *to improve the economic and social outcome of many weather sensitive activities*. The chapter stresses the kinds of calculations which need to be made in order to give us weather sensitive information, e.g. the calculation of an export climate sensitivity index, the compilation of commodity weighted information, of weather forecasting models and of weather/climate sensitivity charts.

We have in this final chapter of the section a rather new angle on how people and environment interact with consequent physical (soil erosion due to prolonged drought), economic (loss of profit) and social (lower incomes, loss of earning power, starvation) impacts — highlighting a see-saw relationship between people and nature which can be assisted by increased knowledge and information.

? When you have studied all the chapters in this section, you should take the last statement as a yardstick and evaluate the issues in each chapter from that point of view. What place does knowledge and information have in a people/environment equation?

18 ENVIRONMENTAL CONFLICTS IN JAPAN: PRODUCTION VERSUS PEOPLE

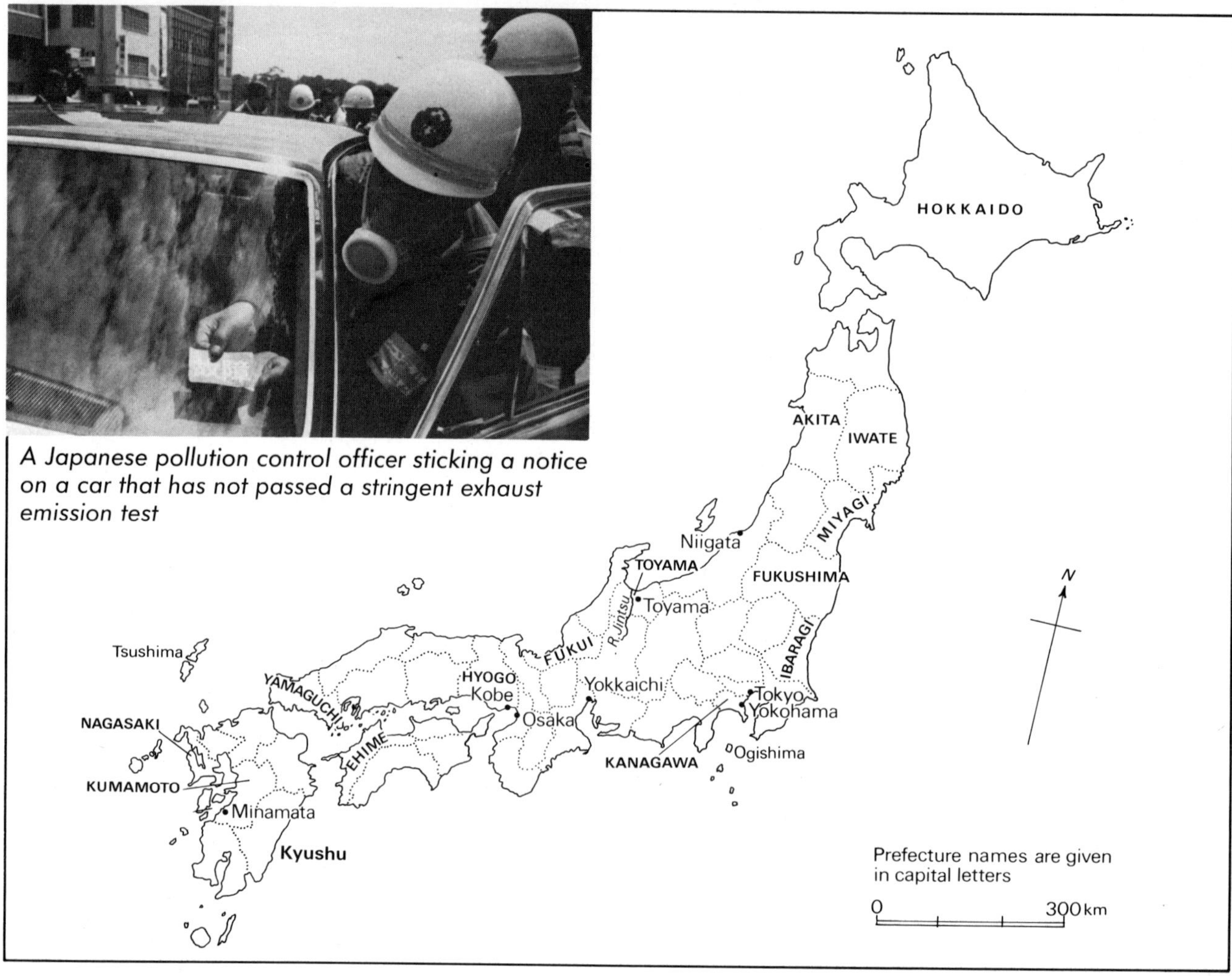

A Japanese pollution control officer sticking a notice on a car that has not passed a stringent exhaust emission test

ENVIRONMENTAL HAZARDS AND CONFLICTS

Geography takes a considerable interest in the environmental hazards which affect people all over the world. Some hazards are natural (for example, typhoons), while others result from human activities (for example, acid rain). In this chapter, we will see how one country, Japan, has created some of its own environmental problems, but has also managed to find some solutions.

During the 1970s, the international press gave a wide airing to Japan's pollution problems, as the following headlines and quotation suggest.

Dr Murata, a medical specialist at Toyama City, has among his patients one middle-aged woman whose whole body is twisted like a corkscrew. Her frail, useless limbs are permanently bent at awkward angles, while her hands are shapeless knots. She is now confined to bed, in constant agony. Her bones have become as brittle as dry twigs, and even a slight cough is enough to break a rib. At the last count, X-rays of her body revealed over 70 bone fractures. Dr Murata says there is absolutely no hope of any cure in this case. Death can only come as a merciful release.

DECLARING WAR ON POLLUTERS

JAPAN – AN OCTOPUS EATING ITS OWN TENTACLES

THE AWFUL COST OF GROWTH

Pollution – industry's dirty word

KŌGAI – PUBLIC ENEMY NUMBER ONE

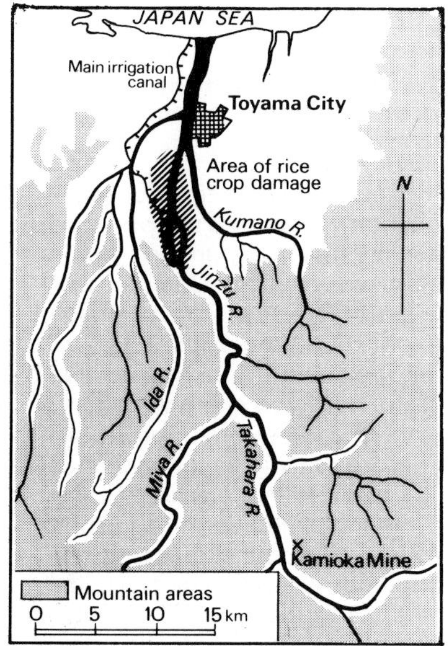

Figure 1 *The Jintsu River Basin*

THE CADMIUM PROBLEM — ITAI-ITAI DISEASE

The nature of cadmium

Cadmium is a heavy metal, closely related to, and occurring naturally with, one of the more common minerals, zinc, which is widely used in electric batteries and in metal plating. Wherever there are zinc mines, there will be traces of cadmium in the surrounding air, soil and water. Near a zinc smelter, for instance, the quantity of cadmium in the air might be as high as $0.55 \mu g/m^3$, whereas in an unpolluted rural area the daily average content would only be $0.001 \mu g/m^3$. Similarly, rivers flowing close to zinc mines are likely to carry traces of cadmium, which may then be deposited further downstream, with unfortunate results.

DISASTER ON THE LOWER JINTSU

The Jintsu River basin, in western Japan (Figure 1), has two notable features. The largest productive zinc mine in Japan is situated in the forested mountains of the upper basin at Kamioka, and the lower valley contains densely populated farmland, mainly producing irrigated rice. It was in this area, near Toyama City, that Dr Murata's patient happened to live. She was just one of hundreds with similar symptoms. As long ago as the 1930s, rural people in this district were complaining of severe, rheumatic-like pains in their joints and back. By 1950, a small number were desperately ill. Their condition became widely known as *itai-itai* (or, literally translated, *ouch-ouch*) disease, because of the acute agony which they suffered. In some cases, the spine seemed to crumble gradually, so that the victim became paralyzed. On X-ray photographs, the bones looked very similar to the soft tissue of the body, indicating that the calcium, which makes up normal bones, had somehow disappeared.

Investigating the problem

The incidence of itai-itai disease in the lower Jintsu valley eventually attracted the attention of scientists and doctors, but it took many years of research before a properly supported answer was pieced together. In 1955, for example, it was suggested that itai-itai was caused by a poor diet of polished rice, which is low in protein and minerals, combined with a history of hard work in the fields. There seemed to be some truth in the hypothesis, since the sufferers, mainly middle-aged women, had indeed lived largely on polished rice, especially during the food shortages of World War II, and had also worked long hours in the flooded rice fields. But so too had millions of other peasant women throughout Japan, and none of these were afflicted with itai-itai. However, by 1961, scientists were closer to the truth when they suggested a link between metallic poisoning and itai-itai disease, since bone analysis of victims always revealed a high content of metals, such as copper, lead, zinc and cadmium.

The findings were presented to the local authority, in this case Toyama Prefecture, and to the mining company. Then, in 1967, local people at last began legal proceedings against Kamioka mine. They were greatly encouraged during the following year, when Cabinet Minister Sonoda, in charge of the Department of Health and Welfare, openly supported them, saying that 'human health should come before industrial profits'. Yet the court case dragged on for four years, until a local judge ruled in favour of the disease victims. Cadmium was the real culprit, with dietary deficiencies being no more than a contributory factor.

By this time, the circumstances of itai-itai were common knowledge throughout Japan. The victims lived mainly in the township of Fuchu-machi (Figure 2), where the irrigated rice fields are surrounded by a maze of natural and man-made waterways. Along these channels, particles of mine waste, including cadmium, were carried, to be deposited at the inlets to the fields. Farmers dug pools in which the unwanted sediments could collect, but all to no avail. The growing rice plants were still able to concentrate the poisonous metal in their roots and grains (Figures 3 and 4). Farm families ate this rice daily, and to make matters worse, some of them drew drinking water from irrigation channels passing directly beneath farmhouse kitchens. Meanwhile, domestic sewage going back onto the fields as fertiliser, helped to maintain the high level of cadmium in the soil (see Figure 3). Traces of cadmium, building up regularly in their bodies, gradually replaced their bone calcium, and the result was itai-itai disease, with its legacy of deformed limbs and pain beyond belief.

? Study Figure 3. Comment on the differences in cadmium content between the Ida and Kumano drainage systems (not linked to Kamioka mine) and the main Jintsu system.

Scientific investigation is rarely satisfied with conclusions drawn from a single location. For this reason, Kobayashi and Hagino, the two Japanese researchers who had first exposed the link between mine wastes and disease, looked for a different area in which to replicate their original study. They soon found one. Until the late 1960s, there was another large zinc mine on the island of Tsushima, off the south-west coast of Japan. Exactly the same pattern appeared as in the lower Jintsu basin, but on a lesser scale. Local crops, such as sweet potato, were blighted by cadmium in the soil, and at Kashine hamlet, for example, several people showed the horrible symptoms of itai-itai. There seemed little further doubt about the relationship of cause and effect.

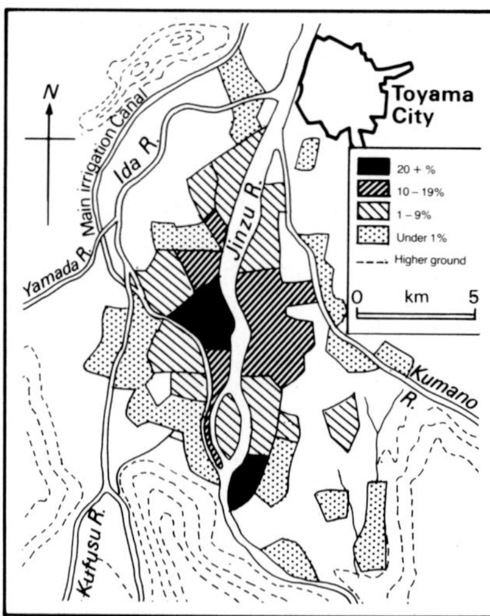

Figure 2 *Percentage of women over 50 years suffering from itai-itai in population of rural townships (1959)*

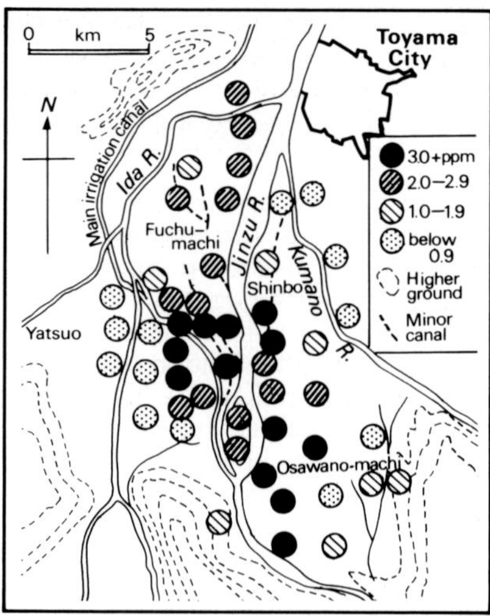

Figure 3 *Amounts of cadmium (parts per million) in surface soil of rice fields (1959)*

Outcomes

Two beneficial results have emerged from this whole sad story. One is the disappearance of itai-itai disease, and the other is the payment of compensation by the mine operators to the victims. There are several factors which explain why the disease has now almost vanished:

1 Locally grown rice now forms a much smaller proportion of the daily diet in Toyama, as elsewhere in Japan. Since the 1950s, living standards in Japan have risen markedly, and there would no longer be any danger of people going short of calcium or vitamin D (both necessary for healthy bones) in their diet, as was once the case. (Cadmium is more likely to be taken into bone tissue when calcium is in short supply.)

2 Rural water supplies have been improved in areas such as Fuchu-machi, and farming families would no longer draw domestic water straight from the irrigation channels.

3 Modern farmers favour chemical fertilisers and would never consider using domestic sewage on their rice fields.

4 Alarmed by adverse criticisms in the world's press and on national television, the Kamioka mine operators (a branch of the gigantic Mitsui combine) have taken steps to reduce the amount of poisonous waste discharged into the Jintsu drainage system.

5 Most important of all, perhaps, is the fact that Japan now has laws governing environmental pollution. Twenty years ago, the public had no such protection.

Payment of compensation to pollution victims had not previously been the rule in Japan, and the notion was clearly unpopular with the major industrial corporations. To be realistic, compensation for damage as serious as that caused by itai-itai has to pay for permanent disablement, or even death. It can therefore run into the equivalent of millions of pounds sterling. Losses on this scale mean reduced profits, and in such circumstances it is not altogether surprising that a guilty company would try to disclaim any responsibility for damaging human health.

?

1 Study Figures 2 and 3. Which two of the rural townships of Fuchu-machi, Shinbo, Osawano and Yatsuo have the highest levels of soil pollution and the highest incidence of itai-itai?

2 Draw a simple diagram to show the chain of cause and effect in itai-itai disease.

3 Study Figure 4. Note also that the following Japanese prefectures contain non-ferrous mines: Hyogo, Ibaragi, Fukui, Toyama and Ehime.
a What are the differences in cadmium content between rice from the prefectures named above and rice from other parts of Japan? Suggest reasons for these differences.
b Compare Toyama Prefecture with Mississippi, the American state most affected by metallic pollution of rice fields.

4 Prepare notes for a debate on one of the following topics:
a No industrial organisation should have the right to pollute the environment, without hindrance or control. (Alternatively, be a devil's advocate and take the opposite view.)
b Every advanced nation has difficulty balancing the benefits gained from technological progress against the losses incurred in environmental terms.

Compare your notes with those from a group of two or three other students, debate the ideas in class or write a short essay about one of the topics.

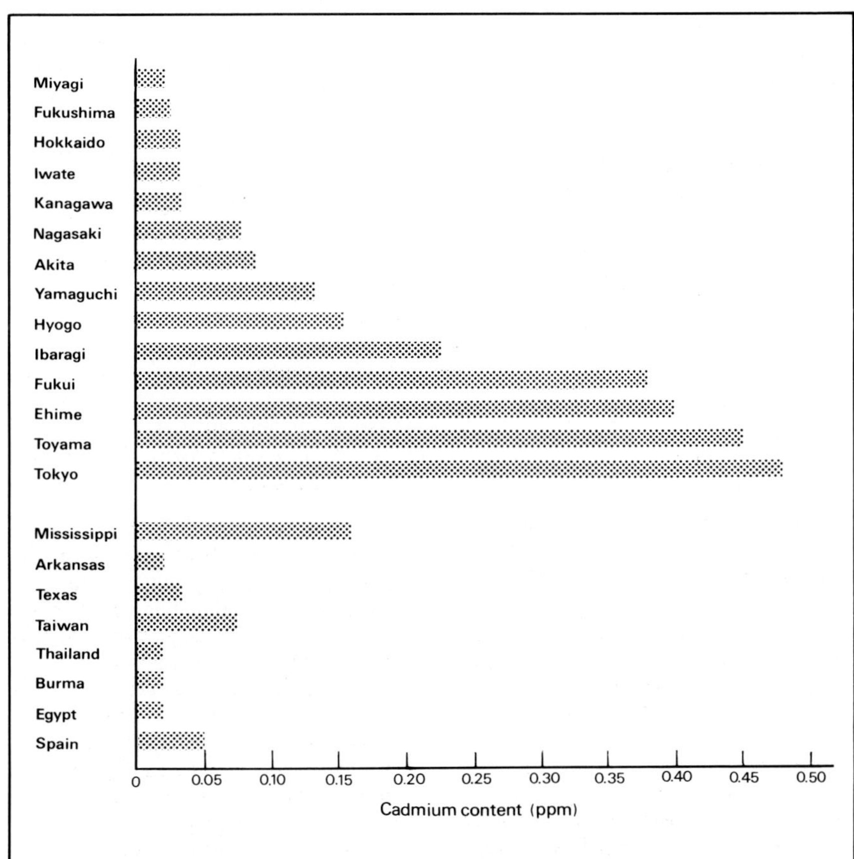

Figure 4 *Cadmium content in rice produced from 14 agricultural experimental stations in selected Japanese prefectures and from other areas (1960)*

To understand why the Japanese allowed their environment to become so badly polluted we have to look beyond the behaviour of individual firms. What we need to consider are the *values* which Japanese society has traditionally held dear. Persisting over the centuries, these values have helped to shape social conduct.

In the case of the Jintsu basin, the old-fashioned values of *patient endurance* (*seishin*) and *respect for authority* seem to have worked to the disadvantage of local people. Itai-itai sufferers and their relatives tended to accept the disease as a misfortune to be bravely endured. This is one reason why they did not rush to sue the mine owners. Another factor is the low value which the Japanese, unlike the Americans for example, place on litigation. Hence, it took years for the victims to begin court action against the mine, which happened to belong to one of Japan's most respected business corporations. Again, the Japanese value of *loyalty* to the nation's interests also stood in the way of justice. Government officials, aware that mineral production is essential to a manufacturing country like Japan, were reluctant to interfere with that production. The mine owners, meanwhile, loyal to their own profit-making ideals, appealed against the 1971 court decision. However, by that time, public opinion in Japan was swinging firmly against polluters. In 1972, a higher court finally ordered substantial payments to be handed over to the itai-itai victims and their families.

THE MERCURY PROBLEM — MINAMATA DISEASE

Kyushu, the most south-westerly of Japan's main islands, has a forested and mountainous interior. Its deeply indented coasts, on the other hand, are fringed with intensive farming, fishing harbours and industrial towns. One of these is Minamata, looking out over the bay of that name. Dotted with islands, Minamata Bay widens northwards into the landlocked Shiranui Sea (see opening map and Figure 5). In the early 1950s, people in this coastal area witnessed some strange happenings. On the surface of the Shiranui Sea, dead fish and octopus were seen floating. Seagulls dropped like stones from the sky, to flutter briefly on the ground before dying. Domestic cats also behaved out of character, by whirling around in a crazy dance and then dropping lifeless. All of this was alarming enough, but worse was yet to come. It was in 1953 that a number of people in Minamata and neighbouring settlements began to show signs of a horrifying disease, never known before in the area. The sufferers had difficulty in hearing, seeing, speaking and in co-ordinating their movements. Within a few months, some were in a state of living death, as this newspaper report details:

> Kumiko M., of Yudo, first became ill in 1956, at the age of six. Within just three months she was totally paralysed. She lay thus for 18 years, never moving a muscle, until she finally died in August 1974, weighing under 20 kilogrammes.

Even newly born infants did not escape. In 1955, for example, seven per cent of all births in the Minamata area showed signs of congenital brain damage.

Investigating the problem

As in the case of itai-itai disease, it took a long time for the Minamata mystery to be properly uncovered. The Japanese spirit of 'seishin' encouraged the victims and their relatives to suffer silently rather than to take action on their own behalf. In any event, the cause had first to be established. One important item of evidence was that people with a particularly high intake of fish in their daily diet seemed to be most at risk.

Then, further evidence appeared in 1959. By far the largest industrial concern in the area was the Chisso Corporation, which produced vinyl and other materials to be used in plastics manufacture. This factory regularly discharged liquid wastes into Minamata Bay. Among the wastes was *mercury*, which had long been under suspicion as a dangerous poison. It was Dr Hosokawa, an employee of Chisso, who carried out a series of animal experiments to determine a definite link between methyl mercury (the organic form of the metal) and damage to the nervous systems of living organisms. Meanwhile, officials of the company were becoming worried. They discouraged any publicity which might help attach blame to Chisso. At the same time, the factory went on pouring its chemical waste into the sea, to build up in the marine ecosystem (see Figure 6). The tragedy is that the

?

1 Study Figure 5 and describe the distribution of disease victims. Set up a number of different hypotheses suggesting the cause or causes. In groups of two or three, pool your ideas and then make up a list.

2 You are employed as a chemist by the Chisso Corporation. You have worked for the firm for 20 years, and you enjoy a good salary with prospects of promotion to chief industrial chemist in the near future. This position carries a very high salary, and you could then afford a large new suburban home. You originally came from Komenotsu, where your relatives still live. Your niece has recently told you of a paralysing disease which has affected her young son. Experimenting secretly in your laboratory, you have hit on the idea that the high mercury content in local food is connected with the disease. However, you know that your company does not want to be involved in any public scandal. What should you do and why?

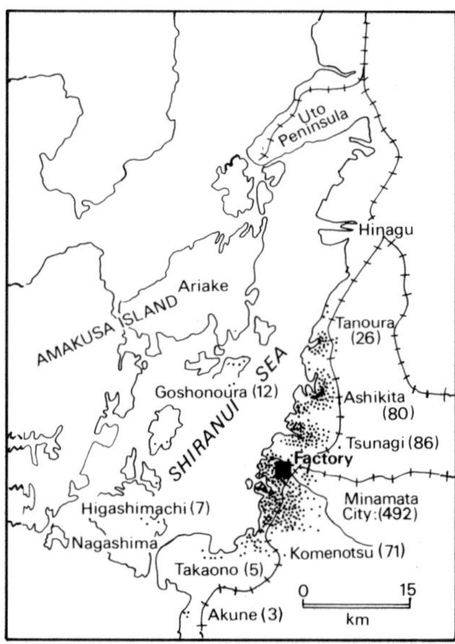

Figure 5 *Location of Minamata disease victims in 1971 (numbers given in brackets)*

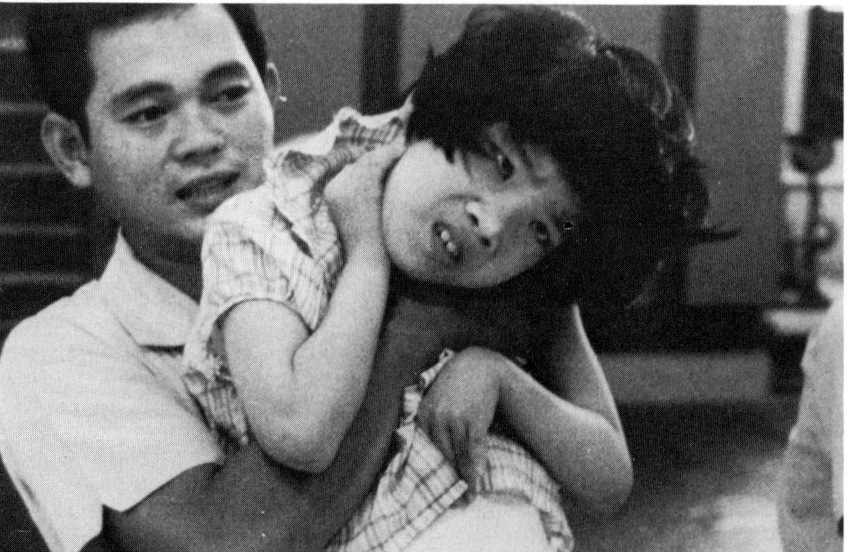

Print 1 *Hisae Morimoto, a 12-year-old victim of Minamata disease at a rehabilitation centre in Minamata*

same conditions continued into the late 1960s. Local people, who had reduced the quantity of fish they ate, somehow got the impression that the crisis was over and then resumed their former eating habits.

Winning compensation

Matters changed when the relatives of Minamata disease victims started looking for outside help. They consulted lawyers, wrote letters to government officials and publicised their cause in newspapers and on television. As often happens when a major grievance is articulated through the mass media, public opinion can be mobilised in favour of the oppressed. In the Minamata case, the Chisso Corporation consistently denied responsibility for as long as possible. Even when the link between mercury waste and human disablement was recognised throughout Japan and beyond, the company insisted that it could never afford to meet the costs of compensation. Yet the verdict was little in doubt, and in 1973, Kumamoto District Court forced Chisso to pay up. A sum equivalent to several million pounds was distributed among 138 afflicted families. In accordance with Japanese custom, company officials also apologised publicly to those they had so badly offended. And thus the curtain descended on the Minamata tragedy. Out of the 100 000 people living around the Shiranui Sea, over a hundred had died from mercury poisoning and something like a further thousand had been affected in some way. However, once the build-up of industrial waste ceased, life in the Minamata waters slowly began to recover. Fish, oysters and crabs, which showed remarkably high levels of mercury in 1959–60, had only a tenth as much in 1973–74. Yet the environment was far from safe even then. Look at Figure 7 to get some idea of the amounts of methyl mercury held in the sludge covering the sea bed beside the chemicals factory. Rather belatedly, in 1978, the government set up a research centre to study mercury related disease. In that year, too, Chisso sought government assistance, complaining that the costs of compensation were forcing the company into bankruptcy.

A point worth remembering is that the cases of cadmium and mercury poisoning which we have been analysing are by no means unique in Japan. During 1965, for instance, there was another serious problem of mercury pollution, near the Showa Denko chemicals plant at Niigata. Several hundred people there became ill or were disabled before the factory was forced to close permanently.

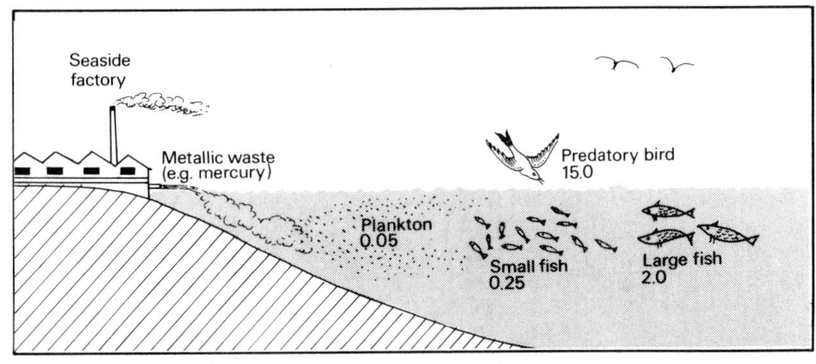

Figure 6 *Build-up of mercury waste (ppm) in the inshore ecosystem*

Print 2 *Supporters of Minamata disease victims outside Kumamoto District Court during the mercury poisoning trial against the Chisso Corporation in 1973*

?

1 Set out in a short paragraph your reaction to an industrial firm's claim that 'effluent containing heavy metals can be disposed of harmlessly into the open ocean, where dilution will take care of it'. Use evidence to support your reaction.

2 List the questions which a local residents' association in Japan might want to ask a chemicals company about to construct a new plant in their area.

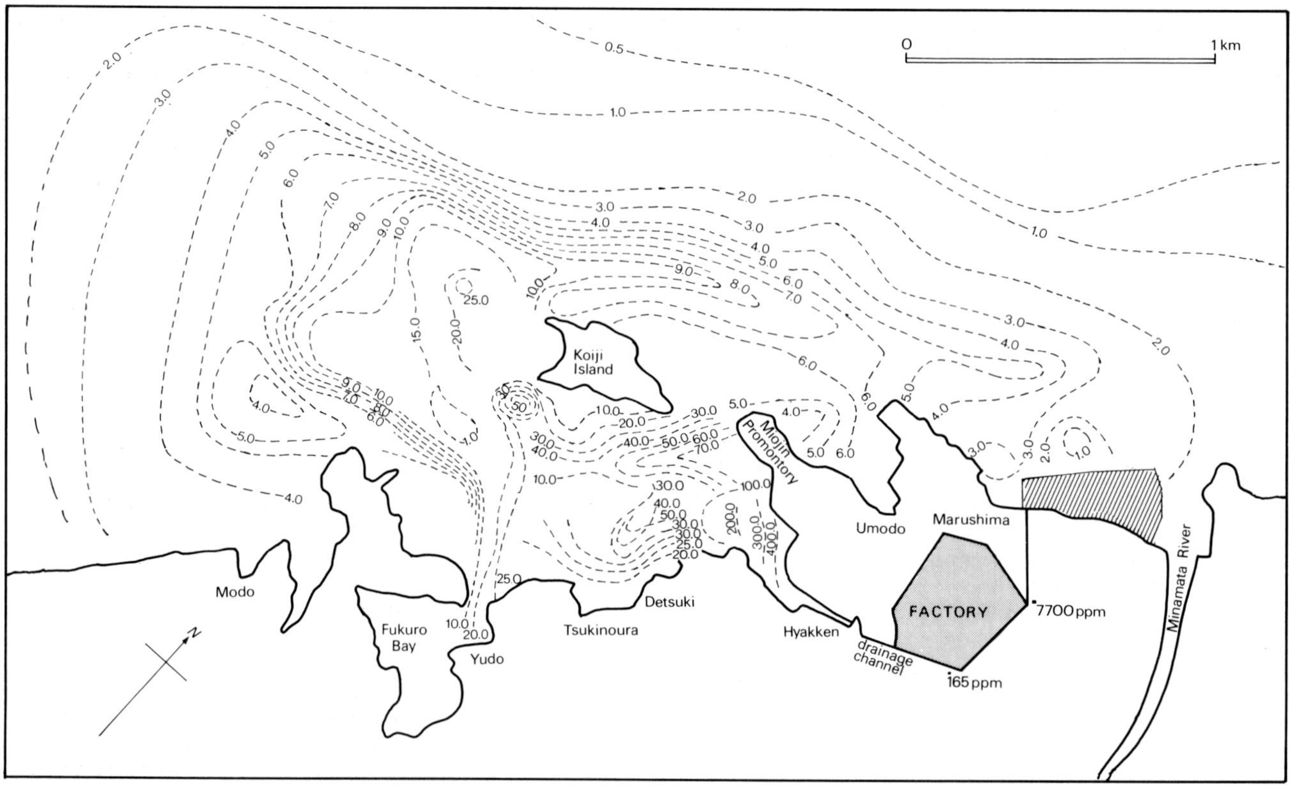

Figure 7 *Mercury content of sludge at the bottom of Minamata Bay. Figures = ppm Hg (dry weight)*
(*Source*: Environmental Agency, 1973)

THE PROBLEM OF AIR POLLUTION

During recent years, anyone visiting a busy Japanese city centre in the morning rush hour would quickly detect two signs of atmospheric pollution. One would be a blue haze, greatly limiting visibility. The other would be a distinctly unpleasant taste in the mouth. In fact, the air inhaled by Japan's urban population has been severely polluted from three main sources: factories, vehicles and power stations.

Factories

As Figure 8 suggests, waste gases and dust particles are given off from many industrial processes. As manufacturing output grew quickly in Japan after the Second World War, the effects of these wastes became even more obvious. One American visitor, Professor Robert Smith of Cornell University, was deeply shaken by the deterioration which he observed. Crossing the Inland Sea in 1951, he noted 'an extraordinarily beautiful body of water, dotted with innumerable islands, providing views of far mountain ranges and great expanses of seascape.' Twenty-five years later, he had a different story to tell. 'The view from the ferry is a nightmare. On some islands, the vegetation has died off completely from the effects of pollutants. Over all hangs a pale sun, sometimes a faint yellow, sometimes a pinkish orange, in whose diffused light objects cast no shadows.' But the polluters did not have things entirely their own way, as the case of *Yokkaichi asthma* clearly shows.

The city of Yokkaichi lies on the sheltered west side of Ise Bay. Japan's biggest oil refining and petrochemicals complex was set up here in 1955, on land on which the Imperial Japanese Navy once had a vast fuel depot. Over the following eight years, other new chemical plants spread across the adjoining

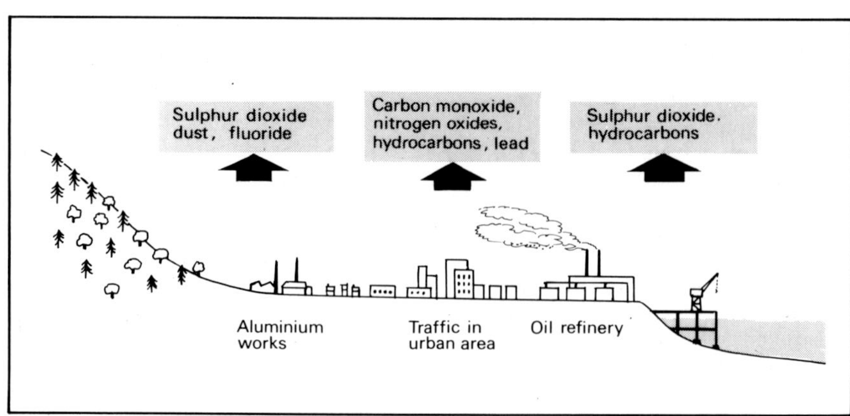

Figure 8 *Some sources of air pollution in Japanese towns and cities*

coastal flats where farmers' grain crops had previously ripened in the summer sun. Between 1955 and 1970, the industrial output of this area grew in value three times as fast as the national average. Yet the companies which were growing rich from the industrial expansion around Yokkaichi were also causing problems for local residents. As the atmosphere became clogged with a fearsome mixture of sulphur dioxide, carbon monoxide and other poisonous gases, many people complained of constant coughing, chest pains and breathing difficulties. Thus the name Yokkaichi asthma entered the Japanese vocabulary. Eventually, a group of sufferers brought a court case against six of the petroleum, chemicals and electricity companies in the vicinity. As at Toyama and Minamata, it took several years for the court to reach a decision, but once again the result was that the guilty firms had to pay compensation.

Vehicles

The major Japanese conurbations (see Figure 11, *page 383*) between them contain one of the world's greatest concentrations of road traffic. On windless days, hundreds of thousands of car exhausts fill the air with emissions of carbon monoxide, nitrogen oxides, traces of lead and hydrocarbons (from unburnt petrol). Sunlight, streaming down on this mixture, activates it to form *photochemical smog*. Because of this, during the 1970s, many urban dwellers developed sore throats, running eyes and respiratory problems. By the 1980s, however, pollution from vehicles was well under control. Indeed, no nation has been as quick as Japan to establish standards for coping with this form of public nuisance. As long ago as 1969, the government first set emission levels for new cars, which would not be allowed to give off more than a tiny amount of waste gases. These standards have been made ever more strict and, although Japanese car makers have met them, few foreign-built cars are designed to cope. Spot checks by the police help ensure that the standards are not exceeded, and another favourable factor is that very few Japanese-owned cars date from the years when higher emission levels were allowed. The policy has worked, and the evidence is that the air in Japanese city centres was clearer in 1980 than it had been in 1970.

Power stations

Where there are dense clusters of homes and factories, as in the cities along Japan's Pacific coast, there is a demand for power. Meeting this demand are a number of large electricity generating stations, many of them fed by imported oil. The waste gases from these stations are another major source of sulphur dioxide, and until the late 1960s electricity generation contributed greatly to the severe atmospheric pollution surrounding Japanese cities. Since that time, a number of factors have combined to reduce the problem:

- The government has encouraged the construction of nuclear power stations, so that heavy petroleum, rich in sulphur, does not now need to be shipped from the Arabian Gulf in such huge quantities. Unfortunately, the nuclear programme itself creates new worries about possible accidents and about the disposal of radioactive wastes. As is often the case in environmental questions, a solution to one problem only leads in turn to new problems.

- Electricity producers have spent money on equipment for de-sulphurising waste gases.

- The simple expedient of building very tall chimneys disperses unwanted gases so that they eventually reach the ground only in small concentrations (though we need to consider where they reach the ground and if they do, in fact, do damage).

However, the most powerful factor of all in controlling environmental damage has been government legislation. This applies to the atmosphere, just as it does to water pollution and noise nuisance, for example.

?

1 Read the newspaper extract as evidence that waste heat from industrial enterprises need not always be wasted.

2a Estimate the approximate cost of installing de-sulphurising equipment in all Japanese thermal power stations. The annual cost would be equivalent to £5 per kilowatt of electricity produced, and the total installed thermal power capacity is 500 000 megawatts (1983). Compare this cost with the losses of about £2000 million caused each year to human health and buildings by sulphur pollution.

b This places a monetary value on human health. Is it morally right to do this? Write an article for your local newspaper on this problem. Then write a letter to the editor by someone who has read your article but takes the opposite view to yours.

3 Imagine that it is 1950 and that you live in an old part of Kobe, close to the port. You have just learned that an electricity generating station is going to be built in a derelict plot close by. You are not happy. Use Figure 13 (*page 384*) for guidance and suggest where the plant could be placed.

4 Acid rain, carried by prevailing westerly winds is now a serious problem in Germany, Scandinavia and the north-east area of North America. It is caused mainly by sulphur dioxide from industrial combustion and by oxides of nitrogen from vehicle exhausts. Suggest some reasons why this problem has not yet been identified on a large-scale in Japan. Bear in mind that Japan has mainly southerly winds in summer and north-easterly winds in winter.

Factory waste to heat a town

From Bernard Thompson in Copenhagen

THE DANISH coastal town of Fredericia is about to become the first community in the world to obtain domestic central heating by using waste heat from an industrial concern.

Under the terms of an agreement with the Danish chemical company, Superfoss, the authority that supplies piped central heating to the town's 50,000 residents, it will save 12,000 tons of oil a year, or half the current consumption. Instead, use will be made of heat generated during the manufacturing process at the company's nearby fertiliser plant.

Local people are, not surprisingly, enthusiastic about the idea. The project will reduce each household central heating bill by at least £100 a year.

Domestic heating in many Danish towns is supplied from a central boiler house operated by the local council. It is expected that the experiment will be repeated in other parts of the country before long, and its potential worldwide as a means of making better use of energy is seen here as enormous.

At present, millions of gallons of hot water produced during the manufacturing process are simply pumped out into the Baltic. Experts who have designed this scheme say that similar heat waste is taking place at chemical plants, oil refineries, and power stations throughout the world.

Superfoss has agreed to supply half of Fredericia's heating needs free of charge. The remainder will be paid for at a price comparable with the current cost of oil. This means that the company will indirectly pay the £3 millions bill to make the necessary changes to the heating installations in the town. On the other hand, the money Superfoss receives for the remainder of the heat supplied will help it to produce cheaper fertilisers, and consequently make its prices more competitive.

(*Source*: *The Guardian*, 29 January 1979)

Table 1 *Some important dates in environmental legislation in Japan*

1967	Basic Law for Environmental Pollution Control
1968	Air Pollution Control Act
1969	Government White Paper on environmental hazards
1970	Amendments to the Basic Law of 1967
1971	Environment Agency established
1973	Open Seas Pollution Control Act National Environment Preservation Law
1974	Pollution Compensation Law

First, we have to note the basic difference between public goods (such as air, sea and wildlife) and private goods. Public goods are not normally traded for profit, whereas private goods are. However, the production of private goods for profit often creates unwanted effects, known as *externalities*. These are costs which are not part of any economic transaction, but are simply imposed on the environment and on the public, who derive no benefit in return. The minamata and itai-itai disasters are examples of externalities, affecting innocent individuals on such a scale that storms of outrage were eventually stirred up. Public authorities react to externalities in two ways: *prevention* and *compulsory compensation*.

THE ANTI-POLLUTION MOVEMENT

The deaths of itai-itai and Minamata disease victims, the foul state of city air and the rapid disappearance of wildlife all combined to convince the Japanese that they had created the most polluted environment in the world. But Japan, to its credit, has dealt with the problem in a vigorous way. There have, of course, always been a few brave individuals to oppose the injustices of pollution. During the early years of mass industrialisation in Japan, for instance, Shozo Tanaka persistently campaigned in the Diet (or parliament) on behalf of miners, peasant farmers and fishermen affected by pollution. But lone voices like his were powerless in the period after 1945, when industrial growth dominated Japanese life. Manufacturing companies were largely free to do as they pleased, because the burden of proof lay entirely on any pollution victim bold enough to lodge a complaint. However, itai-itai and Minamata helped to change this one-sided state of affairs.

The crucial move came in 1967, when the Diet enacted the Basic Law for Environmental Pollution Control. Following that came a series of allied measures (listed in Table 1). For our purpose, the important thing is not simply to list laws and dates, but to see how the environment is protected.

Prevention

Clearly, any economic agency which causes damage to life in its vicinity should be somehow restrained. In most of the industrialised countries, laws for environmental protection have been passed during the past 20 years. Britain, for example, enforced its first Clean Air Act in 1956, while the United States did likewise in 1968. For a long period, Japan did nothing, but when action was taken it was decisive. As Figure 9 shows, the centre piece of Japanese environmental protection is the *Environment Agency*, whose task is to set national standards and to administer the many anti-pollution laws which have now been passed. Daily monitoring of air and water quality is left to the governments of

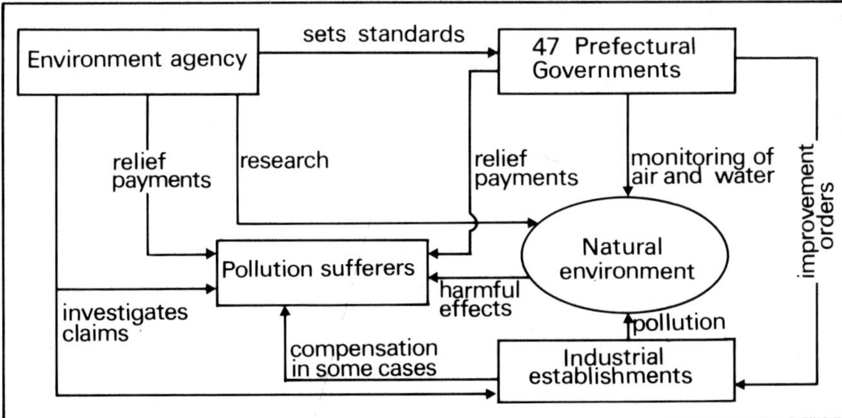

Figure 9 *The system of environmental protection in Japan*

the 47 prefectures, which have all installed expensive metering equipment. The results are sometimes displayed publicly on electronic scoreboards, suggesting that the average Japanese person can make sense of indicators such as 'parts per million of SO_2'. Occasionally, a prefecture will set a local standard which is even stricter than that imposed by the Environment Agency. Kanagawa, for instance, will not allow the slightest trace of cadmium in water, although the nationwide standard permits 0.1 μg/litre.

Compensation

It is one thing to penalise factory owners and car drivers who do not comply with the law. It is quite another to assist those citizens who have suffered as a result of environmental lawbreaking. So it was some years after setting up the Environment Agency that the Japanese Government enacted its compensation laws (see Table 1). To qualify for compensation, an individual must be certified by the Agency as a victim of some pollution related disease. In 1981, for instance, over 70 000 people were in this category, most of them with respiratory illness brought on by air pollution.

It could be argued that compensation, as far as possible, should leave the victims no worse off than they were before pollution affected them. In practice, this ideal situation cannot usually be achieved. It is extremely difficult to put a price on lost health, and impossible to make up for the death of a beloved mother or a child. Nevertheless, Japanese law insists that guilty parties pay up: an idea which has become known as PPP (polluter-pays-principle). On this principle, the Chisso Corporation has been compelled not only to compensate the Minamata victims but also to pay for their continuing medical care.

THE TASK AHEAD

Japan, it can be fairly claimed, has tackled its most obvious forms of pollution. In addition, some Japanese companies have turned the anti-pollution laws into a source of profit. National spending on pollution control was equivalent to two per cent of the country's gross national product during the years 1976–80, and was thus the highest figure for any nation. Those manufacturers, such as Mitsubishi and Hitachi, who produce water treatment and air purifying equipment, along with measuring devices and display units, have found this a profitable business. Their customers, mainly in the steel, chemicals and paper industries, have also been major spenders in *green business* — tree planting and landscaping around industrial areas.

Yet the overall picture is far from satisfactory. For one thing, Japan during the 1960s had allowed environmental pollution to get virtually out of hand. Restoring even some of the damage caused then has turned out to be a lengthy and expensive process. Industrial firms are not unanimous in favouring the present system of environmental legislation. They argue that compulsory anti-pollution spending reduces the funds which a firm can invest in other directions, and may even limit the firm's competitiveness in international markets. Again, these feelings must be set against national satisfaction about the decrease in public complaints concerning pollution. Prefectural governments registered about 80 000 of these during 1974, but the number had fallen to 64 000 by 1981. Environmental scandals on a grand scale were becoming fewer. However, two additional points deserve mention. Although many improvements have been made, there are several huge urban-industrial areas along the Pacific coast of Japan where the environment remains very much under threat. Also, while pollution in its more blatant forms has been checked, the Japanese people are far from happy with their environment in general. Outcries about air and water problems have given way to complaints about noise, inadequate housing and lack of public open space.

?

1 Discuss in groups the effects which the costs of compulsory pollution control might have on a manufacturing firm's profitability, competitiveness and numbers employed.

2 Again in groups, discuss the ways in which each of the following groups could help to reduce industrial pollution: shareholders, trade unions, customers, taxpayers.

3 Make notes on your discussions and decide in each case how feasible and how desirable it would be to take action.

THE PRESSURES OF ECONOMIC GROWTH

In recent decades, the Japanese economy has grown more quickly than that of any other nation, apart from a few small Middle Eastern states. In that period, Japan had been investing heavily in new industrial technology, so that by 1980 it had become a world leader in steel, car manufacture, shipbuilding, electrical goods and optical equipment. As Figure 10 shows, the average annual increase in the *gross national product* (GNP, that is the total value of goods and services produced) has been consistently higher in Japan than in any of its main industrial rivals. In part, this success has been helped by low military spending and by noninvolvement in international problems. In part, too, it has been aided by the Japanese values system. There is, for example, general agreement that *economic growth* is a desirable goal for the nation as a whole. Government, manufacturers and trade unions have co-operated over the years to attain this objective. Another basic value is the loyalty which industrial corporations show to their employees, which workers show to their employers and which almost everyone shows to Japan itself.

Yet, the very success which has made Japan the world's largest producer of so many different manufactured goods has created a series of new problems. Heading the list are environmental pollution, as we have seen, and shortage of space. As the Japanese have become richer, they have had to build vast new steel works, huge car factories and giant chemicals plants. New motorways and rail links have been needed to speed the flow of people and goods. Room for all this expansion has been exceptionally difficult to find, for two pressing reasons. One is the mountainous nature of Japan, which has about 70 per cent of its land surface lying above 500 metres. These uplands are often steeply sloping, with a dense forest cover, and are best left in their natural state. The second factor is that the remaining usable land is already more overcrowded than any other comparable area on Earth. In the international table of population densities, the Japanese figure of 316 persons per square kilometre seems high, but by no means remarkable. However, those other nations such as Belgium and the Netherlands, which have overall densities similar to that of Japan, do not have so much uninhabited upland. Hence, the real density of population in lowland Japan is extreme. The Tokaido Megalopolis (see Figure 11, *page 383*) in particular is quite extraordinary in its merciless congestion. Economic growth has put pressure on the crowded Japanese environment in various other ways. For instance, as the national income per capita has risen, so too has the demand for household equipment and outdoor recreation. Japanese houses, once simple and uncluttered, are now crammed with modern aids to living, and so the urge to acquire a more spacious home is widely felt.

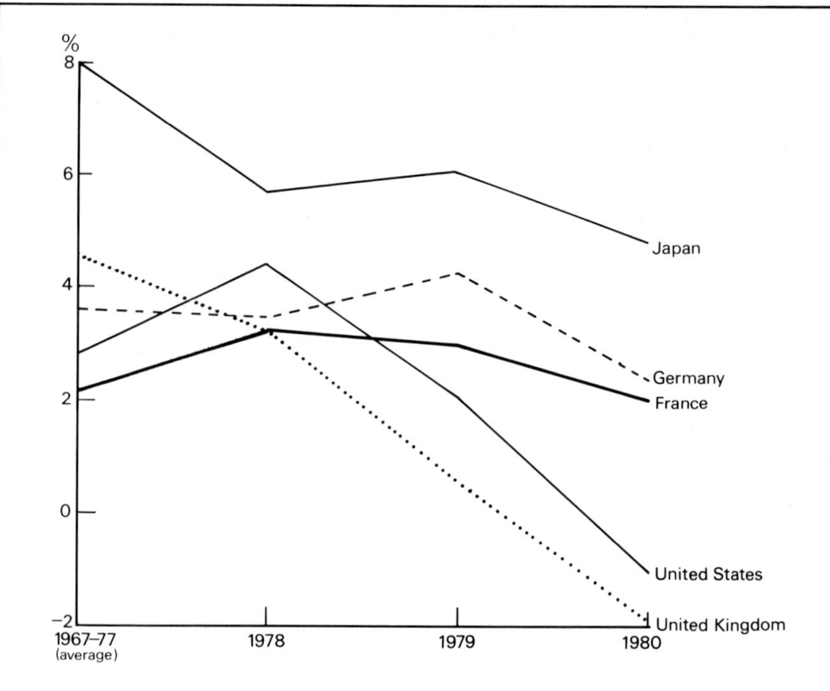

Figure 10 *Real GNP growth in selected OECD countries*
(*Source*: *OECD Economic Outlook*, No. 26, Dec. 1979)

? Before reading the next section, list as many ideas as you can for possible strategies of saving space for living and working. Then compare your list with the one in the next section.

COPING WITH SPACE SHORTAGES

Japan has not been short of strategies (Figure 12) for coping with the ever pressing need for space in which to live and work:

Changes in specialisation
Japanese industry has been investing less in industries which take up a lot of room but have only a low value of output per unit of area, such as shipbuilding, heavy chemicals and textiles. Instead, there has been more emphasis on those industries which combine high value of output with a low demand for space, like microelectronics, robotics, cameras and watches.

Overseas investment
Another way of relieving pressure on lowland Japan is for local companies to invest outside Japan. This has the added advantage of savings on labour costs, which are rarely as high elsewhere as they are in Japan. Beginning in Asia, Japanese firms like Mitsubishi set up branches in Taiwan, Korea and

Singapore, expanding thereafter into North America and Western Europe.

High-rise building

For many years, there were no very tall buildings in urban Japan, for fear of earthquake damage. Tokyo, for instance, had an eight-storey limitation until 1968. This policy has been changed, and tower blocks for offices and housing are now commonly used to make more intensive use of scarce space. A variation on this theme is to build below the surface. The larger Japanese cities now have huge underground shopping plazas, which provide much needed additional space.

Development of new areas

Much of the recent pressure on land in Japan has been concentrated in the three MMAs (Major Metropolitan Areas) shown in Figure 11. Into this zone, the Tokaido Megalopolis, half of the national population is packed. To help remedy the imbalance in population distribution, Prime Minister Tanaka proposed a plan in 1972, *Building A New Japan*. The dispersal of economic activity, he suggested, could be achieved if new sites were developed at the more northern and southern extremities of the country. Although little ever actually came of the Tanaka Plan, the notion of dispersal has influenced the siting of Tokyo's new airport, as we shall see later in this chapter.

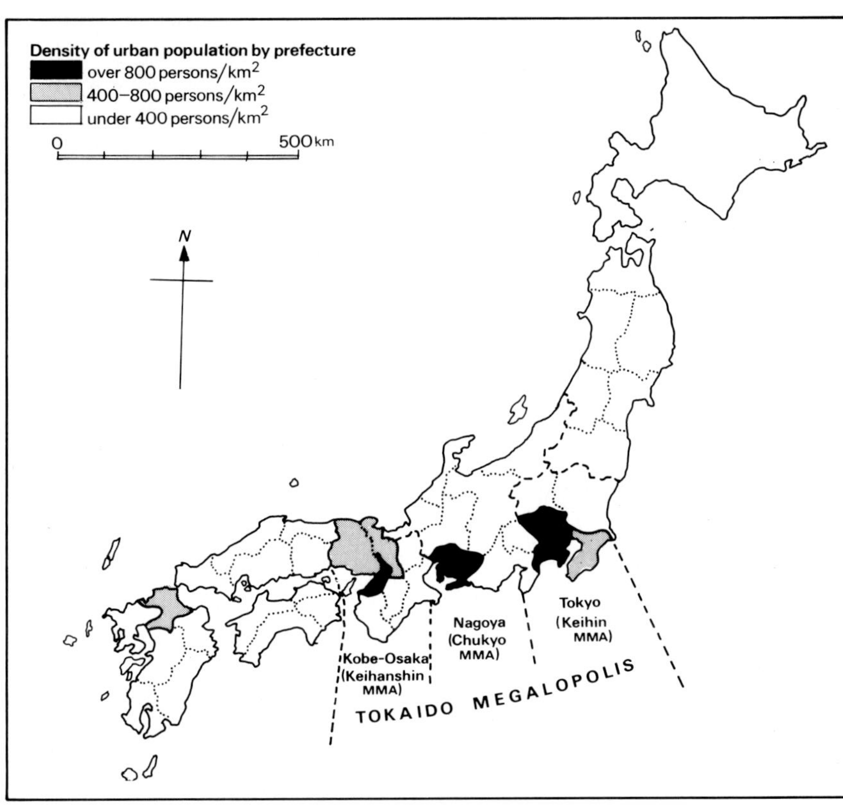

Figure 11 *Major concentrations of urban population in Japan (1980)*

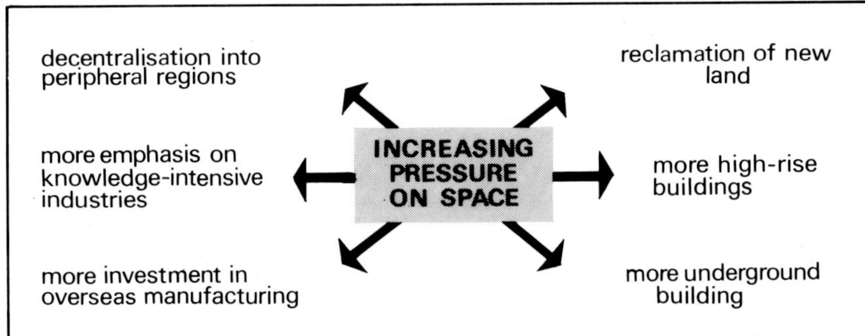

Figure 12 *Ways of coping with pressure on scarce space*

Reclamation

Where the Dutch were once world leaders in claiming new land from the sea, it is the Japanese who currently do more reclamation work than any other nation. Whereas the Netherlands originally reclaimed its polderlands for agriculture, the Japanese want new land for urban uses — housing, transport and industry. Thus, Tokyo and Osaka Bays are ringed about with land artificially created during this century. Other notable examples include the enormous Ogishima steelworks, opened in 1979, and Nagasaki Airport, dating from 1975. The following section will show how reclamation has enabled Kobe to continue successfully as Japan's premier seaport.

In passing and as you are doing the exercise below remember that the strategies outlined above can only be a partial answer to the problems of crowded conditions. Japanese society has also traditionally stressed the importance of adapting culturally to life in restricted spaces. Polite behaviour, simple tastes and mass conformity to group norms are just three obvious examples of this pattern.

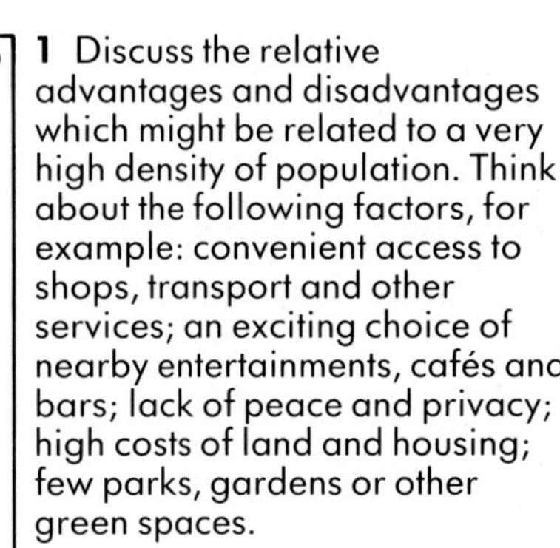

?

1 Discuss the relative advantages and disadvantages which might be related to a very high density of population. Think about the following factors, for example: convenient access to shops, transport and other services; an exciting choice of nearby entertainments, cafés and bars; lack of peace and privacy; high costs of land and housing; few parks, gardens or other green spaces.

2 Shinobu Onoda is a 21 year old girl who has recently left a junior college to work in the

? fashion department of a large Tokyo shopping chain. Junichiro Kirihata, a 71 year old pensioner, now lives alone in his well-kept but old-fashioned farmstead. Consider how these two individuals might differ in their view of current changes in Japanese life. Think, for example, of high consumer spending; many imported products and ideas; a demand for ever higher standards of personal comfort; and a growing belief that leisure is just as important as hard work.

Print 3 *Space is at a premium in Japan's urban areas — A hotel guest emerges from his 'capsule' at the* First Inn, *Tsukubx. The capsule hotel is able to sleep 4 800 people per night*

KOBE

The following case study shows how one particularly crowded area of Japan, the port of Kobe, has come to terms with the problem of limited usable land.

Location of Kobe

The quiet western coast of Japan contrasts with its bustling eastern side, dotted with ocean gateways. Although the country has over a thousand recognised seaports, only 17 of these have a major international role. Two of them are neighbours on Osaka Bay. The older is Osaka, which has been developing since the 1600s, and the other is Kobe, where growth has dated mainly from the 1870s. Yet, despite its later start, it is Kobe which handles double the traffic of its neighbour. Unlike Osaka, where the port is overshadowed by the vast sprawl of the city, Kobe is more of a city organised around a port. Its roads either run parallel to the waterfront or descend from the hills to the sea. At every intersection there is a glimpse of constant maritime activity. The restricted site, with steep hills forcing the city into a coastal strip about 2 kilometres broad and 30 kilometres long, is Kobe's most obvious drawback. But this is more than made up for by the combined advantages of a small tidal range, a bay of deep water well sheltered from the open Pacific, and excellent inland communications. As Figure 13 shows, the proximity of the Hanshin Expressway and the Shinkansen 'bullet' rail link helps to make Kobe a particularly accessible port.

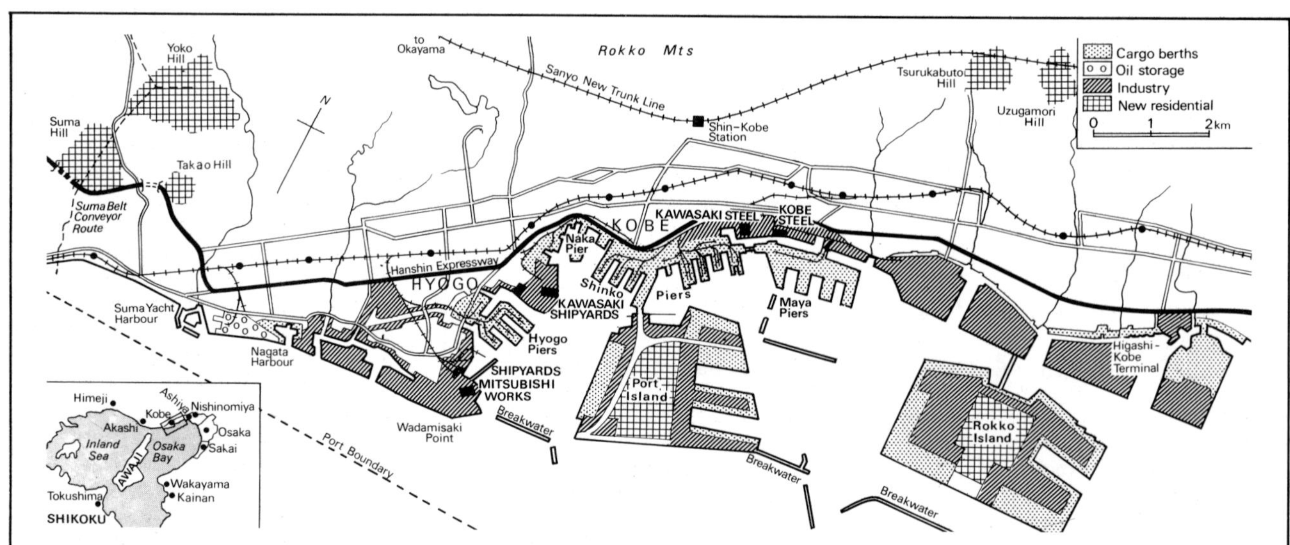

Figure 13 *The port of Kobe*

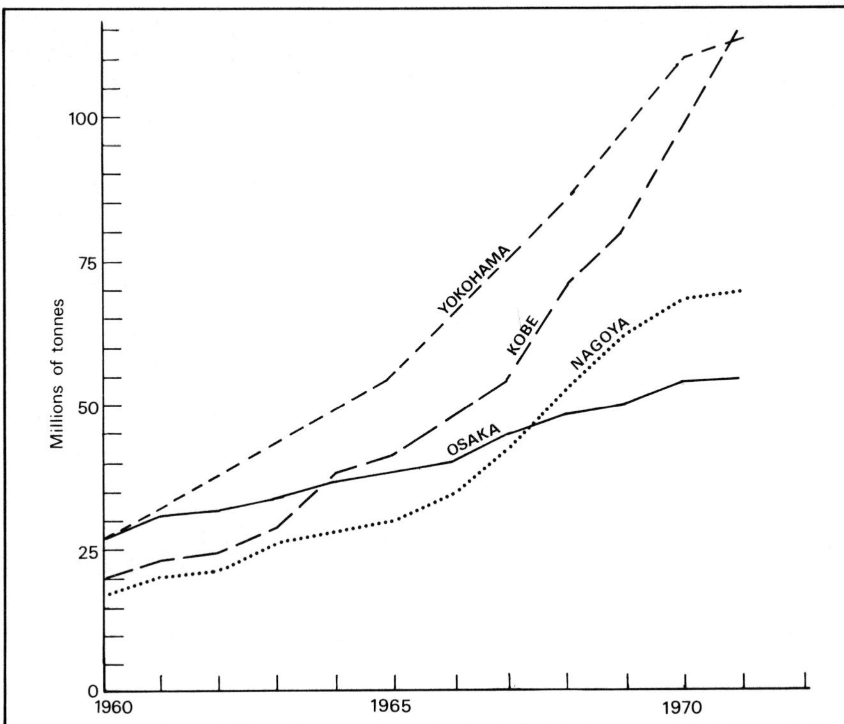

Figure 14 *The growth of seaborne traffic at major Japanese ports, 1960–73*

Growth of the port

When Japan was opened to foreign trade in 1868, after centuries of seclusion, Kobe was designated as an overseas gateway, along with Yokohama and Hakodate. The latter proved too remote to be internationally competitive. Yokohama was severely damaged in the Great Earthquake of 1923, and it has gradually been overtaken by Kobe as the major port (see Figure 14). Throughout its expansion, the port of Kobe has combined *industrial* with *commercial* success. During all that time there has been one continuing local difficulty: shortage of space. The solution has usually been to reclaim new land.

Industrial activity

Kobe has two main ribbons of waterfront manufacturing — one to the west of the city centre and the other to the east. The western side grew first, with the location of shipbuilding and repair yards, such as the Kawasaki Yard (1886) and the Kobe Shipyard and Engine Works (1905). From 1918 onwards, these were supplied with steel from the Kawasaki Steel Corporation's works, built on a now hopelessly congested site near the city centre. Since World War II, the extensive western polders beyond Wadamisaki Point have accommodated various land uses which are dangerous and space-consuming. Examples include an Esso fuel oil storage depot, Osaka Gas Company works and a huge incinerator for urban refuse.

Meanwhile, along the eastern side of Kobe, is a string of modern developments occupying hundreds of hectares of artificial land, created between 1953 and 1970. The largest single unit is the Kobe steel integrated works, beyond which lie engineering and chemical plants, plus food processing — tied to imported vegetable oils, grain and sugar. This intense industrial activity along water frontages is a typical feature of modern Japan. It allows free access to incoming raw materials and outgoing manufactured goods. It also concentrates pressure on every space, which can only expand by new reclamation.

Commercial activity

The rising tonnage of goods handled at Kobe has increased the demand for working space, where ships can berth, barges operate, and container lorries load. At first, simple finger-type piers were built, close to the city centre. More recently, as at Maya and Hyogo Piers, small artificial islands have been created for cargo handling. Even these, however, have proved inadequate. Kobe has been chosen as the major terminal for several container companies, some Japanese (like NYK or Mitsui OSK Lines) and others European (such as the Maersk Line of Copenhagen). It is this commercial expansion, in combination with industrial growth, which has motivated the port authorities to find a radical solution to the congestion problem.

New reclamation

Since 1961, Japan has had five-yearly plans for national port development. Under these, financial aid has helped to create two great man-made islands adjacent to Kobe, providing space for industry, housing, recreation and shipping. *Port Island* was reclaimed between 1966 and 1975, by embanking and infilling. The newly-won space, linked to Kobe city centre by the red painted arch of Kobe Ohashi Bridge, was then devoted to container handling, housing overspill for 20 000 people, business services and public parks. A new automated rail system puts the island within seven minutes travel from central Kobe. *Rokko Island*, begun in 1972, has become another major container terminal, enclosing residential and recreational areas.

Finding material for the islands posed a serious problem, since Kobe, unlike nearby Osaka, did not have sandbanks which could be dredged for the purpose. Instead, the Rokko Mountains, a much eroded block of granite, have been quarried for their sands. Millions of tonnes of material, brought out by fleets of lorries and by an elevated conveyor belt leading down to Suma Harbour, were then ferried out by barge to fill in the island sites. 'Mountains go to the sea' is how one Japanese saying describes the process. The result is that, on the west side of Kobe, Takao and Suma Hills have been torn away to provide sands for Port Island. To the east, Tsurukabuto and Uzugamori Hills have been gouged out to build Rokko Island. This solution to the problem of space shortage did not go ahead without protest. For example, the transport of rock debris in such vast quantities further muddied the murky waters of Osaka Bay, striking another blow at the already weakened inshore fisheries. Levelling the wooded hills also degraded the environment on the fringe of Kobe. Deforestation, never a wise procedure, is especially damaging where atmospheric pollution exists to further weaken the growth of coniferous trees. These aspects of environmental destruction should be set against the argument that the denuded hills have been put to good use. They are now occupied by new housing estates, which are a valuable extension to the crowded city below. Land use conflict around Kobe, however, is likely to continue into the foreseeable future.

?

1 In recent years, Rotterdam has ranked ahead of Kobe, as the world's busiest port. Consider some reasons why this situation would probably continue. Bear in mind the different hinterlands of the two ports, their connections with inland waterways and the scope for further physical development in each case.
2 Study Figure 15. Identify areas where Osaka port seems to have adopted the reclamation solution and list the main uses to which the new land is put.
3 Willingness to adopt up-to-date technology and faith in continuing economic progress are two values prevalent in Japanese society, particularly during the last 30 years. Discuss the importance of these values with reference to the development of Kobe port.

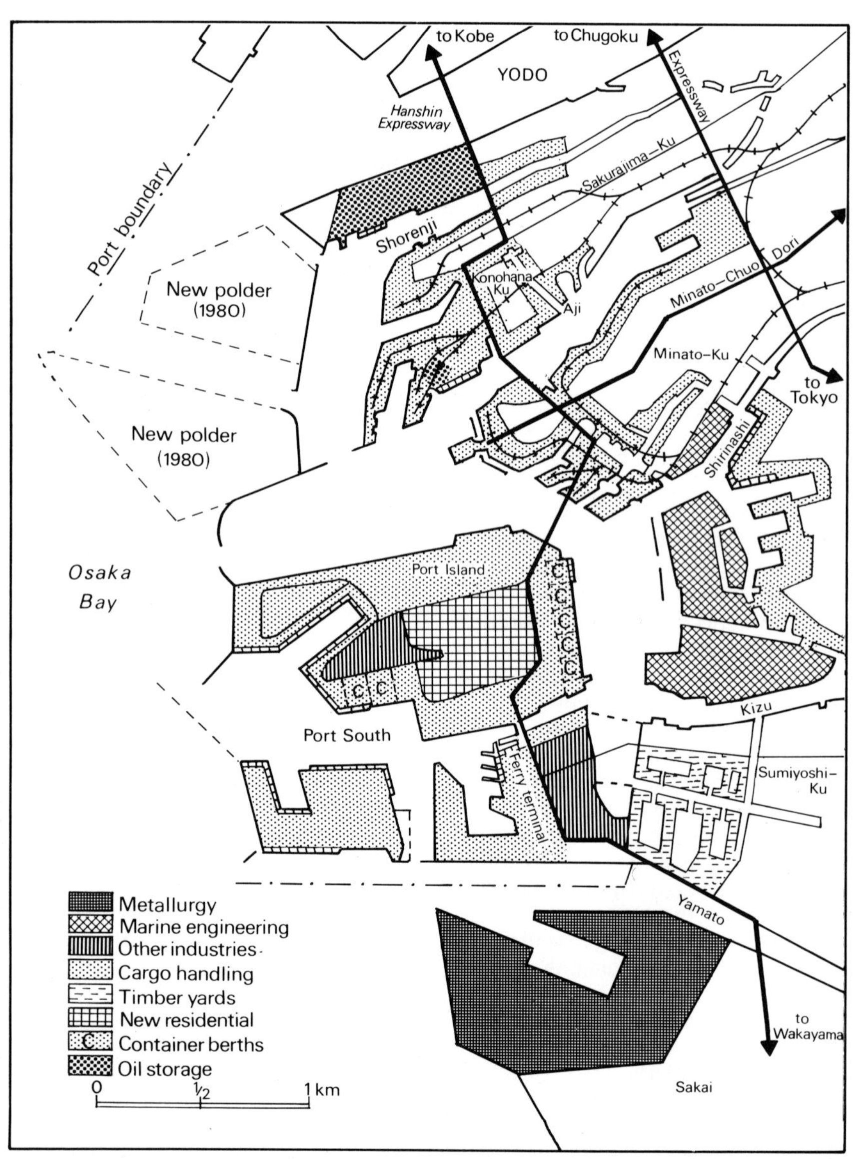

Figure 15 *The port of Osaka*

NARITA AIRPORT

THE NEED FOR A NEW AIRPORT IN JAPAN

We have seen how difficult it is in Japan to find room for space-demanding facilities, such as port extensions or new housing estates. Among the amenities which a modern industrial nation must have is at least one major international airport. Such a facility demands a great deal of space — as a rule, several hundred hectares. Amongst other things, a large airport requires two or three runways, aligned in different directions, and each about 4000 metres long. Its passenger halls, freight and repair sheds, aircraft hangars and car parks all have to be constructed on a vast scale, to cope with the unceasing flood of travellers.

Since 1930, Japan's main door to the air routes of the world has been Haneda Airport, on the south side of Tokyo (Figure 16). The site has certain advantages. Above all, it is well-linked by highway, monorail and conventional railway to the nearby city centre. Despite this, Haneda was clearly unable to deal with the great expansion which took place in Japanese airport traffic during the 1970s (see Table 2). The site is adjacent to an extremely congested area of factories and housing, and the noise of heavy jets taking off and landing is a serious nuisance to thousands of local people. To extend this site, new land would have to be reclaimed from

Tokyo Bay, where the runways would then encroach on the already crowded shipping lanes. Similarly, the packed landward side of Haneda could not be feasibly adapted to accommodate expansion of the airport. At the same time, Japan's part in international trade was inexorably growing. Air connections with commercial partners in Europe, North America and Australasia were therfore in ever greater demand. As Japanese living standards rose, the demand for holidays abroad put still further strain on Haneda's inadequate facilities. The problem could only be solved by building on a new site, which had to be in the general vicinity of Tokyo, the economic decision making centre of Japan.

THE NARITA SITE

The first moves to create a new international airport began in 1966, when a number of government departments jointly selected a site at Narita (Figure 16). Halfway across the Boso Peninsula, Narita is an old temple town, surrounded by farmland and woods. Modern development has made a great impact on this area.

Among the positive features of the airport's construction have been the following (see also Table 3):

Employment

While the airport itself employs many thousands of people in administration, catering, cleaning and baggage handling, it also has a *multiplier effect*. Its existence creates many additional jobs nearby in hotels, car hire and taxi firms, banks and shops.

Space for growth

The site is large enough to handle any growth in air traffic forecast for the remainder of this century. At first, only one main runway was laid out, but a further two are being added during the 1980s, to produce an operating area three times the size of Haneda.

Table 2 *Passengers and cargo transported by Japanese airlines*

	International routes		*Domestic routes*	
Year	*Million passengers*	*Million tonne/km of cargo*	*Million passengers*	*Million tonne/km of cargo*
1965	0.4	64	5.2	17.3
1970	1.6	324	14.7	61.9
1975	2.5	767	25.0	131.8
1980	4.8	1448	40.9	264.3

Table 3 *Impact of a new international airport on its surrounding region*

Positive effects	*Negative effects*
Direct employment	Loss of agricultural land
Indirect employment (hotels, taxis, catering)	Noise and other disturbance to local inhabitants
Airport purchases of goods and services	Cost of additions to the land transport network
New developments in the corridor between airport and main city	Land use restrictions in the surrounding area, because of safety and noise regulations

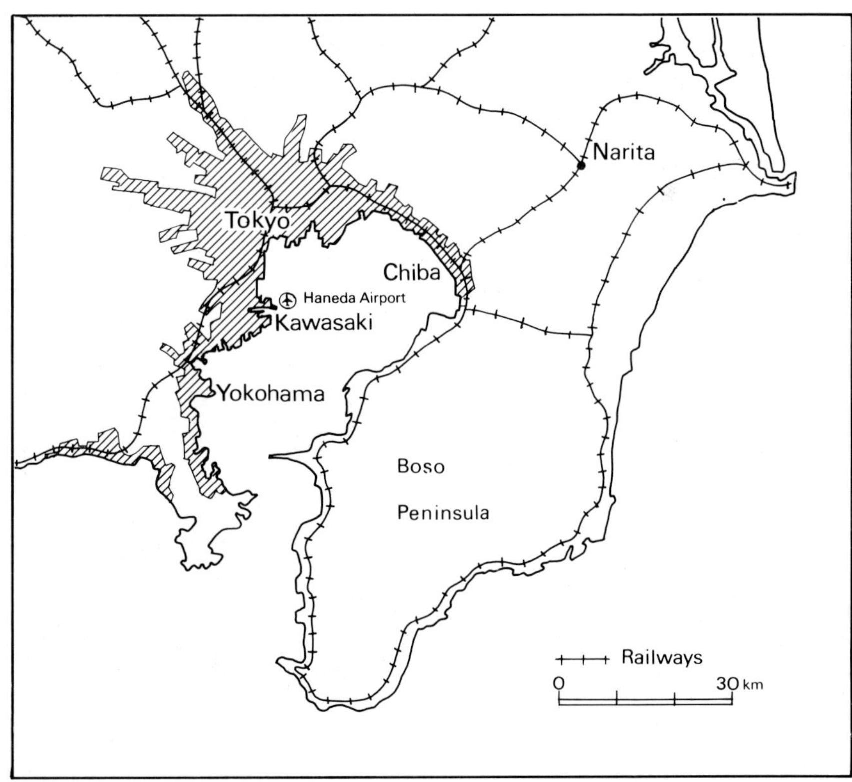

Figure 16 *Location of Narita*

On the other hand, the site has certain drawbacks:

Disturbance
Over 1000 dwellings, lying on or beside the new airport, had to be vacated. Again, hundreds of hectares of farmland had to be sold to the airport authority. Farmers naturally resented the loss of land which, in some cases, had been used by the same families for centuries.

Noise
People who had enjoyed the peace of rural life now had to endure the thunderous din of constant aircraft movements. In this case, improved standards of travel for some mean lowered standards of daily comfort for local inhabitants.

Transport costs
A rural site, 60 kilometres from central Tokyo, has to be made accessible to the city which it serves. Thus, a costly system of high quality expressways and a new railway line have had to be built, linking Narita to Tokyo.

Public protests
While all of these drawbacks could be measured in one way or another, there was one further difficulty which could not. It had to do with the manner in which Narita was chosen. Local people were not consulted, and there was no public inquiry at which the various costs and benefits of the scheme could be aired. Construction work at the site went on during the early 1970s, and Narita might well have been in use by 1975. But it was not to be.

Japan has a history of militant student movements, with leftist revolutionary beliefs. Over the years, student groups have organised public demonstrations over issues such as Japan's relations with the United States, the war in Vietnam and the use of nuclear power. Narita Airport proved to be a major focus for student protesters, who argued that the development was carried out in an undemocratic fashion. Local farmers, furious at the loss of their land, were only too eager to join forces with the students in demonstrating against the building of Narita. For twelve years, on and off, the student–farmer alliance clashed violently with the army of police who had to defend the airport throughout its construction. All over the world, newspapers and television showed ferocious battles between police, using riot shields, clubs and tear gas grenades, and masked protesters, armed with spears or petrol bombs. The climax came in 1978, when the newly built airport control tower was wrecked by demonstrators. By then, both sides had lost heavily in deaths and injuries. Indeed, one German magazine was asking: 'Is this the most dangerous airport in the whole world — surrounded by guerrillas, shrouded in cloud and with only one working runway?'

Yet in the end the authorities won. By erecting miles of stout metal fencing and by digging protective ditches around the airport perimeter, they made the site as safe as possible. Passengers entering the airport are checked again and again, and armed security men are much in evidence. It is only by being a fortress that Narita can function as an airport. Yet the case of Narita is by no means unique. Throughout the richer nations of the West, fierce political arguments will continue to rage over questions such as defence, employment policies, transport planning and social inequality. Wherever groups of people feel that they are not heeded by national governments and other decision-making bodies, there is always a likelihood of organised protest and public disorder.

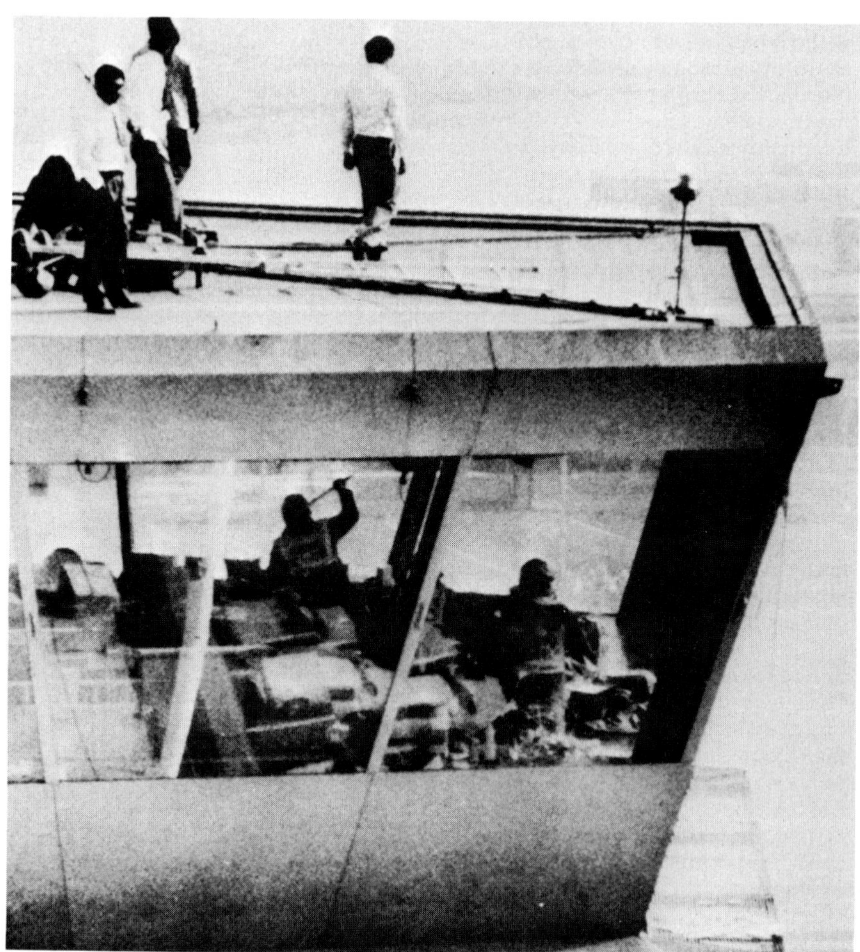

Print 4 *Students smashing equipment in the control tower of Narita airport, during a violent demonstration against the opening of the airport in 1978*

Here are some typical reactions offered by individuals who were affected by the Narita decision.

TOSHIO SAWAMOTO (23) STUDENT ACTIVIST
It's obvious, surely. The capitalist system has always oppressed ordinary people. It offers them paltry wages in return for a lifetime of hard work. So the rich get richer, while the poor just have to suffer the smell and the noise of this monstrosity. Still, we had some good battles.

AKIRA MATSUDA (56) FARMER
This land, you know, where my father and grandfather are buried, has a special place in my heart. Now strangers come and go by the busload. But what do they know about this country of mine, our Emperor, our faith, our festivals? Anyway, what does flying abroad matter to me?

KATSUO SHIMIZU (34) GARAGE OWNER
This is one of the best things that ever happened to me. Otherwise, I'd have had a small country garage and not much else. Now I've got this big place, with 18 employees. Last year I had a 12-day holiday in Hawaii, and we'll probably try California next year.

MICHIKO OTAKE (21) SHOP ASSISTANT
People don't realise how hard it can be to get a decent job out here in the country. Honestly, I don't know what I'd have done if Narita hadn't opened. Most likely I would have had a two-hour trip into Tokyo every day. As it is, my job in the duty-free shop is only 20 minutes from home. Aircraft noise? Doesn't bother me personally.

? Having read these reactions, now work through the following steps.
1 Consider the case for a second new international airport, to be located outside Osaka, where the present facilities are not adequate for the overseas traffic which is forecast.
2 Bearing in mind the reactions already quoted, class members should vote *for* or *against* the Osaka proposal. Individuals should give three reasons for their choice.
3 List the *for* and *against* reasons separately on the blackboard, and select the two most convincing reasons in each list.
4 Voters *for* or *against*, working in two separate groups, should now search for the *values* underlying each reason.
5 Values should now be listed on the blackboard and considered further in class discussion. The implications of each value for the example of airport development should be clarified as far as possible.
6 Individuals should now vote again on the same issue. Those who have changed their minds should explain the reasons behind their thinking.
7 Working in pairs, consider the following possibilities.
a If the vote went in your favour, how would you accommodate the wishes of those who voted differently?
b If the vote went against you, how best could you live with the wishes of the majority?

THE JAPANESE VALUES SYSTEM

It is not possible to understand the human geography of any nation without knowing something of the values system which motivates the people of that country. It would therefore be reasonable to end this chapter on Japan by mentioning five particular values which are powerful forces in social behaviour.

National unity
With one language, no social classes, no large ethnic minorities (apart from Koreans) and no regional separatist movements, Japan holds together more cohesively than most other industrialised nations. One related advantage is that national laws, such as those governing pollution, are almost universally observed.

Stability
With power residing mainly in the hands of the Liberal Democratic Party, the government in Japan can safely pursue long-term policies. Thus, the majority of citizens would appreciate that, when national decisions are taken (as at Narita), there is little likelihood of these being later reversed.

Traditionalism
On all sides there is evidence of a genuine respect for the past. Shinto shrines and Buddhist temples abound, festivals are dutifully observed and old forms of dress, diet and recreation live on, side by side with the new.

Discipline
For centuries, the related qualities of hard work, patient application and respect for authority have been central to Japanese life. Major achievements in industrial growth fit in with this pattern, but so too does the tolerance of unduly high levels of environmental pollution.

Readiness to adapt
Ever since opening its doors to worldwide trade after 1868, Japan has valued technological innovation. Modern technology has been allowed to damage the environment, but there have also been beneficial uses, as in the control of air pollution and the reclamation of new land.

19 MANAGING RECREATIONAL AREAS: THE EFFECTS OF TRAMPLING ON AN ECOSYSTEM

A badly-eroded path on the Pennine Way

WHY MANAGEMENT?

Hampstead Heath, Snowdon's summit, Kynance Cove, Ivinghoe Beacon, Tarn Hows, Box Hill and the Peak District are examples of areas in Britain which have become known as 'tourist honeypots' because their great natural beauty attracts thousands of visitors each year. They all share a common problem — each is being worn away by the people who visit them. The erosion and damage caused by visitors present a problem for several reasons, namely:

a Badly worn paths pose a risk to public safety.
b The natural beauty of the area is marred.
c In special areas valued as Nature Reserves or Sites of Special Scientific Interest, plants and animals are damaged.
d Farming and living conditions for adjacent landowners and residents are made more difficult because of trespass, litter, noise, etc.

? Read through the newspaper article about the Snowdon management scheme, identify the problems and make a list of the possible measures that might be taken to provide solutions to them. Keep a list of the solutions. At the end of the chapter you will be asked to make another list of solutions and to compare them.

THE SNOWDON MANAGEMENT SCHEME

Footpath erosion is a growing threat to some of the more popular walking areas and nowhere is the problem more acute than on Snowdon, the highest mountain in England and Wales which is suffering from the effects of 400 000 pairs of feet climbing to the summit each year.

The Snowdon Management Scheme has been set up in an attempt to improve and maintain the footpath network on Snowdon. Dr Roderick Gritten, Field Research Information Officer for the Snowdonia National Park Authority, outlines the dangers facing Snowdon and the work being done by the Scheme to reconcile the conflicts between conservation, agriculture, recreation and the tourist industry.

About one million people set foot on Snowdon each year and an estimated 400 000 reach the summit. Many of these, expecting a day out away from the bustle of city life, will have been surprised, last summer, to see a mechanical digger lurking by the shores of Llyn Llydaw. And the sight of a dumper truck, laden with gravel, may have been the last straw as it puttered into view by the ruins of the crushing mill on the Miners' Track.

The six paths up Snowdon have become extremely badly eroded and since the spring of 1979 their reconstitution has been the concern of the Snowdon Management Scheme. This is a five-year programme and is run by the Snowdonia National Park and sponsored by the Countryside Commission. Other bodies have indicated their support for the Scheme and the Welsh Development Agency are now also sponsoring it. The Scheme employs, among others, 21 estate workers whose job it is to repair the footpaths so that they may once again be pleasant to walk upon and not the unsightly scars which they are today. The presence of the mechanical digger and dumper truck go a long way to indicate the scale that this footpath erosion has reached.

Footpath erosion is inevitable in popular mountain areas. The boots of walkers quickly tear away the vegetation and thin topsoil, leaving the unstable subsoil to the mercy of the elements. The summit of Snowdon receives 200" of rainfall anually and this quickly washes away the subsoil, leaving boulder-filled gullies which are difficult to walk through. Walkers are only human and prefer, whenever possible, to take the easier path. Compounded with this is the fact that so many visitors are ill-shod. Alternative routes are found on either side of the gullies and the process of erosion starts again. In some places, the footpaths are now nearly 40 feet across. Worst hit are steep sections and low-lying boggy areas.

Erosion on Snowdon is not just confined to the footpaths. With the summit as the inevitable focus of so many people's labours, this area still remains — in the words of Prince Charles, nearly ten years ago — "the highest slum in England and Wales". The Snowdon Management Scheme aims to landscape the summit and negotiations are in progress with the Snowdon Mountain Railway company to help them to improve the appearance of the summit cafe and the surrounding buildings.

The increasing popularity of Snowdon has created other problems beside erosion.

Car parking space at the start of the main footpaths is inadequate. The car park at Pen y Pass is the worst hit and the £1 parking fee, levied in the high season, has done little to discourage motorists from parking there. The Snowdon Sherpa bus service, now in its fourth year, is a joint experimental venture with local bus companies which aims to alleviate these car parking problems. The service allows walkers to park their cars away from the mountain and to catch a bus to the start of their desired path. Climbers are now able to descend by a different route and catch the Sherpa bus back to their cars. Buses run every hour around the perimeter of the mountain during the summer period. The Snowdon Management Scheme also involves increasing the size of the free car park at Nant Peris and Sherpa buses run every half an hour between there and Pen y Pass, at a reduced rate.

Information Programme

Another important aspect of the Scheme is the Information Programme. The majority of the people who climb Snowdon are visitors with little or no experience of mountains who may never attempt to climb another. Snowdon, being the highest mountain in England and Wales, has always had this attraction to the inexperienced — its ascent is the highlight of their stay in Snowdonia. It is the job of the National Park Information Service, under the auspices of the Management Scheme, to inform visitors of the safest way to enjoy the mountain. Information boards will be erected at the start of the footpaths which will provide advice on conditions on the mountain, the sort of terrain to be expected and the equipment walkers need. Four extra Wardens have also been employed with the Scheme to patrol Snowdon and give advice and guidance to the public. Leaflets, Guided Walks and Information Centre exhibitions will be further developed as an aid to informing visitors about mountain safety.

However, there are other problems besides mountain safety, erosion and car parking that have been created by the increasing pressures of recreation on Snowdon. For instance, surprisingly few visitors are aware that most of the land in British National Parks is in fact privately owned. Indeed, most of Snowdon is owned or rented by farmers. Though the land is poor by most agricultural standards, thousands of sheep graze the mountain, providing a livelihood for the farmers concerned. This brings an inevitable conflict between the public and the farmers. Dogs are allowed to run wild, gates are left open and stone walls and fences are damaged. The increasing litter problem on Snowdon also affects farmers in that stock are often injured by broken glass and tins.

There is another conflict on Snowdon. Several of the footpaths run through an important Nature Reserve. When the last of the glaciers retreated from the area 10,000 years ago, a legacy of arctic-alpine plant communities was left behind. Snowdon and a few other nearby upland areas in the National Park represent the southernmost limit of these rare arctic-alpine communities and it is because of them that these areas were designated as National Nature Reserves. Public pressures on the mountain may endanger these important habitats, particularly if visitors stray from the footpaths.

It is the aim of the Snowdon Management Scheme to reconcile the conflicts between recreational, agricultural and nature conservancy interests and the role of the information programme to bring them to the attention of the visiting public.

Once the erosion problem has been tackled and the footpaths modified, drained and graded, then walkers may be encouraged to keep to them. Further erosion should then be minimised provided regular maintenance work is carried out. But the scale of footpath erosion on Snowdon is enormous and it is perhaps unlikely that the Snowdon Management Scheme's five year programme will be adequate to complete the work.

(*Source: The Rucksack*, Vol. 10 No. 1, Winter 1980)

Of course, not all popular beauty spots and nature reserves are severely damaged, but if we want to prevent damage occurring on similar sites, then some attempt must be made to manage sites and visitors in ways that reconcile conflicting interests. How can this be achieved?

MANAGEMENT PLANS

One way of achieving a solution to the problem is through the preparation of an agreed management plan for each site which sets out the aims and objectives for its use and conservation, and provides pratical suggestions about how these objectives can be achieved. In the case of Snowdon's summit, for example (Countryside Commission, 1977), consultants prepared a plan based on three lines of inquiry:

1 The effects on the physical environment of use by visitors.
2 The causes of environmental damage.
3 Proposals for a practical plan to ensure the successful long-term management of the site.

We can use these lines of inquiry to:

1 learn about how ecosystems work;

2 test specific hypotheses about the ability of different ecosystems to resist and recover from damage caused by trampling; and

3 raise discussions about the recreational capacity of different sites.

The purpose of this chapter is to show how studies of recreation sites can be used to improve our understanding of how ecosystems work and to raise issues about site management that reflect people's attitudes and values. As a first step in these studies it is often useful to establish a 'baseline' against which people-induced changes can be observed and measured. It may not always be easy to separate out the effects of the very many factors (both human and physical) that influence how an ecosystem works, but unless some attempt is made to unravel the complex interactions that link together plants, animals and the physical environment, it will be difficult to assess the impact of people-induced changes on an ecosystem. First, therefore, we must go right back to basics and understand how an ecosystem works. In this way we shall clarify which variables are significant and need to be considered.

HOW AN ECOSYSTEM WORKS

An ecosystem is an ordered and highly integrated community of plants and animals together with the environment that influences it. The chalk grassland shown in Figure 1 represents one example. Ecosystems work through the harnessing of the sun's radiant energy by green plants, and the transfer of this stored energy through a food web of grazing and carnivorous animals that live above ground and in the soil. The food web is the means whereby animals gain essential nutrients and energy for growth. The death of plants and animals, and their decay by the decomposer organisms that live in the soil, provide the main means by which nutrients are cycled between the atmosphere, soil and the living world. In this way, living organisms in a particular site and their non-living environment are intimately linked together into a single working whole.

? In pairs discuss Figure 1 and attempt to explain to each other what it says. Redraw the diagram of an ecosystem for the following:
a woodland, **b** moorland, **c** urban park.

The structure of an ecosystem at any one site or at any one point in time is a function of many interacting variables such as climate, soil, topography, and management history. Particular groups of plants and animals only occupy a site that affords them suitable growing conditions. The range of conditions over which a species grows and reproduces is called the tolerance range (Figure 2). No two species have identical tolerance ranges but ranges do overlap. For example bell heather (*Erica cinerea*) and cross-leaved heath (*Erica tetralix*) are closely related plant species and both grow in moorland and heathland communities in Britain (Figure 3). Cross-leaved heath can tolerate much wetter conditions than bell heather so that along a gradient of increasing soil wetness, each becomes dominant at a different point along the gradient, but both species grow together where their ranges overlap.

Soil wetness is only one of the very many factors that influence plant growth. For example, the climatic regime of an area influences the length of the growing season, soil type influences the pH and supply of nutrients available for uptake, and the presence or absence of grazing influences the invasion or suppression of particular species.

The complex interaction of variables can be expressed in the following form (first used by Jenny, 1941):

$$E = f(cl,o,r,p,t...)$$

Where E denotes any ecosystem property, cl the regional climate, o the living organisms present or available, r is the topography, p the parent material of soils, t is the time period over which the ecosystem has developed and the dots represent other unspecified factors such as fire. The 'f' stands for 'a function of'.

People are not included in Jenny's expression but they clearly influence ecosystem properties in very many ways. Deliberate management practices such as cultivation, fertiliser and pesticide application, cutting and burning all affect soil, and influence the structure and workings of an ecosystem favouring some plants at the expense of others. Accidental influences also result from air and water pollution, and from the introduction of alien species. Because few ecosystems have escaped the influence of people,

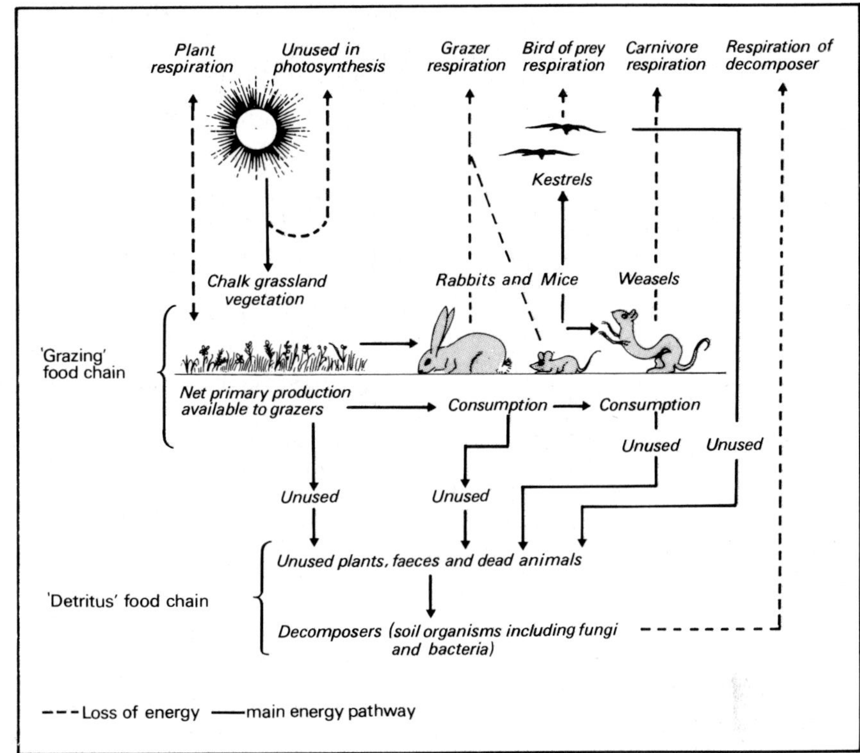

Figure 1 *The processes at work in a chalk ecosystem*

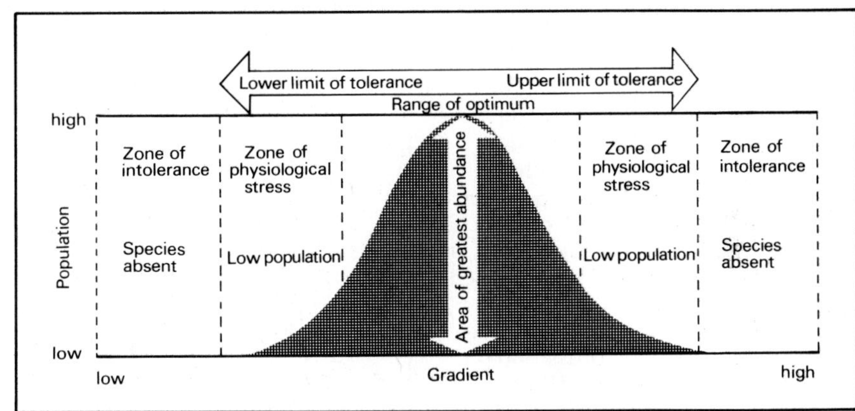

Figure 2 *The tolerance range of a species*

Figure 3 *Bell heather (*Erica cinerea*) and cross-leaved heath (*Erica tetralix*)*

studies which aim to investigate the effects of introducing a new factor, such as trampling, must include in their study programme some analysis of the variability of untrampled areas. The causes of variability will include factors associated with people's activities as well as factors of the physical environment. For example, in an area that is grazed by sheep we would expect to find some plants that are tolerant of trampling. If the sheep were to go, then the tolerant species would be suppressed by other more 'aggressive' plants which could not grow previously. The effects of new paths introduced into grazed and ungrazed grassland on similar soils, slopes and geological substrate will differ, mainly because of the initial composition of the vegetation. Any investigation of the effects of trampling must, therefore, include some attempt to assess what other environmental factors are likely to influence the structure and functioning of the ecosystem. Such 'baseline' investigations form an integral part of the approaches to trampling studies that will be discussed in this chapter.

? Make sure that you understand what is meant by baseline studies and that you can explain their necessity in a scientific investigation.

? This exercise is designed to investigate the influence of several environmental factors on the distribution of two grasses. Figure 4 illustrates some of the results of a random sample drawn from grasslands in the Sheffield region in the north of England. Figures 4c and 4d show the distribution of sites on different substrates, the Magnesian limestone and the Millstone grit respectively. Samples are plotted on a polar-graph which combines information about slope (concentric axes) and aspect (radial axes). At the centre of the diagram, sites are flat and sites on the circumference have a slope of 50°. The three o'clock position on the circumference represents a 50° east-facing slope, the six o'clock position a 50° south-facing slope etc. The pH of the soil at a depth of 10 cm is shown in 10 classes ranging from pH <3.6 (acid) to pH> 7.5 (very alkaline). Figures 4a and 4b plot the occurrence of two species, mat-grass (*Nardus stricta*) and upright brome (*Zerna erecta*) (see also Figure 5). In order to examine the tolerance range of each species and to discover which environmental factors appear to influence species occurrence, construct transparent overlays of the species distributions. Place these overlays onto the polar-graphs and answer the following questions:

What is the relationship between the occurrence of mat grass and upright brome and the following factors:

a geology? **c** aspect?
b pH? **d** slope?

First examine these relationships by eye and then construct histograms of the number of times each plant occurs for given pH classes, slope categories and aspect quadrants. What are the main differences between the two species with respect to tolerance range? Record your results for future reference so that you could write a description of the relationships.

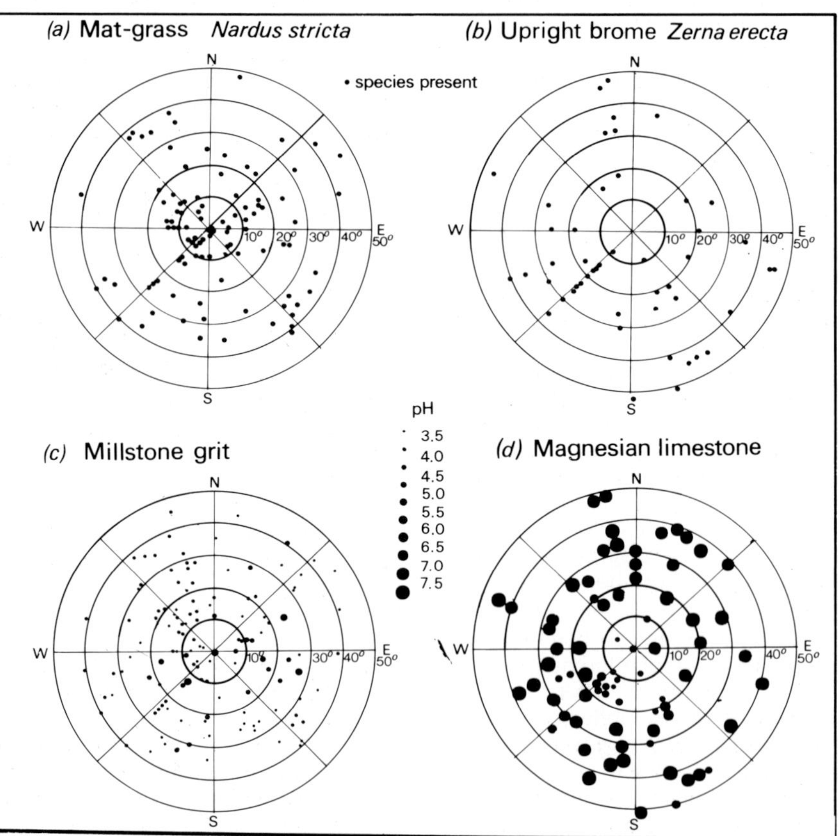

Figure 4 *a Mat-grass (*Nardus stricta*)* **b** *Upright brome (*Zerna erecta*)* **c** *Millstone grit* **d** *Magnesian limestone*

(*Source:* Grime, J P and Lloyd, P S *An Ecological Atlas of Grassland Plant*, 1973)

Figure 5 *Mat-grass (*Nardus stricta*) and upright brome (*Zerna erecta*)*

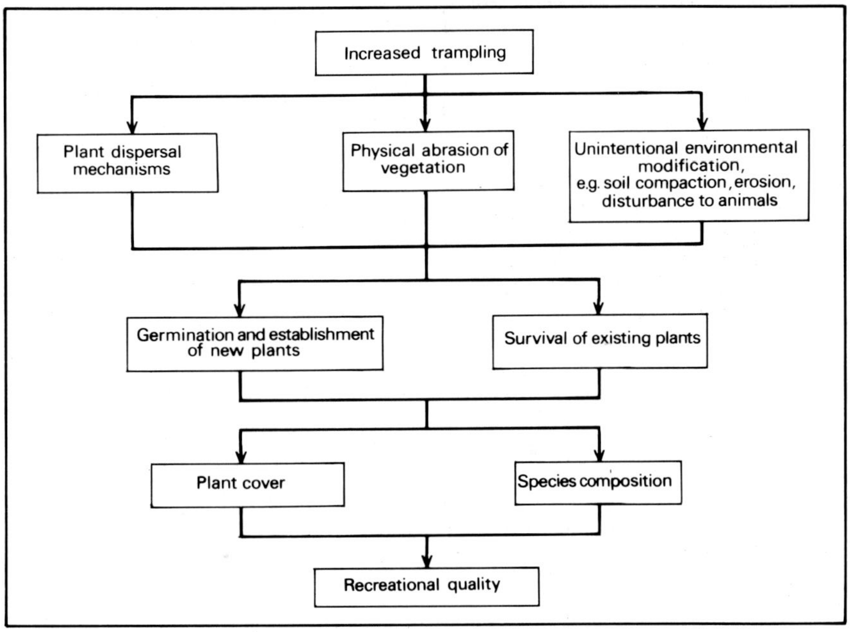

Increased trampling

Plant dispersal mechanisms

Physical abrasion of vegetation

Unintentional environmental modification, e.g. soil compaction, erosion, disturbance to animals

Germination and establishment of new plants

Survival of existing plants

Plant cover

Species composition

Recreational quality

Figure 6 *A model to show the effects of trampling on an ecosystem*

HOW TRAMPLING AFFECTS ECOSYSTEMS

Figure 6 illustrates the many ways in which trampling can affect ecosystems. The diagram emphasises the ways in which a change in one part of the ecosystem initiates a change in another. For example, trampling often leads to a loss in vegetative cover and to compaction of the soil. This is not really a problem on flat sites, but, on steep sloping sites, rapid erosion can follow. That is, for the same amount of use, the impact of trampling is measurably different for different topographical conditions.

The ease with which paths erode is also influenced by the physical properties of the soil on the path. In the absence of the binding effects of roots and organic material, the amount of material likely to be removed from a path through erosion can be determined through a knowledge of the physical structure of the soil. On a path that traverses several different vegetation and soil types, differences in the physical erosion resulting from trampling can be expected to relate to both plant types and soil structure. For example, one study in the Peak District carried out on a path that traversed several upland vegetation and soil types suggested that a critical slope threshold exists for paths on slopes of >14°. Above this threshold, physical erosion on paths was significantly higher than on other shallower slopes, irrespective of soil type. This finding suggests that attempts to improve the resilience of the path to wear by altering the gradient of the path would be more effective than attempts to improve the durability of the surface itself through the use of duck boards, sleepers or gravel dressing, etc. A similar conclusion led to the rerouting of the steep summit path at the top of Cairngorm in Scotland.

A number of studies have been designed to investigate some of the specific relationships suggested in Figure 6. In particular, the response of vegetation to trampling can be studied to illustrate two different properties of vegetation: **1**) its *vulnerability* to trampling pressure, and **2**) its *resilience*. Studies of *vulnerability* emphasise how readily different vegetation types and particular species are damaged by trampling pressure. Studies of *resilience* emphasise properties of the vegetation that survives in trampled areas, and the ability of different vegetation types to recover after trampling pressure is withdrawn. In order to make recommendations about the practical management of popular beauty spots, site managers are likely to require information about *both* the vulnerability of different vegetation types and their ability to recover.

? The interactions depicted in Figure 6 are very complex so examine the diagram carefully, talk it over at some length, and then attempt the following exercise:

As part of a local plan, a local council proposes to open up a newly acquired area of chalk downland to the general public. Use Figure 6 as the basis for a report to be included in the local newspaper, outlining the changes that will take place as a result of increased trampling. Redraw the diagram to show the kinds of processes and changes that are desribed by the existing boxes. Make a list of the other information you would need for a more detailed report.

METHODS OF STUDY

Two main approaches have been used to study the effects of trampling on vegetation and soils. The first includes field studies of existing paths in a variety of physical settings, and the second includes experimental studies conducted in the field and laboratory.

1 Path studies. Numerous studies have examined the detailed pattern of vegetation and soil properties on established paths and adjacent areas, with the objective of relating these patterns to different levels of use. Most of the studies involve the use of *transects* aligned at right angles to the path and along which sample *quadrats* are placed to record vegetation and soil characteristics. Observations are also made about the use of paths and the distribution of users across the path. Pressure pad counters concealed under the turf and visual counts made from a vantage point over set time periods, provide estimates of absolute levels of use. Bayfield's trampleometer is a very simple device for estimating the frequency with which different parts of a patch of ground are used.

The trampleometer consists of a 5 cm oval wire nail with a short length (3–5 cm) of fine wire soldered to the head. Snare wire or enamelled copper transformer wire has proved successful. The pins are inserted in the ground so that the head is flush with the surface and the fine wire projects vertically. Wires bent by walkers are recorded as 'hits' and are straightened for further use. In dense vegetation coloured, plastic-coated wire can ease the task of recording. Pins placed at distances of 10 cm apart can be laid out in a transect across paths and 'hits' recorded over short time intervals. Histograms which show records of 'hits' over a period of time give a clear indication of the extent of trampling across the path and of the zones which receive the heaviest traffic. Data expressed in this way are often suitable for rank correlation techniques.

2 Experimental studies. Field and laboratory experiments have been designed to examine the relationship between trampling and vegetation and soils under controlled conditions. The basic experimental design involves the use of several treatment plots or pot-grown plants and equivalent control plots or plants. In this way the growth and performance of a given vegetation type or plant species can be monitored for different levels of use and under different environmental conditions — flat or steep slopes, high or low fertility etc. Such an approach examines the influence and interactions of environmental variables on plant growth, and 'controls' or 'eliminates' the influence of other variables that are likely to affect the performance of plants.

Both path and experimental studies yield useful and complementary information about the vulnerability and resilience of vegetation to wear by trampling. Some of the findings of each type of study are introduced below.

PATH STUDIES

The zonation of plant communities across a path suggests that some plants are better able to tolerate trampling than others. We can ask four questions about these tolerant plant communities:

1 Are particular plant zonations associated with different levels of use?

2 Are path species invaders or members of the adjacent plant communities from which the path is derived?

3 Do path species exhibit common characteristics?

4 Do different vegetation and soil types differ in their vulnerability to wear?

Print 1 *Grimspound, on Dartmoor in South Devon, is a well-known prehistoric landmark. It is visited all-year round and as a result a distinct path has been worn between the car parking area (to the right of the print) and the summit of the tor. The vegetation on the hillside is heather moor, the soil is a thin peaty topsoil over a subsoil of granitic stones*

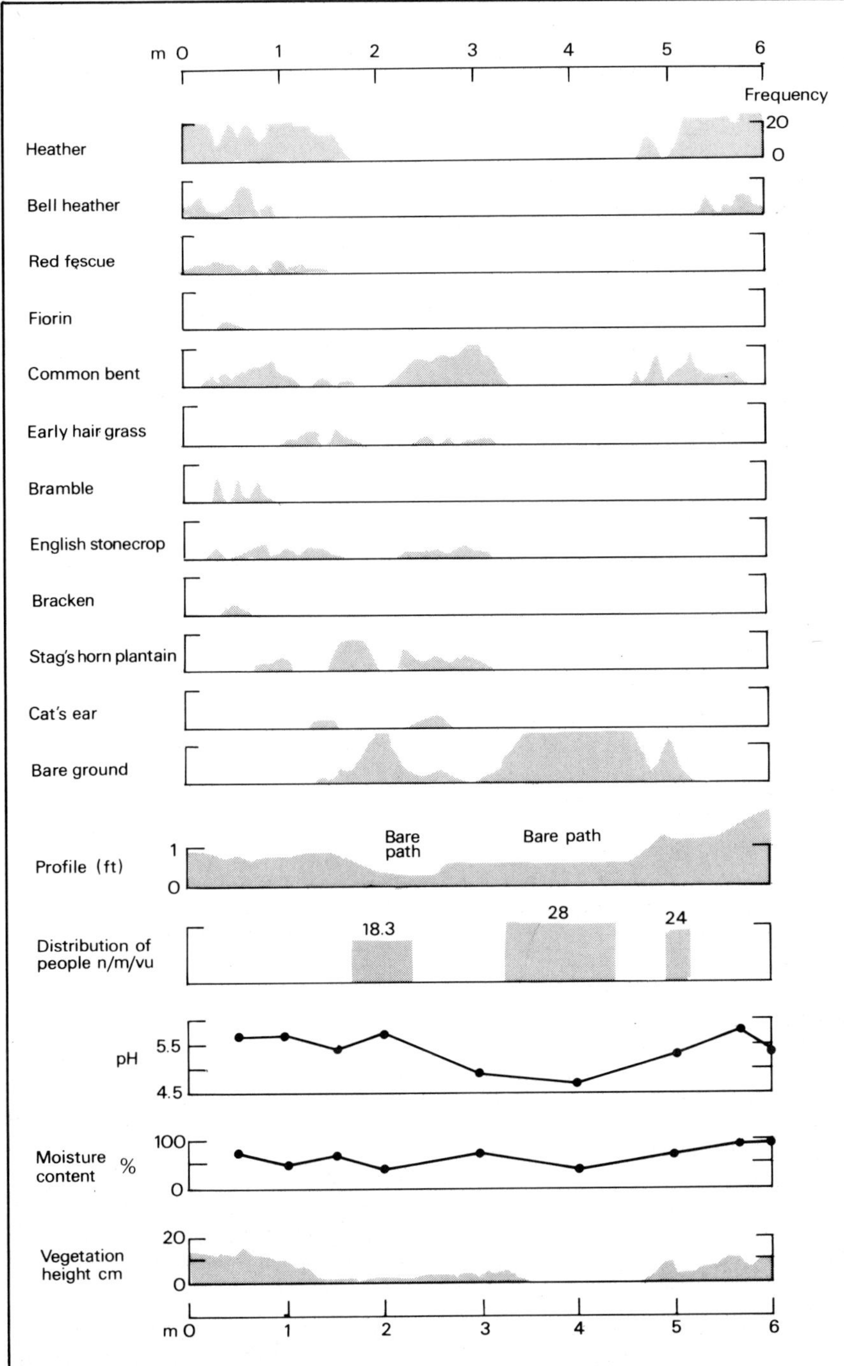

Figure 7 *Histogram showing the results of a study of plant zonation on a coastal path in the Scilly Isles*

(*Source:* Goldsmith, F B *et al*, *The Biological Journal Linn Soc. vol.* 2, 1970)

? Look at the diagram carefully. What distinctive zones can be recognised across the transect? Can you suggest reasons why these zones occur?

Clearly, much of the vegetation zonation appears to be related to the trampling pressure and so too is the penetrability of the soil. A comparison of several transects in the same area yielded the results shown in Table 1, which can be used to test whether or not there is a significant relationship between use and damaged ground. Use a chi-square test to test the significance of this relationship.

1 Are particular plant zonations associated with different levels of use?

Figure 7 illustrates the result of a study undertaken on a coastal path in the Scilly Isles. A series of transects were laid out across the path and at selected distances along the transect several observations were made about the plant community, soils and site conditions. The diagram expresses these data in a histogram form.

Although a significant relationship might be found from this analysis, in proposing that there is a causal relationship, you must be aware that other factors may also exert an influence on the results, for example, distance from the sea and the influence of the sea spray.

2 Are the path species invaders or members of adjacent communities?

By comparing the species composition of adjacent plots with the flora found on a path, an assessment can be made of whether the path species have invaded from distant plant communities or from those nearby. Moreover, by examining the mode of dispersal and spread, some suggestions can be made about how plants arrive on a path. Some seeds will be brought in by people and their domestic animals, but others will be dependent upon other mechanisms of dispersal (wind, mechanical, gravity) and will be found represented in the adjacent communities.

3 Do path species exhibit common characteristics?

Trampling causes mechanical damage to plants through ripping and tearing as well as through compression. As you would expect, certain plant forms are better able to survive on paths than others.

Plants can be grouped together on the basis of their *life-form*. Life-form classifies plants according to the position of the bud or growing point from which new growth is made in relation to soil level. A Danish botanist, Raunkiaer, in 1932 identified the following primary life-form classes:

1 *Phanerophytes* — woody plants with buds more than 25 cm above soil level.
2 *Chamaephytes* — woody or herbaceous plants with buds above the soil surface but below 25 cm.
3 *Hemicryptophytes* — herbs (very rarely woody plants) with buds at soil level.
4 *Geophytes* — herbs with buds below the soil surface.
5 *Helophytes* — marsh plants.
6 *Hydrophytes* — water plants.
7 *Therophytes* — plants which survive as seeds.

For studies of plants associated with paths it is possible to further subdivide the groups, as in Table 2, using a combination of leaf form and arrangement. Using this classification of plants it is possible to examine the percentage of life-forms found among vegetation samples drawn from paths, and from the adjacent community. Many studies have shown that path plants exhibit a preponderance of the Hemicryptophyte, Geophyte and Therophyte groups and that the rosette form of leaf arrangement, such as that found with the plantain and daisy is more frequent on the path community than elsewhere. The advantage that these life-forms have over other forms is found in the protection of the meristematic tissue (the parts of the plant responsible for growth) afforded by the tough, flat leaves and by the soil itself. Other life-forms have buds held above the soil surface and these buds are exposed and readily damaged by trampling (see Figure 8 and the following exercise).

Table 2 *Sub-division of life-form used in Clapham, Tutin and Warburg's* Flora of the British Isles *(2nd edition), 1962*

Phanerophytes — buds >25 cm above soil level
M.M. Mega-mesophanerophytes — from 8 m upwards (trees)
M. Microphanerophytes — 2–8 m (travellers joy)
N. Nanophanerophytes — 25 cm–2 m (butchers broom)

Chamaephytes (Ch.) — buds <25 cm above soil level
Chw. Woody chamaephytes (e.g. thyme, heather)
Chh. Herbaceous chamaephytes (e.g. germander speedwell)
Chc. Cushion plants (e.g. thrift)

Hemicryptophytes (H.) — buds at soil level
Hp. Protohemicryptophytes — with uniformly leafy stems, but basal leaves usually smaller than the rest (e.g. bulbous buttercup, many grasses)
Hs. Semi-rosette hemicryptophytes, with leafy stems but the lower leaves larger than the upper ones and the basal nodes shortened (e.g. foxglove, tormentil)
Hr. Rosette hemicryptophytes, with leafless flowering stems and a basal rosette of leaves (e.g. plantain, silverweed, daisy, dandelion)

Geophytes (G.) — buds below soil surface
Gb. Geophytes with bulbs (field garlic, bluebell)
Gr. Geophytes with buds on roots
Grh. Geophytes with buds on rhizomes (bracken, wood anemone)
Grt. Geophytes with buds on root tubers (spider orchid)
Gt. Geophytes with buds on stem tubers or corms (wild arum)

4 Do different vegetation and soil types differ in their vulnerability to wear?

One study undertaken of the Pennine Way, a long-distance path in the uplands of Britain (Countryside Commission, 1971), examined the extent to which the width of the path and the amount of bare ground exposed on the path varied with different vegetation and soil types, and with different intensities of use. An *index of extent*, was used to describe the vulnerability of different vegetation and soil types to wear. It is a synthetic index which is calculated using the width of the path and the proportion of bare ground:
Index of extent = overall width + bare width + width of undamaged vegetation.

Figure 9, illustrates the features measured on a large number of random points along a section of such a path.

? Table 3 (*page* 400) shows the results from a comparable survey undertaken in similar upland conditions in the Peak District. Carry out the exercise below the table and then calculate the index of extent for each sample and construct a 'wigwam' diagram for each soil type as illustrated in Figure 10. Which soil type is the most vulnerable to wear? Suggest reasons for the differences between soil-types. The causal mechanisms involved might

Table 1

	Number of samples		
	High use	*Low use*	*Total*
High percentage bare ground *< 50 per cent live cover*	90	25	115
Low percentage bare ground *> 50 per cent live cover*	170	115	285
	260	140	400

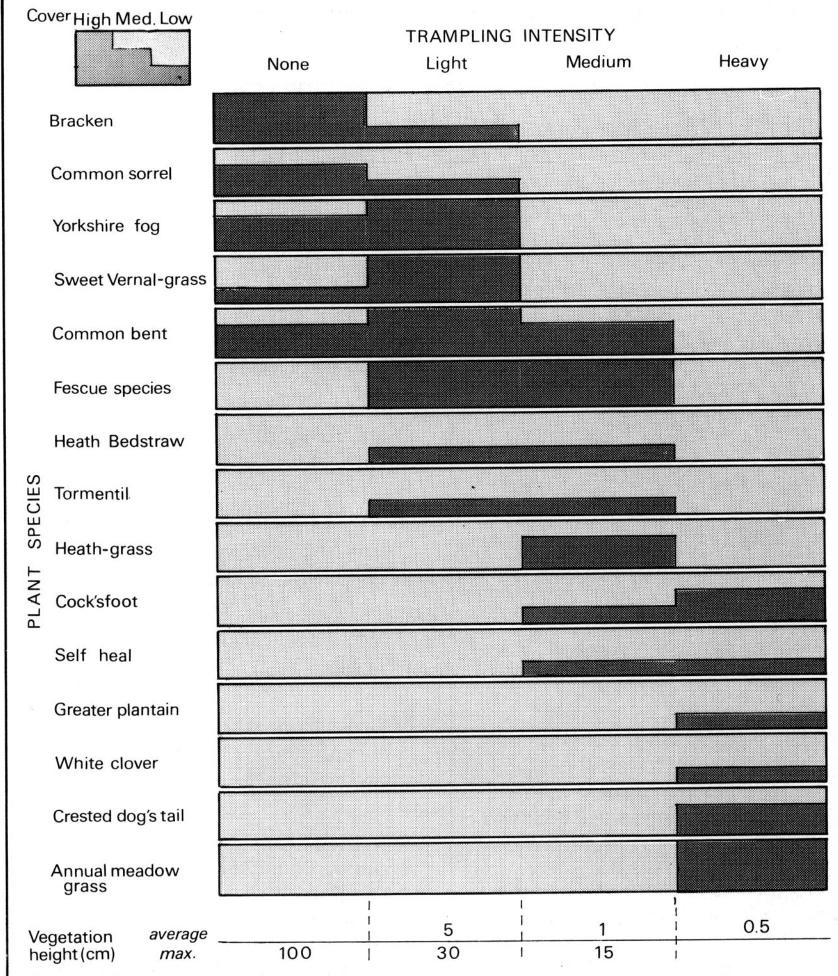

Figure 8 *Vegetation and change on four paths in Tarn Hows as a result of trampling*

? Figure 8 illustrates the kind of graphical presentation of vegetation and change that can be achieved by comparing several paths of known use from a single upland site, Tarn Hows (Countryside Commission, 1978). Use a book such as *Flora of the British Isles* to find out which life-forms are represented by the species shown in Figure 8. Make a list of the life-forms which appear to be the most tolerant of trampling.

? include the role of organic matter as a physical buffer, the moisture status of the two soils, and seasonal differences relating to use.

Conclusions from path studies

The most consistent finding from the numerous studies that have been undertaken on existing paths is that a distinctive path flora is established at moderate levels of trampling. Different species achieve dominance in different habitats but tolerant plants are likely to exhibit similar life-forms. These studies also show that the amount of damage caused by the same amount of use is related to the initial composition of the vegetation. For example, vegetation dominated by bilberry, a dwarf shrub (*Chamaephyte*), is more vulnerable than grassland dominated by mat grass (*Hemicryptophyte*). Other more extensive surveys of this kind have also shown that sloping sites are more readily damaged and eroded than sites on shallow gradients.

Limitations of these studies

One of the main problems associated with these studies is that they make an assumption that the vegetation zones represented on a path are in equilibrium with the present level of trampling and management. On established paths of known history this may be a valid assumption to make, but many paths have multiple uses and have been used for a multitude of purposes in the past. Studies also demand that an estimate of use is made. On some sites, levels of use may be known, on others direct observation may have to be made using particular access points or using pressure pad counters on selected paths. In all events, however, some estimate of use is required before much sense can be made of the empirical data.

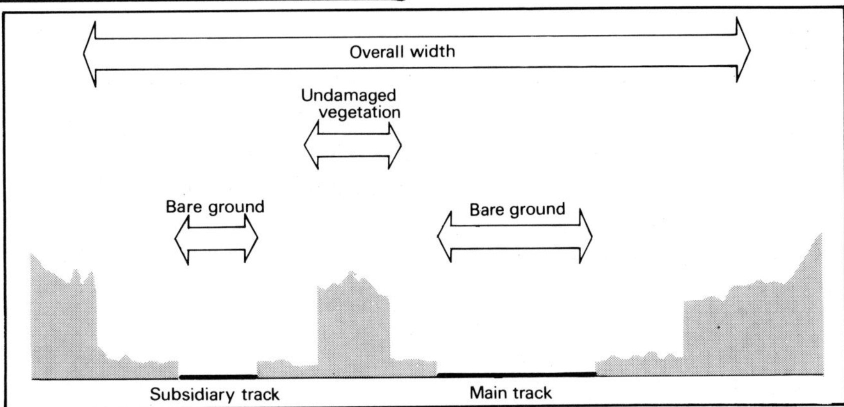

Figure 9 *Features measured at random points on a path to calculate the index of extent*

Table 3 *Footpath extent mean values*

Soil type		*Trampling pressure*	*Width**	*Bare ground**	*Undamaged vegetation**
Podzol	$\bar{x}$	7 852	5.73	0.29	1.34
Podzol	$\bar{x}$	2 000	5.73	2.04	1.48
Podzol	$\bar{x}$	2 000	2.41	3.34	0.05
Groundwater Podzol	$\bar{x}$	7 852	7.51	1.19	0.39
Groundwater Podzol	$\bar{x}$	1 000	3.13	0.17	0.02
Groundwater Podzol	$\bar{x}$	29 082	8.45	4.81	0.65
Peaty mineral	$\bar{x}$	13 289	7.89	3.57	0.63
Peaty mineral	$\bar{x}$	24 809	7.26	3.00	0.07
Peaty mineral	$\bar{x}$	12 028	3.81	2.59	0.55
Blanket peat	$\bar{x}$	27 761	9.60	5.76	0.64
Blanket peat	$\bar{x}$	24 011	11.24	7.34	0.96
Blanket peat	$\bar{x}$	16 380	5.52	3.43	0.68

* in metres

(*Source*: Naylor, M. C., 1977,'The Impact of Recreation on the Vegetation and Soils of Kinderscout', *Discussion Papers in Conservation 16*, University College London)

? Use the data provided in Table 3 to plot the relationship between use and path width and apply the correlation coefficient to test the significance of this relationship. Do your results lend support to the findings of other studies which suggest that a doubling in use leads to a doubling in path width? Carry out a similar exercise to examine the relationship between use and bare ground. Account for the differences between the findings of these two exercises.

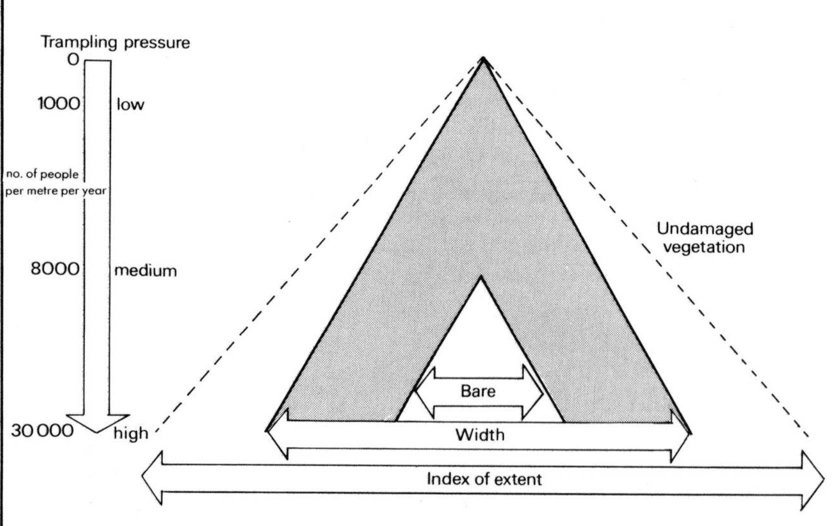

Figure 10 *A wigwam diagram showing the vulnerability to wear of a particular soil type*

? You may be able to identify paths in your local area which could be used for studies similar to those described here. Set out a design for a study in terms of hypothesis to be tested; assumptions made; data collection methods; data processing methods; conclusions; and strengths and weaknesses of the study.

FIELD EXPERIMENTS

A number of field experiments have been designed to study the following questions:

1 What are the effects of different intensities of trampling on vegetation and soil?

2 What are the seasonal effects of trampling?

3 Are different types of user more damaging than others (for example, cars, bicycles, horses and people)?

4 Which plant communities regenerate most quickly after trampling?

Most of these experiments involve the use of field plots in areas of previously untrampled vegetation. The experimental design includes a series of control plots and treatment plots. Control plots provide a baseline against which changes in the treatment plot can be assessed and provide a means of taking into account changes that occur in the plant community as a result of normal conditions of growth. The results of one field experiment designed to assess the cumulative effects of weekly trampling on different ecosystems are found in Table 4. In order to express the results of the trampling treatment on the vegetation cover, we need to express the live cover on the treatment plot as a percentage of that on the control plots. Various aspects of the vulnerability of different vegetation types can be expressed and compared.

Table 4 *Cumulative effects of weekly trampling on different ecosystems*

Species	Percentage Cover: End treatment, Control	Percentage Cover: End treatment, Path	Percentage Cover: End recovery, Control	Percentage Cover: End recovery, Path
Site: Keston Common, heathland on pebble gravels; podzolic soil, pH 4.5				
Wavy hair-grass	2	2	28	16
Heather	43	1	31	–
Moss	–	2	–	–
Litter	55	95	41	76
Bare				8
Site: Redhill Common, grass heath on Lower Greensands; podzolic soil, pH 4.5				
Yorkshire fog	40	4	42	25
Wavy hair-grass	4	7	9	19
Bent	3	1	–	–
Sheeps sorrel	4	1	–	–
Bracken	48	–	49	4
Litter	1	87	–	52
Site: Farthing Downs, annually mown grassland on rendzina soil over chalk, pH 6.5				
Yorkshire fog	16	–	14	–
Cocksfoot	–	1	7	23
Fescue	15	7	5	5
False oats grass	23	2	16	10
Bent	33	21	–	2
Knapweed	2	–	–	1
Thistle	3	–	–	1
Litter	8	69	58	58
Site: Happy Valley, annually mown chalk grassland on clay-with-flints over chalk, pH 7				
False oats grass	19	20	19	29
Cocksfoot	15	12	35	20
Yorkshire fog	14	–	15	–
Bent	23	7	18	17
Buttercup	11	–	–	–
Fescue	6	5	5	–
White clover	4	–	2	–
Dock	2	–	–	1
Dandelion	–	6	–	14
Litter	6	50	6	19
Site: Epsom Common, gleyed,silty clay, pH 6				
Bent	17	26	18	16
Yorkshire fog	21	–	13	12
False oats grass	46	3	8	26
Bedstraw	–	2	–	2
Birch	5	–	–	–
Potentil	1	–	–	–
Litter	10	69	61	44

(*Source*: Harrison, C. M., 1980–81, *Biol. Conserv.* 19)

? Use the data provided in Table 4 (*page 401*) to complete the following tasks:

1 Calculate the total percentage of live cover (excluding litter and bare ground) for controls and path.

2 Express the total live cover on paths as a percentage of live cover on the control. This gives you the *relative cover*.

3 Plot the increase or decrease in relative cover of path at the end of treatment and at the end of the recovery period against pH of soil and suggest reasons for these relationships.

4 Integrate the information for vulnerability (percentage live cover at end of treatment) and resilience (percentage live cover at end of recovery) in a single graph. Does the most vulnerable vegetation recover the most slowly? Explain your answer.

5 Which plants 'decrease' and which 'increase' under trampling? Do they have any common characteristics?

6 Which species are 'invaders'? How do they colonise?

7 Does the performance of selected species, for example, Yorkshire fog, cocksfoot and false oats grass, differ by habitat type? Explain your answer.

LABORATORY EXPERIMENTS

In order to determine precisely how plants respond to trampling and in particular to look at the different ways in which plants respond to repeated abrasion and compression associated with treading, laboratory experiments can be conducted under carefully controlled environmental conditions. Bayfield, for example, has studied the effects of wear using artificial 'feet' made from a block of wood covered with the rubber sole from a walker's boot which is dropped from a known height onto pot plants containing different species. By carefully controlling the environmental conditions under which plants grow, the effects of increased compression on the soil, which may lead to the loss of water-holding capacity and lack of aeration, can be assessed as well as artificially induced changes in nutrient availability, etc.

Comparative studies of different species have shown how tillering (the growing of new shoots) and regrowth of certain species is actually encouraged by moderate levels of trampling but that as trampling pressure increases, regrowth is prevented. Plants die because of the loss of photosynthetic surfaces and the draining of stored food reserves in root tissues. Species with large stores of food held in roots or underground storage organs are at an advantage, but so too are plants that can germinate rapidly from seed. These studies have shown that the reasons why different species are either more vulnerable or more resilient, are complex, and depend very much upon the growth strategy of the plant and the physical and chemical properties of the rooting medium.

Much of the commercial work devoted to breeding resilient strains of grasses for use on sports pitches, for example, has proceeded on the basis of laboratory and field trials of these kinds. As a result, it has been possible to develop commercial seed mixtures that can be used to reinforce a plant community artificially. Most of these plant species, however, depend upon a substrate that is well-drained and fertilised, i.e. optimum growth conditions, and they are not suitable for use in wild areas that are likely to offer sub-optimal conditions for growth. For these reasons, any practical attempt to reinforce the resilience of wild areas is likely to depend upon growth trials undertaken with local species, and under local environmental conditions.

How quickly do different plant communities recover from trampling?

Several studies have begun to examine how quickly worn areas can regenerate again after trampling pressure has stopped. Most of these studies involve long-term monitoring of experimental plots that have been subjected to different types of trampling pressure and then allowed to recover. Other studies have involved the monitoring of worn paths, after use has been withdrawn. The majority emphasise that the ability of vegetation to recover is a function of the amount of damage inflicted, the life-form and growth strategy of plants and the overall suitability of conditions for plant growth at a site, e.g. fertility, soil-moisture, general climatic regime. For example, Bayfield shows that in the exposed and severe environment of an upland moorland on Cairngorm, damage to deer grass

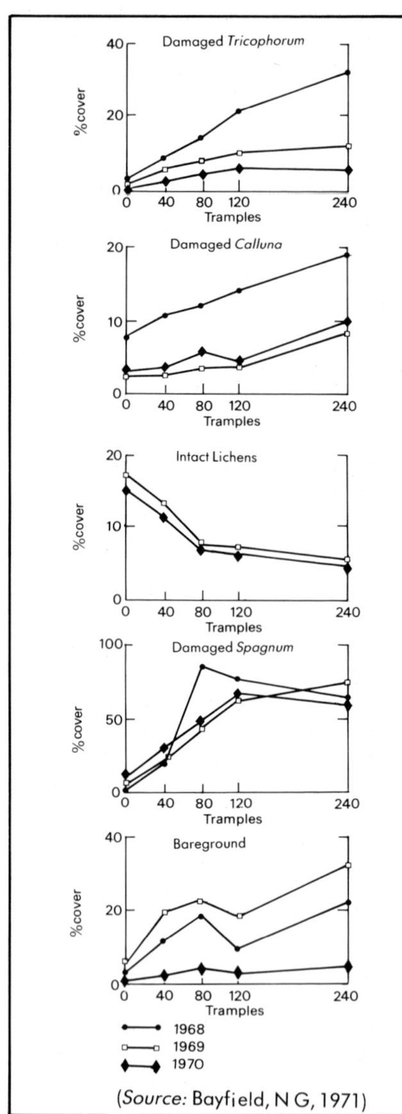

(*Source:* Bayfield, N G, 1971)

Figure 11 *The effects of trampling on the upland moorland communities on Cairngorm and recovery over a period of three years*

(*Tricophorum*) is clearly related to trampling pressure, and reaches a maximum value of 34 per cent for the most heavily trampled plots (Figure 11). Recovery was fairly rapid, probably because of its deciduous habit and basal growing buds. Damage to heather, (*Calluna*) is more severe and included dead, broken, bruised and stripped stems. Recovery of heather was extremely slow and at the end of the monitoring period, an 18 per cent live cover represented a considerable reduction on the 35 per cent before trampling. Bog moss (*Sphagnum*) (see Figure 11) was the most vulnerable of all, because its buds are not protected, and regrowth is a lengthy process under these severe environmental conditions.

By contrast, a study of the recovery of a temporary nature-trail at Ranmore Common on chalk grassland and woodland in lowland England shows how readily this vegetation recovered from a short period of intense trampling. At this site it was almost impossible to tell where the trail had been located after two years of regrowth.

The limitations of experimental studies

The advantage of an experimental approach is that short-term changes in the workings of an ecosystem which result from measured amounts of trampling can be studied in detail. It is difficult, however, to examine long-term changes in the ecosystem that involve adaptive changes in the structure and composition of vegetation. In other words, the arrival and growth of new species that achieve an equilibrium condition, and cause the development of a *new* ecosystem cannot be studied very readily. Many studies have shown that distinctive vegetation is associated with established paths, but the short life of field and laboratory experiments means that we cannot take into account these processes of adaptation. However, the advantage of the experimental approach is that we can control for different environmental variables and hence improve our understanding of the mechanisms involved in the development of vulnerability and resilience to wear by trampling of both vegetation types and particular plant species.

RECREATIONAL CAPACITY

One of the concepts that has been used widely and uncritically in recreation studies is the concept of 'recreation capacity'. Many studies seem to suggest that a site has a readily definable 'capacity' beyond which further use is unacceptable.

? Before reading the next section, attempt to write a short, succinct definition of recreational capacity. When you have finished the section, review your definition and identify which of the four elements of recreational capacity the definition highlights.

The recreational capacity of a site can only be defined once the declared management objectives for that site and the degree of 'naturalness' necessary to preserve it have been determined. In practice, any working definition of recreational capacity is likely to involve four elements:

1 Physical capacity: the maximum number of people or vehicles that can be accommodated by facilities on the site, e.g. car parks, toilets, restaurants, viewpoints, etc.

2 Ecological capacity: the maximum level of recreational use (numbers and activities) that can occur before unacceptable changes take place in the flora, fauna and soil.

3 Economic capacity: the maximum level of recreational use (numbers and activities) that can occur before unacceptable changes take place in commercial activities carried out on the same site, e.g. forestry on woodland sites, water quality of reservoirs, etc.

4 Perceptual capacity: the maximum level of recreational use above which there is a decline in the quality of the recreational experience of the users, e.g. overcrowding, behaviour conflict between riders and walkers, etc.

The recreational capacity of a site is only likely to be agreed upon if these various elements of capacity can be precisely determined and applied. Just how difficult this task can be is exemplified by perceptual capacity. The definition of perceptual capacity depends upon the type of user. Someone seeking peace and tranquillity will have a lower threshold for perceptual capacity than another person who has come to watch and enjoy the activities of others. Likewise, a definition of ecological capacity depends upon some assessment of the ecological 'worth' or 'value' of particular habitats or species. Different groups of people will provide different assessments depending upon their training, knowledge and degree of commitment to natural history.

In practice, recreational capacity can rarely be determined and the preparation of the management plan for an area is likely to involve:

1 a clear statement of what the objectives of management are;

2 an exploration of the various policy and management options that can be taken by management in pursuit of these objectives;

3 a series of discussions which seek to reconcile the views of different interest groups.

Figure 12 illustrates some of the policy and management options that need to be considered when attempting to balance use and capacity, and examples of how some practices have been applied are briefly introduced below.

PRACTICAL MEASURES TO CONTROL TRAMPLING

Limiting access

The main way in which trampling can be controlled is through a limitation on access arrangements. In some cases this may be achieved by restricting the number of access points and the number of parking places. In others, more drastic action may be required and certain paths and tracks may need to be closed and fenced off. Such action would lead to complaints and dissent, especially on public land over which access is regarded as a right that has been hard won; this happened, for example, in the Peak District National Park. Subtle means of controlling use, such as alternative way-marked paths, guided walks and nature trails, are more positive

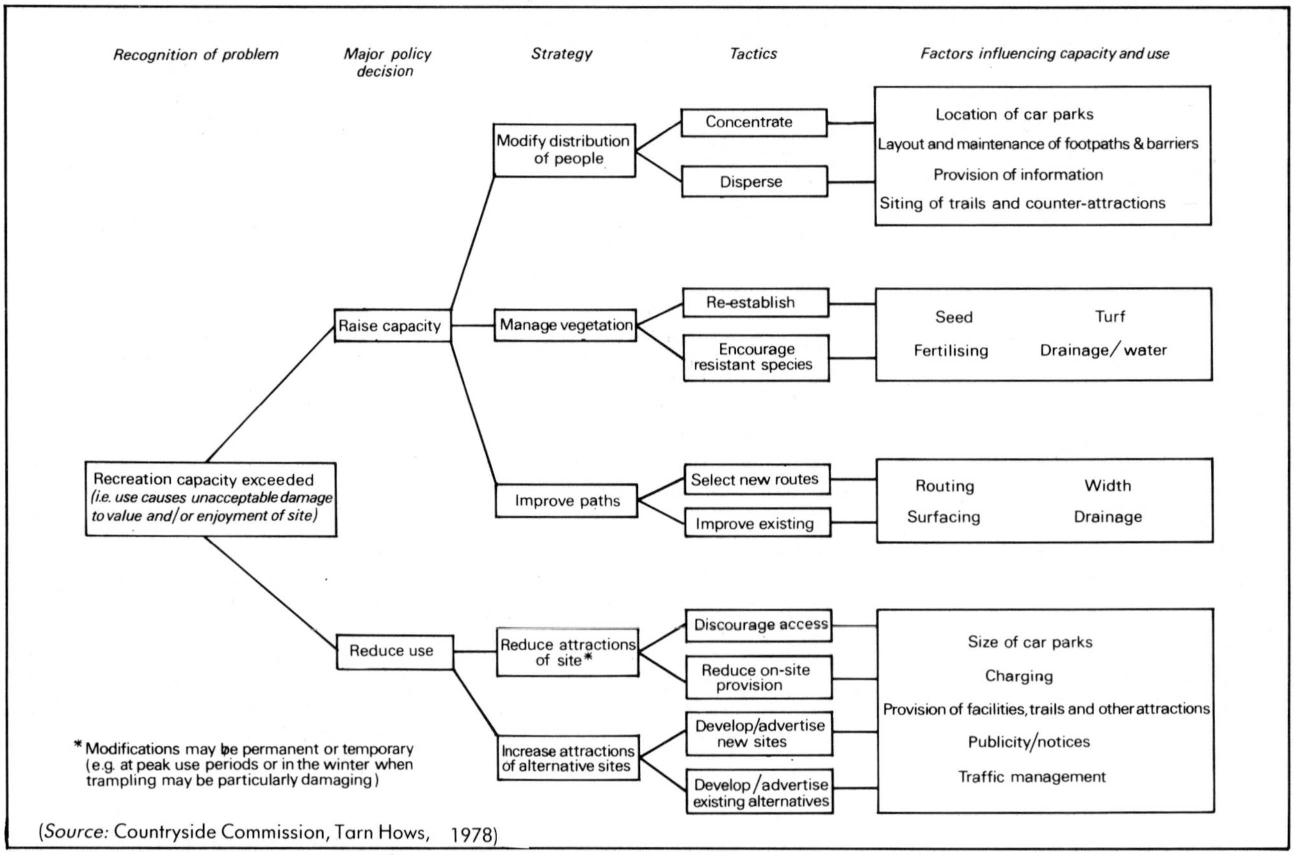

Recognition of problem | *Major policy decision* | *Strategy* | *Tactics* | *Factors influencing capacity and use*

- Recreation capacity exceeded (*i.e. use causes unacceptable damage to value and/or enjoyment of site*)
  - Raise capacity
    - Modify distribution of people
      - Concentrate
      - Disperse
      - Factors: Location of car parks; Layout and maintenance of footpaths & barriers; Provision of information; Siting of trails and counter-attractions
    - Manage vegetation
      - Re-establish
      - Encourage resistant species
      - Factors: Seed; Turf; Fertilising; Drainage/water
    - Improve paths
      - Select new routes
      - Improve existing
      - Factors: Routing; Width; Surfacing; Drainage
  - Reduce use
    - Reduce attractions of site*
      - Discourage access
      - Reduce on-site provision
    - Increase attractions of alternative sites
      - Develop/advertise new sites
      - Develop/advertise existing alternatives
    - Factors: Size of car parks; Charging; Provision of facilities, trails and other attractions; Publicity/notices; Traffic management

* Modifications may be permanent or temporary (e.g. at peak use periods or in the winter when trampling may be particularly damaging)

(*Source:* Countryside Commission, Tarn Hows, 1978)

Figure 12 *Balancing the usage and capacity of a site – some of the possible options*

methods of influencing pressure and directing it to the most tolerant and least vulnerable parts of a site. For example, at Kynance Cove a new route was constructed and people making the climb back from the beach to the cliff top were directed along an inland route away from the sites in which rare plants grew. Similar means have been used on Snowdon, Cairngorm and in the Peak District. Such measures have, however, assumed that damage is merely a function of numbers of people and successful rerouting will need to ensure that a significant number of all users can be permanently rerouted so that any remaining use is consistent with the long-term stability of the sensitive eroded areas. In some cases, the rerouting of the most damaging users at particular times of the year, e.g. horse riders in winter, may be sufficient to promote a desirable outcome without restricting use by other less damaging users such as sightseers.

Reinforcing the habitat

Several measures are open to the site manager under this category, ranging from small-scale intervention, such as reseeding and occasional fertiliser applications, through to progressively more intensive and expensive measures, such as regrading of slopes, step construction, turf laying, the introduction of artificial surfaces such as aggregates, gravel and duckboards, concrete and tarmacadam. The costs of path improvement on Snowdon came to just under £100 000 in 1974 with a projected annual upkeep rate of £13 500. Clearly, each path surface has to be designed with some use, capacity in mind. Any management plan has to consider numbers of people, the durability of the type of surface together with the costs of construction and upkeep.

Manipulating people and surfaces

Because of the prohibitive costs and visual impact of artificial surfaces in countryside locations on the one hand, and on the other, the difficulty of gaining widespread acceptance for the strictly controlled use of open countryside, most management plans are likely to require a dual approach to management. Successful plans such as those carried out in the Peak District, for example, illustrate how a combined approach can be devised and implemented.

Print 2 *Ranger painting a 'Waymark' sign in the Peak District*

Managing the Peak National Park

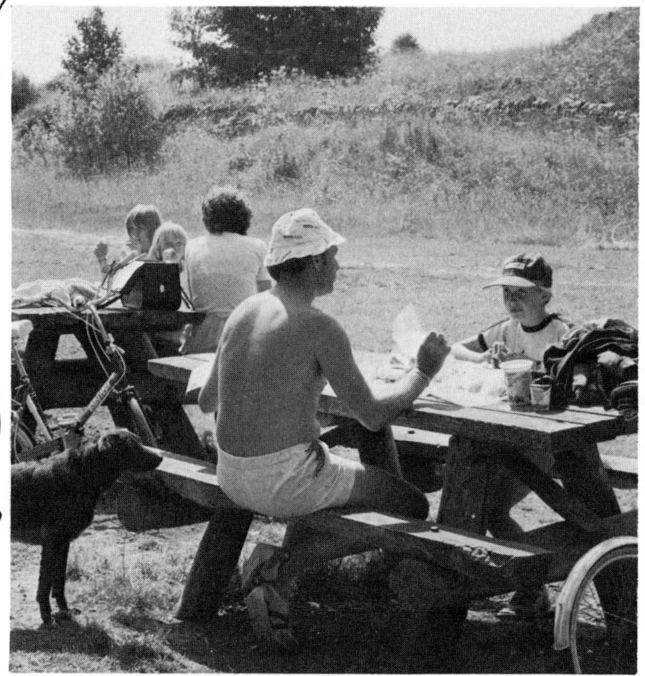

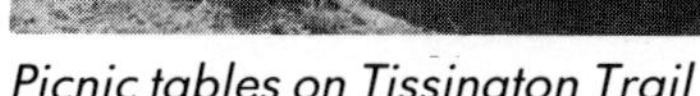

Picnic tables on Tissington Trail

Cycling on High Peak Trail

The Peak is the most urban of the national parks in the sense that it is most closely juxtaposed to great conurbations (Manchester, West Yorkshire, Sheffield) which have traditionally looked to its countryside as a playground. Significantly, the Peak Park was represented at the celebrations in April 1982 of the fiftieth anniversary of the Kinder Scout Mass Trespass, a milestone in the ramblers' struggle for access to open moorland. In the Peak the voluntary conservation bodies, and subsequently the Peak Board itself, belong firmly to the *inclusive* tradition of amenity concern, that of the hikers, cyclists and footpath societies. In other national parks the *exclusive* tradition is dominant, that of the county branches of the Council for the Protection of Rural England and the naturalists' trusts. This tradition holds (though rarely states) that amenity is a fragile good to be preserved despite people rather than for them.

Not surprisingly, the Peak points with pride to its solid achievement in opening some 76 square miles of open moorland to walkers under the type of access agreements provided for in the 1949 National Parks Act. It is doubtful if there is as much access land in all the rest of Britain. But even in the Peak the extension of access has come nearly to a halt in recent years with important stretches of hill country barred by intransigent landowners.

As well as opening the moorlands to the tougher walkers, the Peak Park has planned and spent for those wanting gentler strolls, most notably by turning disused railway lines into paths.

The Board has always taken the view that visitors need to be informed, educated and sometimes — to put it bluntly — *controlled* as well as welcomed with access land. There are radical ramblers who find the well-meant shepherding of the Park rangers, checking that access land bye-laws are obeyed, a touch too near the para-military for their freethinking taste. But the Board's information and interpretative services are the envy of the other nine parks in point of both numbers and quality. And the Peak's two working schemes, whereby cars are banned from a significant mileage of otherwise congested lanes on summer Sundays, put it well ahead of other parks. The elder of these in the Goyt Valley dates back to the early 70s. The later, in the Derwent Valley, was launched more recently. Six and a half miles of road are closed on Sundays and a few more miles of a cul de sac lane are permanently shut. But the local footpaths have been improved, car parks installed and a minibus serves the closed lengths. Cars displaying disabled badges are let through. In a place where 4000 visitors and 600 cars come on a fine Sunday, the transformation is magical. Quiet has been restored. Bike-hire flourishes and the local parish council wants the scheme extended.

National parks have twin duties to conserve the countryside entrusted to them and to promote appropriate recreation in it; achieving these ends is an exacting task.

(*Source:* Christopher Hall, 'Heritage on a Shoe-String', *New Scientist*, 10 June 1982)

? The following role play explores the conflicts and compromises involved in the preparation of an agreed management plan. Before the role play, make lists of the opportunities, constraints and considerations which are open to you or which you must take account of as a manager. Rank your lists in accordance with your earlier list of solutions drawn up at the beginning of the chapter. What have you learned?

Task
Choose an area which is used for informal recreation and which has some areas of semi-natural vegetation in it, such as woodland, heathland, downland or moorland. A common in or near a town, and a country park would be suitable types of site. The more popular the site is, the better. You are the manager of this site and are charged with the task of formulating a management plan for the site. Existing use has caused damage through trampling. At a public meeting various types of evidence are laid before you and different interest groups voice their opinions. After the discussion you will need to present a written report outlining a preferred management plan for the site.

Mechanics
Suggestions for the type of evidence that can be collected and analysed before the exercise, based on field surveys, are listed below:
a diagram of paths in relation to main habitats present (woodland, grassland, heaths, etc.);
b correlation diagram of 'index of extent' of paths and numbers of people using different paths;
c 'wigwam' diagrams to illustrate path development on different vegetation/soils for different levels of use;
d information on use of site by different user groups, either prepared by questionnaire or through simple observation techniques coupled with supportive information from managers.
Suggestions for the type of interest groups represented at the meeting:

- Commons conservators: concerned to preserve the essential countryside character of the site;
- Local residents: concerned to **i** prevent over-use by non-residents' (use correlation diagram), and **ii** to promote views of the lower socio-economic groups who are not represented by traditional user groups;
- Members of British Horse Society: concerned to extend bridleways;
- Member of the Local Naturalists Group: concerned about the extinction of rare plants and animals (use 'wigwam' diagram);
- Ramblers Association: concerned to provide and maintain access for all (use user survey);
- Chairman of local Chamber of Commerce: concerned to see more land released for development.

The Activity
Around the table are:

- the Chairman of the Management Committee;
- the Recreation officer from the local borough concerned about the costs of management;
- a representative from the Traffic and Highways Committee concerned about traffic problems.

1 Divide into groups. Each group represents one interest group and must prepare a script which briefly outlines its case, with the use of supportive evidence where appropriate. Each of the interest groups presents its argument at the public meeting.
2 Now return to your initial role as manager of the site, in charge of formulating a management plan. Your task is to use the material presented during the public meeting as the basis for a written report. This report must outline future policy to be pursued, and must be based on a reasoned criticism of the views of all the parties involved.
3 Compare your own report with those of other members of the class. Do they differ? If so, why? Discuss why you may have arrived at alternative policies and examine the value judgements associated with particular decisions.
4 Do you think differing values and attitudes lead to management conflicts?
5 Finally, return to the list of solutions relating to the Snowdon Management Scheme which you made right at the beginning of the chapter. Are there any alterations you would now make?

20 MANAGING DRAINAGE BASINS

The River Lugg in flood, Hereford

THE RESOURCE WE TAKE FOR GRANTED

In 1976 people in Britain suddenly found themselves worrying about where the next bucket of water was going to come from. The water they had always taken for granted had been switched off. Somebody, unknown and without any great drama, had pulled a lever or turned a wheel in some unremarkable building and the water had stopped flowing. Just like that. For the first time people began to ask how such a fundamental difficulty could have arisen. In fact, the problem was one of resource management. The Water Authorities have only so much money to spend on precautionary measures against drought and they are unable to contend with severe conditions without taking drastic action to reduce consumption.

By contrast, in June 1982, a period of intense rainfall brought havoc to many parts of Britain as stream and river levels rapidly rose causing widespread flooding. Among the affected areas were drainage basins within major cities. In places like London several people, unused to the hazards of floods, were drowned in the fast flowing water. In these places, the natural environments of rivers have been so changed that they behave in a totally different way from that expected in a rural environment. As with drought, the problems of heavy rain have often been intensified by the activities of people.

The most obvious need to conserve water occurs during a drought. But it is by no means the only occasion. As the two articles show, resource management problems can be both sudden and unpredictable, and also long-term and certain.

WATER PLEA

Kent County Council is asking the Government to help in its search for a new water supply. The council says a new source of water must be found by 1988 at the latest, or industrial growth and residential development in the county will be put at risk.

(*Source*: *The Daily Telegraph*, 12 December 1981)

Share bath plea as water runs to waste

Householders in the north west were urged yesterday to share baths and showers, and to use melted snow to flush lavatories in an urgent effort to save water supplies, which might only last another five days.

The shortage, due to burst mains, is worse than the summer drought of 1976 with the worst hit areas being Merseyside, Cheshire and Greater Manchester, where millions of gallons of water are running to waste every hour.

'The situation is very serious indeed,' a spokesman for the North West Water Authority said last night.

The Arctic weather conditions have also hindered repair works, and although maintenance men are working emergency rotas they are having difficulty locating bursts beneath packed ice and thick snow.

? Using the articles, and your impression of basic needs in time of drought, write a short paragraph to illustrate the major reasons for supply difficulties, and some of the realistic methods that could be used to safeguard against them. At this stage you may not know all of the difficulties or solutions, but for the moment simply list your views. These will be similar to those of 'the person in the street'. They will act as a basis for subsequent discussion.

The extract below from *The Daily Telegraph* reveals some aspects of the sudden flood that occurred in the Silk Stream basin of North London in 1982. By contrast, the opening photograph shows the River Lugg in Herefordshire spilling over its banks in a totally rural lowland environment. Using the article and the photograph, write a short paragraph to illustrate the major features that might increase the likelihood of flooding, the area flooded and the danger to life in the two environments. Try also to suggest some ways in which the effects of a flood could be reduced. At this stage you may not know all the answers but for the moment simply list your views. They will act as a basis for subsequent discussion.

Storms and floods leave six dead

At least six young people died as torrential rain, lightning and floods lashed Britain over the weekend.

Three children aged six, nine and 14 were swept away in a stream swollen by the downpour at Burnt Oak, Middlesex.

Two bodies have been recovered by frogmen but the third was still unaccounted for last night.

The father and brother of the nine-year old victim, Stephen Reed, searched for him until they were warned by police that they themselves were in danger of drowning. Police recovered Stephen's body two hours after he fell in.

The three boys were swept away in separate incidents when the normally-placid and shallow Silk Stream in Burnt Oak rose to more than six ft deep.

(*Source*: *The Daily Telegraph*, 28 June 1982)

THE SIZE OF THE PROBLEM

Water supply, river flow and their impact on our environment rarely form part of everyday conversation. Only when there is a flood or drought, or a main pipe bursts and the supply to our homes is cut off, do we pay much attention to this vital life support system. However, if all the water supply and protection apparatus — reservoirs, dams, dykes, pumps, pipes and taps were removed by magic one night, the chances are that much of the population would be dead within a couple of weeks either from dehydration or illness caused by drinking contaminated water. Those that remained would find survival much harder because uncontrolled rivers are an irregular and unreliable source of supply. Flood plains would become increasingly hazardous not only because they would become regularly flooded, but because disease-carrying insects would infest the expanding marshy ground.

We have only to compare the settlements made by ancient civilisations with the site of towns today to see how dependent we are on the control of water in the landscape (Figure 1). But how have we been able to control the environment in this way? What have we discovered about the movement of water on Earth that has enabled successful alteration to the natural regime?

? Examine each of the areas identified by the captions on diagram **b** in Figure 1. Attempt to describe the result of land use changes on the natural environment.

Water is important in other ways. Look at a landscape and you will see the vital role played by water in its formation. A casual glance suggests the river is the main erosive agent, lowering its bed and removing sediment to the sea. However, water flowing through the soils covering hillslopes plays an equally important role.

People have drastically altered many aspects of river flow — they have ploughed up soil and altered many slopes. But can this be done safely? What must be known about the natural environment to prevent alterations causing unforseen disasters?

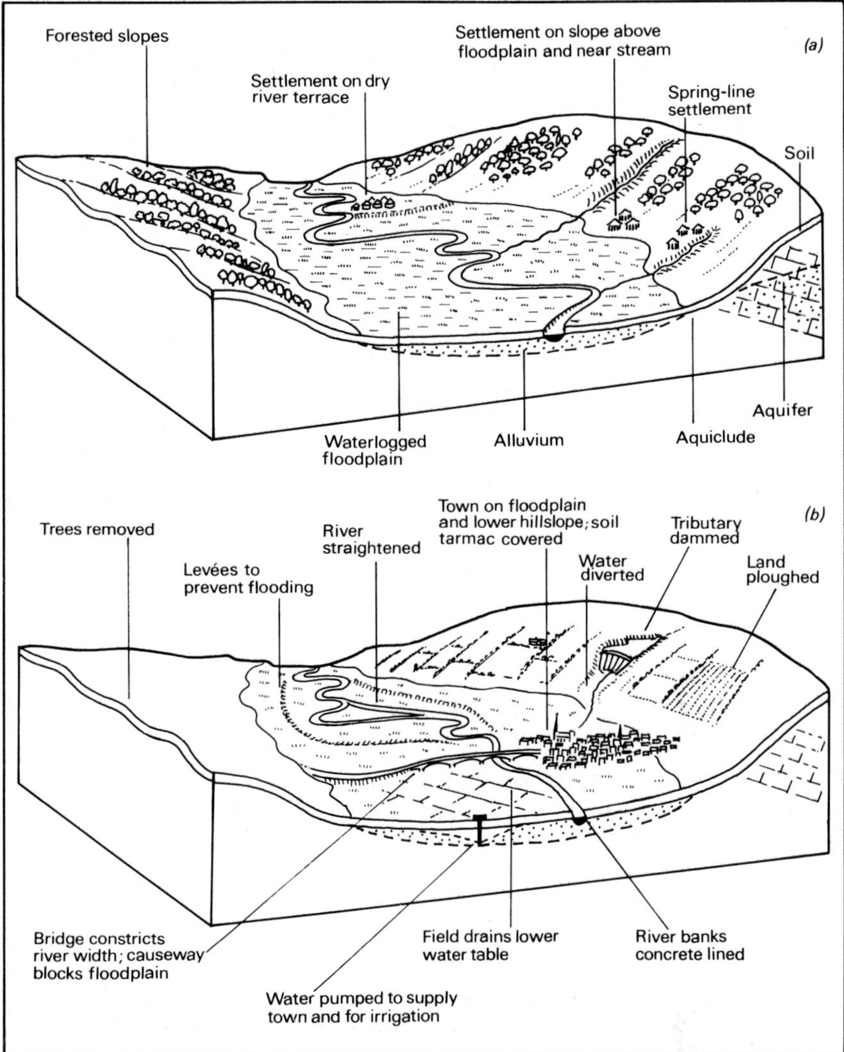

Figure 1 *The past and the present* **a** *Early settlements were closely controlled by the presence of water* **b** *Present settlements occupy sites that are vulnerable to flooding unless rivers are carefully controlled.*

Water, human activity and the natural environment are today bound together by a web of increasing complexity (Figure 2). Clearly, therefore, to be able to identify the nature of people's impact on the drainage basin it is first necessary to understand the behaviour of the natural system.

THE PHYSICAL FRAMEWORK

THE MOVEMENT OF WATER

The drainage basin is a convenient physical unit to work with because it acts as a focus of all water movement. The purpose of a drainage basin is to transfer rain or snow melt to the sea as effectively as possible. This is achieved in a number of ways, involving both surface and subsurface routes (Figure 3). In Britain, it is rare for rainfall intensity to be great enough to exceed the infiltration capacity of the soil, and most rain therefore infiltrates the soil and moves either laterally as *throughflow*, or percolates deeply into a water-bearing rock (aquifer) and then moves as *groundwater*. The rate of subsurface flow depends on the *permeability* of the soil and

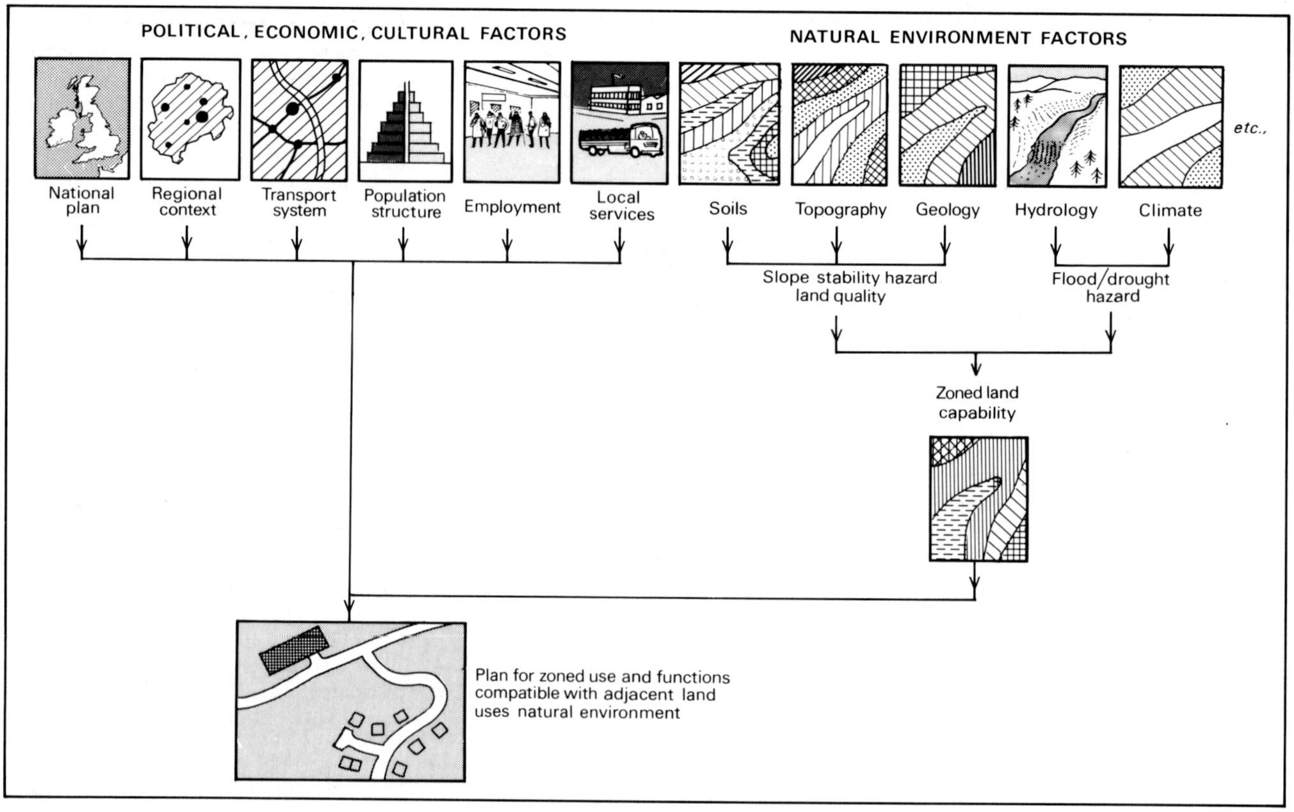

POLITICAL, ECONOMIC, CULTURAL FACTORS
NATURAL ENVIRONMENT FACTORS
National plan
Regional context
Transport system
Population structure
Employment
Local services
Soils
Topography
Geology
Hydrology
Climate
etc.,
Slope stability hazard land quality
Flood/drought hazard
Zoned land capability
Plan for zoned use and functions compatible with adjacent land uses natural environment

Figure 2 *The complex system of water, human activity and the natural environment*

Rainfall
Interception
Infiltration
Direct channel rainfall
Overland flow
Evaporation
Soil moisture storage
Throughflow
Transpiration
Streamflow
Evaporation
Groundwater storage
Groundwater flow
Ocean storage

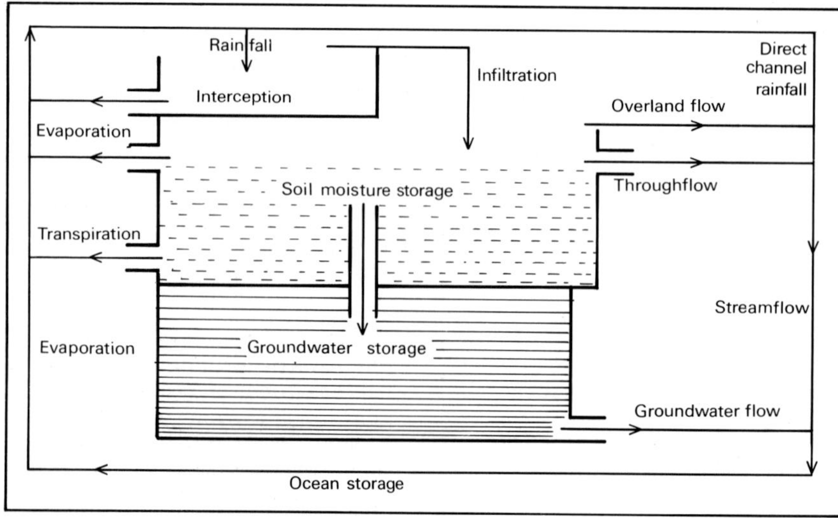

Figure 3 *The hydrological cycle*

rock — that is their ability to allow the passage of water. This depends on the size of the pores in the soil or rock, and to what extent those pores are connected to each other. Fractures running through a rock will serve to increase its permeability. (Remember to distinguish between permeability and *porosity* which is the ability of a rock or soil to hold water in its pores without necessarily allowing the water to move between those pores and through the rock.)

However, in all cases water is driven towards the lowest point in a drainage basin by gravity. Here it appears at the surface to provide a major contribution to streamflow through *channel seepage*. The upper level of saturated material is called the *water table*, and the height of the water table above the stream is called the *hydraulic head*. The rates of throughflow and groundwater flow are therefore directly related to the hydraulic head.

During a prolonged storm, or towards the end of the winter, the accumulation of water in rock and soil may cause the saturated zone to reach the surface. When the water table is coincident with the surface, water flows from the soil or rock. Occasionally the flow is concentrated into a gushing stream, but more usually it simply takes the form of a region of seepage and produces sodden ground (Figure 4). Furthermore, when a soil becomes saturated, rainfall is unable to infiltrate and it runs over the surface. Because the rate of *surface* (overland) flow is very much faster than throughflow or groundwater flow, the lag (delay) between rainfall and water reaching the stream is greatly reduced. Therefore, with the crucial change from throughflow to overland flow, streamflow increases rapidly and

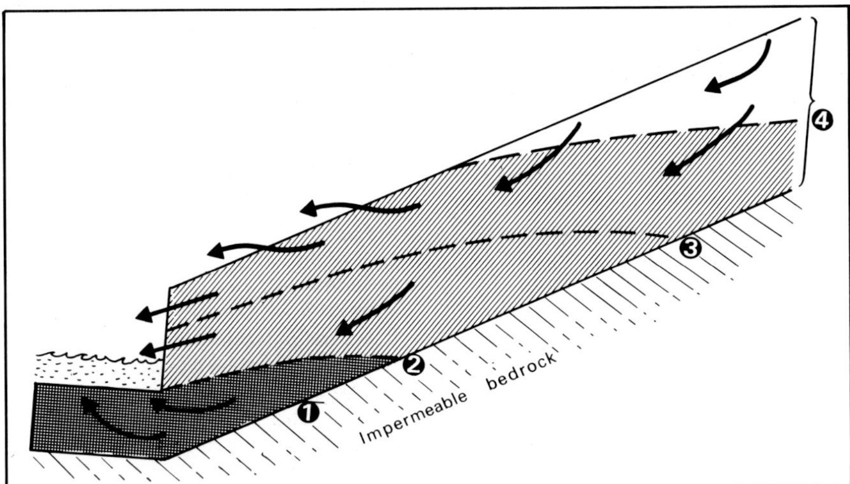

Figure 4 *The principle of stream bank contribution — during a prolonged storm, water flows from upslope causing a rise in the water table from 1 to 2 to 3. This causes a greater flow into the channel and if the water table rises to level 4, water seeps out at the surface as overland flow*

Figure 5 *The drainage basin hydrograph and its relationship to contributing areas*

flooding becomes possible.

The shape of the landscape plays an important part in controlling areas of frequent overland flow. For example, the steeper the slope, the faster throughflow can occur and the smaller the chances of overland flow. On the other hand, where hollows in the surface cause flow to converge, water will be funnelled towards the centre of the hollow, thus increasing the tendency for the soil to become saturated. For this reason most river headwaters begin in hillslope hollows.

Areas within a drainage basin that become saturated during a storm or by the end of winter are called *contributing areas* because they are responsible for the contributions that come together quickly during a storm to provide the peak flow. They are the most crucial areas for generating sufficient water to cause flooding. The relationship between streamflow and time is described by a diagram called a hydrograph (Figure 5).

? The rate of water movement to a stream is closely related to soil properties. The relationship is expressed by Darcy's Law

$$Q = KAH/L$$

where Q is the rate of throughflow; K is the soil permeability; A is the cross-sectional area of the soil; and H/L is the slope gradient.

1 Use this equation to suggest which of the following soil properties would tend to make overland flow more likely:
a coarse texture, **b** fine texture, **c** well-developed structure, **d** poorly developed structure, **e** substantial soil thickness, **f** shallow soil, **g** high initial soil moisture content, **h** low initial soil moisture content.

2 Which landscape properties would tend to make overland flow more likely:
a steep slopes, **b** gentle slopes, **c** hillslope concavities (hollows), **d** hillslope convexities (spurs).
Explain how Darcy's Law is helpful in coming to the right conclusion in each case.

3 Study Figure 5, which shows the changing water contributions within a drainage basin during and after a storm. Notice that the first part of any rainfall is used to wet dry leaves and other surfaces and to replenish the soil moisture deficit caused by drainage since the last storm, evaporation from the soil surface and transpiration from vegetation.

a Describe the development of the contributing areas within the drainage basin and their relationship to the stream hydrograph. Pay particular attention to identifying the chief cause of the hydrograph peak and the tail of the recession limb.

b Now copy Figure 6 and add to it those regions you consider to be most at risk from overland flow and flooding. Mark separately those areas that appear to have the least risk.

SOIL STABILITY

As the preceding section showed, the rate of water movement is closely related to both topography and soil properties, particularly thickness and permeability. However, the soil is not simply a passive element of the landscape, particularly during a major storm. Drainage basin management is therefore also concerned to identify the causes and results of soil instability. Soil instability (soil erosion) is sometimes a major hazard within a basin and can contribute directly to the sediment load of the stream and cause siltation problems to navigable channels and make water purification that much more costly.

Soil development depends on the role of water as a chemical reagent. Water and mineral matter react through the process of hydrolysis (the major component of chemical weathering) to produce secondary products such as clays. Clays behave quite differently from their parent materials when saturated with water.

Water influences soil stability in a number of ways:

1 By flowing over the surface, gradually gathering momentum until it has sufficient energy to entrain particles. In Britain this type of erosion is confined to contributing areas because rainfall intensity alone is not normally sufficient to cause widespread surface runoff and gullying.

2 By saturating the soil and giving the particles a degree of buoyancy so that the intergranular friction is reduced and they can more easily slide or flow. Landslides and debris flows result from local saturation of soils on hillslopes above about 10 per cent.

Vegetation acts to reduce the severity of these processes by bonding the soil with roots, increasing soil permeability and providing a rough surface which keeps the speed of overland flow below the critical value that will cause particle entrainment.

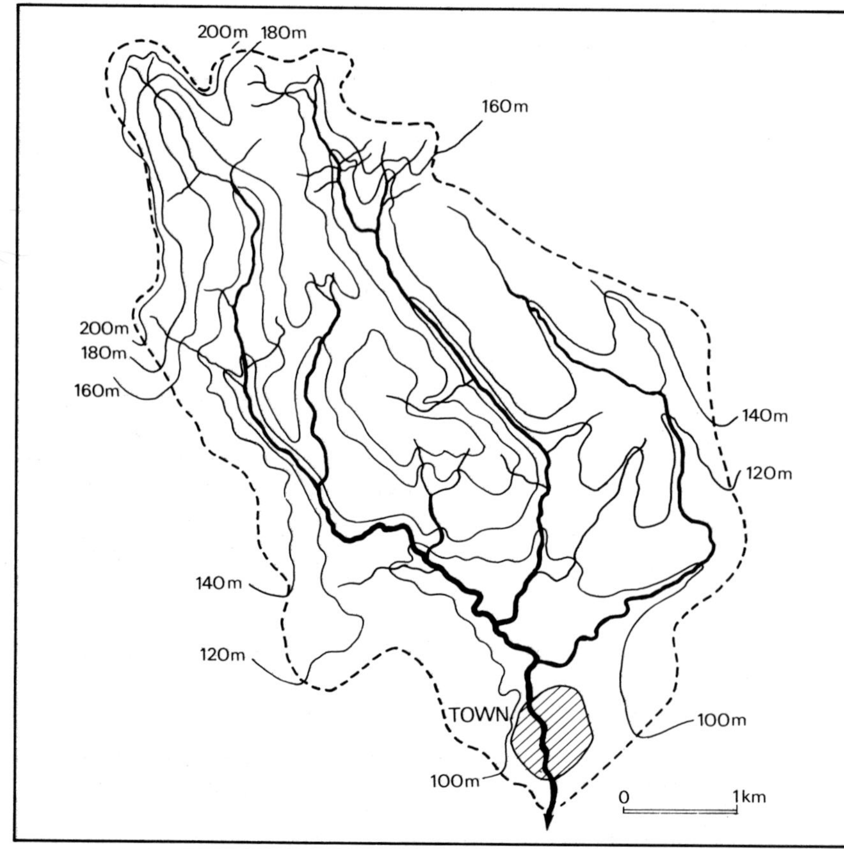

Figure 6 *A drainage basin*

Figure 7 shows the major types of slope element and the associated dominant processes. Develop a pattern of shading that will identify each slope element then, on a copy of the drainage basin map (Figure 6), produce a map of slope stability regions. Explain how your choice has been influenced by the map you drew earlier showing the regions you considered to be most at risk from overland flow and flooding.

RIVER SEDIMENT TRANSPORT

Material delivered to the stream bank by hillslope transport processes must be removed by the stream. Streams usually adopt either meandering or braided channel plans depending on the calibre of the load to be transported and the channel gradient. Most streams in the UK are meandering because they only carry fine debris resulting from bank erosion of chemically weathered material (soil).

Meandering streams are finely tuned to erode material from their beds and banks. However, the time period over which they maintain this tuning is very great. Thus an apparently severe change, such as the break through of a meander bend or the massive scouring that accompanies a major storm, must be seen in the context of long-term change. In a river system, occasional

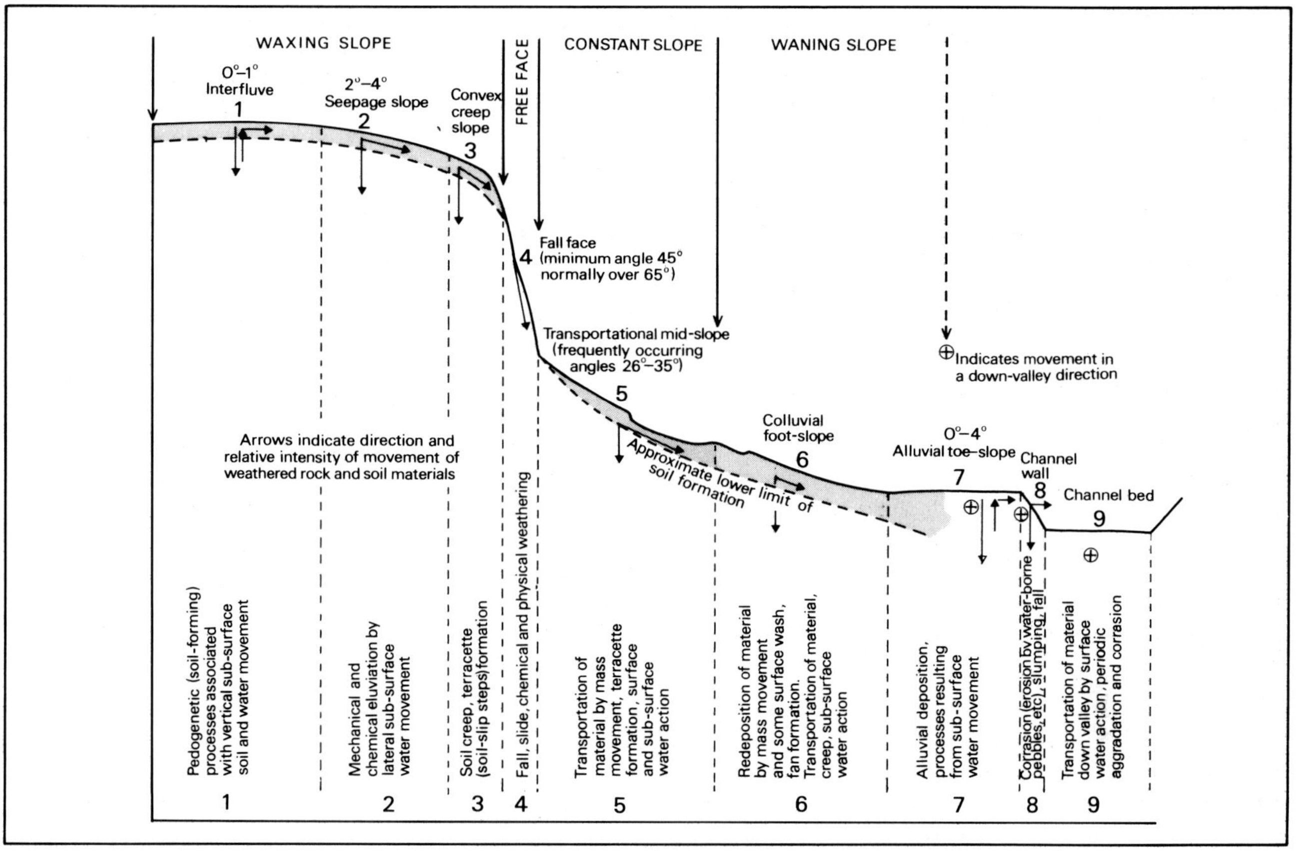

Figure 7 *The major types of slope element and the associated, dominant processes of erosion*

Figure 8 *Two aspects of meander geometry* **a** *Drainage area and meander wavelength* **b** *Channel width and meander wavelength*

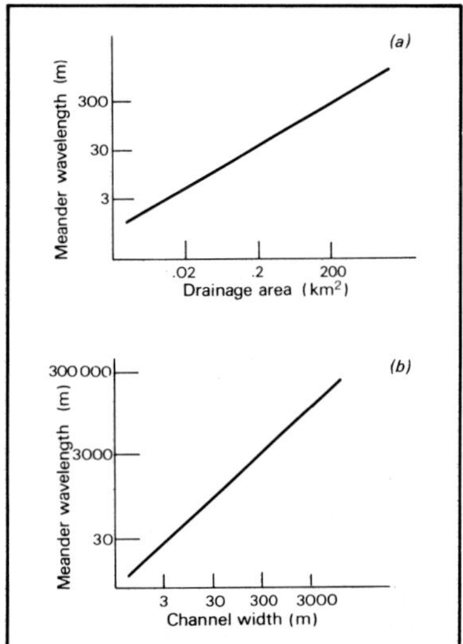

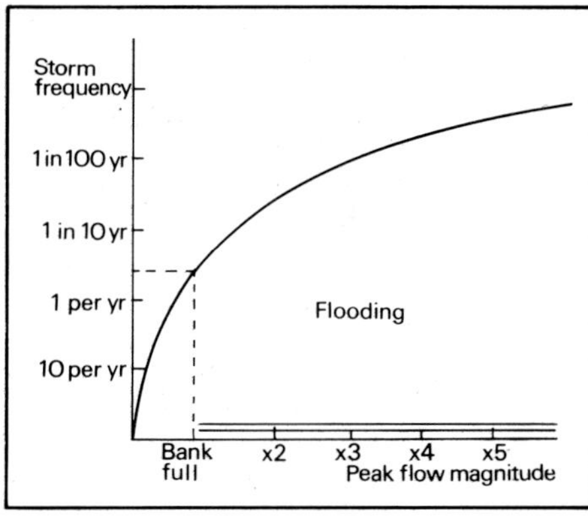

Figure 9 *Strength of storm versus flow magnitude. A storm of the strength that occurs on average every five years will cause this river to burst its banks*

catastrophic changes are always followed by long periods of imperceptible rebuilding. The geometry of a meander system is a reflection of the long-term adjustment (Figure 8).

A stream derives its energy from the mass of water and the channel gradient. This energy is used to overcome friction both internally and against the channel perimeter, to transport loose debris, and to erode new material from the bed and banks. The pattern of stream meanders, and of bed forms such as pools and riffles within the channel are all designed to cause the energy to be spread evenly and to encourage an orderly transfer of sediment from the basin. Stream meanders and bed roughness are particularly important in controlling water velocity, acting as a form of brake.

?

1 Describe the graphs in Figure 8, paying particular attention to the scales on the axes.
2 Explain why it is difficult for a stream to have a meander geometry which is suited to all flow conditions.
3 Using Figure 9 and the information that natural rivers in the UK spill over their banks onto their flood plains on average once every two years, suggest the size of flow that might be most important in determining the geometry of the channel and why rivers are not adjusted to hold the maximum possible flow.

One very clear example of the long-term balance between water flow and stream channel geometry can be illustrated by examining the River Mississippi between 1932 and 1975. This large river has a highly sinuous course, suggesting that the river is only just maintaining a meandering habit and is on the verge of braiding. Many meanders cut off naturally during floods, but others increase in amplitude so that, on average, the river length and degree of sinuosity remains the same. Between 1933 and 1936 many of the meanders were artificially cut off in an attempt to straighten the river and shorten river journeys by barge. The pattern of the river in subsequent years is shown by Figure 10.

? **1** Trace each of the river plans in Figure 10. Then, overlaying each one onto its predecessor, record the changes that took place and the way the engineers responded to them.
2 Measure the channel length for each date using a piece of string. Now calculate the river's sinuosity by dividing the measured length by the straight line distance linking the ends of the river. (A river that flows in a straight path has a sinuosity of 1.0.) Next, plot the sinuosity against time.

? **3** Describe the results shown by graph. What has happened despite the engineers best efforts to prevent it?
4 Why are so many strengthened banks (revetments) and training walls (dykes) now useless?

PEOPLE AND DRAINAGE BASINS

The natural environment has been drastically altered in most parts of the world. In the UK, major changes have stemmed from urbanisation, navigation and farming practices.

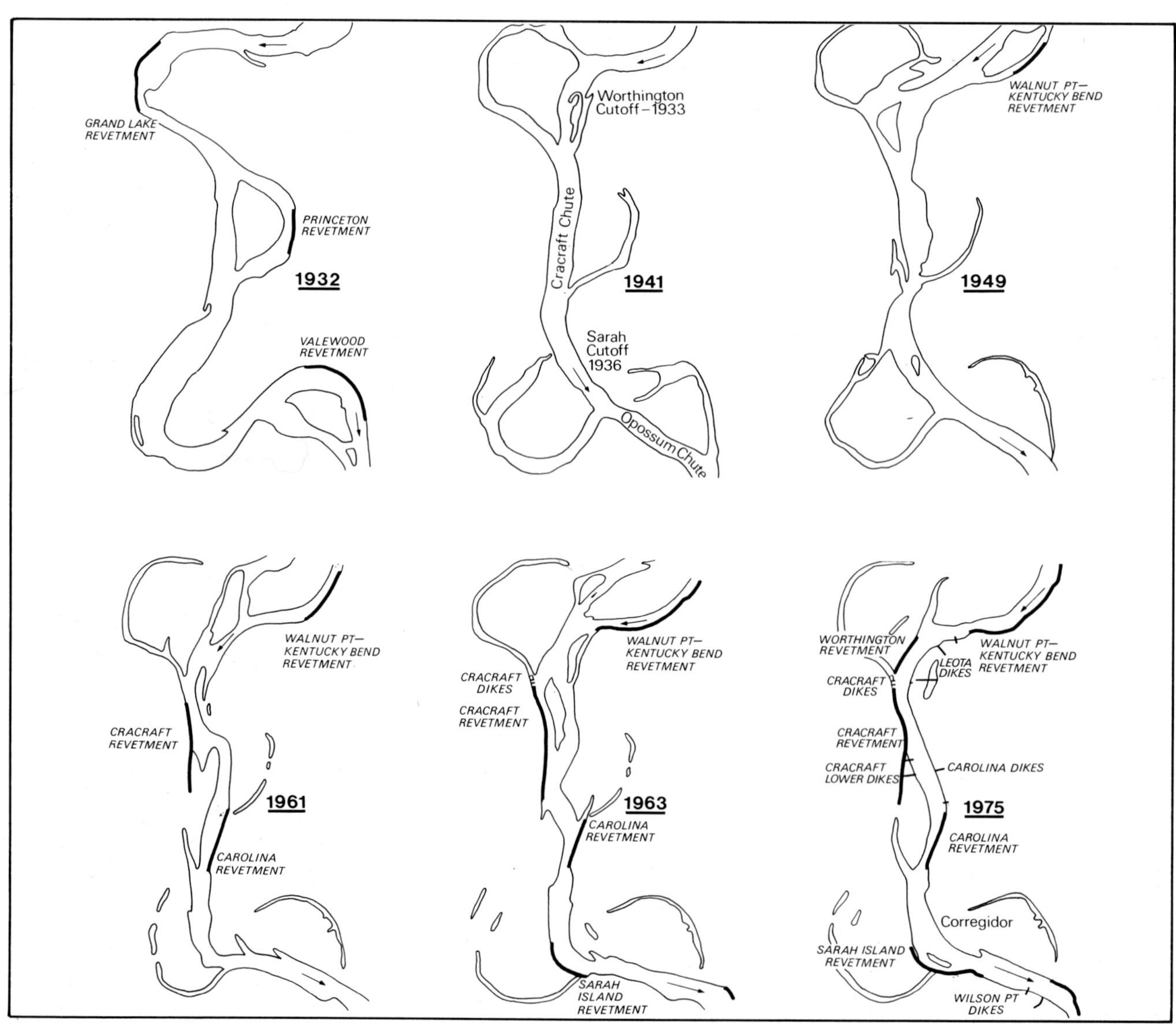

Figure 10 *1932–1975 Development of the Kentucky Bar–Mayersville Reach on the Mississippi*

? **1** Refer to Figures 1 and 3 (*pages 409–10*) and sketch out the hydrological cycle appropriate to:
a An area of complete urbanisation.
b An area of agricultural activity.
2 Critically examine the hydrograph in Figure 5 (*page 411*) and redraw it to illustrate the changes that would accompany both urbanisation and agriculture.
3 Suggest some of the management plans that might follow from the conclusions you made in question 2. Refer to both urban drainage systems and rivers.
4 Now examine Figure 11, which shows the normal progression of sediment concentration from an area with progressive intensification of land use. Describe what might have happened to each section of the graph.
5 Assume the drainage basin in Figure 6 (*page 412*) was entirely used for agricultural purposes. Draw an overlay to your earlier version of Figure 6 and shade the areas that would suffer increased soil erosion, bearing in mind that slopes above about 10° are rarely used for cultivation, but are usually restricted to pasture.
6 Figure 12 shows the drainage basin of the Silk Stream, a minor part of the drainage system in North London. This basin has developed on London clay and therefore has only a small groundwater flow. Nevertheless, in 1929, when the area had 15 per cent urban cover and only a poorly integrated sewer system, the average hydrograph produced by 10 mm of rain in 1 hour had a lag time to peak discharge of 6.6 hours. Explain what happened as the percentage of urban cover increased during the period 1929 to 1942. There has been no change since 1942. Why do you think this is so?
7 In a sub-basin of the Silk Stream there are 12 000 people living in blocks of flats together with a considerable amount of commercial light industry and warehousing. The change in sediment concentration through a storm is shown in Figure 13. Describe the variations of sediment concentration paying particular regard to the time scale involved. What special problems does this distribution pose?

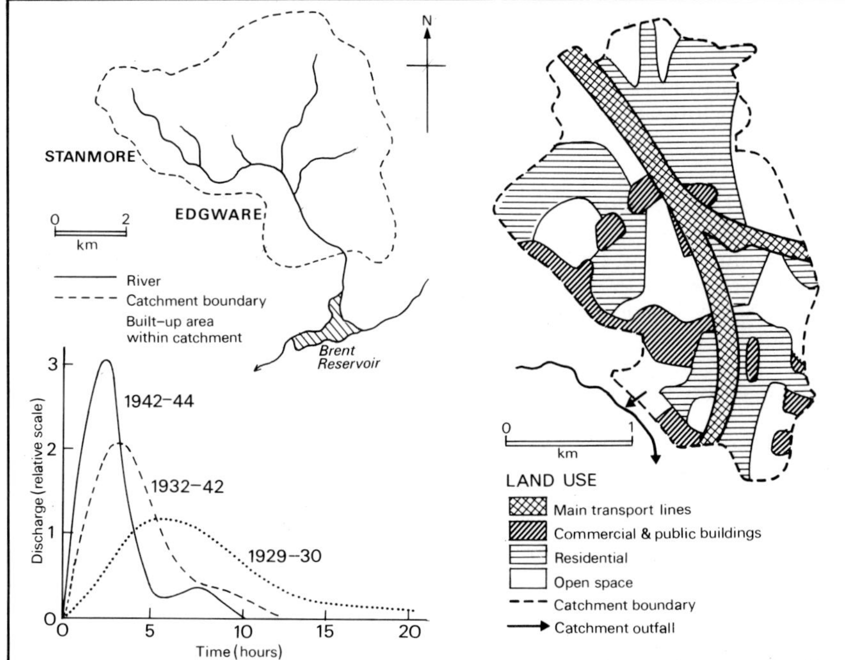

Figure 12 (*above*) *Location map for the Silk Stream, North London.* (*below*) *Progressive changes in the standard response of the Silk Stream to a 10 mm/1 hour storm during urbanisation*
(*Source*: *Teaching Geography*, Geographical Association)

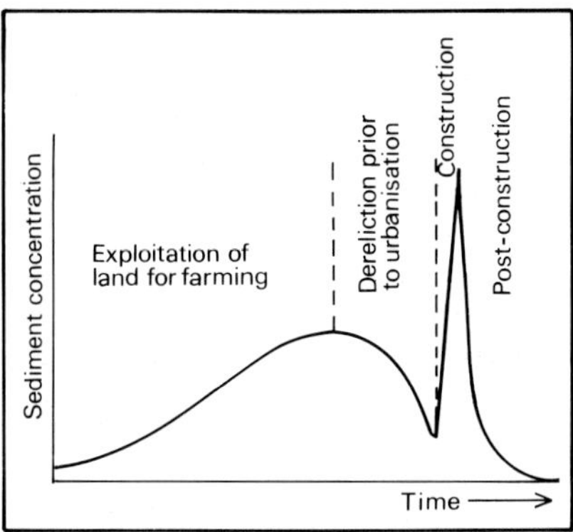

Figure 11 *Long-term changes in sediment yield with progressive intensification of land use*

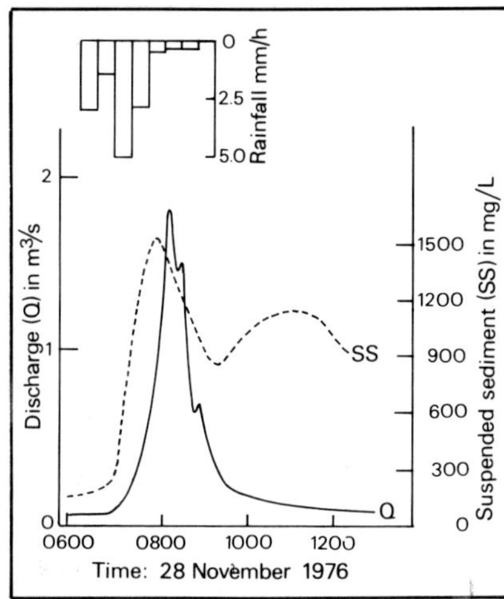

Figure 13 *Hydrograph showing variations in the suspended sediment and biochemical oxygen demand in the Silk Stream during a storm*

FLOOD AND DROUGHT

Many people think of rivers in terms of their flood impact. Rivers are certainly dramatic features of the landscape when in flood, and they may cause considerable loss and damage. However, rivers are also the major source of water for human use, whether it be for the home, in industry or for irrigation. The slow movement of water as throughflow and groundwater flow is particularly valuable in helping provide a constant supply of water because they convert the 'on/off' pattern of rainfall into a much more continuous flow. Nevertheless, without frequent periods of rain to recharge the soil and rock reservoirs, the amount of

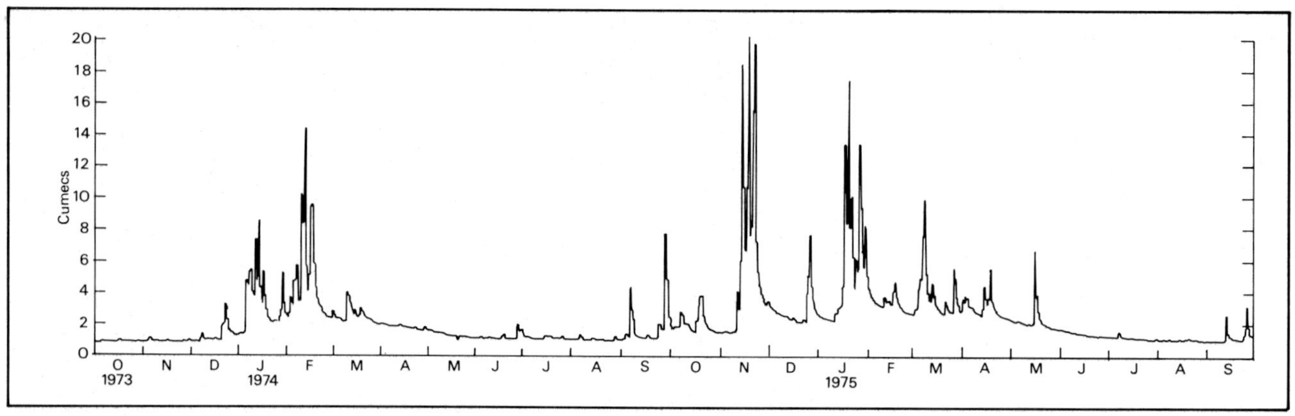

Figure 14 *River Lodden at Sheepbridge — daily mean discharge hydrograph for the water, 1973–75*
(*Source*: B. J. Knapp, *Practical Foundations of Physical Geography*, Allen and Unwin)

water reaching the stream banks will become progressively smaller, tending to a very low value known as the *dry weather flow*. Dry weather flow is the flat part of the hydrograph tail in Figure 5 (*page 411*), and represents something like 10 per cent of the mean river flow.

? **1** Figure 14 shows a two-year hydrograph for a river in the UK. Overlay a piece of tracing paper and copy the hydrograph. Rule a line to show the dry weather flow. Bank full conditions are reached with a flow of 22 cumecs (cubic metres per second). Rule a line to indicate flood conditions.

2 Study the trace between October 1973 and April 1974 and compare it with the period October 1974 to April 1975. Suggest possible causes for the contrast between the two periods? (Consider meteorological reasons and the time of year.)

3 Why should the peaks in November 1974 become progressively higher? (The highest was actually 36 cumecs.)

SUPPLY AND DEMAND

As the hydrological cycle (Figure 3 *page 410*) reveals, by no means all rain reaches streams. A considerable loss is due to the combined effects of transpiration and evaporation (*evapotranspiration*) of which transpiration is the major component. Transpiration is greatest in the summer months when temperatures are high and plants use water both to keep cool and to grow. Most plants have very low transpiration rates at temperatures below 6°C and therefore there is little loss during the winter. Evaporation depends on the absolute moisture content of the air — the drier the air, the more evaporated water can be absorbed. In winter cold air is usually close to its saturation level and, in any case, can hold far less water at a given relative humidity than warm air. For all these reasons, the rate of evapotranspiration is strongly seasonal in character.

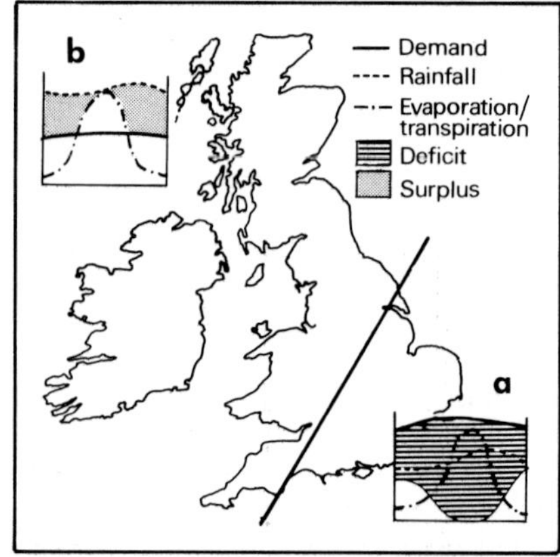

Figure 15 *The water balance for two meteorological stations in the UK*
a *Lowland, populous regions*
b *Upland, sparsely populated regions*

? Figure 15 shows the water balance for two meteorological stations representative of two parts of the UK. One represents lowland and populous regions, and other upland and sparsely populated regions. For each station, the rainfall, evapotranspiration and demand curves are shown. The demand curve is the total requirement of domestic, industrial and agricultural (irrigation) consumers.

1 Describe and explain the form of both water balance diagrams.

2 What conclusions may be drawn that would be of interest to those in charge of the UK's water supply?

3 Using one of the reference sources given at the end of the chapter, set out the suggested strategy for the UK by the year 2000, and explain why so many of the future schemes are for interbasin transfers, rather than for reservoirs.

4 Drainage basin managers can plan to sell their water to other regions as a resource, or they can even sell it abroad. Study the following article, then draw up a water balance for a Middle Eastern country, and explain why water is to be shipped abroad as well as transferred to other regions in the UK.

Welsh push plan to export water

Paul Hoyland reports on a project that could earn £2 m a year

Plans to export millions of gallons of water by oil tanker from Milford Haven in South Wales, to the Middle East and Africa could provide a welcome business for the Welsh Water Authority after a period of changes, economic hardship, and political controversy.

A newly-published brochure in Arabic, French and English extols the virtues of the Welsh scheme, that involves reversing the role of a redundant refinery terminal at Milford Haven to pump water into the crude-carrying tankers and an estimated £2 million a year into the authority.

Mr Roy Webborn, the authority's assistant director of finance and water-export manager, says that the facility will enable a tanker to be loaded within 24 hours with 50 million gallons of water, the equivalent of two days' supply for Cardiff.

The tanker would arrive at Milford Haven with its cargo of oil and turn round and leave with a cargo of water that can be used for irrigation or treated and used as drinking water.

The costs were so attractive that even using tankers exclusively for water was cheaper than the cost of desalinated water, Mr Webborn said.

Huge tanks at the port could store the water that would come from the Llys-y-fran reservoir, which supplies the Milford Haven district.

The Northumbrian Water Authority has similar plans to export water, but the Welsh believe that their ready-made facility at Milford Haven gives them the edge in the trade to win contracts.

It would cost about £100 000 to alter the terminal, but Mr Webborn says the work could be completed within 28 days of a large contract being agreed.

Although the Welsh Water Authority is selling at its normal commercial rate of £1.26 per 1000 gallons, it is willing to negotiate on long-term contracts. 'On that basis I think the potential income for the authority is in the region of at least £2 million,' Mr Webborn said.

The income would be a welcome bonus for the authority. The Welsh Secretary, Mr Nicholas Edwards, has trimmed its board from 35 members to 13, and opposition is mounting against a decision to further reduce the numbers of divisions — cut from 25 to seven in 1979 — to three, with the consequent loss of 330 jobs.

But it is the recession that has dealt the most serious shock to the system. Welsh industry has reduced its spending on water by £15 million a year, wiping out a tenth of the authority's income.

After years of investing in new plant to meet an expected increase in the demand from industry the authority raised its charges by 18 per cent to counter the slump. That provoked a storm of protest.

Welsh nationalists, long aggrieved that consumers in England pay considerably less for water supplied from the same Welsh reservoirs, started a water-rates strike.

The withholding of payments further weakened the authority's finances and culminated in Plaid Cymru's taking the issue to the High Court in a test case due to be heard soon.

As the authority has been prevented by the Government from substantially increasing its charges to the English water authorities it now cannot afford to lose the test case that hinges on whether Welsh consumers should pay more for their water.

(*Source*: *The Guardian*, 3 January 1984)

BASIN MANAGEMENT

From what you have read so far in this chapter it will be clear that many aspects of a drainage basin are interrelated. So it is not surprising to find that when people alter one aspect of the basin to solve a problem, they often inadvertently create problems elsewhere. For example, government grants to farmers to improve the quality of grazing land in the Pennine area of the Yorkshire Ouse basin take the form of assistance to install land drains. Although the drains have reduced waterlogging of the Pennine soils and successfully allowed the land to be upgraded, the drains have had disastrous effects on the lower reaches of the river, providing a much expanded contributing area for stormflow and therefore considerably increasing the flood hazard at York. Thus, in this case the government is giving out money to farmers in one place and having to give out even more elsewhere simply to counter the effects created by the first.

Clearly, river basins need to be managed as units. In the UK the importance of basin management has, in part, been recognised by the establishment of government Water Authorities, each of which exercises control over a major drainage basin.

Basin management involves practices and measures designed to

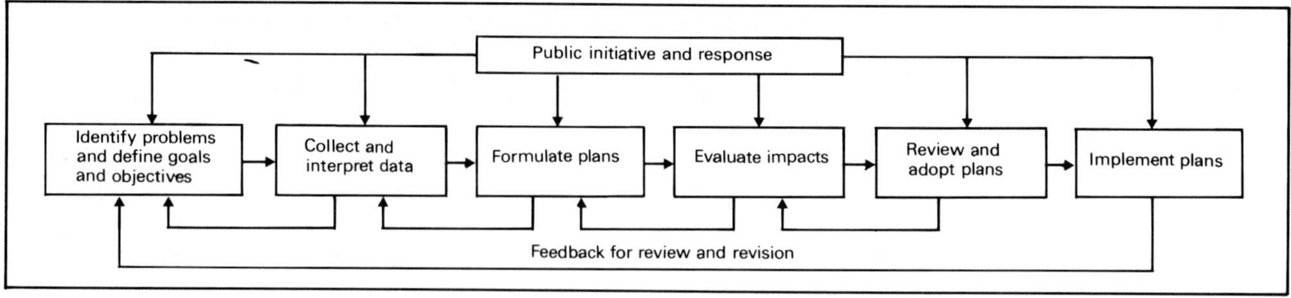

Figure 16 *The planning process*

improve land use, to alleviate drought and flooding, and to reduce erosion and sedimentation. These benefits are achieved through combinations of land treatment and structural measures. The land treatment measures may consist of agricultural practices to extend the vegetation cover, to reduce overland flow and therefore to reduce surface erosion. Flood and drought measures may include dams and dykes. Clearly, the result of each major change in a drainage basin needs to be assessed carefully, and its costs and benefits assessed in the light of its impact on other parts of the basin. This would, ideally, involve a planning process such as that outlined in Figure 16.

DRAINING WETLANDS

Farmers in the Pennine area of the Yorkshire Ouse basin are not the only farmers engaged in land drainage schemes. It is a European-wide trend, as indicated in the extract below. Read the article and decide whether you think the current system needs to be changed.

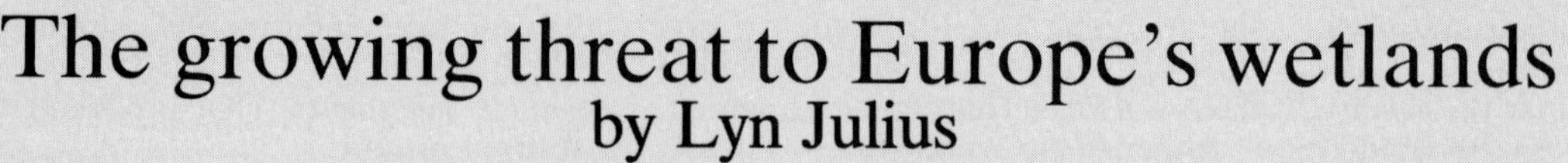

The growing threat to Europe's wetlands
by Lyn Julius

Every year, thousands of migrant birds arrive in the wetlands off western Europe to find that their wintering and breeding grounds have shrunk. Year by year, too, the rich flora and fauna, and the reptiles, otters and amphibians of Europe's marches, bogs, fens, streams, wet meadows, ditches, rivers and lakes all diminish. The major cause: agricultural drainage.

About half of the endangered bird species in the EEC live on wetlands. But if present trends continue, 11 species, including the stork, the corncrake and the shoveler, may disappear altogether outside nature reserves. Of the wetlands themselves, the raised bogs of Britain and Ireland will vanish in the next five years. The Irish turlough, a valuable habitat for many bird species, is already on the point of extinction.

For centuries, drainage was regarded as a benefit, often even a matter of survival. Wetlands were drained to prevent flooding, to control mosquitoes and to add to the food supply. But as the wetlands dwindle, Europeans have been waking up to the need to reserve what is left, and the plant and wildlife which wetlands support, before it is too late.

Not surprisingly, farmers and conservationists are at odds over the issue. The conflict between them is the subject of a new study* which focuses on the impact of drainage in four EEC states: France, Ireland, Britain and the Netherlands. The study is particularly concerned with the effect that the common agricultural policy is having on wetlands. Under its influence, each country has sought to expand its agriculture and to establish an extensive drainage programme.

The Community may pay a small proportion of the cost, and also makes funds available for drainage projects under a series of modernisation directives. On the other hand, French wetlands are still firmly in the hands of the powerful farmers, but conservationists do win the occasional battle. In Ireland, where conservationists are weaker still, the famous peat bogs of Irish geography, with their unique ecosystems, may soon pass into myth, despite repeated initiatives by the European Parliament urging their protection. The turlough, a temporary lake once common in the West of Ireland, is now a rarity.

Ireland is so dependent on farming that drainage is seen as a public good. Irish farmers are determined to improve their land and increase their output in every possible way. As a study puts it, 'drainage is a logical and desirable form of investment and loss of bog, turloughs anad wildlife may seem a small price to pay.'

Most Irish schemes are large-scale arterial schemes, where meandering rivers are straightened out to make them flow more easily. Almost every one is being subsidised by the Community, and more heavily than elsewhere. However, the Commission wants a commitment from Ireland that no measures will be taken that would be incompatible with the environment. It has also commissioned a survey of the environmental impact of one scheme, in County Mayo, which should provide a useful basis for assessing future projects.

After vigorous drainage since Roman times, only a small fragment of Britain's inland bogs, fens and marshland remains; but our long and indented coastline still harbours some important wetlands. In the last 20 years, the effects of drainage have been somewhat tempered by environmental legislation. The Nature Conservancy Council (NCC) actively advises water authorities and negotiates water management agreements; there are government guidelines to encourage more sensitive schemes.

The Community has made small grants to UK drainage schemes, notably in Northern Ireland; EEC aid has also gone towards improving existing drainage in the Fens. But in the Western Isles, where the EEC is funding 40 per cent of a regional development grant, the study warns of the effect of investment and tourist development on the unique sandy grassland, habitat of some of the highest densities of wading birds in Europe.

The Netherlands, with a quarter of its land below sea level and its sophisticated network of dykes, dams and pumps, is the only country whose survival depends on the successful management of water. As in the UK, agriculture is intensive — the CAP is a cornerstone of Dutch prosperity. Also, as in the UK, most drainage projects are improvements of earlier schemes, or land consolidation projects commonly featuring conservation areas.

With the possible exception of Ireland, the study concludes that the availability of Community funds has had little effect on the overall level of drainage. More far-reaching is the impact of high support prices for beef, milk, cereals, etc, which make drainage of neglected meadows, for instance, economically worthwhile.

The study urges a more discriminating drainage policy, based on a more rigorous evaluation of costs and benefits. Wetlands have their uses as centres of peaceful recreation, for fish spawning, for assimilating wastes, and even for flood control. The study argues that halting the drainage of environmentally important sites in the UK would have little impact on farming. Already, in the Netherlands, land is at such a premium that drainage is contemplated only if the return is at least 10 per cent.

**Wetland drainage in Europe: the effects of agricultural policy in four EEC countries.* By David Baldock. Published jointly by the International Institute for Environment and Development and the Institute for European Environmental Policy, 10 Percy Street, London W1P 0DR.

(*Source*: Europe 84)

?

1 Consider the impact of each of the following structures on the drainage basin under both high- and low-flow conditions:
a A reservoir built across an upland tributary.
b A motorway embankment built across a flood plain, but containing a bridge across the river channel.
c Dykes (embankments) along the banks of a river, and designed to keep flood waters away from the town.
d Doubling the size of a town, with consequent extension of its flood water sewer system.
e The straightening of the river within the town, and its encasement in a smooth, concrete-lined channel.

2 In underdeveloped flood-prone areas, non-intensive land uses such as agriculture are usually given high priority.
a Should cultivated or pastureland be adopted for flood-prone land?
b What would be the advantages of keeping a strip of uncultivated land on either side of the river?
c Should the use of hedges and walls and the construction of farm buildings be encouraged or discouraged?
d What would be the impact of the use of fertilisers and pesticides on flood-prone land?
e What are the long-term economic benefits to be gained from allowing the natural flooding of farmland?

3 In developed flood-prone areas, there is a greater problem of land use.
a What might be the advantage of prohibiting all structures from a zone on either side of the river?
b What forms and function of building would suffer least damage if they were located in flood-prone areas?
c Draw a sketch map to indicate the most suitable zones for development of a town sited adjacent to a river. Consider:
i commercial uses,
ii warehousing uses,
iii water-soluble toxic chemical storage,
iv car parks,
v recreational spaces,
vi office use,
vii retailing use,
viii residential use.
How would such a town plan conflict with town plans drawn up on the basis of bid-rent concepts? How might these two patterns be combined to produce an acceptable plan?

THE FLOOD PLAIN

With most people living in towns and cities near to rivers, management of flood plains is of special concern. However, the cost of constructing flood prevention structures, such as headwater reservoirs or extensive embankments, may not justify the expected loss from inundation. Under these circumstances the flood plain needs to be managed in such a way as to keep the flood loss to a minimum. In planning for flood loss reduction, the local and regional land use planners are particularly concerned with defining appropriate land uses for flood-prone areas.

More is at stake in deciding land uses in flood-prone areas than the potential reduction of flood losses. The flood plain is an important environmental and ecological resource meriting attention in its own right. Flood-prone areas often provide wildlife habitats, fertile soils, scenic areas and land suitable for recreation and park uses (Figure 17). Thus the land-use planner must evaluate proposed uses in terms of both the potential flood risk and the beneficial use of the natural attributes of the flood plain. Ideally, a regional plan would provide a framework for more detailed county and city land use in which both urban and non-urban uses are usually considered in detail and attention given to local conditions and problems. A regional plan might recommend open space uses for flood-prone areas; a county or city plan would typically specify the particular open space use, such as grazing or water-related recreation.

? By now you will have gained a greater perspective into the role of water in drainage basin management. Re-examine your answers to the first two exercises in this chapter in the light of the work you have subsequently done. Draw up a list of the problems produced by water that were not in your original answers then combine all the answers, ranking them in the order you think they should be, from the human viewpoint. State clearly the evidence you have used in coming to this choice.

Print 1 *This view of urban developments encroaching on the margins of a flood plain has been taken from within a flood relief channel. For the moment the foreground remains as recreation space because of the flood potential. This is a situation where the conflict between environment and people is at its sharpest*

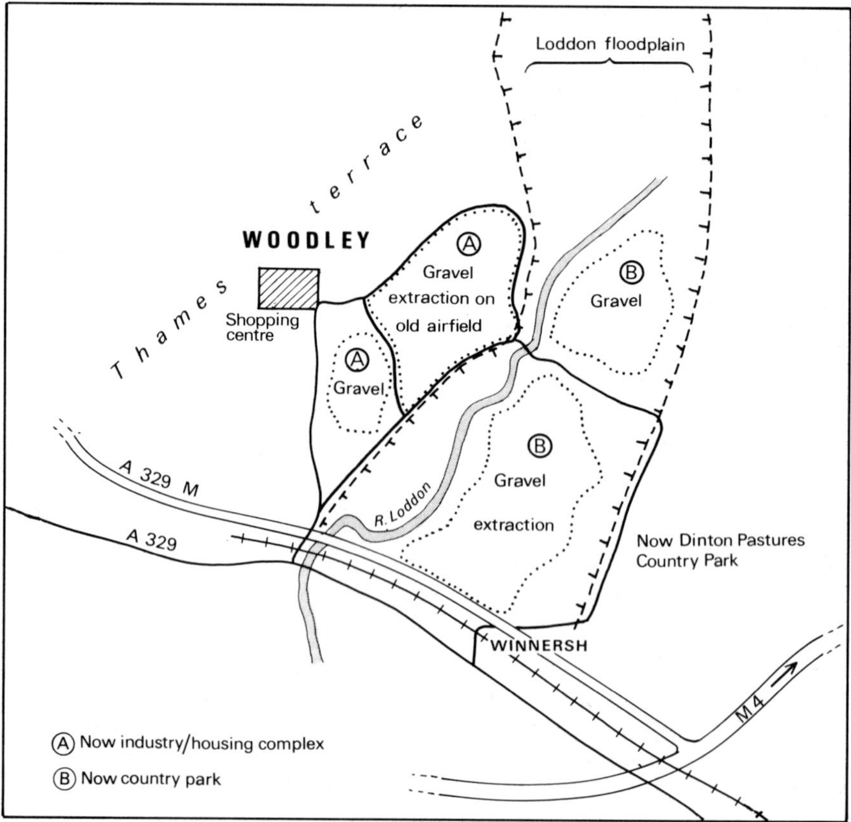

Figure 17 Sequential use of the flood plain of the River Lodden and adjacent areas to maximise benefit to the community

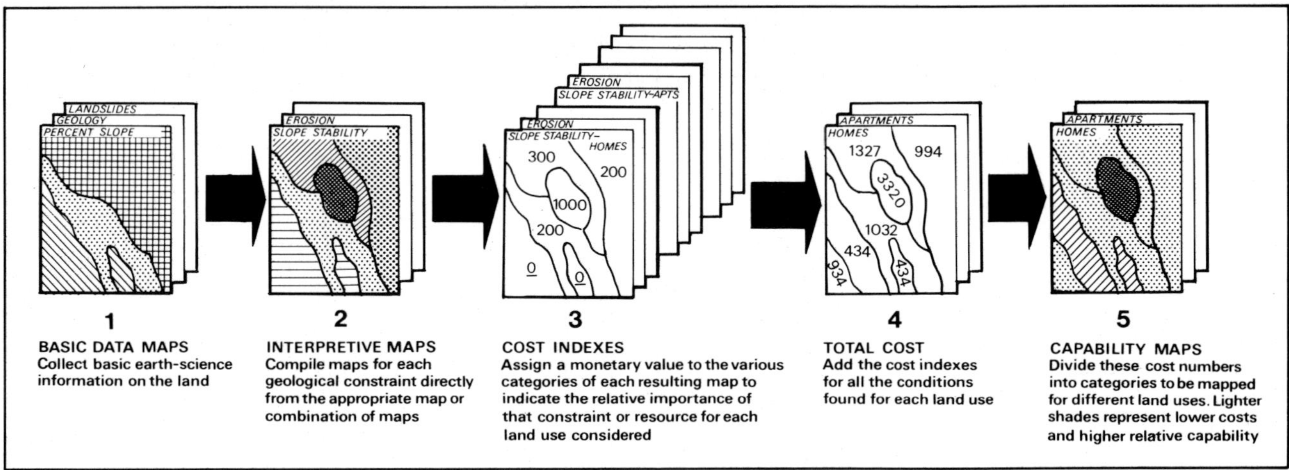

Figure 18 *Five stages in the assessment of land capability within a drainage basin*

EFFECTIVE PLANNING

Despite their efforts in recent years to pay closer attention to natural phenomena, planners have not been totally successful in incorporating earth science information into the planning process, largely because there is ineffective communication between them.

By collecting appropriate basic earth science information, interpreting that information accurately, comparing the hazards, constraints and resources, and using this information together with other environmental and social considerations, planners can help to make better decisions on land use. However, individual maps of factors such as landslide or flooding potential are unsatisfactory because they cannot indicate the relative importance of hazards, constraints or resources, or show whether it is more costly to build on a landslide, a flood plain or a sand and gravel deposit. Furthermore, the importance of a hazard depends on the land use. For example, the cost of correcting for a moderate landslide hazard is different for various types of land use. It is very different say, for low density housing as opposed to high density flats.

One method of assessing land capability within a drainage basin consists of five steps as shown in Figure 18.

1 Define the problems, e.g. flooding, that might result in significant costs for a variety of land uses. This is achieved by examining each natural process separately.
2 Map out the impact (as we have done in earlier sections).
3 Define the social cost (in £) for each type of possible development for each natural process. These costs are:
a design costs **b** potential damage costs **c** opportunity costs, i.e. the revenues that would have resulted from a type of land use and are forgone if the land is used for another purpose.
4 Total all the expected costs for all conditions, the total being used as an indicator of the capability of the land to accommodate each use.
5 Map the totals for use in the planning process. Finally the maps could be digitised and fed into a computer (Figure 19, *page* 422).

IDEAS FOR LOCAL STUDY

Various aspects of the work you have undertaken in this chapter can be tested by local studies.
1 If you live in or near a rural area, it should be possible to find hillslope concavities and to see concentration of water in such concavities and near streams during a prolonged storm. Indirect evidence of zones of high moisture content can be gained by mapping areas where rushes such as *Juncus effusus* grow and where land has been left as permanent pasture.
2 Features of river geometry can be measured directly by surveying techniques, and suspended sediment can be measured simply by dipping a flask into the stream at a known depth at regular intervals. The contents of the flask are then evaporated and the remaining sediment weighed. This technique can be used in urban areas by lowering a weighted flask from a bridge on a piece of string.
3 If the local water authority can be persuaded to provide a map showing the probable limits of flooding every 50 years, then this can be overlain on a map produced by fieldwork showing land uses. The relationship between flood hazard and land use could then be investigated.

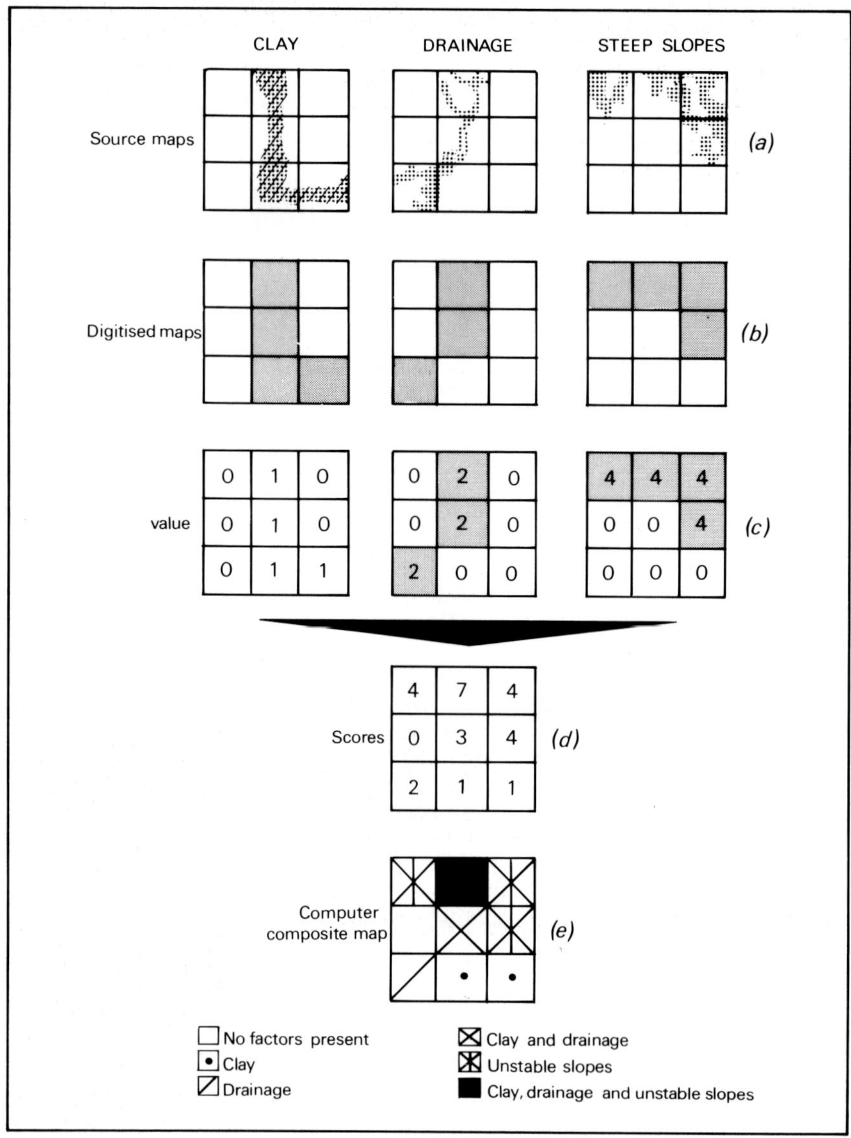

Figure 19 *Computer composite mapping technique for a hypothetical area*

21 THE US MIDWEST: THE PROBLEM OF WATER ALLOCATION

Irrigation canals and aqueducts carry water across the Midwest and California to distribute scarce river resources to fields that would otherwise be parched and unproductive

The insatiable thirst must be quenched

By MICHAEL KEATING

Canadians have smugly assumed that they could never run out of water in a land that holds more than one-seventh of the world's freshwater lakes and rivers.

But North America now faces the need to curb its insatiable thirst; some western regions are using up irreplaceable water sources and are starting to abandon farms, while other areas simply have reached the limit of their growth.

The US high plains states (including New Mexico, Texas, Wyoming, Colorado, Oklahoma and Kansas), once portrayed as dry and dusty, have been turned into lush farm fields by 40 years of tapping the vast underground water source called the Ogallala Aquifer.

Now a growing number of wells in Texas are spitting sand and desperate farmers say only massive water diversions can revive their land. By diversions they mean siphoning water from the Great Lakes or damming Canadian rivers to make them flow south.

Canadian environment officials say large areas of the southern Prairies already face periodic water shortages, a fact that compromises Canada's hope of doubling food production by the end of the century.

'In 10 to 20 years, water may become as valuable as oil', Environment Minister Charles Caccia said in a recent interview.

Last spring, John Carroll, a US specialist in Canadian-US relations, warned that Oklahoma, Kansas, Texas and New Mexico 'will become panicky' as they use up the Ogallala Aquifer. 'They are starting to look north.'

At a recent meeting on boundary waters, a Canadian expert predicted that western US water shortages could become so bad that they will threaten agricultural production and potentially harm relations between the two countries.

Richard Thomas, a Canadian fisheries expert who advises the International Joint Commission, said water shortages could be severe enough to provoke 'the decline or collapse of North American society.' He added that the semi-arid US West is budgeting for between 70 and 200 per cent more water than is available.

In addition, recent US estimates predict air pollution will trap heat near the earth's surface in a greenhouse-like effect that will increase global temperatures and reduce water supplies in many areas.

Mr Thomas said this could reduce the rainfall runoff to some already heavily used western US rivers by between 40 and 70 per cent. This would mean even more US farming areas would dry up and have to be abandoned unless water is imported.

Canadian climate experts say the greenhouse effect will not influence the Prairies as much as some US states, but Canada still will face an upheaval in its agriculture.

Some Canadian water experts say there will be mounting pressure for major diversions within the country, both to increase farm production and to stave off the effects of a climate change.

Water consumption, which shrinks lakes, rivers and underground aquifers, is predicted to increase sharply in North America, breeding conflicts over who gets water in some areas. The United States already consumes four times more Great Lakes Water than Canada.

Water quantity and quality experts warn that it takes decades to develop and build major water-works and that Canada with no water plan, is running out of time to plan for its water needs in the next century.

(*Source*: *Globe & Mail*, 24 January 1984)

THE WATER PROBLEM

'The insatiable thirst must be quenched' is just one of the frequent headlines in North American newspapers which draw attention to the anticipated water shortage over the next decade in the arid south-west and semi-arid Midwest of the United States (see Figure 1). A popular suggested answer to this 'crisis' is to construct long distance aqueducts and divert large volumes of water from Canada which at the moment appear to flow 'uselessly' into Hudson's Bay and the Arctic Ocean.

? Form pairs and make a case for building aqueducts. Take a class vote after a discussion on aqueducts. At the end of the chapter check to see if you have changed your mind. How does the class vote after studying the chapter?

SIMULATING WATER MANAGEMENT DECISIONS

This chapter contains exercises which model the problem at the level of **1** the individual farmer, **2** the regional, and **3** the international scale of government. Through such exercises it is possible to gain greater insight into the detailed intricacies of the problem.

Simulations try to model the 'real world' by reducing the complexities of real situations to a more manageable level but avoiding too great an abstraction. This can be done with varying degrees of sophistication whether mathematical or computer analogue. Most people are familiar with scale physical models of drainage basins and coastlines which allow experimentation in a controlled laboratory environment. Similarly, we are all familiar with popular decision making games which deal with the buying and selling of land. Simple decision making games or simulations can be constructed for modelling the types of decision making activity that characterises the management of water resources. Obviously, in the real world a considerable amount of information is collected and evaluated before

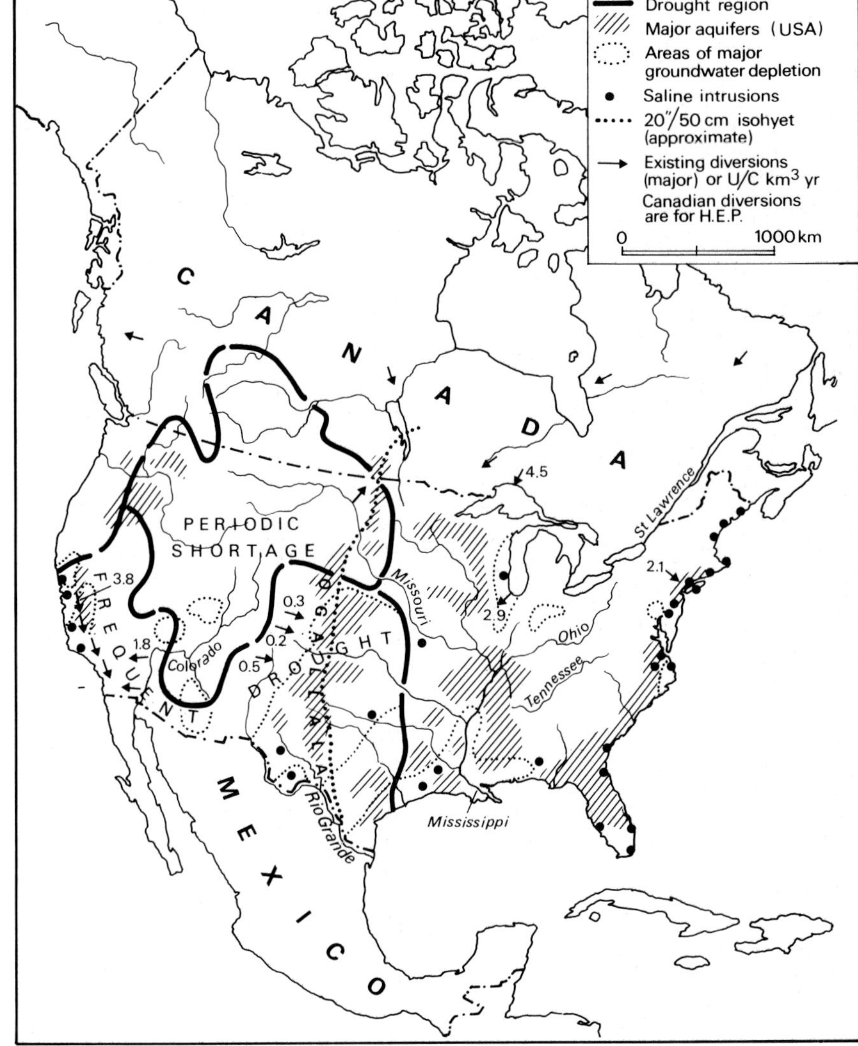

Figure 1 *North America — the water problem*

decision makers, usually well-experienced in such matters, make a final decision. The process ideally draws on a sound conceptual framework and may involve more sophisticated simulations of the type included in this chapter.

In the real world and unlike the physical scale models, the outcomes of a decision making simulation are never exactly determined because the progression of events is associated with uncertainties and probabilities as the simulation unfolds. The value of participating is that players gain insight into the complexities involved and can hopefully gain a measure of skill in coping with similar real world situations which generally have a strong element of indeterminism about them.

THE AMERICAN WEST — THE DROUGHT ISSUE

The issue of drought in the American west is a fascinating mix of geography, history and politics. The 50 cm isohyet which is shown in Figure 1 is a crucial dividing line for agriculture in the US. East of it there is sufficient precipitation for farming, whereas to the west irrigation is generally necessary. While the line is stationary in atlases, in reality its position fluctuates slowly according to the particular climatic phase, bulging eastwards in dry years and westwards in wet phases. The soils in the region are fertile and well-drained while the gently rolling terrain makes it an ideal location for growing particular kinds of crops, notably wheat. Generally farming can flourish in this region if water is available.

Over a century ago, the well-known American explorer, John Wesley Powell submitted a report to the US Congress which, based on his explorations and surveys, argued that settlement should not exceed the natural limits set by the limited water resources. His advice was ignored and the region was thrown open to settlers so that today almost 30 million people live in the Midwest. Unfortunately, many immigrant farmers settled in marginal areas during a 'wet phase' in the late 19th and early 20th centuries. They farmed for a few years but were then forced out during the 'dry phase' of the 1930s. Their plight is vividly captured in John Steinbeck's novel *The Grapes of Wrath*. Despite this adversity, the region was not abandoned because federal and state governments embarked upon water conservation programmes. Better dry farming methods were developed and improved pumping technology enabled the exploitation of groundwater from the Ogallala aquifer (Figure 1). This source has been accumulating water for thousands of years, and in 1950 had an estimated volume of 3.04 billion acre feet and seemed initially to be the answer to the problem. (An acre foot is 1 acre—0.4 hectare—of water, 1 foot—0.30 metre—deep. The USA uses English linear measurements, Canada uses metric.) With annual withdrawal increasing from 7 million to 21 million acre feet in the period 1950–1980, however, it is now becoming apparent that further expansion of both agriculture and urbanisation will lead to serious water shortages within the next two decades and even earlier in the arid south-west part of the region.

What alternatives?

A growing awareness of the problem has stimulated farmers and industrialists to put pressure on water officials to examine various alternatives to replenish and augment the aquifer. This kind of policy has been successful

Print 1 *An abandoned farm in America's Dust Bowl, 1938*

elsewhere, particularly in California and Colorado where dams and aqueducts have been constructed to move water over long distances. Many Americans in the Midwest and south-west have looked north to Canada for a similar answer to the water shortage issue. Several schemes have been proposed which would seem to make this alternative feasible. However, these plans would require large amounts of money. The California scheme has cost $2.5 billion so far and may consume another $11.5 billion if it is ever completed. Considerable opposition would also come from environmentalists who foresee major ecological damage and possibly irreversible change occurring. A Midwest diversion scheme would require the co-ordination and co-operation of several levels of government. Finally, it is not altogether certain that Canadians are in favour of selling their water, whether surplus to their present needs or not. Canadians have become increasingly sensitive to exporting some kinds of resources to the United States and some even feel that exporting water would be the final phase of 'continentalism', a process by which resources would be managed on a United States–Canadian basis (possibly with the involvement of Mexico as well). Critics see this as a means whereby Americans would increasingly place Canadians in an economically subservient position.

? List in as much detail as you can at this stage, the economic, environmental and political arguments against massive water transfer from Canada to the Midwest.

WHAT IS AN OPTIMAL SOLUTION?

In situations which entail expense or time it makes sense to examine the alternatives which may be available and perhaps, if possible, to try one on a trial basis. The phrase 'economic man' has been coined to describe how people maximise their gains and minimise their losses. In real life, people and events are more complicated than this simple model. Studies have shown, for example, that the perceptions and attitudes which mould the making of decisions can vary considerably between water engineers and water users and even between individual farmers, as seen in Table 1. Some may work very hard to 'maximise' while others may be content to work at a less intense level and 'satisfice'. An economist, for example, is interested in ensuring that the best use is made of capital, labour, and other factors contributing to production. An optimal solution, therefore, is one which provides the largest margin of gains over costs. A politician, however, is much more interested in his or her constituency getting what it desires. For him/her the optimal solution is the one which gets him/her re-elected. Some people may think they have a 'right' to cheap water while others may believe it to be over-priced. A political decision may override the rational economic choice of a

Table 1 *Decision Makers: Typical Goals and Possible Solutions*

Decision makers	*Typical goals*	*Possible solution(s)*
Farmer	Grow food and make profits	Increase irrigation; pay at bulk (block) rate
Politician	Stay in power and advance state prosperity	Respond to majority wishes of constituency to maintain votes. Water is an important vote getter
Water engineer	Increase availability of water	Evaluate alternative supply systems; increase supply
Economist	Maximise economic efficiency	Charge by volume of demand; calculate social benefits and costs
Geographer/ Resource planner	Understand/manage resources and environment wisely	As Economist, but also assess environmental impact

particular alternative. The whole question of rights is further complicated by the nature of 'ownership'.

WHO OWNS THE WATER?

Whereas the farmer can plant whatever crop he likes (within reason) because he has the property right to his farm, the situation with water resources is unusual in that water is usually not stationary and is generally regarded as a 'common property'. This means that everyone has access to the resource and it cannot be 'owned' in the usual sense. In North America the situation is more complicated in that while water is generally a common property in the humid east, in the drier west a pattern of 'prior appropriation' gradually became the rule. Where water supplies were limited, the first users claimed that they had exclusive rights and could prevent latecomers from using it. This issue was a common source of friction and lawlessness between miners, ranchers, and farmers before territories became states. By the turn of the century this approach was gradually modified as the emerging states tried to sort out the problems through legislation and the courts. The appropriator had now to apply for a water permit and show that the intended use was not 'contrary to the public interest'. This permit system and its public interest restriction was adopted in 16 of the western states although, curiously, Colorado still maintains common law, self-created appropriations.

The role of government

As a response to a restricted resource, state and provincial governments in the USA and Canada gradually assumed an important licensing and regulatory role in water management, particularly in the western half of the continent. Rivers, however, often both denote and cross state and international boundaries. This means that since the American Constitution gives Congress the power to regulate commerce between states and prevent state laws which may impede it, the federal government has an important say in water management in the United States. It was to enhance commerce and maintain national security that the federal government promoted the multi-state Tennessee Valley Authority hydroelectric power and flood control scheme in the late 1930s and early 1940s. The federal government also asserted its power through decisions of the Supreme Court in declaring that states could not prevent the 'export' of, for example, coal from Pennsylvania or timber from Oregon. However, in the case of water resources, again mainly in the west, upstream states have sought to protect their water from downstream users who might appropriate it, a situation called *area of origin protectionism*.

? Review the terms: prior appropriation, common property, public interest, and area of origin protectionism in relation to water rights.

Examples of water rights issues

As urban and irrigation development grew in the arid south-west states in the 1920s, the states of the upper Colorado River moved to protect what they felt was their long-term interest in the river. Eventually the rights to the use of the river were allocated by the Supreme Court on an *even basis* or fair share based on river length and flow volume in that state, irrespective of whether this was the most efficient way of doing it. Recently, the East Texas Utility Company became interested in moving coal from Wyoming using a water-lubricated slurry pipeline. This required 20 000 acre feet of water from the Little Bighorn River. After lengthy inquiries and negotiations, the project was cancelled because Wyoming politicians opposed the export of the water, not the coal. Similarly, the north-west states have been hostile to a diversion from the Columbia River to the Colorado River Basin.

In all schemes of this kind there are generally some groups in favour and others opposed. Where the water project is within a state it is less complicated, and since proposals usually entail substantial financial outlay and hence taxation, legislators have to put the decision to state voters in the form of a referendum. In the case of Texas, voters in the east out-voted the voters in the west who obviously stood to benefit from a proposed water project. But, in a referendum in California, voters in the south out-voted those in the north to secure the long distance transfer of 'surplus' water from the north to the south. This occurred even though the southern, urban property owners were effectively subsidizing the scheme, i.e. they were paying a higher percentage of the cost through higher property taxes, despite the fact that the agricultural areas were actually going to use a much higher percentage of the water. People were voting for the scheme even though they were not going to benefit from it directly. The Californian decision was also significant in that environmental interest groups began lobbying for the first time on a large scale over the potential ecological and other consequences of the project. Although they were not eventually successful in halting its construction, the opposition did bring to the fore the general question of the desirability of 'mega-projects'.

WHO BENEFITS?

In the face of conflicting demands on water resources, whether at the state, national, or international level, diversion project evaluations tend to identify four main points of concern:

1 Is a planned water development scheme in the best interests of economic efficiency, judged on the basis of the merits of alternative solutions? Major critics of irrigation projects, for example, argue that water is sold too cheaply and that the returns from subsidised farming do not make economic sense, particularly when there is overproduction. Most irrigation schemes in the more arid regions of the United States are heavily subsidised by the Federal government, particularly through very liberal repayment requirements.

Print 2 *Water that would once have gone to the Pacific has been diverted under the Colorado compact agreement to help irrigate the Midwest. The power needed to pump water across the Rockies is gained from HEP schemes such as the Big Thompson*

2 If a diversion project involves large amounts of tax dollars, is it being done fairly so that some regions or groups are not benefiting much more than others?
3 Even if no tax dollars are involved, opponents worry whether diversion will result in other areas enjoying economic growth at the expense of the area of origin of the water.
4 Is the project environmentally sound so that a host of detrimental side effects are not created which may be impossible to rectify?

? Take each of the above points in turn and after discussion in small groups, conclude who should make these decisions, and how, before reading any further.

Resolving each of these issues is at the crux of decision making in water resources management in North America today. The question of equity, or fairness, is decided by the political process and by the courts but, in a complex area such as water rights, it may take many years for such a political or legal decision to be made once it is initiated. The matters of economic efficiency and environmental disruption are treated largely as issues of technical evaluation, and for this two useful procedures have been developed in the USA — cost-benefit analysis and the environmental impact statement.

Cost-benefit analysis
This originated in the USA in government-funded river basin projects. It is an accounting scheme which sets out, theoretically at least, all the costs of a project (such as those relating to dams, channels or labour) and all the benefits (such as reduced flooding, hydro-electric power, increased fishing areas and so on). Depending on the cost of borrowing money to finance the project and its proposed lifetime, a decision can be made according to the ratio of costs to benefits. It can be used to compare different projects or simply to evaluate one project in isolation. Table 2 sets out a typical cost-benefit calculation. If the calculation shows that the ratio of benefits to costs is greater than unity, the project has social desirability. While this approach has become important where governments are paying for projects, there are critics who question the way in which it has been used, claiming it is often used *after* a project has been initiated. When this happens the objective of the evaluation process seems to be one of politicians justifying a decision rather than determining how good it is.

The environmental impact statement
This aid to decision making came as a necessary requirement to any government-funded project, and originated in the USA in the Federal Environment Protection Act of 1969. Proposed projects such as dams, canals, highways, and power stations require federal government departments to consider the environmental consequences of development through a detailed study and specification of:

1 the environmental impact,
2 adverse environmental impacts which may not be avoided,
3 alternatives to the proposed scheme,

Table 2 *Simplified Cost-Benefit Tabulation for Hypothetical Irrigation Scheme (millions $; 20-year period)*

Costs		*Benefits*	
Pipelines and reservoirs	100	Increased agricultural production (at constant prices)	110
Pumping system and energy costs	15	Recreation at reservoirs (fishing, boating, etc.)	20
Loss of wildlife habitat area *A*	10	Increased wildlife habitat area *B*	20
	125		150

Ratio 125:150
Project appears feasible

Note:
It is clear that certain assumptions have to be made in calculating the ratio. In this case, the ratio of benefits to costs is 150:125, or greater than one, and therefore feasible. What criticisms can one level at the assumptions? (e.g. Will food prices stay the same?)

4 short-term use and long-term productivity,
5 any irreversible commitments of resources.

An important Supreme Court decision in 1971 over the proposed Calvert Cliffs power station showed that the Federal Environmental Protection Act (1969) would be enforced. The proponents of the power station had, by the new law, to produce an environmental impact statement as part of the proposal so that the inquiry could decide on its possible consequences. Consequently, the environmental impact statement is now a required input into decision making associated with all projects such as water, energy, and transportation schemes. Although it originated in relation to federal projects in the US, it has subsequently been copied by the federal government in Canada and by many state and provincial governments. Thus, it is impossible to proceed with any proposal without a careful evaluation of the possible impact on the environment.

It can be seen that this kind of technical evaluation goes beyond cost-benefit analysis which focuses on income-generating effects and financial costs. Furthermore, project evaluation is now broadening to encompass social impacts, particularly where native peoples and traditional lifestyles may be affected. The social impact analysis is still evolving and has not diffused to the extent of the environmental impact statement. Despite the complexities of decision making in the water resources area, it is clear that through extensive public debate, legal adjudication and the intensive research of the past three decades, both technical, economic and social, the development of more sophisticated methods of data collection and project evaluation, the likelihood of a water resource project which is clearly unfair, inefficient and environmentally unsound is highly unlikely but, unfortunately, not impossible.

? Now compare your conclusions on the who and how of making decisions with the information outlined in these preceding paragraphs.

WATER AND THE MIDWEST

PHYSICAL FEATURES OF THE MIDWEST

The high plains of the Midwest slope gently south-eastwards to sea level and the lowlands of the Mississippi/Missouri River basin, and in the west they are abruptly terminated by the Rocky Mountains. To the south they are bounded by the Rio Grande River, and to the north they continue into Canada as the Canadian Prairie. The region is drained by south and easterly flowing rivers disbursing into the Gulf of Mexico; in Canada the rivers mainly flow northwards to the Arctic Ocean, often by way of large lake bodies (Figure 2).

The geology of the region is dominated by flat-bedded deposits of Tertiary age material, an unconsolidated remnant of gravel, sand and silt eroded from the Rocky Mountains. This formation provides especially favourable conditions for groundwater to accumulate — the Ogallala Aquifer, in fact, covers more than 354 046 square kilometres.

Coal, oil, and natural gas deposits occur throughout the Midwest region. In the south-east, salt dome intrusions have resulted in gas and oil accumulation. Elsewhere, large reserves of medium to poor quality coal exist. Extensive oil shale deposits occur in the Rocky Mountain region of the USA, while in Canada very large deposits of tar sands occur in Alberta. The processing of oil from these formations would require large amounts of water.

The climate of the Midwest is one of continental conditions with hot

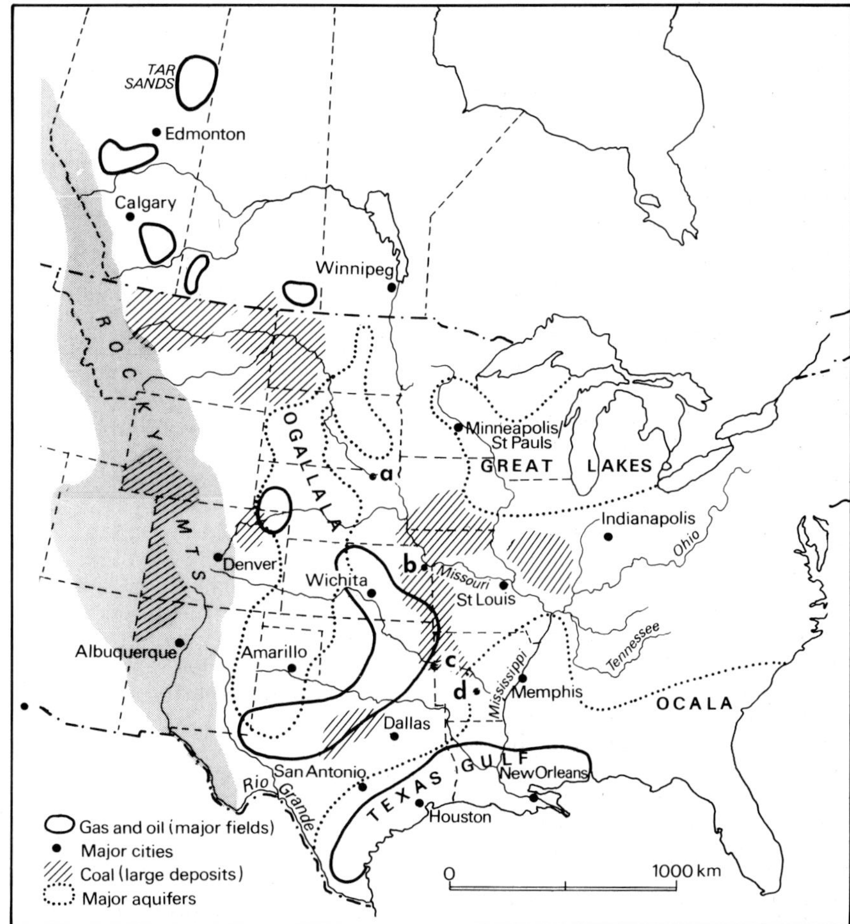

Figure 2 *The Midwest*

summers and cold, dry winters which increase in severity northwards. Precipitation is crucial to plant growth with the 50 cm isohyet being the key zone of moisture demarcation reflecting the rain shadow effect of the Rocky Mountains and hence the need for irrigation. Aridity also increases south-westwards since the region gains little moisture from the Gulf or from the eastward track of the northern hemisphere's cyclonic systems, unlike Canada which has a greater overall surface runoff than the USA. Although the Great Lakes are a huge natural reservoir, both countries are committed to maintaining present lake levels and hence there is a reluctance on the part of states and provinces in that region to export water to other areas.

Print 3 *Irrigation in the Midwest*

WATER FOR AGRICULTURE

Overcoming the problems of semi-arid farming

The settlements in the Midwest which had taken place during a moist phase in the latter part of the 19th century were faced in the late 1920s and early 1930s with severe drought conditions. This crisis was exacerbated by the early farming and ranching methods whereby straight row tillage reduced much of the accumulated moisture in the soil, and without vegetation to hold it, a considerable portion of the top soil blew away in the dust storms. In the ranching areas the land was frequently over-grazed and the hoof-compacted soil deteriorated in the same way. Consequently, the 1930s was a period of dramatic reversal of grassland settlement, with farmers and ranchers abandoning land in the drier parts on both sides of the border. The response to these conditions was both political and technical. Governments eventually intervened through the Conservation Corps and the New Deal in the US and the Prairie Farm Rehabilitation Scheme in Canada. For many farmers both these initiatives were too late; they had already packed up and moved to the far west or to the cities.

For those who stayed, a programme of soil conservation, with sub-surface cultivation, stubble mulching (trash farming) and alternate, contour-aligned crop stripping was encouraged. These measures emphasised minimal disturbance of the top layer of humus to retain moisture and hence permitted farming to continue. Collectively, such adjustments to the realities of farming and ranching in a semi-arid region during the 1930s were as important in the human geography of the Midwest as the original settlement was during the last decades of the 19th century. More significantly, this period changed the attitudes of farmers towards the desire to secure more reliable sources of water, if possible through irrigation. Several irrigation schemes were initiated in the upper Missouri, its tributaries and in Canada, while in the Ogallala Aquifer region, deeper wells were drilled to draw on the groundwater. From the dramatic reversal of the 1930s both agriculture and ranching gradually rebounded.

Modern irrigation

Because the region is well-suited to large-scale, mechanised farming and ranching, agricultural output is enormous. The Ogallala region of the Midwest, which is the area most affected by declining water supplies, has collectively only one per cent of the total US population and six per cent of the total land area, but produces 15 per cent of the nation's wheat, corn, sorghum and cotton. The Canadian Prairies account for 16 per cent of the Canadian population and 80 per cent of that country's agricultural area and have experienced almost continuous economic growth related to the export of wheat and energy, especially in the past 20 years.

The total acreage of irrigated land in the drier southern part of the US Midwest increased from 2.5 million in 1950 to over 14.0 million acres in 1980. Most of this increase was from the Ogallala source where consumption increased from 7 to 21 million acre feet. Similarly, agricultural production grew in the northern Midwest states with increased irrigation from dam and reservoir construction. In the four western provinces of Canada, irrigated acreage expanded from 0.62 million acres in 1950 to 1.5 million acres in 1980, also by extensive dam and reservoir construction. Although total consumption of water has grown, irrigation efficiency has also increased due to improved methods of sprinkling, automatic timing according to soil and atmospheric humidity conditions and less leakage and evaporation.

WATER FOR INDUSTRY

While agriculture is the greatest volume user of water, the plains and prairie region has also surged in economic importance through hydrocarbon discoveries and the subsequent development of an oil and gas industry. Oil and gas booms in Texas, Kansas and in Alberta have had spin-offs in the development of urban growth and related service industries, all of which have increased the demand for water. Industries related to the processing of agricultural products have also grown substantially, such as cattle feed lots and cotton ginning in the south.

As the world and North American supplies of petroleum are gradually depleted, attention is turning to using oil shales as a source of hydrocarbons. Almost 25 000 square miles of Colorado, Utah, and Wyoming have oil shale deposits which are estimated to be able to supply US petroleum needs for a century at least. Present day costs of extraction and refining are high, but, as other supplies dwindle, they could become more attractive. There is also the problem that a daily production of 1 million barrels would require 300 000 acre feet of water, with the Colorado as the likely source. Similarly, in Canada, the large tar sands deposits of Alberta, which have been developed so far on a small scale, consume significant amounts of water during the conversion process.

It is clear, therefore, that the present day agriculture of the plains, which relies on cheap water, may face increasing competition from other regions and other users.

THE EMPTYING OF THE OGALLALA AQUIFER

The steadily increasing utilisation of groundwater in the southern plains has been predicted to result in at least 25 per cent of the resource being tapped by the end of the century. Furthermore, there are areas where the 'overdraft' is already a serious problem. In western Kansas and Nebraska the Ogallala aquifer seems likely to decline in volume by 50 per cent. Whereas that water source was almost 60 feet thick in 1930, it is now less than 8 feet. In Nebraska, half of the irrigation projects in the western part of the state are estimated to be facing a serious decline in water supplies by 1990. These problems are not unique to the southern plains states — the arid south-west, Arizona, New Mexico, and southern California have similar water emergencies. In 1963, the Supreme Court decision over the sharing of the Colorado meant that although Arizona won, it still needed to have the support of California (who lost) to get money to finance water projects. The arid south-west was experiencing an urban boom associated with retirement homes, tourism, leisure, and military development. Many towns and cities such as Albuquerque (New Mexico) grew by more than 20 per cent during the 1970s. Consequently, the south-west states decided to look for water elsewhere rather than keep squabbling amongst themselves. One source has been the Columbia River — but the political leaders of the north-west states have been vehemently opposed. Consequently, it can be seen that the semi-arid parts of the plains states face competition from the even more arid states to the south for available sources of water within the US.

? Read through this section again and make a list of competing regions and users of water. Refer to Figures 1 and 2 and your atlas if necessary.

SIMULATION 1: DECISION MAKING AT THE GRASS ROOTS — THE PLAINS FARMER

SCENARIO

The greatest part of the water consumed in the Plains is used for agriculture; specific irrigation requirements vary according to local soil, climatological, drainage conditions and the crops being grown. The typical irrigation requirement is approximately 1.5 acre feet per acre per annum.

The farmer is the decision maker and has to evaluate many factors when deciding on:

1 Which crops to select and plant.

2 How much and when to irrigate or how to improve irrigation.

3 How to cope with the prospect of drought.

External factors also play a part, such as:

1 Crop prices and demand for agricultural products.

2 Government decisions to build irrigation canals.

3 Changes in energy costs.

4 Changes in technology.

Clearly, there are a range of approaches to coping with drought hazard over which one assumes the farmer may exercise a measure of choice. Insofar as external factors are concerned, the farmer may have little or no control and can only respond in a limited manner. Furthermore, there are those events which are uncertain and may change, moving the direction of simulation in unplanned ways. If a farmer is reasonably successful in decision making, the farm should prosper; if not, the enterprise may fail.

SIMULATION PROCEDURE

Farms are 1000 acres in size. Rounds simulate one year of farming operations. Participants assume the role of farmers.

STEP 1. Locate the farm Using a random number table allocate participants to a regional location in the High Plains area according to the co-ordinates of eastings and northings in Figure 3. Thus, a random number of 22 would give a location at square, easting 2, northing 2. Three main zones of decreasing water stress are delineated: **A B C**

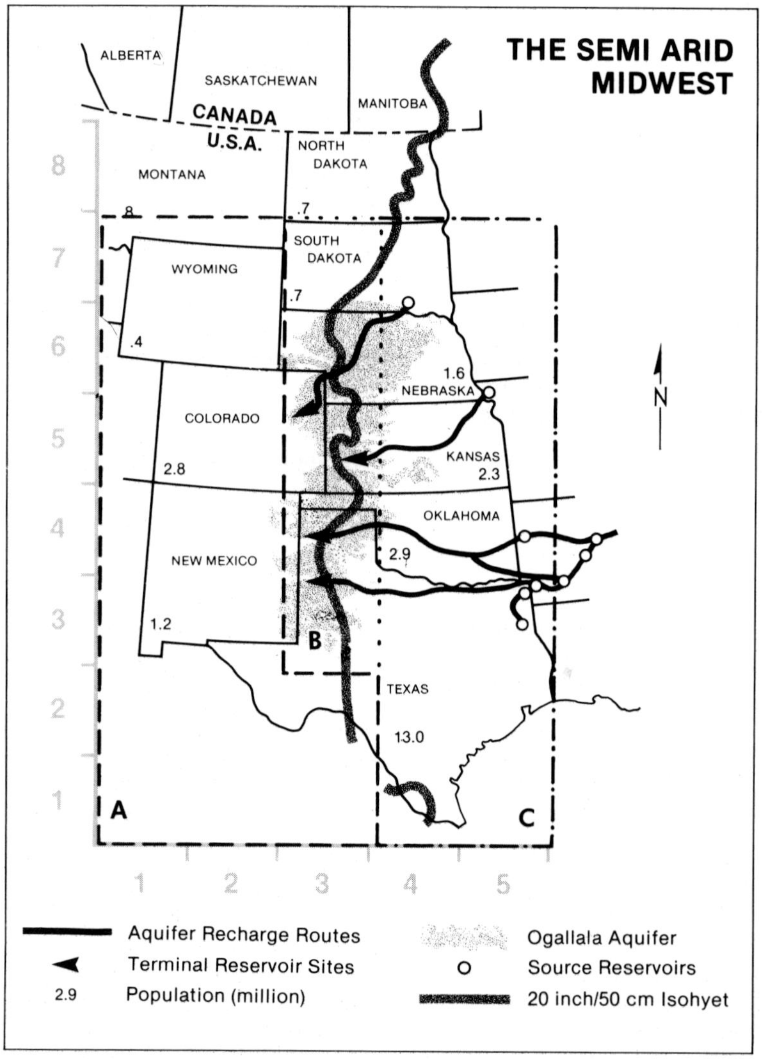

Figure 3 *The semi-arid Midwest*

STEP 2. Examine crop choice Table 3, *page 433*, shows the main crops grown and their expected yield per acre (late 1970s yields). All players start as dry land farmers. When sufficient capital has been saved one can switch to irrigation. This table can be excluded for the first or early rounds (why?).

STEP 3. Environment Using a random number table, determine the environmental effect on harvest for year 1 (2,3, etc.), at the appropriate row–column intersection in Table 4.

STEP 4. Crop Price Using a random number table determine price effect on returns for each round in Table 5.

ACCOUNTING

Participants keep an account per year of costs and profits. The approximate operational or input cost per hectare is $50 for dry farming and $100 for irrigated land. Irrigated land needs a capital reserve of $5 000 before start up is possible. Lifetime for irrigation is five years Zone **A**, 10 years Zone **B**, 20 years Zone **C**.
Simulation can commence with participants:

REVIEW

1 What kinds of strategies did players develop to cope with poor crops or low prices?
2 Do some players emerge as gamblers and others as cautious conservatives?
3 Did strategies change as experience grew?

1 having the necessary initial cash for one year's input costs;
2 being able to borrow the amount by an advance against the expected harvest; and
3 being able to borrow to obtain irrigation equipment.

Table 3

Water stress zone	Dry farming A	B	C	Irrigated A	B	C
Wheat	15	25	30	30	40	45
Sorghum	20	35	55	95	80	80
Corn	—	—	—	115	115	115

Crop value (per bushel): Wheat $2.75, Corn $2.10, Sorghum $2.00.
Yields: in bushels per acre

Table 4

Growing conditions	Very poor	Poor	Average	Good	Very good
i Dry Farming	−40%	−20%	+5%	+20%	+30%
ii Irrigated	−10%	−5%	+10%	+25%	+35%
iii Numbers	0,1	2,3	4,5,6	7,8	9

Table 5

	Low	Average	High
Wheat	−10%	+5%	+40%
Corn	−20%	+5%	+30%
Sorghum	−40%	+2%	+20%
Numbers:	0,1,3	3,4,5,6	7,8,9

ALTERNATIVE STRATEGIES FOR THE MIDWEST

Given that there is an impending water shortage, what are the alternatives available to the Midwest? Before even examining them, some critics have argued that the money could be spent more effectively by increasing food production elsewhere in the US or Canada, or through aid to the developing countries. Generally this viewpoint, although proposed by many environmentalists, is not viewed with much favour in the Midwest. After 150 years of development and conquering the hazards of the region, the ethos is still that of 'development' and a 'well-watered earth'. Consequently, the response has been to examine alternative means of increasing the availability of water: a supply-side solution.

? Look at the viewpoints expressed here and in Chapter 13. Given the wider context provided in that chapter, set up a debate on the proposition that the money could be spent more effectively through aid to developing countries.

Various studies to solve the problem have been undertaken in all the south-west states which have experienced water scarcity, and by many of the city governments there too. In addition, the Federal Economic Development Commission (FEDC) has been active in examining the problem. This burst of activity has been a response, not only to perceived future shortages, but also to actual droughts in the early 1970s and early 1980s (and also because there was a moratorium by Congress during 1968–1978 on the study of major inter-basin transfers).

THE HIGH PLAINS STUDY

The various studies have identified, on a preliminary basis, projected water deficits, potential impacts and the kinds of alternatives open to decision makers in, for example, the High Plains region of the Ogallala Aquifer. The High Plains Study (1981) pointed out a projected decline in irrigation capability by 1990 of 40 per cent, using optimistic assumptions, and a 60 per cent decline using conservative assumptions. Among the possible alternative courses of action were the following:

1 Voluntary conservation
This could make possible a small increase in production (1 per cent) through more efficient use and the introduction of drought resistant crops.

2 Mandatory conservation
This would lead to a decline of $1.04 billion in the regional economy (i.e. simply shutting off water use).

3 Augmentation
Additions to the water supply might be feasible but this is generally unproven on a large scale.

4 Intrastate transfers
So far major intrastate diversions have only been studied in Oklahoma and could divert water from the eastern part of the state. The northern diversion would cost $5.3 billion over 30 years.

Nebraska also investigated this option and found that because of the cost, a re-examination of state policy was necessary before a plan could be drawn up. Colorado is still studying the possibility of diverting from the South Platte River to the High Plains. The other states generally feel that there is more water to be diverted.

5 Interstate imports and interbasin transfers

The report identified four possible routes for interstate transfers drawing on the Missouri River at **a** Fort Randall and **b** St Joseph, and the Arkansas River at **c** Van Buren and **d** Pine Bluff (Figure 2).

6 A combination of conservation and importation

The importation would recharge the aquifer and thus stabilise production and also increase surface supplied irrigation. Gross benefits were estimated to be: Colorado — \$65–\$115 per acre foot; Kansas — \$43–\$111; Nebraska — \$3–\$284; Oklahoma — \$73–\$93; Texas — \$45–\$246; New Mexico — \$103. In such calculations, as has already been discussed earlier in relation to cost-benefit studies, certain assumptions have to be made about interest rates and the period for writing off the cost of the projects, and the cost of energy for pumping water uphill.

Despite the strong opposition to such developments (both the Carter and Reagan administrations have balked at the high costs involved), a strong lobby of various interest groups still pressures for water projects (see the article below). Although no inter-basin diversion exists to convey water to a state lying entirely outside the basin of origin, the importance of the Arizona versus California court decision in 1963 was that Congress *could* legislate major inter-basin transfer. Furthermore, other court rulings have determined that no state or coalition of states could veto a large scale inter-basin transfer believed to be in the overall national interest. Consequently, there is no constitutional barrier to a group of states, and/or Congress, planning and undertaking a long distance water project.

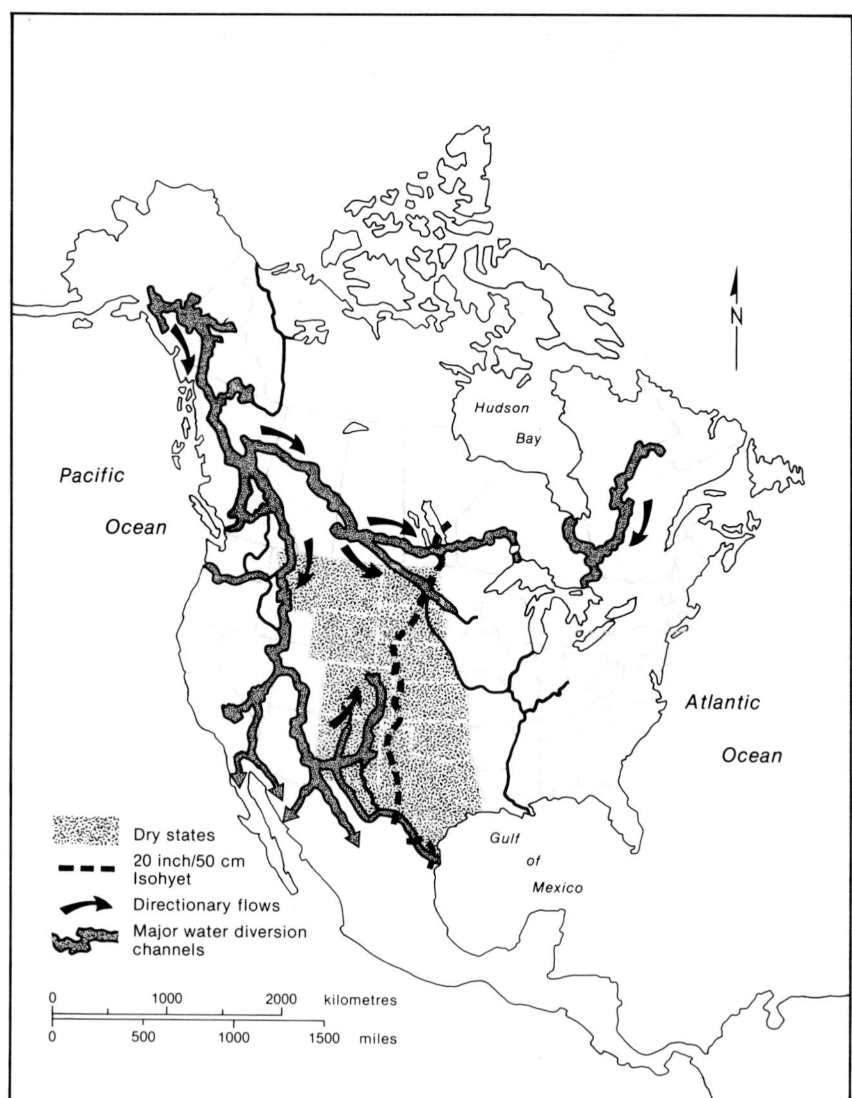

Figure 4 *The North American water and power alliance scheme*

WATER DIVERSIONS FROM CANADA

There have been several proposals for longer transfers of water to the south-west and southern High Plains, principally from Canada. The attraction is that on an average annual basis, Canada's rivers discharge in total approximately 3.2 million cubic feet per second. No major international transfer of water presently exists and there is no legal basis for an American claim to Canadian water. This has not inhibited several American and Canadian engineers from drawing up plans to move water south from the north flowing rivers (Figure 4).

Perhaps the best known and most controversial is that proposed by the Ralph Parsons Company, the North American Water and Power Alliance (NAWAPA) scheme described in the article below. The cost was estimated in 1966 at more than \$100 billion and it would take at least 20 years to build. Not only would this drown major mineral deposits and forest land, but it would be located in an earthquake-prone zone. Understandably, in the face of such anticipated costs and fear of the unknown consequences of the scheme, many Canadians are vehemently opposed. Although some alternative plans such as the Grand and Kuiper schemes originated in Canada (see Figure 7 *page 439*), they are more modest in scale and have probably less dramatic environmental impact.

Appeal Made to Import Water

Canals Sought to Save Ogallala Aquifer

WASHINGTON — Outgoing Texas Gov. William Clements appealed to the Reagan administration on Thursday for federal funds to develop a plan to import millions of gallons of water, at a cost of billions of dollars, to six western states.

Clements made the appeal as chairman of the High Plains Study Council, which has just completed a four-year, $6 million study of the six-state area served by the Ogallala Aquifer.

In presenting the study to Commerce Secretary Malcolm Baldridge, Clements raised the prospect of canals stretching as far as Canada to bring water into Texas, Oklahoma, Colorado, New Mexico, Kansas and Nebraska.

However, his vision contrasted sharply with a cover letter to the study signed by all six governors on the council. They concluded that importation of water into the agriculturally rich High Plains isn't financially feasible.

The letter said the cost of importing water would far exceed the ability of customers to pay and concluded water conservation programmes are the answer.

The area gets most of its water from the Ogallala Aquifer, an underground supply being used more quickly than it is being replaced. The 170 000 irrigation wells tapping the aquifer support a thriving agricultural industry that produces 15 per cent of the value of the nation's wheat, corn, sorghum and cotton.

Brig. Gen. Hugh Robinson of the Army Corps of Engineers, which co-operated in the study, said, 'There isn't sufficient water available in the area to continue farming as it is now, without some importation.'

Clements said he raised the possibility of importing water from Canada during his meeting with Baldridge.

However, Monte Pascoe, director of Colorado's Department of Natural Resources who worked with the council, said the council had agreed the solution was to change present farming techniques and crops to conserve water.

The council specifically decided against seeking Canadian water some time ago after the Canadian Embassy said Canada needed its own water, Pascoe added.

He said the costs of importing water from other states, if they were willing to part with it 'are fantastic. It just boggles the mind.

'It seemed far more practical to concentrate on conservation techniques,' Pascoe said.

'They've been mining water in west Texas a lot longer than we have and don't have the same conservation techniques.' Pascoe said, explaining why Clements was more interested in importing water than the other governors.

'Conservation is the only answer,' added Wayne Wyatt, director of the High Plains Underground Water District in an interview in Lubbock, Texas, after the report was released.

Wyatt said on Thursday that officials have discovered 1 billion acre feet of extra water in the Ogallala Aquifer 'that we didn't know was there.'

(*Source*: *D.P.* 1 July 1983)

NAWAPA — A VIABLE SOLUTION TO THE WATER PROBLEM?

During 1980, Americans used 235.5 billion gallons of fresh water every day. Nearly half of this amount, or 105.2 billion gallons, was not replaced by snow or rainfall. To make up the difference, great underground reservoirs known as aquifers are gradually being tapped for their ancient water supplies. The public is alarmed by reports of 'the browning of America.'

This is the atmosphere which has given a fresh lease on life to enormous water transfer projects such as that dreamed up in California in 1964 by the huge engineering firm, the Ralph M. Parsons Co. Called the North American Water and Power Alliance, this gigantic project with a price tag of more than $200 billion would bring water from the Yukon down to Mexico.

The entire Rocky Mountain Trench in B.C. would become one enormous reservoir 800 kilometres long which would obviously mean kissing goodbye to cities such as Prince George, Cranbrook and Golden. The logistics of the project are breathtaking — it would ultimately produce 70 million kilowatts of electricity, divert 160 million acre feet of water per year, and would require the cooperation of three nations, 36 states, seven provinces, one territory and thousands of cities and local governments.

NAWAPA has always been laughed off as an engineers' pipedream but they are deadly earnest these days at the Pasadena headquarters of the Parsons organisation.

'We are very encouraged by the new administration of President Reagan and we've been talking to many politicians about NAWAPA,' said Parsons' director of corporate relations, Dorn S. Dicker. 'We think there's a possible regeneration of interest in our project. We feel the climate is very ripe for this type of thing.'

(*Source*: *The Weekend Sun*, Vancouver 4 April 1981)

Whether such dreams ever materialize is one thing; another and more immediate problem is that many streams and rivers cross the international boundary and both countries share the Great Lakes so that arguments over water rights have been almost inevitable. Conflict first arose when ranchers in Montana began diverting flow from the Milk River in 1889. In Canada, diversion work to irrigate land near Lethbridge affected users downstream in Montana. It was clear that adverse impacts would emerge in both countries and that the traditional friendship between them would be upset. Subsequently, under the aegis of the Boundary Waters Treaty (1909), the International Joint Commission (IJC) was set up to investigate such problems and to recommend action to the two governments. Composed of equal representation from Canada and the USA, the Commission has looked at disputes over the Milk, Saskatchewan, Souris and Poplar Rivers in the Plains area as well as many others. These issues were not properly resolved until the principle of 'apportionment' was devised in 1921. This principle is an agreed and binding share of water volume which varies from river to river according to the flow in any given year during the irrigation season. In its application, and lately through its responsibility over pollution monitoring, the Commission has performed important roles both in maintaining good relations between the two countries and applying sound water management ideas. Clearly, if any plans to undertake long distance, international transfers of water were seriously considered by the two governments, the IJC would be at the centre of negotiation and decision making.

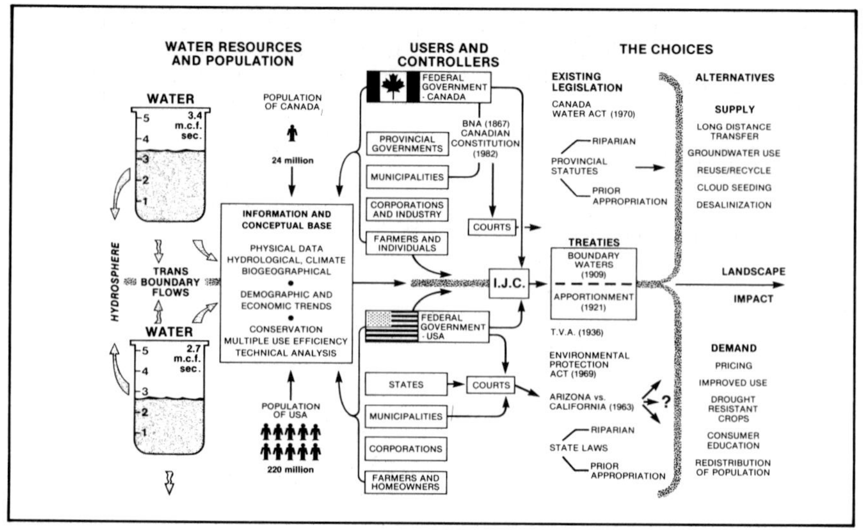

Figure 5 *Water resources and population*

The Organisation of water management decision-making

How the general organisation of water management decision making exists today is shown in a simplified format in Figure 5. The diagram is split two ways into Canadian and American halves, top and bottom, and three major components — the resource, the users and the choices, left to right. Ideally, Mexico, as the other North American nation sharing similar water supply and trans-boundary problems, ought to be included, but it is beyond the scope of this chapter to do so.

Looking from left to right, the first section, *Water Resources and Population*, draws attention to the great differences in population and water resources between the two countries. Managing water effectively requires basic information on supplies — literally how much is there and trends in use, such as regional population change. However, the data can only be used effectively if certain principles and concepts of water resource use are established and adhered to, and if suitable technical analysis is performed.

The central section, *Users and Controllers*, picks out the main groups of users and the levels of government control, all of whom are water resource managers in one way or another. Generally, the USA has more groups and a more complex and sophisticated decision system than Canada, a function of its greater development and regional variation. The authority to decide over water rights is specified in each country's constitution and subsequent legislation, but is then subject to interpretation by the courts of law. The two inserts, IJC and TREATIES, show that a procedure for managing disputes between the two countries is in place and functions effectively.

The third section, *The Choices*, points out that legislation has been enacted in both countries at different levels of government to specify the rights in and procedures for water use. In both countries the two concepts of riparian and prior appropriation underlie the body of legislation. The first originated where supplies were plentiful; it is based in English common law and generally applies to humid eastern North America. Prior appropriation evolved where supplies were limited; it evolved in western North America on the rangelands and in gold mining areas and became the common pattern of water rights in the drier west. Choices also encompass alternatives for action and can vary between supply (e.g. transfer) and demand (e.g. pricing) methods.

One or a combination of the alternatives in Figure 5 may be chosen in either country or by particular states or provinces. The *actual* choice is going to be related to the degree of environmental stress (literally how likely or bad is the drought?); the functioning of the decision making web at a particular point in time (e.g. how much money is allocated to the government and what legislation exists); and how familiar the decision maker is with the range available. In short, the management of water resources is complex and exact solutions are impossible to predict, although the resource geographer can indicate which is the most reasonable alternative. Whatever the chain of events and responses, whether no action or more irrigation, the impact will appear and affect the landscape whether dried grassland or green, irrigated fields.

The generalisation of Figure 5 must now be matched with the detailed reality of decision making at three scales of geographical significance. Faced with the prospect that local water supplies may be insufficient to sustain past levels of agricultural production, what can a local farmer do? This relates to Simulation 1 on *page 431*. Recognising that their political survival may be at stake, what should a provincial or state legislator recommend? Believing that the aims of their own interest group are laudable, what should the lobbyist or protester support? This relates to Simulation 2 below. Beyond this, at the national and international levels, what should central government officials and politicians recommend? This relates to Simulation 3 on *page 438*.

SIMULATION 2: REGIONAL TRANSFER OF WATER

SCENARIO

Several regional, interbasin transfers of water, wholly within state or provincial jurisdictions, have been proposed in the last two decades or, in the case of California, are under construction. In the Province of Alberta, a scheme to divert water southwards by a series of canals, tunnels and holding reservoirs, the Prairie Rivers Improvement Management and Evaluation Plan (PRIME) has been proposed and discussed for several years (Figure 6).

Diversion downplayed

EDMONTON (CP) — Henry Kroeger hasn't abandoned his dream of water diversion but now says it will be a long time before anyone has to think of funnelling northern Alberta rivers to the dry south.

The newly formed, nine-member Alberta Water Commission does not have water transfer as a goal, said Kroeger, the former transportation minister who is now chairman of the commission and a Progressive Conservative member of the legislature for Chinook.

The commission's job, he said in an interview, will be to identify the province's dryland and wetland areas and to find ways of correcting the imbalances.

'Long before you talk about diversion in a major sense you have to look at storage, conservation and more efficient ways of water management,' Kroeger said.

'Somewhere down the road, interbasin water transfer will have to be looked at. But that's an awful long way down the road.'

Kroeger said he did not want to discuss water transfer because of widespread criticism he received for supporting the concept.

He said, however, that diversion of rivers would be the way to change a balance that sees 87 per cent of Alberta's water flow through the sparsely populated north of the province while only 13 per cent flows through the heavily agricultural south.

'The only other alternative is to move people up north,' Kroeger said. 'We need more development in the north, but people aren't likely to take to the climate.'

The commission will have $300 000 a year for the next five years. Its budget does not include provision for major studies.

Alberta NDP Leader Grant Notley has said he supports the principle of a water commission but is worried about the appointment of Kroeger and of assistant deputy environment minister Peter Menychuk, another proponent of interbasin transfer.

Their presence on the commission fits too well with the secret public relations scheme envisaged in government documents leaked last year, Notley said.

Those documents stressed the need to interest the public in water diversion.

Notley said he would like three conditions attached to the commission's role:

— Clear mechanisms for and guarantees of substantial public input into commission deliberations.

— A commitment by Premier Peter Lougheed that all the commission's reports and studies be made public.

— The immediate ruling out by the commission of any possibility of interbasin transfer of water.

(*Source*: *Calgary Herald*, 27 December 1982)

In this simulation, the major interest groups, of the kind which can sway the government's decision, come to the forefront and argue their relative merits. This is accomplished through the procedure of a public inquiry where participants in the simulation, representing the key interests, present their views and attempt to influence the 'strategic' decision.

SIMULATION PROCEDURE

STEP 1. Allocation of participants to one of five groups: the Provincial Government Board of Inquiry (who will make the decision); Oil Companies; Farmers; Industrialists; Environmentalists.

STEP 2. Groups prepare positions on PRIME, and the Board of Inquiry decides on the structure of the public inquiry.

The Government has to choose the degree of influence the inquiry will have by selecting the appropriate rung of S. Arnstein's 'Ladder of Participation'.

Citizen Control
Delegated Power
Partnership
Placation
Consulting
Informing
Therapy
Manipulation

If the rung is too low, the groups may overreact and feel misled; if too high, the government may feel it's no longer governing. Public consultation techniques fall into five broad areas:

1 start-up (brochures about the meeting);

2 collecting information,

3 mutual education (workshops, open house);
4 public preference (the meeting);
5 decision (presenting the decision);
Which is/are the best technique(s) to choose?

Albertans have tended to elect conservative governments not prone to radical change. With large reserves from oil revenues, the government may favour the long-term investment value of the project, especially as oil will decline and the economy will focus again on agriculture. Who is chosen to sit on the inquiry is crucial for the government.

Oil Companies have to decide whether to support or oppose the plan since substantial volumes of water are needed for processing the Tar Sands. (At full volume, the greater part of the Athabasca River would be appropriated, mainly for effluent disposal.) The companies may have conflicting goals. On the one hand they need to protect water requirements for tar sands processing. On the other, the value of land which they own or control in the south will increase if water is available.

Farmers are located mainly in the south half of the province and are keen to get further supplies of water.

STEP 3. The Public Inquiry takes place and each group presents their viewpoint and attempts to refute the other group's arguments.

STEP 4. The Government Board of Inquiry decides whether to:
1 proceed with the plan;
2 modify it in some way;
3 abandon the plan;

Review
1 Which group seemed to sway the decision and why?
3 Environmentalists often argue that the system is against them; does it appear so in this simulation? If so, why? If not, why not in this case?
3 What forms of communication took place in the simulation? Explain their significance. Did coalitions form?

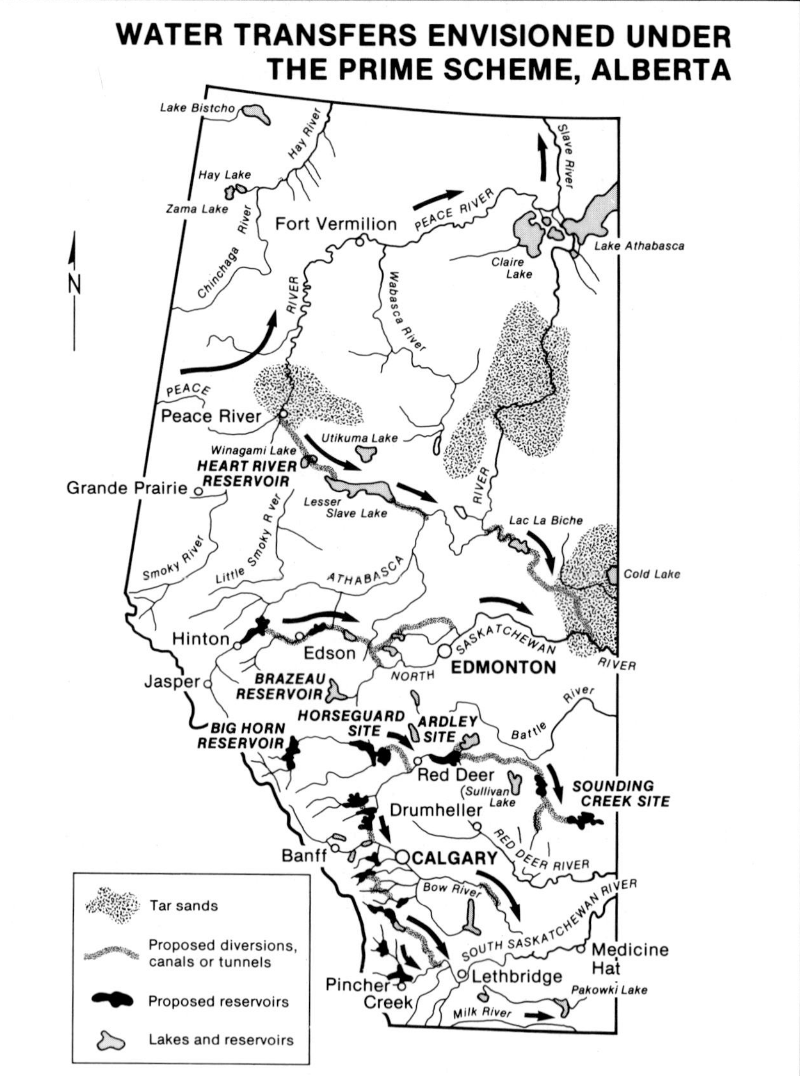

Figure 6 *Water transfers envisioned under the Prime Scheme, Alberta*

SIMULATION 3: INTERNATIONAL TRANSFER
SCENARIO
As the arid south-west and semi-arid Midwest US continue to grow in urban settlement and irrigation needs, there is increasing pressure to find additional sources of water. This perhaps means long distance transfer from Canada as the only large volume source (Figure 7). Any proposal which affects both countries would require substantial finance, careful evaluation, agreement between the parties and consent from their citizens. Assuming that the need does arise, the International Joint Commission will be at the focus of decision making. As a quasi-judicial body, it has, over a period of 70 years, dealt with over 100 issues and had 80 per cent of its recommendations accepted by both governments.

In this simulation, have players assume the role of major interest groups who each prepare briefs which summarise their argument for or against the proposal. These are then presented to a player or group acting as the IJC which then decides on the issue and presents a recommendation to the USA and Canadian governments.

Experts fear border rows from shortage

For over two decades Canadian policy has firmly rejected water diversions and the US Congress has refused to pay for the expensive studies needed to trigger such works.

But the planners are simply biding their time.

In a recent interview Thomas Kierans, a St John's engineer was still promoting his 1959 Grand Canal scheme. He has an $84-billion plan to dike the 150-kilometre-wide mouth of James Bay, turning it into a fresh-water lake.

Although the United States has yet to make a formal request for water, Canadian officials are acting as if it was imminent.

'The United States is withdrawing ground water faster than it is recharging (the aquifers)' said William Mountain, federal assistant deputy minister of inland waters.

'There may be a hell of a lot less water in the interior basins,' he said, adding that a drying up of the west would produce a social upheaval like that of the dustbowl period of the 1930s.

Last year the federal Government told the Canada-US International Joint Commission on boundary waters that it has a long-standing opposition to unilateral increases in diversions. 'Large-scale diversions would play havoc with the ecological and economic balance of the Great Lakes basin,' Federal Environment Minister Charles Caccia said.

Ontario Natural Resources Minister Alan Pope said a proposal to siphon 280 cubic metres a second from Lake Superior would drop Lakes Huron, Michigan and Erie by as much as 15 centimetres, creating an 'almost catastrophic effect.'

At a meeting of Great Lakes premiers and governors in Indianapolis last November, Ontario Premier William Davis and Quebec Vice-Premier Jacques-Yvan Morin found allies when the governors of Michigan, Ohio, Wisconsin, Indiana and Illinois agreed to oppose diversions outside their borders.

William Ruckelshaus, head of the US Environmental Protection Agency, said at another meeting last fall that there will be no US attempts to divert Great Lakes water without Canadian agreement.

Canada is relying for protection on the Boundary Waters Treaty of 1909 an agreement which binds it and the United States to agree on any diversions from shared waters.

But a growing number of water experts inside and outside of governments are calling for public discussion of how Canada will deal with North American water issues.

'Down the road we need a national strategy on diversions and the export of water,' Mr Mountain said. The provinces control most of Canada's water and he said it must be a truly national, not just federal position.

(*Source*: *The Globe and Mail*, 25 January 1984)

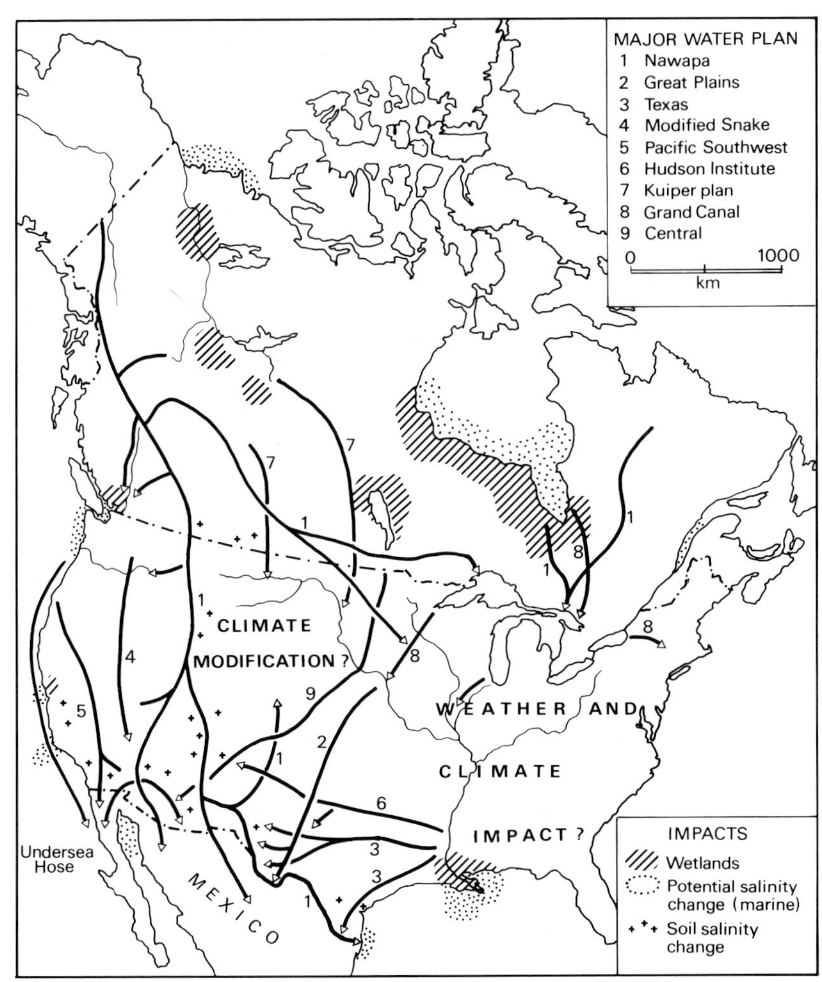

Figure 7 *International water transfers — proposed routes and potential impacts*

SIMULATION PROCEDURE

STEP 1. Allocation of players to groups (5)

1 International Joint Commission

2 United States engineering companies

3 Canadian economists and geographers

4 Western lobby

5 Environmentalists

STEP 2. Groups prepare briefs or summary arguments.

IJC In preparation for reviewing the proposals, it will have to decide:
1 whether both countries have equal rights to the total resources of the Continent;
2 the basis of apportionment (even or some other ratio?);
3 whether the river basin should be the basis of water resource planning and allocation or state boundary;
4 the order of importance of efficiency, equity, environment OR nationality.

US engineering companies A conglomerate of companies who have devised a long distance water transfer system. More than one route may be proposed and construction in phases may be emphasised (use Figure 2). Stress the growing problem of world hunger and North American efficiency in food production. Emphasis can be placed on the job creation and spin-off benefits.

Canadian economists and geographers will review costs and benefits in terms of export revenues, displacement costs, loss of future use of water. The economists may stress the importance of alternatives as encouraged by higher prices for water (e.g. greater care with evaporation and leakage). Geographers stress range of alternatives and attempt to estimate human and landscape impact of proposals.

Western lobby An alliance of western states, corporations and farm associations who are determined to increase their water supply. Their brief can focus on:
1 national economic development;
2 reliability and productivity of irrigated agriculture;
3 provision of jobs in the Midwest and south-west;
4 people having 'a right' to sufficient water.

Environmentalists An umbrella organisation of environmental interest groups such as the Sierra Club who oppose 'mega-projects' and want 'small is beautiful'. They can stress the long-term ecological implications of the project such as loss of wetland in the north, possible increased salinity of the Arctic Ocean and threats to wildlife (the transfer of possible ecological misfits between river basins).

STEP 3. Briefs are presented to the International Joint Commission and a recommendation is made.

PROSPECTS FOR THE FUTURE

The availability of cheap water is approaching a crisis in the Midwest region of the USA because the greatest user, agriculture, relies on a source which is diminishing through overuse, particularly in the south and west. The political history of the continent has seen two countries evolve with long east to west boundaries which cut across the broad natural biomes and major topographic features that trend north to south. Canada has large water resources and a small population, one-tenth that of its neighbour. To many in the United States, the solution of the crisis seems to be a matter of moving the surplus to the deficit. This attitude, partly due to the belief in technology and the 'supply side' answers of the plainsmen, is also fostered by the long history of trade and friendship between the two nations: the political boundary seems insignificant.

However, to the voters of the north-west States and Canada, these attitudes seem irrational in being expensive, unfair and environmentally unsound. Environmentalists and social scientists interested in water management point out that other alternatives may be cheaper and less disruptive. The 1970s may have been the decade when the large 'mega-project' proposals were finally tamed. The next two decades will be crucial for the sound management of water resources particularly in the semi-arid and arid regions. It is unlikely that the struggle between the belief in technology with its 'supply side' emphasis and the principles of efficiency, equity and conservation is over.

? Use what you have learned in this chapter to add another paragraph to 'Assault on the Earth' *page 366*.

22 WEATHER IMPACTS AND WEATHER MANAGEMENT

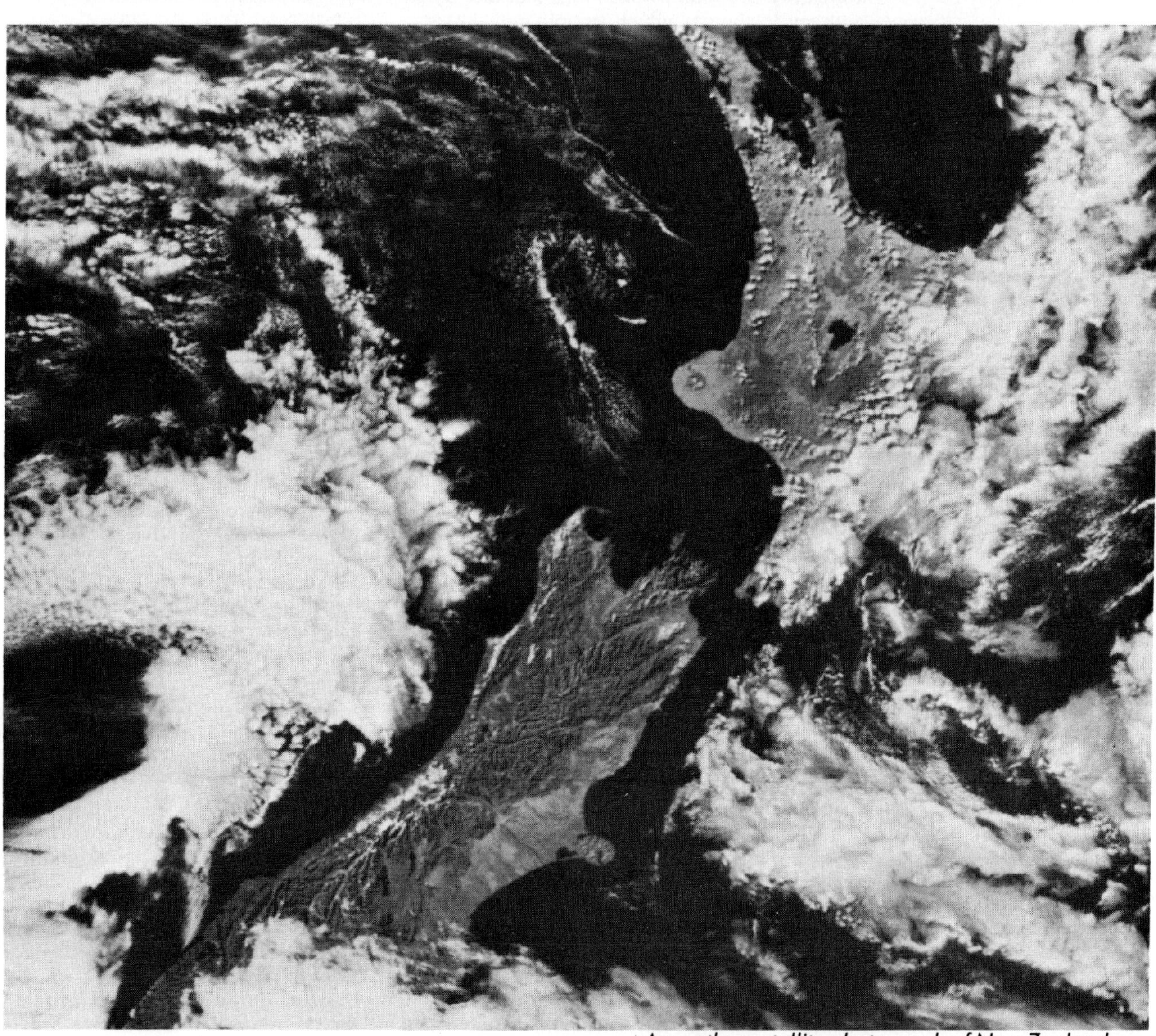

A weather satellite photograph of New Zealand

WEATHER SENSITIVITY — THE NEED FOR INFORMATION

While there is not enough evidence to conclude that the world's climate is deteriorating, it is clear that climate has become more variable. This means that this year's bumper crop may soon become a vague memory.
(*Source: Business Week*, August 1976)

The World's Climate Unpredictable

Record rains and floods soaked some areas while droughts parched others, with potentially serious social, economic and political effects.
(*Source: Time*, August 1976)

The reporting of weather forecasts on television, radio and in newspapers has grown to meet an increasing need from a wide variety of people and organisations. National and international businesses, agriculturalists, transport planners and sports event planners are among those who need information about future weather conditions. 'Climate sensitivity' (the impact of weather and climate on people and their activities), with its social and political overtones, is now well-recognised.

It is important to realise, however, that most of the applied aspects of meteorology and climatology have been, and in many instances still are, concerned with the interaction between the atmosphere and physical or biological processes, such as the effect of carbon dioxide on the climate. Little if any attention has been given to the human aspects of the subject. In geography, this has meant that meteorology and climatology have, in most cases, been considered exclusively as a part of physical geography, and not (as they should be) as part of geography as a whole, especially economic and human geography.

It is now well-established that a much more overall concept of *applied* meteorology must be used, involving the relationships between the atmosphere and people as decision makers. Indeed, it is clear that in many situations the impact of the weather on physical, biological, and economic processes and activities is influenced both by the weather events that occur and by the choices which people may make.

? List the people and organisations mentioned in the first paragraph. Alongside, suggest the choices they may make given the occurrence of a week's severe frost.

WEATHER INFORMATION AND TIME-SCALES

The key factor to note is the concept of weather information, which may be made up of a number of separate items. It may, for example, consist of:
1 the weather existing at the present moment,
2 the weather that is expected to exist at a specified time in the future, *and/or*
3 analysis and interpretation of the records of weather that have existed in the past, including the recent past.

All three items are of value, but the actual weather information required is generally determined by the kind of problems confronted, and the associated decision making. For example, an analysis of past weather is useful for assisting in planning the location and design of many types of facilities, such as irrigation schemes, hydroelectric dams, and business enterprises. On the other hand, immediately available and very recent weather information is essential for activities such as marketing, agricultural production forecasts, and the analysis of business trends, whereas present weather and short-period forecasts of future weather are most useful in making immediate operating decisions, such as flight planning for a commercial airline, the scheduling of irrigation, or estimating domestic energy demand.

? Think about and discuss in groups how information on **1** present, **2** future and, **3** past weather would be useful in predicting the weather for a sporting event you are interested in.

WEATHER AS A RESOURCE

Meteorological research has identified three important concepts:
1 That the atmosphere is an important natural resource which may be perceived, tapped, modified, despoiled, or ignored, and whose availability may be forecast.
2 That information of past weather and climate, present weather and future weather and climate is also an important resource.
3 That given an understanding of physical–biological–sociological interactions with the atmosphere, and given sufficient information about atmospheric events, people can, at times, use their management abilities to improve the economic and social outcome of many weather-sensitive activities.

These concepts have been identified and developed in the last 20 years but it is true to say that there are still very few case studies which give the cost in dollars or pounds of a specific weather or climatic event.

THE CANTERBURY DROUGHT

In the Canterbury area of New Zealand (Figure 1), where 'fat-lamb' and wheat farming is the dominant agricultural industry, droughts occur from time to time. The most 'recent' drought started in July 1980. In parts of Canterbury a very serious drought also occurred in the 1984–85 season. It was especially critical in South Canterbury and the neighbouring North Otago areas.

The primary cause of the very dry conditions relates to a prevalence of strong westerly winds with correspondingly wetter than usual conditions on the western side of the South Island of New Zealand and drier than usual conditions in the area east of the main mountain range, the Southern Alps. In addition, during the period of the

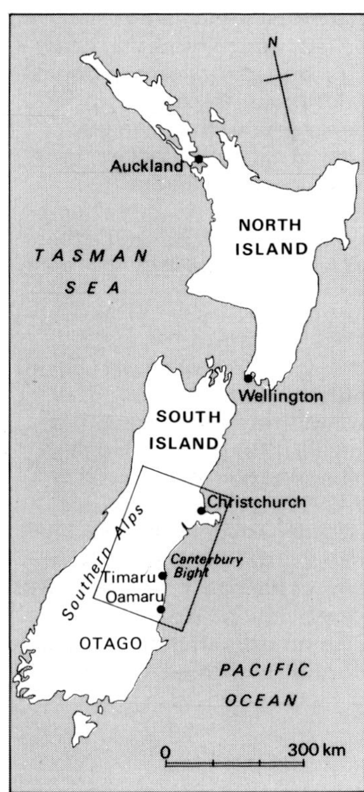

Figure 1 *Map of New Zealand showing location of Canterbury area*

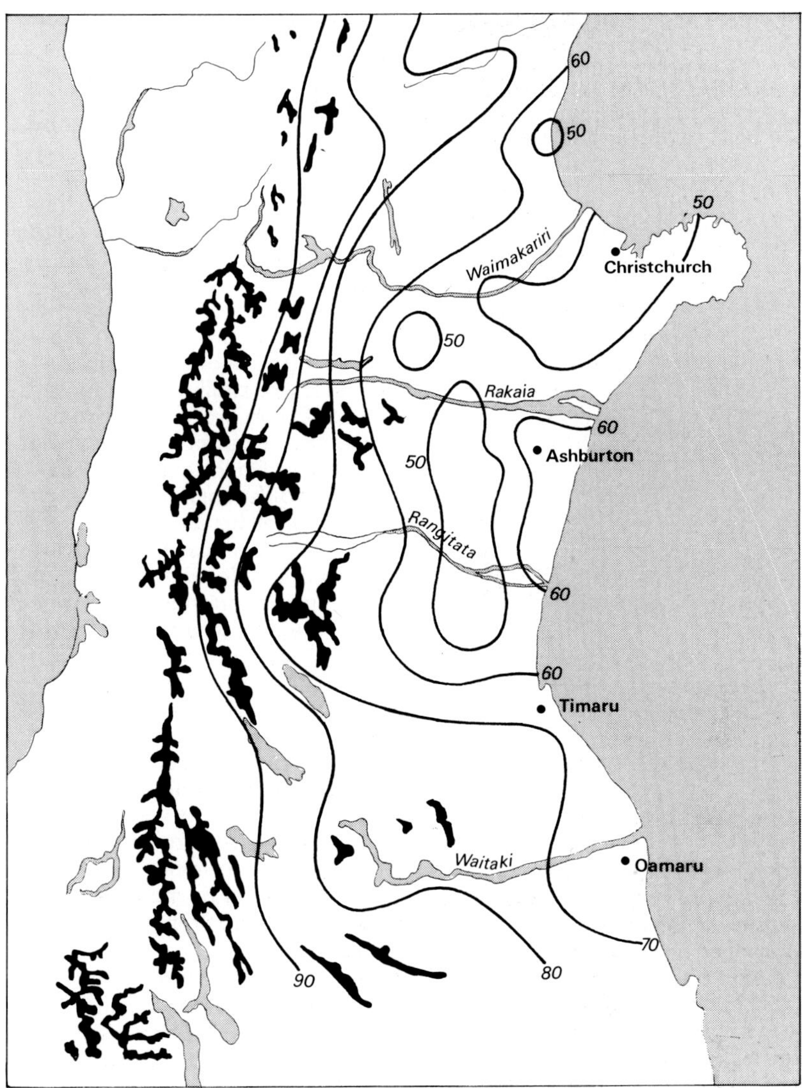

Figure 2 *November 1981 to August 1982 (inclusive) rainfalls in Canterbury/North Otago expressed as a percentage of the 1941–70 norm for the period*

drought, few depressions have formed in the central and northern Tasman Sea and even fewer have moved south-east to the area east of South Island. Pressure patterns and ocean temperatures in various parts of the tropics, particularly in the Darwin and Tahiti areas, probably contributed to the stronger than usual westerly flow.

? Examine Figure 2 carefully, then copy it and shade in those areas most severely affected. How would you describe the general pattern of drought severity?

THE AGRICULTURAL IMPACT

To the concerned farmer, as well as the concerned community in which the farmer lives, two questions may well arise. Firstly, does long-range weather forecasting provide any answers? Secondly, is weather modification (specifically rain-making) possible? Regrettably the answer to the first question is 'No', except under very unusual circumstances which may permit a forecast of trends to be made some time ahead. Similarly, at least under drought conditions in Canterbury, the answer to the second question regarding weather modification is also 'No'. Indeed, although a considerable amount of work has been done in both subjects there are, at present, few positive results which could lead to providing any real help to farming communities such as those in Canterbury.

The individual farmer, then, has to make his own decisions with appropriate advice from agricultural and financial advisors, and from those who can analyse and interpret climatological records. The decisions made by one Canterbury sheep farmer in 1982–83 are detailed below.

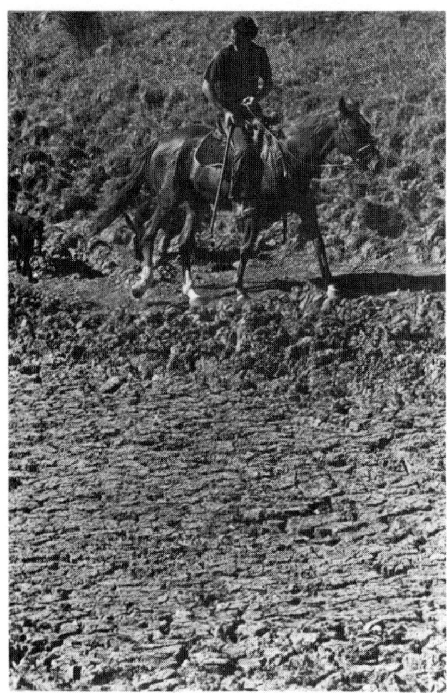

Print 1 *Drought in the Otago area of New Zealand*

Decision making on a Canterbury farm

The farm has 210 hectares, and the income comes from raising sheep. In January 1982, the total number of sheep on the farm was 2500. By this time, conditions were very bad as little rain had fallen since early June 1980. In the absence of long-range rainfall forecasts, there was no indication of how much longer the drought might continue.

The farmer chose to spend $30 000 in providing supplementary feed. This included spending $17 000 on barley which was brought in as grain feed to carry the ewes over the winter (June, July, August). The farmer also chose to spend $6000 in order to graze sheep on a nearby farm. The balance of $7000 was spent on 'invisible' costs such as selling lambs at lower than normal prices. The owners of this particular farm were established farmers and hence were able to afford to spend the additional $30 000, but a farmer in a less fortunate situation may not have been able to carry such additional costs. In this case, however, the decision to buy supplementary feed allowed the owners to close off most of the farm over the winter and, in September 1982, the pastures on the farm were in reasonably good condition, and the ewes which had been fed on the barley (at a cost of $17 000) went into lambing in good condition.

Many questions arise from the decisions which were made on this specific farm. They include, whether it was wise to spend $17 000 on supplementary feed (e.g. if it had rained from February and so produced ideal grass growth, would the $17 000 have been a total loss?), whether it was really necessary to spend $6000 to graze sheep on a nearby farm (presumably the owner of this farm with the extra pasture had made different decisions — possibly to sell his sheep and lambs at low prices because he could not afford any supplementary feed), and whether irrigation should have been used. (In reality irrigation water was not available, and estimates of $150 000 for a full-scale irrigation scheme would obviously have to be considered on a long-term basis, and certainly not in the middle of a drought.) These questions are very real, and in a drought situation anywhere, similar sorts of questions will arise. The key factors are that decisions have to be made — even the decision to do nothing.

? **1** Set out in table form the expenditure made by the farmer. Write your own commentary on the decisions he made.
2 Consider the same case if the owner had no available money to spend. What options would then be open to him?

WEATHER SENSITIVITY OF NATIONS

Research on the economic value of weather is only useful if it affects the way decisions are made by managers of weather sensitive processes. A continuing need exists to assess and present the impact of weather in terms of production figures, costs, or similar measures, which can be used *directly* by decision makers, including economists, agriculturalists, planners, transportation companies, and politicians. The ranking of nations in terms of their climatic sensitivity is a difficult concept in weather economics and in the analysis which follows, one attempt at ranking what may be called the 'export climate' of nations has been made.

? How would you attempt such an analysis? Consider, for example, the value of exports from the USA in 1979 and the value of exports from India in 1976–77 as given in Tables 1 and 2. How would you rate each of the commodities exported in terms of their climatic sensitivity to each nation? Do the commodities listed provide the information that you need?

Table 1 *Exports from the United States (1979)*

Commodity	*Value (US$ million)*
Meat and meat preparations	1 127
Dairy products and eggs	161
Fish	1 027
Wheat and wheat flour	5 491
Rice	850
Corn and other grains	8 109
Fruit, nuts, vegetables	2 126
Animal feed stuffs	2 317
Inedible crude materials	20 755
Synthetic rubber	579
Raw cotton	2 198
Coal	3 394
Petroleum and products	1 916
Animal and vegetable oils	1 845
Chemicals	17 306
Machinery and transport equipment	70 491
Other goods	28 879

(*Source*: *Whitaker's Almanac*, 1981)

Table 2 *Exports from India (1976–77)*

Commodity	*Value (Rs. million)*
Iron and steel	3 872
Cotton manufacturing	3 312
Iron ore and concentrates	2 982
Fresh fruit and nuts	1 151
Vegetable oils	486
Leather	2 635
Jute manufactures	1 974
Coffee	1 140
Tea	2 929
Tobacco	968

(*Source: Whitaker's Almanac*, 1981)

CALCULATING AN 'EXPORT CLIMATE SENSITIVITY INDEX'

In the analysis which follows, the weather sensitivity of 133 nations was measured in terms of their exports by considering the proportion of the exports of each country in five broad classes: food/beverages, non-food agriculture, fuels/minerals/metals, machinery equipment, and other manufactured goods. These components were weighted by factors of 10, 7, 2, 1 and 3 respectively to give what could be called an 'export climate sensitivity index'.

The weightings of 10, 7, 2, 1, and 3 are based on the estimate of each component's sensitivity to weather or climatic factors, and obviously must be looked at as first guesses only. In effect, a total weight of 23 was allocated (10 + 7 + 2 + 1 + 3), of which 10/23 (or about 44 per cent) is the weight allocated to the obviously climatic sensitive food/beverages export sector, whereas only 1/23 (or about 4 per cent) was allocated to the relatively low climatic sensitive machinery equipment export sector.

For example, Fiji, with mid-1970 exponent values of 75 per cent for food/beverages exports, 1 per cent for non-food agriculture, 14 per cent for fuels/minerals/metals, 3 per cent for machinery equipment, and 7 per cent for other manufactured goods, has an 'export climate sensitivity index' of 809 (that is 75 × 10, plus 1 × 7, plus 14 × 2, plus 3 × 1, plus 7 × 3), whereas the USA has a comparable index of 361 (made up of 19 × 10, plus 5 × 7, plus 6 × 2, plus 43 × 1, plus 29 × 3).

Values for other countries vary from 1000 for Gambia to 198 for Japan. Some of the major agricultural exporting countries had relatively high rankings (for example, Argentina came 31st, New Zealand 43rd, Australia 61st, United States 83rd), whereas several of the oil exporting countries had relatively low rankings (for example, Venezuela 126th, and Saudi Arabia 130th). A shortened list of the rankings of 40 countries' indexes is given in Table 3.

It must be emphasised that these weights are based on a *general* econo-climatic appraisal and other weightings may well be more appropriate. Naturally, the weightings will vary considerably from region to region, but if a nation by nation comparison, or ranking, is to be made (as has been done in this case) there are advantages in using the same weighting system for each country. By so doing, a useful climatic sensitivity ranking is obtained, which may of course be quite easily calculated and updated each year. If this is done, trends in the world 'rankings' of 'export climates' may be assessed.

Obviously the ranking and the 'export climate sensitivity index' will vary according to the specific year being studied, as the composition of exports from some countries varies considerably from year to year (in many cases these are related to weather or climate). The index will also be different if another set of weightings is used, but it is considered that any reasonably similar weighting (based on an appropriate econo-climatic assessment) would give a ranking of nations not too dissimilar to that provided by this analysis.

COMMODITY WEIGHTED WEATHER INFORMATION

People in pursuit of useful and profitable enterprises, must make many weather related decisions. Such decisions are made for many purposes, and are based, in part if not in whole, on the type of weather information available. Therefore, when weather information is used in decision making in connection with some weather sensitive activity, particularly on a national scale, it is essential that the information be in a form appropriate to the problem and the area concerned. One way to do this is to provide decision makers with commodity weighted information about weather and climate.

For example, a forest industry (whether regional or national) is basically interested in the weather and climate conditions pertaining to the forested areas, a wool industry in the weather and climate in the sheep areas, and an electricity industry in the weather pertaining to the populated areas with respect to consumption, but also in other areas with respect to energy sources and transmission routes. Despite this almost obvious need for commodity weighting, it has been traditional for most national meteorological

?

1 You are the budget director of an international finance company. Several requests have been received from countries seeking financial aid to assist their agricultural development. Using the information contained in this section (including Table 3) assess how you would assign priorities for financial assistance. You should be aware that the need for assistance, and the ability to repay the loan, are both important.

2 In the light of the matters discussed in this section, how would you rank, from a climate sensitive viewpoint, the key agricultural exporting countries of Australia, Argentina, New Zealand, USA, USSR and Canada? Three of these countries are in the Southern Hemisphere and three are in the Northern Hemisphere. Is this important?

Table 3 *Sensitivity ranking of export climates for 40 selected countries*

Country	*Exports (%)* Food, beverages	Non-food agriculture	Fuels, minerals, metals	Machinery, equipment	Other manufacture	*Export-climate sensitivity index*
Weight	*10*	*7*	*2*	*1*	*3*	
Gambia	100	0	0	0	0	1000
Cameroon	73	19	5	1	3	883
Iceland	77	1	16	1	5	825
Sri Lanka	65	17	5	0	13	818
Argentina	69	6	1	9	15	788
Kenya	65	5	18	2	10	753
Thailand	59	12	9	3	17	746
New Zealand	46	29	6	5	14	722
Brazil	62	2	10	12	14	708
Uruguay	45	21	0	2	31	692
Cyprus	47	1	7	8	37	610
Peru	44	6	49	0	1	583
Australia	31	13	33	5	18	526
Greece	33	3	14	5	45	519
India	32	3	9	6	50	515
Mexico	33	9	29	7	22	461
South Africa	22	10	27	7	34	453
Portugal	16	10	4	15	56	421
Israel	15	4	2	10	70	402
Spain	21	2	6	26	45	397
Jamaica	20	0	71	2	6	362
United States	19	5	6	43	27	361
Korea	11	2	2	17	68	349
Singapore	10	14	32	24	20	346
Poland	11	2	13	14	60	344
Canada	12	12	26	33	18	343
France	15	3	6	37	39	337
Hong Kong	4	2	1	16	77	303
Austria	4	7	5	28	56	295
Italy	8	2	7	34	50	292
USSR	8	10	25	22	35	288
United Kingdom	7	2	10	37	44	273
Sweden	3	11	6	44	36	271
Switzerland	4	1	3	33	58	260
Nigeria	6	0	93	0	1	249
Germany, F D. Republic	5	1	5	48	41	238
Algeria	2	0	97	0	1	217
Venezuela	1	0	98	0	1	209
Saudi Arabia	0	0	100	0	0	200
Japan	1	1	1	56	41	198

(*Source*: Compiled from mid-1970 export data published by The World Bank)

Table 4 *Computations involved in the calculation of weighted rainfall indices in the South Canterbury, North Otago region of New Zealand (data for March, 1969)*

Climatological station	*Rainfall: percentage of normal*	*Land area*		*Human population*		*Sheep population*		*Dairy cows in milk*	
		A[1]	*B*[2]	*A*	*B*	*A*	*B*	*A*	*B*
Lake Coleridge	54	1.1	59.4	0.1	5.4	1.7	•	0.0	•
Ashburton	39	1.2	46.8	0.8	31.2	2.4	•	0.2	•
Lake Tekapo	97	2.7	261.9	0.1	9.7	1.3	•	0.0	•
Timaru Aerodrome	68	0.8	54.4	0.4	27.2	1.1	•	0.3	•
Timaru	48	0.3	14.4	1.2	57.6	0.7	•	0.1	•
Waimate	40	1.4	56.0	0.3	12.0	1.9	•	0.1	•
Tara Hills	89	1.2	106.8	0.1	8.9	0.7	•	0.0	•
Oamaru	74	1.2	88.8	0.9	66.6	1.0	•	0.1	•
South Canterbury/ North Otago Region	—	9.9	685.5	3.9	218.6	632.2			
Weighted Index (*B*/*A*)	—	$\frac{685.5}{9.9}$	= 70	$\frac{218.6}{3.9}$	= 56	$\frac{632.2}{10.8}$	= 59	——	= 56

A[1] = Percentage of the New Zealand total of the economic parameter located in the geographical county and associated climatological station(s)
B[2] = (Column *A*) × (the rainfall at the climatological station expressed as a percentage of normal)

services to publish weather and climatic information for *places* usually only on a routine monthly or, in some cases, weekly basis. In some cases an attempt is made to map this information, but a considerable deficiency still exists in that the basic meteorological point data cannot, even when mapped, be used in association with economic data which is generally related to areas. The weather and climate information has to be adjusted to fit the available economic data. One way to do this is to use economic data collected on a county basis, and this method has been used in New Zealand for many years.

COMMODITY WEIGHTED INFORMATION IN NEW ZEALAND

Weather and climatic indices weighted according to the distribution of 100 economic and agricultural parameters are currently analysed by the New Zealand Meteorological Service for several areas of New Zealand, and for New Zealand as a whole. The actual weightings used in each analysis are based on the contribution of the 'geographical county' to the New Zealand total 'population' or 'area'.

A specific example of the calculation of weighted rainfall indices for the Canterbury/North Otago area of New Zealand for March 1969 is given in Table 4. In this example, eight climatological stations are listed (each being said to be linked to a specific geographical county), together with the weights (or percentage of the New Zealand total of the various economic parameters shown) 'located' in each county. The rainfall expressed as a percentage of normal for each of the climatological stations is also shown. The detailed computations necessary to determine the weighted (regional) index for land area (that is 70 per cent), and human population (that is 56 per cent) are shown.

Partial computations are also given for the sheep population weight, and for dairy cows in milk. In order to understand the techniques used for computing weighted rainfall indices properly, complete Table 4.

?

1 What weightings of weather and climate data would you use for your country?

2 What would you consider to be the five most important weightings for, say, Algeria, Australia, Argentina and Austria?

3 Do you think the weightings should be updated every year?

4 How would you use the weather weightings to help decision makers?

FORECASTING AGRICULTURAL PRODUCTION

Research into climate and agricultural production usually emphasises the relationship between climatic variability and, either the productivity of human-altered environments such as a cornfield, or the productivity of 'natural' environments such as grasslands for sheep or cattle. The economic and, in many cases, the political importance of large-scale weather fluctuations must also be emphasised, as they contribute greatly to variability in global food production.

AGRO-CLIMATOLOGICAL MODELS

The relationship between agricultural production and climate is usually assessed through the development of an agro-climatological model. Such models attempt to mirror the physical processes that cause variations in agricultural production (such as soil moisture availability), but in many cases a 'simple' statistical relationship is established linking, say, variations in rainfall with variations in wheat production. These 'statistical' models do not usually explain environmental processes, but the relationships obtained from such models do condense complex cause and effect linkages into simpler associations which are much more readily understood and computed.

Weather based agricultural production models have been in operation in New Zealand for several years. In these models, the weighted number of days of soil moisture deficit (weighted by the distribution of sheep, cows, cattle) was found to be a convenient way of condensing a vast amount of weather information. The number of days of rainfall deficit is calculated from the balance between the daily rainfall data and estimates of the daily potential evapo-transpiration (the return of water from the earth to the air by evaporation from the soil and transpiration from plants).

The weighted number of days of deficit for New Zealand for the 35 seasons 1949/50 to 1983/84, using three different weightings, are shown in Table 5. This data is for New Zealand as a whole, and can therefore be said to represent the conditions experienced by the 'national' sheep farm, the 'national' dairy farm, and the 'national' beef cattle farm.

? Examine Table 5 and comment on the variability of the New Zealand pastoral climate from 1949/50 to1983/84. Isolate, in particular, the drought seasons of the 1970s, and compare the 1970s as a whole with the 1960s.

Table 5 *Weighted number of days of water deficit in New Zealand (weighted by distribution of sheep, dairy cows and cattle)*

	June–May		
Season	*Sheep*	*Cows*	*Cattle*
1949–50	41.3	40.3	43.3
1950–51	20.1	13.1	15.3
1951–52	34.6	29.4	31.2
1952–53	16.3	12.7	14.0
1953–54	39.2	36.9	14.1
1954–55	38.2	35.0	35.9
1955–56	42.9	20.8	33.2
1956–57	26.7	23.0	28.7
1957–58	20.7	19.5	29.9
1958–59	22.8	4.7	13.9
1959–60	34.2	13.7	27.0
1960–61	19.5	19.0	17.0
1961–62	35.3	22.4	32.0
1962–63	29.8	23.4	28.3
1963–64	42.4	26.4	39.2
1964–65	18.9	4.8	12.7
1965–66	19.3	4.6	14.1
1966–67	26.7	5.8	14.9
1967–68	32.3	29.1	35.2
1968–69	32.7	12.8	24.1
1969–70	35.2	34.4	31.3
1970–71	39.0	16.4	26.8
1971–72	31.0	19.8	24.5
1972–73	59.7	45.8	52.8
1973–74	40.5	35.0	37.3
1974–75	30.4	17.1	25.0
1975–76	37.0	20.8	25.3
1976–77	27.2	24.1	22.8
1977–78	55.2	47.6	51.6
1978–79	27.2	20.1	27.5
1979–80	10.7	2.9	8.3
1980–81	37.6	21.0	28.0
1981–82	38.1	20.7	30.4
1982–83	43.6	39.6	49.9
1983–84	19.1	6.7	14.7

A MODEL FOR NEW ZEALAND DAIRY FARMING

An important factor in this model is the desirability of providing dairy production *forecasts* with sufficient lead-time to enable appropriate decisions (transportation, marketing etc.) to be taken. A summary of the 'lead-time' dairy production forecasts made for April 1978 is given in Table 6. This shows that in the 1977–78 season, the dramatic drop in production in April 1978 in comparison with April 1977 was well-anticipated. It shows that, assuming there was no further rain on the 'National Dairy Farm' in January (specifically after January 9 1978, the date of the first prediction), the April 1978 New Zealand dairy production was first predicted to decrease by 30 per cent, compared with the April 1977 production. Similar progressive forecasts were made as shown and, as noted at the foot of Table 6, the *actual* difference in production between April 1978 and April 1977 of −54 per cent (which was actually not known until about 20 May 1978), was reasonably accurately forecast with lead-times of up to 15 weeks.

Such weather-based forecasts are of direct benefit to decision makers within the dairy industry, and if correctly applied, could lead to a considerably more efficient system 'beyond the farm gate'. For example, in the case of the New Zealand Dairy Board, the forecasts of production given in Table 6 provided the Board with considerable advance notice of the quantity of milk fat (specifically, milk fat converted into butter, cheese, dried milk etc.) available for customers outside New Zealand. Similarly, the forecasts gave valuable lead-time to the several shipping companies in New Zealand on the quantity of milk fat likely to be available for export.

Table 6 *Predictions of New Zealand Dairy Production for April 1978*

Date of prediction (1978)	*Model**	*Assumption*	*Predicted production differences*** %	*\$ million*
Jan. 9	J	No further rain in Jan.	−30	−12
Jan. 31	J	None	−35	−14
Feb. 1	J+F	Normal Feb.	−19	− 7
Feb. 1	J+F	As Feb. 1977	−27	−11
Feb. 1	J+F	No rain in Feb./ Temp. +1½°	−54	−21
Feb. 1	J+F	No days of deficit/ Temp. −1½°	−14	−55
Feb. 15	J+F	No further rain in Feb.	−52	−20
Feb. 15	J+F	No further days of deficit	−31	−12
Feb. 28	J+F	None	−45	−18
March 1	J+F+M	Normal March	−32	−13
March 13	J+F+M	No further rain in March	−69	−27
March 31	J+F+M	None	−62	−24

*J = model used January weather data only
F = February
M = March
**April 1978 compared with April 1977
(*Note*: The actual difference was −54 per cent, 'worth' \$21 million at 1978 prices)

Print 2 *Sheep farming in New Zealand is a highly climate sensitive occupation — the quality of wool and meat products varies greatly with the state of the pasture and a period of drought can lead to great hardship*

THE NATIONAL CLIMATIC SENSITIVITY OF NEW ZEALAND

Although, when looking at an economy, the trained econo-climatic eye may be able to differentiate quickly between those sectors of an economy which are more sensitive to weather and climate than others, it is not always easy for those not so trained. Indeed, it appears that it is sometimes very difficult for politicians, economists, and meteorologists to ascertain what really are the weather and climate sensitive sectors of an area.

Table 7 illustrates, for selected sectors of the New Zealand economy, an econo-climatic view. An 'econo-climatic view' first involves an overall view of an economy, and second a detailed look at its key elements. Then, and only then, should the climatic sensitivity be considered. In the specific case cited for New Zealand, the items noted as being climate sensitive are listed in the order in which the various components of the economy appear in the *New Zealand Official Year Book*.

A few examples will indicate the typical step-by-step process one has to go through in assessing 'national climatic sensitivity'.

Table 7 *Weather/climate sensitivity — the New Zealand example*

Railways	*Expenditure ($ million)*	*%*
Gross expenditure	404	
of which FUEL was	26	6

Shipping	*Units*
Container traffic – Unloaded	107 000
– Loaded	113 000

Roads	*Expenditure ($ million)*	*%*
Highway maintenance	51	30
Highway construction	36	21
Grants to local authorities	70	41

Air *Air NZ*	*Million*
International: Passenger/km flown	4 430
International: Cargo/tonne/km flown	152
International: Total revenue (tonne/km)	573

Energy *Total Energy*	*Petajoules*[1]	*%*
Imported oil	178	50
Primary electricity	72	22
Coal	55	15
Natural gas	37	10
Indigenous oil	14	4

[1] Petajoule = 10^{15} joules

Energy Users	*%*
Industry	39
Transport	34
Households	16
Commerce	11

Electric power generation	*%*
Hydro	86
Geothermal	5
Other	9

Agriculture *Gross Agricultural Production*	*$ million*	*%*
Wool	851	19
Dairy produce	688	15
Cattle	668	15
Sheep and lambs	565	13
Agricultural services	199	4
Crops and seeds	184	4
Fruit	130	3
Vegetables	121	3
Poultry and eggs	112	2

Subsidies from Public Funds

	Adverse events relief ($ million)	*Fertiliser transport subsidy ($ million)*
1970/71	3.5	7.1
1971–72	0.3	9.0
1972–73	0.2	12.5
1973–74	1.0	11.6
1974–75	0.2	8.2
1975–76	0.3	9.2
1976–77	0.4	12.7
1977–78	0.2	16.1
1978–79	6.1	23.4
1979–80	0.3	28.7

Gross Domestic Product *Market Production Groups**	*Value ($ million)*	*%*
Agriculture	2 378	11
Food manufacture	1 266	6
Transport/Storage	1 199	6
Construction	934	5
Electricity/Gas/Water	700	3
Wood product's manufacture	342	2
Forestry	176	1

*Weather/climate related

(*Note*: All data are from the N Z Official Year Book (1981), and unless otherwise stated refer to the most recent period available. The percentages shown are the percentage of the relevant totals.)

Transport

This is obviously an important component of the economy, but it is not transport as a whole that should be considered but its various sectors, and specifically the climate sensitive aspects of the sector. For example, the gross expenditure on railways in New Zealand in 1981 was $404 million, including $26 million (6 per cent) spent on fuel. In New Zealand, the fuel is usually petroleum, which must be imported. It could then be argued that it is the fuel element of railway transport which is weather or climate sensitive, on the basis that any reduction in fuel oil used as a result of better weather and climate information for rail transportation, will be a benefit, through a reduction in imports, to the nation. While it may be argued that other aspects of the railways expenditure are weather and climate sensitive (and this may in fact be the case in other countries), a careful analysis of the situation shows that few measurable sectors of the railway operations in New Zealand are weather and climate sensitive, the exception being this expenditure on fuel.

In the case of shipping, key weather/climate factors are the loading and unloading of containers. In New Zealand, the number of such containers moved exceeds 200 000 in a year, and most contain weather and climate sensitive commodities such as dairy produce, meat, wool or fruit products. The optimal movement of these mainly perishable commodities is crucial to New Zealand's export competitiveness and thus it is logical to include container movements in the list of weather and climatic sensitivities, because they reflect very clearly the end product of a fundamentally agricultural nation.

Agriculture

In a country like New Zealand, the agricultural sector has a very high 'national weather sensitivity' profile. Wool accounts for 19 per cent of the gross value of New Zealand's agricultural production, compared with 3 per cent for vegetables. Thus, irrespective of the fact that some aspects of the vegetable sector (such as the transport of the vegetables to markets) are much more weather-sensitive than many aspects of the wool industry (such as the effects of severe frost on the quality of wool), the fact remains that in terms of monetary value the wool sector of New Zealand agriculture is seven times more important than the vegetable sector. Similarly, 62 per cent of New Zealand's agricultural income is from pastoral products (that is wool, dairy products, and meat) and it is evident that the key climatic factors will be very much related to the state of the nation's pastures. In other countries, there could be a quite different picture, especially where field crops or horticultural-type crops are the principal agricultural products.

Government subsidies
A third aspect worthy of note is the item on subsidies from public funds. Two subsidies are listed — the first concerns adverse events (essentially drought conditions), the second is a fertiliser transport subsidy. This latter subsidy is, in several ways, highly sensitive to weather and climate, since it is usually given *after* a relatively poor climatic (and income) season, in order to encourage farmers to fertilise their pastures during the following season. The natural response by farmers following a 'poor' season is for less money to be spent on fertilisers. The government fertiliser subsidy is therefore a means of smoothing the irregularities in the application of fertiliser to New Zealand's pastures. Since these irregularities result from fluctuations in the farmers' incomes caused by climatic variations, the amount of subsidy is directly related to climate.

Energy
In this sector, a key factor in New Zealand is the heavy reliance on imported oil, which supplies 50 per cent of the country's total energy requirement. A further 22 per cent of New Zealand's total energy is met by electricity, of which 86 per cent is produced through hydroelectric generation. However, this proportion is highly weather/climate related as, for example, in summer there is less water available for generating hydroelectric power than in winter. More importantly, in cold winters, where more electricity is required, more non-hydroelectricity has to be generated; that is the proportion of hydroelectricity as a percentage of total electricity is lower.

The total energy requirements of New Zealand are therefore related to two main factors: the need to use imported oil, and the availability of relatively cheap hydroelectricity. The first factor depends to some extent on the amount of water available for hydroelectric power production, and the general overall state of the New Zealand economy. The second factor involves:
a the adequate supply of water during the critical summer and autumn periods in the hydroelectric dams, and,
b the severity of the winter which dictates what proportion of electricity needs to be generated by more expensive oil and gas.

Gross domestic product
Of New Zealand's gross domestic product, 11 per cent comes from the agricultural sector, and 6 per cent from food manufacturing. Other important weather and climate related market production groups include transport and storage 6 per cent, construction 5 per cent, energy 3 per cent, and wood products and forestry 3 per cent. These components comprise one-third of New Zealand's gross domestic product, and it is evident that these components are clearly more sensitive to weather and climate than other sectors.

Print 3 *Container cargo in Auckland Harbour waiting to be shipped*

In order for you to understand how national climatic sensitivity works in your own country, or another of your choice, obtain a copy of the *Official Year Book* or its equivalent. A *Year Book* attempts to describe a nation in terms of its resources and gives specific information, such as how various parts of the economy contribute to the Gross National Product (for example, the comparative importance of mining and agriculture, or perhaps more specifically coal mining and corn production). To understand the *Year Book* you must first ask the question: 'What makes the country like it is?'. That is, you must have an appreciation of what the important sectors are; how one sector compares with another; and how the primary, secondary and tertiary sectors work together. Then, you must ask the question: 'Does the weather and climate affect any of these sectors?'

In many cases it may be easier to eliminate those sectors of the country that do not appear to be influenced by weather and climate, such as education, the steel industry, insurance, television, egg production, crime rates, political elections, or salaries paid to public servants. But it is not that simple, and it is quite possible that the insurance industry in one country may be very weather/climate sensitive, and indeed it may well be much more weather/climate sensitive than, say, that country's dairy production, or energy consumption. It is important therefore, to understand what makes a specific country 'tick', and then use appropriate 'econo-climatic' thinking to at least separate the various sectors into high, moderate, and low sensitivity to weather and climate.

Throughout this chapter the interrelationships between the physical and the human/economic world have been stressed; indeed it is evident that climate is probably far more humanly and economically orientated than physically orientated. This is well-demonstrated in many references to weather and climate given by the press, radio and television — think of droughts in Ethiopia and Australia, tornados in the USA, floods affecting coffee production in Brazil, blizzards affecting transporation, abnormal death totals, and increased energy consumption.

You should now be able to appreciate and evaluate the points raised in the section below. Choose two exercises and complete them.

1 *'During the 1980s both weather and climate fluctuations will produce major economic, social, and political consequences. Some of these consequences will be relatively short lived (although very dramatic at the time) such as the record low temperatures which affected several parts of the United States during January 1982, while other weather/climate events will be much more significant such as those reported by the official Soviet news agency Tass on 28 January 1982 that "... 1981 was one of the most difficult years for the economic development of the USSR". Tass commented that extremely unfavourable weather — a drought of rare intensity and duration — adversely affected not only agricultural output but also a number of industries running on agricultural raw materials.'*

Note that the word 'will' is used in the first sentence of the extract above. In the light of what has been discussed in this chapter, do you consider that 'will' is the correct word, or should it be 'may'? What might cause the 'will' to become 'may' if this chapter were being written in the 1990s? Write a paragraph setting out your reasoning.

2 *'The world's vulnerability to both weather and climate fluctuations has undoubtedly grown as the world population has increased and the use of available resources has become more intense.'*

In discussion in small groups, prepare the outline of an essay which would examine the validity of this sentence. You should include what you have learned in other chapters in the book, other geography lessons and your general knowledge and common sense.

3 *'One of the consequences of the weather/economic mix, particularly on a national and international scale, has been the increasing demand for improved techniques for measuring the sensitivity of weather and climate on economic activities. This has been due to the increasing sensitivity and impact that food production has on world political and economic events, the increasing cost of energy and its implications for heating, cooling, etc. and the national and international concern expressed in several areas that people — because they are not judiciously utilising their climatic resources — are living beyond their "climatic income".'*

Consider one enterprise in any country in the world. Is that enterprise being run beyond the country's climatic income?

4 The Ministers of Finance/Agriculture/Energy of three different countries (assume one is a well-developed western-type country, the second a relatively fragile island community in the South Pacific, and the third a well-developed agricultural country but with serious balance of payment problems), come to the World Meteorological Organisation (WMO) for advice on how weather and climate information could be better used in each of their countries. What would WMO need to know in order to assist these countries, and what kind of answers to each of the three countries do you think WMO would give?

FURTHER READING

1 POPULATION IN EUROPE: TRENDS AND ISSUES

Hall, R. and Ogden, P. (1983) *Europe's Population in the 1970s and 1980s*, Department of Geography, Queen Mary College, University of London, Special Publication No. 4.

Hall, R. (1984) 'Changing fertility in the developing world and its impact on global population growth,' *Geography*, Jan. 1984 pp. 19–27.

Verduin, J. A. and Verduin-Muller, H. S. (1982) *On European Population: Demographic Developments*, Geographical Institute, State University, Utrecht (Department of Geography for Education).

2 LATIN AMERICAN MIGRANTS: A TALE OF THREE CITIES

Butterworth, D. and Chance, J. (1981) *Latin American urbanisation*, Cambridge University Press.

Gilbert, A. (1974) *Latin American development: a geographical perspective*, Penguin.

Gilbert, A. (5 March 1982) 'The poor man's lot', *The Guardian*.

Gilbert, A. and Gugler, J. (1982) *Urbanisation in the Third World*, Oxford University Press.

Lloyd, P. (1979) *Slums of hope? Shanty towns of the Third World*, Penguin.

Mangin, W. (1967) Latin American squatter settlements: a problem and a solution, *Latin American Research Review 2*, pp 65–98.

3 HOUSING IN LUSAKA: AN EXAMPLE OF SELF-HELP

Adamson, P. (1982) The Gardens, *New Internationalist*, March 1982 issue.

Dwyer, D. J. (1975) *People and Housing in Third World Cities*, Longman

Geographical Magazine Habitat — a series of articles on Third World housing, May 1976 issue. *Also* What is the City but the People? May 1978 issue.

Phillips, D. R. and Yeh, A. G. O. (1983) Changing attitudes to housing provision: BLISS in the Philippines, *Geography*, 68, 37–40 *also* Filipinos Help Themselves to Housing, *Geographical Magazine* Vol. LV, No. 3 pp. 154–159.

Martin, R. (1976) Lusaka Squatters are Licensed, *Geographical Magazine* vol. XLVIII, No. 8, pp. 475–477.

Martin, R. (1976) Lusaka squatters are licensed *Geographical Magazine* May vol. XLVIII, No. 8 p. 475–477.

Murison, H. S. & Lea, J. P. (1979) *Housing in Third World Cities*, Macmillan Press.

New Internationalist Report on meeting basic needs — articles on low cost housing in Third World countries and a report on housing in Britain, August 1976 issue. *Also* Streets paved with gold and other articles on Third World cities, June 1978 issue.

Third World Research Reports (referred to in the text)

Jackman, M. E. (1973) *Recent Population Movements in Zambia*, Manchester University Press for the Institute of African Studies, University of Zambia.

National Housing Authority (1975) *Self-Help in Action: A Study of Site and Service in Zambia*, National Housing Authority Research Study: No. 2. Lusaka, N.H.A.

Schlyter, A. and Schlyter, T. (1979) *George — the Development of a Squatter Settlement in Lusaka, Zambia*, Stockholm, Swedish Council for Building Research (Byggforskningsradet).

Schlyter, A. and Rakodi, C. (1981) *Upgrading a Squatter Environment*, Stockholm, Swedish Council for Building Research (Byggforskningsradet).

Note: Third World research reports may not be easily accessible except in specialised university libraries.

Information on Britain

BEE (Bulletin for Environmental Education) this monthly publication contains a wealth of information on inner city housing problems in Britain.

Herbert, D. T.and Smith, D. M. (eds.) (1979) *Social Problems and the City*: Geographical Perspectives, Oxford University Press.

Map Extracts

Stanford International Map Centre, 12–24 Long Acre, London WC2 E9LP, produce map extracts for certain countries in the Third World. An extract is available of Lusaka, Rep. of Zambia at a scale of 1:50 000 (parts of Sheets 1528 A3 and 1528 A4), which would be useful for further work on Lusaka.

4 THE AMERICAN CITY TODAY

Castells, (1983) *The City and the Grassroots*, Arnold.

Hall, P. (1977) *The World Cities*, Weidenfeld & Nicholson.

Johnston, R. J. (1982) *The American Urban System*, Longman.

Kirby, A. M. and Lambert D. M. (1984) *Space and Society*, Longman.

Palm, R. (1981) *The American City*, Oxford University Press.

Thomson, M. (1978) *Great Cities and their Traffic*, Penguin.

Acknowledgement

This chapter was produced within the Program of Political and Economic Change, University of Colorado at Boulder.

5 FOOD SUPPLIES, AGRICULTURAL PRODUCTION AND LAND REFORM: THE CASE OF PERU

Blakemore, H. and Smith, L. (1983) *Latin America: geographical perspectives*, Methuen.

The Report of the Independent Commission on International Development Issues under the Chairmanship of Willy Brandt, (1980) *North–South: a programme for survival*, Pan Books, London.

Thorp, R. and Bertram, G. (1978) *Peru, 1890–1977, growth and*

policy in an open economy, Macmillan.
UN Food and Agriculture Organisation, *Production Yearbook, 1972 and 1980*, Rome, 1973 and 1981.
World Bank, (1981) *Peru, major development policy issues and recommendations*, World Bank, Washington.

6 US AGRICULTURE: THE PROBLEM OF OVERPRODUCTION

Alexander, J. W. and Gibson, L. J. (1979) *Economic Geography* 2nd Edn. Prentice Hall, New Jersey.
Paterson, J. H. (1979) *North America* 6th Edn. Oxford University Press, New York.

The best way to keep up with developments in American agriculture is to read *The Economist* each week.

Statistics are available in the *Statistical Abstract of the United States* and *Agricultural Statistics*, both published annually.

7 GOVERNMENT POLICY AND THE RURAL ENVIRONMENT IN MEXICO

Arizpe, Lourdes (1980) 'Cultural change and ethnicity in rural Mexico' in David A. Preston, (ed.) *Environment, Society and Rural Change in Latin America*, Wiley
Redclift, Michael (1980) Agrarian Populism in Mexico — the 'Via Campesina' *The Journal of Peasant Studies* 7, no. 4.
Revel-Mouroz (1980) 'Mexican colonisation experience in the humid tropics' in Preston, D. (ed.) *Environment, Society and Rural Change in Latin America*, Wiley.

8 PEASANTS, THE ENVIRONMENT AND THE COMMON AGRICULTURAL POLICY

Fennell, R. (1979) *The Common Agricultural Policy of the European Community.* Granada.
Franklin, S. H. (1969) *The European Peasantry.* Methuen.
Hallett, G. (1981) *The Economics of Agricultural Policy* (2nd Edition) Basil Blackwell.
Marsh, J. S. and Swanney, P. J. (1980) *Agriculture and the European Community.* Allen and Unwin.
Tracy, M. (1982) *Agriculture in Western Europe: Challenge and Response 1880–1980.* 2nd Edition. Granada.
The Economist 'Europe's green and expensive land', November 1980 pp. 51–64; 'Down on the farm', October 23, 1982, pp. 54–55.

9 CHINA: CAN SELF SUFFICIENCY BE ACHIEVED?

Buchanan, K. (1970) *The Chinese People and the Chinese Earth*, Bell.
Buchanan, K. (1970) *The Transformation of the Chinese Earth*, Bell.
Tregear, T. R. (1980) *China, A Geographical Survey*, Hodder and Stoughton.
Tuan Yi-fu (1970) *The World's Landscapes: China*, Longman. Historical and contemporary evolutions of China's landscape.

Yearbooks and journals

Beijing Review (Beijing, weekly) International and internal development including agriculture are covered with authority.
China Quarterly (SOAS, London) Scholarly, often with useful detailed articles.
FAO (Rome) Annual reports with essential raw data plus various quarterly publications.
The Asia Yearbook (Hong Kong, Far Eastern Economic Review) is most useful in keeping up to date with this region. China is heavily featured.
The Geographical Magazine (London, monthly) Articles on China appear frequently. For instance, June 1982 'Less Land for Chinese Farmers' is particularly good. October 1980 'The Northeast is Transformed' includes a relevant agricultural survey.
United Nations Demographic Yearbook (New York) for essential population data.

10 LESOTHO: POPULATION PRESSURE AND RESOURCE DETERIORATION

Bawden, M. G. and Carroll, D. M. (1968) *The Land Resources of Lesotho*, Directorate of Overseas Surveys, Tolworth UK.
Jilbert, J. *Population Change in Lesotho 1966–1976; some Implications for National Planning. Mohlomi*; Journal of Southern African Historical Studies, Vol. 1 1976.
Smits, L. G. A. The Distribution of the Population in Lesotho and some Implications for Economic Development, *Lesotho Notes and Records No. 7*, 1968.

Acknowledgements The author would like to express his thanks to the following individuals: Gerard Schmitz and Lucas Smits, Professors in Geography at the National University of Lesotho, who were sources of many of the ideas which appear in this chapter.

Roger Porkess, Head of Mathematics at Denstone College, for computer analysis of population data in this chapter.

11 BRAZIL: REGIONAL PLANNING AND INEQUALITIES

Ambio; in journal of the human environment. Recommended articles include 'Parks and Biological reserves in the Brazilian Amazon, Vol. XI, No. 5 (1982), pp. 309–314, 'Depletion of Tropical Rain Forests', Vol. XII, No. 2 (1983), pp. 67–71. 'Ouro Prêto: Brazil's Monument Town', Vol. XII, No. 3–4 (1983), pp. 213–215, 'The Tragedy of our Tropical Rain Forests, Vol. XII, No. 5 (1983), pp. 252–254.
Brazilian Embassy (1979) *Brazil, a Geography*, Brazilian Embassy, London.
Cole, J. P. (1981) *The Development Gap*, Wiley, London.
Dickinson, J. P. (1982) *Brazil*, The World Landscapes Series, Longman, London.
Latin American Bureau Special Brief (1982) *Brazil state and struggle*, Latin American Bureau Ltd., London.

National Westminster Bank (1981) *Brazil*, An economic report published by National Westminster Bank, August 1981, London.

The Observer Magazine (1979) 'The Rape of the Amazon', *The Observer Magazine*, 22 April 1979, London.

12 INDIA: PLANNING AND PRIORITIES

Bradnock, R. W. (1984) *Agricultural Change in South Asia*, John Murray.

Bradnock, R. W. (1984) *Urbanisation in India*, John Murray.

Farmer, B. H. (1983) *An Introduction to South Asia*, Methuen.

Johnson, B. L. C. (1979) *India*, Heinemann.

Spate, O. H. K. and Learmonth, A. T. A. (1967) *India and Pakistan: a General and Regional Geography* (3rd ed.), Methuen.

13 NIGERIA: CHOICES IN AGRICULTURAL DEVELOPMENT

Adams W. M. (1983) 'Green revolution — by order', *Geographical Magazine*, August 1983 pp. 405–11.

Ajaegbu H. I. (1976) *Urban and Rural Development in Nigeria*, Heinemann.

Kirke-Green A. and Rimmer D. (1981) *Nigeria Since 1970: A Political and Economic Outline*, Hodder and Stoughton.

Ogoyuntibo J. S., Areola O. O. and Filani M. (1978) *A Geography of Nigerian Development*, Heinemann (Nigeria).

Richardson P. (1985) *Indigenous Agricultural Revolution: the Ecology of Food Production in West Africa*, Hutchinson University Press.

Acknowledgements The author would like to thank Dick Grove, who introduced him to Africa, Graham Chapman who introduced him to gaming simulation, and the many who helped him in his visits to Nigeria, especially Alan Bird, Jacob Bala, Jack Griffiths and Mike Mortimore.

14 THE IMPACT OF INDUSTRIAL CHANGE: SOME EUROPEAN EXAMPLES

Dunford, M. and Perrons, D. (1983) *The Arena of Capital*, Macmillan.

Eatwell, J. (1982) *Whatever Happened to Britain? The Economics of Decline*, Duckworth.

Hedges, N. and Beynon, H. (1982) *Born to Work, Images of Factory Life*, Pluto Press.

Martin, R. and Rowthorn, R. (eds) (1985) *Deindustrialisation and the British Space Economy*, Macmillan.

Townsend, A. (1983) *The Impact of Recession on Industry, Employment and the Regions, 1976–1981*, Croom Helm.

15 CHANGING TRADE PATTERNS: NEWLY INDUSTRIALISING COUNTRIES IN SOUTH-EAST ASIA

Balaffa, (1981) *Newly Industrialising Countries in the World Economy*, Pergamon, New York.

Barclay's Review (February 1982) — Two very readable articles on the NICs and Asian NICs.

Saunders, C. (ed) (1981) *The Political Economy of New and Old Industrial Countries*, Butterworths.

16 ZAMBIA:A ONE-MINERAL ECONOMY?

Maunder, P. (ed.) (1982) *Case studies in Development Economics* (Case 5 on Zambia), Heinemann.

Rostow, W.W. (1960) *The Stages of Economic Growth* Cambridge University Press.

Chisholm, M. (1982) *Modern World Development, A geographical perspective* Hutchinson.

Fanon, F. (1967) *The Wretched of the Earth*, Penguin Books.

17 THE NUCLEAR POWER DEBATE IN THE UK

Bunyard, P. (1981) *Nuclear Britain*, New English Library.

Elliott, D. et al. (1978) *The Politics of Nuclear Power*, Pluto Press.

Foley, G. (1976) *The Energy Question*, Penguin.

Hoyle, F. (1977) *Energy or Extinction — the case for nuclear energy*, Heinemann.

Nuclear Energy Questions Study Pack, Information Service on Energy, 2 Forth Street, Edinburgh.

Wild, M. (ed) (1983) *Energy in the 80s*, Longman.

18 ENVIRONMENTAL CONFLICTS IN JAPAN: PRODUCTION VERSUS PEOPLE

Dalyell, T. (24th Feb. 1983) *New Scientist* 'Acid rain erodes our credibility'.Japan Pictorial, Vol. 2, No. 1, 1979, p. 13–20. (Available from: Japan Information Centre, 9 Grovesnor Square, London W1.)

Kobe City Government (1979) *Port of Kobe*. (Available from: Port of Kobe Authority, UK Office, Plantation House, 31–35 Fenchurch St, London EC3M 3DX.)

MacDonald, D. (1985) *A Geography of Modern Japan*, Paul Norbury Publications.

19 MANAGING RECREATIONAL AREAS: THE EFFECTS OF TRAMPLING ON AN ECOSYSTEM

The Countryside Commission, Cheltenham publishes many useful surveys on subjects such as the Pennine Way, Snowdon and Kynance Cove.

Bayfield, N.G. (1979) 'Recovery of Four Montane Heath Communities on Cairngorm, Scotland, from Disturbance by Trampling', *Biological Conservation* 15, pp. 165–79.

Burden, R.F. and Randerson, P.F. (1972) 'Quantitative Studies on the Effects of Human Trampling on Vegetation as an Aid to the Management of Semi-Natural Areas', *Journal of Applied Ecology* 9, pp. 439–57.

Clapham, A.R., Tutin, T.G. and Warburg, E.F. (1962); *Flora of the British Isles*, (2nd Edition) Cambridge University Press.

Harrison, C.M. (1980–1) 'Recovery of Lowland Grassland and

Heathland in Southern England from Disturbance by Seasonal Trampling', *Biological Conservation* 19, pp. 119–30.

20 MANAGING DRAINAGE BASINS

Clark, M. and Small, J. (1982) *Slopes and Weathering*, CUP.
Goudie, A. (1981) *The Human Impact*, Blackwell
Knapp, B. J. (1979) *Elements of Geographical Hydrology*, Allen and Unwin.
Knapp, B. J. (1986) *Systematic Geography*, Allen and Unwin.
Knapp, B. J. and Child, S. (1979) Updating geomorphology: hydrological effects of man's activities *Teaching Geography* (5)2 p. 78–82.
Smith, D. I. and Stopp, P. (1978) *The River Basin*, CUP.

21 THE US MIDWEST: THE PROBLEM OF WATER ALLOCATION

Chorley, R. (ed.) (1969) *Introduction to Geographical Hydrology* — essays on river basins and general hydrology, Methuen.
Foster, H. and Sewell, W.R.D. (1981) *Water: The Emerging Crisis in Canada*, Lormier, Toronto.
Newsweek, February 23, 1981, 'The Browning of America'.
Overman, M. (1968) *Water*, Open University Press.
Norbeck, E. (1976) *Changing Japan*, Holt Rinehart and Winston.
O.E.C.D. (1975) *Airports and the Environment*.
O.E.C.D. (1976) *Water Management in Japan*.

22 WEATHER IMPACTS AND WEATHER MANAGEMENT

Dorrnkamp, J.C., Gregory, K.J. and Brown, A.S. (1980) *Atlas of Drought in Britain*, Institute of British Geographers.
Kates, R.W. (ed.) (1984) *Climate Impact Assessment: Studies of the Interaction of Climate and Society*, SCOPE, Wiley, Toronto.
Taylor, J.A. (ed.) (1974) *Climatic Resources and Economic Activity*, David and Charles.
Maunder, W. J. (1970) *The Value of the Weather*, Methuen.

AUTHORS

Dr W M Adams, *Assistant lecturer in Geography, University of Cambridge*

Dr Michael Dunford, *Lecturer in Human Geography, School of European Studies, University of Sussex*

Professor S H Franklin, *Professor of Geography, Victoria University of Wellington, New Zealand*

Dr Alan Gilbert, *Reader in Geography, Institute of Latin American Studies, University College London*

Dr Carolyn M Harrison, *Lecturer in Geography, University College London*

Dr Margaret E Harrison, *Lecturer, The School of Geography and Geology, The College of St Paul and St Mary, Cheltenham*

John Huckle, *Head of Geography, Bedford College of Higher Education*

John Jilbert, *Head of Geography, Denstone College, Nr Uttoxeter*

Professor Andrew Kirby, *Department of Geography, University of Colorado, USA*

Dr Brian J Knapp, *Head of Geography, Leighton Park School, Reading*

Dr W J Maunder, *Meteorological Service, Wellington, New Zealand*

Donald MacDonald, *Head of Geography, Jordanhill College of Education, Glasgow*

Raymond Pask, *Geography teacher, Melbourne High School, Melbourne, Australia*

Les Potts, *Senior Lecturer, Department of Business Studies, Aylesbury College*

Professor Michael Redclift, *Lecturer in Latin American Rural Sociology, Department of Environmental Studies, Wye College, (University of London)*

Professor Derrick Sewell, *University of Victoria, Canada*

Dr Frances Slater, *Senior Lecturer in Education, University of London Institute of Education*

Dr Clifford T Smith, *formerly Director of Centre for Latin American Studies, University of Liverpool*

Dr P T H Unwin, *Lecturer in Geography, Royal Holloway and Bedford New College, University of London*

Dr Jan Verduin, *Lecturer in Human Geography, University of Utrecht, Netherlands*

Professor Henriette Verduin-Muller, *Lecturer in Human Geography, University of Utrecht, Netherlands*

Dr Peter Ward, *Lecturer in Geography, Fitzwilliam College, Cambridge*

John Wareing, *Principal Lecturer in Geography, Polytechnic of North London*

Meryl Welsh, *formerly Geography teacher at Strode's College, Egham*

Professor Colin J B Wood, *Associate Professor, University of Victoria, Canada*

INDEX

Cities, states and companies appear as main entries, and not under their countries. Alphabetical arrangement is letter by letter. Most detailed entries can be found under individual countries.

ACKNOWLEDGEMENTS

The publishers would like to thank the following for permission to reproduce photographs:
Bill Adams, p 276; *Ardea London/ David and Katie Urry*, p 418; *Associated Press*, pp 251, 376, 377, 388; *Aycliffe and Peterlee Development Corporation*, p 302 (top right); *Barnaby's Picture Library*, pp 16, 306, 307; *BBC Hulton Picture Library*, p 66; *Martin Bond*, p 351; *Brazilian Embassy*, p 212 (top); *Cambridge University Collection*, p 407; *Camerapix Hutchison*, pp 38 (top left), 202, 266, 273, 278, 330; *Camera Press, p 21; J. Allan Cash Ltd.*, p 217; *Centre for World Development Education/World Bank/E. Huffman*, pp 43, 46, 52; — *G. Franchini*, pp 120, 127 — *ILO*, pp 84, 100, 129 — *Hilda Bijur*, p 136 — *Mattioli*, p 204 — *Y. Hadar*, p 269; *Christian Aid*, p 288; *Crown Copyright*, p 199; *Susan I. Cunningham*, pp 225, 230; *Earthscan/Mark Edwards*, pp 41, 220, 247, 253 — *Marcos Santilli*, pp 207, 219, 223; *Earthscape/ B. Knapp*, pp 26, 396, 420, 423, 428; *European Commission*, p 140; *Farmers Weekly*, p 138; *Ford Motor Co. Ltd.*, p 288; *Peter Fraenkel*, p 328, 334; *Friends of the Earth/Graeme Montgomery*, p 352; *German National Tourist Board*, p 153; *Alan Gilbert and Peter Ward*, pp 38 (bottom right), 39; *Gisborne Herald*, NZ, p 444; *Sally and Richard Greenhill*, pp 65, 67 (top right), 73, 112, 125, 164, 311, 312, 318, 319, 321; *Greenpeace*, p 353; *The Guardian*, p 362; *Richard Hadley*, pp 24, 34 (top left), 36; *Tom Hanley*, pp 322, 324; *Jon Jacobson — Photographer*, p 102; *John Jilbert*, pp 186, 196, 198, 201, 205; *Andrew Kirby*, pp 70, 74, 75; *LDDC*, p 77; *Tony Morrison, South American Pictures*, pp 88, 91, 92, 93; *V. Mueller*, p 20; *Ian Murphy*, p 337; *National Coal Board*, p 304; *National Union of Journalists*, p 291; *Network/Mike Abrahams*, p 7; *New Zealand Meteorological Service*, p 441; *Oxfam*, pp 237, 242, 250; *Raymond Pask*, pp 157, 158, 160, 163, 168, 171, 173; *Peak Park Joint Planning Board*, pp 390, 404, 405; *Michael Redclift*, p 132; *RSPB/ Michael Richards*, p 357; *Tony Sacks*, p 341; *Thomas Schlyter*, p 56; *Clifford Smith*, pp 90, 96, 98; *Frank Spooner Pictures*, pp 34 (bottom right), 302 (bottom left); *Studio Jon Ltd*, p 418; *John Topham Picture Library*, pp 144, 372; *United States Department of Agriculture — Doug Wilson*, p 107 — *Earl Wilson*, p 108 — *Tim McCabe*, p 115 — *Jay Beavers*, p 116; *Vision International/ M. Mann*, pp 449, 451; *A. C. Waltham*, p. 430; *Zefa*, p 358.
Front cover photograph reproduced by permission of Camerapix Hutchison